AF766578

Handbook of Numerical Analysis
Volume 20

Processing, Analyzing and Learning of Images, Shapes, and Forms: Part 2

Handbook of Numerical Analysis

Volume 20

Processing, Analyzing and Learning of Images, Shapes, and Forms: Part 2

Edited by

Ron Kimmel
Department of Computer Science
Technion - Israel Institute of Technology
Haifa, Israel

Xue-Cheng Tai
Hong Kong Baptist University, Department of Mathematics,
Kowloon Tong, Hong Kong;
University of Bergen, Department of Mathematics,
Bergen, Norway

ELSEVIER

N·H

North-Holland

An imprint of Elsevier

North-Holland is an imprint of Elsevier
Radarweg 29, PO Box 211, 1000 AE Amsterdam, The Netherlands
The Boulevard, Langford Lane, Kidlington, Oxford OX5 1GB, United Kingdom

Copyright © **2019** Elsevier B.V. All rights reserved.

No part of this publication may be reproduced or transmitted in any form or by any means, electronic or mechanical, including photocopying, recording, or any information storage and retrieval system, without permission in writing from the publisher. Details on how to seek permission, further information about the Publisher's permissions policies and our arrangements with organizations such as the Copyright Clearance Center and the Copyright Licensing Agency, can be found at our website: www.elsevier.com/permissions.

This book and the individual contributions contained in it are protected under copyright by the Publisher (other than as may be noted herein).

Notices

Knowledge and best practice in this field are constantly changing. As new research and experience broaden our understanding, changes in research methods, professional practices, or medical treatment may become necessary.

Practitioners and researchers must always rely on their own experience and knowledge in evaluating and using any information, methods, compounds, or experiments described herein. In using such information or methods they should be mindful of their own safety and the safety of others, including parties for whom they have a professional responsibility.

To the fullest extent of the law, neither the Publisher nor the authors, contributors, or editors, assume any liability for any injury and/or damage to persons or property as a matter of products liability, negligence or otherwise, or from any use or operation of any methods, products, instructions, or ideas contained in the material herein.

ISBN: 978-0-444-64140-3
ISSN: 1570-8659

For information on all North-Holland publications
visit our website at https://www.elsevier.com/

Publisher: Zoe Kruze
Acquisition Editor: Sam Mahfoudh
Editorial Project Manager: Joanna Collett
Production Project Manager: James Selvam
Cover Designer: Alan Studholme

Typeset by SPi Global, India

ELSEVIER Book Aid International
Working together to grow libraries in developing countries
www.elsevier.com • www.bookaid.org

Contents

Contributors

Numbers in Parentheses indicate the pages on which the author's contributions begin.

Yonathan Aflalo (273), Alibaba Research, Haifa, Israel

Omri Azencot (117), Math Department, University of California, Los Angeles, CA, United States

Martin Bauer (613), Department of Mathematics, Florida State University, Tallahassee, FL, United States

Mirela Ben-Chen (117), Computer Science Department, Technion - Israel Institute of Technology, Haifa, Israel

R. Bergmann (553), Research Group Numerical Mathematics (Partial Differential Equations), Faculty of Mathematics, TU Chemnitz, Chemnitz, Germany

Andrea L. Bertozzi (503), Department of Mathematics, University of California, Los Angeles, CA, United States

Nicolas Charon (613), Department of Applied Mathematics and Statistics, Johns Hopkins University, Baltimore, MD, United States

Da Chen (233), University Paris Dauphine, PSL Research University, CEREMADE, CNRS, UMR 7534; Centre Hospitalier National d'Ophtalmologie des Quinze-Vingts, Paris, France

Ke Chen (579), Department of Mathematical Sciences, University of Liverpool, Liverpool, United Kingdom

Laurent D. Cohen (233), University Paris Dauphine, PSL Research University, CEREMADE, CNRS, UMR 7534, Paris, France

Kimon Drakopoulos (149), Marshall School of Business, University of Southern California, Los Angeles, CA, United States

Roland Glowinski (191), Department of Mathematics, University of Houston, Houston, TX, United States; Department of Mathematics, Hong Kong Baptist University, Kowloon Tong, Hong Kong

Oshri Halimi (273), Electrical Engineering Department, Technion - Israel Institute of Technology, Haifa, Israel

Michael Hintermüller (437), Humboldt-Universität zu Berlin; Weierstrass Institute for Applied Analysis and Stochastics (WIAS), Berlin, Germany

Sung Ha Kang (533), School of Mathematics, Georgia Institute of Technology, Atlanta, GA, United States

Hariprasad Kannan (351), Inria Saclay, Palaiseau, France

Ron Kimmel (273), Computer Science Department, Technion - Israel Institute of Technology, Haifa, Israel

Nikos Kolotouros (149), Department of Computer and Information Science, University of Pennsylvania, Philadelphia, PA, United States

Nikos Komodakis (351), Ecole des Ponts ParisTech, Champs-sur-Marne, France

Rongjie Lai (315), Department of Mathematics, Rensselaer Polytechnic Institute, Troy, NY, United States

F. Laus (553), Image Analysis Group, Department of Mathematics, TU Kaiserslautern, Kaiserslautern, Germany

Roy R. Lederman (1), Department of Statistics and Data Science, Yale University, New Haven, CT, United States

Gi-Ren Liu (1), Department of Mathematics, National Chen-Kung University, Tainan, Taiwan

Lok Ming Lui (579), Department of Mathematics, Chinese University of Hong Kong, Sha Tin, Hong Kong, China

Shousheng Luo (191), Beijing Computational Science Research Center, Beijing; School of Mathematics and Statistics, Data Analysis Technology Lab, Institute of Applied Mathematics, Henan University, Kaifeng, Henan Province, PR China

Petros Maragos (149), School of Electrical and Computer Engineering, National Technical University of Athens, Athens, Greece

Ekaterina Merkurjev (503), Department of Mathematics and CMSE (Computational Mathematics, Science, and Engineering), Michigan State University, East Lansing, MI, United States

Jan Modersitzki (579), Institute of Mathematics and Image Computing, University of Lübeck, Lübeck, Germany

Kostas Papafitsoros (437), Weierstrass Institute for Applied Analysis and Stochastics (WIAS), Berlin, Germany

Nikos Paragios (351), CentraleSupelec, Gif-sur-Yvette, France

J. Persch (553), Image Analysis Group, Department of Mathematics, TU Kaiserslautern, Kaiserslautern, Germany

Dan Raviv (273), School of Electrical Engineering, Faculty of Engineering, Tel Aviv University, Tel Aviv-Yafo, Israel

Shoham Sabach (401), Faculty of Industrial Engineering and Managment, The Technion, Haifa, Israel

Christos Sakaridis (149), Department of Information Technology and Electrical Engineering, ETH Zürich, Zurich, Switzerland

Jie Shen (647), Department of Mathematics, Purdue University, West Lafayette, IN, United States

Tal Shnitzer (1), Viterbi Faculty of Electrical Engineering, Technion - Israel Institute of Technology, Haifa, Israel

Justin Solomon (41), Department of Electrical Engineering and Computer Science, Massachusetts Institute of Technology, Cambridge, MA, United States

G. Steidl (553), Image Analysis Group, Department of Mathematics, TU Kaiserslautern; Fraunhofer ITWM, Kaiserslautern, Germany

Xue-Cheng Tai (191, 533), Hong Kong Baptist University, Department of Mathematics, Kowloon Tong, Hong Kong; University of Bergen, Department of Mathematics, Bergen, Norway

Ronen Talmon (1), Viterbi Faculty of Electrical Engineering, Technion - Israel Institute of Technology, Haifa, Israel

Marc Teboulle (401), School of Mathematical Sciences, Tel-Aviv University, Tel-Aviv-Yafo, Israel

Yu Wang (41), Department of Electrical Engineering and Computer Science, Massachusetts Institute of Technology, Cambridge, MA, United States

Hau-Tieng Wu (1), Department of Mathematics; Department of Statistical Science, Duke University, Durham, NC, United States

Laurent Younes (613), Department of Applied Mathematics and Statistics, Johns Hopkins University, Baltimore, MD, United States

Hongkai Zhao (315), Department of Mathematics, University of California, Irvine, CA, United States

Wei Zhu (533), Department of Mathematics, University of Alabama, Tuscaloosa, AL, United States

Preface

This is the second of two volumes about mathematical tools for analysing, processing, and synthesizing shapes and images. These two volumes include topics that range from variation methods, learning, understanding, and exploring geometric structures, surfaces, and low-dimensional manifolds, including modelling, interface problems, numerical methods, and optimization techniques. Experts in different aspects of these fields introduce their recent philosophy, theory, and resulting techniques. It ranges from a novel attempt to put deep learning within the framework of compressed sensing and sparse models, reconstruction of low rank matrices, shifting into learning geometry, shape representation that has the potential to migrate geometry analysis into that of deep learning, and pure geometric problems dealt in a novel, yet axiomatic manner. While the focus of the first volume was about modelling problems with considerations on geometric structures of images, this second volume continues the discussion on modelling and computational problems with geometric flavours, involving metric and spectral geometries for matching surfaces and quantifying the distortions of shapes, optimal control of curves as shape boundaries, and manifold learning. Computationally efficient and more importantly—practical axiomatic methods, are included like a novel active contour method, and an operator splitting algorithm. In other chapters, experts from different fields provide their introductions to neural networks and deep learning exploring the relationship between the obvious image processing and the more challenging aspect of this field, that is, machine learning of geometric structures.

Ron Kimmel
Xue-Cheng Tai

Chapter 1

Diffusion operators for multimodal data analysis

Tal Shnitzer[a], Roy R. Lederman[b], Gi-Ren Liu[c], Ronen Talmon[a,*] and Hau-Tieng Wu[d,e]

[a]*Viterbi Faculty of Electrical Engineering, Technion - Israel Institute of Technology, Haifa, Israel*
[b]*Department of Statistics and Data Science, Yale University, New Haven, CT, United States*
[c]*Department of Mathematics, National Chen-Kung University, Tainan, Taiwan*
[d]*Department of Mathematics, Duke University, Durham, NC, United States*
[e]*Department of Statistical Science, Duke University, Durham, NC, United States*
[*]*Corresponding author: e-mail: ronen@ef.technion.ac.il*

Chapter Outline

Abstract

In this chapter, we present a Manifold Learning viewpoint on the analysis of data arising from multiple modalities. We assume that the high-dimensional multimodal data lie on underlying low-dimensional manifolds and devise a new data-driven representation that accommodates this inherent structure. Based on diffusion geometry, we present three composite operators, facilitating different aspects of fusion of information from different modalities in different settings. These operators are shown to recover the common structures and the differences between modalities in terms of their intrinsic geometry and allow for the construction of data-driven representations which capture these

Handbook of Numerical Analysis, Vol. 20. https://doi.org/10.1016/bs.hna.2019.07.008

© 2019 Elsevier B.V. All rights reserved.

characteristics. The properties of these operators are demonstrated in four applications: recovery of the common variable in two camera views, shape analysis, foetal heart rate identification and sleep dynamics assessment.

Keywords: Manifold Learning, Diffusion Maps, Multimodal data, Sensor fusion, Shape differences, Common variable, Alternating Diffusion

AMS Classification Codes: 62-07, 62M15, 57R40, 57M50

1 Introduction

The analysis of data arising from multiple modalities presents various opportunities to gain perspective on different aspects of the measured phenomena. In some cases, a common phenomenon is measured by different modalities, and many sources of variability may be introduced due to modality-specific properties and irrelevant parameters. In these cases, capturing the common variables to all modalities is of great interest. In other cases, differences between modalities can embody important information as well. For example, in two-lead electrocardiogram (ECG) recordings of pregnant women, when both leads are placed on the abdomen, the differences between the leads could capture information related to the foetal heart activity (as demonstrated in Shnitzer et al. (2019)).

The work presented in this chapter addresses multimodal data analysis from a Manifold Learning viewpoint. It relies on concepts from Diffusion Maps (Coifman and Lafon, 2006), a particular Manifold Learning technique, which assumes that the (possibly) high-dimensional data lie on an underlying lower-dimensional manifold. Given such a dataset, Diffusion Maps provides a data-driven embedding into a Euclidean space, where the Euclidean distances between the embedded coordinates approximate the diffusion distances on the manifold. However, Diffusion Maps, as well as other Manifold Learning techniques (e.g., Laplacian Eigenmaps (Belkin and Niyogi, 2003) and ISOMAP (Tenenbaum et al., 2000)), incorporate all variables from all the modalities, in case the data are multimodal. The task of combining several modalities and recovering the common components of multimodal data using Manifold Learning has been addressed in various studies. Two Manifold Learning methods for sensor fusion are presented in Davenport et al. (2010) and Keller et al. (2010), where measurements from different sensors are combined by either concatenating the measurements from the different sensors (Davenport et al., 2010), or by first applying Diffusion Maps to each sensor separately and then concatenating the embedded coordinates (Keller et al., 2010). However, such methods aggregate all the underlying sources of variability and do not distinguish between common and sensor-specific variables. Several other studies have addressed spectral clustering of multiview data by combining kernels representing the different views in various ways (De Sa, 2005; Kumar and Daumé, 2011; Lindenbaum et al., 2015; Zhou and Burges, 2007), e.g., multiplication of affinity matrices in De Sa (2005) and in Lindenbaum et al. (2015). In another work (Wang et al., 2012), an iterative framework for metric fusion is presented based on multiplication of normalized affinity kernels. In Froyland (2015) and Salhov et al. (2016), dynamical systems are addressed, in which multiple time frames are treated as different views. It is shown there that combining diffusion kernels from different time frames can recover coherent sets (Froyland, 2015) or intrinsic parameters relating to the geometry of the system (Salhov et al., 2016). Other works addressing the recovery of a common variable include the classical Canonical Correlation Analysis (CCA) (Hotelling, 1936) and its more recent nonlinear kernel extensions (Bach and Jordan, 2002; Lai and Fyfe, 2000). In addition, a few studies have addressed the recovery of differences between modalities. For example, in Rustamov et al. (2013), an operator for recovering shape differences is proposed, which provides a mapping between shapes that distorts functions only in the locations of the shape differences.

Following this line of work, we review here three composite operators, first presented in Lederman and Talmon (2018), Shnitzer et al. (2019) and Talmon and Wu (2019), that are based on concepts from Diffusion Maps, and facilitate the integration of high-dimensional data measured by multiple modalities in a data-driven manner. These operators are shown to recover the common structures and the differences between the different modalities and provide new and meaningful representations for multimodal data.

The chapter is organized as follows. In Section 2 we present a brief description of one version of Diffusion Maps as a reference for Sections 3 and 4. In Section 3, we present the Alternating Diffusion (AD) operator (Lederman and Talmon, 2018) and demonstrate that it recovers the common variable in multimodal data. In Section 4, two composite diffusion operators from Shnitzer et al. (2019) are presented, which are shown to recover the common structures as well as the differences between different modalities. In Section 5, we show three examples to applications.

2 Preliminaries: Diffusion Maps

The work presented in this chapter utilizes operators which are based on Diffusion Maps (Coifman and Lafon, 2006). In the following section we briefly describe the derivation of the Diffusion Maps operator.

Consider a dataset, $\{s_i\}_{i=1}^N$, which is assumed to be sampled from some underlying Riemannian manifold, denoted by $\mathcal{M}$, with a metric g. The following pairwise affinity kernel is constructed from the data:

$$W_{i,j} = \exp\left(-\frac{d_g(s_i, s_j)^2}{\epsilon^2}\right), \tag{1}$$

where d_g denotes a distance function between the observations induced by the metric g, and $\epsilon > 0$ is the kernel scale.

The kernel is then normalized and the "backward" operator **Q** is defined as follows:

$$Q_{i,j} = \frac{W_{i,j}}{d_i}, \tag{2}$$

where $d_i = \sum_j W_{i,j}$. See other normalization alternatives and their implications in Coifman and Lafon (2006). The operator **Q** can be interpreted as the transition probability matrix of a Markov chain defined on the data. Note that a corresponding "forward" operator can be defined by dividing $W_{i,j}$ by d_j.

Based on the constructed Markov chain, the following diffusion distance is defined:

$$D_T(s,r) = \int_{\mathcal{M}} \frac{1}{\pi(u)} (p_T(s, u) - p_T(r, u))^2 d\mu(u), \tag{3}$$

where $p_T(s, u)$ is the transition probability of reaching s from u in T time steps, π denotes the stationary distribution of the transition probability matrix of the Markov chain (**Q**) and μ represents the distribution of the points on $\mathcal{M}$.

The diffusion distance describes similarity between points based on the probability of transition along paths connecting the points, so that highly connected points are more similar. This distance measure takes into account the geometry of the data. For example, data clusters and "bottlenecks" are captured by the diffusion distance (Coifman and Lafon, 2006, Figure 1; Talmon et al., 2013, Figure 3).

Denote the eigendecomposition of the operator **Q** by $\mathbf{Q}\psi_\ell = \lambda_\ell \psi_\ell$. In Coifman and Lafon (2006) it was shown that the diffusion distance between the data points can be approximated using a subset of m eigenvectors and eigenvalues of **Q**, as follows:

$$D_T(s_i, s_j) \approx \left(\sum_{\ell=1}^{m} \lambda_\ell^{2T} \left(\psi_\ell(s_i) - \psi_\ell(s_j)\right)^2 \right)^{\frac{1}{2}}. \tag{4}$$

A new representation of the dataset is then defined by the following embedding, termed Diffusion Maps:

$$s_i \mapsto \left[\lambda_1^T \psi_1(s_i), \lambda_2^T \psi_2(s_i), \ldots, \lambda_m^T \psi_m(s_i)\right]. \tag{5}$$

Note that the right hand side of (4) describes the Euclidean distance between the points in this embedded space, and therefore, the diffusion distances in the data domain can be approximated by the Euclidean distances in this new embedded space.

This interpretation depicts that the embedded coordinates obtained by the Diffusion Maps algorithm provide a new representation of the data in a Euclidean space (commonly low-dimensional), such that the Euclidean distances capture meaningful geometrical properties of the data.

3 Alternating Diffusion

3.1 Problem formulation: Metric spaces and probabilistic setting

In this section we review the formulation of the common variable problem as presented in Lederman and Talmon (2018). The formal description is followed by an illustrative example in the next section. Suppose that $X \in \mathcal{X}$, $Y \in \mathcal{Y}$ and $Z \in \mathcal{Z}$ are hidden random variables, where $\mathcal{X}$, $\mathcal{Y}$ and $\mathcal{Z}$ are (possibly) high-dimensional spaces, with the density $(X, Y, Z) \sim \pi_{x,y,z}(X, Y, Z)$, such that given X, the variables Y and Z are independent, i.e.,

$$\pi_{x,y,z}(X, Y, Z) = \pi_x(X)\pi_{y|x}(Y|X)\pi_{z|x}(Z|X). \tag{6}$$

While we do not have direct access to X, Y or Z, we observe the random variables $S^{(1)} \in \mathcal{S}^{(1)}$ and $S^{(2)} \in \mathcal{S}^{(2)}$ that are functions of these variables:

$$S^{(1)} = g(X, Y), \tag{7}$$

$$S^{(2)} = h(X, Z), \tag{8}$$

where g and h are some bilipschitz maps.

The given dataset consists of N observations $\{(s_i^{(1)}, s_i^{(2)})\}$, $i = 1, \ldots, N$, where the ith pair of observed variables $(s_i^{(1)}, s_i^{(2)})$ is driven by a realization of the hidden random variables (x_i, y_i, z_i).

In this setup, we consider tasks of unsupervised learning. One classic problem would be to recover a parametrization of the space of the hidden variables (X, Y, Z) (up to ambiguities), for example, by concatenating the observations $S^{(1)} \in \mathcal{S}^{(1)}$ and $S^{(2)} \in \mathcal{S}^{(2)}$ treating them as a single observation which is a function of X, Y and Z.

In this section, we would like to recover a parametrization $f(S^{(1)}, S^{(2)})$ of *only* X (up to a bilipschitz map), which "ignores" the variables Y and Z (see illustrative example in the next section). One interpretation of this operation, discussed in Shaham and Lederman (2018), is that f recovers equivalence classes, such that if two realizations i and j share the same value of X (i.e., $x_i = x_j$, but, in general $y_i \neq y_j$ and $z_i \neq z_j$), then $f(s_i^{(1)}, s_i^{(2)}) = f(s_j^{(1)}, s_j^{(2)})$, whereas if $x_i \neq x_j$ then $f(s_i^{(1)}, s_i^{(2)}) \neq f(s_j^{(1)}, s_j^{(2)})$. The algorithm presented in Section 3.3 obtains a parametrization which approximates these properties.

Remark 1. The problem has a natural extension to the case of a larger number of hidden variables and a larger number of observed modalities $(S^{(1)}, S^{(2)}, S^{(3)}, \ldots)$. For simplicity, we discuss the case of two observed modalities.

3.2 Illustrative example

In general, the variables X, Y and Z can each be high-dimensional, and the maps g and h might be unknown and involved. To illustrate the problem, we present a simple example which is easy to interpret visually.

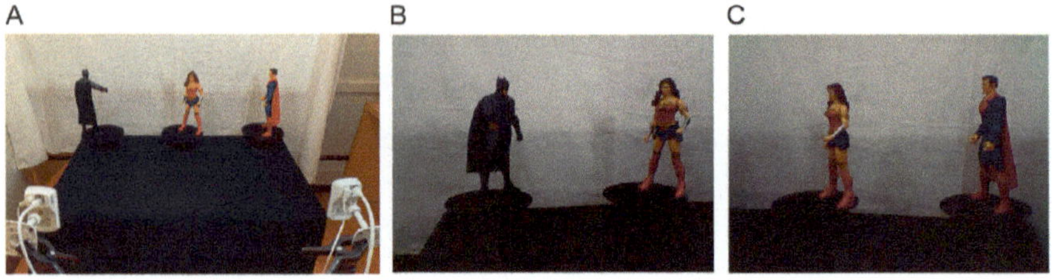

FIG. 1 Experiment setup: (A) Three spinning figures, two cameras. (B) and (C) Simultaneous views from Camera 1 and Camera 2.

Consider the experimental setup depicted in Fig. 1, which consists of three action figures, a wonder woman figure (WF), a batman figure (BF) and a superman figure (SF), on spinning tables. Two cameras produce pairs of *simultaneous* snapshots; Camera 1 observes WF and BF, whereas Camera 2 observes WF and SF. In this example, X is the orientation of WF, Y is the orientation of BF, and Z is the orientation of SF, $S^{(1)}$ is the image taken by Camera 1 (a function of the orientations of WF and BF, i.e., X and Y) and $S^{(2)}$ is the image taken by Camera 2 (a function of the orientations of WF and SF, i.e., X and Z). The spinning tables spin at different speeds and we ignore the temporal order in which the images are taken; so our dataset contains pairs of simultaneous images, with X, Y and Z approximately satisfying (6). Obviously, the images are susceptible to additional noise in practice.

The variable X is special in this setup, because the images in both cameras depend on this variable, whereas, for example, we can get identical images in Camera 2 even if the value of Y changes (as long as the values of X and Z are identical). However, WF does not look the same when viewed from the two different cameras. In fact, in general, the variable X can be manifested in very different ways: X can look like WF from one direction and like Aquaman from another direction. The measurements can even be of different modalities (Camera 2 can be replaced by some sound recording, which is unfortunately less convenient to demonstrate in a written book chapter). The relations between the hidden variables and the measurements can be more abstract.

The parametrization of X that we would like to obtain would ignore the variables Y and Z. To illustrate the properties of such parametrization, we begin with a one variable example in Fig. 2. In this example, we assume that we know how to crop the images so that we can obtain images of WF as viewed from Camera 1. In this case, we would take the images, run the Diffusion Maps algorithm as described in Section 2, and map every image to a 2-dimensional point in Fig. 2B. We observe that the 2-dimensional diffusion map of the images is approximately a circle, capturing the cyclic nature of the variable X (up to some bilipschitz mapping and noise). In the three-figures common variable problem, we cannot assume that we know how to crop the images, but we would like to obtain a result with similar properties.

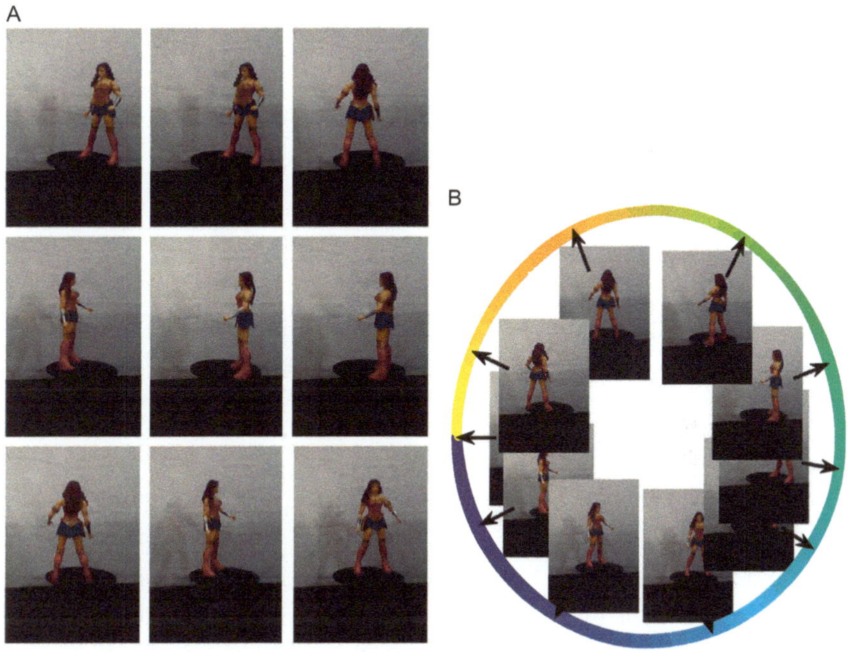

FIG. 2 One object example: if we could crop the images and extract the common object as viewed from Camera 1, we would get images like those in (A). Application of Diffusion Maps to these images yields the embedding in (B). Alternating Diffusion does not assume that we know enough about the data to crop the images, or even that there exists a simple projection operation with these properties. Our goal would be to obtain similar results without such prior knowledge about the problem.

Moreover, in general, such a cropping might not be possible; the pixels in the observations could consist of a combination of the hidden variables. In other words, we would like to obtain a parametrization that is analogous to the parametrization obtained if we had measurements that were only a function of X, ignoring Y and Z. We will see that the key to this analysis is having pairs of images that are taken simultaneously.

To understand the difficulty in extracting X from the data, let us first analyze the data from a single camera, ignoring the other camera. Diffusion Maps of the images from Camera 1 (ignoring the images from Camera 2) are presented in Fig. 3. The true value of X for the image represented by each point is illustrated by the colour assigned to each point. The key property to observe in this figure is that images with very similar values of X (i.e., similar colours) are mapped to very distant points. This means that these maps do not ignore Y, and do not identify the proper equivalence classes. Note that in some cases, X is sufficiently dominant to stand out in this map, but this is not the case in general.

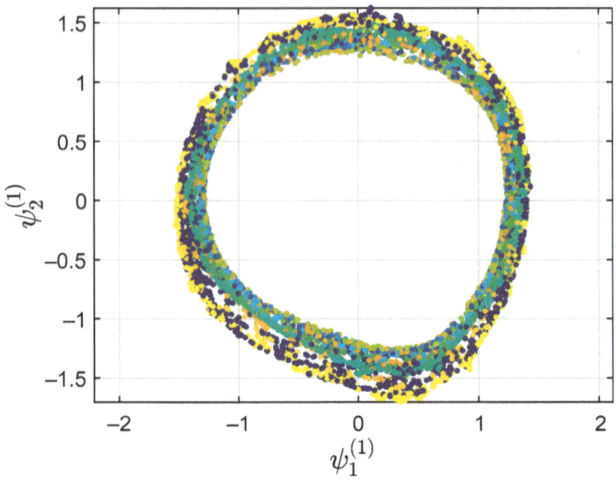

FIG. 3 Diffusion Maps embedding of the data from Camera 1. The colour corresponds to the true value of X of the image corresponding to each point.

Various Manifold Learning methods and spectral graph methods can be very loosely interpreted as comparing sets of neighbours (i.e., nodes reached in a small number of hops): if two nodes in the graph have similar sets of neighbours, then the nodes are considered similar. To illustrate the idea behind AD, we consider *indirect neighbours*, as depicted in Fig. 4. We start with the single realization—a pair of simultaneous images at the top of the tree, and consider its two nearest neighbours *based only on Camera 1*; these two nearest neighbours (pairs of images) are presented in the second row of Fig. 4. Naturally, the images from Camera 1 in the second row are very similar to the image from Camera 1 in the root, because we chose the most similar images. In particular, WF (the variable X) is in roughly the same state in all the images. However, we note that the images from Camera 2 are very different: SF is in a different state in each image. In other words, in our first step from the root to the second row, we obtained a set of pairs of images that all have roughly the same values of X and Y, but going from the root to the second row, we lost track of Z (SF). Next, we consider each of the two pairs in the second row and find their nearest neighbours *based only on Camera 2*. The resulting set of pairs is presented in the third row of the tree. We note that each node in the third row is very similar to its parent in Camera 2, so that WF (X) and SF (Z) are in a similar state to those in the parent node; but BF (Y) is now in a very different state. Therefore, the pairs of images in the third row, in the second row and in the root share the same state of WF (X), but the pairs in the third row exhibit very different states of BF (Y) and SF (Z). We refer to the pairs in the third row (and beyond) as the indirect neighbours of the root. We can continue the process of finding further indirect neighbours by

FIG. 4 Illustration of Alternating Diffusion using a tree of indirect neighbours (see text for details).

alternating between distances in Camera 1 and distances in Camera 2. After several iterations, we obtain a set of pairs of images that all exhibit a very similar state of WF, but lose track of BF and SF. Comparing the sets of pairs obtained for different roots reveals how similar these roots are in the state of WF (X), almost as if the other figures were not there.

The AD algorithm (see Section 3.3) uses a slightly different process of considering different probabilities of hopping to different images based on their similarities, but it is based on the same idea of propagation that alternates between distances from different modalities, essentially "forgetting" the original state of Y and Z.

3.3 Algorithm

The algorithm begins by applying a simplified version of Diffusion Maps, separately to the two subsets of observations $\{s_i^{(1)}\}_{i=1}^N$ and $\{s_i^{(2)}\}_{i=1}^N$. For each subset $\ell = 1, 2$, we construct the affinity matrix $\mathbf{W}^{(\ell)} \in \mathbb{R}^{N \times N}$, given by

$$W_{j,k}^{(\ell)} = \exp\left(-\frac{\|s_j^{(\ell)} - s_k^{(\ell)}\|^2}{\epsilon_\ell^2}\right), \tag{9}$$

where $\epsilon_\ell > 0$ is a scale parameter and $\|\|$ is a norm that could be non-Euclidean and observation specific. By dividing the elements of each column by its sum, we construct the column stochastic diffusion operators $\mathbf{P}^{(1)}$ and $\mathbf{P}^{(2)}$. Note that these operators are the "forward" operators, in contrast to the "backward" diffusion operator defined in (2).

From each sample, a diffusion flow is defined by alternating applications of $\mathbf{P}^{(1)}$ and $\mathbf{P}^{(2)}$. Specifically, the flow from the ith sample is defined as a sequence of vectors $\{v_t\}_{t=0}^{\infty}$ such that

$$v_{t+1} = \begin{cases} \mathbf{P}^{(1)} v_t, & t = 2m \\ \mathbf{P}^{(2)} v_t, & t = 2m+1 \end{cases} \tag{10}$$

for every integer $m \geq 0$, where the initial vector $v_0 = (0, 0, \ldots, 0, 1, 0, \ldots)^T$ is a vector of dimensionality N, of all zeros, except for the ith position.

It follows that the even steps of the propagation defined in (10) can be recast as

$$v_{2m} = \mathbf{P}^m v_0, \tag{11}$$

where $\mathbf{P}$ is termed the *AD operator*, given by

$$\mathbf{P} = \mathbf{P}^{(2)} \mathbf{P}^{(1)}. \tag{12}$$

The *AD distance* between sample i and sample j is defined as the following Euclidean distance

$$d_t(i, j) = \|v_t - u_t\|_2, \tag{13}$$

ALGORITHM 1 Alternating Diffusion

Input: Two corresponding discrete datasets $\{s_k^{(1)}\}_{k=1}^N$ and $\{s_\ell^{(2)}\}_{\ell=1}^N$
Output: The AD operator **P** and AD distances $d_t(k, \ell)$.

1. Construct the affinity matrices of the two datasets
$$W_{k,\ell}^{(1)} = \exp\left(-\frac{\| s_k^{(1)} - s_\ell^{(1)} \|^2}{\epsilon_1^2}\right),\quad W_{k,\ell}^{(2)} = \exp\left(-\frac{\| s_k^{(2)} - s_\ell^{(2)} \|^2}{\epsilon_2^2}\right), \tag{14}$$
where $\epsilon_1 > 0$ and $\epsilon_2 > 0$ are the kernel bandwidths, commonly set to the median of the distances.
2. Based on the affinity matrices, the "forward" operator (the column stochastic diffusion kernel) is constructed as follows
$$\mathbf{P}^{(i)} = \mathbf{W}^{(i)}\left(\mathbf{D}^{(i)}\right)^{-1}, \tag{15}$$
where $i = 1, 2$ and $\mathbf{D}^{(i)}$ is a diagonal matrix with $d_{k,k} = \sum_{\ell=1}^N W_{k,\ell}^{(i)}$ on its diagonal.
3. The AD operator **P** is constructed according to
$$\mathbf{P} = \mathbf{P}^{(2)}\mathbf{P}^{(1)}. \tag{16}$$
4. The AD distances are defined for $t = 2m$, where m is a positive integer, as the Euclidean distances between the columns of $\mathbf{P}^m$, that is
$$d_t(k, \ell) = \| \mathbf{P}^m[:, k] - \mathbf{P}^m[:, \ell] \|_2, \tag{17}$$
where $k, \ell = 1, \ldots, N$.

where v_t and u_t are vectors in the propagation sequences (defined in (10)) from the sample i and the sample j, respectively. Note that $\mathbf{P}^m v_0$ is the ith column of $\mathbf{P}^m$ and that $\mathbf{P}^m u_0$ is the jth column of $\mathbf{P}^m$. Consequently, the AD distance (13) can be computed using the columns of the matrix $\mathbf{P}^m$.

In Lederman and Talmon (2018), it was shown that the AD operator **P** is related to an "effective" diffusion operator that extracts the structure of the common variable X and ignores the observation-specific variables Y and Z. It was further shown that the AD distance (13) is related to the classical diffusion distance (Coifman and Lafon, 2006) computed from data where the common variable is the only source of variability. In the sequel, we will briefly describe the analysis leading to the above results.

To gain further intuition about AD, let us revisit first the construction of the classical Diffusion Maps (Coifman and Lafon, 2006). The application of Diffusion Maps to the subset $\{s_i^{(1)}\}_{i=1}^N$ constructs a Markov chain defined on the graph $G^{(1)} = (V, E^{(1)})$. Each of the N samples is represented by one node in $V = \{1, \ldots, N\}$. The weight of the edge $e_{ij}^{(1)} \in E^{(1)}$ between node i and node j is determined by $W_{i,j}^{(1)}$. Consequently, the diffusion operator $\mathbf{P}^{(1)}$

can be viewed as the transition probability matrix on this graph, where $P^{(1)}_{i,j}$ is the probability of transition to node i from node j in a single step.

Similarly, a separate application to the subset $\{s_i^{(2)}\}_{i=1}^N$ constructs a Markov chain with the transition probability matrix $\mathbf{P}^{(2)}$ on the graph $G^{(2)} = (V, E^{(2)})$. In the end, we have two graphs with the same set of nodes V and two different sets of weighted edges $E^{(1)}$ and $E^{(2)}$, inducing two different transition probability matrices $\mathbf{P}^{(1)}$ and $\mathbf{P}^{(2)}$, which are determined by the distances in each subset of observations.

The procedure of AD, defined in (10), constructs a Markov chain on the same set of nodes V, similarly to the two classical applications of Diffusion Maps above. The difference is that the transition probabilities are defined as follows: the transition probability in odd steps is $\mathbf{P}^{(1)}$ and the transition probability in even steps is $\mathbf{P}^{(2)}$. Consequently, two consecutive steps can be considered as a Markov chain on a new, directed, graph $G = (V, E)$, where the transition probabilities matrix is $\mathbf{P}$.

To illustrate AD, let us revisit the toy problem described in Section 3.2, and specifically, the tree depicted in Fig. 4. Suppose the pair of images at the root of the tree is the ith observation, represented by a vector v_0 of all zeros but the ith entry which equals 1. The first step of AD is the application of $\mathbf{P}^{(1)}$ to v_0, that is, $v_1 = \mathbf{P}^{(1)}v_0$. This application can be viewed as a "soft" nearest neighbours search, where the neighbours are indicated by the nonzero entries of v_1. Specifically, the two pairs corresponding to the two indices with the largest values at v_1 are depicted at the second row of Fig. 4. This soft nearest neighbours search of $v_1 = \mathbf{P}^{(1)}v_0$ is an alternative interpretation to the classical diffusion interpretation, in which the "mass" concentrated solely at the ith entry in v_0 is diffused to other entries in v_1. At the second step of AD, $\mathbf{P}^{(2)}$ is applied to v_1, yielding $v_2 = \mathbf{P}^{(2)}v_1 = \mathbf{P}^{(2)}\mathbf{P}^{(1)}v_0$. This step can be seen as another soft neighbours search, but this time neighbours of the neighbours identified at the first step; or in other words, indirect (soft) neighbours of the pair at the root. These indirect neighbours are conveyed by the nonzero entries of v_2, where the neighbours that attain maximal values are depicted at the third row of Fig. 4. In the end, after t alternating applications of $\mathbf{P}^{(1)}$ and $\mathbf{P}^{(2)}$, as described in the beginning of this section, by traditional nearest neighbours search, the nonzeros entries of v_t indicate pairs of images in which the common figure (WF) remains coherent, that is, at a similar orientation, whereas the two observation-specific figures (BF and SF) are in noncoherent arbitrary orientations. Comparing such vectors v_t obtained for different roots reveals how similar these roots are only in terms of the state of WF (X), ignoring the other figures (Y and Z).

3.4 Common manifold interpretation

In this section we will discuss AD in simplified special settings. Suppose there exists a common structure underlying two observations, which is modelled by a low-dimensional *common manifold*, and suppose that each observation introduces various deformations and interferences, which are modelled by other irrelevant nuisance structures. Mathematically, let $\mathcal{M}$ denote a d-dim Riemannian manifold with the metric tensor $g^{(i)}$, which depends on the observation $i = 1, 2$, and let $d_{g^{(i)}}$ denote the distance function on $\mathcal{M}$ associated with $g^{(i)}$. Allowing the metrics to be different, i.e., $g^{(1)} \neq g^{(2)}$ accounts for possible discrepancies and deformations between the two observations of the common manifold $\mathcal{M}$. Let $\mathcal{N}_i$ be a compact metric space with the distance function $d_{\mathcal{N}_i}$, which models the observation-specific factors. Let $\mathcal{S}_i$ denote the *observable* metric space with the distance function $d_{\mathcal{S}_i}$, which models the space of the accessible data by the ith observation. Assume that

$$s^{(i)} : \mathcal{M} \times \mathcal{N}_1 \times \mathcal{N}_2 \to \mathcal{S}_i \tag{18}$$

is a smooth isometric embedding of $\mathcal{M} \times \mathcal{N}_i$ into $\mathcal{S}_i$. The observation-specific variables are modelled by the fact that $s^{(1)}$ ignores $\mathcal{N}_2$ and that $s^{(2)}$ ignores $\mathcal{N}_1$. In other words, each observation embodies two structures: a deformation of the common manifold of interest $\mathcal{M}$ and an interference component. Note that these are latent structures which are not directly accessible.

The observations model (18) implies that for $(x, y, z), (x', y', z') \in \mathcal{M} \times \mathcal{N}_1 \times \mathcal{N}_2$, where $x, x' \in \mathcal{M}$, $y, y' \in \mathcal{N}_1$ and $z, z' \in \mathcal{N}_2$, we have

$$\begin{aligned} d_{\mathcal{S}_1}(s^{(1)}(x, y, z), s^{(1)}(x', y', z'))^2 &= d_{g^{(1)}}(x, x')^2 + d_{\mathcal{N}_1}(y, y')^2 \\ d_{\mathcal{S}_2}(s^{(2)}(x, y, z), s^{(2)}(x', y', z'))^2 &= d_{g^{(2)}}(x, x')^2 + d_{\mathcal{N}_2}(z, z')^2. \end{aligned} \tag{19}$$

In practice, this metric decomposition is unavailable as well, and we have access only to the observable distance functions $d_{\mathcal{S}_i}$.

The discrete datasets of observations are modelled by realizations of random variables defined on the manifold structure. Formally, let $(\Omega, \mathcal{F}, P)$ be a probability space, where Ω is the event space, $\mathcal{F}$ is the sigma algebra on Ω, and P is a probability measure defined on $\mathcal{F}$. Consider a random vector $S : (\Omega, \mathcal{F}, P) \to \mathcal{M} \times \mathcal{N}_1 \times \mathcal{N}_2$. The dataset of the ith observation is modelled by realization of the random vector $S_i = s^{(i)} \circ S$. Let $\nu := \nu_{\mathcal{M} \times \mathcal{N}_1 \times \mathcal{N}_2} = S_* P$ denote the induced probability measure on $\mathcal{M} \times \mathcal{N}_1 \times \mathcal{N}_2$. By the assumption that conditional on $\mathcal{M}$ the observation-specific variables are independent, we have

$$\nu(x, y, z) = \nu_{\mathcal{M}}(x) \nu_{\mathcal{N}_1|\mathcal{M}}(y|x) \nu_{\mathcal{N}_2|\mathcal{M}}(z|x), \tag{20}$$

where $\nu_{\mathcal{M}}(x)$ is the marginal distribution on $\mathcal{M}$, and $\nu_{\mathcal{N}_i|\mathcal{M}}(\cdot|x)$ is the conditional distribution on $\mathcal{N}_i$, such that

$$\int_{\mathcal{N}_i} \mathrm{d}\nu_{\mathcal{N}_i|\mathcal{M}}(\cdot|x) = 1. \tag{21}$$

Based on the common manifold model, we apply the AD scheme to analyze data collected simultaneously from two observations, possibly of different modalities.

3.4.1 Observable operators

Definition 1 (Observable diffusion kernels).
Let $\tilde{P}^{(i)} \in C([0, \infty))$, $i = 1, 2$, be two kernels that decay sufficiently fast and are associated with the two observations. Define

$$\tilde{P}_\epsilon^{(i)}(o, o') := \tilde{P}^{(i)}\left(\frac{d_{S_i}(s^{(i)}(o), s^{(i)}(o'))}{\sqrt{\epsilon}}\right) \tag{22}$$

$$P_\epsilon^{(i)}(o, o') := \frac{\tilde{P}_\epsilon^{(i)}(o, o')}{\int_{\mathcal{M}\times\mathcal{N}_1\times\mathcal{N}_2} \tilde{P}_\epsilon^{(i)}(o, o'')\mathrm{d}\nu(o'')}, \tag{23}$$

where $o = (x, y, z)$ and $o' = (x', y', z')$.

In other words, the observable diffusion kernel $P_\epsilon^{(i)}$ is a *diffusion kernel with bandwidth* ϵ associated with the observable metric measure space $\mathcal{S}_i = s^{(i)}(\mathcal{M}\times\mathcal{N}_1\times\mathcal{N}_2)$. While the bandwidth may depend on the observation depending on the chosen metric g, for notational simplicity, we assume that the kernels share the same ϵ value. Note that $P_\epsilon^{(i)}$ is normalized, such that

$$\int_{\mathcal{M}\times\mathcal{N}_1\times\mathcal{N}_2} P_\epsilon^{(i)}(o, o')\mathrm{d}\nu(o') = 1.$$

which circumvents unwanted nonintrinsic quantities (Coifman and Lafon, 2006).
Definition 2 (Observable diffusion operators).
For a function $f \in C(\mathcal{M}\times\mathcal{N}_1\times\mathcal{N}_2)$, let $D^{(i)} : C(\mathcal{M}\times\mathcal{N}_1\times\mathcal{N}_2) \to C(\mathcal{M}\times\mathcal{N}_1\times\mathcal{N}_2)$ denote the diffusion operator on the ith observation, defined by

$$(D^{(i)}f)(o) := \int_{\mathcal{M}\times\mathcal{N}_1\times\mathcal{N}_2} P_\epsilon^{(i)}(o, o')f(o')\mathrm{d}\nu(o'), \tag{24}$$

where $o = (x, y, z)$ and $o' = (x', y', z')$.

Based on the observable kernels and operators, we define next the *AD* kernel and operator, which can be computed directly from the observations.
Definition 3 (Alternating Diffusion kernel and operator).
Let P_ϵ be the AD kernel, defined by

$$P_\epsilon(o, o'') := \int_{\mathcal{M}\times\mathcal{N}_1\times\mathcal{N}_2} P_\epsilon^{(2)}(o, o')P_\epsilon^{(1)}(o', o'')\mathrm{d}\nu(o').$$

where $o = (x, y, z)$, $o' = (x', y', z')$, and $o'' = (x'', y'', z'')$.

The AD operator is defined by

$$(Df)(o) := \int_{\mathcal{M}\times\mathcal{N}_1\times\mathcal{N}_2} P_\epsilon(o, o'')f(o'')\mathrm{d}\nu(o''). \tag{25}$$

Note that by definition, we have

$$D := D^{(2)}D^{(1)}, \tag{26}$$

3.4.2 *Effective operators*

To show that in effect AD extracts only the common variables, we define an auxiliary operator that marginalizes the (latent) observation-specific variables.

Definition 4 (Marginalization operator).
Let $\mathcal{E}: C(\mathcal{M}\times\mathcal{N}_1\times\mathcal{N}_2)\to C(\mathcal{M})$ be a marginalization operator, which is defined by

$$(\mathcal{E}f)(x) := \int_{\mathcal{N}_1\times\mathcal{N}_2} f(x,y,z)\mathrm{d}\nu_{\mathcal{N}_1|\mathcal{M}}(y|x)\mathrm{d}\nu_{\mathcal{N}_2|\mathcal{M}}(z|x). \tag{27}$$

The marginalization operator $\mathcal{E}$ is a bounded linear operator, which evaluates the marginal distribution of the observations. On the one hand, it performs the desirable operation—attenuating the observation-specific variables while preserving the common variables. On the other hand, it cannot be computed directly from the observations, since we do not have access to the hidden manifold structure and to the latent variables. In the sequel, we show that AD, which is computed from the observations, attains an equivalent operation.

We observe that $\tilde{P}_\epsilon^{(1)}((x, y, z),(x', y', z'))$, defined in (22) is not a function of z and z' because $s^{(1)}$ only sees $\mathcal{M}\times\mathcal{N}_1$:

$$\begin{aligned}\tilde{P}_\epsilon^{(1)}((x, y, z),(x', y', z')) &= \tilde{P}_\epsilon^{(i)}\left(\frac{d_{\mathcal{S}_1}(s^{(1)}(x, y, z), s^{(1)}(x', y', z'))}{\sqrt{\epsilon}}\right)\\ &= \tilde{P}_\epsilon^{(i)}\left(\frac{d_{g^{(1)}}(x, x')^2 + d_{\mathcal{N}_2}(y, y')^2}{\sqrt{\epsilon}}\right).\end{aligned} \tag{28}$$

It follows that $P_\epsilon^{(1)}$, defined in (23) is not a function of z. So, with a mild abuse of notation we can write

$$P_\epsilon^{(1)}((x, y, z),(x', y', z')) = P_\epsilon^{(1)}((x, y),(x', y')). \tag{29}$$

For similar reasons, $P_\epsilon^{(2)}$ is not a function of y, and therefore:

$$P_\epsilon^{(2)}((x, y, z),(x', y', z')) = P_\epsilon^{(2)}((x, z),(x', z')). \tag{30}$$

Consequently, $\left(D^{(1)}f\right)(x, y, z)$, defined in (24) is constant in z (for a given x, y), and $\left(D^{(2)}f\right)(x, y, z)$ is constant in y.

Next, we define the *effective* counterparts of the diffusion kernels and operators.

Definition 5 (Effective diffusion kernel and operator).
Let $P_\epsilon^{(e_i)}(x, x')$ be the effective diffusion kernel of the ith observation define on $\mathcal{M}$ by

$$P_\epsilon^{(e_1)}(x, x') := \int_{\mathcal{N}_1}\int_{\mathcal{N}_1} P_\epsilon^{(1)}((x, y), (x', y'))\mathrm{d}\nu_{\mathcal{N}_1|\mathcal{M}}(y|x)\mathrm{d}\nu_{\mathcal{N}_1|\mathcal{M}}(y'|x'), \quad (31)$$

$$P_\epsilon^{(e_2)}(x, x') := \int_{\mathcal{N}_2}\int_{\mathcal{N}_2} P_\epsilon^{(2)}((x, z), (x', z'))\mathrm{d}\nu_{\mathcal{N}_2|\mathcal{M}}(z|x)\mathrm{d}\nu_{\mathcal{N}_2|\mathcal{M}}(z'|x'). \quad (32)$$

and let $D^{(e_i)}$ be the effective diffusion operator of the ith sensor, defined by

$$D^{(e_i)}\mathcal{E}f(x) := \int_{\mathcal{M}} P_\epsilon^{(e_i)}(x, x')(\mathcal{E}f)(x')\mathrm{d}\nu_{\mathcal{M}}(x'), \quad (33)$$

where $f \in C(\mathcal{M} \times \mathcal{N}_1 \times \mathcal{N}_2)$.

Definition 6 (Effective AD kernel and operator).
Let $P_\epsilon^{(e)}$ be the effective AD kernel, defined by

$$P_\epsilon^{(e)}(x, x'') := \int_{\mathcal{M}} P_\epsilon^{(e_2)}(x, x')P_\epsilon^{(e_1)}(x', x'')\mathrm{d}\nu_{\mathcal{M}}(x'). \quad (34)$$

and let $D^{(e)}$ be the effective AD operator, defined by

$$D^{(e)} := D^{(e_2)}D^{(e_1)}. \quad (35)$$

The main result in this analysis claims that after marginalization, AD applied directly to the two observations is, in effect, equivalent to a diffusion operator applied to the hidden common manifold.

Theorem 1. *If $f \in C(\mathcal{M} \times \mathcal{N}_1 \times \mathcal{N}_2)$ is constant on $\mathcal{N}_i$, then*

$$\mathcal{E}[D^{(i)}f] = D^{(e_i)}\mathcal{E}f, \quad (36)$$

for $i = 1, 2$. Furthermore, if f is constant on $\mathcal{N}_1$, we have

$$\mathcal{E}[Df] = D^{(e)}\mathcal{E}f. \quad (37)$$

For the proof and more details, see Talmon and Wu (2019).

In other words, the marginalization operator and the AD operator commute in these settings, in the sense described above. However, importantly, neither the marginalization operator $\mathcal{E}$ nor the effective AD operator $D^{(e)}$ can be directly computed from data.

Consider the eigenvalue decomposition of the observable AD operator $D\phi = \lambda\phi$, where ϕ is an eigenfunction of D associated with the eigenvalue λ.

Then, by the commutativity and the linearity of $\mathcal{E}$, we have for the setting of Theorem 1:

$$D^{(e)}\mathcal{E}\phi = \mathcal{E}D\phi = \mathcal{E}\lambda\phi = \lambda\mathcal{E}\phi. \tag{38}$$

This entails that if λ is an eigenvalue of D with the eigenspace $E_\lambda(D)$, then λ is an eigenvalue of $D^{(e)}$ with the eigenspace $E_\lambda(D^{(e)}) = \{\mathcal{E}\phi | \phi \in E_\lambda(D)\}$.

In practice, we only have access to the eigenfunction $\phi \in E_\lambda(D)$ rather than $\mathcal{E}\phi \in E_\lambda(D^{(e)})$. Yet, ϕ carries information about $\mathcal{E}\phi$. Since D and $D^{(e)}$ are both compact operators, the dimensions of their corresponding eigenspaces $E_\lambda(D)$ and $E_\lambda(D^{(e)})$ are both finite, and for each eigenfunction of $D^{(e)}$ associated with λ, there exists at least one eigenfunction of D associated with λ.

The common manifold framework described above provides additional insight about AD and its spectral interpretation. Next, it will be used to informally describe the main result concerning AD distances (13), presented in Lederman and Talmon (2018). In the next section, we will revisit the spectral analysis of the AD operator and introduce two modified AD operators with much improved spectral properties and theoretical guarantees.

In the context of the current continuous setting, the main result about AD, presented in Lederman and Talmon (2018, Theorem 5), can be informally stated as follows. For observations with the hidden variables $(x, y, z), (x', y', z') \in \mathcal{M} \times \mathcal{N}_1 \times \mathcal{N}_2$, $\delta_{x,y,z}$ a dirac delta at (x, y, z) and δ_x a dirac delta at x,

$$\|D^{t+1}\delta_{x,y,z} - D^{t+1}\delta_{x',y',z'}\|_2 \approx \|(D^{(e)})^t\mathcal{E}\delta_x - (D^{(e)})^t\mathcal{E}\delta_{x'}\|_Q, \tag{39}$$

where t is a positive integer and $\|\|_Q$ is an appropriate norm that appears naturally during the derivation. For the formal derivation and more details, see Lederman and Talmon (2018).

The result in (39), which is the continuous counterpart of the AD distance defined in (13), implies that the AD distance (left-hand side) depends only the common variable x and ignores the observation-specific variables y and z.

We note that the results of AD depend on the order of the kernels consisting the AD kernel; this dependence is manifested in the norm $\|\|_Q$ in (39). This is a natural consequence of not having access to a single objective metric for $\mathcal{M}$ in the most general settings. This aspect will be further discussed in Section 4.

4 Self-adjoint operators for recovering hidden components

In this section, we review an extension of the AD operator, presented in Lederman and Talmon (2018) and Talmon and Wu (2019). The extension is described in a modified setup, consisting two (observable) manifolds with a diffeomorphism between them. The main difference between the two versions is that the diffeomorphism assumed in this section implies that there is a map between the two observable manifolds, whereas in the preceding sections it is assumed that such a map does not exist, because there are nuisance parameters, Y and Z, in addition to X. The extension concerns two aspects of the AD operator presented in Section 3. First, the order of the kernel composition is important and influences the resulting embedding, as described in Lederman and Talmon (2018) and Talmon and Wu (2019). Second, the AD operator is not guaranteed to have the convenient spectral properties of a self-adjoint operator when there is only a finite number of sampling points. These aspects are addressed in Shnitzer et al. (2019) where a new self-adjoint operator is introduced.

In addition, we discuss an antiself-adjoint operator presented in Shnitzer et al. (2019), which, in contrast to the AD operator, is shown to recover differences, instead of the common manifold. This section is organized as follows. In Section 4.1 we describe the general setting and formulation of the problem. In Section 4.2 we present the two operators and the main theoretical results from Shnitzer et al. (2019). In Section 4.3 we describe the algorithm and the properties of the operators in a discrete setting.

4.1 Problem formulation

Consider two different modalities or sensors, where each modality measures data lying on a different manifold. The two modalities are assumed to contain both common properties (structures) and sensor-specific properties, and our goal is to recover and characterize these common structures and the differences between the two manifolds. More concretely, given two diffeomorphic d-dimensional Riemannian manifolds without a boundary, denoted by $\mathcal{M}^{(1)}$ and $\mathcal{M}^{(2)}$, with corresponding metrics, denoted by $g^{(1)}$ and $g^{(2)}$, respectively, and the diffeomorphism $\phi : \mathcal{M}^{(1)} \mapsto \mathcal{M}^{(2)}$, we define the following structures:

$$\Omega_\alpha = \left\{ s^{(1)} \in \mathcal{M}^{(1)} : \ \nabla\phi|_x = \alpha \mathrm{I} \right\} \subset \mathcal{M}^{(1)} \tag{40}$$

$$\Omega_\alpha^c = \mathcal{M}^{(1)} \backslash \mathring{\Omega}_\alpha, \tag{41}$$

where $\nabla\phi|_x$ is represented as a $d \times d$ matrix by choosing proper orthonormal bases at $T_{s^{(1)}}\mathcal{M}^{(1)}$ and $T_{\phi(s^{(1)})}\mathcal{M}^{(2)}$, $\alpha > 0$ denotes scaling, I denotes a $d \times d$ identity matrix and $\mathring{\Omega}_\alpha$ denotes the interior of Ω_α.

Given measurement pairs from the two manifolds, $\left(s^{(1)}, s^{(2)}\right)$ such that $s^{(1)} \in \mathcal{M}^{(1)}$ and $s^{(2)} \in \mathcal{M}^{(2)}$, our goal is to identify and represent Ω_α and Ω_α^c, i.e., the similar structures (up to a scaling α) and the difference structures, respectively, in a data-driven manner.

For example, consider the two shapes in Fig. 5, which presents a sphere (plot (A)) and a scaled and deformed sphere (plot (B)). In this example,

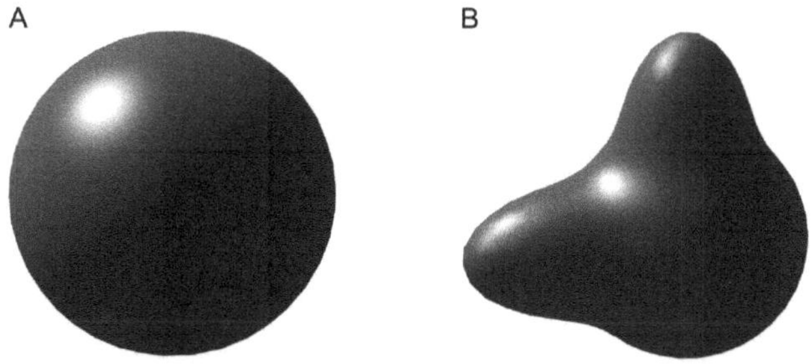

FIG. 5 Two shapes, a sphere (A) and a sphere with two "bumps" (B), which differ by a nonisometric deformation and scaling.

the common structures, Ω_α, are the locations on the sphere which are not deformed in plot (B) and the differences, Ω_α^c, are the deformed locations (the two "bumps").

4.2 Operator definition and analysis

We define the new operators as follows. Based on the given data pairs, for each manifold $\mathcal{M}^{(\ell)}$, $\ell = 1, 2$, we construct the following affinity kernel using the distance function on the manifold, $d_{g^{(\ell)}}$:

$$w_{\epsilon_\ell}^{(\ell)}(s, s') = \exp\left(-\frac{d_{g^{(\ell)}}(s, s')^2}{\epsilon_\ell^2}\right), \tag{42}$$

where $s, s' \in \mathcal{M}^{(\ell)}$, $\epsilon_\ell > 0$ is the kernel bandwidth and $\ell = 1, 2$.

Two "forward" and "backward" kernels are then defined for each manifold according to

$$p_{\epsilon_\ell}^{(\ell)}(s, s') = \frac{w_{\epsilon_\ell}^{(\ell)}(s, s')}{d_{\epsilon_\ell}^{(\ell)}(s')} \tag{43}$$

$$q_{\epsilon_\ell}^{(\ell)}(s, s') = \frac{w_{\epsilon_\ell}^{(\ell)}(s, s')}{d_{\epsilon_\ell}^{(\ell)}(s)}, \tag{44}$$

where $d_{\epsilon_\ell}^{(\ell)}(s) = \int w_{\epsilon_\ell}^{(\ell)}(s, s')\, \mu^{(\ell)}(s')\, dV^{(\ell)}(s')$, $V^{(\ell)}$ is the volume measure induced by the metric $g^{(\ell)}$, and $\mu^{(\ell)}$ is the density function of the points on $\mathcal{M}^{(\ell)}$.

Using these kernels, the following operators are constructed for any function $f \in C^\infty\left(\mathcal{M}^{(\ell)}\right)$:

$$P_{\epsilon_\ell}^{(\ell)} f(s) = \int p_{\epsilon_\ell}^{(\ell)}(s, s') f(s')\, \mu(s')\, dV(s') \tag{45}$$

$$Q_{\epsilon_\ell}^{(\ell)} f(s) = \int q_{\epsilon_\ell}^{(\ell)}(s, s') f(s')\, \mu(s')\, dV(s'). \tag{46}$$

The operators for recovering the common structures and the differences are then constructed as follows, for any function $f \in C^\infty\left(\mathcal{M}^{(1)}\right)$:

$$S_{\epsilon_1,\epsilon_2} f(s) = \frac{1}{2}\left(\phi^* Q^{(2)}_{\epsilon_2} (\phi^*)^{-1} P^{(1)}_{\epsilon_1} + Q^{(1)}_{\epsilon_1} \phi^* P^{(2)}_{\epsilon_2} (\phi^*)^{-1}\right) f(s) \tag{47}$$

$$A_{\epsilon_1,\epsilon_2} f(s) = \frac{1}{2}\left(\phi^* Q^{(2)}_{\epsilon_2} (\phi^*)^{-1} P^{(1)}_{\epsilon_1} - Q^{(1)}_{\epsilon_1} \phi^* P^{(2)}_{\epsilon_2} (\phi^*)^{-1}\right) f(s), \tag{48}$$

where $s \in \mathcal{M}^{(1)}$, $\phi^* : C^\infty(\mathcal{M}^{(2)}) \to C^\infty(\mathcal{M}^{(1)})$ denotes the pullback operator from $\mathcal{M}^{(2)}$ to $\mathcal{M}^{(1)}$, i.e., $(\phi^* g)(s) = g(\phi(s))$, $g \in C^\infty(\mathcal{M}^{(2)})$, and $(\phi^*)^{-1}$ denotes the pullback from $\mathcal{M}^{(1)}$ to $\mathcal{M}^{(2)}$, which inverts ϕ^*, since ϕ^* is assumed to be diffeomorphic.

The asymptotic expansions of $S_{\epsilon_1,\epsilon_2}$ and $A_{\epsilon_1,\epsilon_2}$ are presented in the following proposition.

Proposition 1. *When $\epsilon_1, \epsilon_2 > 0$ are sufficiently small and $\mu^{(1)}$ is smooth enough, the asymptotic expansions of the operators $S_{\epsilon_1,\epsilon_2}$ and $A_{\epsilon_1,\epsilon_2}$ are given by*

$$\begin{aligned} S_{\epsilon_1,\epsilon_2} f(s) =& f(s) - \epsilon_1^2 \Delta^{(1)} f(s) - \epsilon_2^2 \phi^* \Delta^{(2)} (\phi^*)^{-1} f(s) \\ & - \frac{\epsilon_2^2}{2}\left(\phi^* \frac{2\nabla^{(2)}(\phi^*)^{-1} f \cdot \nabla^{(2)} \mu^{(2)}}{\mu^{(2)}}(s) - f\phi^* \frac{\Delta^{(2)}\mu^{(2)}}{\mu^{(2)}}(s)\right) \\ & - \frac{\epsilon_1^2}{2}\left(\frac{2\nabla^{(1)} f \cdot \nabla^{(1)}\mu^{(1)}}{\mu^{(1)}}(s) - \frac{f\Delta^{(1)}\mu^{(1)}}{\mu^{(1)}}(s)\right) + O(\epsilon_1^4 + \epsilon_2^4) \end{aligned} \tag{49}$$

$$\begin{aligned} A_{\epsilon_1,\epsilon_2} f(x) =& \frac{\epsilon_1^2}{2}\left(\frac{2\nabla^{(1)} f \cdot \nabla^{(1)}\mu^{(1)}}{\mu^{(1)}}(s) + \frac{f\Delta^{(1)}\mu^{(1)}}{\mu^{(1)}}(s)\right) \\ & - \frac{\epsilon_2^2}{2}\left(\phi^* \frac{2\nabla^{(2)}(\phi^*)^{-1} f \cdot \nabla^{(2)} \mu^{(2)}}{\mu^{(2)}}(s) + f\phi^* \frac{\Delta^{(2)}\mu^{(2)}}{\mu^{(2)}}(s)\right) + O(\epsilon_1^4 + \epsilon_2^4). \end{aligned} \tag{50}$$

Note that for better readability, we omitted two manifold specific constants in the asymptotic expansions of the operators, multiplying the zero-order and second-order terms. These constants are omitted in the following proposition as well.

For comparison, we present the asymptotic expansion of the AD operator in the problem setting described in Section 4.1.

Proposition 2. *When $\epsilon_1, \epsilon_2 > 0$ are sufficiently small, the asymptotic expansion of the AD operator is given by*

$$\begin{aligned} P_{\epsilon_1,\epsilon_2} f(s) =& f(s) - \epsilon_1^2 \Delta^{(1)} f(s) - \epsilon_2^2 \phi^* \Delta^{(2)} (\phi^*)^{-1} f(s) \\ & - \epsilon_1^2 \frac{2\nabla^{(1)} f \cdot \nabla^{(1)}\mu^{(1)}}{\mu^{(1)}}(s) - \epsilon_2^2 \phi^* \frac{2\nabla^{(2)}(\phi^*)^{-1} f \cdot \nabla^{(2)} \mu^{(2)}}{\mu^{(2)}}(s) + O(\epsilon^4). \end{aligned} \tag{51}$$

The proofs of Propositions 1 and 2, and an analysis of a special case where the densities $\mu^{(1)}$ and/or $\mu^{(2)}$ are uniform can be found in Shnitzer et al. (2019).

Note that the first terms in the asymptotic expansions of the operator $S_{\epsilon_1,\epsilon_2}$ and the AD operator, in (49) and (51), respectively, are similar, and comprise a sum of two Laplace–Beltrami operators, corresponding to the two manifolds, and the function itself, $f(s)$. The difference between the two operators is in the remaining elements of the asymptotic expansion. The asymptotic expansion of $S_{\epsilon_1,\epsilon_2}$ contains components which relate to the two manifolds in a symmetric manner leading to a self-adjoint operator, in contrast to the AD operator (for a proof that the AD operator is not self-adjoint see the supplementary material in Shnitzer et al. (2019).

Proposition 3. *Suppose* $\epsilon_2 = \alpha\epsilon_1$ *for some* $\alpha > 0$. *The operators* $A_\alpha : C^\infty\left(\mathcal{M}^{(1)}\right) \mapsto C^\infty\left(\mathcal{M}^{(1)}\right)$ *and* $S_\alpha : C^\infty\left(\mathcal{M}^{(1)}\right) \mapsto C^\infty\left(\mathcal{M}^{(1)}\right)$ *are antiself-adjoint and self-adjoint, respectively, where* $A_\alpha = \lim_{\epsilon_1\to 0} A_{\epsilon_1,\alpha\epsilon_1}/\epsilon_1^2$ *and* $S_\alpha = \lim_{\epsilon_1\to 0} S_{\epsilon_1,\alpha\epsilon_1}/\epsilon_1^2$.

For the proof, see Shnitzer et al. (2019).

When the density functions of the two manifolds are uniform (e.g., when $\mu^{(1)}$ is uniform and the diffeomorphism is volume preserving), the asymptotic expansions of the operator $S_{\epsilon_1,\epsilon_2}$ and the AD operator are reduced to the sum of the two Laplace–Beltrami operators corresponding to the two manifolds. This asymptotic expansion coincides with the analysis presented in Froyland (2015), which proposed an operator that is analogous to the AD operator and to $S_{\epsilon_1,\epsilon_2}$. There, the operator is applied to dynamical systems for recovering coherent sets, which can be viewed as the common structures in different time frames. The similarity of the asymptotic expansions of the operators, presented here and in Froyland (2015), strengthens our claim that the operator $S_{\epsilon_1,\epsilon_2}$ recovers the common structures of the two manifolds.

In addition, in the case of uniform densities, the asymptotic expansion of the operator $A_{\epsilon_1,\epsilon_2}$ contains only fourth-order terms and above since the second-order terms in (50) vanish. This indicates that the operator $A_{\epsilon_1,\epsilon_2}$ depends mostly on the diffeomorphism and the density functions of the two manifolds. Moreover, the following proposition holds

Proposition 4. *Given* $f \in C^\infty\left(\mathcal{M}^{(1)}\right)$, *if* $\operatorname{supp} f \subset \mathring{\Omega}_\alpha$, *then* $A_\alpha f(s) = 0$.

Note that a direct result of this proposition is that the eigenfunctions of the operator A_α, corresponding to nonzero eigenvalues, are supported on Ω_α^c, i.e., the differences between the two manifolds, since $A_\alpha \psi_n^A = \lambda_n^A \psi_n^A$, where ψ_n^A and λ_n^A denote the eigenfunctions and eigenvalues of the operator A_α, respectively. For the proof, see Shnitzer et al. (2019).

4.3 Discrete setting

Given pairs of samples from two discrete datasets, $\{s_k^{(1)}, s_k^{(2)}\}_{k=1}^N$, which are samples of the two manifolds $\mathcal{M}^{(1)}$ and $\mathcal{M}^{(2)}$, the construction of the discrete counterparts of the operators S_α and A_α, defined in (47) and (48), is detailed in Algorithm 2. In addition, two new representations for the two datasets can be obtained from these discrete operators, which capture the common structures and the difference structures.

ALGORITHM 2 Construction of the operators A and S in the discrete setting

Input: Two corresponding discrete datasets $\{s_k^{(1)}\}_{k=1}^N$ and $\{s_\ell^{(2)}\}_{\ell=1}^N$

Output: The operators **A** and **S** along with a set of coordinates $\{\phi_j^A(k)\}_{j=1}^m$ and $\{\phi_j^S(k)\}_{j=1}^m$, $k = 1, \ldots, N$, representing the two datasets and highlighting their differences and their common structures, respectively.

1. Construct the affinity matrices of the two datasets

$$W_{k,\ell}^{(1)} = \exp\left(-\frac{\|s_k^{(1)}, s_\ell^{(1)}\|^2}{\epsilon_1^2}\right), \quad W_{k,\ell}^{(2)} = \exp\left(-\frac{\|s_k^{(2)}, s_\ell^{(2)}\|^2}{\epsilon_2^2}\right) \tag{52}$$

   where ϵ_1 and ϵ_2 are the kernel bandwidths, commonly set to the median of the distances.

2. Based on the affinity matrices, the column stochastic and row stochastic matrices are constructed as follows

$$\mathbf{P}^{(i)} = \mathbf{W}^{(i)}\left(\mathbf{D}^{(i)}\right)^{-1}, \quad \mathbf{Q}^{(i)} = \left(\mathbf{D}^{(i)}\right)^{-1}\mathbf{W}^{(i)} \tag{53}$$

   where $i = 1, 2$ and $\mathbf{D}^{(i)}$ is a diagonal matrix with $d_{k,k} = \sum_{\ell=1}^N W_{k,\ell}^{(i)}$ on its diagonal.

3. The discrete operators **A** and **S** are constructed according to

$$\mathbf{A} = \mathbf{Q}^{(1)}\mathbf{P}^{(2)} - \mathbf{Q}^{(2)}\mathbf{P}^{(1)} \tag{54}$$

$$\mathbf{S} = \mathbf{Q}^{(1)}\mathbf{P}^{(2)} + \mathbf{Q}^{(2)}\mathbf{P}^{(1)} \tag{55}$$

4. To obtain a new representation for the data, the eigenvalue decompositions of **A** and **S** are computed, $\mathbf{A}\psi_n^A = \lambda_n^A\psi_n^A$ and $\mathbf{S}\psi_n^S = \lambda_n^S\psi_n^S$, respectively. The eigenvalue decomposition of **S** results in a set of N real eigenvectors with real eigenvalues and the decomposition of **A** results in a set of N complex eigenvectors with purely imaginary eigenvalues. For **A**, the first $m/2$ eigenvectors, corresponding to the largest nonconjugate eigenvalues (in absolute value) are taken and the new representation for each pair of points, $\{s_k^{(1)}, s_k^{(2)}\}$, is constructed using the real and imaginary parts of these eigenvectors. For **S**, the new representation is constructed based on the first m eigenvectors corresponding to the largest eigenvalues.

$$\phi^A(k) = \left[\mathrm{real}\{\psi_{1:m/2}^A(k)\}, \mathrm{imag}\{\psi_{1:m/2}^A(k)\}\right], \; \phi^S(k) = \left[\psi_{1:m}^S(k)\right] \tag{56}$$

   where $k = 1, \ldots, N$.

In the discrete setting, the constructed operators **S** and **A** are symmetric and antisymmetric, respectively, and the spectral theorem holds. Therefore, two new orthogonal representations for the common structures in the datasets and for the differences between the two datasets are constructed based on the eigenvectors of the two operators, as described in Algorithm 2, clause 4. Note that since the operator **A** is antisymmetric, its eigenvectors are complex. It was shown in Shnitzer et al. (2019) that the real and imaginary parts of these eigenvectors are orthogonal and, therefore, are used as separate coordinates in the new representation. Additional motivation for these embedding choices appears in Shnitzer et al. (2019).

To summarize this section we cite the following proposition, which states that the discrete operator **A** is related to the differences between the two datasets.

Proposition 5. *Define* $V_{\Omega_\alpha^c} = \{i \mid s_i^{(1)} \in \Omega_\alpha^c\}$, *the discrete operator* **A** has the following properties:

1. $A_{i,j} \neq 0$ *only when* $i \in V_{\Omega_\alpha^c}$ *or* $j \in V_{\Omega_\alpha^c}$.
2. *The rank of* **A** is bounded by $2|V_{\Omega_\alpha^c}|$.

For the proof, see Shnitzer et al. (2019).

5 Applications

In this section, we apply the operators **A** and **S** to three problems: shape analysis, noninvasive foetal heart rate detection and sleep dynamics visualization and annotation. We demonstrate that the operator **A** recovers the differences between modalities and can be used to construct a new meaningful representation that emphasizes these differences, in Sections 5.1 and 5.2. The properties of the operators **S** and **A** are demonstrated in the sleep dynamics application, in Section 5.3. An additional application has been presented in Section 3.2.

5.1 Shape analysis

Consider two shapes, a sphere and a deformed sphere, presented in Fig. 5A and B, respectively. The deformation between the two shapes consists of a nonisometric deformation (two "bumps") and scaling, which was added to emphasize that our method is not sensitive to scale differences. We apply the operator **A** to these shapes and show that it recovers the differences between the two shapes.

We construct the operator **A** according to Algorithm 2, where the points on the sphere (Fig. 5A) are denoted by $s_k^{(1)}$ with $k = 1 \ldots N$ and the points on the deformed sphere (Fig. 5B) are denoted by $s_k^{(2)}$ with $k = 1 \ldots N$ and $N = 1922$. Here, the affinity matrices in Algorithm 2 were constructed using the geodesic distances on the two shapes and ϵ_1 and ϵ_2 were empirically set to be the median

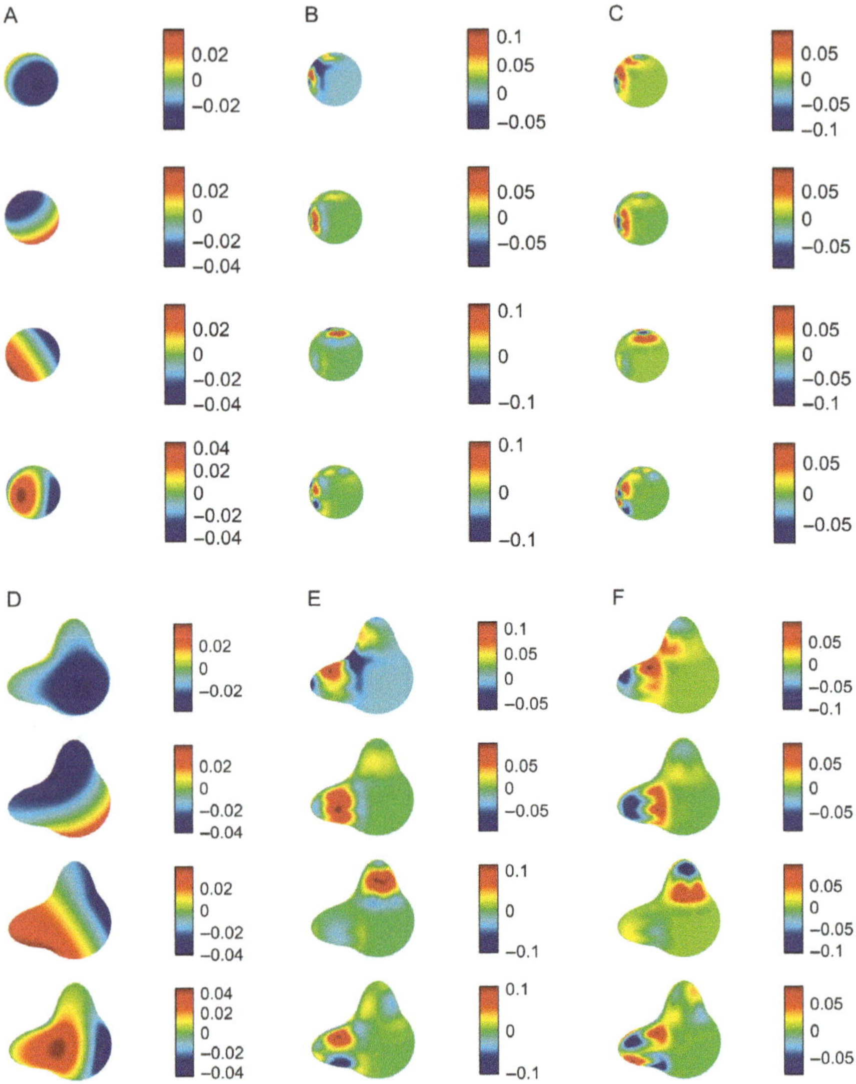

FIG. 6 The sphere, plots (A)–(C), and the deformed sphere, plots (D)–(F), coloured according to the first four eigenvectors of the Alternating Diffusion operator in plots (A) and (D), according to the real parts of the first four eigenvectors of **A** in plots (B) and (E) and coloured according to the imaginary parts of the first four eigenvectors of **A** in plots (C) and (F).

of the distances in each shape divided by 10. The first four eigenvectors of **A**, corresponding to the largest eigenvalues (in absolute value) are presented in Fig. 6. In this figure, panels (B) and (E) present the sphere and the deformed sphere, respectively, coloured according to the real part of the first four eigenvectors of the operator **A**. Panels (C) and (F) present the sphere and the deformed sphere, respectively, coloured according to the imaginary part of these eigenvectors. These plots depict that the eigenvectors of the operator **A** are supported on the differences between the two shapes, i.e., the two "bumps", and that they approximately take the form of local spherical harmonics in these areas.

For comparison, we calculate the first four eigenvectors (nontrivial), corresponding to the largest eigenvalues, of the AD operator. The operator and its eigenvalue decomposition were constructed according to Algorithm 1, where the affinity matrices were constructed using the geodesic distances on the shapes and the kernel scales were empirically set to be the median of the distances divided by 2. Fig. 6, panels (A) and (D) present the sphere and the deformed sphere, respectively, coloured according to the first four eigenvectors of the AD operator. These figures depict that, in contrast to the operator **A**, the eigenvectors of the AD operator capture the common manifold, the sphere, and take the form of standard spherical harmonics on the entire shape. As expected, the eigenvectors of the AD operator attenuate the sensor-specific factors (the two "bumps") whereas the eigenvectors of the operator A emphasize these differences.

5.2 Foetal heart rate recovery

To further demonstrate the properties and capabilities of the operator **A**, we consider a real application of noninvasive foetal heart rate monitoring. In our setting, the foetal heart rate is measured using two electrocardiograph (ECG) leads placed on the maternal abdomen. These electrodes capture both the foetal and the maternal heart activities and the goal is to isolate the foetal heart activity from the two abdominal leads. This task can be challenging since the maternal heart activity is stronger and often masks the foetal heart activity.

From the physiological properties, our basic assumption is that a different view of the foetal heart activity is captured by each electrode, since the source of the signal is close to the two maternal abdominal leads. Conversely, we assume that the same view of the maternal heart activity is captured by the two electrodes, since the source of the signal is far from the two abdominal leads. Therefore, in the context of this chapter, the common component of the two ECG recordings is the maternal heart activity and the differences between the two ECG recordings are embodied in the foetal heart activity.

Based on these assumptions, in order to isolate the foetal heart activity we apply the operator **A** and construct a new representation, which captures the differences between the two sensors. We demonstrate this approach using the publicly available dataset of trans-abdominal maternal ECG recordings in 2013 PhysioNet/Computing in Cardiology Challenge. This dataset includes simultaneous recordings from four ECG leads placed on the maternal abdomen, sampled at 1000 Hz. In our analysis, we use only two ECG leads, 1 and 4, which were chosen based on the results presented in Li et al. (2017).

The data were first preprocessed using a low pass filter (with a cut-off frequency of 100 Hz), a median filter for trend-removal (with a window size of 101 samples) and was then down-sampled to 100 Hz to reduce the amount of samples. Each ECG recording in the dataset contains 60 s, resulting in 6000 samples.

The operator **A** was constructed based on the two ECG recordings according to Algorithm 2. The dataset elements, $s_k^{(1)}$ and $s_\ell^{(2)}$ in Algorithm 2, were set to be overlapping windows of eight samples, i.e., $s_k^{(1)} = [r_{k+1}^{(1)}, \ldots, r_{k+8}^{(1)}]$ and $s_\ell^{(2)} = [r_{\ell+1}^{(2)}, \ldots, r_{\ell+8}^{(2)}]$, where $r_i^{(1)}$ denotes one sample from the first ECG recording and $r_i^{(2)}$ denotes one sample from the second ECG recording. The affinity matrices were calculated using the Euclidean distances between the dataset elements and the kernel scales were set to be the medians of the distances. The operator **A** was then constructed based on the normalized affinity matrices, as described in Algorithm 2. The first 20 eigenvectors, corresponding to the largest eigenvalues (in absolute value), were computed. Using the real and imaginary parts of these eigenvectors, a new representation for the ECG recordings was obtained, $\{\phi_j^A(k)\}_{j=1}^{40}$, $k = 1, \ldots, 6000$. In order to visualize and evaluate the capabilities of the new representation in extracting the foetal heart rate, we use the deshape short time Fourier transform (dsSTFT) (Lin et al., 2018). This transform creates a spectrogram in which dominant frequencies of oscillatory signals are emphasized by masking the multiples associated with the nonsinusoidal oscillatory patterns. We apply the dsSTFT to each representation vector $\phi_j^A \in \mathbb{R}^{6000}$ separately, resulting in 40 spectrograms, and then take the pixel-wise median over all 40 spectrograms. This results in a new spectrogram representing the dominant frequencies in the eigenvectors of the operator **A**.

The median dsSTFT spectrogram for subject $a01$ in the 2013 PhysioNet/Computing in Cardiology Challenge dataset is presented in Fig. 7 plots (C) and (D). This spectrogram contains two dominant frequency lines (denoted by darker pixels). The dark lines located close to 1 Hz, denoted by a red arrow in plot (C), correspond to the maternal heart rate. The line in frequency range 2–3 Hz, denoted by a blue arrow in plot (C), corresponds to the foetal heart rate. As a reference, in plot (D), we mark the true foetal heart rate on the spectrogram by a dashed blue line. For comparison, we applied the dsSTFT to the two ECG recordings (separately) after the preprocessing stage. The resulting spectrograms corresponding to the two recordings are presented in Fig. 7 plots

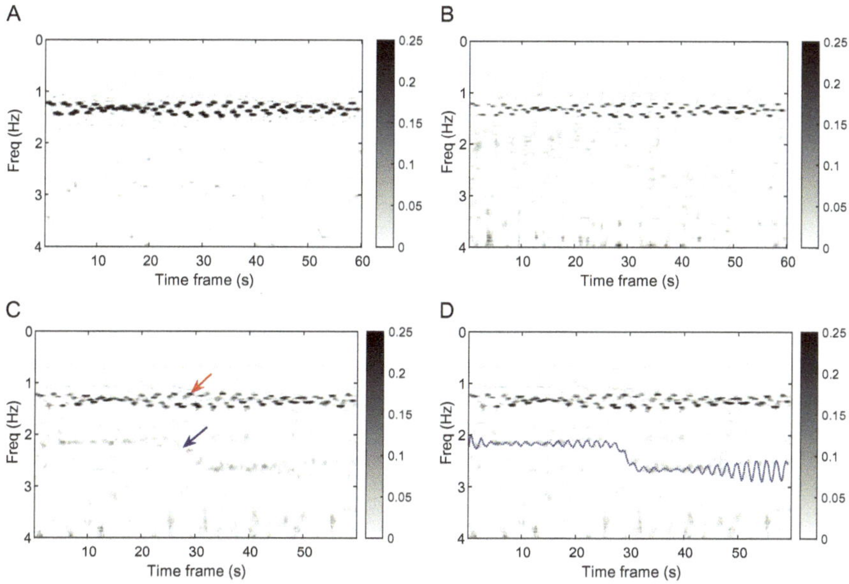

FIG. 7 Spectrograms calculated by the dsSTFT for: (A) abdominal ECG channel 1, (B) abdominal ECG channel 2, (C) and (D) the new representation, $\phi^A(k)$, constructed based on the operator **A**. In plot (C), the maternal heart rate in denoted by a *red arrow* and the foetal heart rate is denoted by a *blue arrow*. The ground-truth foetal heart rate is presented in plot (D) by a *dashed blue line*.

(A) and (B). Fig. 7 depicts that the operator **A** indeed extracts and emphasizes the foetal heart rate based on the two channels, even when it is mostly unobservable in each channel separately.

We note that meaningful results were obtained for other subjects in the dataset as well and that here, for brevity, we demonstrated the capabilities of the operator **A** on one of the subjects. A more extensive analysis and error rates on the entire dataset are available in Shnitzer et al. (2019).

5.3 Sleep dynamics assessment

Sleep is a universal recurring dynamical activity in mammals that is closely related to health. According to the American Academy of Sleep Medicine (AASM) (Berry et al., 2012; Iber et al., 2007), sleep dynamics can be divided into two broad stages, the rapid eye movement (REM) and the nonrapid eye movement (NREM). The NREM stage is further divided into stages N1, N2 and N3. Polysomnography (PSG) is the gold-standard for sleep dynamics assessment and includes recordings of several different signals, such as electroencephalogram (EEG) signals and electrocardiogram signals (ECG). Sleep experts can provide sleep stage annotations for clinical use based on the EEG signals. In this example, we apply the operators **S** and **A** from Section 4 to EEG sleep recordings for sleep dynamics visualization and to extract features for automatic sleep annotation.

The algorithm is composed of two steps. First, we apply a modern nonlinear time-frequency analysis tool, the synchrosqueezing transform (SST) (Daubechies et al., 2011), to extract spectral features from two simultaneously recorded EEG signals. The spectral features are then used to construct a new representation based on the operators **A** and **S** according to Algorithm 2. The new representation is then utilized both for visualization purpose and to establish a prediction model, using a hidden Markov model (HMM) (Fraser, 2008), for automatic sleep annotation. For details on the spectral features extraction step, see Liu et al. (2018).

To evaluate the proposed algorithm, we use a publicly available database—the Sleep-EDF Database [Expanded] from Physionet (Goldberger et al., 2000). The database contains two subsets (marked as SC* and ST*). Both subsets contain simultaneous recordings of two EEG channels, Fpz-Cz and Pz-Oz, sampled at 100 Hz. The first subset, SC*, is acquired from healthy subjects without any sleep-related medication, and consists of recordings from 10 males and 10 females of ages 25–34. For 19 of the 20 subjects, the database includes two recordings per subject of approximately 20 h, and the last subject has only a single recording. The EEG signals were recorded during two subsequent day-night periods at the subjects' homes. Since there are long periods of wakefulness at the beginning and end of the recordings, in Supratak et al. (2017) the wakefulness periods were truncated such that only 30 min of wakefulness were included before and after the sleep periods. For a fair comparison, here we truncated the data in a similar manner. The second subset, ST*, was obtained in a 1994 study of the Temazepam drug effects on subjects with mild insomnia. This subset contains recordings of 7 males and 15 females, who had mild insomnia. Since this dataset was originally used for studying the effects of Temazepam, the EEG signals were recorded in a hospital during two nights, one of which was after Temazepam intake and the other after administration of a placebo. Only the placebo recordings are available in Goldberger et al. (2000). For both SC* and ST* sets, 30 second epochs of EEG data were annotated and divided into six classes: Awake, REM, N1, N2, N3 and N4. Epochs in which movement was detected or the sleep stage was unknown were excluded. In addition, epochs labelled by N4 were relabelled to N3 according to the AASM standard. For more details on the database, we refer the reader to https://www.physionet.org/physiobank/database/sleep-edfx/.

In our analysis we follow the state-of-the-art papers (Liu et al., 2018; Supratak et al., 2017) and apply the HMM prediction model with leave-one-subject-out cross-validation (CV), where the validation set and training set contain different subjects. In this scheme, the main challenge comes from the intersubject variability, but it is closely related to the real-world scenario—how to predict the sleep dynamics of a new subject based on a given annotated database.

Denote the unnormalized confusion matrix resulting from the leave-one-subject-out CV by $\mathbf{M} \in \mathbb{R}^{5\times 5}$, where M_{ij} represents the number of expert-assigned i-class epochs, which were predicted to be in the j-class, where $i, j = 1, \ldots, 5$. In the following results, the precision (PR_p), recall (RE_p), and F1-score ($\mathrm{F1}_p$) of the pth class, where $p = 1, \ldots, 5$, are computed, respectively, through

$$\mathrm{PR}_p = \frac{M_{pp}}{\sum_{p=1}^{5} M_{pq}}, \quad \mathrm{RE}_p = \frac{M_{pp}}{\sum_{q=1}^{5} M_{pq}}, \quad \mathrm{F1}_p = \frac{2\mathrm{PR}_p \cdot \mathrm{RE}_p}{\mathrm{PR}_p + \mathrm{RE}_p}. \tag{57}$$

The overall accuracy (ACC), macro F1 score (Macro-F1) and kappa (κ) coefficient are computed, respectively, through

$$\mathrm{ACC} = \frac{\sum_{p=1}^{5} M_{pp}}{\sum_{p,q=1}^{5} M_{pq}}, \quad \text{Macro-F1} = \frac{1}{5}\sum_{p=1}^{5} \mathrm{F1}_p, \quad \kappa = \frac{\mathrm{ACC} - \mathrm{EA}}{1 - \mathrm{EA}}, \tag{58}$$

where

$$\mathrm{EA} = \frac{\sum_{p=1}^{5}\left(\sum_{q=1}^{5} M_{pq}\right) \times \left(\sum_{q=1}^{5} M_{qp}\right)}{\left(\sum_{p,q=1}^{5} M_{pq}\right)^2}. \tag{59}$$

5.3.1 Visualization

We start by visualizing the intrinsic sleep features of the 20 subjects from the SC* dataset using the new representations obtained from operators **S** and **A**, where the representation vectors are denoted by ϕ_i^S and ϕ_i^A for **S** and **A**, respectively, according to Algorithm 2, clause 4. Fig. 8 presents a visualization of the representation based on the operator **S**, which captures the common structures

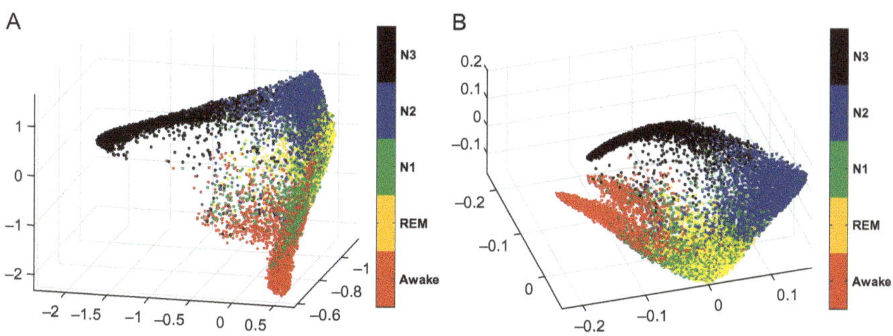

FIG. 8 Visualization of intrinsic sleep features (from two EEG channels) based on the operator **S**, for 20 different subjects from the SC* dataset. Plot (A) presents representation vectors ϕ_1^S, ϕ_2^S and ϕ_3^S and plot (B) presents representation vectors ϕ_2^S, ϕ_3^S and ϕ_4^S.

of the two EEG channels. Clearly, Awake, REM, N2 and N3 stages are well clustered in this new representation, whereas samples from the N1 class are mixed with other sleep stages. Moreover, the samples are organized in a circular structure, where different sleep stages are located in different regions of the circle and follow the usual sleep dynamics pattern.

Fig. 9 presents a visualization of the representation based on the operator **A**. In contrast to the plots in Fig. 8, here the samples are less "structured," but a clustering according to the sleep stages can still be seen. Since the clustering effect is visible for both operators, we use both representations in the automatic annotation task.

The ST* dataset of subjects with mild insomnia is visualized in Figs. 10 and 11. Fig. 10 presents a visualization of the representation based on the operator **S** and Fig. 11 presents the representation based on the operator **A** for this dataset. In both figures, the Awake, REM, N2 and N3 sleep stages are still well clustered. However, compared with the normal subjects in the SC* dataset, the separation and the circular structure (for the operator **S**) are less clear.

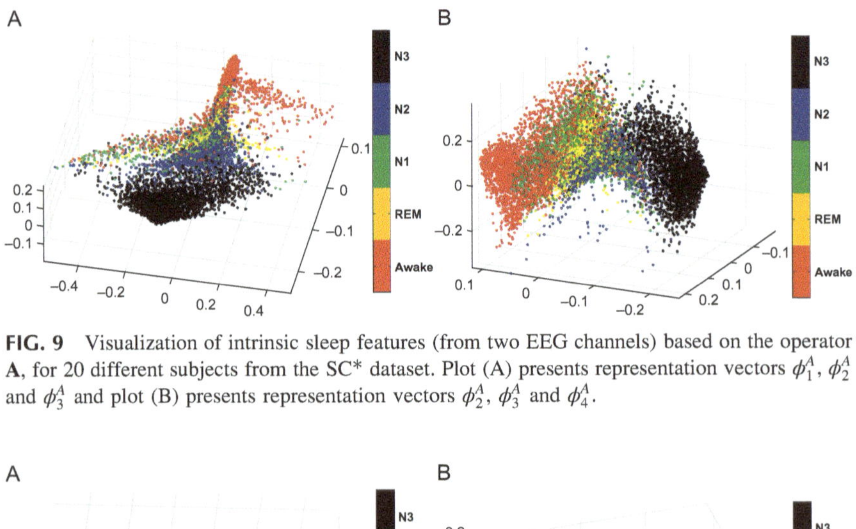

FIG. 9 Visualization of intrinsic sleep features (from two EEG channels) based on the operator **A**, for 20 different subjects from the SC* dataset. Plot (A) presents representation vectors ϕ_1^A, ϕ_2^A and ϕ_3^A and plot (B) presents representation vectors ϕ_2^A, ϕ_3^A and ϕ_4^A.

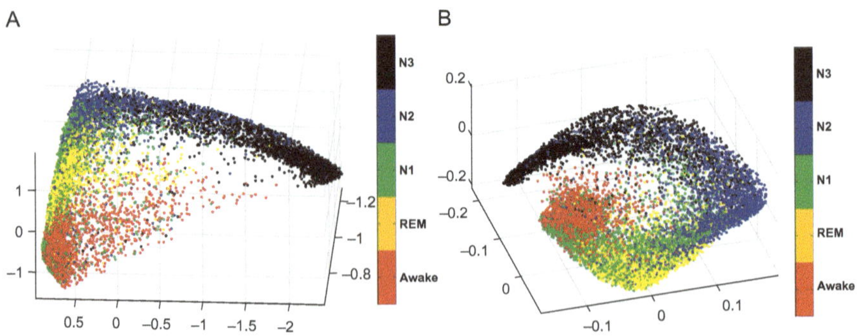

FIG. 10 Visualization of intrinsic sleep features (from two EEG channels) based on the operator **S**, for 22 different subjects from the ST* dataset. Plot (A) presents representation vectors ϕ_1^S, ϕ_2^S and ϕ_3^S and plot (B) presents representation vectors ϕ_2^S, ϕ_3^S and ϕ_4^S.

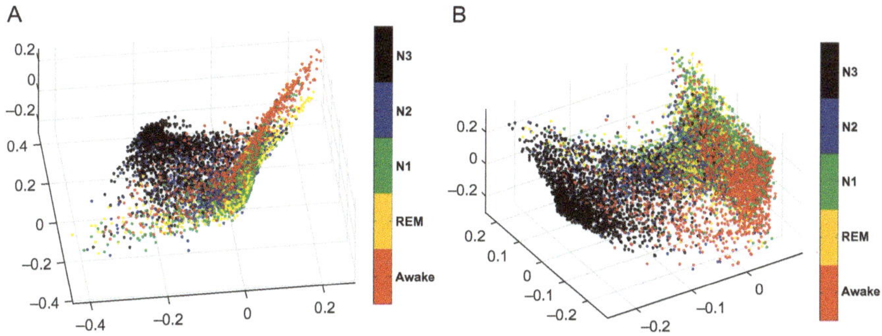

FIG. 11 Visualization of intrinsic sleep features (from two EEG channels) based on the operator **A**, for 22 different subjects from the ST* dataset. Plot (A) presents representation vectors ϕ_1^A, ϕ_2^A and ϕ_3^A and plot (B) presents representation vectors ϕ_2^A, ϕ_3^A and ϕ_4^A.

5.3.2 Automatic annotation

Using the representations obtained by the operators **S** and **A**, three HMM models for automatic sleep annotation are trained, one based on both operator representations, i.e., ϕ_i^S and ϕ_i^A, one based on the representation vectors constructed from **S**, i.e., ϕ_i^S, and one based on the representation vectors constructed from **A**, i.e., ϕ_i^A.

In both considered datasets the labelled epochs are imbalances, with 42.4% epochs labelled as N2 and only 6.6% epochs labelled N1. We follow Liu et al. (2018) and Supratak et al. (2017), which handle the imbalanced data by setting the number of epochs per-stage per-recording *in the training data* (consisting of 19 subjects for SC* and 21 subjects for ST*) to be equal to the number of epochs in the smallest sleep stage, N1.

The average confusion matrices for the SC* dataset (averaged over 20 subjects) obtained by the three HMM models, are presented in Tables 1–3, for representations based on **S** and **A**, based on **S** and based on **A**, respectively.

In comparison with leading recent sleep automatic annotation solutions (Supratak et al., 2017) the presented results for all three HMM models are comparable. Since we use two channels and Supratak et al. (2017) only uses a single channel, seemingly the results in Supratak et al. (2017) are better. However, note that our approach is unsupervised in nature, whereas Supratak et al. (2017) is supervised. In addition, the obtained accuracy results are comparable to the results reported in Liu et al. (2018), which proposed a similar algorithm to the algorithm presented here but with the AD operator (instead of **S** and **A**) and applied it to the two EEG channels from the SC* dataset.

For the ST* dataset, the average confusion matrices (over 22 subjects) obtained by the three HMM models are presented in Tables 4–6, for representations based on **S** and **A**, based on **S** and based on **A**, respectively. The obtained accuracy results in this dataset are comparable with Liu et al. (2018).

TABLE 1 Performance of the new representations obtained using both operators, S and A, evaluated by 20-fold leave-one-subject-out cross-validation on EEG channels Fpz-Cz and Pz-Oz from the SC* dataset.

	Predicted					Per-class metrics		
	Awake	*REM*	*N1*	*N2*	*N3*	*PR*	*RE*	*F1*
Awake (19%)	6962 (**88%**)	148 (**2%**)	621 (**8%**)	160 (**2%**)	36 (**0%**)	92	88	90
REM (18%)	86 (**1%**)	6727 (**87%**)	402 (**5%**)	502 (**7%**)	3 (**0%**)	76	87	81
N1 (7%)	418 (**15%**)	761 (**27%**)	1111 (**40%**)	501 (**18%**)	13 (**0%**)	38	40	39
N2 (42%)	112 (**1%**)	1211 (**7%**)	741 (**4%**)	14643 (**82%**)	1092 (**6%**)	89	82	85
N3 (14%)	22 (**0%**)	0 (**0%**)	20 (**0%**)	671 (**12%**)	4990 (**88%**)	81	88	84

The overall accuracy equals 82.08%, the macro F1 score equals 75.93% and Cohen's kappa equals 75.65%. The bold numbers in brackets indicate the percentile of each predicted class out of the total samples in the true class.

TABLE 2 Performance of the new representation obtained based on S, evaluated by 20-fold leave-one-subject-out cross-validation on EEG channels Fpz-Cz and Pz-Oz from the SC* dataset.

	Predicted					Per-class metrics		
	Awake	*REM*	*N1*	*N2*	*N3*	*PR*	*RE*	*F1*
Awake (19%)	6777 (**85%**)	116 (**1%**)	871 (**11%**)	133 (**2%**)	30 (**0%**)	91	85	88
REM (18%)	87 (**1%**)	6600 (**86%**)	573 (**7%**)	457 (**6%**)	0 (**0%**)	77	86	81
N1 (7%)	435 (**16%**)	707 (**25%**)	1190 (**42%**)	460 (**16%**)	12 (**0%**)	34	42	38
N2 (42%)	114 (**1%**)	1148 (**6%**)	889 (**5%**)	14655 (**82%**)	993 (**6%**)	89	82	86
N3 (14%)	41 (**1%**)	0 (**0%**)	18 (**0%**)	732 (**13%**)	4912 (**86%**)	83	86	84

The overall accuracy equals 81.37%, the macro F1 score equals 75.32% and Cohen's kappa equals 74.74%. The bold numbers in brackets indicate the percentile of each predicted class out of the total samples in the true class.

TABLE 3 Performance of the new representation obtained based on A, evaluated by 20-fold leave-one-subject-out cross-validation on EEG channels Fpz-Cz and Pz-Oz from the SC* dataset.

	Predicted					Per-class metrics		
	Awake	*REM*	*N1*	*N2*	*N3*	*PR*	*RE*	*F1*
Awake (19%)	6878 (**87%**)	170 (**2%**)	678 (**9%**)	152 (**2%**)	49 (**1%**)	92	87	89
REM (18%)	89 (**1%**)	6459 (**83%**)	628 (**3%**)	539 (**7%**)	2 (**0%**)	75	84	79
N1 (7%)	395 (**14%**)	769 (**27%**)	1147 (**41%**)	475 (**17%**)	18 (**1%**)	34	41	37
N2 (42%)	105 (**0%**)	1266 (**7%**)	878 (**5%**)	14369 (**81%**)	1181 (**7%**)	89	81	85
N3 (14%)	24 (**0%**)	0 (**0%**)	32 (**1%**)	604 (**11%**)	5043 (**88%**)	80	88	84

The overall accuracy equals 80.79%, the macro F1 score equals 74.80% and Cohen's kappa equals 74.01%. The bold numbers in brackets indicate the percentile of each predicted class out of the total samples in the true class.

TABLE 4 Performance of the new representations obtained using both operators, S and A, evaluated by 22-fold leave-one-subject-out cross-validation on EEG channels Fpz-Cz and Pz-Oz from the ST* dataset.

	Predicted					Per-class metrics		
	Awake	*REM*	*N1*	*N2*	*N3*	*PR*	*RE*	*F1*
Awake (11%)	2036 (**89%**)	58 (**3%**)	126 (**6%**)	49 (**2%**)	15 (**1%**)	72	89	80
REM (20%)	42 (**1%**)	3478 (**84%**)	61 (**1%**)	536 (**13%**)	14 (**0%**)	78	84	81
N1 (10%)	551 (**27%**)	399 (**20%**)	611 (**30%**)	477 (**23%**)	6 (**0%**)	57	30	39
N2 (45%)	165 (**2%**)	543 (**6%**)	269 (**3%**)	7481 (**79%**)	1035 (**11%**)	81	79	80
N3 (15%)	25 (**1%**)	2 (**0%**)	4 (**0%**)	678 (**22%**)	2436 (**77%**)	69	77	73

The overall accuracy equals 76.04%, the macro F1 score equals 70.60%, and Cohen's kappa equals 66.61%. The bold numbers in brackets indicate the percentile of each predicted class out of the total samples in the true class.

TABLE 5 Performance of the new representation obtained based on S, evaluated by 22-fold leave-one-subject-out cross-validation on EEG channels Fpz-Cz and Pz-Oz from the ST* dataset.

	Predicted					Per-class metrics		
	Awake	*REM*	*N1*	*N2*	*N3*	*PR*	*RE*	*F1*
Awake (11%)	2026 (**89%**)	50 (**2%**)	162 (**7%**)	41 (**2%**)	5 (**0%**)	73	89	80
REM (20%)	43 (**1%**)	3417 (**83%**)	138 (**3%**)	521 (**13%**)	12 (**0%**)	80	83	81
N1 (10%)	550 (**27%**)	370 (**18%**)	658 (**32%**)	460 (**23%**)	6 (**0%**)	50	32	39
N2 (45%)	133 (**1%**)	454 (**5%**)	340 (**4%**)	7627 (**80%**)	939 (**10%**)	82	80	81
N3 (15%)	29 (**1%**)	1 (**0%**)	7 (**0%**)	690 (**22%**)	2418 (**77%**)	72	77	74

The overall accuracy equals 76.53%, the macro F1 score equals 71.11%, and Cohen's kappa equals 67.26%. The bold numbers in brackets indicate the percentile of each predicted class out of the total samples in the true class.

TABLE 6 Performance of the new representation obtained based on A, evaluated by 22-fold leave-one-subject-out cross-validation on EEG channels Fpz-Cz and Pz-Oz from the ST* dataset.

	Predicted					Per-class metrics		
	Awake	*REM*	*N1*	*N2*	*N3*	*PR*	*RE*	*F1*
Awake (11%)	1987 (**87%**)	64 (**3%**)	158 (**7%**)	49 (**2%**)	26 (**1%**)	71	87	78
REM (20%)	98 (**2%**)	3352 (**81%**)	120 (**3%**)	540 (**13%**)	21 (**1%**)	76	81	78
N1 (10%)	540 (**26%**)	387 (**19%**)	654 (**32%**)	458 (**22%**)	5 (**0%**)	52	32	40
N2 (45%)	156 (**2%**)	616 (**6%**)	309 (**3%**)	7331 (**77%**)	1081 (**11%**)	81	77	79
N3 (15%)	19 (**1%**)	2 (**0%**)	11 (**0%**)	692 (**22%**)	2421 (**77%**)	68	77	72

The overall accuracy equals 74.63%, the macro F1 score equals 69.50%, and Cohen's kappa equals 64.78%. The bold numbers in brackets indicate the percentile of each predicted class out of the total samples in the true class.

Note that the prediction performance of N3 in this dataset is lower than the prediction performance in the SC* dataset, and that a significant portion of N3 is mis-classified as N2.

Evidently, the approach presented in this chapter for unsupervised, agnostic, latent component analysis of multimodal data using diffusion operators has proven useful in the task of sleep stage identification. Specifically, without prior knowledge, based solely on the signal recordings, the presented diffusion operators give rise to new data representations, which facilitate accurate automatic sleep annotation, comparable to the state of the art. In addition, this application demonstrates the importance of such latent component analysis. Particularly, we observe that both the common as well as the observation-specific components of the two input EEG channels carry relevant information.

References

Bach, F.R., Jordan, M.I., 2002. Kernel independent component analysis. J. Mach. Learn. Res. 3, 1–48.

Belkin, M., Niyogi, P., 2003. Laplacian eigenmaps for dimensionality reduction and data representation. Neural Comput. 15 (6), 1373–1396.

Berry, R.B., Budhiraja, R., Gottlieb, D.J., Gozal, D., Iber, C., Kapur, V.K., Marcus, C.L., Mehra, R., Parthasarathy, S., Quan, S.F., et al., 2012. Rules for scoring respiratory events in sleep: update of the 2007 AASM manual for the scoring of sleep and associated events. J. Clin. Sleep Med. 8 (05), 597–619.

Coifman, R.R., Lafon, S., 2006. Diffusion maps. Appl. Comput. Harmon. Anal. 21, 5–30.

Daubechies, I., Lu, J., Wu, H.-T., 2011. Synchrosqueezed wavelet transforms: an empirical mode decomposition-like tool. Appl. Comput. Harmon. Anal. 30 (2), 243–261.

Davenport, M.A., Hegde, C., Duarte, M.F., Baraniuk, R.G., 2010. Joint manifolds for data fusion. IEEE Trans. Image Process. 19 (10), 2580–2594.

De Sa, V. R., 2005. Spectral clustering with two views. In: ICML Workshop on Learning with Multiple Views, pp. 20–27.

Fraser, A.M., 2008. Hidden Markov Models and Dynamical Systems. vol. 107 Siam.

Froyland, G., 2015. Dynamic isoperimetry and the geometry of Lagrangian coherent structures. Nonlinearity 28 (10), 3587.

Goldberger, A.L., Amaral, L.A., Glass, L., Hausdorff, J.M., Ivanov, P.C., Mark, R.G., Mietus, J.E., Moody, G.B., Peng, C.-K., Stanley, H.E., 2000. PhysioBank, PhysioToolkit, and PhysioNet: components of a new research resource for complex physiologic signals. Circulation 101 (23), e215–e220.

Hotelling, H., 1936. Relations between two sets of variates. Biometrika 28 (3/4), 321–377.

Iber, C., Ancoli-Israel, S., Chesson, A.L., Quan, S., 2007. The AASM Manual for the Scoring of Sleep and Associated Events: Rules, Terminology and Technical Specifications. American Academy of Sleep Medicine, Westchester, IL.

Keller, Y., Coifman, R.R., Lafon, S., Zucker, S.W., 2010. Audio-visual group recognition using diffusion maps. IEEE Trans. Signal Process. 58 (1), 403–413.

Kumar, A., Daumé, H., 2011. A co-training approach for multi-view spectral clustering. In: Proceedings of the 28th International Conference on Machine Learning (ICML-11), pp. 393–400.

Lai, P.L., Fyfe, C., 2000. Kernel and nonlinear canonical correlation analysis. Int. J. Neural Syst. 10 (05), 365–377.

Lederman, R.R., Talmon, R., 2018. Learning the geometry of common latent variables using alternating-diffusion. Appl. Comput. Harmon. Anal. 44 (3), 509–536.

Li, R., Frasch, M.G., Wu, H.-T., 2017. Efficient fetal-maternal ECG signal separation from two channel maternal abdominal ECG via diffusion-based channel selection. Front. Physiol. 8, 277.

Lin, C.-Y., Su, L., Wu, H.-T., 2018. Wave-shape function analysis. J. Fourier Anal. Appl. 24 (2), 451–505.

Lindenbaum, O., Yeredor, A., Salhov, M., Averbuch, A., 2015. Multiview diffusion maps. arXiv preprint arXiv:1508.05550.

Liu, G.-R., Lo, Y.-L., Sheu, Y.-C., Wu, H.-T., 2018. Diffuse to fuse EEG spectra-intrinsic geometry of sleep dynamics for classification. arXiv preprint arXiv:1803.01710.

Rustamov, R.M., Ovsjanikov, M., Azencot, O., Ben-Chen, M., Chazal, F., Guibas, L., 2013. Map-based exploration of intrinsic shape differences and variability. ACM Trans. Graph. (TOG) 32 (4), 72.

Salhov, M., Lindenbaum, O., Silberschatz, A., Shkolnisky, Y., Averbuch, A., 2016. Multi-view kernel consensus for data analysis and signal processing. arXiv preprint arXiv:1606.08819.

Shaham, U., Lederman, R.R., 2018. Learning by coincidence: Siamese networks and common variable learning. Pattern Recogn. 74, 52–63.

Shnitzer, T., Ben-Chen, M., Guibas, L., Talmon, R., Wu, H.-T., 2019. Recovering hidden components in multimodal data with composite diffusion operators. SIAM J. Math. Data Sci. (accepted).

Supratak, A., Dong, H., Wu, C., Guo, Y., 2017. DeepSleepNet: a model for automatic sleep stage scoring based on raw single-channel EEG. IEEE Trans. Neural Syst. Rehabil. Eng. 25 (11), 1998–2008.

Talmon, R., Wu, H.-T., 2019. Latent common manifold learning with alternating diffusion: analysis and applications. Appl. Comput. Harmon. Anal. 47 (3), 848–892.

Talmon, R., Cohen, I., Gannot, S., Coifman, R.R., 2013. Diffusion maps for signal processing: a deeper look at manifold-learning techniques based on kernels and graphs. IEEE Signal Process. Mag. 30 (4), 75–86.

Tenenbaum, J.B., De Silva, V., Langford, J.C., 2000. A global geometric framework for nonlinear dimensionality reduction. Science 290 (5500), 2319–2323.

Wang, B., Jiang, J., Wang, W., Zhou, Z.-H., Tu, Z., 2012. Unsupervised metric fusion by cross diffusion. In: 2012 IEEE Conference on Computer Vision and Pattern Recognition (CVPR), pp. 2997–3004.

Zhou, D., Burges, C. J. C., 2007. Spectral clustering and transductive learning with multiple views. In: Proceedings of the 24th International Conference on Machine Learning, pp. 1159–1166.

Chapter 2

Intrinsic and extrinsic operators for shape analysis

Yu Wang* and Justin Solomon

Department of Electrical Engineering and Computer Science, Massachusetts Institute of Technology, Cambridge, MA, United States

*Corresponding author: e-mail: wangyu9@mit.edu

Chapter Outline

Handbook of Numerical Analysis, Vol. 20. https://doi.org/10.1016/bs.hna.2019.08.003

© 2019 Elsevier B.V. All rights reserved.

Abstract

Geometric operators are common objects in surface-based shape analysis and geometry processing. While the intrinsic Laplace–Beltrami operator has been a ubiquitous choice thanks to its intuitive and often desirable properties, it fails to capture the spatial embedding of a shape because it discards extrinsic information; furthermore, it is not always sensitive to the geometric features relevant for a given shape analysis task. To address these challenges, several alternative operators for shape analysis have been proposed in recent work, with an emphasis on operators sensitive to extrinsic features. Many operators appearing in previous work on other problems also encode aspects of extrinsic geometry and are potentially suitable for shape analysis. In this survey, we unify discussion of operators for shape analysis, highlighting key theoretical properties as well as their numerical discretizations. Additionally, we provide numerical experiments on model tasks in the operator-based shape analysis pipeline, including computation of descriptors, distances, and segmentations, to demonstrate the effect of using different operators on the qualitative behaviour of algorithms in this space.

Keywords: Shape analysis, Geometric operator, Linear operator, Spectral geometry, Laplace–Beltrami, Extrinsic geometry, Intrinsic geometry

AMS Classification Codes: 65M70, 58J50, 35P05

1 Introduction

Two viewpoints typify classical measurements for analyzing surface geometry. The *extrinsic* approach considers a surface as it is embedded in three dimensions, relying on distances in the underlying Euclidean space $\mathbb{R}^3$. The *intrinsic* approach, in contrast, restricts consideration to quantities that can be measured without leaving the surface, such as geodesic distances, decoupling surface geometry from its embedding. Historically, the intrinsic approach played a central role in the development of the modern theory of geometry. Largely a byproduct of its prevalence in the mathematical literature as well as a preference for isometry invariance in certain applications, this perspective has become pervasive in computer graphics, geometry processing, and machine learning, leading to many successful algorithms. In practice, that intrinsic methods are inherently invariant to isometry contributes to their popularity: Real-world objects often deform in a nearly isometric fashion, and

FIG. 1 Two *isometric* cubes with an inward or outward bump.

intrinsic methods by design do not distinguish poses under these transformations. This is often a desirable feature for applications like shape classification.

The intrinsic approach, however, employs an incomplete description of a surface, in the sense that the embedding as well as critical cues for humans' perception of shape cannot be recovered from intrinsic measurements alone. As a simple illustration, Fig. 1 shows cubes with inward or outward bumps; these models are different extrinsically but identical intrinsically. As an extreme example, all origami models are intrinsically equivalent to flat sheets of paper. Hence, shape analysis pipelines that desire to use or learn from all possible geometric cues require reconsideration of their intrinsic building blocks, discarding extrinsic information only if it is truly irrelevant for a given task.

A critical example of the intrinsic–extrinsic divide appears in the operator-based approach to geometry processing. *Linear operators*, such as the intrinsic Laplace–Beltrami operator (or Laplacian for short), play a central role in geometry processing and shape analysis, applied to tasks including segmentation, retrieval, parameterization, correspondence, local descriptors, deformation, physical simulation, and deep learning. In particular, *Laplacian-based shape analysis* has achieved remarkable success and popularity for most, if not all, of these tasks. By removing extrinsic information, however, the purely intrinsic surface Laplacian operator can lead to undesirable behaviour in shape analysis algorithms. For example, Laplacian-based algorithms cannot distinguish between the two cubes in Fig. 1.

Analyzing shapes using alternative operators has attracted recent interest (Liu et al., 2017; Wang et al., 2018b), avoiding the loss of information associated to purely intrinsic methods. These operators can capture aspects of extrinsic geometry and can be sensitive to features other than those typically detected by the Laplacian (Wu and Levine, 1997; Liu and Zhang, 2007; Raviv et al., 2010; Hildebrandt et al., 2012; Au et al., 2012; Wang et al., 2014; Andreux et al., 2014; Ye et al., 2018). While the Laplacian is the "default" choice in many shape analysis methods, replacing it with another operator can immediately yield algorithms with different—and often improved—

behaviour. This substitution provides geometric insight that contrasts from that derived using Laplacian-based constructions but yields additional information about shape. For this reason, this survey compares relevant operators, including the Laplacian, the Dirac operator, boundary integral operators, and others, in a unified framework. We provide experiments to highlight similarities and differences among the many options.

2 Preliminaries

2.1 Extrinsic and intrinsic geometry

We briefly review basic concepts and theoretical constructions for analyzing intrinsic and extrinsic geometry in the paragraphs below; interested readers can refer to texts in differential geometry (Frankel, 2011; Lee, 2013; Do Carmo, 2016) for expanded mathematical discussion and to Botsch et al. (2010) for a practical guide to applied geometry.

2.1.1 Intrinsic geometry

Quantities that can be measured without leaving a surface $\mathcal{M}$ are considered intrinsic. The core intrinsic measurement on a surface is its *first fundamental form* (or *metric tensor*), which provides a means of measuring lengths and angles of vectors in the tangent space. Typically, this tensor per point on the surface is expressed as a matrix $g(\mathbf{x}) \in \mathbb{R}^{2\times 2}$ giving the inner products of basis vectors for the tangent space at $\mathbf{x} \in \mathcal{M}$. We denote entries in the metric tensor as g_{ij}, where $i, j \in \{1, 2\}$, and we use $(g^{ij}) = (g^{-1})_{ij}$ to denote the inverse metric. Intrinsic quantities like geodesic distances are fully determined by the metric of a surface.

The metric allows us to perform calculations about a surface abstractly, without its embedding in $\mathbb{R}^3$. Associating each point with a metric tensor g is already sufficient to define a rich family of mathematical objects. Intrinsic operators like the Laplacian, as we will see later in Section 5.2, can be defined using the metric g only, remaining invariant if the embedding changes so long as the metric g remains the same.

A crowning result of classical differential geometry is that *Gaussian curvature*, which compares the circumference of an infinitesimal geodesic circle to that of a circle in the plane with the same area, is intrinsic. The better-known definition of Gaussian curvature as $K(\mathbf{x}) = \kappa_1(\mathbf{x})\kappa_2(\mathbf{x})$, i.e., the product of the two *principal curvatures* $\kappa_1(\mathbf{x})$, $\kappa_2(\mathbf{x})$ at a point $\mathbf{x} \in \mathcal{M}$, involves the eigenvalues of the extrinsic second fundamental form (below).

2.1.2 Extrinsic geometry

Extrinsic geometry considers a surface as an object embedded in $\mathbb{R}^3$. In classical differential geometry, when taking this view, it is typical to treat a surface locally using a parameterization $\mathbf{r}(u, v) : \mathbb{R} \times \mathbb{R} \to \mathbb{R}^3$ and then apply the analytical tools developed in multivariable calculus to $\mathbf{r}(\cdot, \cdot)$. In addition to the first fundamental form, this embedding enables measurement of the *second fundamental form*, a 2×2 matrix giving the rate of change of the surface normal; the eigenvalues of the second fundamental form are the principal curvatures $\kappa_1(\mathbf{x})$, $\kappa_2(\mathbf{x})$ mentioned above. The mean curvature at a point $\mathbf{x} \in \mathcal{M}$, defined as average of the two principal curvatures $H(\mathbf{x}) = (\kappa_1(\mathbf{x}) + \kappa_2(\mathbf{x}))/2$, is an extrinsic quantity.

2.2 Operators and spectra

The operator-based approach to shape analysis constructs an operator (discretely, a matrix) that acts on functions over the surface; for example, the Laplacian roughly maps functions to their second derivatives. Then, the eigenvalues (or, *spectrum*) of the operator and its eigenfunctions encode geometric features. This representation is convenient for a number of reasons. Whereas the list of vertex positions of a triangle mesh is only known up to permutation, rigid motion, and potentially deformation, operators easily can be designed whose list of eigenvalues are invariant to these irrelevant changes.

For many choices of operators including the Laplacian (Zeng et al., 2012; Boscaini et al., 2015; Chern et al., 2018), the Hessian operator (Stein et al., 2018), and the Dirichlet-to-Neumann operator (Wang et al., 2018b), the operator encodes *all* information about the shape up to a class of transformations, in the sense that one can recover an embedding from the operator and/or its spectrum (up to isometry or rigid transformation, depending on the operator). In this case, the operator can be thought as an alternative representation of the shape, providing a *lossless* means of computing geometric information for algorithms downstream. For this reason, it is fairly common for shape analysis methods to require only access to the operator constructed from a shape rather than access to the shape itself.

Spectral methods extract geometric information directly from operator eigenvalues and eigenfunctions (Zhang et al., 2010a). As we will see later, spectral methods naturally separate features at different scales and extract global features about a shape, making them suitable for design of robust algorithms.

3 Theoretical aspects and numerical analysis

In this section, we present a brief introduction to the theory of linear operators and their applicability to shape analysis; we also introduce general techniques for their discretization.

3.1 Basics of linear operators

Let $\mathcal{M}$ be a smooth surface possibly with boundary $\partial\mathcal{M}$, and let $\mathcal{L}^2(\mathcal{M})$ be the space of square (Lebesgue) integrable functions. A linear operator is a map $\mathcal{A} : \mathcal{L}^2(\mathcal{M}) \to \mathcal{L}^2(\mathcal{M})$ taking in one function on the surface and returning another function, such that $\mathcal{A}(u+v) = \mathcal{A}u + \mathcal{A}v$ and $\mathcal{A}(c \cdot u) = c \cdot \mathcal{A}u$ for $c \in \mathbb{R}$.

For our purposes, we can simply think of the function space $\mathcal{L}^2(\mathcal{M})$ as analogous to the vector space $\mathbb{R}^n$ and the set of operators $\mathcal{L}^2(\mathcal{M}) \to \mathcal{L}^2(\mathcal{M})$ as analogous to the set of matrices $\mathbb{R}^{n\times n}$, where n is the number of vertices in a triangle mesh. A complete, rigorous treatment of operators involves functional analysis and linear operator theory, which is beyond the scope of this paper; we refer interested readers to Brezis (2010) for discussion.

3.1.1 Equivalent strong and weak forms

A linear operator and the PDEs associated with it usually can be expressed using an equivalent integral equation. Let us start with the Laplace equation with Neumann boundary conditions on $\mathcal{M}$:

$$\begin{cases} -\Delta_{\mathcal{M}} u(\mathbf{x}) = f(\mathbf{x}) & \mathbf{x} \in \mathcal{M}\backslash\partial\mathcal{M} \\ \nabla_{\mathbf{n}(\partial\mathcal{M})} u(\mathbf{x}) = g(\mathbf{x}) & \mathbf{x} \in \partial\mathcal{M} \end{cases} \tag{1}$$

in which $\Delta_{\mathcal{M}}$ is the Laplacian operator on the curved surface, and $\nabla_{\mathbf{n}(\partial\mathcal{M})}$ is the normal derivative operator, satisfying $\nabla_{\mathbf{n}(\partial\mathcal{M})} u(\mathbf{x}) = \mathbf{n}(\mathbf{x}) \cdot \nabla_{\mathcal{M}} u(\mathbf{x})$ where $\nabla_{\mathcal{M}}$ is the intrinsic gradient operator and $\mathbf{n}(\mathbf{x})$ is the normal to the boundary at $\mathbf{x} \in \partial\mathcal{M}$. For the purpose of discussion in this section, readers may simply think of the special case where $\mathcal{M} \subseteq \mathbb{R}^2$, i.e. a planar domain in $\mathbb{R}^2$, for which $\Delta_{\mathcal{M}}$ becomes $\Delta_{\mathbb{R}^2} = \frac{\partial^2}{\partial x^2} + \frac{\partial^2}{\partial y^2}$ and $\nabla_{\mathcal{M}}$ becomes the usual gradient operator in $\mathbb{R}^2$. The general definition of $\Delta_{\mathcal{M}}$ is postponed to Section 5.2, although our discussion below holds in the general case when $\mathcal{M}$ becomes a curved surface. To simplify notation, we omit the subscript in $\Delta_{\mathcal{M}}$, $\nabla_{\mathcal{M}}$ and $\nabla_{\mathbf{n}(\partial\mathcal{M})}$ when these operators are used within integral $\int_{\mathcal{M}}$, $\int_{\mathcal{M}}$ and $\int_{\partial\mathcal{M}}$, respectively.

Eq. (1) gives a PDE that is known as the *strong form* of the Laplace equation. If we integrate both sides against an arbitrary function $v(\mathbf{x})$ and sum the results, we obtain the *weak form* of the same problem:

$$-\int_{\mathcal{M}} v(\mathbf{x})\Delta u(\mathbf{x}) + \int_{\partial\mathcal{M}} v(\mathbf{x})\nabla_{\mathbf{n}} u(\mathbf{x}) = \int_{\mathcal{M}} v(\mathbf{x}) f(\mathbf{x}) + \int_{\partial\mathcal{M}} v(\mathbf{x}) g(\mathbf{x}), \quad \forall v(\mathbf{x}). \tag{2}$$

In this formula, the function $v(\mathbf{x})$ is known as a *test function*. Under appropriate assumptions about the function spaces containing $u(\mathbf{x})$ and $v(\mathbf{x})$, one can show that the weak form and the strong form are solved by the same $u(\mathbf{x})$. Roughly speaking, when $v(\mathbf{x})$ is chosen as a delta function, the strong form is recovered: The relationship involving $f(\mathbf{x})$ appears when the delta function is in the interior $\mathcal{M}\backslash\partial\mathcal{M}$, and the relationship involving $g(\mathbf{x})$ appears when the delta function is on the boundary $\partial\mathcal{M}$. Using Green's first identity,

$$\int_{\mathcal{M}} \nabla u(\mathbf{x}) \cdot \nabla v(\mathbf{x}) = -\int_{\mathcal{M}} v(\mathbf{x})\Delta u(\mathbf{x}) + \int_{\partial\mathcal{M}} v(\mathbf{x})\nabla_{\mathbf{n}} u(\mathbf{x}). \tag{3}$$

The left side of the weak form in Eq. (3) gives a symmetric *bilinear form* $\int_{\mathcal{M}} \nabla u(\mathbf{x}) \cdot \nabla v(\mathbf{x})$, which takes two input functions u, v and outputs a scalar. When $v = u$, the bilinear form becomes $\int_{\mathcal{M}} \|\nabla u(\mathbf{x})\|_2^2$, the *Dirichlet energy* measuring the smoothness of u.

Our discussion so far introduces an equivalence between a PDE (1) and an integral equation (2) specified by a bilinear form $\int_{\mathcal{M}} \nabla u(\mathbf{x}) \cdot \nabla v(\mathbf{x})$. This equivalence can be generalized to a large range of PDEs with natural boundary conditions.

Recall that a bilinear form $a(u, v)$ is linear in the arguments u, v, i.e., $a(u_1 + u_2, v) = a(u_1, v) + a(u_2, v)$ and $a(c \cdot u, v) = c \cdot a(u, v)\ \forall c \in \mathbb{R}$, and analogously for the second argument v. Bilinear forms and linear operators generalize matrix operations in finite-dimensional spaces to infinite-dimensional spaces. Hence, we can make definitions for continuous operators similar to their counterparts in linear algebra, including the following key classes of bilinear operators:

- *Self-adjoint*: We call a bilinear form self-adjoint if $\forall u, v : a(u, v) = a(v, u)$. This generalizes the definition of symmetric matrices $\mathbf{A}$, for which $\mathbf{A} = \mathbf{A}^\top$.
- *Positive semidefinite*: We call a self-adjoint bilinear form positive semidefinite (p.s.d.) if we have $a(u, u) \geq 0\ \forall u$; the form is positive definite (p.d.) if $a(u, u) = 0$ implies $u \equiv 0$. This generalizes the definition of a p.s.d. matrix $\mathbf{A}$, for which $\mathbf{u}^\top \mathbf{A} \mathbf{u} \geq 0\ \forall \mathbf{u} \in \mathbb{R}^n$, and a p.d. matrix $\mathbf{A}$, which additionally satisfies the condition that $\mathbf{u}^\top \mathbf{A} \mathbf{u} = 0$ implies $\mathbf{u} \equiv \mathbf{0}$.

Our discussion of the Laplacian follows a template common in functional analysis, in which linear operators are converted to bilinear forms; the linear operator is the strong form and the bilinear form is the weak form. In particular, given a linear operator $\mathcal{A}: \mathcal{L}^2(\mathcal{M}) \mapsto \mathcal{L}^2(\mathcal{M})$ we can define a corresponding bilinear form $a(u, v) : \mathcal{L}^2(\mathcal{M}) \times \mathcal{L}^2(\mathcal{M}) \to \mathbb{R}$ as $a(u, v) = \langle u, \mathcal{A}v \rangle_{\mathcal{M}}$, where $u, v \in \mathcal{L}^2(\mathcal{M})$. In the reverse direction, a linear operator can be recovered from a bilinear form, e.g., by applying the Riesz representation theorem (Brezis, 2010). Such equivalence is trivial in finite-dimensional spaces: Both bilinear forms and linear maps are given by matrices.

This equivalence allows us to equate the problem $\mathcal{A}u = f$ with a weak form

$$a(u,v) = \langle f, v \rangle_{\mathcal{M}} \quad \forall v \tag{4}$$

where $\langle u,v \rangle_{\mathcal{M}} := \int_{\mathcal{M}} u(\mathbf{x}) v(\mathbf{x})$ is the inner product on $\mathcal{L}^2(\mathcal{M})$, under mild assumptions that u, v, f come from appropriate function spaces—usually a Sobolev space, consisting of functions whose low-order partial derivatives have bounded $\mathcal{L}^2$ integrals. Returning to the Laplacian operator $\Delta_{\mathcal{M}}$, its corresponding bilinear form is $a(u,v) = \int_{\mathcal{M}} \nabla u(\mathbf{x}) \cdot \nabla v(\mathbf{x})$; note that fewer derivatives are required of u, v to define the bilinear form than to define the linear operator.

3.1.2 Variational formulation

Given functions $f(\mathbf{x})$ and $g(\mathbf{x})$, consider the variational problem for a function $u(\mathbf{x})$:

$$\min_{u(\cdot)} \int_{\mathcal{M}} \left(\|\nabla u(\mathbf{x})\|_2^2 - u(\mathbf{x})f(\mathbf{x})\right) - \int_{\partial\mathcal{M}} u(\mathbf{x})g(\mathbf{x}). \tag{5}$$

Using variational calculus, one can show that problems (1) and (5) are solved by the same function $u(\mathbf{x})$. This relationship justifies calling the Neumann condition $\nabla_{\mathbf{n}} u(\mathbf{x}) = g(\mathbf{x})$ in (1) by its alternative name, the *natural* boundary condition, since it arises from the variational problem (5) without an explicit boundary constraint.

3.1.3 Eigenvalue problem

The eigenvalue problem associated to operator $\mathcal{A}: \mathcal{H} \to \mathcal{H}$ is defined as follows:

$$\mathcal{A}\phi = \lambda\phi, \tag{6}$$

where $\lambda \in \mathbb{R}$ is known as an eigenvalue and $\phi(\cdot)$ is its corresponding eigenfunction. The *spectral theorem* states that in the most common case, namely when $\mathcal{A}$ is a compact self-adjoint operator and $\mathcal{H}$ is a separable Hilbert space (Zhu, 2007), there are countably many eigenvalues and corresponding eigenfunctions. We mainly consider this case in our survey, and hence we use $\{\lambda_i\}_{i=0}^{\infty}$ and $\{\phi_i(\mathbf{x})\}_{i=0}^{\infty}$ to denote the sets of eigenvalues and corresponding eigenfunctions of $\mathcal{A}$, respectively, sorted in ascending order such that $\lambda_0 \leq \lambda_1 \leq \lambda_2 \leq \cdots$.

Known as the *Courant–Fischer min-max theorem*, the strong form (6) can be converted to an equivalent weak form by finding saddle points of the optimization problem

$$\begin{cases} \min\limits_{\phi(\cdot)} & a(\phi,\phi) \\ \text{s.t.} & \langle \phi,\phi \rangle_{\mathcal{M}} = 1. \end{cases} \tag{7}$$

Assuming $a(\cdot,\cdot)$ is symmetric, we can follow the convention that $\{\phi_i(\mathbf{x})\}_{i=0}^{\infty}$ are orthonormal: Eigenfunctions corresponding to different λ's must be orthogonal, and applying Gram–Schmidt orthogonalization to eigenfunctions with the same λ, followed by normalization, ensures that $\{\phi_i(\mathbf{x})\}_{i=0}^{\infty}$ are orthonormal. A consequence of the spectral theorem, for many choices of operators $\mathcal{A}$, is that the ϕ_i's form a *complete* orthonormal basis[a]; in classical mathematics, the completeness of the Laplacian is a consequence of the Sturm–Liouville decomposition (Chavel, 1984; Rosenberg, 1997). Laplacian

[a]Completeness requires that any function in $\mathcal{L}^2(\mathcal{M})$ can be approximated arbitrarily closely by linear combinations of $\{\phi_i(\mathbf{x})\}_{i=0}^{\infty}$.

eigenfunctions are also known as *manifold harmonics*. When the surface is a sphere, the Laplacian eigenfunctions are called *spherical harmonics*.

The *spectrum* of an operator, $\{\lambda_i\}_{i=0}^{\infty}$, is the generalization of eigenvalues of a matrix. This spectral decomposition of $\mathcal{A}$, as we will see later, extracts information about $\mathcal{M}$, from large- to small-scale.

3.2 PDEs and Green's functions

A few classic PDEs written in terms of the Laplacian are widely used in shape analysis and provide physical intuition for behaviour of algorithms in spectral geometry. As we will see later, solutions to these problems closely relate to the spectrum of the Laplacian. For simplicity, starting from this section in the paper we only consider closed surfaces (i.e., ones without boundaries), unless noted otherwise and in which case boundary conditions will be explicitly discussed.

3.2.1 Laplace equation

The Laplacian admits a *Green's function*, also known as the *fundamental solution*, given by a function $G(\mathbf{x}, \mathbf{y})$ satisfying

$$\Delta_{\mathbf{x}} G(\mathbf{x}, \mathbf{y}) = \delta_{\mathbf{y}}(\mathbf{x}), \tag{8}$$

where $\Delta_{\mathbf{x}}$ is still the operator $\Delta_{\mathcal{M}}$ with the notation chosen to emphasize that it operates on $\mathbf{x}$ and $\delta_{\mathbf{y}}(\cdot)$ denotes the Dirac δ-function centred at $\mathbf{y}$. Given G, up to an additive constant function, the solution to the Laplace equation $\Delta_{\mathcal{M}} u(\mathbf{x}) = f(\mathbf{x})$ for "smooth enough" $f(\cdot)$ can be written in closed-form as

$$u(\mathbf{x}) = \int_{\mathcal{M}} G(\mathbf{x}, \mathbf{y}) f(\mathbf{y}) d\mathbf{y}.$$

In other words, the linear operator $f(\cdot) \mapsto \int_{\mathcal{M}} G(\mathbf{x}, \mathbf{y}) f(\mathbf{y}) d\mathbf{y}$ based on the Green's function "inverts" the original Laplacian operator. The Green's function of a linear operator plays an analogous role to the pseudo-inverse of a (finite-dimensional) matrix: Recall that for a matrix $\mathbf{A} \in \mathbb{R}^{n \times n}$ whose null space contains only constant vectors, the solution to the linear system $\mathbf{Au} = \mathbf{f}$ is $\mathbf{u} = \mathbf{A}^{\dagger}\mathbf{f}$ up to an additive constant, when the right-hand side $\mathbf{f} \in \mathbb{R}^{n \times 1}$ sums to zero; here $\mathbf{A}^{\dagger}$ is the (Moore–Penrose) pseudo-inverse of $\mathbf{A}$ that satisfies the relationship $\mathbf{A} \cdot \mathbf{A}^{\dagger} = \mathbf{I} - \frac{1}{n}\mathbf{1}^{\top}\mathbf{1}$, similar to the definition in Eq. (8).

Given the spectral decomposition of $\Delta_{\mathcal{M}}$ defined in Section 3.1.3, we have $(\Delta_{\mathcal{M}} u)(\mathbf{x}) = \int_{\mathcal{M}} \sum_{k=0}^{\infty} \lambda_k \phi_k(\mathbf{x}) \phi_k(\mathbf{y}) u(\mathbf{y})\, d\mathbf{y}$. This expression first decomposes $f(\cdot)$ into an orthonormal basis, and multiplies each component with the corresponding eigenvalue: high-frequency components are amplified by a factor of λ_k, so $\Delta_{\mathcal{M}}$ can be thought as a high-pass filter in the signal processing view. The following spectral expansion gives an explicit formula for the Green's function, which can be understood as the pseudo-inverse of the operator $\Delta_{\mathcal{M}}$:

$$G(\mathbf{x}, \mathbf{y}) = \sum_{k=n_0}^{\infty} \frac{1}{\lambda_k} \phi_k(\mathbf{x}) \phi_k(\mathbf{y}). \tag{9}$$

Here, n_0 is the number of zero eigenvalues of $\Delta_{\mathcal{M}}$; $n_0 = 1$ for simply connected surfaces. The corresponding linear operator $f(\cdot) \mapsto \int_{\mathcal{M}} G(\mathbf{x}, \mathbf{y}) f(\mathbf{y}) d\mathbf{y}$, conversely, is a low-pass filter, in the sense that low-frequency components are exaggerated and emphasized by the multiplicative factor $1/\lambda_k$.

3.2.2 Heat equation

The heat equation (also known as the diffusion equation) describes a time-varying evolution of a function $u(\mathbf{x}, t)$ given its initial distribution $u(\mathbf{x}, 0)$. Physically, this PDE is used to determine the spatial distribution of temperature on a conductive surface after it diffuses for time t:

$$\text{(Heat equation)} \quad \frac{\partial}{\partial t} u(\mathbf{x}, t) = \Delta_{\mathbf{x}} u(\mathbf{x}, t) \quad \text{s.t.} \quad u(\mathbf{x}, 0) = f(\mathbf{x}). \tag{10}$$

The *heat kernel* $k_t(x, y)$ specifies the transition density function from $\mathbf{x}$ to $\mathbf{y}$ over a time period t:

$$k_t(\mathbf{x}, \mathbf{y}) = \sum_{i=0}^{\infty} e^{-\lambda_i t} \phi_i(\mathbf{x}) \phi_i(\mathbf{y}). \tag{11}$$

Fig. 2 visualizes (one column of) the heat kernel in log scale, i.e., $\log k_t(\mathbf{x}, \cdot)$, where $\mathbf{x}$ locates at the dot in the centre of the isolines. $k_t(\mathbf{x}, \cdot)$ corresponds the

FIG. 2 Heat kernel in log scale, $\log k_t(\mathbf{x}, \cdot)$.

temperature distribution at time t, when the initial distribution is $\delta_{\mathbf{x}}(\cdot)$, a Dirac δ-function centred at $\mathbf{x}$. It is not a coincidence that $\log k_t(\mathbf{x}, \cdot)$ looks like $d_g(\mathbf{x}, \cdot)$, the geodesic distance to $\mathbf{x}$: a connection between them is given in Section 3.4.

Using heat kernel, the solution to the heat equation can be written as

$$u(\mathbf{x}, t) = \int_{\mathcal{M}} k_t(\mathbf{x}, \mathbf{y}) f(\mathbf{y}) d\mathbf{y}. \tag{12}$$

Substituting (11), this expression for u first decomposes $f(\cdot)$ into an orthonormal basis associated to the PDE, and then each component decays independently at a rate associated to its eigenvalue. The heat kernel can be understood similarly to Eq. (9): the linear operator $f(\cdot) \mapsto \int_{\mathcal{M}} k_t(\mathbf{x}, \mathbf{y}) f(\mathbf{y}) d\mathbf{y}$ act as the exponential of $\Delta_{\mathcal{M}}$. If we interpret $u(\mathbf{x}, t)$ as a probability density function, then $k_t(\mathbf{x}, \mathbf{y})$ in (12) provides the transition density function for Brownian motion (Knight et al., 1981). Heat diffusion is a Markov process: $k_{t+s}(\mathbf{x}, \mathbf{y}) = \int_{\mathcal{M}} k_t(\mathbf{x}, \mathbf{z}) k_s(\mathbf{y}, \mathbf{z}) d\mathbf{z}$, reflecting the fact that heat starting at $\mathbf{x}$ has to diffuse to $\mathbf{y}$ via some $\mathbf{z}$ in a "memory-less" fashion. These facts are starting points to prove convergence of some discrete approximations to the Laplace–Beltrami operator (Coifman and Lafon, 2006).

3.2.3 Wave and Schrödinger equations

Wave propagation is governed by the wave equation, in which $u(\mathbf{x}, t)$ measures the magnitude of displacement at $\mathbf{x}$ at time t:

$$\text{(Wave equation)} \quad \frac{\partial^2}{\partial t^2} u(\mathbf{x}, t) = \Delta_{\mathbf{x}} u(\mathbf{x}, t) \ \text{ s.t. } \ u(\mathbf{x}, 0) = f(\mathbf{x}) \ \text{ and } \ \frac{\partial}{\partial t} u(\mathbf{x}, 0) = g(\mathbf{x}). \tag{13}$$

The wave equation, as a second-order differential equation, also requires initial conditions on $\frac{\partial}{\partial t} u(\mathbf{x}, 0)$ to yield a unique solution $u(\cdot, \cdot)$, making the discussion complicated; instead, here we examine a similar first-order PDE.

Closely related to the wave equation, the *Schrödinger equation* in Eq. (14) prescribes the evolution of a quantum particle via the wave function $u(\mathbf{x}, t) : \mathcal{L}^2(\mathcal{M}) \times \mathbb{R}^+ \to \mathbb{C}$ such that $|u(\mathbf{x}, t)|^2$ is interpreted as the particle's probability density function (p.d.f.) at location $\mathbf{x}$ and time t:

$$\text{(Schrödinger equation)} \quad i \frac{\partial}{\partial t} u(\mathbf{x}, t) = -\Delta_{\mathbf{x}} u(\mathbf{x}, t) \ \text{ s.t. } \ u(\mathbf{x}, 0) = f(\mathbf{x}) \tag{14}$$

where $i = \sqrt{-1}$. This equation is later used to define the wave kernel signature (see Section 4.3) and the Hamiltonian/Schrödinger operator (see Section 5.10).

3.2.4 Generalization

The PDEs above can be generalized by replacing the operator $-\Delta$ with a general p.s.d. operator $\mathcal{A}$:

$$\text{(Poisson equation)} \qquad \mathcal{A}u(\mathbf{x}) = f(\mathbf{x}) \tag{15}$$

$$\text{(Heat equation)} \qquad \frac{\partial}{\partial t}u(\mathbf{x}, t) = -\mathcal{A}u(\mathbf{x}, t) \ \text{ s.t. } \ u(\mathbf{x}, 0) = f(\mathbf{x}) \tag{16}$$

$$\text{(Schrödinger equation)} \qquad i\frac{\partial}{\partial t}u(\mathbf{x}, t) = \mathcal{A}u(\mathbf{x}, t) \ \text{ s.t. } \ u(\mathbf{x},0) = f(\mathbf{x}) \tag{17}$$

Solutions to these PDEs can be written in the unified form $u(\mathbf{x}) = \int_{\mathcal{M}} G_t(\mathbf{x}, \mathbf{y}) f(\mathbf{y}) \mathrm{d}\mathbf{y}$, where

$$G_t(\mathbf{x}, \mathbf{y}) = \begin{cases} \sum_{k=n_0}^{\infty} \frac{1}{\lambda_k}\phi_k(\mathbf{x})\phi_k(\mathbf{y}) & \text{Poisson equation} \\ \sum_{k=0}^{\infty} e^{-\lambda_k t}\phi_k(\mathbf{x})\phi_k(\mathbf{y}) & \text{Heat equation} \\ \sum_{k=0}^{\infty} e^{i\lambda_k t}\phi_k(\mathbf{x})\phi_k(\mathbf{y}) & \text{Schrödinger equation} \end{cases}$$

is the Green's function for each PDE and $\{\phi_k(\cdot)\}_{k=0}^{\infty}$ and $\{\lambda_k(\cdot)\}_{k=0}^{\infty}$ are eigenfunctions and eigenvalues of the operator $\mathcal{A}$, such that $\lambda_0 = 0, \ldots, \lambda_{n_0} = 0, \lambda_{n_0+1} > 0$. For fixed t, the map $f(\cdot) \mapsto \int_{\mathcal{M}} G_t(\mathbf{x}, \mathbf{y}) f(\mathbf{y}) \mathrm{d}\mathbf{y}$ is also a linear operator, which can be thought as the pseudo-inverse or exponential of $\mathcal{A}$, i.e., $\mathcal{A}^{\dagger}$, $e^{-t\mathcal{A}}$, $e^{it\mathcal{A}}$, respectively, for these PDEs. Loosely speaking, solving these three PDEs amounts to applying the three operators to $f(\cdot)$. For additional details, we refer the reader to Evans (1998).

When $\mathcal{A} = -\Delta_{\mathcal{M}}$, the exponential operator becomes the Laplacian heat kernel $e^{t\Delta_{\mathcal{M}}}$. This relation allows us to compute the heat kernel from the Laplacian operator, bypassing the difficulty of discretizing the "exponentiated" operator directly. When $\mathcal{A}$ is a covariant derivative operator or connection Laplacian operator, the exponentiated operator becomes an operator of parallel transport along vector field or geodesics, respectively, with applications to vector field analysis and processing (Azencot et al., 2015; Sharp et al., 2018) and manifold learning (Singer and Wu, 2012; Wu, 2017).

3.3 Operator derivation and discretization

An operator associated to a smooth surface has to be discretized to be used in the computational setting; there are many techniques for operator discretization that are relevant to this task. On regular grids, differentials can be approximated by finite differences, in a consistent manner such that the approximation converges to the differential quantity as the grid resolution increases (Atkinson and Han, 2005). On irregular domains like meshes, many discretization methods are possible, including the finite element method (FEM) and the finite volume method (FVM).

The goal of operator discretization or construction is to obtain a matrix **A**, accompanied with a *mass matrix* **M**, such that the continuous equation $\mathcal{A}u = f$ is discretized as the linear equation $\mathbf{Au} = \mathbf{Mf}$, where **u**, **f** are vectors discretizing $u(\cdot)$ and $f(\cdot)$, e.g., **u**, **f** can be vectors storing the values of $u(\cdot)$ and $f(\cdot)$ at mesh vertices, respectively. In the scenario that function $u(\cdot)$ is given, intuitively the action of an operator $\mathcal{A}$, i.e., applying $\mathcal{A}$ to a scalar function $u \in \mathcal{L}^2(\mathcal{M})$ to yield $f \in \mathcal{L}^2(\mathcal{M})$, is approximated by matrix multiplication: $\mathbf{f} = \mathbf{M}^{-1}\mathbf{Au}$; ideally **f** should converge to the function $f(\cdot)$ under mesh refinement. In the other scenario when function $f(\cdot)$ is given, usually the major case of interest, ideally $\mathbf{u} = \mathbf{A}^{-1}\mathbf{Mf}$ should converge to the continuous solution $u(\cdot)$ under mesh refinement. The weak form corresponding to $\mathcal{A}$ is approximated by the discrete quadratic form: $a(u,v) \approx \mathbf{u}^{\top}\mathbf{Av}$; again **u**, **v** are discretization of functions $u(\cdot)$, $v(\cdot)$. We summarize the typical approaches used to obtain the matrices **A** and **M** below.

3.3.1 Finite and boundary element methods

The *finite element method* (FEM) and *boundary element method* (BEM) (Steinbach, 2007; Śmigaj et al., 2015) are frameworks for discretizing operators in a convergent fashion. The idea of FEM and BEM is to consider the weak form (4), which has to hold for every test function v, and to restrict u, v and f to finite-dimensional linear subspaces $\{\phi_i(\mathbf{x})\}_{i=1}^n$ and $\{\psi_i(\mathbf{x})\}_{i=1}^m$. This finite-dimensional approximation can be written $u = \sum_{i=1}^n \mathbf{u}_i\phi_i$, $f = \sum_{i=1}^m \mathbf{f}_i\psi_i$, where $\mathbf{u} \in \mathbb{R}^n, \mathbf{f} \in \mathbb{R}^m$ stack the coefficients of u and f in the basis. Allowing the test function v to be any basis function in $\{\psi_i(\mathbf{x})\}_{i=1}^m$, Eq. (4) becomes a finite-dimensional linear system $\mathbf{Au} = \mathbf{Mf}$, where

$$\mathbf{A} \in \mathbb{R}^{m\times n} : \mathbf{A}_{ij} = a(\psi_i, \phi_j) = \langle \psi_i, \mathcal{A}\phi_j \rangle_{\mathcal{M}} \tag{18}$$

$$\mathbf{M} \in \mathbb{R}^{m\times m} : \mathbf{M}_{ij} = \langle \psi_i, \psi_j \rangle_{\mathcal{M}}. \tag{19}$$

The most commonly used basis is the piecewise-linear basis, i.e., the "hat functions" on a triangle mesh, as well as the piecewise-constant functions, specifying per-vertex and per-triangle data, respectively. Fig. 3 shows an example of piecewise-linear basis on a triangle mesh.

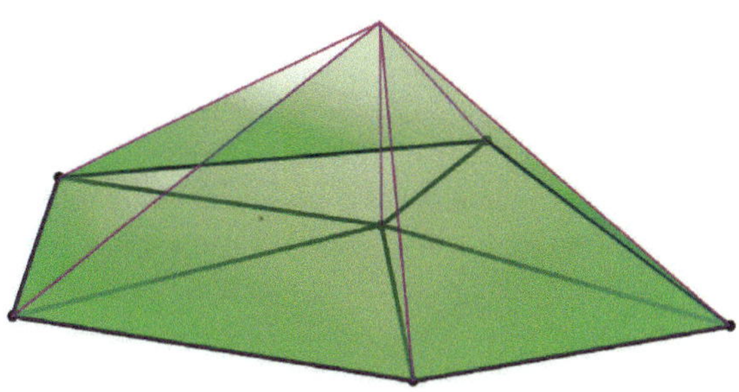

FIG. 3 The "hat" basis function corresponding to the vertex in the centre.

The *Galerkin method* refers to the case that $\{\phi_i(\mathbf{x})\}_{i=1}^n$ and $\{\psi_i(\mathbf{x})\}_{i=1}^m$ are chosen to be the same basis, so $\mathbf{A}$ becomes a p.s.d. matrix. FEM mostly applies to differential operators whose action is local, leading to a sparse matrix $\mathbf{A}$ with sparsity specified by the vertex adjacency. BEM follows the same procedure above, but applies to boundary integral operators instead; this leads to differences in how the inner product $a(\psi_i, \phi_j)$ is evaluated; we postpone the discussion of BEM until it is used in Section 5.17.

3.3.2 Discrete exterior calculus (DEC)

DEC (Desbrun et al., 2005) provides an alternative framework to derive operators, building discrete operators operating on per-element quantities directly. DEC gives discrete equivalents of the exterior derivative, Hodge star, and other operators used as building blocks to construct more complicated differential operators on a surface. Integrated forms and operators are defined analogously to their continuous counterparts but on mesh elements; they are typically designed so that discrete versions of important theorems hold exactly, such as the divergence theorem or other specializations of the generalized Stokes' theorem for differential forms (Frankel, 2011). Sometimes the operator derived using DEC coincides with the one derived by linear FEM; the cotangent Laplacian is such an example (see Section 5.2).

3.3.3 Mixed methods and compositions

Complicated geometric operators often can be derived as the composition of "simpler" operators, like differential operators of lower orders, reducing the difficulty of discretization. As examples, the bi-Laplacian applies the Laplacian twice (Botsch and Kobbelt, 2004; Jacobson et al., 2011); the Hessian operator is composed of the gradient operator and the matrix divergence operator (Stein et al., 2018); and in Wang et al. (2018b), the Dirichlet-to-Neumann operator is constructed as the composition of a few boundary integral operators that are straightforward to discretize. Such composition usually can be justified by the mixed finite element method.

3.3.4 Graph affinity Laplacian

For a graph with n vertices connected by edges, its graph Laplacian $\mathbf{L} \in \mathbb{R}^{n\times n}$, is a matrix such that $\mathbf{L}_{ij} \neq 0$ for $i \neq j$ if and only if vertices i, j are adjacent and $\mathbf{L}_{ii} = -\sum_{j\neq i} \mathbf{L}_{ij}$. Spectral analyses of graph Laplacians have led to remarkable success for graph partitioning and image segmentation (Shi and Malik, 2000). Entries in the graph Laplacian are weights measuring the affinity or similarity between vertices; a larger edge weight $\mathbf{L}_{ij}$ indicates a stronger bond between vertex i and j such that, e.g., in graph partitioning the edge ij should be less likely to be cut.

Triangle meshes can be viewed as graphs, and hence graph-based methods can be applied to meshes. Moreover, the discrete operators obtained using other methods, e.g., the cotangent Laplacian (Section 5.2), can be thought as particular versions of the graph Laplacian with local-geometry-aware weights that typically depend on angles and vertex areas in the triangle mesh. With this connection between discrete operators and graph Laplacians, sometimes the discrete operator matrix $\mathbf{A}$ can be constructed heuristically, as long as matrix entry $\mathbf{L}_{ij}$ provides some measure of affinity between vertex i and vertex j.

3.3.5 Variants and generalizations

The domain $\mathcal{D}$ and image $\mathcal{I}$ of a general linear operator $\mathcal{A} : \mathcal{D} \to \mathcal{I}$ are not necessarily the same function space, in which case the discrete operator becomes a rectangular matrix $\mathbb{R}^{m \times n}$ rather than a square matrix. As an example, the gradient operator (see, e.g., Tong et al., 2003) maps a scalar function to a vector field, which live in different function spaces. We may also obtain a rectangular matrix due to the choice of basis: In Eq. (18), if $\{\phi_i(\mathbf{x})\}_{i=1}^{n}$ and $\{\psi_i(\mathbf{x})\}_{i=1}^{m}$ are chosen as spaces of piecewise linear and piecewise constant functions, respectively, then we obtain an operator of size $\mathbb{R}^{f \times n}$, i.e., #faces $\times$ #vertices. The function space can go beyond the set of real-valued functions. For example, the Dirac operator (Liu et al., 2017) maps quaternion-valued functions $\mathcal{L}^2(\mathcal{M}; \mathbb{H})$ to $\mathcal{L}^2(\mathcal{M}; \mathbb{H})$, and the operator becomes a quaternion-valued matrix in $\mathbb{H}^{m \times n}$. Operators can also act on vector fields or forms rather than scalar fields, e.g., divergence and curl operators (Tong et al., 2003).

3.3.6 Operator properties and desiderata

When discretizing an operator, it is desirable that certain algebraic properties of the continuous operator transfer to the discrete operator. In this case, theorems and properties of some continuous functions can be preserved in a discrete sense. As an example, applying the Laplacian to a constant function yields zero function: $\Delta_{\mathbf{x}} 1(\mathbf{x}) = 0(\mathbf{x})$. Accordingly, for a discrete Laplacian $\widetilde{\mathbf{L}}$, we would like $\widetilde{\mathbf{L}}\mathbf{1} = \mathbf{0}$. One may ask for a discretization of a discrete operator that preserves as many of the properties of the smooth operator as possible. It is usually not possible for a discrete operator to enjoy all possible discrete analogues of smooth properties; see Wardetzky et al. (2007) for a "no free lunch" theorem in the case of discrete Laplacians.

3.4 Operators and geometry

Geometric PDEs (Taylor, 2011, 2013) defined on $\mathcal{M}$ as well as geometric operators like $\Delta_{\mathcal{M}}$ encode and reveal information about the geometry and topology of $\mathcal{M}$. Several methods in shape analysis and geometry processing are based on this observation.

Geometric analysis is a branch of mathematics that applies analytical tools to operators and PDEs for purposes of studying the underlying geometry.

It also connects the *global geometry* of a surface with its *topology*. As an example, elliptic differential operators on $\mathcal{M}$ satisfy the Atiyah-Singer index theorem (Booss, 2012), which relates analytical properties of the operator to the topology of $\mathcal{M}$. As a special case, the familiar Gauss–Bonnet theorem equates the total Gaussian curvature $\int_{\mathcal{M}} K(\mathbf{x})$ on $\mathcal{M}$ with a topological quantity—the Euler characteristic—distinguishing spheres, torii, and other surfaces with varying genus. Equipped with the tools to study geometry and topology quantitatively, Perelman's proof of the Poincaré conjecture (Perelman, 2002, 2003a,b) is among the most remarkable achievements of geometric analysis.

Spectral geometry studies the relationship between spectral properties of operators and the underlying geometry. As an example, the (Laplacian) heat kernel $k_t(\mathbf{x}, \mathbf{y})$ is related to the geodesic distance $d_g(\cdot, \cdot)$ via the following formula obtained by Varadhan (1967):

$$\lim_{t \to 0} t \log k_t(\mathbf{x}, \mathbf{y}) = -\frac{1}{4} d_g^2(\mathbf{x}, \mathbf{y})$$

In graphics, this relation is exploited for efficient geodesic distance computation by the *heat method* (Crane et al., 2013b).

Spectrum asymptotics provide another example relating an operator with the geometry. Weyl's Law gives the following asymptotic formula for Dirichlet eigenvalues of Laplacian on a domain $\Omega \subseteq \mathbb{R}^d$:

$$\lambda_j \sim \frac{(2\pi)^2}{(\mathrm{vol}(\Omega)|B(d)|)^{2/d}} j^{2/d}, \quad j \to \infty$$

where $|B(d)|$ is the volume of the unit ball in $\mathbb{R}^d$, and vol(Ω) is the volume of Ω. See Chavel (1984), Rosenberg (1997), Zelditch (2009), and Craioveanu et al. (2013) for Weyl's law and other results relating the Laplacian spectrum to geometric quantities. The recent survey (Henrot, 2017) summarizes progress in spectral geometry with connections to shape optimization, which studies problems such as determining the shape that maximizes or minimizes λ_j for certain j.

3.5 Inverse problems

While an operator and its spectrum can be deterministically computed on a given surface $\mathcal{M}$, *inverse problems* study what information about $\mathcal{M}$ can be recovered and extracted from an operator or its spectrum.

3.5.1 *Shape-from-operator*

A central inverse problem for operator-based geometry is to ask, given a particular operator, whether or not we can recover the shape, possibly up to rigid motion or isometry. Below we list some examples drawn from the smooth and discrete geometry literatures:

- *Laplacian*: In continuous theory, the Laplacian determines the metric of a surface; accordingly, one can recover edge lengths and vertex coordinates (up to isometry) from the discrete Laplacian operator (Zeng et al., 2012; Boscaini et al., 2015; Corman et al., 2017a; Chern et al., 2018).
- The *single layer potential* is another example of an operator for which the inverse problem above is well-posed, since it encodes the inverse distance between pairs of vertices; multidimensional scaling can recover the embedding from distances, up to rigid motion, see, e.g., Williams (2002).
- The *Dirichlet-to-Neumann* operator determines the first and second fundamental forms of the underlying surface, which in turn determines its geometry up to rigid motion due to the Bonnet theorem (Wang et al., 2018b).
- *Dirac*: Although this property is not explicitly mentioned in the geometry processing literature, it is also straightforward to see that a mesh can be recovered up to translation from the strong form of discrete relative Dirac operator (Crane et al., 2011; Liu et al., 2017; Kostrikov et al., 2018), which stacks edge vectors in an edge-vertex adjacency graph.

Algorithms have been proposed to recover shape from forms (Wang et al., 2012), metrics (Chern et al., 2018), and operators (Boscaini et al., 2015; Corman et al., 2017a), although many of these techniques rely on nonlinear methods without guarantee of success.

3.5.2 Shape-from-spectrum

A more difficult question is whether or not it is possible to uniquely recover a shape from an operator's eigenvalues only, again up to rigid motion or isometry. This is a famous problem in mathematics: "Can one hear the shape of a drum?," as posed by Kac (1966). This question is also of practical interest: Eigenvalues are used as "ShapeDNA," global shape signatures, and "shape2vector" features, using various operators (Reuter et al., 2006; Liu et al., 2017; Wang et al., 2018b), and hence the inverse problem is equivalent to asking whether these signatures are unique/invertible.

For the Laplacian operator, in the general setting the answer is negative: there exist isospectral surfaces (Gordon et al., 1992a, b). However, if the shape space is restricted to convex planar regions with analytic boundary, the answer is positive (Zelditch, 2000). Furthermore, it remains an open question for non-convex analytic planar domains; also, for the case that the shape is a closed 2-manifold or 2-manifold with analytical boundary, which attracts the most practical interest, the question is still open. The recent work (Cosmo et al., 2019) tackles this inverse problem using differentiable eigensolvers, leading to empirical successes recovering surfaces from their Laplacian spectra.

Few results are known for operators other than Laplacian. For the intrinsic Dirac operator, examples of Dirac-isospectral tori are given in Ammann and Bär (1998). For Dirichlet-to-Neumann operator (to the best of our knowledge), whether "hearing the shape" for 3D surfaces is possible is still an open challenge. A recent result indicates that the 2D polyline can be recovered from its Steklov eigenvalues (Levitin et al., 2017).

4 Spectral shape analysis and applications

In this section, we discuss how practical algorithms can benefit from geometric operators and their eigendecompositions. Before we look into particular methods, we first present a few ideas that motivate the use of geometric operators and their eigendecompositions in computational tasks.

4.1 Spectral analysis: From Euclidean space to manifold

4.1.1 A signal processing viewpoint

Spectral methods provide bases to represent geometric signals and associated data. Particularly, eigenfunctions $\{\phi_i\}_{i=0}^{\infty}$ of an appropriate operator (see Section 3.1.3) generalize the Fourier basis 1 − periodic functions $\mathbb{T}(1) = \{f \in \mathcal{L}^2(\mathbb{R}) : f(x) = f(x+1)\}$, i.e., $\{\sin(2\pi nx)\}_{n=1}^{\infty} \cup \{\cos(2\pi nx)\}_{n=0}^{\infty}$, to a curved manifold $\mathcal{M}$. This provides a basis to represent geometric signals, allowing us to generalize traditional signal processing and image analysis for data associated to points in $\mathbb{R}^n$, to data on a geometric domain. Fig. 4 shows an example of reconstructing geometric signals—in particular mesh coordinates—using its Laplacian eigenfunctions. Typically, the Laplacian eigenfunctions are used as the generalization of Fourier bases, but eigenfunctions of any p.s.d. operator $\mathcal{A}$ that form a complete basis can be used as well. Applying this idea to problems in graphics dates back to *curve Fourier descriptors* (Zahn and Roskies, 1972) and *surface signal processing* (via Laplacian) (Taubin, 1995).

4.1.2 Multiresolution and hierarchical representation

Spectral methods enable multiresolution analysis on meshes. The multiresolution approach processes information and features at different scales separately. Inspired by the success of multiresolution methods in signal processing and image analysis, the importance of multiresolution analysis on meshes has been recognized since the pioneering work of Eck et al. (1995) and Kobbelt (1997).

Guskov et al. (1999) generalize image-based Laplacian pyramids and wavelets to meshes, based on mesh decimation and upsampling procedures. As an alternative, Mahadevan (2007) and Zhong and Qin (2014) construct wavelets out of Laplacian eigenfunctions. This spectral approach has a natural advantage: It separates geometric signals at increasing frequencies, making it

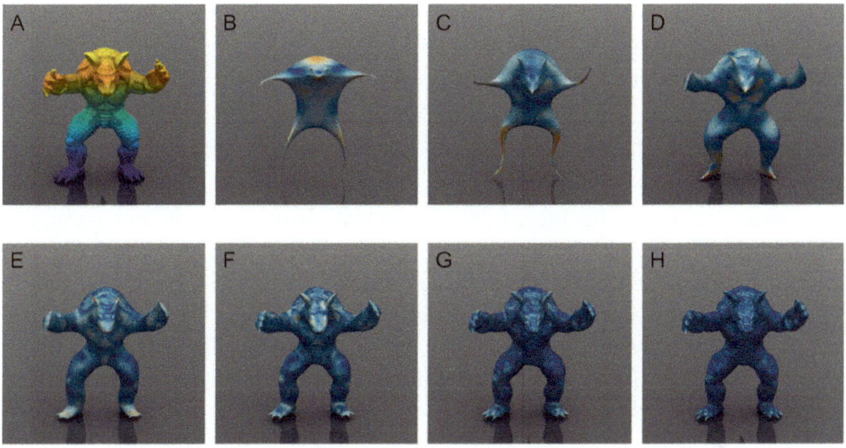

FIG. 4 Reconstructing mesh coordinates using increasing numbers of eigenfunctions. The colour encodes the error of reconstructed coordinates. (A) Rest pose; (B) 10; (C) 20; (D) 40: (E) 100; (F) 200; (G) 400; (H) 800.

easy to apply multiresolution analysis to geometric data. The multiresolution consideration is also influential in designing practical algorithms: A recurrent strategy we will see in this section is to apply the multiscale organization of a spectral basis to obtain a *coarse-to-fine* hierarchy, an idea that appears in hierarchical segmentation (De Goes et al., 2008; Huang et al., 2009) and approximation of multiscale descriptors (Vaxman et al., 2010).

4.1.3 Spectral shape analysis and processing

Motivated by the facts above, geometric operators and their spectral decompositions provide a versatile toolbox for mesh processing, known as *spectral mesh processing*. Zhang et al. (2010a, b) survey this topic, focusing on Laplacian-based methods. Although most methods in spectral mesh processing were originally introduced for Laplacian, many remain valid if we simply substitute the Laplacian with an alternative operator, e.g., any of the operators discussed in Raviv et al. (2011b), Hildebrandt et al. (2012), Aflalo et al. (2013), Liu et al. (2017), Wang et al. (2018b), and Ye et al. (2018).

4.2 Spectral data analysis

Spectral shape analysis is strongly influenced by developments and advances in spectral data analysis, particularly *spectral graph theory* (Chung and Graham, 1997) and *manifold learning* (Cayton, 2005). These areas study the geometric structures of graphs and data manifolds, usually sharing similar considerations with problems in analyzing 3D surfaces. As we will see in the section, many spectral or operator-based methods in shape analysis can be thought as adaptions or variants of prior counterparts in these areas; as an example, Rustamov (2009) points out an equivalence between the Laplacian mesh editing and semisupervised learning. For this reason, we highlight relevant methods in spectral data analysis.

4.2.1 Spectral graph theory with applications

The analogy between spectral shape analysis and spectral graph theory has been explored since the inception of spectral shape analysis and geometry processing (Gotsman, 2003; Gotsman et al., 2003). Indeed, many discrete geometric operators can be thought as special instances of the graph Laplacian or adjacency matrix, with geometry-aware entries, i.e., weights determined by geometry of the underlying polygonal mesh like edge lengths or angles. From this view, discrete mesh analysis might be considered a *specialization* of graph analysis.

Earlier work in (spectral) shape analysis uses various versions of graph Laplacians constructed on the mesh graph (Taubin, 1995; Liu and Zhang, 2007), largely inspired by applications of spectral graph theory to image analysis and computer vision (Shi and Malik, 2000). While graph combinatorial measurements such as the average node degree are of interest for graph analysis, this information is irrelevant to geometry processing, reflecting only, e.g., which polygon mesh representation is used or the meshing algorithms generated the data. To alleviate sensitivity to graph combinatorics, graph Laplacians, as well as their spectra (Levy, 2006), were quickly replaced by the cotangent Laplacian, which converges to the smooth Laplace–Beltrami operator and exhibits more robust behaviour (Dyer et al., 2007).

Many concepts and algorithms used for mesh processing and analysis are exactly same as those for graphs. As an example, the *Fiedler vector* $\phi_1(\mathbf{x})$ is the eigenvector corresponding to the second-smallest eigenvalue of the graph Laplacian, which has been used for graph partitioning (Fiedler, 1975; Spielman and Teng, 1996). *Fiedler theory* reveals the number of zero Laplacian eigenvalues equals to the number of connected components in a graph; an identical statement holds for manifolds.

4.2.2 Manifold learning and data science

Manifold learning explores geometric structure in clouds of data points. Techniques in this area employ spectral methods for analyzing data manifolds, under the assumption that a dataset is an empirical sample from a low-dimensional manifold on which the data resides. Many of these methods have been adapted to analyze meshes or are incorporated in mesh processing pipelines, as we will see in Section 4.4.

Two classic examples from the early literature in manifold learning are *Iso-Map* (Tenenbaum et al., 2000) and *LLE* (*locally linear embedding*) (Roweis and Saul, 2000), which use spectral methods to find embeddings of data that preserve global and local distances within a set of data points, respectively. IsoMap uses *multidimensional scaling* (MDS) to embed data points, so that Euclidean distances in the embedded space preserve, as much as possible, the geodesic distances from the original manifold. Prior to IsoMap, Schwartz et al. (1989) propose to flatten a curved surface by minimizing a related distance-based distortion metric, giving a similar solution to what they referred to as the mapmaker's problem. LLE uses *k-nearest neighbours* (kNN) to build a graph over the data points, computes the (row-wise) normalized graph Laplacian $\mathbf{L}$ (which is asymmetric) and its singular value decomposition (SVD) $\mathbf{L} = \mathbf{U}\Sigma^{\frac{1}{2}}\mathbf{V}^{\top}$, and finally outputs $\mathbf{V}\Sigma^{\frac{1}{2}}$ as the embedding.

Laplacian LLE Belkin and Niyogi (2003) is an variant improving LLE by considering the nonnormalized graph Laplacian, which is symmetric and whose spectrum provably converges to that of the Laplace–Beltrami operator and applying an eigen-decomposition rather than the SVD; this is identical to how the Laplacian typically is used in shape analysis on meshes. These methods inspire mesh parameterization methods, as we will see in Section 4.4. *Hessian LLE* (Donoho and Grimes, 2003) replaces the Dirichlet energy used in Laplacian LLE with the Hessian energy (whose formula is given later in Eq. (32). There are various followups including *local tangent space alignment* (*LTSA*) (Zhang and Zha, 2004). LTSA first constructs the approximation of tangent space at each data point, and then aligns all tangent spaces, yielding a global parameterization of the underlying manifold. Later Rosman et al. (2010) take the topological properties of the embedding into consideration. With spirit similar to LTSA, *spectral affine-kernel embeddings (SAKE)* (Budninskiy et al., 2017) introduce the multi-Laplacian quadratic form $\mathbf{Q}$, which is assembled from local operators $\mathbf{L}_i$ whose null space only contain locally affine functions around each point i, and replaces the Laplacian matrix previously used in manifold learning with this new matrix $\mathbf{Q}$. Recent *parallel transport unfolding* (*PTU*) (Budninskiy et al., 2019) replaces the geodesic distances obtained via Dijkstra paths as used in IsoMap with geodesic distances obtained via parallel transport, removing the limitation that Isomap can handle only geodesically convex sampled domains. PTU can generate a quasi-isometric, low-dimensional embedding of a manifold with arbitrary topology.

Diffusion maps (Coifman and Lafon, 2006) draw connections between the row-normalized graph Laplacian $\mathbf{P}$ (defined as follows) and Markov chains. They relate $\mathbf{P}$ to anisotropic diffusion, proving that $\mathbf{P}$ converges in a pointwise fashion to the Laplace–Beltrami operator plus a term depending on the pointwise sampling density, when $t \mapsto 0^+$. In particular, $\mathbf{P}$ is defined by taking

$$\mathbf{W}_{ij} = \exp\left(-\frac{d(\mathbf{x}_i, \mathbf{x}_j)^2}{t}\right) \quad \text{and} \quad \mathbf{P} = \mathbf{D}^{-1}\mathbf{W},$$

where $d_i = \sum_{j=1}^{n} \mathbf{W}_{ij}$ and $\mathbf{D} = \text{diag}(d_i)$, i.e., $\mathbf{D}$ is a matrix that stacks row sums from $\mathbf{W}$ along the diagonal. After defining the operator $\mathbf{P}$, we can compute the *diffusion embedding*

$$\left(e^{-\lambda_1 t}\phi_1(\mathbf{x}), e^{-\lambda_2 t}\phi_2(\mathbf{x}), \ldots, e^{-\lambda_k t}\phi_k(\mathbf{x})\right) \tag{20}$$

for clustering data points, an embedding designed to map similar points on the manifold also close to each other in the embedded space, while being robust and insensitive to noise.

The $\mathbf{P}$ used above is an example of a *point cloud Laplacian* operator. For applications like diffusion maps, various results on the convergence of different versions of point cloud Laplacians are derived in Belkin and Niyogi (2005), Hein et al. (2005), Belkin et al. (2008, 2009), Liang and Zhao (2013), and Shi and Sun (2017).

There are related methods in machine learning. Multiple methods (Shi and Malik, 2000; Meila and Shi, 2001; Ng et al., 2002) employ eigenfunctions of a Laplacian or affinity matrix for spectral clustering and segmentation on data points. Harmonic interpolation solves a Laplacian system for label propagation in semisupervised learning (Zhu et al., 2003). *Manifold regularization* (Belkin et al., 2006) proposes to add Dirichlet energy as a regularization term to the loss function in classification, yielding Laplacian support vector machines (LapSVM).

4.3 Spectral analysis: Point embedding, signature, and geometric descriptors

Rather than general clouds of data points, our focus in this survey is spectral analysis of geometric data expressed as a mesh. To this end, in this section we provide a few problems, including segmentation, point signature computation, and distance computation, whose solutions often explicitly involve eigenfunctions of a geometric operator, closely related to the methods we saw in Section 4.2.

4.3.1 Spectral embedding and segmentation

Methods for mesh vertex clustering (i.e., segmentation) follow patterns similar to those in clustering data manifolds.

As an example, the *global point signature* (*GPS*) (Rustamov, 2007) considers the following high-dimensional embedding of points $\mathbf{x}$ on a triangle mesh surface based on the eigenfunctions and eigenvalues of the Laplacian:

$$\left(\frac{\phi_1(\mathbf{x})}{\sqrt{\lambda_1}}, \frac{\phi_2(\mathbf{x})}{\sqrt{\lambda_2}}, \ldots, \frac{\phi_k(\mathbf{x})}{\sqrt{\lambda_k}}\right) \tag{21}$$

This embedding is isometry-invariant, thanks to the use of Laplacian eigenfunctions. Rustamov (2007) applies standard k-means algorithms to the embedded vertices, yielding an isometry-invariant segmentation. Fig. 5 illustrates the first

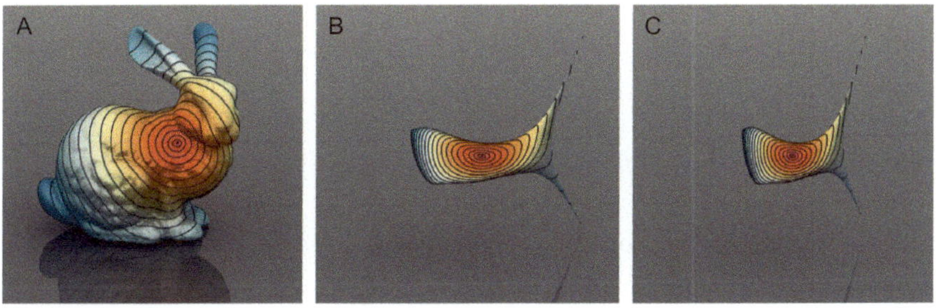

FIG. 5 The Stanford bunny model with a log heat kernel $\log k_t(\mathbf{x}, \cdot)$ is shown in (A). The same bunny model but replacing the x, y, z coordinates of a vertex $\mathbf{x}$ with (B) the harmonic embedding $\left(\frac{\phi_1(\mathbf{x})}{\sqrt{\lambda_1}}, \frac{\phi_2(\mathbf{x})}{\sqrt{\lambda_2}}, \frac{\phi_3(\mathbf{x})}{\sqrt{\lambda_3}}\right)$ or (C) the diffusion embedding $\left(e^{-\lambda_1 t}\phi_1(\mathbf{x}), e^{-\lambda_2 t}\phi_2(\mathbf{x}), e^{-\lambda_3 t}\phi_3(\mathbf{x})\right)$.

few dimensions of the diffusion and harmonic embeddings, respectively. Later methods derive analogues of this embedding for different operators, with success for segmentation (Hildebrandt et al., 2012; Liu et al., 2017; Wang et al., 2018b).

4.3.2 *Spectral distances*

Computing the distance between any pair of points on the surface is a fundamental task for shape analysis. While the geodesic distance is a natural choice, it is sensitive to noise and small topology changes; moreover, computing full pairwise geodesic distances is expensive. Spectral distances including diffusion distance (Coifman and Lafon, 2006) and biharmonic distance (Lipman et al., 2010) are alternatives with favourable robustness properties.

The diffusion distance $d_D(\cdot, \cdot)$ and biharmonic distance $d_B(\cdot, \cdot)$ between points on surface $\mathbf{x}, \mathbf{y} \in \mathcal{M}$ are defined as L^2 distances between the corresponding spectral embeddings:

$$d_D(\mathbf{x}, \mathbf{y})^2 = \sum_{i=1}^{\infty} e^{-2t\lambda_i}(\phi_i(\mathbf{x}) - \phi_i(\mathbf{y}))^2, \quad d_B(\mathbf{x}, \mathbf{y})^2 = \sum_{i=1}^{\infty} \frac{1}{\lambda_i^2}(\phi_i(\mathbf{x}) - \phi_i(\mathbf{y}))^2.$$

The following equalities for d_D and d_B provide interpretations of these distances: The distance between $\mathbf{x}$ and $\mathbf{y}$ is the integrated difference between the Green's functions or heat kernel functions at $\mathbf{x}$ and $\mathbf{y}$:

$$d_D(\mathbf{x}, \mathbf{y})^2 \equiv k_{2t}(\mathbf{x}, \mathbf{x}) + k_{2t}(\mathbf{y}, \mathbf{y}) - 2k_{2t}(\mathbf{x}, \mathbf{y}) \equiv \int_{\mathcal{M}} (k_t(\mathbf{x}, \mathbf{z}) - k_t(\mathbf{y}, \mathbf{z}))^2 \mathbf{dz},$$

$$d_B(\mathbf{x}, \mathbf{y})^2 \equiv G_B(\mathbf{x}, \mathbf{x}) + G_B(\mathbf{y}, \mathbf{y}) - 2G_B(\mathbf{x}, \mathbf{y}) \equiv \int_{\mathcal{M}} (G(\mathbf{x}, \mathbf{z}) - G(\mathbf{y}, \mathbf{z}))^2 \mathbf{dz},$$

where $G_B(\mathbf{x}, \mathbf{y}) = \sum_{k=1}^{\infty} \frac{1}{\lambda_k^2}\phi_k(\mathbf{x})\phi_k(\mathbf{y})$ is the biharmonic Green's function. Note diffusion distance is not a true metric, since $d_D(\mathbf{x}, \mathbf{y}) = 0$ does not necessarily imply $\mathbf{x} = \mathbf{y}$; this issue is addressed by the biharmonic distance (Lipman et al., 2010).

The commute-time distance or harmonic distance (Qiu and Hancock, 2007) is popular in graph analysis, due to connections to random walks on graph:

$$d_H(\mathbf{x},\mathbf{y})^2 = \sum_{i=1}^{\infty} \frac{1}{\lambda_i}(\phi_i(\mathbf{x}) - \phi_i(\mathbf{y}))^2 \equiv G(\mathbf{x},\mathbf{x}) + G(\mathbf{y},\mathbf{y}) - 2G(\mathbf{x},\mathbf{y}).$$

However, it cannot be defined on continuous surfaces since $G(\mathbf{x},\mathbf{y})|_{\mathbf{y}=\mathbf{x}}$ is singular.

4.3.3 Point signatures and descriptors

The goal of shape descriptors or signatures is to compute a vector per point $\mathbf{x}$ on a surface, such that the vector summarizes and encodes local (and sometimes global) geometry near $\mathbf{x}$. After discretization, these descriptors are expressed as per-vertex feature vectors that can be used for higher-level tasks such as shape matching or correspondence:

- The *global point signature* (GPS) (Rustamov, 2007) uses the embedding in Eq. (21) as a signature, which can be ambiguous due to rotations within eigenspaces:

$$\mathrm{GPS}(\mathbf{x}) : \mathcal{M} \to \mathbb{R}^K, \quad \mathrm{GPS}(\mathbf{x}) := \left(\frac{\phi_1(\mathbf{x})}{\sqrt{\lambda_1}}, \frac{\phi_2(\mathbf{x})}{\sqrt{\lambda_2}}, \frac{\phi_3(\mathbf{x})}{\sqrt{\lambda_3}} \right). \tag{22}$$

- The *heat kernel signature* (*HKS*) (Sun et al., 2009) uses the diagonal of the heat kernel as a signature, summarizing the geometry near a point $\mathbf{x}$:

$$\mathrm{HKS}(\mathbf{x},t) : \mathcal{M} \times \mathbb{R}^+ \to \mathbb{R}, \quad \mathrm{HKS}(\mathbf{x},t) := k_t(\mathbf{x},\mathbf{x}) \equiv \sum_{i=0}^{\infty} e^{-\lambda_i t}\phi_i(\mathbf{x})^2. \tag{23}$$

  The HKS is not subject to the rotation ambiguity of the global point signature (GPS). Additionally, an "informative property" shows that for any $\mathbf{x}$, HKS $(\mathbf{x},\cdot)$ determines $\mathcal{M}$ up to isometry. The HKS has been extended using operators beyond the Laplacian, leading to the affine invariant HKS (Raviv et al., 2011a), the modified Dirichlet HKS (Hildebrandt et al., 2012), the Steklov HKS (Wang et al., 2018b), and the Dirac HKS (Liu et al., 2017). Fig. 6 illustrates a typical heat kernel signature with a fixed small time, computed using the Steklov eigenfunction, combining Sun et al. (2009) and Wang et al. (2018b): we see the signature roughly aligns with the local curvature.

  Developed in parallel to the HKS, the *auto diffusion function* (*ADF*) (Gebal et al., 2009) proposes to consider the time-varying scalar field $\mathrm{ADF}_t(\mathbf{x}) := k_{t/\lambda_1}(\mathbf{x},\mathbf{x})$, which is a time scaled version of HKS, for segmentation and skeletonization.
- The *scale invariant heat kernel signature* (*SI-HKS*) (Bronstein and Kokkinos, 2010) modifies the HKS to factor out the effect of scale using operations in the spatial and frequency domains.

- The *wave kernel signature* (*WKS*) (Aubry et al., 2011) is derived from the Schrödinger equation, Eq. (14), whose solutions can be written in the form $u(\mathbf{x},t)=\sum_{k=0}^{\infty} e^{i\lambda_k t}\phi_k(\mathbf{x})f_E(\lambda_k)$. In particular, the WKS is defined as follows:

$$\mathrm{WKS}(\mathbf{x},E):\mathcal{M}\times\mathbb{R}^+\to\mathbb{R},\quad \mathrm{WKS}(\mathbf{x},E):=\lim_{T\to\infty}\int_0^T \|u(\mathbf{x},t)\|^2\equiv\sum_{k=0}^{\infty} f_E^2(\lambda_k)\phi_k(\mathbf{x})^2,$$

where $f_E^2(\cdot)$ is chosen as a log-normal distribution function with parameter E:

$$f_E^2(x)=\left[\sum_k e^{\frac{-(\log E-\log\lambda_k)^2}{2\sigma^2}}\right]^{-1} e^{\frac{-(\log E-\log x)^2}{2\sigma^2}}.$$

The WKS can be interpreted as the average (over time) probability density at $\mathbf{x}$. Compared to the HKS, the WKS is more discriminative to geometric features and information at small scales, usually leading to more accurate shape matching.
- From a signal processing perspective, the HKS and WKS apply different one-parameter filter functions (parameterized by t or E) to λ, i.e., $e^{-\lambda t}$ and $f_E^2(\lambda)$, respectively. *Optimal spectral descriptors* (Litman and Bronstein, 2014) learn this filter function rather than using a pre-designed function.
- The *discrete-time evolution process* (*DEP*) descriptor (Melzi et al., 2018a) considers a discrete-time process rather than a continuous one as in HKS and WKS. The discrete-time process is fully specified by a generic pairwise relation function $A_\delta(\mathbf{x},\mathbf{y})$ depending on a scale parameter δ, similar to the heat kernel $k_t(\mathbf{x},\mathbf{y})$ which depends on a time parameter t in HKS. Given an initial function $f^{(0)}(\cdot)$, this iterative evolution is very similar to the continuous diffusion in Eq. (12), with time t replaced with discrete step index $l\in\mathbb{Z}$:

$$f_\delta^{(l)}(\mathbf{x})=\int_{\mathcal{M}} A_\delta(\mathbf{x},\mathbf{y})f_\delta^{(l-1)}(\mathbf{y})d\mathbf{y}.$$

The DEP point signature is defined as follows:

$$\mathrm{DEP}(\mathbf{x},\{1,2,\ldots,K\}):\mathcal{M}\times\{1,2,\ldots,K\}\to\mathbb{R},\quad \mathrm{DEP}(\mathbf{x},k):=s_{\delta_k}(\mathbf{x}).$$

in which the score function $s_\delta(\mathbf{x})$ is defined as

$$s_\delta(\mathbf{x}):=\sum_{l=1}^{\infty} r_\delta^l f_\delta^{(l)}(\mathbf{x}),$$

where r_δ is a constant that depends on δ.
- The *intrinsic shape context* (*ISC*) descriptor (Kokkinos et al., 2012) builds a local geodesic chart system for each $\mathbf{x}\in\mathcal{M}$ and uses the average of some

FIG. 6 The heat kernel signature computed using the Steklov eigenfunctions.

per-vertex descriptor $I(\mathbf{x})$ over intrinsic angular bins and radial bins within the chart. This corresponds to diffusion in the chart; Geodesic CNNs (GCNN) (Masci et al., 2015) apply similar operations in learnable settings.

4.4 Shape analysis and geometry processing

That operators and their spectra encode geometric information has been explored by a variety of applications in geometry processing, shape analysis, and computer graphics. Early work utilizing geometric operators include Reuter et al. (2005), which proposes to use eigenvalues of the continuous Laplace–Beltrami operator as a shape representation, and Levy (2006) which considers the Laplacian eigenfunctions as a means to understand geometry, with applications to surface registration, segmentation and parameterization. Since then a large number of methods—mainly based on the Laplacian—has been developed.

Geometric operators encode geometric information, from which topological properties can also be inferred. We first look at segmentation, skeletonization, and quadrangulation, which leverage the fact that topological structures can be extracted from level sets of functions derived from eigenfunctions and eigenvalues. In particular, the zero level-set of an eigenfunction is called the *nodal set*, which appears in many methods. Geometric operators also encode information such as symmetry. Particularly symmetric patterns in eigenfunctions and derived descriptors reveal symmetry in the shape. Ovsjanikov et al. (2008) embed shape vertices to their point-wise spectral signature and reduces the problem of intrinsic symmetry detection to determining Euclidean symmetries in the signature space. Thomas and Natarajan (2014) propose a multiscale symmetry detection method by examining contours of level sets of first few eigenfunctions.

4.4.1 Segmentation

Liu and Zhang (2004) apply spectral clustering methods from other fields to mesh segmentation; their method applies the normalized cuts (Shi and Malik, 2000) to a point cloud Laplacian. Extending this work, Liu and Zhang (2007) embed meshes to 2D using the first two nonzero eigenfunctions of Laplacian and a related geometric operator, i.e., mapping $\mathbf{x} \in \mathcal{M} \to (\phi_1(\mathbf{x}), \phi_2(\mathbf{x})) \in \mathbb{R}^2$, and apply contour analysis to the projected planar shape for segmentation. De Goes et al. (2008) use diffusion distances to build a coarse-to-fine hierarchy of segments for articulated bodies. Huang et al. (2009) use eigenvectors of the Hessian of a nonlinear deformation energy instead of Laplacian, for hierarchically decomposing deformable shapes. Finally, Reuter et al. (2009) use the Laplacian nodal sets for shape segmentation.

Related, *spectral surface reconstruction* (Kolluri et al., 2004) applies a variant of spectral graph partitioning to a Delaunay tetrahedralization of a point cloud, to identify inside and outside tetrahedra, and outputs triangular faces adjacent to both inside and outside tetrahedra.

4.4.2 Skeletonization (and segmentation)

Skeletonization might be considered as a "dual" problem to segmentation. As an example of this duality, Shi et al. (2008) consider the Fiedler vector of an anisotropic Laplacian, i.e., the scalar function $\phi_1(\mathbf{x})$, and constructs its *Reeb graph*, intuitively a graph whose edges pass orthogonally to the isolines of $\phi_1(\mathbf{x})$ and whose vertices correspond to regions where the isolines have a particular connectivity; in this method, the vertices of the Reeb graph roughly correspond to segments on the surface determined by the Fiedler vector. Similarly, Patane et al. (2008) identify critical points (saddle and extreme points) of Laplacian eigenfunctions, apply isoline analysis to these points, and connect them to extract a Reeb graph, from which shape segmentations and skeletonizations can be then obtained.

Nodal sets of the Fiedler vector can be unstable. To address this issue, the auto diffusion function (Gebal et al., 2009) uses diffusion to define a time-varying scalar field $\text{ADF}_t(\mathbf{x}) := k_{t/\lambda_1}(\mathbf{x}, \mathbf{x})$ (see Section 4.3), evaluates $\text{ADF}_t(\mathbf{x})$ at multiple t values, extracts Reeb graphs for each of them, and finally outputs a skeleton combining these Reeb graphs. More broadly, the Reeb graph extracts a topological skeleton for $\mathcal{M}$, and hence can be used as a (graph-valued) shape descriptor.

4.4.3 Parameterization and remeshing

Laplacian eigenfunctions demonstrate patterns that can be exploited and extracted for surface parameterization and quadrangulation. The most famous relevant result, *Courant's theorem* states that the number of partitions by the nodal set for the k-th Laplacian eigenfunction $\phi_k(\cdot)$ is less or equal to k (Courant, 1923).

The *Morse–Smale complex* associated to a real-valued function $f : \mathcal{M} \rightarrow \mathbb{R}$ is a cellular decomposition of $\mathcal{M}$ into quadrangular patches. The edges of the complex form a graph on $\mathcal{M}$ whose vertices are critical points of f and whose edges roughly align with directions of the gradient $\nabla_{\mathcal{M}} f$. Laplacian eigenfunctions enjoy several properties that make them strong candidates as the generating function f. In particular, their minima and maxima appear alternately, and critical points are evenly distributed with respect to geodesic distance.

Spectral surface quadrangulation (Dong et al., 2006) generates an initial Morse–Smale complex from $\phi_k(\mathbf{x})$, where k is the expected number of nodal sets; in practice, k roughly equals the eigenfunction index thanks to Courant's theorem above (usually $k < 100$). Computation of the complex is followed by topological noise removal, patch boundary adjustment, and refinement. This yields a smooth parameterization that can be used to generate a quadrilateral mesh. Examples of this technique and comparison to other quad meshing algorithms are provided in the survey (Bommes et al., 2013).

Spherical parameterization is the task of mapping genus-0 meshes to the unit sphere. Gotsman et al. (2003) tackle this problem by optimizing for a graph Laplacian matrix on the mesh graph such that its spectral embedding $\mathbf{x} \in \mathcal{M} \rightarrow (\phi_0(\mathbf{x}), \phi_1(\mathbf{x}), \phi_2(\mathbf{x})) \in \mathbb{S}^3$ becomes a valid spherical parameterization.

Iso-charts (Zhou et al., 2004) apply an IsoMap-like procedure for mesh parameterization and spectral clustering, as basic steps in a pipeline to decompose a mesh into low-stretch atlases/charts. Based on the key observation that geodesic distance distortion is closely related to stretch distortion induced by parameterization, IsoMap is used to generate the initial parameterization. A similar idea appears in the earlier work (Zigelman et al., 2002), which applies MDS to preserve the surface geodesic distances and uses the first two dimensions of the optimal MDS embedding as a surface parameterization, leading to a texture map with low distance distortion.

The spectral embedding $(\mathbf{u}, \mathbf{v}) = (\phi_1(\mathbf{x}), \phi_2(\mathbf{x}))$ can be used as a parameterization with fairly low distortion (Liu and Zhang, 2007). *Spectral conformal parameterization* (Mullen et al., 2008) uses a generalized Fiedler vector of the *conformal energy* matrix $\mathbf{C}$, defined via

$$\begin{bmatrix} \mathbf{u} \\ \mathbf{v} \end{bmatrix}^{\mathsf{T}} \mathbf{C} \begin{bmatrix} \mathbf{u} \\ \mathbf{v} \end{bmatrix} := \mathbf{u}^{\mathsf{T}} \mathbf{L} \mathbf{u} + \mathbf{v}^{\mathsf{T}} \mathbf{L} \mathbf{v} - \frac{1}{2} \sum_{e_{ij} \in \partial \mathcal{M}} (\mathbf{u}_i \mathbf{v}_j - \mathbf{u}_j \mathbf{v}_i),$$

where $\mathbf{L}$ is the cotangent Laplacian and e_{ij} is an edge along the boundary. The third term in this energy equals the total area of the parameterization.

An earlier method coined *intrinsic parameterization* (Desbrun et al., 2002) first proposes this energy and solves a related linear system.

Relatedly, eigenfunctions have been used for frame field design, tangential vector and n-vector field processing on surfaces (Azencot et al., 2015, 2017; Brandt et al., 2017, 2018).

4.4.4 Data compression

Since the ϕ_i's generalize the Fourier basis, it is straightforward to use $\{\phi_i\}_{i=0}^k$ to compress geometric data and signals. Spectral compression of mesh geometry (Karni and Gotsman, 2000) projects vertex (x, y, z) coordinate functions onto $\{\phi_i(\cdot)\}_{i=0}^k$. Ben-Chen and Gotsman (2005) present theoretical results roughly showing that spectral compression using the Laplacian basis is optimal in the mean-square error sense, for a random mesh whose vertex positions are drawn from certain probability distributions. Aflalo et al. (2015) prove the optimality of the Laplacian eigenfunctions for approximating functions with bounded variation on surfaces.

4.4.5 Mesh smoothing and fairing

As an example of generalizing signal processing to meshes, the *spectral filtering* methods proposed in Vallet and Lévy (2008) apply low-pass, high-pass, and band-exaggeration filters to eigenvalues, leading to effects of mesh smoothing, details sharpening, and band-feature enhancement, respectively. Relatedly, locally amplifying surface area according to its Gaussian curvature produces natural exaggeration effects (Sela et al., 2015).

4.4.6 Feature extraction

Descriptors like heat and wave kernel signatures in Section 4.3 provide powerful tools to find keypoints that can be candidates for landmark correspondences. Similarly, spectral information can be used to detect salient points and regions on a surface. For example, spectral mesh saliency detection (Song et al., 2014) generalizes image spectral saliency detection (Hou and Zhang, 2007) to meshes: Simply speaking, instead of applying a filter function to eigenvalues as in spectral filtering, the method essentially applies a filter to the log of eigenvalues, to amplify and emphasize the low-frequency part of the spectrum.

4.4.7 Shape matching and correspondence

Ovsjanikov et al. (2010) apply the HKS to match feature points for the correspondence problem. The influential framework of *functional maps* (Ovsjanikov et al., 2012) directly computes a map F between eigenfunctions on two domains $\mathcal{M}$ and $\mathcal{N}$, i.e., $F : \mathcal{L}^2(\mathcal{M}) \to \mathcal{L}^2(\mathcal{N})$. Many methods have been proposed to improve this framework and extend it to alternative applications, such as the partial shape correspondence (Rodolà et al., 2017). We refer to the recent course notes (Ovsjanikov et al., 2016) for an extensive survey.

4.4.8 Shape differences

Shape differences provide a framework to compare two shapes that are put into correspondence using a functional map (Rustamov et al., 2013). Given two geometric domains $\mathcal{M}$ and $\mathcal{N}$ as well as a functional map $F:\mathcal{L}^2(\mathcal{M})\rightarrow\mathcal{L}^2(\mathcal{N})$, the shape difference operator $D:\mathcal{L}^2(\mathcal{M})\rightarrow\mathcal{L}^2(\mathcal{M})$ measures the pointwise distortion induced by F on $\mathcal{M}$. In particular, D modifies any function $f\in\mathcal{L}^2(\mathcal{M})$ linearly, i.e., the modification yields $Df\in\mathcal{L}^2(\mathcal{M})$, such that the bilinear form is best preserved under the map F, in the sense that $\forall g\in\mathcal{L}^2(\mathcal{M}):|a_{\mathcal{M}}(g,Df)-a_{\mathcal{N}}(Fg,Ff)|$ is minimized. Here $a(\cdot,\cdot)$ is a positive semidefinite bilinear form (i.e., inner product), which can be the Dirichlet energy (Laplacian), the area form (identity operator) (Rustamov et al., 2013), or the DtN/Steklov operator in weak form (Wang et al., 2018b). The resulting operator $D=\mathcal{A}_{\mathcal{M}}^{\dagger}F^{\top}\mathcal{A}_{\mathcal{N}}F$ reflects how the local surface geometry of $\mathcal{M}$ and $\mathcal{N}$ differ under map F, where $\mathcal{A}_{\mathcal{M}},\mathcal{A}_{\mathcal{N}}$ is the operator $\mathcal{A}$ operating on the domain $\mathcal{M},\mathcal{N}$, respectively, operator $\mathcal{A}$ is the strong form that corresponds to the bilinear form $a(\cdot,\cdot)$ with $a(u,v)=\langle u,\mathcal{A}v\rangle_{\mathcal{M}}$ for $u,v\in\mathcal{L}^2(\mathcal{M})$, and the superscript $\cdot^{\dagger}$ denotes the pseudo-inverse of an operator. Spectral basis reduction can be applied to $\mathcal{A}.$ and F, decreasing the computational cost.

The shape difference framework provides methods for shape analogy tasks and shape dataset exploration. Combined with a shape-from-Laplacian routine like those proposed in Boscaini et al. (2015), Corman et al. (2017a), and Chern et al. (2018), it leads to applications including style and deformation transfer, and computation of an "intrinsic average" of a few shapes by "averaging" their Laplacian eigenspaces. Recent work (Huang et al., 2019) proposes a learning-based framework to reconstruct surfaces from shape difference operators.

4.4.9 Shape retrieval

Reuter et al. (2006) advocate using the vector of Laplacian eigenvalues as "ShapeDNA," an ID for shape retrieval. In "Shape Google" (Bronstein et al., 2011), descriptors including heat kernel signatures and scale invariant heat kernel signatures are collectively used to construct a "bag of features" whose distributions are used to represent a shape. Lian et al. (2013) provide a comprehensive comparison of shape retrieval algorithms including many spectral methods.

Self-functional maps (Halimi and Kimmel, 2018) provide an alternative shape representation/signature that can be used for retrieving shapes. Conceptually similar to the construction of shape difference operators, this framework studies the interaction between two different metric spaces on the same domain. It uses functional maps between spaces spanned by the regular and the scale invariant Laplacian eigenfunctions (see Section 5.5) as the shape signature.

4.4.10 Geometric (deep) learning

Geometric operators have also been applied to geometric deep learning (Bronstein et al., 2017), in architectures based on convolutional neural networks (CNNs). *Geodesic CNN* (*GCNN*) (Masci et al., 2015) and *anisotropic CNN* (*ACNN*) (Boscaini et al., 2015) consider diffusion and anisotropic diffusion operators as alternative models to planar 2D convolutions on a curved surface. *Surface networks* (Kostrikov et al., 2018) use the Laplacian and Dirac operators in lieu of the graph adjacency matrix used in graph neural networks (Scarselli et al., 2009). The recent work (Wang et al., 2019) further proposes a framework for learning a family of discretized operators on meshes, including the Laplacian and Dirac operators.

Instead of applying operators in spatial domain, *spectral graph neural networks* (Bruna et al., 2014; Defferrard et al., 2016) apply the graph Laplacian operator in the spectral domain. *Synchronized spectral CNN* (Yi et al., 2016) applies and extends this approach to 3D meshes, with a setup similar to functional maps. *Deep functional maps* (Litany et al., 2017) learn to refine the point-wise SHOT descriptor (Tombari et al., 2010) to be used in a differentiable functional map layer, minimizing the geodesic error from the ground-truth correspondence. Later unsupervised versions (Roufosse and Ovsjanikov, 2018; Halimi et al., 2019) remove the need of the ground-truth: Halimi et al. (2019) employ a geodesic distortion loss, and Roufosse and Ovsjanikov (2018) define an unsupervised loss function in spectral domain, combined with the functional map framework.

While the deep learning methods above have gained recent popularity, the application of spectral data to learning from geometry emerged prior to geometric deep learning. For example, Aflalo et al. (2011) learn a diffusion kernel for shape retrieval, and *optimal spectral descriptors* (Litman and Bronstein, 2014) learn spectral filter coefficients in a signal processing fashion.

4.4.11 Structure and vibration analysis

In mechanics, the eigenfunctions of the stiffness matrix (Hessian) of some nonlinear energy—which become the Laplacian eigenfunctions for a certain energy—correspond to principal vibration modes to which the shape resonates most, under external forces and perturbations without damping. The corresponding eigenvalues are known as the intrinsic frequencies for the shape. Beyond the Laplacian, the biharmonic equation, having been considered in geometric modelling (Botsch and Kobbelt, 2004), is another common linearized elastic model, whose eigenfunctions represent modes of vibration of a thin elastic plate.

4.4.12 Acoustics

The eigenvalues of the stiffness matrix correspond to the characteristic frequencies of the sound made by the shape, appearing in a large number of works in sound simulation (van de Doel and Pai, 1996, 1998; O'Brien et al., 2002; Chadwick et al., 2009; Ren et al., 2013; Langlois et al., 2014). This is the forward problem of computing eigenvalues from the shape. Next we talk about its inverse problem, that is, designing a shape making sound of certain frequencies.

4.4.13 Shape optimization

Shape optimization searches the optimal shape that minimizes a given objective functional—which frequently involve eigenvalues and eigenfunctions—from a class of shapes. In the inverse acoustic problem of instrument design, eigenvalues are directly used as part of an objective function for shape optimization (Yoo et al., 2006; Yu et al., 2010; Bharaj et al., 2015). In shape analysis, Cosmo et al. (2019) propose *isospectralization*, optimizing and deforming shapes to match a given spectrum, as a prepossessing step for better correspondence. See Henrot (2017) for recent theoretical progress in spectral geometry with connections to shape optimization.

4.4.14 Reduced simulation

Constraining shape deformation to be within the span of top eigenfunctions of some operator reduces the degrees of freedom for physical simulation significantly, achieving interactive speeds for simulating deformations of large meshes. Using eigenfunctions of the stiffness matrix associated to a physical system is known as *modal analysis* (Pentland and Williams, 1989; Hauser et al., 2003; Barbič and James, 2005; Hildebrandt et al., 2011); Laplacian eigenfunctions are often used for reduced deformation, when the deformation energy is not specified (Rong et al., 2008). The scalar- and vector-valued Laplacian eigenfunctions are also used for reduced fluid simulation (Liu et al., 2015).

4.5 Other aspects of spectral shape analysis

4.5.1 Localized bases

Laplacian eigenfunctions, as the generalization of the Fourier basis to manifolds, are globally supported, which can be a limitation in many applications. Reduced simulation is an example: deformation, restricted as a linear combination of eigenfunctions, must be global, leading to counter-intuitive visual effects. Hence, some works promote locality and sparsity in a spectral-type basis. For instance, *compressed manifold harmonics* (*CMH*), originating in scientific computing (Ozoliņš et al., 2013), are introduced to graphics in Neumann et al. (2014). These methods add an additional term, the L^1 norm on eigenfunctions $\phi_i(\cdot)$, to the objective in the original eigenvalue problem (see Eq. 7), to promote sparsity in the solution. Relatedly, Bronstein et al. (2016) study how to appropriately and consistently discretize the L^1-norm on meshes. *Localized manifold harmonics* (*LMH*) (Melzi et al., 2018b) propose to modify the standard Laplacian, yielding a new operator whose eigendecomposition leads to localized orthogonal bases. *Multiscale diffusion wavelets* (Mahadevan, 2007) construct hierarchical wavelet trees using Laplacian eigenfunctions.

4.5.2 Bases for shape collections

When multiple poses of a shape are available, methods like the *singular value decomposition (SVD)* or *proper orthogonal decomposition (POD)* provide modes that are frequently used for modal analysis in engineering sciences, see, e.g., Kunisch and Volkwein (2002). Data-driven methods, like those used in tensor factorization on shape frames, animation compression (Alexa and Müller, 2000), and inverse skinning (Kry et al., 2002), also involve similar techniques; see the recent survey (Jacobson et al., 2014) for an extensive review. Relatedly, given a collection of 3D shapes, *fuzzy correspondence* (Kim et al., 2012) proposes a framework for robustly computing point correspondence. It first constructs an approximate shape-wise correspondence matrix $\mathbf{C} \in \mathbb{R}^{ns \times ns}$, where s is the number of shapes and n the number of vertices in each shape, and then applies spectral analysis to $\mathbf{C}$. The similarity between two vertices is defined as the distance between their spectral (diffusion) embedding of $\mathbf{C}$.

When there are multiple nonisometric shapes, Laplacian eigenfunctions are not compatible across shapes. To address this issue, *coupled quasi-harmonic bases* (Kovnatsky et al., 2013) study the joint diagonalization problem of Laplacians computed on shape collections. *Laplace–Beltrami basis pursuit* (Schonsheck et al., 2018) aligns the eigenfunctions between two shapes, with applications to nonisometric shape registration. It optimizes for a conformal deformation of the source shape, such that the Laplacian eigenfunctions of the deformed shape match those on the target shape.

4.6 Numerical aspects

Most of the applications above rely on matrix approximations of an operator designed to capture quantities on a smooth surface. Machinery from numerical analysis is needed to justify the discrete approximations of smooth quantities. Algorithms that are well-founded numerically tend to have stronger invariance to remeshing and other common confounding factors in geometry processing.

4.6.1 Convergence

When $\mathcal{M}$ is a flat domain in $\mathbb{R}^n$, the mathematical theory of FEM establishes convergence of the linear problem $\mathbf{Au} = \mathbf{Mf}$, showing that $\mathbf{u}$ converges to $u(\cdot)$ under suitable assumptions (Brenner and Scott, 2007). Similarly, the continuous eigenvalue problem is approximated by the generalized eigenvalue problem of the discretized matrix system

$$\mathbf{A}\boldsymbol{\phi} = \lambda\mathbf{M}\boldsymbol{\phi} \quad \text{s.t.}\ \boldsymbol{\phi}^{\mathsf{T}}\mathbf{M}\boldsymbol{\phi} = \mathbf{I}. \tag{24}$$

The discrete spectrum $\{\lambda_i, \boldsymbol{\phi}_i\}_{i=0}^{k}$ usually converges to the continuous one, proved on a case-by-case basis using similar techniques (Sun and Zhou, 2016).

Surprisingly, for a curved surface $\mathcal{M}$, convergence analysis of linear problems involving the Laplacian only recently has been established (Holst, 2001; Dziuk and Elliott, 2013). A rigorous analysis of FEM on manifolds inevitably requires theorems and techniques involving geometric analysis on curved surfaces, machinery that is not necessary for FEM on flat domains.

4.6.2 Eigensolvers

Most methods like FEM yield sparse linear systems whose eigenvalue problems need to be solved. Recall that linear systems can be solved using either direct or iterative algorithms; Botsch et al. (2005) include pointers to many sparse linear solvers and as well as the corresponding preconditioners for systems arising in mesh processing. Similarly, for the eigenvalue problem, direct or iterative algorithms can be employed. Direct solvers output all eigenvectors and eigenvalues, while iterative solvers return the first few ones. In most applications, only the first few eigenfunctions/eigenvalues are needed, in which case iterative solvers are favoured.

Iterative eigensolvers, including shifted power method, orthogonal projection methods, and Krylov subspace methods (e.g., Arnoldi method and Lanczos algorithm), further combined with filtering and restarting techniques and the use of a preconditioner, have a large number of variants (Saad, 2011). The state-of-the-art methods, including the implicitly restarted Arnoldi method, are widely available in sparse linear algebra packages including ARPACK (Lehoucq et al., 1998). Some methods, particularly BEM for integral-based operators, lead to dense linear systems. They can be solved by iterative algorithms like LOBPCG (locally optimal block preconditioned conjugate gradient) (Knyazev, 2001), which only asks for an oracle that implements the matrix-vector product, with implementations widely available.

4.6.3 Accelerations and preconditioners

Iterative eigensolvers can be accelerated using *spectral shift* and *preconditioning*. In computer graphics, since Dong et al. (2006) a shift eigenvalue solver is frequently used, with a properly chosen value of spectral shift σ. σ prescribes the maximum frequency (eigenvalue) of interest and separates the eigenvectors out of the range $[0, \sigma)$, efficiently accelerating the convergence.

The use of a preconditioner can greatly accelerate the convergence of an iterative linear or eigen-solver. Preconditioners proposed first for linear problems can be used for eigensystems as well, including those based on the incomplete Cholesky/LU factorization, the multigrid methods (Golub and Van Loan, 2012), or more general black-box algebraic multigrid methods. Inspired by the breakthroughs in theoretical computer science of linear time graph preconditioners (Spielman and Teng, 2004), in Krishnan et al. (2013) a multilevel preconditioning scheme is efficiently implemented for the Laplacian, with linear time and memory complexity, successful applied to mesh and image editing.

For dense BEM systems, efficient approximations can be obtained by employing the *hierarchical matrix* (Börm et al., 2003) or *fast multipole* (Coifman et al., 1993) methods.

4.6.4 Multiresolution, subsampling, and approximation methods

There are multiple methods to approximate eigenfunctions. Since low-frequency eigenfunctions are stable to sampling rate, one approach to accelerate computations is to reduce the mesh size. This can be done by either explicitly maintaining a smaller downsampled mesh, or implicitly using algebraic prolongation and restriction operators like ones in multigrid method (Brenner and Scott, 2007).

The Nyström subsampling method (Williams and Seeger, 2001; Drineas and Mahoney, 2005) samples columns from the operator matrix. Liu et al. (2006) apply the Nyström subsampling method to efficiently approximate eigenvectors; farthest point sampling is recommended over random point sampling for better accuracy. Multiresolution techniques are used in Dong et al. (2006) and Wang et al. (2018b) for Laplacian and Steklov eigenvalue problems, respectively. Wang et al. (2018a) apply low-rank estimation through the Nyström method to the kernel matrix in kernelized functional maps. Vaxman et al. (2010) introduce a multiscale approach to evaluate heat kernels hierarchically.

The recent work (Nasikun et al., 2018) proposes a fast approximation scheme for the Laplacian eigenfunctions. It first constructs a subspace, and then solves the eigenvalue problem in variational form (Eq. 7) but restricting the solution within the subspace. The subspace, whose construction is inspired by skinning weights for shape deformation, is fast to compute and capable to well approximate low-frequency functions.

4.6.5 Robustness

Dyer et al. (2007) empirically study the robustness of the Laplacian eigenvalues when evaluated on different discretizations of the same object, showing superiority of the mesh Laplacian over the graph Laplacian. Exploring the fact that lowest eigenvalues are robust to sampling, the recent *spectral coarsening* (Liu et al., 2019) algorithm constructs and optimizes a sparse operator matrix, of a much smaller size than the mesh size, whose lowest eigenvalues well preserve those of the original geometric operator. It first applies a combinatorial coarsening procedure to select key nodes, and then uses the nodes to construct a graph Laplacian. Entries of this Laplacian are determined by minimizing the error of multiplying the Laplacian to the ground truth top k eigenfunctions.

4.6.6 *Padé approximation of heat kernel and distance*

For some problems, it is possible to avoid computing the sequence $\{\phi_i, \lambda_i\}_{i=0}^k$ altogether. As an example, we have seen that the matrix exponential $e^{-t\mathbf{L}}$ and reciprocal $1/\mathbf{L}^p$ are frequently used in operator-based geometry processing, and it suffices to approximate these matrix functions directly. To this end, Patané (2015, 2017) explores the idea of using the Padé approximant of matrix exponentials and reciprocals. Generically speaking, the Padé approximant of a function is a ratio of two polynomials $P(\cdot)$, $Q(\cdot)$ whose coefficients can be derived using a Taylor expansion. For instance, the following expression gives the Padé approximation of the matrix exponential:

$$e^{-t\mathbf{L}} := \sum_{i=0}^{\infty} \frac{(-t\mathbf{L})^i}{i!} \approx \frac{P(\mathbf{L})}{Q(\mathbf{L})} = c_0\mathbf{I} + \sum_{i=1}^{k} \frac{c_i}{\mathbf{L} + s_i\mathbf{I}}$$

we have the last equality, as the result of algebraic factoring of the polynomials. Applying this operator reduces to solving a few linear systems rather than solving an eigenvalue problem. Patanè (2017) reviews Laplacian kernels and distances including numerical aspects of fast approximation.

5 Relevant geometric operators

In this section, we will discuss choices of the geometric operator $\mathcal{A}$ to perform spectral shape analysis. Relevant geometric operators roughly can be categorized as coming from two approaches. The first approach considers a shape as a thin shell specified by the surface; all intrinsic and some extrinsic methods fall in this category. The second approach considers a shape as a 3D solid whose boundary is specified by the surface. In this section, we first discuss methods using the first approach, and then those using the second approach.

5.1 Identity operator, area form, and mass matrix

When discretizing a general linear operator $\mathcal{A} : \mathcal{L}^2(\mathcal{M}) \to \mathcal{L}^2(\mathcal{M})$ or equivalently the bilinear form $a(\cdot, \cdot) : \mathcal{L}^2(\mathcal{M}) \times \mathcal{L}^2(\mathcal{M}) \to \mathbb{R}$, FEM outputs the final discrete operator in the form $\mathbf{M}^{-1}\mathbf{A}$, where $\mathbf{M}$ is the mass matrix.

As an example, when $\mathcal{A}$ is the *identity operator*, the corresponding bilinear form is $a(u, v) = \int_{\mathcal{M}} uv$. If we let $u, v = 1_{\mathcal{R}}$, i.e., the constant function supported on region $\mathcal{R} \in \mathcal{M}$, then $a(u, v)$ evaluates to the area of $\mathcal{R}$. For this reason, $a(u, v) = \int_{\mathcal{M}} uv$ is also frequently referred to as the *area form*. For this simple identity operator, we have $\mathbf{A} = \mathbf{M}$, yielding the discrete operator

$\mathbf{M}^{-1}\mathbf{A} = \mathbf{I}$; in other words, the mass matrix $\mathbf{M}$ is the discrete weak form of the identity operator. $\mathbf{M}$ is also a discrete area form, in that $\mathbf{1}^{\mathsf{T}}\mathbf{M}\mathbf{1}$ evaluates to the total area of a triangle mesh.

Despite its simplicity, the area form already reveals interesting information about the shape. The shape difference framework (Rustamov et al., 2013) introduces an area-based shape difference, which measures the area distortion between two surfaces under a functional map using measurements derived from the area form.

5.1.1 Discretization: Full mass matrix

Following exactly the FEM, we have $\mathbf{M}_{ij} = \langle \phi_i, \phi_j \rangle_{\mathcal{M}}$. Using a piecewise linear basis function on a triangle mesh, $\mathbf{M}$ evaluates to the following expression:

$$\mathbf{M}_{ij} = \langle \phi_i, \phi_j \rangle_{\mathcal{M}} = \sum_{T \in \mathcal{T}(i) \cap \mathcal{T}(j)} \int_{\mathcal{M}} \phi_i \phi_j = \sum_{T \in \mathcal{T}(i) \cap \mathcal{T}(j)} \begin{cases} \frac{1}{6}\,\text{area}(T) & \text{if } i = j \\ \frac{1}{12}\,\text{area}(T) & \text{otherwise,} \end{cases}$$

where $\mathcal{T}(i)$ is the set of triangles adjacent to vertex i. Such $\mathbf{M}$ is referred to as the *full mass matrix*, which has the same sparsity pattern as the graph Laplacian induced by the mesh graph.

5.1.2 Discretization: Lumped mass matrix

The full mass matrix $\mathbf{M}$ is a sparse but nondiagonal matrix, making it expensive to invert $\mathbf{M}$ or solve the linear system $\mathbf{Mx} = \mathbf{b}$, where $\mathbf{b}$ is, e.g., the mean curvature normal in Laplacian smoothing (Sorkine et al., 2004). In practice, instead of using this exact full mass matrix, a *lumped* (diagonalized) mass matrix is usually used instead. This substitution can be justified, in the sense that the asymptotic convergence rate remains the same if a lumped mass matrix is used (Ciarlet, 2002). The diagonal entries of a lumped mass matrix associate each vertex i with some area A_i surrounding it, such that $\mathbf{M}_{ii} = A_i$. Popular area dividing schemes lead to the barycentric or Voronoi mass matrices.

For any triangle T in the mesh, the *barycentric mass matrix* $\mathbf{M}^b$ simply associates $\frac{1}{3}\text{Area}(T)$ to each of T's incident vertex, as shown in Fig. 7A

$$\mathbf{M}^b_{ii} = \sum_{T \in \mathcal{T}(i)} \frac{1}{3}\text{area}(T).$$

The *Voronoi mass matrix* $\mathbf{M}^v$ sets $\mathbf{M}^v_{ii}$ as the Voronoi area around vertex i, i.e., the area of a neighbourhood in the Voronoi diagram as shown in Fig. 7B. In Meyer et al. (2003), the Voronoi mass matrix is recommended, since it leads to a provably tight error bound when used to compute the mean curvature; a modified Voronoi scheme is proposed to handle negative areas due to obtuse triangles.

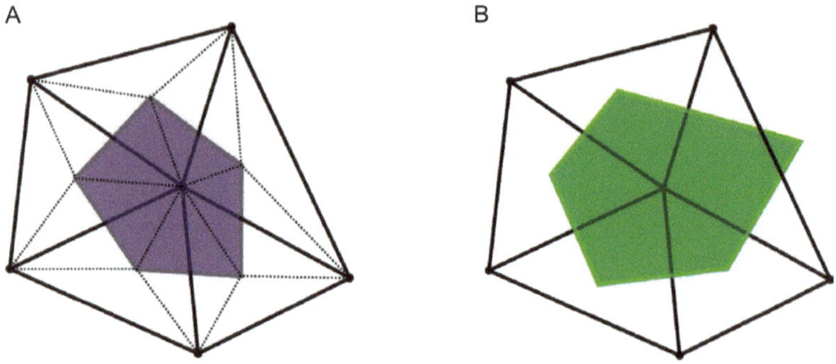

FIG. 7 The barycentric area (A) and the Voronoi area (B) associated to the vertex in the centre.

5.2 Laplace–Beltrami (intrinsic Laplacian)

In the extrinsic 2-dimensional space (flat planar domain), the Laplacian operator is defined as: $\Delta_{\mathbb{R}^2} = \frac{\partial^2}{\partial x^2} + \frac{\partial^2}{\partial y^2}$. The manifold Laplacian (Laplace–Beltrami) generalizes this definition to 2-manifolds, including curved surfaces, as follows:

$$\Delta_{\mathcal{M}} = \nabla_{\mathcal{M}} \cdot \nabla_{\mathcal{M}} = \frac{1}{\sqrt{|\det g|}} \partial_i \left(\sqrt{|\det g|} g^{ij} \partial_j \right) \tag{25}$$

where g_{ij} is the metric tensor and $(g^{ij}) = (g_{ij})^{-1}$ is the inverse metric. This definition of Laplacian only involves the metric of $\mathcal{M}$; in other words, the Laplacian is an intrinsic operator, remaining invariant under isometric transformations which does not change the metric g.

5.2.1 Harmonic functions

Harmonic functions, i.e., function $u(\cdot)$ such that $\Delta_{\mathcal{M}} u = 0$, enjoy a long list of special properties, including the *mean value theorem*, which roughly states that $u(\mathbf{x})$ equals to the average value of $u(\cdot)$ evaluated in any neighbourhood of $\mathbf{x}$, and consequently the *maximum principle*, which states that extrema of $u(\cdot)$ must lie on the boundary $\partial\mathcal{M}$. Additionally, harmonic functions are C^∞. This is a remarkable fact: although the definition of harmonic function only involves second-order derivatives, the condition $\Delta_{\mathcal{M}} u = 0$ is so strong that $u(\cdot)$ has to be infinitely differentiable (i.e., smooth).

These properties contribute to the wide applicability of the Laplacian operator. In mathematics and physics, the Laplace equation (1) is particularly common, describing a wide range of physical phenomena including static electric potential fields, steady-state fluid flow, equilibrium states of diffusion processes and temperature distributions, and others. These models have been introduced to graphics and geometry processing, appearing in a large number of early works: Laplacian mesh editing (Sorkine et al., 2004; Sorkine, 2005),

Poisson mesh editing (Yu et al., 2004), Poisson reconstruction (Kazhdan et al., 2006), Harmonic coordinates (Joshi et al., 2007), Green coordinates (Lipman et al., 2008), mesh deformation and fairing (Botsch and Kobbelt, 2004), and so on. Solutions to the Laplacian system are robust to noises and sampling bias, favourable in practical applications.

5.2.2 Discretization

On flat or curved triangle meshes, the Laplacian operator has the following celebrated cotangent formula (Pinkall and Polthier, 1993):

$$\mathbf{L}_{ij} = \begin{cases} \frac{1}{2}\left(\cot\alpha_{ij} + \cot\beta_{ij}\right) & \text{if } \{i,j\} \in \mathbf{E} \\ -\sum_{j \neq i} \mathbf{L}_{ij} & \text{if } i = j \\ 0 & \text{otherwise,} \end{cases} \tag{26}$$

where α_{ij}, β_{ij} are the two angles opposite to edge ij, as labelled in Fig. 8.

The contangent formula in Eq. (26) can be derived in many ways. Using FEM, this formula can be derived by evaluating $\mathbf{L}_{ij}$ as the inner product between gradients of hat basis functions $\mathbf{L}_{ij} = -\langle \nabla_{\mathcal{M}}\phi_i, \nabla_{\mathcal{M}}\phi_j \rangle_{\mathcal{M}}$, where ϕ_i is piecewise linear basis (hat basis) centred at vertex i. Evaluating this integral is a standard computation using calculus; the exact steps of calculation can be found in, e.g., Zhang et al. (2010a). An alternative derivation of the formula Eq. (26) using the finite volume method is given in Botsch et al. (2010, Chapter 3.3). A derivation using DEC can be found in, e.g., Hirani (2003).

The discrete approximants can be understood in the strong sense that $\Delta_{\mathcal{M}} u \approx \mathbf{M}^{-1}\mathbf{L}\mathbf{u}$ or in the weak sense that $\int_{\mathcal{M}} (\nabla_{\mathcal{M}} u)^2 \approx -\mathbf{u}^{\mathsf{T}}\mathbf{L}\mathbf{u}$, where $\mathbf{M}$ is

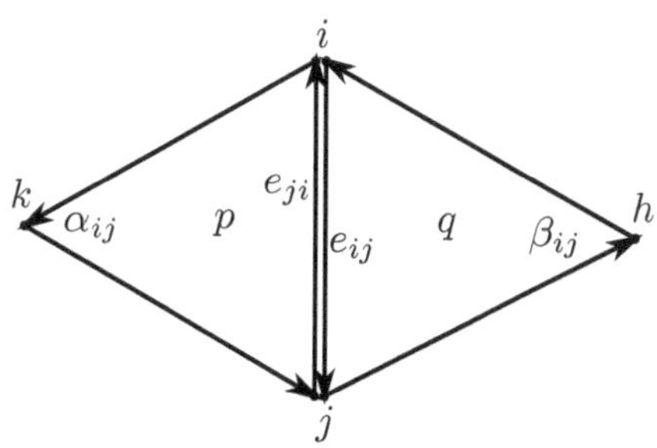

FIG. 8 An illustration of notations used in this paper. p, q are two adjacent triangles. Triangle p consists of vertices ikj, and triangle q consists of vertices ijh. α_{ij} and β_{ij} are the two angles opposite to the undirected edge ij in the two triangles, respectively. $e_{ij} = -e_{ji}$ is a pair of opposite directed edges ij and ji.

the mass matrix. Using an lumped mass matrix leads to the following discrete definition of $\Delta_{\mathcal{M}}$ on a triangulated mesh:

$$\Delta_{\mathcal{M}} u(v_i) \approx \frac{1}{\mathbf{M}_{ii}} \sum_{v_j \in \mathcal{N}_1(i)} \frac{1}{2} \left(\cot \alpha_{ij} + \cot \beta_{ij} \right) \left(\mathbf{u}_j - \mathbf{u}_i \right)$$

where $\mathcal{N}_1(i)$ is the set of 1-ring neighbours adjacent to vertex i.

5.3 Combinatorial and graph Laplacians

The signal processing approach to geometry processing was first introduced in the seminal work (Taubin, 1995), generalizing classical discrete Fourier analysis to discrete surface signal analysis. Prior to the popularity of the cotangent Laplacian, the uniform Laplacian (Tutte Laplacian) or combinatorial Laplacian (Zhang, 2004) was used instead. The uniform Laplacian (Taubin, 1995) is a graph Laplacian with uniform weights:

$$(\mathbf{L}\mathbf{u})_i := \frac{1}{|\mathcal{N}_1(i)|} \sum_{v_j \in \mathcal{N}_1(i)} \left(\mathbf{u}_j - \mathbf{u}_i \right).$$

This purely combinatorial and topological definition fails to converge to a continuous operator that captures the geometry of the surface. This discrepancy, in practical applications, can lead to undesirable behaviour; as an example, the resulting parameterization can have large distortions as observed in Zhou et al. (2004). This issue is resolved by the geometry-aware Laplacian, which behaves more consistently compared to the graph Laplacian; a qualitative comparison of spectral behaviours can be found in, e.g., Sorkine (2006) and Dong et al. (2006). For this reason, graphics methods based on the graph Laplacian have been replaced by those using the cotangent Laplacian. Despite the drawbacks, eigenfunctions of graph Laplacians, favoured for their simplicity, have been applied to a variety of geometric deep learning methods recently such as the spectral mesh CNN (Yi et al., 2016).

5.4 Restricted Laplacian

The restricted Laplacian (Chuang et al., 2009) also discretizes the Laplace–Beltrami operator, but instead of using piecewise linear hat basis $\phi_i(\cdot)$, it uses a different basis $b_i(\cdot)$ defined in the extrinsic 3D space:

$$\mathbf{L}_{ij} = \langle \nabla_{\mathcal{M}} b_i, \nabla_{\mathcal{M}} b_j \rangle_{\mathcal{M}}.$$

In particular, the basis $b_i(\cdot)$ is chosen as spatial polynomials (obtained as product of second-order B-spline basis) supported on nearby voxels. Although the basis is defined extrinsically, the resulting operator is close to the cotangent Laplacian in Eq. (26) and converges to the same intrinsic Laplacian under mesh refinement. A major advantage of this restricted Laplacian is its robustness to mesh tessellation: the cotangent Laplacian has undesirable numerics and behaviour on meshes with low-quality tessellation since the function $\cot\theta \to \infty$ as $\theta \to 0$.

5.5 Scale invariant Laplacian

In Eq. (25), the Laplacian can be defined given any metric g, not necessarily the natural metric induced by the embedded space $\mathbb{R}^3$. Following this observation, Aflalo et al. (2013) propose to use an alternative metric $\tilde{g}$, the scale and isometry invariant metric, which scales the original metric by $\tilde{g}_{ij} = |K| g_{ij}$, where K is the Gaussian curvature. This modified metric factors out the effect of scaling, allowing scale invariant analysis. This new metric tensor induces a new Laplacian $\Delta_{\tilde{g}}$, using the same definition in Eq. (25) but replacing g with $\tilde{g}$. The new Laplacian $\Delta_{\tilde{g}}$ and its derived shape descriptors, by design, are scale invariant. Prior work (Bronstein and Kokkinos, 2010) develops scale invariant descriptors directly, without introducing a scale invariant metric. The scale invariant metric is also introduced for volumetric domains (Raviv and Raskar, 2015).

The scale invariant Laplacian $\Delta_{\tilde{g}}$ is discretized as $\mathbf{K}^{-1}\mathbf{M}^{-1}\mathbf{L}$, where $\mathbf{M}, \mathbf{L}$ are the same (lumped) mass matrix and cotangent matrix used for intrinsic Laplacian, and $\mathbf{K}$ is a diagonal matrix such that $\mathbf{K}_{ii} = \sqrt{K_i^2 + \epsilon^2}$, where K_i is the discrete Gaussian curvature at vertex i. K_i can be approximated as $K_i = \frac{2\pi - \sum_j \gamma_j^i}{\frac{1}{3} A_i}$, where γ_j^i is an angle adjacent to vertex i, and A_i again is the area of vertex i.

Self-functional maps (Halimi and Kimmel, 2018) propose to use functional maps between eigenfunctions of $\Delta_{\tilde{g}}$ and Δ_g as a shape signature.

5.6 Affine and equi-affine invariant Laplacian

The sequence of works (Raviv et al., 2011b, 2014a; Spagnuolo et al., 2012; Raviv and Kimmel, 2015) consider the equi-affine and affine invariant geometry for analyzing shapes under nonrigid transformations. *Affine invariance* means that surfaces are considered the same under affine transformations, i.e., linear transformations $\mathbf{x} \mapsto \mathbf{Ax} + \mathbf{b}$, including squeezing and shearing. *Equi-affine invariance* means that surfaces are considered the same under affine transformations that preserve volumes, i.e., $\det(\mathbf{A}) = 1$. Particularly, an equi-affine invariant metric is first defined in Raviv et al. (2011b), on top of which an *equi-affine invariant Laplacian* can be constructed, again using the same definition in Eq. (25). The later Raviv and Kimmel (2015) obtain an affine invariant metric and the derived *affine invariant Laplacian*, by applying the technique of scale invariant metric construction (Aflalo et al., 2013) to the equi-affine invariant metric.

Raviv et al. (2011a) apply this insight to a combined geometry-colour space, constructing shape descriptors fusing geometric and photometric information. Raviv et al. (2014b) apply the affine invariant approach to medical imaging, evaluating local contractions of soft tissues.

5.7 Anisotropic Laplacian

One implicit assumption in deriving the (regular) Laplacian $\Delta_{\mathcal{M}}$ is the isotropic assumption, that within a local chart (tangent plane) the model assumes there is no difference along any direction. Consequently, the diffusion equation associated with $\Delta_{\mathcal{M}}$ assumes that the heat diffuses at the same speed in all directions, as in Fig. 2. This assumption can be removed, leading to a more general model.

In the flat domain $\mathbb{R}^2$, removing isotropy means that $\Delta_{\mathbb{R}^2} = \frac{\partial^2}{\partial x^2} + \frac{\partial^2}{\partial y^2}$ is replaced by the more general formula $\Delta^A_{\mathbb{R}^2} = \sum_{i,j=1}^2 \frac{\partial}{\partial x_j}\left[a_{ji}(\mathbf{x})\frac{\partial}{\partial x_i}\right]$, where $a_{ji}(\mathbf{x})$ is a function of $\mathbf{x}$. On a curved surface, the formula becomes the expression

$$\Delta^A_{\mathcal{M}} = \nabla_{\mathcal{M}} \cdot (A(\mathbf{x})\nabla_{\mathcal{M}})$$

where $A(\mathbf{x}) \in \mathbb{R}^{2\times 2}$ is the thermal conductivity tensor (matrix) that operates on tangent vectors. $\Delta^A_{\mathcal{M}}$ is the *anisotropic Laplacian* operator; the additional degrees of freedom to choose a tensor field $A(\mathbf{x})$ allows the anisotropic Laplacian to be *directionally sensitive.*

Andreux et al. (2014) propose to use an anisotropic Laplacian for shape analysis, surpassing methods based on the usual isotropic Laplacian by a large margin on tasks including segmentation and region detection. *Anisotropic diffusion descriptors* (Boscaini et al., 2016b) consider spectral feature descriptors derived by the anisotropic Laplacian, and propose to learn anisotropic kernels. Learnable anisotropic kernels have been further applied to geometric deep learning to construct *anisotropic convolutional neural networks (ACNN)* (Boscaini et al., 2016a). A more restrictive anisotropic Laplacian has been proposed earlier in Shi et al. (2008) to extract Reeb graphs and skeletons; this version of operator amounts to setting $a_{12}(\mathbf{x}) = a_{21}(\mathbf{x}) = 0$ and $a_{1,1}(\mathbf{x}) = a_{2,2}(\mathbf{x})$ in the general form.

5.7.1 *Discretization*

The anisotropic diffusion kernel $A(\mathbf{x})$ needs to be prescribed in a pointwise fashion, usually in a local coordinate system; Andreux et al. (2014) use principal curvature directions for this purpose, and Boscaini et al. (2016b) propose an alternative kernel parameterization independent of the extrinsic curvature. Following Andreux et al. (2014), we assume an orthonormal frame field is given $\hat{\mathbf{u}}, \hat{\mathbf{v}} \in \mathbb{R}^3$, which can be, e.g., the directions of principal curvatures. The thermal conductivity tensor $A(\mathbf{x})$ is given per-triangle as $\mathbf{U}_{ijk} \in \mathbb{R}^{3\times 3}$, operating w.r.t. the frame field $\hat{\mathbf{u}}, \hat{\mathbf{v}}$. Let $\hat{\mathbf{e}}_{ij} \in \mathbb{R}^3$ denote the normalized edge vector from vertex i to vertex j. Define $\mathbf{H}$ as the *shear matrix*, which depends on $\mathbf{U}$ and $A(\mathbf{x})$ and encodes anisotropic scaling up to an orthogonal basis change; for an exact formula for $\mathbf{H}$, we refer to the original paper (Andreux et al., 2014). Then, we can write

$$\mathbf{L}_{ij} = \begin{cases} \frac{1}{2}\left(\frac{\langle \hat{\mathbf{e}}_{kj}, \hat{\mathbf{e}}_{ki}\rangle_{\mathbf{H}}}{\sin\alpha_{ij}} + \frac{\langle \hat{\mathbf{e}}_{hj}, \hat{\mathbf{e}}_{hi}\rangle_{\mathbf{H}}}{\sin\beta_{ij}}\right) & \text{if } \{i,j\} \in \mathbf{E} \\ -\sum_{k\neq i} \mathbf{L}_{ik} & \text{if } i=j \\ 0 & \text{otherwise} \end{cases} \tag{27}$$

where $\langle\cdot,\cdot\rangle_{\mathbf{H}}$ is the $\mathbf{H}$-weighted inner product, α_{ij}, β_{ij} are the two angles opposite to edge ij, and k, h are opposite vertices on which α_{ij}, β_{ij} are located, respectively, as labelled in Fig. 8. In the isotropic case, we have $\mathbf{H} = \mathbf{I}$.

5.8 Hessian and normal-restricted Hessian: A family of linearized energies

The Hessian matrix of any nonlinear deformation energy at the rest pose can be used as an alternative to the Laplacian. This Hessian is guaranteed to be p.s.d., since the deformation energy is minimized at the rest pose. Discussed in Zorin (2005) and Hildebrandt et al. (2010), this observation provides an approach to systematically derive new operators by linearizing any nonlinear deformation energy. Analyzing eigenmodes of the Hessian is known as *modal analysis*, a technique that has been applied to reduced physical simulation (Barbič and James, 2005), shape segmentation (Huang et al., 2009), and shape analysis (Hildebrandt et al., 2010).

The resulting Hessian is a matrix in $\mathbb{R}^{3n\times 3n}$ for a mesh with n vertices, and its eigenfunctions in $\mathbb{R}^{3n}$ can be thought of as principal velocity fields along which the energy varies. Analogous to the regular heat kernel signature introduced as the trace of heat kernel of Laplacian, Hildebrandt et al. (2010) introduce a vector version named the *vibration signature*, which uses the principal velocity eigenmodes. Similarly, a vector-based *vibration distance* is introduced using a similar definition to the diffusion distance.

Hildebrandt et al. (2010) also propose restricting the velocity to the normal direction, leading to a matrix in $\mathbb{R}^{n\times n}$.

5.9 Modified Dirichlet energy

Applying the Hessian approach to the thin shell energy leads to the *modified Dirichlet energy* (Hildebrandt et al., 2010, 2012). The modified Dirichlet energy is the sum of the Dirichlet energy $\int_{\mathcal{M}}\frac{1}{2}\|\nabla_{\mathcal{M}}u(\mathbf{x})\|_2^2$ and an additional term

$$\int_{\mathcal{M}}\frac{1}{2}(\kappa_1^2(\mathbf{x})+\kappa_2^2(\mathbf{x}))u(\mathbf{x})^2 \equiv \int_{\mathcal{M}}\left(H(\mathbf{x})^2 - 1/2K(\mathbf{x})\right)u(\mathbf{x})^2$$

where $\kappa_1(\mathbf{x})$ and $\kappa_2(\mathbf{x})$ are the two principal curvatures at $\mathbf{x}$. In geometric modelling, when $u(\mathbf{x}) = 1(\mathbf{x})$ the additional term is often referred to as the total curvature functional and is closely related to the Willmore energy (Zorin, 2005), that is, $\int_{\mathcal{M}} \frac{1}{4}(\kappa_1(\mathbf{x}) - \kappa_2(\mathbf{x}))^2 \equiv \int_{\mathcal{M}} \left(H(\mathbf{x})^2 - K(\mathbf{x})\right)$.

The additional term is extrinsic, penalizing $u(\cdot)$ in regions with large extrinsic bending. Applying the spectral surface quadrangulation method of Dong et al. (2006), but instead using these new eigenfunctions, yields quadrangulations that better align to the extrinsic curvature of the surface. The descriptors and distances derived using the modified Dirichlet energy are sensitive to extrinsic curvatures.

5.10 Hamiltonian operator and Schrödinger operator

The *Hamiltonian operator* (Choukroun et al., 2018b) $\mathcal{H}$, or the *Schrödinger operator* (Choukroun et al., 2018a), is an elliptic operator of the form

$$\mathcal{H}f = -\Delta_{\mathcal{M}} f + \mathcal{V}f, \tag{28}$$

where $f \in \mathcal{L}^2(\mathcal{M})$, and $\mathcal{V} : \mathcal{L}^2(\mathcal{M}) \to \mathcal{L}^2(\mathcal{M})$ is the *diagonal operator* defined as $[\mathcal{V}f](\mathbf{x}) = V(\mathbf{x})f(\mathbf{x})$, in which $V : \mathcal{M} \to \mathbb{R}$ is a prescribed real-valued scalar function.

The weak form of the Hamiltonian operator amounts to modifying the Dirichlet energy by adding the term

$$\int_{\mathcal{M}} V(\mathbf{x})u(\mathbf{x})^2.$$

This additional term penalizes $u(\cdot)$ differently on different regions on $\mathcal{M}$. In particular, using the variational definition of the eigenvalue problem, the lowest eigenfunctions will avoid regions where $V(\mathbf{x})$ is large and will instead concentrate on regions with low potential. Choukroun et al. (2018a) study the problem of designing potential functions.

The PDE associated with the operator, the *Schrödinger equation*, can be viewed as replacing the Laplacian in the wave equation with the Hamiltonian operator. The Schrödinger equation in quantum mechanics prescribes the probability wave density function $\Psi(x, t)$ of a particle with mass m under potential field V:

$$i\hbar \frac{\partial \Psi}{\partial t} = \frac{-\hbar^2}{2m} \Delta_{\mathcal{M}} \Psi + \mathcal{V}\Psi, \tag{29}$$

where $\hbar$ is the Planck's constant and $|\Psi(\mathbf{x}, t)|^2$ is interpreted as the probability density distribution.

5.11 Curvature Laplacian

Among the earliest work applying a spectral approach to shape analysis, Liu and Zhang (2007) introduce a graph Laplacian $\mathbf{L}$ called *curvature Laplacian.* $\mathbf{L} := \mathbf{D} - \mathbf{W}$, where $\mathbf{D}$ is a matrix whose diagonal stacks summations of rows in the weight matrix $\mathbf{W}$, i.e, $d_i = \sum_{j=1}^{n} \mathbf{W}_{ij}, \mathbf{D} = \text{diag}(d_i)$, and $\mathbf{W}$ is defined as

$$\mathbf{W}_{ij} = \begin{cases} \left(\left|\kappa_i^{min}\right| + \left|\kappa_j^{min}\right|\right) \frac{|\mathbf{e}_{ij}|}{h} |\langle \frac{\mathbf{e}_{ij}}{|\mathbf{e}_{ij}|}, \mathbf{z}\rangle| & \text{if } \kappa_i^{min} < 0 \text{ or } \kappa_j^{min} < 0 \\ \varepsilon & \text{otherwise,} \end{cases}$$

where κ_i^{min} is the minimal principal curvature at vertex i, $\mathbf{e}_{ij}$ is the (unnormalized) edge vector of edge ij (as labelled in Fig. 8), h is the average edge length in the mesh, and $\mathbf{z}$ is the normalized average of the directions for curvatures κ_i^{min} and κ_j^{min}. When $\kappa_i^{min}, \kappa_j^{min} \geq 0$, which implies *local convexity*, $\mathbf{W}_{ij}$ is set to a small value ϵ merely to maintain mesh connectivity. An edge ij will have a large weight if vertex i or j has a negative minimal principal curvature direction roughly aligning with $\mathbf{e}_{ij}$; this implies that locally the mesh is bent roughly along edge ij, so the edge ij should not be cut in a spectral embedding using this operator. The spectral embedding of this Laplacian, combined with contour analysis, results in segmentation that is aware of the extrinsic bending.

5.12 Concavity-aware Laplacian

With the observation that vertices on concave seams are often part of potential segmentation boundaries, Au et al. (2012) and Wang et al. (2014) propose a *concavity-aware Laplacian* of the following form:

$$\mathbf{L}_{ij} = \begin{cases} \frac{w_{ij}}{\sum_{\{i,k\} \in \mathbf{E}} w_{ik}} & \text{if } \{i,j\} \in \mathbf{E} \\ -1 & \text{if } i = j \\ 0 & \text{otherwise,} \end{cases}$$

where $\mathbf{E}$ is the set of edges, and w_{ij} is a concavity-sensitive weights:

$$w_{ij} = \frac{e_{ij}\beta}{|K_i| + |K_j| + \epsilon}.$$

In this expression, K_i is the Gaussian curvature at vertex i, and $\beta = 0.01$ if either i or j is a concave vertex and $\beta = 1$ otherwise. The vertex concavity factor β is determined locally by using the inner product between edge vector directions and normal difference, or more robustly via PCA within the vertex's 1-ring neighbourhood; to further improve robustness, this operator can be constructed on a smoothed mesh.

Au et al. (2012) define solutions to linear systems involving the concavity-aware Laplacian to be *concavity-aware fields*, whose isolines whom align with concave seams and serve as candidate cuts. Wang et al. (2014) use the eigenfunctions of the concavity-aware Laplacian in a framework involving hierarchical spectral analysis and isoline-based boundary detection.

5.13 Extrinsic and relative Dirac operators

The Dirac operator D, named after Paul Dirac, is a square root of the negative semidefinite Laplacian operator, whose definition thus has to resort to number systems with imaginary parts. In 1D we have $D = \sqrt{-1}\frac{\partial}{\partial x}$ and $D^2 = -\frac{\partial^2}{\partial x^2} = \Delta_x$; this definition can be generalized to a curved surface. Dirac operators are defined to operate on spinors on abstract manifolds, and in particular on a 3D surface, spinors can be understood as a quaternion-valued fields, suggesting the notation $D : \mathcal{L}^2(\mathcal{M}; \mathbb{H}) \to \mathcal{L}^2(\mathcal{M}; \mathbb{H})$. There are multiple Dirac operators D; we discuss the *extrinsic Dirac operator* (Kamberov et al., 1996) first.

5.13.1 *Extrinsic Dirac operator* $\mathrm{D_f}$

Identifying $\mathbb{R}^3$ with $\mathrm{Im}(\mathbb{H})$, a 3D surface can be defined as a map $f : \mathcal{M} \to \mathbb{H}$, i.e., from a point on the (abstract) surface $\mathcal{M}$ to $\mathbb{R}^3$. The differential of the map, $df : T\mathcal{M} \to \mathbb{H}$, maps a tangent vector to $\mathbb{R}^3$ or $\mathrm{Im}(\mathbb{H})$.

Consider a surface deformation that applies a local conformal transformation (i.e., rotation and scaling) to each tangent plane, using a smooth quaternion field $\phi(\cdot)$. In quaternion language, this procedure can be written as a pointwise modification on df to $\widetilde{df} = \overline{\phi} \cdot df \cdot \phi$. For $\widetilde{df}$ to be integrable, i.e., consistent in the sense that the modification remains the differential of a surface, the following Dirac equation (or *integrability condition*) has to hold:

$$D_f \phi = \rho \phi \tag{30}$$

where $\rho : \mathcal{M} \to \mathbb{R}$ is a real-valued function that controls local distortion of the curvature, and D_f is the extrinsic Dirac operator defined as

$$D_f \phi := -\frac{\mathrm{d}f \wedge \mathrm{d}\phi}{|\mathrm{d}f|^2}.$$

Here, $|\mathrm{d}f|^2$ is the squared length element that corresponds to the unsigned area element for surface f. Crane et al. (2011) propose to use the Dirac equation (30) to compute the conformal transformation $\phi(\cdot)$ of a triangle mesh, with prescribed function $\rho(\cdot)$ encoding curvature information. This work proposes the following discrete version of the extrinsic Dirac operator $\mathbf{D} \in \mathbb{H}^{|F| \times |V|}$. $\mathbf{D}_{pi}$ is nonzero as long as vertex i is within triangle p, given by the expression

$$\mathbf{D}_{pi} = -\frac{1}{2A_p} e_p^{(i)},$$

where A_p is the area of triangle p, and $e_p^{(i)}$ is the opposing edge vector of vertex i within triangle p in quaternion value. The right-hand rule has been assumed, such that edge vectors are oriented counter-clockwise around each triangle. Numerically, a quaternion-valued matrix can be equivalently written as a real-valued matrix by expanding each entry into a 4×4 block. The extrinsic Dirac operator also appears in conformal surface flows (Crane et al., 2013a), with applications to surface fairing. *Surface networks* (Kostrikov et al., 2018) use this version of Dirac operator in deep neural networks, to capture extrinsic geometric information.

5.13.2 *Relative Dirac operator* $\mathrm{D_N}$

As mentioned earlier, D_f can be thought of as a square root of the Laplacian. This phrase can be understood via the relation

$$D_f^2\phi = \Delta\phi + \frac{\mathrm{d}N \wedge \mathrm{d}\phi}{|\mathrm{d}f|^2}$$

where N is the Gauss map. Liu et al. (2017) make the observation that the additional term

$$D_N\phi := D_{f,N}\phi = -\frac{\mathrm{d}N \wedge \mathrm{d}\phi}{|\mathrm{d}f|^2}$$

defined as the *relative Dirac operator* of a single surface f, is a purely extrinsic operator. The closely related relative Dirac operator *between two surfaces* f_1, f_2 with correspondence is defined as

$$D_{f_1,f_2}\phi := -\frac{\mathrm{d}f_2 \wedge \mathrm{d}\phi}{|\mathrm{d}f_1|^2}.$$

With this theory in place, Liu et al. (2017) use the following one-parameter family of operators to linearly blend the Laplacian and the relative Dirac operator

$$L(\tau) := (1-\tau)\Delta + \tau D_N.$$

Since the relative Dirac operator vanishes on purely flat regions, this blending with the Laplacian injects ellipticity (positive definiteness) that constrains the large kernel of the operator and stabilizes the lowest eigenfunctions.

By considering ϕ as a face-based function, Ye et al. (2018) define an alternative discrete extrinsic Dirac operator $\mathbf{D}_E \in \mathbb{H}^{|F|\times|F|}$. We denote the extrinsic Dirac operator as D_E, which comes from the same continuous operator as D_f; this alternative notation is merely to reflect that a different discretization will be used. For each triangle p,

$$(\mathbf{D}_E\boldsymbol{\phi})_p = \frac{1}{2}\sum_q (2H_{pq} + e_{pq})(\boldsymbol{\phi}_q - \boldsymbol{\phi}_p)$$

where $\boldsymbol{\phi} \in \mathbb{H}^{|F| \times 1}$, the sum $\sum_q \cdot$ is over all the three triangles adjacent to triangle p, H_{pq} is a quaternion whose real part equals to the integrated mean curvature H_{pq} over the edge, and e_{pq} is a quaternion whose imaginary part equals to the edge vector in $\mathbb{R}^3$; unspecified real or imaginary parts are always zero. Here the integrated mean curvature is $H_{pq} := \frac{1}{2}|e_{pq}| \tan\frac{\theta_{pq}}{2}$, where θ_{pq} is the bending angle at edge pq.

This version of the extrinsic Dirac operator connects to a discrete spin transformation, resulting in improved numerical performance. It also leads to a discrete intrinsic Dirac operator derived in a unified framework, discussed as follows.

5.14 Intrinsic Dirac operator D_I

The recent work (Ye et al., 2018) also proposes to use an *intrinsic Dirac operator*, whose original intrinsic definition requires the languages of spinors and spin manifolds. An equivalent definition of the intrinsic Dirac operator D_I is as follows:

$$D_I = D_f + H$$

where H is the mean curvature (as a diagonal operator). Although it can hardly be seen from this formula, H cancels out the extrinsic part of D_f, yielding an intrinsic operator D_I that only depends on surface metric.

Following the same notation in the previous subsection, an alternative discrete intrinsic Dirac operator $\mathbf{D}_I \in \mathbb{H}^{|F| \times |F|}$ on a mesh can be defined as

$$(\mathbf{D}_I \boldsymbol{\phi})_p = \frac{1}{2} \sum_q (2H_{pq} + e_{pq}) \cos\frac{\theta_{pq}}{2} \boldsymbol{\phi}_q.$$

The cosine factor plays the role of a correction, ensuring exact covariance of the discrete operator to isometric deformations, providing a key theoretical property analogous to the continuous counterpart (Ye et al., 2018). Empirically this discretization improves performance on relevant applications. See Hoffmann and Ye (2018) for more theoretical justifications of this version of discrete Dirac operators, using a definition of the discrete spin structure. The definition of the intrinsic Dirac operator explicitly involves the spin structure, making it more discriminative than the intrinsic Laplacian, e.g., in distinguishing two approximately isometric tori with different spin structures as illustrated in Ye et al. (2018, Fig. 5).

5.15 Volumetric (extrinsic) Laplacian

So far we have discussed operators taking the thin shell approach; now we move to operators that consider shapes as the boundaries of solids. Namely, starting from this subsection, we consider a volumetric domain Ω whose boundary is specified by a surface $\mathcal{M} = \partial\Omega$.

The Laplace equation with Neumann boundary conditions is defined as follows:

$$\begin{cases} -\Delta_{\mathbb{R}^3} u(\mathbf{x}) = 0 & \mathbf{x} \in \Omega \\ \dfrac{\partial}{\partial n} u(\mathbf{x}) = 0 & \mathbf{x} \in \partial\Omega \end{cases} \tag{31}$$

where $\Delta_{\mathbb{R}^3} = \frac{\partial^2}{\partial x^2} + \frac{\partial^2}{\partial y^2} + \frac{\partial^2}{\partial z^2}$ is the (extrinsically defined) Laplacian operator in $\mathbb{R}^3$. Critical information about the surface is encoded by parts corresponding to the boundary condition. The volumetric heat kernel signature (Raviv et al., 2010; Wang and Wang, 2015; Patané, 2015) considers the heat kernel of $\Delta_{\mathbb{R}^3}$ and applies it to analyze solid shapes.

This volumetric Laplacian can also be discretized by a cotangent matrix $\mathbf{L}$: $\mathbf{L}_{ij}$ is nonzero if i, j are adjacent vertices, and the weights $\mathbf{L}_{ij}$ are cotangents of dihedral angles opposite to edges within the same tetrahedron; see, e.g., Jacobson (2013, Section 2.1) for an explicit formula and derivation.

5.16 Hessian energy

The Hessian LLE (Donoho and Grimes, 2003) substitutes the Hessian in place of the Laplacian used in LLE, enjoying several theoretical guarantees. For a volumetric domain Ω given by a tetrahedral mesh, Stein et al. (2018) consider and discretize the Hessian energy defined as:

$$E_{\mathbf{H}^2}(u) := \frac{1}{2}\int_\Omega \|\mathbf{H}(u(\mathbf{x}))\|_F^2 = \frac{1}{2}\sum_{i,j}\int_\Omega \left(\frac{\partial^2}{\partial \mathbf{x}_i \partial \mathbf{x}_j} u(\mathbf{x})\right)^2 \tag{32}$$

where $\mathbf{H}(u(\mathbf{x})) \in \mathbb{R}^{3\times3}$ is the symmetric matrix consisting of second-order partial derivatives. Spectral analysis on the Hessian energy can be thought as an application of Hessian LLE to mesh data. An interesting property of the Hessian energy is that its null space contains only affine functions, i.e., $\mathbf{1}$, $\mathbf{x}$, $\mathbf{y}$, $\mathbf{z}$. Stein et al. (2018) discretize the Hessian energy as a composition of the gradient operator and a matrix divergence operator, using a mixed FEM approach. As pointed out in their paper, such discretized operator sometimes includes spurious modes in its kernel; this issue seems to be empirically avoided by the followup work (Stein et al., 2019). Furthermore, Stein et al. (2019) generalize the (planar) Hessian energy to a curved surface, with an additional term correctly taking into account the effect of intrinsic curvature.

The bi-Laplacian energy, used for a variety of tasks (Botsch and Kobbelt, 2004; Zorin, 2005; Jacobson et al., 2010, 2011), is closely related to the Hessian energy; they both correspond to the same PDE (biharmonic equation $\Delta_{\mathbb{R}^3}^2 u = 0$) but with different boundary conditions. Fisher et al. (2007) present a DEC Laplacian with modified boundary conditions, and the resulting bi-Laplacian energy (Wang et al., 2015) also includes affine functions in the kernel, similarly to the Hessian energy.

5.17 Single layer potential operator and kernel method

Starting from this subsection, we look at operators defined using boundary integrals. Such operators can be discretized using boundary element methods (BEM).

One example of a boundary integral operator is the *single layer potential operator* $\mathcal{V}: H^{-1/2}(\Gamma) \to H^{1/2}(\Gamma)$, which is defined on surface Γ as

$$[\mathcal{V}u](\mathbf{x}) := \int_\Gamma G(\mathbf{x}, \mathbf{y})u(\mathbf{y})\mathrm{d}\Gamma(\mathbf{y}),$$

where $G(\mathbf{x}, \mathbf{y}) := \frac{1}{4\pi}\frac{1}{|\mathbf{x}-\mathbf{y}|}$ is the fundamental solution of the Laplace equation.

Wu and Levine (1997) simulate the shape as an electric conductor and exploit the resulting charge distribution for shape segmentation. This amounts to solving a linear system consisting of (the inverse of) the single layer potential operator. The charge density function for an electric conductor with constant electrical potential is $[\mathcal{V}^{-1}1](\mathbf{x})$, where $1(\cdot)$ is the constant function. It is well known in physics that this charge density function is highly correlated with the shape concavity: the charge tends to vanish at most concave regions and accumulate at sharp convexities. Based on this key observation, Wu and Levine (1997) use the charge density function as a segmentation clue: it first detects highly concave seams by tracing local minima charge density, and then cuts along these seams to segment the shape.

5.17.1 Boundary element method (BEM)

The single layer potential is an example of an integral operator for which vertices globally interact with each other, unlike differential operators whose actions are local. In Section 3.3 we presented a framework applicable to both FEM and BEM; following the discussion there, BEM is a similar technique to FEM, with the difference that the inner product $\langle \psi_i, \phi_j \rangle_a$—which now involves boundary integrals—will be evaluated with different techniques. In the case of the single layer potential operator, the corresponding weak form is

$$\begin{aligned} a(u,v) &= \int_\Gamma \int_\Gamma u(\mathbf{x})G(\mathbf{x}, \mathbf{y})v(\mathbf{y})\mathrm{d}\Gamma(\mathbf{x})\mathrm{d}\Gamma(\mathbf{y}) \\ &\approx \sum_{\mathbf{x}_i} \sum_{\mathbf{y}_i} w(\mathbf{x}_i, \mathbf{y}_j)u(\mathbf{x}_i)G(\mathbf{x}_i, \mathbf{y}_j)v(\mathbf{y}_j), \,\mathrm{Area}(\mathbf{x}_i)\mathrm{Area}(\mathbf{y}_j), \end{aligned}$$

where $\mathbf{x}_i$, $\mathbf{y}_j$ are the quadrature points, $\mathrm{AREA}(\mathbf{x}_i)$, $\mathrm{AREA}(\mathbf{y}_j)$ are the associated areas, and $w(\mathbf{x}_i, \mathbf{y}_j)$ is the corresponding quadrature weight. Since $G(\mathbf{x}, \mathbf{y})|_{\mathbf{y}=\mathbf{x}}$ is singular, special case needs to be taken when designing the quadrature scheme to properly handle the integrable singularity in the inverse distance kernel (Steinbach, 2007).

The strong form of the single layer operator is $\mathbf{V} \in \mathbb{R}^{n \times n}$, given by $\mathbf{V}_{ij} = \int_\Gamma \int_\Gamma \phi_i(\mathbf{x})G(\mathbf{x}, \mathbf{y})\phi_j(\mathbf{y})\mathrm{d}\Gamma(\mathbf{x})\mathrm{d}\Gamma(\mathbf{y})$, an integrated version of piecewise inverse distances. $\mathbf{V}$ is a dense matrix, which usually admits a low rank approximation for efficient computations; details can be found in Steinbach (2007). Getting back to the method of Wu and Levine (1997), after discretization, the resulting charge density is $\mathbf{V}^{-1}\mathbf{M1}$, where $\mathbf{1} \in \mathbb{R}^n$ is a constant vector.

More generally, it is possible to replace the kernel $\frac{1}{|\mathbf{x}-\mathbf{y}|}$ with $\frac{1}{|\mathbf{x}-\mathbf{y}|^p}$ or Gaussian $e^{-\frac{1}{t}|\mathbf{x}-\mathbf{y}|^2}$: this amounts to applying the kernel method in machine learning for vertex clustering (Shawe-Taylor et al., 2004).

5.18 Dirichlet-to-Neumann operator (Poincaré–Steklov operator)

Consider the Laplace equation with Dirichlet boundary conditions:

$$\begin{cases} \Delta_{\mathbb{R}^3} g(\mathbf{x}) = 0 & \mathbf{x} \in \Omega \\ g(\mathbf{x}) = u(\mathbf{x}) & \mathbf{x} \in \Gamma = \partial\Omega. \end{cases} \tag{33}$$

The *Dirichlet-to-Neumann operator (DtN)* $\mathcal{S} : \mathcal{L}^2(\Gamma) \to \mathcal{L}^2(\Gamma)$ is defined as the map $u \mapsto g_n = \frac{\partial}{\partial n} g(\Gamma)$. In physics, this operator models small vertical oscillations of an ideal fluid in a container. The eigenvalue problem of the DtN operator is known as the *Steklov eigenvalue problem*. Like the Laplace–Beltrami spectrum encoding intrinsic geometry, the Steklov spectrum encodes extrinsic geometry information: the Steklov heat kernel admits an asymptotic expansion whose coefficients are determined by the mean and the Gaussian curvatures $H(\mathbf{x})$ and $K(\mathbf{x})$ (Polterovich and Sher, 2015). The Steklov spectral geometry also enjoys many other properties (Girouard and Polterovich, 2017).

The DtN operator can be written as the composition of boundary integral operators:

$$\mathcal{S} = \mathcal{V}^{-1}\left(\frac{1}{2}\mathcal{I} + \mathcal{K}\right) = \mathcal{H} + \left(\frac{1}{2}\mathcal{I} + \mathcal{K}^{\mathrm{T}}\right)\mathcal{V}^{-1}\left(\frac{1}{2}\mathcal{I} + \mathcal{K}\right)$$

where the double layer potential $\mathcal{K} : H^{1/2}(\Gamma) \to H^{1/2}(\Gamma)$ and hypersingular operator $\mathcal{H} : H^{1/2}(\Gamma) \to H^{-1/2}(\Gamma)$ are defined as:

$$[\mathcal{K}\phi](\mathbf{x}) := \int_\Gamma \frac{\partial G(\mathbf{x},\mathbf{y})}{\partial n(\mathbf{y})} \phi(\mathbf{y}) d\Gamma(\mathbf{y}), \quad [\mathcal{H}\phi](\mathbf{x}) := -\int_\Gamma \frac{\partial^2 G(\mathbf{x},\mathbf{y})}{\partial n(\mathbf{x})\partial n(\mathbf{y})} \phi(\mathbf{y}) d\Gamma(\mathbf{y}).$$

To discretize the operator $\mathcal{S}$, it suffices to discretize these boundary operators $\mathcal{V}, \mathcal{K}, \mathcal{H}$. Although these operators are defined on the boundary, another remarkable property of $\mathcal{S}$ is that its weak form corresponds to the volumetric Dirichlet energy on the harmonic extension of a surface function into the interior. That is, the corresponding strong form is $a(u,v) = \int_\Omega \nabla\mathcal{E}v \cdot \nabla\mathcal{E}u \equiv \int_\mathcal{M} u\mathcal{S}v$, where $\mathcal{E}u$ is the smooth (harmonic) extension of u into the interior, i.e., $\mathcal{E}u$ is the solution to Eq. (33).

Wang et al. (2018b) use the operator $\mathcal{S}$ for shape analysis, as an extrinsic and volumetric alternative to the Laplace–Beltrami operator. It successfully applies this operator to a large number of tasks, including point signature computation, volume-aware distances, segmentation, and extrinsic shape differences, while being robust to topological changes and noise that affects Laplacian-based computations.

5.19 Other extrinsic methods

There are many other methods capturing various aspects of extrinsic geometry, which may not necessarily define extrinsic operators explicitly.

The early spectral method (Pauly and Gross, 2001) decomposes meshes into patches, and for each patch applies discrete Fourier transform (DFT) and inverse DFT on extrinsic voxel grids. The distribution of distances (D2) (Osada et al., 2002) and the signature of histograms of orientations (SHOT) (Tombari et al., 2010) descriptors are extrinsically defined geometric histograms. Spherical harmonics (Kazhdan et al., 2006) are basis functions in the extrinsic 3D space, independent to the shape to be analyzed. Wang et al. (2012) define the discrete first and second fundamental forms as edge lengths and dihedral angles, and proposes a linear method to reconstruct the surface from the two discrete fundamental forms. Corman et al. (2017b) introduce multiple shape difference operators including those encode extrinsic geometry. Rustamov (2011) interpolates intrinsic Laplacian eigenfunctions into the interior of a surface using barycentric coordinates, which carry extrinsic information, and builds descriptors using the interpolants.

6 Summary and experiments

The geometric operators in this paper roughly fall into two categories: operators that view shapes as thin shells and operators that view shapes as solids. There are also other patterns and common wisdom about these operators, summarized as follows.

Relevance of curvature. A common machinery occurring in multiple methods is to explicitly bake curvatures into a thin shell operator. Indeed, the modified Dirichlet energy (Hildebrandt et al., 2012) directly adds the scalar function $\frac{1}{2}(\kappa_1^2(\mathbf{x})+\kappa_2^2(\mathbf{x}))=H(\mathbf{x})^2-K(\mathbf{x})$ as a diagonal operator; the scale invariant Laplacian (Aflalo et al., 2013) uses a multiplicative factor, the inverse Gaussian curvature $K(\mathbf{x})^{-1}$, to cancel out the scaling effect in metric g; the curvature Laplacian (Liu and Zhang, 2007) is defined using the minimal principal curvature as well as its direction; the concavity-aware Laplacian (Au et al., 2012) constructs a discrete Laplacian whose weights are inverse Gaussian curvatures; and the extrinsic Dirac operator equals to the sum of the mean curvature $H(\mathbf{x})$ (as a diagonal operator) and the intrinsic Dirac operator (Ye et al., 2018).

Diagonal operator and operator blending. Both the modified Dirichlet energy (Hildebrandt et al., 2012) and the Hamiltonian operator (Choukroun et al., 2018b) construct a diagonal operator from a scalar (potential) function, such that the two methods coincide if we set the potential function to be $V(\mathbf{x}) = \frac{1}{2}(\kappa_1^2(\mathbf{x}) + \kappa_2^2(\mathbf{x}))$. Dirac operators (Liu et al., 2017) use infinite well boundary conditions, which also amounts to adding a large potential function—as a diagonal operator—only at boundaries. Any nonnegative scalar function can be converted into a p.s.d. diagonal operator, and it would be interesting to investigate alternative scalar functions rather than ones based on curvatures.

Also, all these methods effectively blend two eigen spaces by linearly adding one operator—usually the Laplacian—to another operator. The blended operator has a set of "compromised" eigenfunctions adapting to both operators, yields low energies measured by (the weak form of) both operators: take the Hamiltonian operator as an example, the Laplacian, or the Dirichlet energy, favours eigenfunctions that are smooth, while the potential energy $\int_{\mathcal{M}} \frac{1}{2} V(\mathbf{x}) u(\mathbf{x})^2$ encourages eigenfunctions to avoid regions with large $V(\mathbf{x})$, so the blended Hamiltonian operator has eigenfunctions that are smooth except in areas with very large $V(\mathbf{x})$. To sum up, operator blending provides a means of balancing the considerations of both operators.

Extrinsic or intrinsic. We already discuss the pros and cons of being an extrinsic or intrinsic operator in Section 1. Table 1 summarizes this key property of geometric operators.

6.1 Experiments

In the rest of this section, we provide experiments comparing the spectral properties of a few representative operators. First, we show the eigenfunctions of these operators, which can reveal rich information about geometry. As eigenfunctions are not unique due to sign ambiguities and rotations within eigenspaces formed by repeating eigenvalues, we then visualize quantities derived from eigenfunctions including heat kernel signatures and spectral distances, which remain invariant w.r.t. rotations within eigenspaces.

6.2 Eigenfunctions

First, we show the eigenfunctions computed on a cube with an outward bump in Fig. 9, comparing them with those computed on a cube with an inward bump in Fig. 10. Conforming with theoretical predictions, we see all intrinsic operators demonstrate identical patterns on both cubes, while the extrinsic operators behave differently on the two cubes. It is also interesting to see distinguishing behaviours among these operators. All intrinsic eigenfunctions, as well as extrinsic ones, are aware of corners of the cube, which are singularities of the Gaussian curvature. This is particularly the case for the scale

TABLE 1 Summary of intrinsic and extrinsic operators.

Laplacian (Pinkall and Polthier, 1993)	$\Delta_{\mathcal{M}}$	Intrinsic
Scale invariant (Aflalo et al., 2013)	$\Delta_{\mathcal{M}}^{(si)}$	Intrinsic
Anisotropic Laplacian (Andreux et al., 2014)	$\Delta_{\mathcal{M}}^{A}$	Either
Modified Dirichlet (Hildebrandt et al., 2012)	$\Delta_{\mathcal{M}}^{\kappa}$	Extrinsic
Relative Dirac (Liu et al., 2017)	D_R	Extrinsic
Extrinsic Dirac (Ye et al., 2018)	D_E	Extrinsic
Intrinsic Dirac (Ye et al., 2018)	D_I	Intrinsic
DtN/Steklov (Wang et al., 2018b)	$\mathcal{S}$	Extrinsic
Volumetric Laplacian	$\Delta_{\mathbb{R}^3}$	Extrinsic

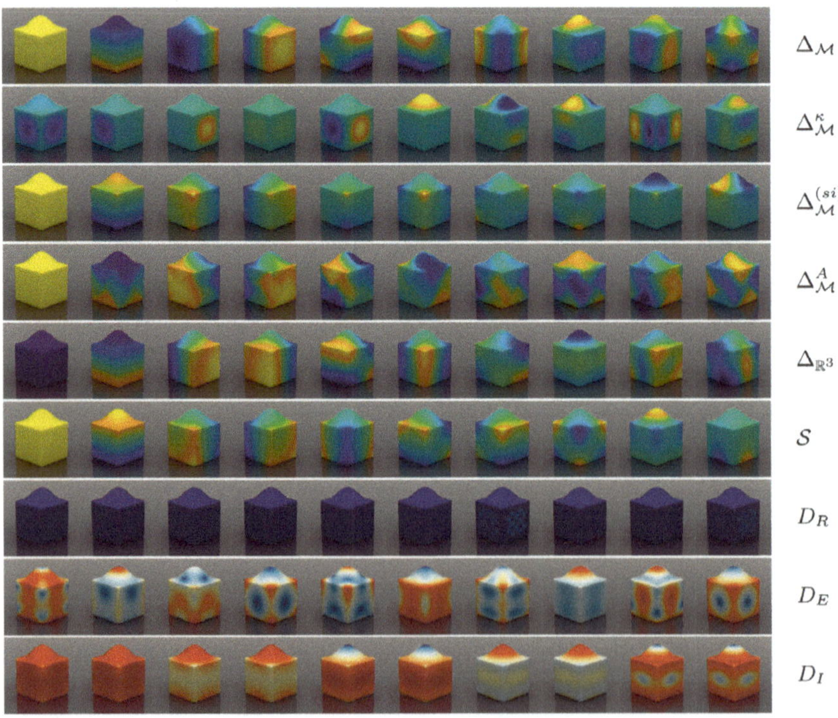

FIG. 9 First few eigenfunctions computed on the cube with an outward bump. For the Dirac operators, norms of the (quaternion valued) eigenfunctions are visualized.

FIG. 10 First few eigenfunctions computed on the cube with an inward bump. For the Dirac operators, norms of the (quaternion valued) eigenfunctions are visualized.

invariant Laplacian: since the Gaussian curvature approaches infinity at the corners, the corresponding eigenfunctions also have singular patterns at these corners. The anisotropic Laplacian demonstrates "vortious patterns" in its eigenfunctions, thanks to the explicit use of an anisotropic kernel.

In addition to corners, all extrinsic operators are sensitive to edges, where the mean curvature becomes large. For the modified Dirichlet energy, the additional term $\int_{\mathcal{M}} \frac{1}{2}(\kappa_1^2(\mathbf{x}) + \kappa_2^2(\mathbf{x}))u(\mathbf{x})^2$ penalizes function $u(\cdot)$ at edges and corners where $(\kappa_1^2(\mathbf{x}) + \kappa_2^2(\mathbf{x}))$ becomes very large, making eigenfunctions vanish at edges and effectively decomposing the cube into six disconnected pieces separated by edges. Similarly, the extrinsic Dirac operator, as the sum of the intrinsic Dirac operator with the mean curvature, also has eigenfunctions whose patterns are separated by edges. The eigenfunctions of the DtN operator also put a particular emphasis on edges of the cube. The relative Dirac operator resonates most in extrinsically flat regions on the shape.

Additional exemplar shapes are provided in Figs. 11 and 12, in which we illustrate eigenfunctions on a bowl with an inward or outward bumps. Compared to the cube in which the mean curvature changes in an extreme fashion, the bowls have smooth curvature varying to a lesser extent across the shape. Consequently, the bowl's Laplacian eigenfunctions become more similar to

FIG. 11 Eigenfunctions computed on a bowl with an inward bump.

eigenfunctions of the modified Dirichlet and the scale invariant Laplacian. The relative Dirac operator vanishes on purely flat regions in the cubes, obstructing a visual comparison; the bowls provide examples that do not contain such regions, and hence we see the relative Dirac eigenfunctions have patterns aligning to ridges of the bumps, resonating most at nearly flat plateaus.

Fig. 13 visualizes eigenfunctions computed on the gargoyle model, at a few representative frequencies. For all operators, eigenfunctions with larger eigenvalues have denser nodal sets. The gargoyle model also has a relative flat region at the bottom, making relative Dirac eigenfunctions concentrate in the bottom.

Since visually interpreting and extracting information from the eigenfunctions may not always be easy, we now start to look at quantities derived from eigenfunctions and eigenvalues.

6.3 Heat kernel signatures

We compute heat kernel signatures (HKS) using the definition in Eq. (23), using either a small time parameter $t = \frac{4 \log 10}{10\lambda_{120}}$, or a large time parameter

FIG. 12 Eigenfunctions computed on a bowl with an outward bump.

$t = \frac{4 \log 10}{\lambda_5}$. We set $k = 120$, i.e., using the first 120 eigenfunctions and eigenvalues. The resulting heat kernel signatures are shown in Fig. 14. For HKS with a small time parameter, the operators commonly put emphasis on regions where there is large Gaussian curvature, such as the horns in the gargoyle model or the fingertips in the hand model. The effect of adding the principal curvature term in the modified Dirichlet energy $\Delta^{\kappa}_{\mathcal{M}}$ is most visible in the gargoyle model: the HKS has a large value at edges of wings in the model, which are places with large mean curvatures. Such places with large mean curvatures are also promoted by the Steklov and volumetric Laplacian HKS with a small time parameter, but less significantly as in the modified Dirichlet HKS.

6.4 Segmentation

We evaluate the results of spectral segmentation on the Princeton segmentation dataset (Chen et al., 2009) and provide visual comparisons of the results obtained by different operators. We first compute the harmonic embedding $\left(\frac{\phi_1(\mathbf{x})}{\sqrt{\lambda_1}}, \frac{\phi_2(\mathbf{x})}{\sqrt{\lambda_2}}, \ldots, \frac{\phi_k(\mathbf{x})}{\sqrt{\lambda_k}}\right)$, as introduced in Eq. (21), and then apply standard

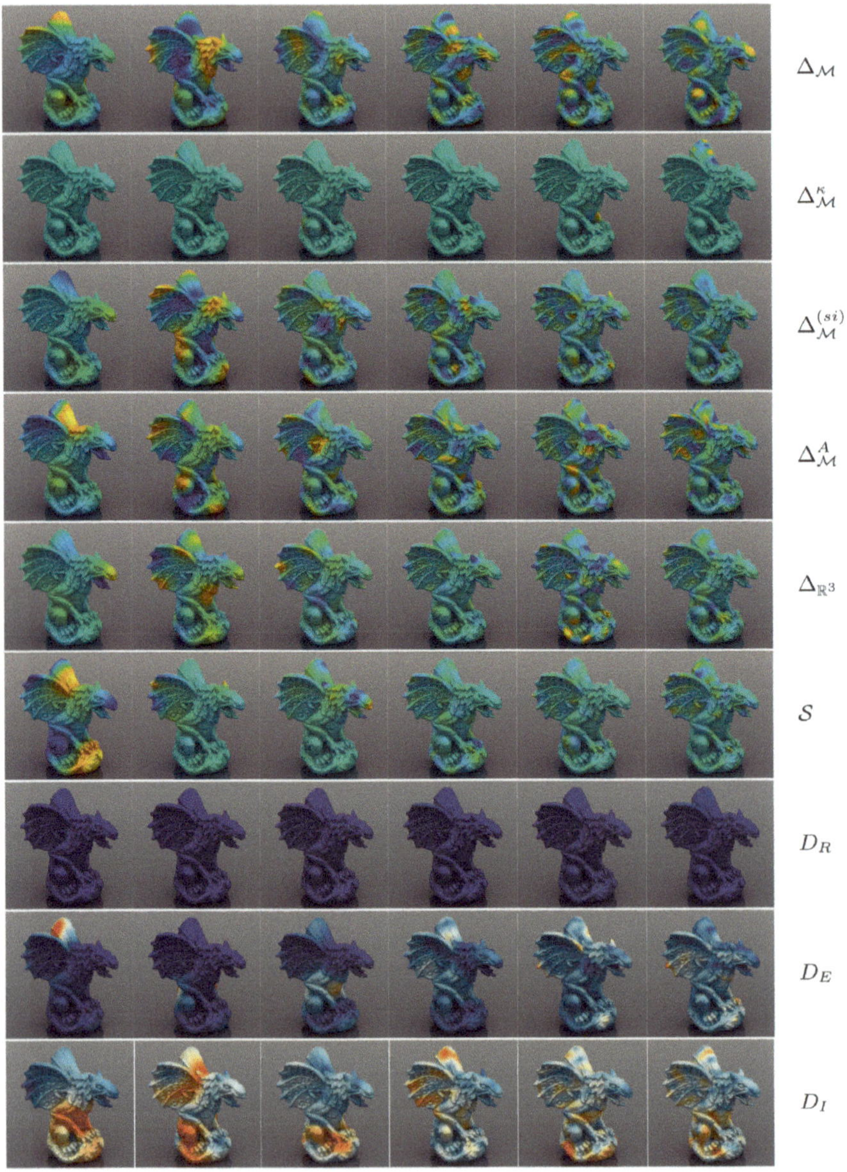

FIG. 13 The 5th, 10th, 20th, 40th, 80th, 100th, 120th eigenfunctions computed on the gargoyle model.

clustering algorithms to the points embedding in $\mathbb{R}^k$. We simply run k-means clustering, with a user-specified number of clusters. For the Dirac operators with quaternion-valued eigenfunctions, we use $\left(\frac{|\phi_1(\mathbf{x})|}{\sqrt{\lambda_1}}, \frac{|\phi_2(\mathbf{x})|}{\sqrt{\lambda_2}}, \ldots, \frac{|\phi_k(\mathbf{x})|}{\sqrt{\lambda_k}}\right)$ following (Ye et al., 2018). We choose $k = 100$ in all experiments.

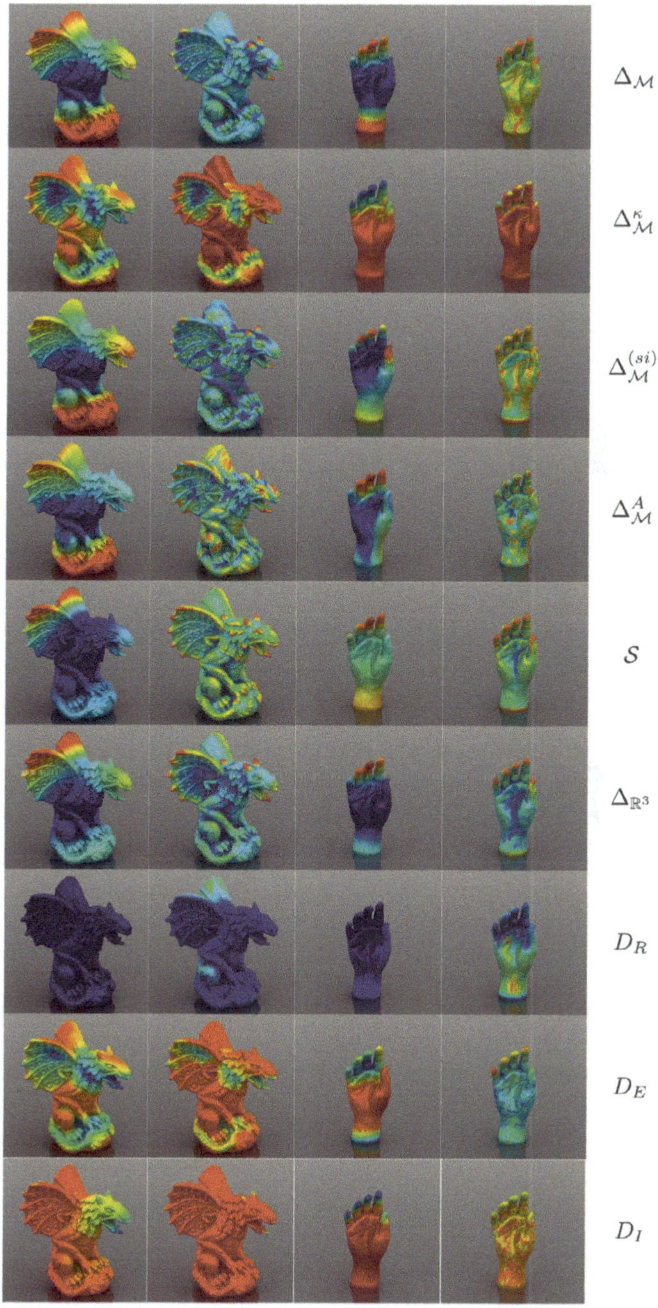

FIG. 14 Heat kernel signatures in log scale, computed using either a small time (*odd rows*) or a large time (*even rows*) parameter.

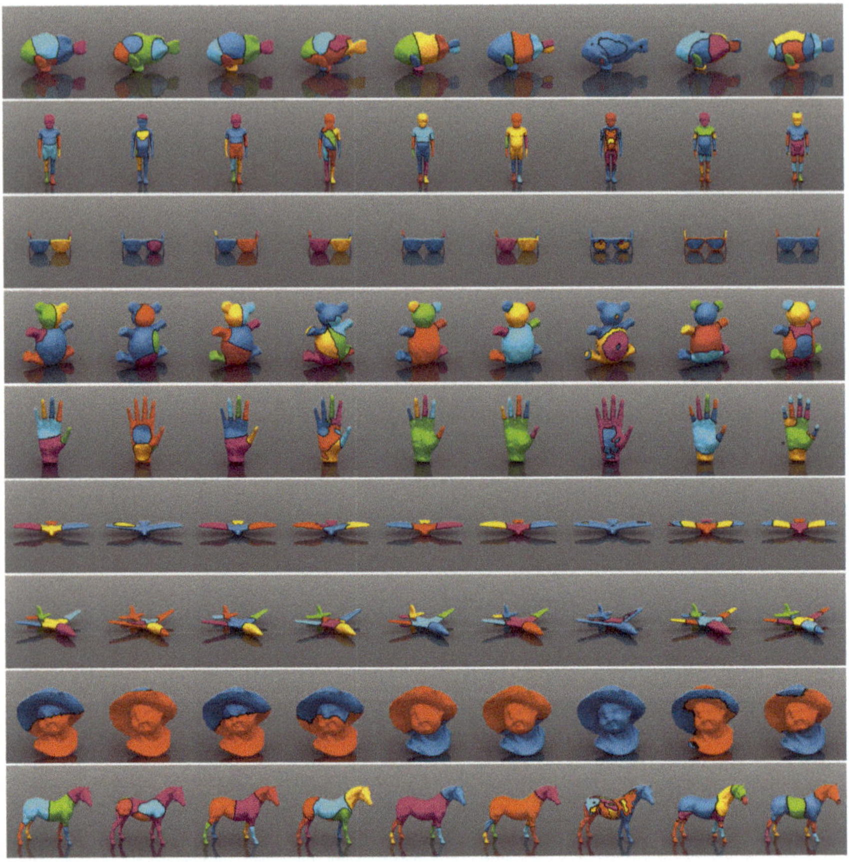

$\Delta_{\mathcal{M}}$ $\Delta_{\mathcal{M}}^{\kappa}$ $\Delta_{\mathcal{M}}^{(si)}$ $\Delta_{\mathcal{M}}^{A}$ $\mathcal{S}$ $\Delta_{\mathbb{R}^3}$ D_R D_E D_I

FIG. 15 A comparison of segmentation results based on different operators.

Fig. 15 shows results of segmentation obtained by following this simple approach. We see that the Laplacian tends to produce segmentation avoiding regions with large Gaussian curvatures, such as ears in the teddy bear model. Volume-based operators tend to decompose shapes into volumetric parts, and the other extrinsic operators including modified Dirichlet energy and Dirac operators cut flat regions from the rest of the shape. An interesting example is the intrinsic Dirac operator, which tends to gather regions with similar geometry (e.g., all fingertips in the hand model) into the same component, different from other intrinsic operators. The relative Dirac operator segments the nearly flat plateaus from other regions; in Liu et al. (2017) a more sophisticated clustering algorithm other than the k-means is used to extract a segmentation.

6.5 Distance or dissimilarity

In Figs. 16 and 17, we show the diffusion distance and the "bioperator" distance, which are generalizations of the diffusion and biharmonic distance for

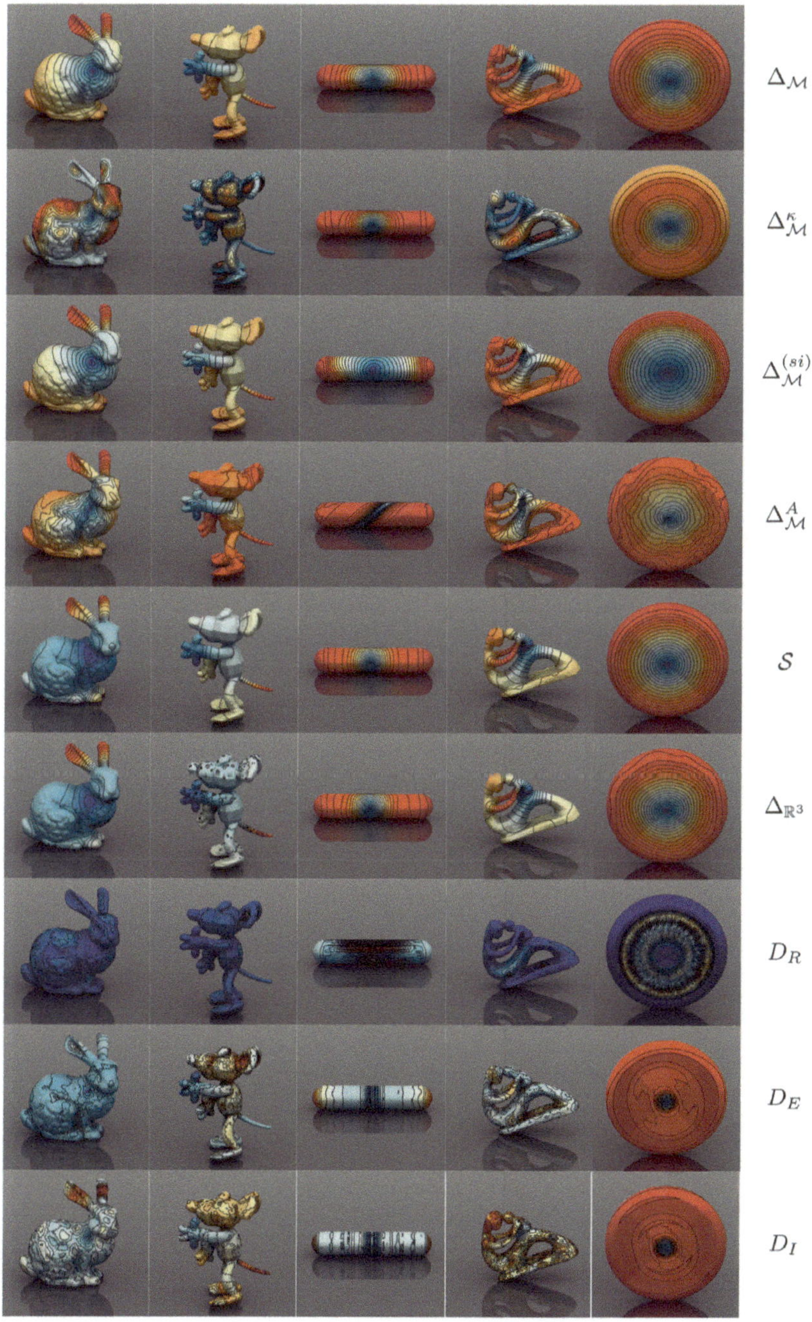

FIG. 16 Diffusion distances computed on a variety of operators.

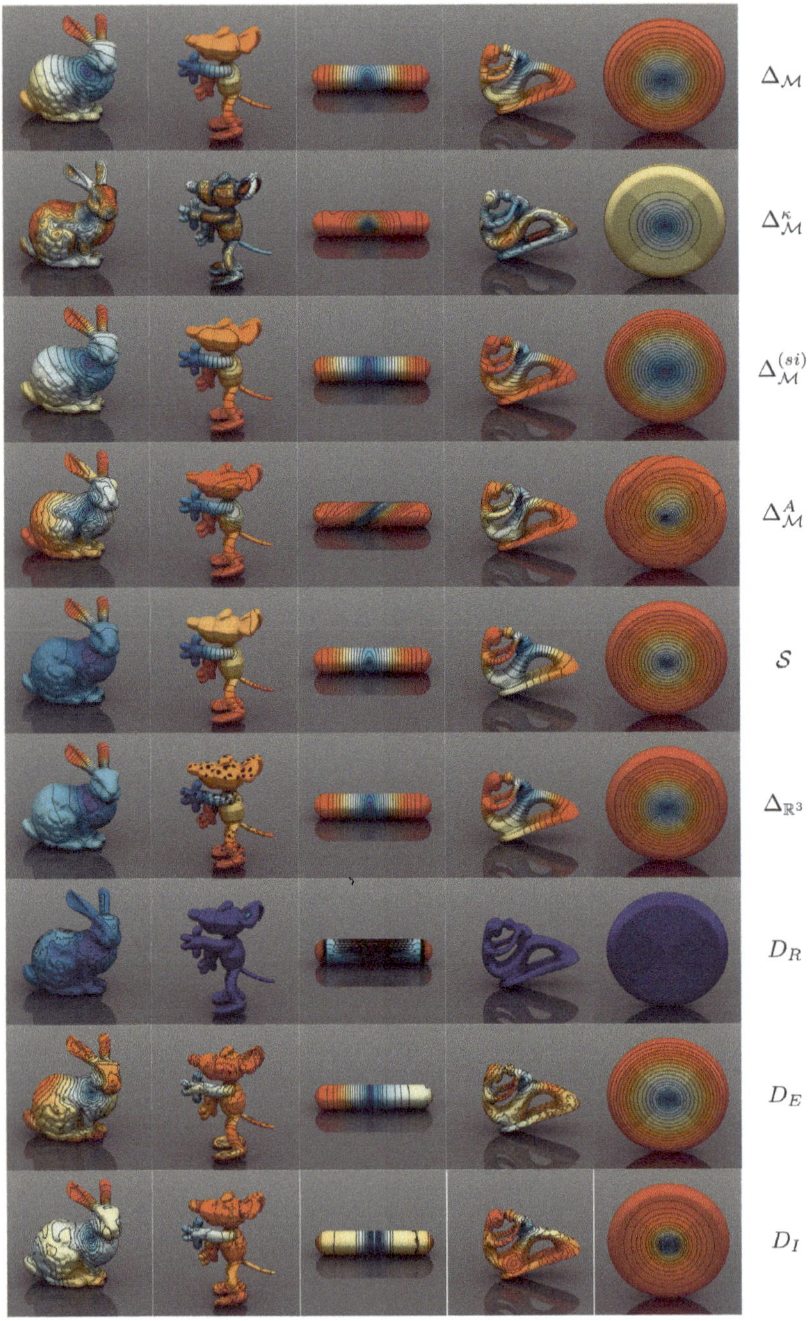

FIG. 17 Bi-"operator" distances computed on a variety of operators.

the Laplacian (Coifman and Lafon, 2006; Lipman et al., 2010) to other operators. Though the generalizations may not make perfect sense as distance measures for other operators, these distances can always be interpreted as dissimilarities of spectral embeddings between points on the surface. Indeed we see that distances in the same category, i.e., Laplacian-like operators (the Laplacian and the scale invariant Laplacian) or volumetric operators (the volumetric Laplacian and the Steklov), tend to have similar patterns. In terms of differences, we can see that the scale invariant bidistance is less evenly distributed in the geodesic sense, compared to the one for Laplacian, as a consequence of using a modified metric. For the anisotropic diffusion distance, as expected the resulting distances copy the anisotropic nature induced by the eigenfunctions. For the volumetric distances like ones based on the volumetric Laplacian or the DtN/Steklov operator, we see eigenfunctions can have local minimum on the surface. This is the correct behaviour due to the volume-geodesic nature of the distances.

7 Conclusion and future work

7.1 Summary

We have seen a few key distinguishing features among these geometric operators, either theoretically or empirically:

- *Laplacian and relatives.* Beyond the Laplacian, the scale invariant Laplacian, the equi-affine and affine invariant Laplacians, and the conformal Laplacian can be thought as the (regular) Laplacian on a "ghost manifold." That is, given an input surface, we first create an abstract manifold that is not necessarily embeddable by deforming the input surface, and then we compute the Laplacian on that manifold. Eigenfunctions of the new Laplacian, are pull backs of the regular Laplacian eigenfunctions, computed on the abstract manifold, onto the original surface.
- *Volumetric operators.* The DtN and volumetric Laplacian are defined on the volumetric domains. Although Wang et al. (2018b) also propose a generalization of DtN to open surfaces, all experiments in this survey are performed on closed surface with a volumetric interior. Accordingly, the eigenfunctions have similar volume-aware patterns.
- *Dirac operators.* The definitions of Dirac operators involve the spin structure of a surface, making them unique from other operators. The spin structure allows the intrinsic Dirac to detect topological information to which Laplacians are insensitive (Ye et al., 2018).
- *Anisotropic operators.* Most operators we survey implicitly use the isotropic assumption. The anisotropic operator is uniquely characterized by an alternative choice, and its eigenfunctions, descriptors, and segmentation all demonstrate direction-sensitive patterns.

7.2 Future work

The diverse and versatile set of operators discussed in this paper suggests possible directions for future work, presenting mathematical, algorithmic, and application-oriented challenges.

7.2.1 Better understanding of existing operators

Most known theoretical results in spectral geometry are for the Laplacian operator and relatives, and theoretical results on other operators are not as prevalently available. While the Laplacian may be simply replaced by other operators, identifying most relevant substitutions for certain applications with theoretical justification provides many promising opportunities for future work.

7.2.2 Synthesizing new operators

We have seen that different operators capture varying aspects of intrinsic and extrinsic geometry. Additively blending operators and multiplicatively composing operators provide a means of constructing new operators from existing ones. It would be interesting to create new operators by combining existing ones inside or outside this survey, and to better characterize the various means of composition. Some techniques used in one operator potentially can be applied to other operators, such as introducing scale invariant metrics, to derive an even larger collection of operators.

7.2.3 Task-dependent and learnable operators

Identifying the most relevant operator for a certain task is another promising direction. While this may not be easy, in practice it is also possible to take a "bag-of-operators" approach, similar to the bag of descriptors method proposed in Bronstein et al. (2011) to further improve the performance of shape analysis algorithms.

When exemplar data or outputs are available, the most effective operators can be learned. We have seen multiple algorithms that add learnable degrees of freedom to operators, and the emerging field of geometric deep learning has inspired considerable attention in this direction (Bronstein et al., 2017).

Acknowledgements

We would like to thank Sebastian Claici and Paul Zhang for proofreading, as well as the anonymous reviewers for their insightful comments. The authors acknowledge the generous support of Army Research Office grant W911NF-12-R-0011, National Science Foundation grant IIS-1838071, Air Force Office of Scientific Research award FA9550-19-1-0319, the MIT–IBM Watson AI Laboratory, the Toyota–CSAIL Joint Research Center, and a gift from Adobe Systems. Part of this work was performed while the first author was visiting the Institute for Pure and Applied Mathematics (IPAM), supported by the National Science Foundation grant DMS-1440415. Any opinions, findings, and conclusions or recommendations expressed in this material are those of the authors and do not necessarily reflect the views of these organizations.

References

Aflalo, Y., Bronstein, A.M., Bronstein, M.M., Kimmel, R., 2011. Deformable shape retrieval by learning diffusion kernels. In: International Conference on Scale Space and Variational Methods in Computer Vision, pp. 689–700.

Aflalo, Y., Kimmel, R., Raviv, D., 2013. Scale invariant geometry for nonrigid shapes. SIAM J. Imaging Sci. 6 (3), 1579–1597.

Aflalo, Y., Brezis, H., Kimmel, R., 2015. On the optimality of shape and data representation in the spectral domain. SIAM J. Imaging Sci. 8 (2), 1141–1160.

Alexa, M., Müller, W., 2000. Representing animations by principal components. Comput. Graph. Forum 19 (3), 411–418.

Ammann, B., Bär, C., 1998. The Dirac operator on nilmanifolds and collapsing circle bundles. Ann. Glob. Anal. Geom. 16 (3), 221–253.

Andreux, M., Rodola, E., Aubry, M., Cremers, D., 2014. Anisotropic Laplace-Beltrami operators for shape analysis. In: European Conference on Computer Vision, pp. 299–312.

Atkinson, K., Han, W., 2005. Theoretical numerical analysis. vol. 39. Springer.

Au, O.K.-C., Zheng, Y., Chen, M., Xu, P., Tai, C.-L., 2012. Mesh segmentation with concavity-aware fields. IEEE Trans. Vis. Comput. Graph. 18 (7), 1125–1134.

Aubry, M., Schlickewei, U., Cremers, D., 2011. The wave kernel signature: a quantum mechanical approach to shape analysis. In: Proc. ICCV Workshops, 1626–1633.

Azencot, O., Ovsjanikov, M., Chazal, F., Ben-Chen, M., 2015. Discrete derivatives of vector fields on surfaces—an operator approach. Trans. Graph. 34 (3), 29.

Azencot, O., Corman, E., Ben-Chen, M., Ovsjanikov, M., 2017. Consistent functional cross field design for mesh quadrangulation. ACM Trans. Graph. 36 (4), 92.

Barbič, J., James, D.L., 2005. Real-time subspace integration for St. Venant-Kirchhoff deformable models. 24 (3), 982–990.

Belkin, M., Niyogi, P., 2003. Laplacian eigenmaps for dimensionality reduction and data representation. Neural Comput. 15 (6), 1373–1396.

Belkin, M., Niyogi, P., 2005. Towards a theoretical foundation for Laplacian-based manifold methods. In: International Conference on Computational Learning Theory, pp. 486–500.

Belkin, M., Niyogi, P., Sindhwani, V., 2006. Manifold regularization: a geometric framework for learning from labeled and unlabeled examples. J. Mach. Learn. Res. 7, 2399–2434.

Belkin, M., Sun, J., Wang, Y., 2008. Discrete Laplace operator on meshed surfaces. In: Proc. SODA, pp. 278–287.

Belkin, M., Sun, J., Wang, Y., 2009. Constructing Laplace operator from point clouds in $\mathbb{R}^d$. In: Proc. SODA, pp. 1031–1040.

Ben-Chen, M., Gotsman, C., 2005. On the optimality of spectral compression of mesh data. ACM Trans. Graph. 24 (1), 60–80.

Bharaj, G., Levin, D.I.W., Tompkin, J., Fei, Y., Pfister, H., Matusik, W., Zheng, C., 2015. Computational design of metallophone contact sounds. ACM Trans. Graph. 34 (6), 223.

Bommes, D., Lévy, B., Pietroni, N., Puppo, E., Silva, C., Tarini, M., Zorin, D., 2013. Quad-mesh generation and processing: a survey. Comput. Graph. Forum 32 (6), 51–76.

Booss, B., 2012. Topology and analysis: the Atiyah-Singer index formula and gauge-theoretic physics. Springer Science & Business Media.

Börm, S., Grasedyck, L., Hackbusch, W., 2003. Hierarchical Matrices. Max Planck Institute.

Boscaini, D., Eynard, D., Kourounis, D., Bronstein, M.M., 2015. Shape-from-operator: recovering shapes from intrinsic operators. Comput. Graph. Forum 34, 265–274.

Boscaini, D., Masci, J., Rodolà, E., Bronstein, M., 2016a. Learning shape correspondence with anisotropic convolutional neural networks. In: Advances in Neural Information Processing Systems, pp. 3189–3197.

Boscaini, D., Masci, J., Rodolà, E., Bronstein, M.M., Cremers, D., 2016b. Anisotropic diffusion descriptors. Comput. Graph. Forum 35 (2), 431–441.

Botsch, M., Kobbelt, L., 2004. An intuitive framework for real-time freeform modeling. ACM Trans. Graph. 23 (3), 630–634.

Botsch, M., Bommes, D., Kobbelt, L., 2005. Efficient linear system solvers for mesh processing. In: Mathematics of Surfaces XI, Springer, pp. 62–83.

Botsch, M., Kobbelt, L., Pauly, M., Alliez, P., Lévy, B., 2010. Polygon Mesh Processing. AK Peters/CRC Press.

Brandt, C., Scandolo, L., Eisemann, E., Hildebrandt, K., 2017. Spectral processing of tangential vector fields. Comput. Graph. Forum 36 (6), 338–353.

Brandt, C., Scandolo, L., Eisemann, E., Hildebrandt, K., 2018. Modeling n-symmetry vector fields using higher-order energies. ACM Trans. Graph. 37 (2), 18.

Brenner, S., Scott, R., 2007. The Mathematical Theory of Finite Element Methods. vol. 15, Springer Science & Business Media.

Brezis, H., 2010. Functional Analysis, Sobolev Spaces and Partial Differential Equations. Springer Science & Business Media.

Bronstein, M.M., Kokkinos, I., 2010. Scale-invariant heat kernel signatures for non-rigid shape recognition. In: 2010 IEEE Computer Society Conference on Computer Vision and Pattern Recognition, pp. 1704–1711.

Bronstein, A.M., Bronstein, M.M., Guibas, L.J., Ovsjanikov, M., 2011. Shape Google: geometric words and expressions for invariant shape retrieval. Trans. Graph. 30 (1), 1.

Bronstein, A., Choukroun, Y., Kimmel, R., Sela, M., 2016. Consistent discretization and minimization of the L1 norm on manifolds. In: 2016 Fourth International Conference on 3D Vision (3DV), pp. 435–440.

Bronstein, M.M., Bruna, J., LeCun, Y., Szlam, A., Vandergheynst, P., 2017. Geometric deep learning: going beyond Euclidean data. IEEE Signal Process. Mag. 34 (4), 18–42.

Bruna, J., Zaremba, W., Szlam, A., Lecun, Y., 2014. Spectral networks and locally connected networks on graphs. In: International Conference on Learning Representations (ICLR2014), CBLS, April, 2014.

Budninskiy, M., Liu, B., Tong, Y., Desbrun, M., 2017. Spectral affine-kernel embeddings. Comput. Graph. Forum 36 (5), 117–129.

Budninskiy, M., Yin, G., Feng, L., Tong, Y., Desbrun, M., 2019. Parallel transport unfolding: a connection-based manifold learning approach. SIAM J. Appl. Algebr. Geom. 3 (2), 266–291.

Cayton, L., 2005. Algorithms for manifold learning. Univ. of California at San Diego Tech. Rep 12 (1-17), 1.

Chadwick, J.N., An, S.S., James, D.L., 2009. Harmonic shells: a practical nonlinear sound model for near-rigid thin shells. ACM Trans. Graph. 28 (5), 119–121.

Chavel, I., 1984. Eigenvalues in Riemannian geometry. vol. 115. Academic Press.

Chen, X., Golovinskiy, A., Funkhouser, T., 2009. A benchmark for 3D mesh segmentation. Trans. Graph. 28 (3), 73.

Chern, A., Knöppel, F., Pinkall, U., Schröder, P., 2018. Shape from metric. ACM Trans. Graph. 37 (4), 63.

Choukroun, Y., Pai, G., Kimmel, R., 2018a. Sparse approximation of 3D meshes using the spectral geometry of the Hamiltonian operator. J. Math. Imaging Vision 60 (6), 941–952.

Choukroun, Y., Shtern, A., Bronstein, A.M., Kimmel, R., 2018b. Hamiltonian operator for spectral shape analysis. IEEE Trans. Vis. Comput. Graph.

Chuang, M., Luo, L., Brown, B.J., Rusinkiewicz, S., Kazhdan, M., 2009. Estimating the Laplace-Beltrami operator by restricting 3D functions. Comput. Graph. Forum 28 (5), 1475–1484.

Chung, F. R. K., Graham, F.C., 1997. Spectral Graph Theory, vol. 92. American Mathematical Soc.

Ciarlet, P.G., 2002. The Finite Element Method for Elliptic Problems. vol. 40. SIAM.

Coifman, R.R., Lafon, S., 2006. Diffusion maps. Appl. Comput. Harmonic Anal. 21 (1), 5–30.

Coifman, R., Rokhlin, V., Wandzura, S., 1993. The fast multipole method for the wave equation: a pedestrian prescription. IEEE Antennas Propag. Mag. 35 (3), 7–12.

Corman, E., Solomon, J., Ben-Chen, M., Guibas, L., Ovsjanikov, M., 2017a. Functional characterization of intrinsic and extrinsic geometry. ACM Trans. Graph. 36 (2), 14.

Corman, E., Solomon, J., Ben-Chen, M., Guibas, L., Ovsjanikov, M., 2017b. Functional characterization of intrinsic and extrinsic geometry. Trans. Graph. 36 (2), 14:1–14:17.

Cosmo, L., Panine, M., Rampini, A., Ovsjanikov, M., Bronstein, M.M., Rodolà, E., 2019. Isospectralization, or how to hear shape, style, and correspondence. In: Proceedings of the IEEE Conference on Computer Vision and Pattern Recognition, pp. 7529–7538.

Courant, R., 1923. Ein allgemeiner satzt zur theorie der eigenfunktionen selbsadjungierter differentialausdrücke. Nachrichten von der Gesellschaft der Wissenschaften zu Göttingen, Mathematisch-Physikalische Klasse 1923, 81–84.

Craioveanu, M.-E., Puta, M., Rassias, T., 2013. Old and New Aspects in Spectral Geometry. vol. 534. Springer Science & Business Media.

Crane, K., Pinkall, U., Schröder, P., 2011. Spin transformations of discrete surfaces. ACM Trans. Graph. 30 (4), 104.

Crane, K., Pinkall, U., Schröder, P., 2013a. Robust fairing via conformal curvature flow. ACM Trans. Graph. 32 (4), 61.

Crane, K., Weischedel, C., Wardetzky, M., 2013b. Geodesics in heat: a new approach to computing distance based on heat flow. Trans. Graph. 32 (5), 152.

De Goes, F., Goldenstein, S., Velho, L., 2008. A hierarchical segmentation of articulated bodies. Comput. Graph. Forum 27 (5), 1349–1356.

Defferrard, M., Bresson, X., Vandergheynst, P., 2016. Convolutional neural networks on graphs with fast localized spectral filtering. In: Advances in Neural Information Processing Systems, pp. 3844–3852.

Desbrun, M., Meyer, M., Alliez, P., 2002. Intrinsic parameterizations of surface meshes. Comput. Graph. Forum 21 (3), 209–218.

Desbrun, M., Hirani, A.N., Leok, M., Marsden, J.E., 2005. Discrete exterior calculus. arXiv math/0508341.

Do Carmo, M. P., 2016. Differential Geometry of Curves and Surfaces: Revised and Updated Second Edition. Courier Dover Publications.

Dong, S., Bremer, P.-T., Garland, M., Pascucci, V., Hart, J.C., 2006. Spectral surface quadrangulation. ACM Trans. Graph. 25 (3), 1057–1066.

Donoho, D.L., Grimes, C., 2003. Hessian eigenmaps: locally linear embedding techniques for high-dimensional data. Proc. Natl. Acad. Sci. 100 (10), 5591–5596.

Drineas, P., Mahoney, M.W., 2005. On the nyström method for approximating a Gram matrix for improved kernel-based learning. J. Mach. Learn. Res. 6, 2153–2175.

Dyer, R., Zhang, H., Möller, T., Clements, A., 2007. An investigation of the spectral robustness of mesh Laplacians. Simon Fraser University.

Dziuk, G., Elliott, C.M., 2013. Finite element methods for surface PDEs. Acta Numer. 22, 289–396.

Eck, M., DeRose, T., Duchamp, T., Hoppe, H., Lounsbery, M., Stuetzle, W., 1995. Multiresolution analysis of arbitrary meshes. In: Siggraph, vol. 95, pp. 173–182.

Evans, L.C., 1998. Partial differential equations. Grad. Stud. Math. 19 (4), 7.

Fiedler, M., 1975. A property of eigenvectors of nonnegative symmetric matrices and its application to graph theory. Czechoslov. Math. J. 25 (4), 619–633.

Fisher, M., Schröder, P., Desbrun, M., Hoppe, H., 2007. Design of tangent vector fields. ACM Trans. Graph. 26 (3), 56.

Frankel, T., 2011. The Geometry of Physics: An Introduction. Cambridge University Press.

Gebal, K., Bærentzen, J.A., Aanæs, H., Larsen, R., 2009. Shape analysis using the auto diffusion function. Comput. Graph. Forum 28 (5), 1405–1413.

Girouard, A., Polterovich, I., 2017. Spectral geometry of the Steklov problem. J. Spectral Theory 7 (2), 321–359.

Golub, G.H., Van Loan, C. F., 2012. Matrix Computations. vol. 3. JHU Press.

Gordon, C., Webb, D., Wolpert, S., 1992a. Isospectral plane domains and surfaces via Riemannian orbifolds. Inven. Math. 110 (1), 1–22.

Gordon, C., Webb, D.L., Wolpert, S., 1992b. One cannot hear the shape of a drum. Bull. Am. Math. Soc. 27 (1), 134–138.

Gotsman, C., 2003. On graph partitioning, spectral analysis, and digital mesh processing. In: 2003 Shape Modeling International, pp. 165–171.

Gotsman, C., Gu, X., Sheffer, A., 2003. Fundamentals of spherical parameterization for 3D meshes. ACM Trans. Graph. 22 (3), 358–363.

Guskov, I., Sweldens, W., Schröder, P., 1999. Multiresolution signal processing for meshes. In: Proceedings of the 26th Annual Conference on Computer Graphics and Interactive Techniques, pp. 325–334.

Halimi, O., Kimmel, R., 2018. Self functional maps. In: 2018 International Conference on 3D Vision (3DV), pp. 710–718.

Halimi, O., Litany, O., Rodola, E., Bronstein, A.M., Kimmel, R., 2019. Unsupervised learning of dense shape correspondence. In: Proceedings of the IEEE Conference on Computer Vision and Pattern Recognition, pp. 4370–4379.

Hauser, K.K., Shen, C., O'Brien, J.F., 2003. Interactive deformation using modal analysis with constraints. In: Graphics Interface, vol. 3, pp. 16–17.

Hein, M., Audibert, J.-Y., Von Luxburg, U., 2005. From graphs to manifolds-weak and strong pointwise consistency of graph Laplacians. In: International Conference on Computational Learning Theory, pp. 470–485.

Henrot, A., 2017. Shape Optimization and Spectral Theory. De Gruyter Open. ISBN: 9783110550856. https://books.google.com/books?id=XXvNAQAACAAJ.

Hildebrandt, K., Schulz, C., von Tycowicz, C., Polthier, K., 2010. Eigenmodes of surface energies for shape analysis. In: GMP, pp. 296–314.

Hildebrandt, K., Schulz, C., Tycowicz, C.V., Polthier, K., 2011. Interactive surface modeling using modal analysis. ACM Trans. Graph.(TOG) 30 (5), 119.

Hildebrandt, K., Schulz, C., von Tycowicz, C., Polthier, K., 2012. Modal shape analysis beyond Laplacian. Comput. Aided Geom. Des. 29 (5), 204–218.

Hirani, A. N., 2003. Discrete Exterior Calculus (Ph.D. thesis). California Institute of Technology.

Hoffmann, T., Ye, Z., 2018. A discrete extrinsic and intrinsic Dirac operator. arXiv preprint arXiv:1802.06278.

Holst, M., 2001. Adaptive numerical treatment of elliptic systems on manifolds. Adv. Comput. Math. 15 (1–4), 139–191.

Hou, X., Zhang, L., 2007. Saliency detection: a spectral residual approach. In: 2007 IEEE Conference on Computer Vision and Pattern Recognition, pp. 1–8.

Huang, Q.-X., Wicke, M., Adams, B., Guibas, L., 2009. Shape decomposition using modal analysis. Comput. Graph. Forum 28 (2), 407–416.

Huang, R., Rakotosaona, M.-J., Achlioptas, P., Guibas, L., Ovsjanikov, M., 2019. Operatornet: recovering 3D shapes from difference operators. arXiv preprint arXiv:1904.10754.

Jacobson, A., 2013. Algorithms and Interfaces for Real-Time Deformation of 2D and 3d Shapes (Ph.D. thesis). ETH Zurich.

Jacobson, A., Tosun, E., Sorkine, O., Zorin, D., 2010. Mixed finite elements for variational surface modeling. Comput. Graph. Forum 29 (5), 1565–1574.

Jacobson, A., Baran, I., Popovic, J., Sorkine, O., 2011. Bounded biharmonic weights for real-time deformation. ACM Trans. Graph. 30 (4), 78–81.

Jacobson, A., Deng, Z., Kavan, L., Lewis, J.P., 2014. Skinning: real-time shape deformation. In: ACM SIGGRAPH 2014 Courses, pp. 24.

Joshi, P., Meyer, M., DeRose, T., Green, B., Sanocki, T., 2007. Harmonic coordinates for character articulation. ACM Trans. Graph. 26 (3), 71.

Kac, M., 1966. Can one hear the shape of a drum? Am. Math. Mon. 73 (4), 1–23.

Kamberov, G., Pedit, F., Pinkall, U., 1996. Bonnet pairs and isothermic surfaces. arXiv preprint dg-ga/9610006.

Karni, Z., Gotsman, C., 2000. Spectral compression of mesh geometry. In: Proceedings of the 27th Annual Conference on Computer Graphics and Interactive Techniques, pp. 279–286.

Kazhdan, M., Bolitho, M., Hoppe, H., 2006. Poisson surface reconstruction. In: Proceedings of the Fourth Eurographics Symposium on Geometry Processing. vol. 7.

Kim, V.G., Li, W., Mitra, N.J., DiVerdi, S., Funkhouser, T.A., 2012. Exploring collections of 3d models using fuzzy correspondences. ACM Trans. Graph. 31 (4), 54.

Knight, F.B., et al., 1981. Essentials of Brownian Motion and Diffusion. vol. 18. American Mathematical Soc.

Knyazev, A.V., 2001. Toward the optimal preconditioned eigensolver: locally optimal block preconditioned conjugate gradient method. SIAM J. Sci. Comput. 23 (2), 517–541.

Kobbelt, L., 1997. Discrete fairing. In: Proceedings of the Seventh IMA Conference on the Mathematics of Surfaces, vol. 97, pp. 101–131.

Kokkinos, I., Bronstein, M.M., Litman, R., Bronstein, A.M., 2012. Intrinsic shape context descriptors for deformable shapes. In: Proc. CVPR, pp. 159–166.

Kolluri, R., Shewchuk, J.R., O'Brien, J.F., 2004. Spectral surface reconstruction from noisy point clouds. In: Proceedings of the 2004 Eurographics/ACM SIGGRAPH Symposium on Geometry Processing, pp. 11–21.

Kostrikov, I., Jiang, Z., Panozzo, D., Zorin, D., Joan, B., 2018. Surface networks. In: 2018 IEEE Conference on Computer Vision and Pattern Recognition, CVPR 2018.

Kovnatsky, A., Bronstein, M.M., Bronstein, A.M., Glashoff, K., Kimmel, R., 2013. Coupled quasi-harmonic bases. Comput. Graph. Forum 32 (2, Pt. 4), 439–448.

Krishnan, D., Fattal, R., Szeliski, R., 2013. Efficient preconditioning of Laplacian matrices for computer graphics. ACM Trans. Graph. 32 (4), 142.

Kry, P.G., James, D.L., Pai, D.K., 2002. Eigenskin: real time large deformation character skinning in hardware. In: Proceedings of the 2002 ACM SIGGRAPH/Eurographics Symposium on Computer Animation, pp. 153–159.

Kunisch, K., Volkwein, S., 2002. Galerkin proper orthogonal decomposition methods for a general equation in fluid dynamics. SIAM J. Numer. Anal. 40 (2), 492–515.

Langlois, T.R., An, S.S., Jin, K.K., James, D.L., 2014. Eigenmode compression for modal sound models. ACM Trans. Graph. 33 (4), 40.

Lee, J.M., 2013. Smooth manifolds. In: Introduction to Smooth Manifolds. Springer, pp. 1–31.

Lehoucq, R.B., Sorensen, D.C., Yang, C., 1998. ARPACK users' guide: solution of large-scale eigenvalue problems with implicitly restarted Arnoldi methods. vol. 6. Siam.

Levitin, M., Parnovski, L., Polterovich, I., Sher, D.A., 2017. Sloshing, Steklov and corners I: asymptotics of sloshing eigenvalues. arXiv:1709.01891.

Levy, B., 2006. Laplace-Beltrami eigenfunctions towards an algorithm that "understands" geometry. In: IEEE International Conference on Shape Modeling and Applications 2006 (SMI'06). pp. 13.

Lian, Z., Godil, A., Bustos, B., Daoudi, M., Hermans, J., Kawamura, S., Kurita, Y., Lavoué, G., Van Nguyen, H., Ohbuchi, R., et al., 2013. A comparison of methods for non-rigid 3D shape retrieval. Pattern Recogn. 46 (1), 449–461.

Liang, J., Zhao, H., 2013. Solving partial differential equations on point clouds. SIAM J. Sci. Comput. 35 (3), A1461–A1486.

Lipman, Y., Levin, D., Cohen-Or, D., 2008. Green coordinates. ACM Trans. Graph. 27 (3), 78.

Lipman, Y., Rustamov, R.M., Funkhouser, T.A., 2010. Biharmonic distance. Trans. Graph. 29 (3), 27.

Litany, O., Remez, T., Rodolà, E., Bronstein, A.M., Bronstein, M.M., 2017. Deep functional maps: structured prediction for dense shape correspondence. In: ICCV, pp. 5660–5668.

Litman, R., Bronstein, A.M., 2014. Learning spectral descriptors for deformable shape correspondence. IEEE Trans. Pattern Anal. Mach. Intell. 36 (1), 171–180.

Liu, R., Zhang, H., 2004. Segmentation of 3D meshes through spectral clustering. In: 12th Pacific Conference on Computer Graphics and Applications, 2004. PG 2004. Proceedings., pp. 298–305.

Liu, R., Zhang, H., 2007. Mesh segmentation via spectral embedding and contour analysis. Comput. Graph. Forum 26 (3), 385–394.

Liu, R., Jain, V., Zhang, H., 2006. Sub-sampling for efficient spectral mesh processing. In: Computer Graphics International Conference, pp. 172–184.

Liu, B., Mason, G., Hodgson, J., Tong, Y., Desbrun, M., 2015. Model-reduced variational fluid simulation. ACM Trans. Graph. (TOG) 34 (6), 244.

Liu, H.-T.D., Jacobson, A., Crane, K., 2017. A Dirac operator for extrinsic shape analysis. Comput. Graph. Forum 36, 139–149.

Liu, H.-T.D., Jacobson, A., Ovsjanikov, M., 2019. Spectral coarsening for geometric operators. In: ACM Transactions on Graphics (Proceedings of SIGGRAPH 2019, to appear).

Mahadevan, S., 2007. Adaptive mesh compression in 3D computer graphics using multiscale manifold learning. In: Proceedings of the 24th International Conference on Machine Learning, pp. 585–592.

Masci, J., Boscaini, D., Bronstein, M., Vandergheynst, P., 2015. Geodesic convolutional neural networks on Riemannian manifolds. In: Proceedings of the IEEE International Conference on Computer Vision Workshops, pp. 37–45.

Meila, M., Shi, J., 2001. Learning segmentation by random walks. In: Advances in Neural Information Processing Systems, pp. 873–879.

Melzi, S., Ovsjanikov, M., Roffo, G., Cristani, M., Castellani, U., 2018a. Discrete time evolution process descriptor for shape analysis and matching. ACM Trans. Graph. 37 (1), 4.

Melzi, S., Rodolà, E., Castellani, U., Bronstein, M.M., 2018b. Localized manifold harmonics for spectral shape analysis. Comput. Graph. Forum 37 (6), 20–34.

Meyer, M., Desbrun, M., Schröder, P., Barr, A.H., 2003. Discrete differential-geometry operators for triangulated 2-manifolds. In: Visualization and Mathematics III, Springer, pp. 35–57.

Mullen, P., Tong, Y., Alliez, P., Desbrun, M., 2008. Spectral conformal parameterization. Comput. Graph. Forum 27 (5), 1487–1494.

Nasikun, A., Brandt, C., Hildebrandt, K., 2018. Fast approximation of Laplace-Beltrami eigenproblems. Comput. Graph. Forum 37 (5), 121–134.

Neumann, T., Varanasi, K., Theobalt, C., Magnor, M., Wacker, M., 2014. Compressed manifold modes for mesh processing. Comput. Graph. Forum 33 (5), 35–44.

Ng, A.Y., Jordan, M.I., Weiss, Y., 2002. On spectral clustering: analysis and an algorithm. In: Advances in Neural Information Processing Systems, pp. 849–856.

O'Brien, J.F., Shen, C., Gatchalian, C.M., 2002. Synthesizing sounds from rigid-body simulations. In: Proceedings of the 2002 ACM SIGGRAPH/Eurographics Symposium on Computer Animation, pp. 175–181.

Osada, R., Funkhouser, T., Chazelle, B., Dobkin, D., 2002. Shape distributions. Trans. Graph. 21 (4), 807–832.

Ovsjanikov, M., Sun, J., Guibas, L., 2008. Global intrinsic symmetries of shapes. Comput. Graph. Forum 27 (5), 1341–1348.

Ovsjanikov, M., Mérigot, Q., Mémoli, F., Guibas, L., 2010. One point isometric matching with the heat kernel. Comput. Graph. Forum 29 (5), 1555–1564.

Ovsjanikov, M., Ben-Chen, M., Solomon, J., Butscher, A., Guibas, L., 2012. Functional maps: a flexible representation of maps between shapes. Trans. Graph. 31 (4), 30.

Ovsjanikov, M., Corman, E., Bronstein, M., Rodolà, E., Ben-Chen, M., Guibas, L., Chazal, F., Bronstein, A., 2016. Computing and processing correspondences with functional maps. In: SIGGRAPH ASIA 2016 Courses, pp. 9.

Ozoliņš, V., Lai, R., Caflisch, R., Osher, S., 2013. Compressed modes for variational problems in mathematics and physics. Proc. Natl. Acad. Sci. 110 (46), 18368–18373.

Patané, G., 2015. Volumetric heat kernel: Padé-Chebyshev approximation, convergence, and computation. Comput. Graph. 46, 64–71.

Patané, G., 2017. Accurate and efficient computation of Laplacian spectral distances and kernels. Comput. Graph. Forum 36 (1), 184–196.

Patanè, G., 2017. An introduction to Laplacian spectral distances and kernels: theory, computation, and applications. Synthesis Lectures on Visual Computing: Computer Graphics, Animation, Computational Photography, and Imaging 9 (2), 1–139.

Patane, G., Spagnuolo, M., Falcidieno, B., 2008. Reeb graph computation based on a minimal contouring. In: 2008 IEEE International Conference on Shape Modeling and Applications, pp. 73–82.

Pauly, M., Gross, M., 2001. Spectral processing of point-sampled geometry. In: Proceedings of the 28th Annual Conference on Computer Graphics and Interactive Techniques, pp. 379–386.

Pentland, A., Williams, J., 1989. Good Vibrations: Modal Dynamics for Graphics and Animation. Vision and Modeling Group, Media Laboratory, Massachusetts Institute of Technology.

Perelman, G., 2002. The entropy formula for the Ricci flow and its geometric applications. arXiv preprint math/0211159.

Perelman, G., 2003a. Finite extinction time for the solutions to the ricci flow on certain three-manifolds. arXiv preprint math. DG/0307245.

Perelman, G., 2003b. Ricci flow with surgery on three-manifolds, arxiv: math. D6/0303109.

Pinkall, U., Polthier, K., 1993. Computing discrete minimal surfaces and their conjugates. Exp. Math. 2 (1), 15–36.

Polterovich, I., Sher, D.A., 2015. Heat invariants of the Steklov problem. J. Geom. Anal. 25 (2), 924–950.

Qiu, H., Hancock, E.R., 2007. Clustering and embedding using commute times. IEEE Trans. Pattern Anal. Mach. Intell. 29 (11), 1873–1890.

Raviv, D., Kimmel, R., 2015. Affine invariant geometry for non-rigid shapes. Int. J. Comput. Vis. 111 (1), 1–11.

Raviv, D., Raskar, R., 2015. Scale invariant metrics of volumetric datasets. SIAM J. Imaging Sci. 8 (1), 403–425.

Raviv, D., Bronstein, M.M., Bronstein, A.M., Kimmel, R., 2010. Volumetric heat kernel signatures. In: Proc. 3DOR, pp. 39–44.

Raviv, D., Bronstein, A.M., Bronstein, M.M., Kimmel, R., Sochen, N., 2011a. Affine-invariant geodesic geometry of deformable 3D shapes. Computers & Graphics 35 (3), 692–697.

Raviv, D., Bronstein, M.M., Bronstein, A.M., Kimmel, R., Sochen, N., 2011b. Affine-invariant diffusion geometry for the analysis of deformable 3D shapes. In: CVPR 2011, pp. 2361–2367.

Raviv, D., Bronstein, A.M., Bronstein, M.M., Waisman, D., Sochen, N., Kimmel, R., 2014a. Equi-affine invariant geometry for shape analysis. J. Math. Imaging Vision 50 (1-2), 144–163.

Raviv, D., Lessick, J., Raskar, R., 2014b. Evaluating local contractions from large deformations using affine invariant spectral geometry. In: International Workshop on Statistical Atlases and Computational Models of the Heart, pp. 147–157.

Ren, Z., Yeh, H., Lin, M.C., 2013. Example-guided physically based modal sound synthesis. ACM Trans. Graph. 32 (1), 1.

Reuter, M., Wolter, F.-E., Peinecke, N., 2005. Laplace-spectra as fingerprints for shape matching. In: Proceedings of the 2005 ACM symposium on Solid and Physical Modeling ACM, pp. 101–106.

Reuter, M., Wolter, F.-E., Peinecke, N., 2006. Laplace-Beltrami spectra as 'shape-DNA' of surfaces and solids. Computer-Aided Design 38 (4), 342–366.

Reuter, M., Biasotti, S., Giorgi, D., Patanè, G., Spagnuolo, M., 2009. Discrete Laplace-Beltrami operators for shape analysis and segmentation. Comput. Graph. 33 (3), 381–390.

Rodolà, E., Cosmo, L., Bronstein, M.M., Torsello, A., Cremers, D., 2017. Partial functional correspondence. Comput. Graph. Forum 36 (1), 222–236.

Rong, G., Cao, Y., Guo, X., 2008. Spectral mesh deformation. Vis. Comput. 24 (7–9), 787–796.

Rosenberg, S., 1997. The Laplacian on a Riemannian Manifold: An Introduction to Analysis on Manifolds. vol. 31. Cambridge University Press.

Rosman, G., Bronstein, M.M., Bronstein, A.M., Kimmel, R., 2010. Nonlinear dimensionality reduction by topologically constrained isometric embedding. Int. J. Comput. Vis. 89 (1), 56–68.

Roufosse, J.-M., Ovsjanikov, M., 2018. Unsupervised deep learning for structured shape matching. arXiv preprint arXiv:1812.03794.

Roweis, S.T., Saul, L.K., 2000. Nonlinear dimensionality reduction by locally linear embedding. Science 290 (5500), 2323–2326.

Rustamov, R.M., 2007. Laplace-Beltrami eigenfunctions for deformation invariant shape representation. In: Proceedings of the Fifth Eurographics Symposium on Geometry Processing, pp. 225–233.

Rustamov, R.M., 2009. On mesh editing, manifold learning, and diffusion wavelets. In: IMA International Conference on Mathematics of Surfaces, pp. 307–321.

Rustamov, R.M., 2011. Interpolated eigenfunctions for volumetric shape processing. Vis. Comput. 27 (11), 951–961.

Rustamov, R.M., Ovsjanikov, M., Azencot, O., Ben-Chen, M., Chazal, F., Guibas, L., 2013. Map-based exploration of intrinsic shape differences and variability. Trans. Graph. 32 (4), 72.

Saad, Y., 2011. Numerical methods for large eigenvalue problems: revised edition. vol. 66. Siam.

Scarselli, F., Gori, M., Tsoi, A.C., Hagenbuchner, M., Monfardini, G., 2009. The graph neural network model. IEEE Trans. Neural Netw. 20 (1), 61–80.

Schonsheck, S.C., Bronstein, M.M., Lai, R., 2018. Nonisometric surface registration via conformal Laplace-Beltrami basis pursuit. arXiv preprint arXiv:1809.07399.

Schwartz, E.L., Shaw, A., Wolfson, E., 1989. A numerical solution to the generalized mapmaker. IEEE Trans. Pattern Anal. Mach. Intell. (9), 1005–1008.

Sela, M., Aflalo, Y., Kimmel, R., 2015. Computational caricaturization of surfaces. Comput. Vis. Image Understanding 141, 1–17.

Sharp, N., Soliman, Y., Crane, K., 2018. The vector heat method. arXiv preprint arXiv:1805.09170.

Shawe-Taylor, J., Cristianini, N., et al., 2004. Kernel Methods for Pattern Analysis. Cambridge University Press.

Shi, J., Malik, J., 2000. Normalized cuts and image segmentation. IEEE Trans. Pattern Anal. Mach. Intell. 22 (8), 888–905.

Shi, Z., Sun, J., 2017. Convergence of the point integral method for Laplace-Beltrami equation on point cloud. Res. Math. Sci. 4 (1), 22.

Shi, Y., Lai, R., Krishna, S., Sicotte, N., Dinov, I., Toga, A.W., 2008. Anisotropic Laplace-Beltrami eigenmaps: bridging Reeb graphs and skeletons. In: 2008 IEEE Computer Society Conference on Computer Vision and Pattern Recognition Workshops, pp. 1–7.

Singer, A., Wu, H.-T., 2012. Vector diffusion maps and the connection Laplacian. Commun. Pure Appl. Math. 65 (8), 1067–1144.

Śmigaj, W., Betcke, T., Arridge, S., Phillips, J., Schweiger, M., 2015. Solving boundary integral problems with BEM++. Trans. Math. Softw. 41 (2), 6.

Song, R., Liu, Y., Martin, R.R., Rosin, P.L., 2014. Mesh saliency via spectral processing. ACM Trans. Graph. 33 (1), 6.

Sorkine, O., 2005. Laplacian mesh processing. In: Eurographics State of the Art Report.

Sorkine, O., 2006. Differential representations for mesh processing. Comput. Graph. Forum 25 (4), 789–807.

Sorkine, O., Cohen-Or, D., Lipman, Y., Alexa, M., Rössl, C., Seidel, H.-P., 2004. Laplacian surface editing. In: Proc. SGP, pp. 175–184.

Spagnuolo, M., Bronstein, M., Bronstein, A., Ferreira, A., 2012. Affine-Invariant Photometric Heat Kernel Signatures. The Eurographics Association, pp. 39–46.

Spielman, D.A., Teng, S.-H., 1996. Spectral partitioning works: Planar graphs and finite element meshes. In: Proceedings of 37th Conference on Foundations of Computer Science, pp. 96–105.

Spielman, D.A., Teng, S.-H., 2004. Nearly-linear time algorithms for graph partitioning, graph sparsification, and solving linear systems, vol. 4. In: Proceedings of the STOC, pp. 4.

Stein, O., Grinspun, E., Wardetzky, M., Jacobson, A., 2018. Natural boundary conditions for smoothing in geometry processing. ACM Trans. Graph. 37 (2), 23.

Stein, O., Jacobson, A., Wardetzky, M., Grinspun, E., 2019. A smoothness energy without boundary distortion for curved surfaces. arXiv preprint arXiv:1905.09777.

Steinbach, O., 2007. Numerical Approximation Methods for Elliptic Boundary Value Problems: Finite and Boundary Elements. Springer.

Sun, J., Zhou, A., 2016. Finite Element Methods for Eigenvalue Problems. Chapman and Hall/CRC.

Sun, J., Ovsjanikov, M., Guibas, L., 2009. A concise and provably informative multi-scale signature based on heat diffusion. Comput. Graph. Forum 28 (5), 1383–1392.

Taubin, G., 1995. A signal processing approach to fair surface design. In: Proceedings of the 22nd Annual Conference on Computer Graphics and Interactive Techniques, pp. 351–358.

Taylor, M.E., 2011. Partial Differential Equations I: Basic Theory (Applied Mathematical Sciences). Springer New York.

Taylor, M., 2013. Partial Differential Equations II: Qualitative Studies of Linear Equations. vol. 116. Springer Science & Business Media.

Tenenbaum, J.B., De Silva, V., Langford, J.C., 2000. A global geometric framework for nonlinear dimensionality reduction. Science 290 (5500), 2319–2323.

Thomas, D.M., Natarajan, V., 2014. Multiscale symmetry detection in scalar fields by clustering contours. IEEE Trans. Vis. Comput. Graph. 20 (12), 2427–2436.

Tombari, F., Salti, S., Di Stefano, L., 2010. Unique signatures of histograms for local surface description. In: Proc. ECCV, pp. 356–369.

Tong, Y., Lombeyda, S., Hirani, A.N., Desbrun, M., 2003. Discrete multiscale vector field decomposition. ACM Trans. Graph. 22 (3), 445–452.

Vallet, B., Lévy, B., 2008. Spectral geometry processing with manifold harmonics. Comput. Graph. Forum 27 (2), 251–260.

van de Doel, K., Pai, D.K., 1996. Synthesis of Shape Dependent Sounds With Physical Modeling. Georgia Institute of Technology.

van de Doel, K., Pai, D.K., 1998. The sounds of physical shapes. Presence 7 (4), 382–395.

Varadhan, S.R.S., 1967. On the behavior of the fundamental solution of the heat equation with variable coefficients. Commun. Pure Appl. Math. 20 (2), 431–455.

Vaxman, A., Ben-Chen, M., Gotsman, C., 2010. A multi-resolution approach to heat kernels on discrete surfaces. ACM Trans. Graph. 29 (4), 121.

Wang, G., Wang, Y., 2015. Multi-scale heat kernel based volumetric morphology signature. In: International Conference on Medical Image Computing and Computer-Assisted Intervention, pp. 751–759.

Wang, Y., Liu, B., Tong, Y., 2012. Linear surface reconstruction from discrete fundamental forms on triangle meshes. Comput. Graph. Forum 31 (8), 2277–2287.

Wang, H., Lu, T., Au, O.K.-C., Tai, C.-L., 2014. Spectral 3D mesh segmentation with a novel single segmentation field. Graph. Models 76 (5), 440–456.

Wang, Y., Jacobson, A., Barbič, J., Kavan, L., 2015. Linear subspace design for real-time shape deformation. ACM Trans. Graph. 34 (4), 57.

Wang, L., Gehre, A., Bronstein, M.M., Solomon, J., 2018a. Kernel functional maps. Comput. Graph. Forum 37 (5), 27–36.

Wang, Y., Ben-Chen, M., Polterovich, I., Solomon, J., 2018b. Steklov spectral geometry for extrinsic shape analysis. ACM Trans. Graph. 37 (7), 21.

Wang, Y., Kim, V., Bronstein, M., Solomon, J., 2019. Learning geometric operators on meshes. In: 7th International Conference on Learning Representations (ICLR 2019) Workshop on Representation Learning on Graphs and Manifolds.

Wardetzky, M., Mathur, S., Kälberer, F., Grinspun, E., 2007. Discrete Laplace operators: no free lunch. In: Symposium on Geometry Processing, pp. 33–37.

Williams, C.K.I., 2002. On a connection between kernel PCA and metric multidimensional scaling. Mach. Learn. 46 (1–3), 11–19.

Williams, C. K. I., Seeger, M., 2001. Using the nyström method to speed up kernel machines. In: Advances in Neural Information Processing Systems, pp. 682–688.

Wu, H.-T., 2017. Embedding Riemannian manifolds by the heat kernel of the connection Laplacian. Adv. Math. 304, 1055–1079.

Wu, K., Levine, M.D., 1997. 3D part segmentation using simulated electrical charge distributions. IEEE Trans. Pattern Anal. Mach. Intell. 19 (11), 1223–1235.

Ye, Z., Diamanti, O., Tang, C., Guibas, L., Hoffmann, T., 2018. A unified discrete framework for intrinsic and extrinsic Dirac operators for geometry processing. Comput. Graph. Forum 37 (5), 93–106.

Yi, L., Su, H., Guo, X., Guibas, L., 2016. Syncspeccnn: Synchronized spectral CNN for 3D shape segmentation. arXiv preprint arXiv:1612.00606.

Yoo, H.H., Cho, J.E., Chung, J., 2006. Modal analysis and shape optimization of rotating cantilever beams. J. Sound Vib. 290 (1-2), 223–241.

Yu, Y., Zhou, K., Xu, D., Shi, X., Bao, H., Guo, B., Shum, H.-Y., 2004. Mesh editing with Poisson-based gradient field manipulation. ACM Trans. Graph. 23 (3), 644–651.

Yu, Y., Jang, I.G., Kim, I.K., Kwak, B.M., 2010. Nodal line optimization and its application to violin top plate design. J. Sound Vib. 329 (22), 4785–4796.

Zahn, C.T., Roskies, R.Z., 1972. Fourier descriptors for plane closed curves. IEEE Trans. Comput. 100 (3), 269–281.

Zelditch, S., 2000. Spectral determination of analytic bi-axisymmetric plane domains. Geom. Funct. Anal. GAFA 10 (3), 628–677.

Zelditch, S., 2009. Local and global analysis of eigenfunctions. arXiv preprint arXiv:0903.3420.

Zeng, W., Guo, R., Luo, F., Gu, X., 2012. Discrete heat kernel determines discrete Riemannian metric. Graph. Models 74 (4), 121–129.

Zhang, H., 2004. Discrete combinatorial Laplacian operators for digital geometry processing. In: Proceedings of SIAM Conference on Geometric Design and Computing. Nashboro Press, pp. 575–592.

Zhang, Z., Zha, H., 2004. Principal manifolds and nonlinear dimensionality reduction via tangent space alignment. SIAM J. Sci. Comput. 26 (1), 313–338.

Zhang, H., Van Kaick, O., Dyer, R., 2010a. Spectral mesh processing. Comput. Graph. Forum 29 (6), 1865–1894.

Zhang, H., Van Kaick, O., Dyer, R., 2010b. Spectral mesh processing. Comput. Graph. Forum 29 (6), 1865–1894.

Zhong, M., Qin, H., 2014. Sparse approximation of 3D shapes via spectral graph wavelets. Vis. Comput. 30 (6–8), 751–761.

Zhou, K., Synder, J., Guo, B., Shum, H.-Y., 2004. Iso-charts: stretch-driven mesh parameterization using spectral analysis. In: Proceedings of the 2004 Eurographics/ACM SIGGRAPH Symposium on Geometry Processing, pp. 45–54.

Zhu, K., 2007. Operator Theory in Function Spaces, 138. American Mathematical Soc.

Zhu, X., Ghahramani, Z., Lafferty, J.D., 2003. Semi-supervised learning using Gaussian fields and harmonic functions. In: Proceedings of the 20th International Conference on Machine Learning (ICML-03), pp. 912–919.

Zigelman, G., Kimmel, R., Kiryati, N., 2002. Texture mapping using surface flattening via multidimensional scaling. IEEE Trans. Visualization Comput. Graph. 8 (2), 198–207.

Zorin, D., 2005. Curvature-based energy for simulation and variational modeling. In: International Conference on Shape Modeling and Applications 2005 (SMI'05), pp. 196–204.

Chapter 3

Operator-based representations of discrete tangent vector fields

Mirela Ben-Chen[a,*] and Omri Azencot[b]
[a]*Computer Science Department, Technion - Israel Institute of Technology, Haifa, Israel*
[b]*Math Department, University of California, Los Angeles, CA, United States*
[*]*Corresponding author: e-mail: mirela@cs.technion.ac.il*

Chapter Outline

Abstract

Tangent vector fields are prominent objects in geometry processing and are used in a myriad of applications, e.g., texture synthesis, fluid simulation and polygonal remeshing. It is, however, challenging to represent, analyze and synthesize tangent vector fields on discrete surfaces such as triangle meshes. The goal of this chapter is to review the use of *linear operators* for these tasks. We introduce the formulation, show the advantages and demonstrate the unique applications such a representation makes possible.

Handbook of Numerical Analysis, Vol. 20. https://doi.org/10.1016/bs.hna.2019.07.007
© 2019 Elsevier B.V. All rights reserved.

Keywords: Geometry processing, Tangent vector fields, Linear operators, Discrete differential geometry

AMS Classification Codes: 65Dxx, 68Uxx

List of key symbols

Smooth

$\mathcal{M}$	a smooth, orientable and compact surface embedded in $\mathbb{R}^3$
$\mathcal{X}$	surface embedding $\mathcal{M} \to \mathbb{R}^3$
$\mathscr{f}, \mathscr{g}, \mathscr{h}$	real-valued functions $\mathcal{M} \to \mathbb{R}$
$\mathscr{u}, \mathscr{v}$	tangent vector fields $\mathcal{M} \to T\mathcal{M}$
$\langle \mathscr{u}, \mathscr{v} \rangle$	pointwise inner product of two tangent vector fields, given by the Riemannian metric on $\mathcal{M}$
$\nabla \mathscr{f}$	gradient vector field of $\mathscr{f}$
$\nabla \cdot \mathscr{v}$	divergence function of $\mathscr{v}$
Δ	Laplace–Beltrami operator
$\mathcal{D}_{\mathscr{v}}$	functional vector field (FVF) corresponding to $\mathscr{v}$
$\overline{\mathcal{D}}_{\mathscr{f}}$	dual FVF corresponding to $\mathscr{f}$
$\widetilde{\mathcal{D}}_{\mathscr{v}}$	divergence-based FVF corresponding to $\mathscr{v}$
$\varphi_{\mathscr{v},t}$	the flow map of the vector field $\mathscr{v}$ for time $t \in \mathbb{R}$, $\varphi_{\mathscr{v},t} : \mathcal{M} \to \mathcal{M}$
$\phi_{\mathscr{v},t}$	the functional flow map of the vector field $\mathscr{v}$ for time $t \in \mathbb{R}$

Discrete

$\boldsymbol{M}$	a triangle mesh, where the vertices, edges and faces are denoted by $(\mathcal{V}, \mathcal{E}, \mathcal{F})$, respectively
$\boldsymbol{n}$	number of vertices $\lvert\mathcal{V}\rvert$
$\boldsymbol{m}$	number of faces $\lvert\mathcal{F}\rvert$
$\boldsymbol{X}$	mesh embedding $X \in \mathbb{R}^{n\times 3}$
$\boldsymbol{f}, \boldsymbol{g}, \boldsymbol{h}$	piecewise linear functions $f, g, h \in \mathbb{R}^n$
$\boldsymbol{u}, \boldsymbol{v}$	piecewise constant tangent vector fields $u, v \in \mathbb{R}^{3m}$
grad	gradient operator grad $\in \mathbb{R}^{3m\times n}$
div	divergence operator div $\in \mathbb{R}^{n\times 3m}$
$\boldsymbol{A}_{\mathcal{V}}$	vector of vertex areas $A_{\mathcal{V}} \in \mathbb{R}^n$
$\boldsymbol{A}_{\mathcal{F}}$	vector of face areas $A_{\mathcal{F}} \in \mathbb{R}^m$
$\boldsymbol{I}_{\mathcal{V}}^{\mathcal{F}}$	interpolation matrix from faces to vertices $I_{\mathcal{V}}^{\mathcal{F}} \in \mathbb{R}^{n\times m}$
$\boldsymbol{I}_{\mathcal{F}}^{\mathcal{V}}$	interpolation matrix from vertices to faces $I_{\mathcal{F}}^{\mathcal{V}} \in \mathbb{R}^{m\times n}$
$\boldsymbol{L}$	Laplace–Beltrami operator $L \in \mathbb{R}^{n\times n}$
$\boldsymbol{D}_{v}^{\mathcal{F}}$	rectangular FVF corresponding to v, $D_{v}^{\mathcal{F}} \in \mathbb{R}^{m\times n}$
$\boldsymbol{D}_{v}^{\mathcal{V}}$	square FVF corresponding to v, $D_{v}^{\mathcal{V}} \in \mathbb{R}^{n\times n}$
$\overline{\boldsymbol{D}}_{f}^{\mathcal{F}}$	dual to rectangular FVF corresponding to f, $\overline{D}_{f}^{\mathcal{F}} \in \mathbb{R}^{m\times 3m}$

$\overline{\boldsymbol{D}}_f^{\mathcal{V}}$	dual to square FVF corresponding to f, $\overline{D}_f^{\mathcal{V}} \in \mathbb{R}^{n\times 3m}$
$\widetilde{\boldsymbol{D}}_v^{\mathcal{F}}$	rectangular divergence-based FVF corresponding to v, $\widetilde{D}_v^{\mathcal{F}} \in \mathbb{R}^{n\times m}$
$\widetilde{\boldsymbol{D}}_v^{\mathcal{V}}$	square divergence-based FVF corresponding to v, $\widetilde{D}_v^{\mathcal{V}} \in \mathbb{R}^{n\times n}$
$\boldsymbol{\Phi}_{v,t}$	the functional flow map of the vector field v for time $t\in\mathbb{R}$, $\Phi_{v,t} \in \mathbb{R}^{n\times n}$

1 Introduction

Tangent vector fields are geometric objects that represent *directional* information tangent to a surface. Examples include: velocity fields, curvature directions and tangents to the parameter lines of a surface parameterization.

On smooth surfaces, a tangent vector field is often defined as a *smooth* assignment of a tangent vector to each point on the surface. In geometry processing and computer graphics, surfaces are often represented using triangle meshes. Thus, on a triangle mesh it is natural to represent a discrete tangent vector field using a *piecewise constant* (PC) representation, namely, as a constant tangent vector per triangle. However, formulating and enforcing smoothness of PC vector fields is challenging, since they are piecewise constant and therefore nondifferentiable. Some approaches to tackle this problem include using finite differences (Crane et al., 2010), using a higher order polynomial discretization (Knöppel et al., 2013; Zhang et al., 2006), and using discrete 1-forms (Desbrun et al., 2005).

In this chapter, we advocate a different approach, first suggested by Azencot et al. (2013). Specifically, we consider the *action* of tangent vector fields on scalar functions, by treating tangent vector fields as *directional derivative* operators. These operators are linear, and, in the smooth case, completely encode the corresponding vector fields. The smooth theory involving these operators, also known as *derivations* (Morita, 2001, Chapter 1.4), is quite vast, as this operator-based point of view is classical in differential geometry.

In the discrete setting, however, this discretization is novel, and considerably simplifies many vector field related tasks, for both synthesis and analysis. This is due to two main properties of this scheme: (1) the only objects that are differentiated are *scalar functions*, and (2) the linear operator can be represented as a compact matrix in a reduced spectral basis. Consider, for example, the problem of joint design of vector fields on two corresponding surfaces, or on an intrinsically symmetric surface $\mathcal{M}$ with a self-correspondence $\varphi:\mathcal{M}\to\mathcal{M}$. A tangent vector field $\mathcal{v}$ is said to be *consistent* with φ if $d\varphi(\mathcal{v}(q))=\mathcal{v}(\varphi(q))$ for any $q\in\mathcal{M}$. This definition requires the *differential* of the map, which in a discrete setting on a triangle mesh is difficult to obtain. However, if we use the *operator* point of view, we can formulate the consistency by $\langle \mathcal{v}, \nabla(f\circ\varphi)\rangle = \langle \mathcal{v}, \nabla f\rangle\circ\varphi$ for any scalar function $f:\mathcal{M}\to\mathbb{R}$. With this formulation, there is no need for the map differential, which considerably simplifies the implementation on discrete surfaces. Fig. 1 shows an example of a nearly symmetric quadrangular mesh generated using this approach (Azencot et al., 2017).

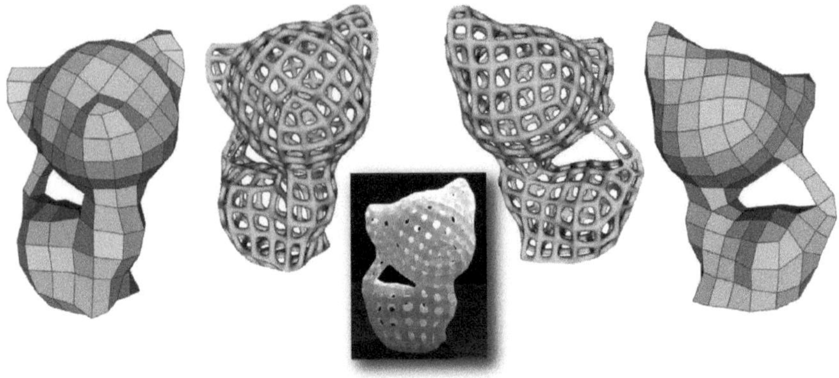

FIG. 1 A nearly symmetric quadrangular mesh generated using the operator-based representation of vector fields (Azencot et al., 2017) (*left* and *right*). The symmetry enables the generation of a very low resolution model, which can be fabricated using additive manufacturing, by thickening the quad mesh edges (*left centre* and *right centre*). The *centre image* is a photograph of the final printed model.

Due to these two properties, the operator-based approach to tangent vector fields enables many additional applications, such as vector field design (Azencot et al., 2013), function transport (Azencot et al., 2013, 2014b, 2015b, 2016) and vector field transport (Azencot et al., 2015a, 2018).

1.1 Organization

The rest of the chapter is organized as follows. In Section 2 we present the formulation of operator-based tangent vector fields in the smooth setting. We then proceed to describe the discrete representation, its properties and some applications in Section 3, which is based on the paper that first introduced the topic (Azencot et al., 2013). Then, in Section 4 we present a new, divergence-based discretization of the smooth setting, which allows for a more accurate representation of composite operators, such as the *Lie bracket* operator, and we present a few applications. We conclude with a brief outlook and some open questions in Section 5.

2 Smooth functional vector fields

In this section we provide the definition of *functional vector fields*, or the operator representation of tangent vector fields, in the smooth setting. For clarity of the exposition we avoid fully rigorous definitions and refer the interested reader to the excellent book by Morita (2001) for a full treatment.

2.1 Notation

In the following, $\mathcal{M}$ is a smooth, orientable, compact and embedded surface, and $f, g, h : \mathcal{M} \to \mathbb{R}$ are C^∞ real-valued functions on it. In addition, u, v are tangent vector fields on $\mathcal{M}$, and $\langle v, u\rangle$ is a function $\mathcal{M} \to \mathbb{R}$ such that $\langle v, u\rangle(p) = \langle v(p), u(p)\rangle_p$, for any point $p \in \mathcal{M}$. Here, the inner product $\langle \cdot, \cdot \rangle_p$ is given by the Riemannian metric of $\mathcal{M}$ at p.

2.2 Directional derivative of functions

A tangent vector field v on $\mathcal{M}$ is given by a smooth assignment of a tangent vector $v(p)$ to every point $p \in \mathcal{M}$. Such a vector field can be used to compute the *directional derivative* of a smooth real-valued function $f : \mathcal{M} \to \mathbb{R}$. The directional derivative function $g : \mathcal{M} \to \mathbb{R}$ is given by:

$$g = \langle v, \nabla f\rangle, \tag{1}$$

where ∇f is the gradient vector field of f.

Thus, given v, we can compute the directional derivative of any function f. Intriguingly, *the opposite is also true*. If two vector fields agree on the directional derivative of *any* arbitrary function f, then the two vector fields are identical (Morita, 2001, Prop. 1.39). Hence, we can *represent* a vector field uniquely by its directional derivative action, as is formalized in the next definition.

2.3 Functional vector fields

Definition 1. Given a tangent vector field v on $\mathcal{M}$, the corresponding *functional vector field* operator $\mathcal{D}_v$ is given by:

$$\mathcal{D}_v(f) = \langle v, \nabla f\rangle. \tag{2}$$

The functional vector field operator $\mathcal{D}_v$ is linear in the function it acts on, due to the linearity of the gradient. Hence,

$$\mathcal{D}_v(af + bg) = a\mathcal{D}_v(f) + b\mathcal{D}_v(g), \tag{3}$$

for any two functions $f, g : \mathcal{M} \to \mathbb{R}$ and any two constants $a, b \in \mathbb{R}$.

We further define the *dual* functional vector field operator, which switches the role of the function and the vector field.

Definition 2. Given a real-valued function $f : \mathcal{M} \to \mathbb{R}$, the corresponding *dual functional vector field* operator $\overline{\mathcal{D}}_f$ is given by:

$$\overline{\mathcal{D}}_f(v) = \mathcal{D}_v(f) = \langle v, \nabla f\rangle. \tag{4}$$

Due to the linearity of the inner product, the dual operator is linear in its vector field argument, with respect to both functions and constants. Thus,

$$\overline{\mathcal{D}}_h(fv + gu) = f\overline{\mathcal{D}}_h(v) + g\overline{\mathcal{D}}_h(u), \tag{5}$$

for any two tangent vector fields u, v, and functions $h, f, g : \mathcal{M} \to \mathbb{R}$.

As discussed previously, a functional vector field operator uniquely encodes its corresponding vector field. If the surface is *embedded* in $\mathbb{R}^3$, the vector field can be easily extracted form the corresponding operator using Lemma 1.

Lemma 1. *Let v be a tangent vector field on $\mathcal{M}$, and let $\mathcal{D}_v$ be its corresponding functional vector field operator. Further, denote the embedding of $\mathcal{M}$ by $\mathcal{X}:\mathcal{M}\rightarrow\mathbb{R}^3$, where $\mathcal{X}$ is given by the three coordinate functions $\mathcal{X}=(x_1, x_2, x_3)$. Then:*

$$\mathcal{D}_v(x_i)=v_i, \quad i\in(1,\ldots,3), \tag{6}$$

where v_i are the coordinate functions of v.

Proof. One approach to compute the gradient of a function $\mathcal{M}\rightarrow\mathbb{R}$ is to extend it to a function over $\mathbb{R}^3$, take its Euclidean gradient and project the result on $T_p\mathcal{M}$, the tangent plane to $\mathcal{M}$ at p (Absil et al., 2009, Eq. (3.37)). Let (x_1, x_2, x_3) denote the canonical coordinates of $\mathbb{R}^3$. By the definition of $\mathcal{X}$ we have that $x_i(p)=x_i(p)$, for all the points $p\in\mathcal{M}$, and therefore x_i is the *extension to* $\mathbb{R}^3$ of x_i. Hence, we have that

$$(\nabla x_i)(p)=[(\nabla x_i)(p)]_{T_p\mathcal{M}}=\left[\sum_{j=1}^{3}\frac{\partial x_i}{\partial x_j}(p)\hat{x}_j\right]_{T_p\mathcal{M}},$$

where $[\cdot]_{T_p\mathcal{M}}$ denotes projection on $T_p\mathcal{M}$ and $(\hat{x}_1, \hat{x}_2, \hat{x}_3)$ denote the canonical directions in $\mathbb{R}^3$.

Note that for all $p\in\mathcal{M}$, $(\frac{\partial x_i}{\partial x_j})(p)=\delta_{ij}$, i.e., it equals 1 when $i=j$ and 0 otherwise. Thus, we have

$$\langle v, \nabla x_i\rangle(p)=\left\langle v(p), \left[\sum_{j=1}^{3}\delta_{ij}\hat{x}_j\right]_{T_p\mathcal{M}}\right\rangle_p=\left\langle v(p), \sum_{j=1}^{3}\delta_{ij}\hat{x}_j\right\rangle, \tag{7}$$

where the second equality holds since v is tangent to $\mathcal{M}$, and we assumed that the Riemannian metric is inherited from the standard inner product in $\mathbb{R}^3$.

Finally, by definition, v is given in coordinates by $v(p)=\sum_{j=1}^{3}v_j(p)\hat{x}_j$. Therefore:

$$(\mathcal{D}_v(x_i))(p)=\langle v, \nabla x_i\rangle(p)=\sum_{j=1}^{3}\delta_{ij}v_j(p)=v_i(p), \tag{8}$$

which completes the proof. □

2.4 Flow maps

Tangent vector fields are closely related to *self-maps* of $\mathcal{M}$. Consider a particle at a point $p\in\mathcal{M}$ (see Fig. 2). Given a tangent vector field v, we can ask

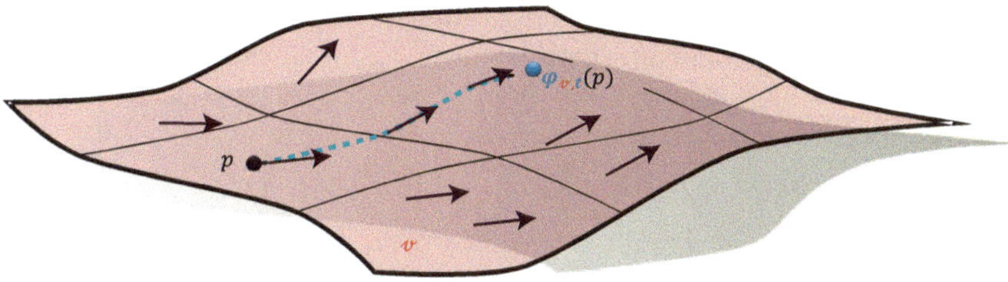

FIG. 2 Illustration of a tangent vector field v and its flow $\varphi_{v,t}$.

where will this particle be after time t, if its velocity is given by v. Such a particle will trace a *flow line* $\gamma_v(p)(t),\ t \in \mathbb{R}$ of v on $\mathcal{M}$, that starts at p for $t = 0$. This flow line is also known as an *integral curve*, and it fulfils:

$$\frac{d}{dt}\gamma_v(p)(t) = v(\gamma_v(p)(t)), \quad \gamma_v(p)(0) = p, \quad \forall t \in \mathbb{R}. \tag{9}$$

Similarly, the paths that *all* the points on $\mathcal{M}$ trace, for every $t \in \mathbb{R}$, define a map, in fact a *diffeomorphism*, from $\mathcal{M}$ to itself, known as the *flow map*.
Definition 3. The *flow map* of v is denoted by $\varphi_{v,t} : \mathcal{M} \to \mathcal{M}$, and is given by $\varphi_{v,t}(p) = \gamma_v(p)(t)$, for all $p \in \mathcal{M},\ t \in \mathbb{R}$.

For compact surfaces, the flow map is guaranteed to exist for any vector field v and any $t \in \mathbb{R}$ (Morita, 2001, Thm 1.45). Furthermore, the flow maps have additional structure, since (Morita, 2001, Prop. 1.42):

$$\varphi_{v,t} \circ \varphi_{v,s} = \varphi_{v,t+s}, \quad \forall t, s \in \mathbb{R}. \tag{10}$$

The group of diffeomorphisms $\{\varphi_{v,t} \mid t \in \mathbb{R}\}$, with the group operation given by composition, is known as the *one parameter group of transformations* generated by v. By applying Eq. (9) to all points $p \in \mathcal{M}$, we get that the flow maps fulfil:

$$\frac{d}{dt}\varphi_{v,t} = v(\varphi_{v,t}), \quad \varphi_{v,0} = Id. \tag{11}$$

Finally, combining Eqs. (10) and (11), it is straightforward to see that the *inverse flow map* is given by $\varphi_{v,t}^{-1} = \varphi_{v,-t}$.

2.5 Functional flow maps

Functional maps (Ovsjanikov et al., 2012) have played an important role in many geometry processing applications. Given a diffeomorphism $\mathcal{T}_{12} : \mathcal{M}_1 \to \mathcal{M}_2$, the corresponding functional map takes functions on $\mathcal{M}_1$ to functions on $\mathcal{M}_2$. The functional map is given by composition with the inverse of $\mathcal{T}_{12}$ and can be used to transport functions between surfaces, even without the knowledge of the corresponding pointwise map.

In the context of flow maps, we can think of $\varphi_{v,t}$ as the transporter of quantities, and define the corresponding *functional flow map*.

Definition 4. The *functional flow map* of v, denoted by $\phi_{v,t}$, takes functions on $\mathcal{M}$ to functions on $\mathcal{M}$, and is given by:

$$\phi_{v,t}(f)(p) = f(\varphi_{v,-t}(p)), \ \ \forall f : \mathcal{M} \to \mathbb{R}, p \in \mathcal{M}, t \in \mathbb{R}. \tag{12}$$

See Fig. 3 for an example of a function f transported on the flow of a vector field v for two values of t, denoted by t_1, t_2. Consider, for example, the value of the function $\phi_{v,t_1}(f)$ at the dark blue point $\varphi_{v,t_1}(p)$ (Fig. 3, centre right). By Definition 4 this value is $\phi_{v,t_1}(f)(\varphi_{v,t_1}(p)) = f(\varphi_{v,-t_1}(\varphi_{v,t_1}(p))) = f(p)$. Hence, this value is equal to the value of the function f (Fig. 3, centre left) at the light blue point p.

Functional flow maps have some interesting properties. First, note that the functional flow maps fulfil the same composition rules as the flow maps.

Lemma 2. *Let $\{\phi_{v,t} \mid t \in \mathbb{R}\}$ be the group of functional flow maps corresponding to the one parameter group of transformations $\varphi_{v,t}$ generated by the tangent vector field v. Then the following holds for any function $f : \mathcal{M} \to \mathbb{R}$:*

$$\phi_{v,t}(\phi_{v,s} f) = \phi_{v,t+s}(f), \ \ \forall t, s \in \mathbb{R}, \tag{13}$$

and in addition, $\phi_{v,0} f = f$.

Proof. By definition we have for any point $p \in \mathcal{M}$:

$$\phi_{v,t}(\phi_{v,s} f)(p) = (\phi_{v,s} f)(\varphi_{v,-t}(p)) = f(\varphi_{v,-s}(\varphi_{v,-t}(p))). \tag{14}$$

By Eq. (10) we get

$$f(\varphi_{v,-s}(\varphi_{v,-t}(p))) = f(\varphi_{v,-(t+s)}(p)) = \phi_{v,t+s}(f)(p). \tag{15}$$

Finally, $\phi_{v,0}(f)(p) = f(\varphi_{v,0}(p)) = f(p)$, due to Eq. (11), which completes the proof. □

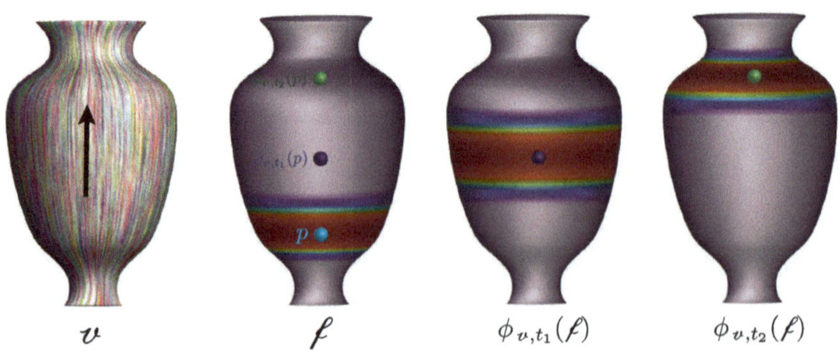

FIG. 3 Illustration of the transportation of a function f on the flow of a vector field v (visualized using Line Integral Convolution), for two values of $t = t_1, t_2$.

Furthermore, the functional flow map is closely related to the functional vector field operator by differentiation with respect to t, as follows.

Lemma 3. *Let* $\{\phi_{v,t} \mid t \in \mathbb{R}\}$*be the group of functional flow maps corresponding to the one parameter group of transformations* $\varphi_{v,t}$ *generated by the tangent vector field* v*. Then the following holds for any function* $f : \mathcal{M} \to \mathbb{R}$ *(Morita, 2001, Eq. (2.28)):*

$$\frac{d}{dt}\phi_{v,t}(f)_{|t=0} = -\mathcal{D}_v(f). \tag{16}$$

We repeat the proof here, for completeness, and using our notations and definitions so far.

Proof. By the definition of $\phi_{v,t}$ and $\varphi_{v,t}$ we have

$$\begin{aligned}
\frac{d}{dt}\phi_{v,t}(f)(p) &= \frac{d}{dt}f(\varphi_{v,-t}(p)) \\
&= \left\langle (\nabla f)(\varphi_{v,-t}(p)), \frac{d}{dt}\varphi_{v,-t}(p) \right\rangle && \text{Multivariate chain rule} \\
&= \langle (\nabla f)(\varphi_{v,-t}(p)), -v(\varphi_{v,-t}(p)) \rangle && \text{Eq.(11)} \\
&= -(\mathcal{D}_v f)(\varphi_{v,-t}(p)) && \text{Definition 1,}
\end{aligned} \tag{17}$$

and since $\varphi_{v,0} = Id$, the result follows. □

Finally, when both are considered as operators on functions, the functional flow map and the functional vector field operators *commute*, as formalized in the following Lemma.

Lemma 4. *Let* $\{\phi_{v,t} \mid t \in \mathbb{R}\}$ *be the group of functional flow maps corresponding to the one parameter group of transformations* $\varphi_{v,t}$ *generated by the tangent vector field* v*, and let* $\mathcal{D}_v$ *be the corresponding functional vector field operator. Then, for any function* $f : \mathcal{M} \to \mathbb{R}$*, and for any* $t \in \mathbb{R}$ *we have*

$$\mathcal{D}_v(\phi_{v,t} f) = \phi_{v,t}(\mathcal{D}_v f). \tag{18}$$

Proof.

$$\begin{aligned}
\mathcal{D}_v(\phi_{v,t} f) &= -\frac{d}{ds}\phi_{v,s}(\phi_{v,t} f)_{|s=0} && \text{Lemma 3} \\
&= -\frac{d}{ds}\phi_{v,s+t} f_{|s=0} && \text{Lemma 2} \\
&= -\frac{d}{ds}\phi_{v,t}(\phi_{v,s} f)_{|s=0} && \text{Lemma 2} \\
&= \phi_{v,t}\left(-\frac{d}{ds}\phi_{v,s} f\right)_{|s=0} && \phi_{v,t} \text{ independent of } s \\
&= \phi_{v,t}(\mathcal{D}_v f) && \text{Lemma 3}
\end{aligned} \tag{19}$$

□

These properties of the functional flow maps allow us to derive a useful PDE for the family of functions generated by the functional flow maps.

Lemma 5. *Let $\{\phi_{v,t} \mid t \in \mathbb{R}\}$ be the group of functional flow maps corresponding to the one parameter group of transformations $\varphi_{v,t}$ generated by the tangent vector field v. Let $f_0 : \mathcal{M} \to \mathbb{R}$, and define the time-varying function $f : \mathbb{R} \times \mathcal{M} \to \mathbb{R}$ by $f(t) = \phi_{v,t}(f_0)$. Then the following holds*

$$\frac{d}{dt} f(t) = -\mathcal{D}_v f(t), \quad f(0) = f_0, \quad \forall t \in \mathbb{R}. \tag{20}$$

Proof.

$$\begin{aligned}
\frac{d}{dt} f(t) &= \frac{d}{dt} \phi_{v,t}(f_0) && \text{definition of } f(t) \\
&= \frac{d}{ds}\Big|_{s=0} \phi_{v,t+s}(f_0) && \\
&= \frac{d}{ds}\Big|_{s=0} \phi_{v,s}\phi_{v,t}(f_0) && \text{Lemma 2} \\
&= \frac{d}{ds}\Big|_{s=0} \phi_{v,s} f(t) && \text{definition of } f(t) \\
&= -\mathcal{D}_v f(t) && \text{Lemma 3.}
\end{aligned} \tag{21}$$

□

Functional flow maps enable us to work with *functions* and *linear operators* as the main object of interest instead of *points* and *maps*. On smooth surfaces, the two approaches are similar. However, when considering optimization problems on discretized surfaces, represented by triangle meshes, there are inherent advantages to using functions.

First, the discretization of the space of functions is well studied, and methods such as finite elements are very effective at this task. Second, the space of functions on a surface forms a *linear vector space*, which can be effectively represented using a basis. Maps and points, on the other hand, are more difficult to deal with, due to the nonlinear and nonconvex constraint which requires that each point is mapped to a point that lies *on the surface*.

2.6 Lie bracket

Tangent vector fields have an additional important structure, denoted by the *Lie bracket*. Given two tangent vector fields v, w, their Lie bracket $[v, w]$ is a tangent vector field, given by (Morita, 2001, Eq.(1.9)):

$$\mathcal{D}_{[v,w]} f = \mathcal{D}_v(\mathcal{D}_w f) - \mathcal{D}_w(\mathcal{D}_v f). \tag{22}$$

When $\mathcal{M}$ is an embedded surface we can use Lemma 1 to get a coordinate expression for the Lie bracket.

Lemma 6. *Let $\mathcal{M}$ be an embedded surface. If v, w are two tangent vector fields, then the coordinate functions of $[v, w]$ are*

$$[v, w]_i = \mathcal{D}_v(w_i) - \mathcal{D}_w(v_i), \quad i \in [1, \ldots, 3], \tag{23}$$

where v_i, w_i are the coordinate functions of v, w, respectively: $v = \sum_{i=1}^{3} v_i \hat{x}_i$, and $w = \sum_{i=1}^{3} w_i \hat{x}_i$.

Proof. Let $\mathcal{X}:\mathcal{M}\rightarrow\mathbb{R}^3$ be the embedding functions of $\mathcal{M}$, $\mathcal{X}=(x_1,x_2,x_3)$. Then, according to Lemma 1, we have

$$[v,w]_i=\mathcal{D}_{[v,w]}(x_i)=\mathcal{D}_v(\mathcal{D}_w x_i)-\mathcal{D}_w(\mathcal{D}_v x_i)=\mathcal{D}_v(w_i)-\mathcal{D}_w(v_i). \quad (24)$$

□

The Lie bracket defines a special relation between vector fields, known as *commutation*. Two vector fields v,w commute if $[v,w]=0$, and this holds if and only if their flow maps commute (Lee, 2012, Thm. 9.44), namely:

$$\varphi_{v,s}\circ\varphi_{w,t}=\varphi_{w,t}\circ\varphi_{v,s},\quad \forall s,t\in\mathbb{R}. \quad (25)$$

Fig. 4 demonstrates the relationship between the bracket of two vector fields u, v and the resulting flows. We consider the two vector fields on the bunny (left), and the frog (right). The centre figures show the flow line of v starting from a point p for time t (blue) followed by the flow line of u (red), and then backward on v (green) and backward on u (black), all for time t. For the bunny, the two vector fields commute and so do their flows, "closing" the rectangle. For the frog, the vector fields do not commute, resulting in a mismatch of the original point p and the final point, leading to interesting flow behaviour on the right.

In the context of applications, the Lie bracket is important because of its relation to finding local coordinates, or a *local parameterization*, on $\mathcal{M}$. In applications such as curvature-aligned quadrangular remeshing (Bommes et al., 2009), two tangent vector fields v,w are given, and we would like to find a local coordinate chart $\sigma:\mathcal{U}\rightarrow\mathbb{R}^2$ for a subdomain $\mathcal{U}\subset\mathcal{M}$, such that if σ is given by the two functions (s_v,s_w), then the vector fields v,w are mapped by $d\sigma$ to the coordinate directions $\frac{\partial}{\partial s_v},\frac{\partial}{\partial s_w}$, respectively.

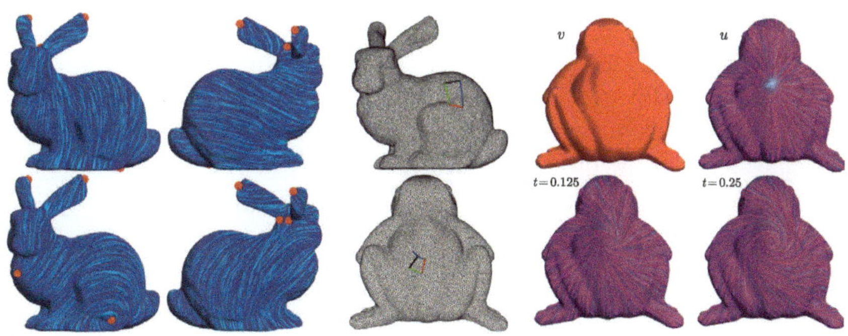

FIG. 4 *Left*: The vector field shown at the bottom row yields a zero bracket with respect to the vector field shown at the top row. *Centre*: The Lie bracket of two vector fields u, v encodes the failure of their flows to commute. We use the vector fields shown on the left and on the right for each shape, correspondingly. *Right*: Two noncommuting vector fields u, v (*top*) and the transport of u over the differential of the flow of v for time t (*bottom*). *Images on the* left *and* right *are reused with permission from Azencot, O., Vantzos, O., Ben-Chen, M., 2016. Advection-Based Function Matching on Surfaces. Comput. Graph. Forum 35, 55–64.*

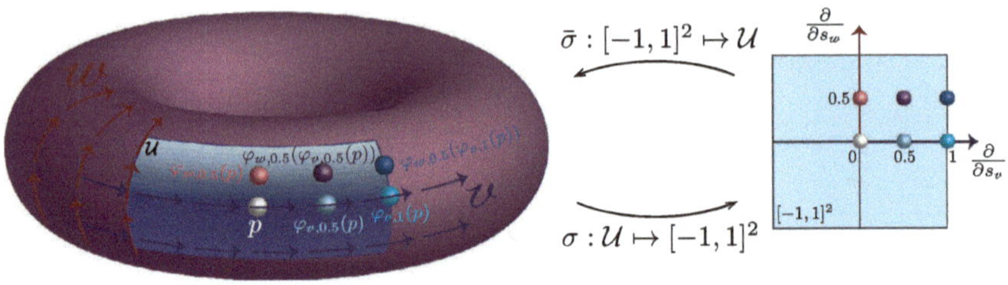

FIG. 5 Illustration of the construction of $\overline{\sigma} : [-1,1]^2 \mapsto \mathcal{U}$, the inverse of a local chart σ, using the flows of two commuting vector fields v, w in the neighbourhood of a point p.

An important result (Lee, 2012, Thm. 9.46) is that if the vector fields v, w *commute*, then such a parameterization indeed exists. Given two linearly independent commuting tangent vector fields v, w on an open subset $\mathcal{U} \subset \mathcal{M}$, we can construct the local coordinate chart as follows (see Fig. 5). Given a point $p \in \mathcal{U}$, there exists an $\epsilon > 0$ such that the map $\overline{\sigma} : (-\epsilon, \epsilon)^2 \to \mathcal{U}$, given by $\overline{\sigma}(s_v, s_w) = (\varphi_{w,s_w}(\varphi_{v,s_v}(p)))$, is a diffeomorphism, and its differential maps the coordinate directions $\frac{\partial}{\partial s_v}, \frac{\partial}{\partial s_w}$ to the tangent vector fields v, w, respectively. Then, the required coordinate chart is given by $\sigma = \overline{\sigma}^{-1}$.

3 Discrete functional vector fields

In this section, we describe a discretization of the concepts that were presented in the previous section. The discretization approach follows the original presentation of the method (Azencot et al., 2013). In this case, some (but not all) of the properties from the smooth setting carry over to the discrete setting.

3.1 Notation

In the following, M is an orientable, closed and embedded triangle mesh, where $M = (\mathcal{V}, \mathcal{E}, \mathcal{F})$ are the sets of vertices, edges and faces, respectively, and we denote by $n = |\mathcal{V}|$ and $m = |\mathcal{F}|$ the number of vertices and faces. The embedding of the mesh in $\mathbb{R}^3$ is denoted by X.

3.1.1 Functions and vector fields

We use finite elements to represent discrete scalar functions and tangent vector fields. Specifically, we use conforming piecewise linear (PL) elements for functions, and piecewise constant (PC) elements for vector fields. Thus, $f, g, h : \mathcal{V} \to \mathbb{R}$ are column vectors of length n that represent the degrees of freedom of PL functions, and $u, v : \mathcal{F} \to \mathbb{R}^3$ are column vectors of length $3m$ that represent PC vector fields. We also occasionally use piecewise constant functions, which we denote by $\mathrm{f}, \mathrm{g}, \mathrm{h} : \mathcal{F} \to \mathbb{R}$, and they are column vectors of length m.

3.1.2 Rotations

We denote by $\mathcal{J}_t \in \mathbb{R}^{3\times 3}$ the counter-clockwise rotation by $\pi/2$ in the plane of the face $t \in \mathcal{F}$, and by $\mathcal{J} \in \mathbb{R}^{3m\times 3m}$ the block diagonal matrix that encodes the rotations for all the faces.

3.1.3 Vectors to matrices

We use the bracket $[\cdot]$ operator to convert vectors in $\mathbb{R}^n$ and $\mathbb{R}^m$ to diagonal matrices in $\mathbb{R}^{n\times n}$, $\mathbb{R}^{m\times m}$ and $\mathbb{R}^{3m\times 3m}$, respectively (replicating each entry 3 times for the latter). The dimensions of the resulting matrix are inferred from the context.

We additionally require an operator that encodes the face-wise multiplication of a vector field by a piecewise constant function, which we denote by $[\cdot]_\bullet \in \mathbb{R}^{3m\times m}$. Specifically, if $w = [v]_\bullet \mathrm{f}$, then $w(t) = v(t)\mathrm{f}(t) \in \mathbb{R}^3, \forall t \in \mathcal{F}$. The transpose of this operator, $[\cdot]_\bullet^T \in \mathbb{R}^{m\times 3m}$ evaluates face-wise inner products, i.e., if $\mathrm{f} = [v]_\bullet^T w$, then $\mathrm{f}(t) = \langle v(t), w(t)\rangle \in \mathbb{R}$.

Note that these operators have the following properties:

1. $[\cdot]$ is symmetric w.r.t its input and its argument, i.e., $[f]g = [g]f$.
2. $[\cdot]_\bullet$ is linear in its argument, i.e., $[v + w]_\bullet = [v]_\bullet + [w]_\bullet$.
3. $[\cdot]_\bullet^T$ is symmetric w.r.t its input and its argument, i.e., $[v]_\bullet^T w = [w]_\bullet^T v$.
4. $[v]_\bullet^T [w]_\bullet = [[v]_\bullet^T w] \in \mathbb{R}^{m\times m}$.
5. $[v]_\bullet [w]_\bullet^T \in \mathbb{R}^{3m\times 3m}$ is a block diagonal matrix, with blocks of size 3×3, that encodes for every face t the outer product $v(t)w(t)^T \in \mathbb{R}^{3\times 3}$.
6. $[w]_\bullet \mathrm{f} = [\mathrm{f}]w \in \mathbb{R}^{3m}$.

3.1.4 Areas and inner products

The vector of triangle areas and vertex areas are denoted by $A_\mathcal{F} \in \mathbb{R}^m$ and $A_\mathcal{V} \in \mathbb{R}^n$, respectively. We take the vertex area as 1/3 of the total area of its adjacent triangles. The L_2 inner product between two PL functions f, g, two PC functions f, g and two vector fields u, v are given by:

$$\langle f, g\rangle_{L_2} = f^T [A_\mathcal{V}] g, \quad \langle \mathrm{f}, \mathrm{g}\rangle_{L_2} = \mathrm{f}^T [A_\mathcal{F}] \mathrm{g}, \quad \langle u, v\rangle_{L_2} = u^T [A_\mathcal{F}] v. \tag{26}$$

3.1.5 Interpolation

The interpolation matrix $I_\mathcal{V}^\mathcal{F} \in \mathbb{R}^{n\times m}$ interpolates quantities from faces to the vertices, i.e., $I_\mathcal{V}^\mathcal{F}(i,j) = \frac{A_\mathcal{F}(j)}{3A_\mathcal{V}(i)}$, iff vertex i belongs to face j and 0 otherwise. Similarly, $I_\mathcal{F}^\mathcal{V}$ interpolates data from vertices to faces and is defined by $I_\mathcal{F}^\mathcal{V} = [A_\mathcal{F}]^{-1} (I_\mathcal{V}^\mathcal{F})^T [A_\mathcal{V}]$ such that $\langle \mathrm{f}, I_\mathcal{F}^\mathcal{V} g\rangle_{L_2} = \langle I_\mathcal{V}^\mathcal{F} \mathrm{f}, g\rangle_{L_2}$ holds.

3.1.6 Differential operators

We additionally require discrete gradient grad $\in \mathbb{R}^{3m\times n}$ and divergence div $\in \mathbb{R}^{n\times 3m}$ operators. Specifically we use

$$(\text{grad} f)(t) = \frac{1}{2A_{\mathcal{F}}(t)} \sum_{v_i \in t} f_{v_i} \mathcal{J}_t e_{tv_i}, \quad t \in \mathcal{F}, \tag{27}$$

where e_{tv_i} is the directed edge opposite to the vertex v_i in the triangle $t = (v_1, v_2, v_3) \in \mathcal{F}, v_i \in \mathcal{V}$ (see Fig. 6). Alternatively, this operator can be written as:

$$\text{grad} = \frac{1}{2}[A_{\mathcal{F}}]^{-1} \mathcal{J} X_{\mathcal{E}}, \tag{28}$$

where $X_{\mathcal{E}} \in \mathbb{R}^{3m\times n}$ is a matrix that encodes the edge vectors of M. Specifically, $X_{\mathcal{E}}(t, v_i) = e_{tv_i} = X(v_{i+2}) - X(v_{i+1}) \in \mathbb{R}^3$, $i \in 1,\ldots,3$.

The discrete divergence is given in terms of the gradient operator:

$$\text{div} = -[A_{\mathcal{V}}]^{-1} \text{grad}^T [A_{\mathcal{F}}]. \tag{29}$$

Note that these operators are identical to the standard ones in geometry processing, see, e.g., Botsch et al. (2010, Chapter 3). We would like to note two important properties that follow directly from the discrete definitions of functions, vector fields, L_2 inner products and differential operators that we use.

3.1.7 Divergence free vector fields

In the smooth setting, for all $v = \mathcal{J}\nabla f$ we have that $\nabla \cdot v = 0$. The analogous property holds when using the discrete operators, as was shown by Polthier and Preuss (2003), i.e., let $v = \mathcal{J}\,\text{grad} f$ then div $v = 0$.

3.1.8 Discrete integration by parts

In the smooth setting, for a closed surface, the following holds (Petersen et al., 2006, pp. 382):

$$\langle v, \nabla f\rangle_{L_2} + \langle \nabla \cdot v, f\rangle_{L_2} = \int_{\mathcal{M}} \langle v, \nabla f\rangle \, da + \int_{\mathcal{M}} f(\nabla \cdot v)\, da = \int_{\mathcal{M}} \nabla \cdot (f v) = 0. \tag{30}$$

Analogously, using our definitions, we have Lemma 7. It was proved by Azencot et al. (2013), and we repeat the proof here using our notation.

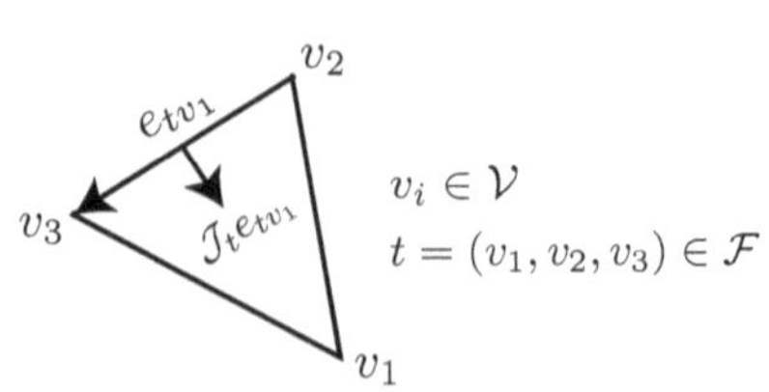

FIG. 6 Illustration of the computation of the gradient operator in Eq. (27).

Lemma 7. *For a closed triangle mesh M, a function f and a vector field v, the following holds*

$$\langle v, \operatorname{grad} f\rangle_{L_2} + \langle \operatorname{div} v, f\rangle_{L_2} = v^T[A_{\mathcal{F}}]\ \operatorname{grad} f + f^T[A_{\mathcal{V}}]\ \operatorname{div} v = 0. \tag{31}$$

Proof. The proof follows directly from the definition of grad in Eq. (29):

$$\begin{aligned} f^T[A_{\mathcal{V}}]\operatorname{div} v &= -f^T[A_{\mathcal{V}}][A_{\mathcal{V}}]^{-1}\operatorname{grad}^T[A_{\mathcal{F}}]v \\ &= -(\operatorname{grad} f)^T[A_{\mathcal{F}}]v = -v^T[A_{\mathcal{F}}]\operatorname{grad} f, \end{aligned} \tag{32}$$

where in the last step we used the fact that a scalar is the transpose of itself, and that $[A_{\mathcal{F}}]$ is diagonal and thus symmetric. □

Note that we did *not* use the product rule $\nabla \cdot (\mathscr{f}v) = \langle v, \nabla \mathscr{f}\rangle + \mathscr{f}(\nabla \cdot v)$, which was used to show integration by parts in the smooth case. This is a much stronger condition, since it is defined pointwise, and in fact it *does not* hold for our definition of discrete differential operators.

3.2 Directional derivative of functions

Analogous to the smooth case, we can define the directional derivative of a PL function $f \in \mathbb{R}^n$ in the direction of a PC vector field $v \in \mathbb{R}^{3m}$. Unlike the smooth case, the resulting function is *not* as smooth as f, but is instead *piecewise constant*. The directional derivative $\mathrm{g} : \mathcal{F} \rightarrow \mathbb{R}$ is given by:

$$\mathrm{g}(t) = \langle v(t), (\operatorname{grad} f)(t)\rangle, \quad \forall t \in \mathcal{F}, \tag{33}$$

or equivalently:

$$\mathrm{g} = [v]_{\bullet}^{T} \operatorname{grad} f \in \mathbb{R}^m. \tag{34}$$

In the discrete case, the directional derivative function g is *not* in the same space as the input function f, and therefore the linear operator that encodes directional derivatives is *rectangular*. In the following, we use an interpolation approach to address this issue, whereas in the next section we consider a different discretization.

3.3 Functional vector fields

Definition 5. Given a tangent vector field v on M, we define *two* corresponding *functional vector field* operators:

$$D_v^{\mathcal{F}} = [v]_{\bullet}^{T} \operatorname{grad} \in \mathbb{R}^{m\times n}, \tag{35}$$

$$D_v^{\mathcal{V}} = I_{\mathcal{V}}^{\mathcal{F}} D_v^{\mathcal{F}} \in \mathbb{R}^{n\times n}. \tag{36}$$

The operator $D_v^{\mathcal{F}}$ does not use interpolation, and thus retains all the information about v. However, this operator is *rectangular*, and therefore cannot be used as is in settings which require *square* matrices, e.g., for flow maps. In such settings, we will use $D_v^{\mathcal{V}}$.

In the discrete case, the functional vector field operators are simply *matrices*, and therefore linear in their function argument with respect to constants.

Lemma 8. *The sparsity structure of $D_v^{\mathcal{F}}$ and $D_f^{\mathcal{V}}$ are given by the following.*

1. *If the entry (t,i), $t \in \mathcal{F}, i \in \mathcal{V}$ in $D_v^{\mathcal{F}}$ is nonzero, then $i \in t$, i.e., the vertex i belongs to the triangle t.*
2. *If the entry $(i,j), i,j \in \mathcal{V}$ in $D_v^{\mathcal{V}}$ is nonzero, then either $i = j$ or $(i,j) \in \mathcal{E}$, i.e., (i, j) is an edge in M.*

Proof.

1. Let $f_i \in \mathbb{R}^n$ be a PL function such that $f_i(j) = 1$ if and only if $i = j$ and 0 otherwise, and let $g_i = D_v^{\mathcal{F}} f_i$. If the entry (t, i) is nonzero, then $g_i(t)$ is nonzero. By the definition of $D_v^{\mathcal{F}}$, we have that $g_i(t) = \langle v(t), (\text{grad } f_i)(t) \rangle$. Since grad$f_i$ is nonzero only for triangles t which are neighbours of the vertex i, the result follows.
2. If the entry (i, j) is not zero, then there should exist an entry (i, t) in $I_{\mathcal{V}}^{\mathcal{F}}$ and an entry (t, j) in $D_v^{\mathcal{F}}$ which are not zero, for some triangle $t \in \mathcal{F}$. By the definition of $I_{\mathcal{V}}^{\mathcal{F}}$ and the first part of the proof, this implies that both i and j belong to the triangle t. This is possible only if $i = j$ or $(i,j) \in \mathcal{E}$.

□

Definition 6. Given a function $f : \mathcal{V} \to \mathbb{R}$, the corresponding *dual functional vector field* operators are given by:

$$\overline{D}_f^{\mathcal{F}} = [\text{grad } f]_\bullet^T \in \mathbb{R}^{m \times 3m}, \tag{37}$$

$$\overline{D}_f^{\mathcal{V}} = I_{\mathcal{V}}^{\mathcal{F}} \overline{D}_f^{\mathcal{F}} \in \mathbb{R}^{n \times 3m}. \tag{38}$$

3.3.1 Duality

As in the smooth case, we have that $\overline{D}_f^{\mathcal{F}}(v) = D_v^{\mathcal{F}}(f)$, since $[\text{grad } f]_\bullet^T v = [v]_\bullet^T \text{grad } f$, and similarly for $\overline{D}_f^{\mathcal{V}}$.

3.3.2 Linearity

Linearity with respect to constants follows since the operators are matrices. Linearity with respect to functions, however, is more subtle in the discrete case, and holds only for $\overline{D}_f^{\mathcal{F}}$. We have that:

$$\overline{D}_h^{\mathcal{F}}([v]_\bullet \mathrm{f} + [u]_\bullet \mathrm{g}) = [\text{grad } h]_\bullet^T [v]_\bullet \mathrm{f} + [\text{grad} h]_\bullet^T [u]_\bullet \mathrm{g} = [\overline{D}_h^{\mathcal{F}} v]\mathrm{f} + [\overline{D}_h^{\mathcal{F}} u]\mathrm{g}, \tag{39}$$

for any two tangent vector fields u, v, and functions $h : \mathcal{V} \to \mathbb{R}$, $\mathrm{f}, \mathrm{g} : \mathcal{F} \to \mathbb{R}$.

3.3.3 Reconstruction

In the discrete case, in general, full reconstruction given the embedding X is possible only for $D_v^{\mathcal{F}}$. We have

Lemma 9. *Let* v *be a tangent vector field on* M*, and let* $D_v^{\mathcal{F}}$ *be its corresponding functional vector field operator. Further, denote the embedding of* M *by* $X:\mathcal{V} \rightarrow \mathbb{R}^3$*, where* X *is given by the three coordinate functions* $X = (x_1, x_2, x_3)$*. Then:*

$$D_v^{\mathcal{F}}(x_i) = \mathrm{v}_i, \quad i \in (1,\ldots,3), \tag{40}$$

where $\mathrm{v}_i \in \mathbb{R}^m$ *are the coordinate functions of* v*.*

Proof. Let $\mathcal{M}_t$ denote the plane of the triangle t, and extend $v(t)$ to a constant vector field $\mathcal{v}_t$ on that plane. Since $\mathcal{M}_t$ is a smooth surface, and $\mathcal{v}_t$ a smooth vector field, we can apply Lemma 1, and deduce that $\mathcal{D}_{\mathcal{v}_t}(x_i) = \mathcal{v}_{ti}$. Since $\mathcal{v}_t$ is constant, $\mathcal{D}_{\mathcal{v}_t}(x_i)(p) = D_v^{\mathcal{F}}(x_i)(t) = \mathcal{v}_{ti}(p) = \mathrm{v}_i(t)$, $\forall t \in \mathcal{F}, p \in \mathcal{M}_t$. □

3.4 Flow maps

Carrying the flow map definition presented in Definition 3 to the discrete case introduces some challenges, due to the discontinuity of piecewise constant vector fields at mesh edges. In fact, tracing vector fields robustly on a triangle mesh is a separate and important research topic, see, e.g., the work by Ray and Sokolov (2014) and citations within. We therefore avoid defining discrete flow maps, and define directly discrete *functional* flow maps.

3.5 Functional flow maps

In Lemma 5, we have seen that given a one parameter group of transformations $\{\varphi_{\mathcal{v},t} \mid t \in \mathbb{R}\}$ generated by the tangent vector field $\mathcal{v}$, its corresponding group of functional flow maps $\phi_{\mathcal{v},t}$, and a function f_0, then the following holds

$$\frac{d}{dt} f(t) = -\mathcal{D}_{\mathcal{v}} f(t), \quad f(0) = f_0, \quad \forall t \in \mathbb{R}, \tag{41}$$

where $f(t) = \phi_{\mathcal{v},t}(f_0)$.

In the discrete setting, we mimic functional flow maps, by defining operators that generate solutions to the analogous discrete equation.

Definition 7. Let v be a tangent vector field on M, and $D_v^{\mathcal{V}}$ its corresponding functional vector field. The *discrete functional flow map* of v is given by:

$$\Phi_{v,t} = \exp(-tD_v^{\mathcal{V}}) \in \mathbb{R}^{n \times n}, \quad \forall t \in \mathbb{R}, \tag{42}$$

where exp is the matrix exponential.

Lemma 10. *Let $\{\Phi_{v,t} \mid t \in \mathbb{R}\}$ be as in Eq. (42), for the tangent vector field v. Let $f_0 : \mathcal{V} \to \mathbb{R}$, and define the time-varying function $f : \mathbb{R} \times \mathcal{V} \to \mathbb{R}$ by $f(t) = \Phi_{v,t} f_0$. Then the following holds*

$$\frac{d}{dt} f(t) = -D_v^{\mathcal{V}} f(t), \quad f(0) = f_0, \quad \forall t \in \mathbb{R}. \tag{43}$$

Proof. The system of linear differential equations in Eq. (43) is a matrix ODE, which is known to have a unique solution (Godunov, 1997, Thm 1), given by (Hochbruck and Ostermann, 2010, Eq. (1.4)) $f(t) = \exp(-tD_v^{\mathcal{V}}) f_0 = \Phi_{v,t} f_0$. □

Discrete functional flow maps allow us to use flow-related properties *without explicitly computing the discrete flow map*. Furthermore, all the operations are not geometric, but algebraic, thus we can leverage various algebraic tools such as spectral analysis, representation in a reduced basis, etc. While computing the matrix exponential is challenging numerically (Moler and Van Loan, 2003), there exist efficient methods for computing *the product of the matrix exponential with a vector* (Al-Mohy and Higham, 2011), which is the only computation that we need in order to compute the functional flow of a given function f_0. The flow in Fig. 3, for example, was generated using this approach.

Using this definition, additional properties of functional flow maps carry over to the discrete case, due to the properties of the matrix exponential.

Lemma 11. *Let $\{\Phi_{v,t} \mid t \in \mathbb{R}\}$ be the group of functional flow maps generated by the tangent vector field v. Then the following hold*

1. *Identity at 0: $\Phi_{v,0} = Id$.*
2. *Composition: $\Phi_{v,t}\Phi_{v,s} = \Phi_{v,t+s}, \quad \forall t,s \in \mathbb{R}$.*
3. *Derivative at 0: $\frac{d}{dt}\Phi_{v,t|t=0} = -D_v^{\mathcal{V}}$.*
4. *Commutativity: $D_v^{\mathcal{V}}\Phi_{v,t} = \Phi_{v,t}D_v^{\mathcal{V}}$.*

Proof. All the properties follow directly from $\Phi_{v,t} = e^{-tD_v^{\mathcal{V}}}$.

1. Trivial since $e^0 = Id$.
2. Follows from $e^{-tD_v^{\mathcal{V}}} e^{-sD_v^{\mathcal{V}}} = e^{-(t+s)D_v^{\mathcal{V}}}$.
3. We have that $e^{-tA} = Id - tA + \frac{1}{2}t^2A^2 + O(t^3)$, thus $\frac{d}{dt} e^{-tA}_{|t=0} = -A$.
4. We have that $e^{-tA} = \sum_{k=0}^{\infty} \frac{1}{k!}(-tA)^k$, therefore $Ae^{-tA} = \sum_{k=0}^{\infty} \frac{1}{k!}(-t)^k A^{k+1} = e^{-tA}A$.

□

In addition to the properties we have shown in the smooth case, additional properties are easy to show in the discrete finite dimensional case. For example, we have the following:

Lemma 12. *Let $\{\Phi_{v,t} \mid t \in \mathbb{R}\}$ be the group of functional flow maps generated by the tangent vector field v, and let $O \in \mathbb{R}^{n \times n}$ be a real matrix. Then the following holds*

$$D_v^{\mathcal{V}} O = O D_v^{\mathcal{V}} \quad \Leftrightarrow \quad \Phi_{v,t} O = O \Phi_{v,t} \quad \forall t \in \mathbb{R}. \tag{44}$$

Proof.

$\Rightarrow$ If $D_v^{\mathcal{V}} O = O D_v^{\mathcal{V}}$ then it is easy to see by induction on k that for all $k \geq 0$ we have $(D_v^{\mathcal{V}})^k O = O (D_v^{\mathcal{V}})^k$. Hence, O also commutes with polynomials in $D_v^{\mathcal{V}}$ and the result follows.

$\Leftarrow$ Taking the derivative with respect to t on both sides at $t = 0$, we have

$$\begin{aligned} \frac{d}{dt}\Phi_{v,t} O_{|t=0} &= \frac{d}{dt} O \Phi_{v,t|t=0} \quad \forall t \in \mathbb{R} \\ \frac{d}{dt}\Phi_{v,t|t=0} O &= O \frac{d}{dt}\Phi_{v,t|t=0} \quad \forall t \in \mathbb{R} \qquad O \text{ does not depend on } t \\ -D_v^{\mathcal{V}} O &= -O D_v^{\mathcal{V}} \quad \forall t \in \mathbb{R} \qquad \text{Lemma } 11(3) \end{aligned} \tag{45}$$

□

3.6 Lie bracket

Analogously to the smooth case, we define the Lie bracket operator of two tangent vector fields u, v on M as the commutator of their *square* functional vector field operators:

$$D_{[v,w]}^{\mathcal{V}} = D_v^{\mathcal{V}} D_w^{\mathcal{V}} - D_w^{\mathcal{V}} D_v^{\mathcal{V}}. \tag{46}$$

As in the smooth case, we have the relationship between the functional flow maps and the Lie bracket operator.

Lemma 13. *The functional flow maps of two vector fields commute if and only if their Lie bracket vanishes:*

$$\Phi_{v,t}\Phi_{w,s} = \Phi_{w,s}\Phi_{v,t}, \ \forall s,t \in \mathbb{R} \quad \Leftrightarrow \quad D_{[v,w]}^{\mathcal{V}} = 0. \tag{47}$$

Proof. First, set $O = \Phi_{w,s}$, and apply Lemma 12, to get that:

$$\Phi_{v,t}\Phi_{w,s} = \Phi_{w,s}\Phi_{v,t}, \ \forall s,t \in \mathbb{R} \quad \Leftrightarrow \quad D_v^{\mathcal{V}}\Phi_{w,s} = \Phi_{w,s} D_v^{\mathcal{V}}, \ \forall s \in \mathbb{R}. \tag{48}$$

Now, set $O = D_v^{\mathcal{V}}$, and apply Lemma 12 again, yielding:

$$D_v^{\mathcal{V}}\Phi_{w,s} = \Phi_{w,s} D_v^{\mathcal{V}}, \ \forall s \in \mathbb{R} \quad \Leftrightarrow \quad D_v^{\mathcal{V}} D_w^{\mathcal{V}} = D_w^{\mathcal{V}} D_v^{\mathcal{V}}. \tag{49}$$

□

Unfortunately, unlike in the smooth case, the operator $D_{[v,w]}^{\mathcal{V}}$ is *not* a functional vector field. This is easy to see when considering the sparsity structure of $D_v^{\mathcal{V}}$. As shown in Lemma 8, $D_v^{\mathcal{V}}$ has a 1-ring neighbourhood sparsity structure. When two such matrices are multiplied, the resulting matrix has a 2-ring neighbourhood sparsity structure, and therefore, cannot in general be a functional vector field. In the smooth case, these second-order dependencies cancel out, however, in the discrete case they do not.

To extract a vector field from $D_{[v,w]}^{\mathcal{V}}$ it is still possible to use the embedding and compute $u = D_{[v,w]}^{\mathcal{V}} X$, where $u \in \mathbb{R}^{n \times 3}$. However, this yields a vector field on the *vertices*, which is only approximately tangent to the input surface. This approach was used by Azencot et al. (2018) to numerically simulate the EPDiff flow on a triangulated mesh.

4 Divergence-based functional vector fields

As we have seen, discrete functional vector fields are quite powerful, allowing us to represent complicated geometric objects such as the flow map, using simple algebraic operations on matrices, such as computing the matrix exponential.

There are, however, some properties of the smooth case that do not carry over nicely to the discrete case. One such example is the Lie bracket described in Section 3.6, which fails to yield a discrete functional vector filed as in the smooth case. In this section, we propose an *additional discretization* of the functional vector field operator, which, combined with the previous one, allows us to alleviate some of the inconsistencies in the discrete case.

4.1 Smooth DFVF

A well-known result from Riemannian geometry (see, e.g., Petersen et al., 2006, pp. 58), the product rule for the product of functions and vector fields, is given by:

$$f \nabla \cdot v + \langle \nabla f, v \rangle = \nabla \cdot (f v), \tag{50}$$

where this equation holds for any point $p \in \mathcal{M}$. Thus, instead of directly working with gradients of functions, we define the following.

Definition 8. Given a tangent vector field v on $\mathcal{M}$, the corresponding *divergence-based functional vector field* (DFVF) operator $\widetilde{\mathcal{D}}_v$ is given by:

$$\widetilde{\mathcal{D}}_v(f) = \nabla \cdot (f v) - f \nabla \cdot v. \tag{51}$$

In the smooth case, the definitions in Eqs. (2) and (51) are equivalent, i.e., $\mathcal{D}_v(f) = \widetilde{\mathcal{D}}_v(f)$ for all functions f. In the discrete case, however, this choice leads to different discrete operators.

4.2 Discrete DFVF

Definition 9. Given a tangent vector field v on M, the corresponding *divergence-based functional vector field* (DFVF) operators are given by:

$$\begin{aligned} \widetilde{D}_v^{\mathcal{F}} &= \operatorname{div}[v]_{\bullet} - [\operatorname{div} v] I_{\mathcal{V}}^{\mathcal{F}} \in \mathbb{R}^{n \times m}, \\ \widetilde{D}_v^{\mathcal{V}} &= \operatorname{div}[v]_{\bullet} I_{\mathcal{F}}^{\mathcal{V}} - [\operatorname{div} v] \in \mathbb{R}^{n \times n}. \end{aligned} \tag{52}$$

The corresponding dual operators are given by:

$$\begin{aligned} \overline{\widetilde{D}_{\mathrm{f}}^{\mathcal{F}}} &= \operatorname{div}[\mathrm{f}] - [I_{\mathcal{V}}^{\mathcal{F}} \mathrm{f}] \operatorname{div} \in \mathbb{R}^{n \times 3m}, \\ \overline{\widetilde{D}_{f}^{\mathcal{V}}} &= \operatorname{div}[I_{\mathcal{F}}^{\mathcal{V}} f] - [f] \operatorname{div} \in \mathbb{R}^{n \times 3m}, \end{aligned} \tag{53}$$

and we have $\widetilde{D}_v^{\mathcal{F}} \mathrm{f} = \overline{\widetilde{D}_{\mathrm{f}}^{\mathcal{F}}} v$ and $\widetilde{D}_v^{\mathcal{V}} f = \overline{\widetilde{D}_{f}^{\mathcal{V}}} v$.

The main advantage of the divergence-based rectangular operator is that it can be multiplied with the rectangular functional vector field operator that we have introduced earlier, to yield a better discretization of the Lie bracket operator.

4.3 Mixed Lie bracket operator

Definition 10. The *mixed Lie bracket operator* of two tangent vector fields v, w is given by:

$$D^{\mathcal{F}}_{[v,w]} = \widetilde{D}^{\mathcal{F}}_{v} D^{\mathcal{F}}_{w} - \widetilde{D}^{\mathcal{F}}_{w} D^{\mathcal{F}}_{v} \in \mathbb{R}^{n \times n}. \tag{54}$$

The mixed lie bracket operator uses both the discrete divergence and discrete gradient operators, to yield a better discretization, that has some interesting properties.

Lemma 14. *The operator $D^{\mathcal{F}}_{[v,w]}$ can be written as:*

$$D^{\mathcal{F}}_{[v,w]} = \operatorname{div} B(v,w) \operatorname{grad} - [\operatorname{div} v] D^{\mathcal{V}}_{w} + [\operatorname{div} w] D^{\mathcal{V}}_{v} \ , \tag{55}$$

where $B(v,w) = [v]_{\bullet}[w]^{T}_{\bullet} - [w]_{\bullet}[v]^{T}_{\bullet}$ is a $3m \times 3m$ block diagonal matrix, encoding the outer products $v(t)w(t)^{T} - w(t)v(t)^{T}$ per face $t \in \mathcal{F}$.

Proof.

$$\begin{aligned}
D^{\mathcal{F}}_{[v,w]} &= (\operatorname{div}[v]_{\bullet} - [\operatorname{div} v] I^{\mathcal{F}}_{\mathcal{V}})[w]^{T}_{\bullet} \operatorname{grad} - (\operatorname{div}[w]_{\bullet} - [\operatorname{div} w] I^{\mathcal{F}}_{\mathcal{V}})[v]^{T}_{\bullet} \operatorname{grad} \\
&= \operatorname{div}[v]_{\bullet}[w]^{T}_{\bullet} \operatorname{grad} - [\operatorname{div} v] I^{\mathcal{F}}_{\mathcal{V}}[w]^{T}_{\bullet} \operatorname{grad} - \operatorname{div}[w]_{\bullet}[v]^{T}_{\bullet} \operatorname{grad} + [\operatorname{div} w] I^{\mathcal{F}}_{\mathcal{V}}[v]^{T}_{\bullet} \operatorname{grad} \\
&= \operatorname{div}([v]_{\bullet}[w]^{T}_{\bullet} - [w]_{\bullet}[v]^{T}_{\bullet}) \operatorname{grad} - [\operatorname{div} v] I^{\mathcal{F}}_{\mathcal{V}}[w]^{T}_{\bullet} \operatorname{grad} + [\operatorname{div} w] I^{\mathcal{F}}_{\mathcal{V}}[v]^{T}_{\bullet} \operatorname{grad} \\
&= \operatorname{div}([v]_{\bullet}[w]^{T}_{\bullet} - [w]_{\bullet}[v]^{T}_{\bullet}) \operatorname{grad} - [\operatorname{div} v] D^{\mathcal{V}}_{w} + [\operatorname{div} w] D^{\mathcal{V}}_{v} \\
&= \operatorname{div} B(v,w) \operatorname{grad} - [\operatorname{div} v] D^{\mathcal{V}}_{w} + [\operatorname{div} w] D^{\mathcal{V}}_{v}.
\end{aligned} \tag{56}$$

□

We can further simplify the operator B as follows.

Lemma 15. *Let $B(v, w)$ be the operator defined in Lemma 14. Then the following holds*

$$B(v,w) = [[v]^{T}_{\bullet} \mathcal{J} w] \mathcal{J}. \tag{57}$$

Proof. By the definition in Lemma 14 we have that $B(v, w)$ at the triangle $t \in \mathcal{F}$ depends only on the values of v, w at t. Let $\xi_i \in \mathbb{R}^3$, $i \in 1,\ldots,3$ be an orthonormal frame at the triangle t, such that ξ_3 is in the normal direction, and set $\mathcal{B} = (\xi_1 \ \ \xi_2 \ \ \xi_3)$. Note, that the rotation by $\pi/2$ in the triangle t is given by

$\mathcal{J}_t = \mathcal{B}\begin{pmatrix} 0 & -1 & 0 \\ 1 & 0 & 0 \\ 0 & 0 & 0 \end{pmatrix}\mathcal{B}^T$ Now, let $v(t) = \mathcal{B}\begin{pmatrix} a \\ b \\ 0 \end{pmatrix}, w(t) = \mathcal{B}\begin{pmatrix} c \\ d \\ 0 \end{pmatrix}$ be the representations of $v(t)$, $w(t)$ in this basis, respectively. Then $B(v, w)(t)$ given by

$$\begin{aligned}
B(v,w)(t) &= \mathcal{B}\left(\begin{pmatrix} a \\ b \\ 0 \end{pmatrix}(c \;\; d \;\; 0) - \begin{pmatrix} c \\ d \\ 0 \end{pmatrix}(a \;\; b \;\; 0)\right)\mathcal{B}^T \\
&= \mathcal{B}\begin{pmatrix} 0 & ad-bc & 0 \\ bc-ad & 0 & 0 \\ 0 & 0 & 0 \end{pmatrix}\mathcal{B}^T \\
&= (bc-ad)\mathcal{B}\begin{pmatrix} 0 & -1 & 0 \\ 1 & 0 & 0 \\ 0 & 0 & 0 \end{pmatrix}\mathcal{B}^T \\
&= (a \;\; b \;\; 0)\begin{pmatrix} 0 & -1 & 0 \\ 1 & 0 & 0 \\ 0 & 0 & 0 \end{pmatrix}\begin{pmatrix} c \\ d \\ 0 \end{pmatrix}\mathcal{J}_t \\
&= (a \;\; b \;\; 0)\mathcal{B}^T\mathcal{B}\begin{pmatrix} 0 & -1 & 0 \\ 1 & 0 & 0 \\ 0 & 0 & 0 \end{pmatrix}\mathcal{B}^T\mathcal{B}\begin{pmatrix} c \\ d \\ 0 \end{pmatrix}\mathcal{J}_t \\
&= v(t)^T\mathcal{J}_t w(t)\mathcal{J}_t \\
&= \langle v(t), \mathcal{J}_t w(t)\rangle\mathcal{J}_t.
\end{aligned} \tag{58}$$

Since this holds for every triangle t, the result follows by the definitions of the $[\cdot]_{\bullet}^{T}$ and $[\cdot]$ operators. □

4.3.1 *Sparsity structure*

Using Lemma 14, we can deduce the sparsity structure of the mixed Lie bracket operator.

Lemma 16. *The mixed Lie bracket operator $D^{\mathcal{F}}_{[v,w]}$ has a 1-ring sparsity structure. I.e, if the entry (i, j) is nonzero, then either $i = j$ or $(i,j) \in \mathcal{E}$.*

Proof. The first component, up to diagonal mass matrices and constants, has the structure of $\text{grad}^T B(v, w)$ grad. As we have shown earlier, grad has a nonzero entry (t, i) only if $i \in t$. The matrix B has no effect on the sparsity structure, since it is block diagonal, and does not have mixed entries for two different triangles. Hence, if (i, j) is nonzero in the first component, then there exists a triangle t such that $i \in t$ and $j \in t$, therefore either $i = j$ or $(i, j) \in \mathcal{E}$. Therefore, the first component has a 1-ring sparsity structure. The last two components are formed by the multiplication of a diagonal matrix with a 1-ring sparse matrix, according to Lemma 8, and therefore are also 1-ring sparse. Hence, $D^{\mathcal{F}}_{[v,w]}$ is the sum of 1-ring sparse matrices, and the result follows. □

The sparsity structure of the Lie bracket operator is of great importance in practice. For example, when applying the operator to a smooth function (e.g., the first few eigenfunctions of the Laplace–Beltrami operator), the mixed Lie bracket operator $D^{\mathcal{F}}_{[v,u]}$ yields a smooth function, whereas the operator from Azencot et al. (2013), $D^{\mathcal{V}}_{[v,u]}$, exhibits visible oscillations (see Fig. 7).

Furthermore, one of the most useful applications of the functional vector field operator is transporting functions on flows of vector fields using the functional flow map, e.g., as was done in Azencot et al. (2014a, 2015b) for fluid simulation. This computation is done, as described in Definition 7, by evaluating $f(t) = e^{-tD_v} f_0$. Thus, powers of D_v are involved in the computation, and the wrong sparsity structure can generate considerable discretization noise during function transport, as can be seen in Fig. 8.

4.3.2 The Lie bracket of divergence-free vector fields

Consider the special case when the smooth vector fields u, v are gradients of functions rotated in the tangent plane by $\pi/2$, namely $v = J\nabla f, u = J\nabla g$.

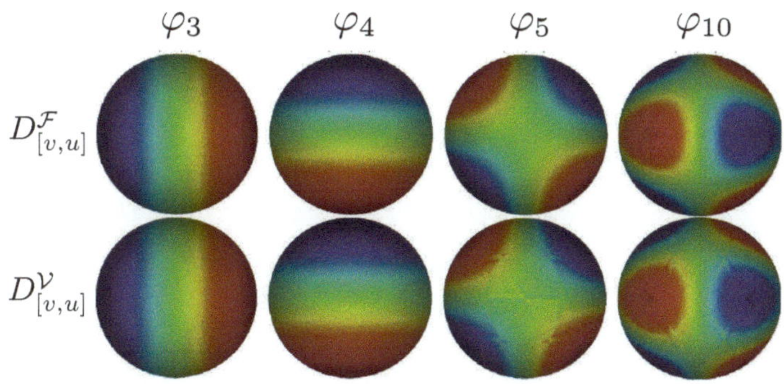

FIG. 7 Applying the Lie bracket operator $D^{\mathcal{V}}_{[v,u]}$ to the ith eigenfunction φ_i of the Laplace–Beltrami operator leads to visible oscillations (*bottom*), whereas when using the mixed Lie bracket $D^{\mathcal{F}}_{[v,u]}$ the results are smoother (*top*).

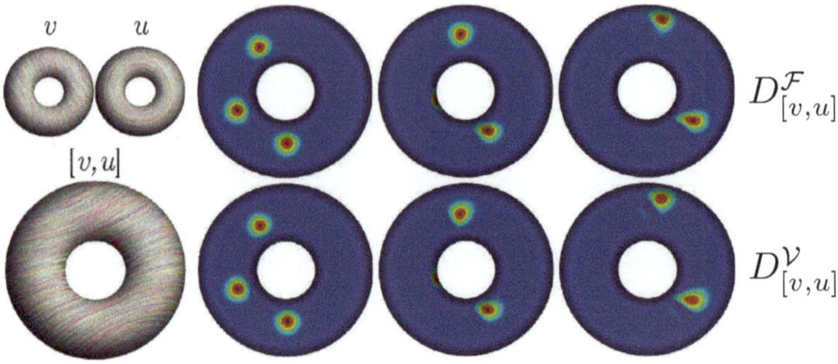

FIG. 8 Transporting a function using the bracket operator $D^{\mathcal{V}}_{[v,u]}$ leads to more discretization error (*bottom*) compared to computing transport using the mixed operator $D^{\mathcal{F}}_{[v,u]}$ (*top*).

Then, it is well known (Arnold, 1989, Appendix 2) that $[\mathscr{v},\mathscr{u}] = J\nabla\{\mathscr{f},\mathscr{g}\}$, where $\{\mathscr{f},\mathscr{g}\} = \langle \mathcal{J}\nabla\mathscr{f}, \nabla\mathscr{g}\rangle$ is known as the *Poisson bracket* of the functions $\mathscr{f},\mathscr{g}$. Hence in the smooth case, the following holds

$$\mathcal{D}_{[\mathscr{v},\mathscr{u}]}\mathscr{h} = \langle \nabla\mathscr{h}, \mathcal{J}\nabla\{\mathscr{f},\mathscr{g}\}\rangle = -\widetilde{\mathcal{D}}_{\mathcal{J}\nabla\mathscr{h}}\{\mathscr{f},\mathscr{g}\}, \tag{59}$$

where the first equality is by the definition of $\mathcal{D}$, and the second by multiplying both sides of the inner product by $\mathcal{J}$ and using the definition again.

In the discrete setting using our operators, we have the following result.

Theorem 1. *Let* $v = \mathcal{J}\,\mathrm{grad}\,f, u = \mathcal{J}\,\mathrm{grad}\,g$, *and define* $\{f,g\} = \langle \mathcal{J}\,\mathrm{grad}\,f, \mathrm{grad}\,g\rangle$, *analogously to the smooth case. Then the following holds*

$$D^{\mathcal{F}}_{[v,u]}h = -\widetilde{D}^{\mathcal{F}}_{\mathcal{J}\mathrm{grad}\,h}\{f,g\}. \tag{60}$$

Proof. First, note that div v = div u = 0, as was discussed in Section 3.1.7. Hence, the last two addends in Eq. (55) are zero. Thus, from Lemma 15, for any function h we have

$$\begin{aligned}
D^{\mathcal{F}}_{[v,u]}h &= \mathrm{div}\,([[v]^T_{\bullet}\mathcal{J}u]\mathcal{J})\,\mathrm{grad}\,h \\
&= \mathrm{div}\,([[\mathcal{J}\,\mathrm{grad}\,f]^T_{\bullet}\mathcal{J}\mathcal{J}\,\mathrm{grad}\,g]\mathcal{J})\,\mathrm{grad}\,h \\
&= -\,\mathrm{div}\,[[\mathcal{J}\,\mathrm{grad}\,f]^T_{\bullet}\,\mathrm{grad}\,g]\mathcal{J}\,\mathrm{grad}\,h \\
&= -\,\mathrm{div}\,[\{f,g\}]\mathcal{J}\,\mathrm{grad}\,h \\
&= -\,\mathrm{div}\,[\mathcal{J}\,\mathrm{grad}\,h]_{\bullet}\{f,g\} \\
&= -\widetilde{D}^{\mathcal{F}}_{\mathcal{J}\mathrm{grad}\,h}\{f,g\},
\end{aligned} \tag{61}$$

where in the step before last we used Property 6 of the bracket operators (see Section 3.1), and in the last step we used again the fact that div $\mathcal{J}$ grad $h = 0$. □

4.3.3 Extraction of the commutator vector field

Given the operator $D^{\mathcal{F}}_{[v,w]}$, it is possible to extract a corresponding vector field by taking $u = D^{\mathcal{F}}_{[v,w]} X \in \mathbb{R}^{n\times 3}$. While this procedure results in a vector field on the *vertices*, which is not guaranteed to be tangent to the surface, in practice this procedure seems to yield a satisfactory result. We show in Fig. 9 the robustness this extraction to the triangulation of the mesh. Given two fields on a smooth mesh, we compute their bracket and show its reconstruction in the top row. The bottom row shows the same computation, but on a low resolution uniform sampling of the original mesh, which contains many skinny triangles. Note that the resulting vector fields are similar in both cases.

4.3.4 Limitation: Loss of group structure

While the sparsity structure of $D^{\mathcal{F}}_{[v,u]}$ is the correct one, we do not have a result linking it back to a corresponding vector field w, such that $D^{\mathcal{V}}_{w} = D^{\mathcal{F}}_{[v,u]}$. Furthermore, the result from Lemma 13 that links a vanishing Lie bracket to the commutation of the functional flows also does not hold, since the separate flows are no longer well defined. It remains an open question whether one can design discrete functional vector fields, and a corresponding Lie bracket operator, such that both the group structure, and the closure under bracket properties hold.

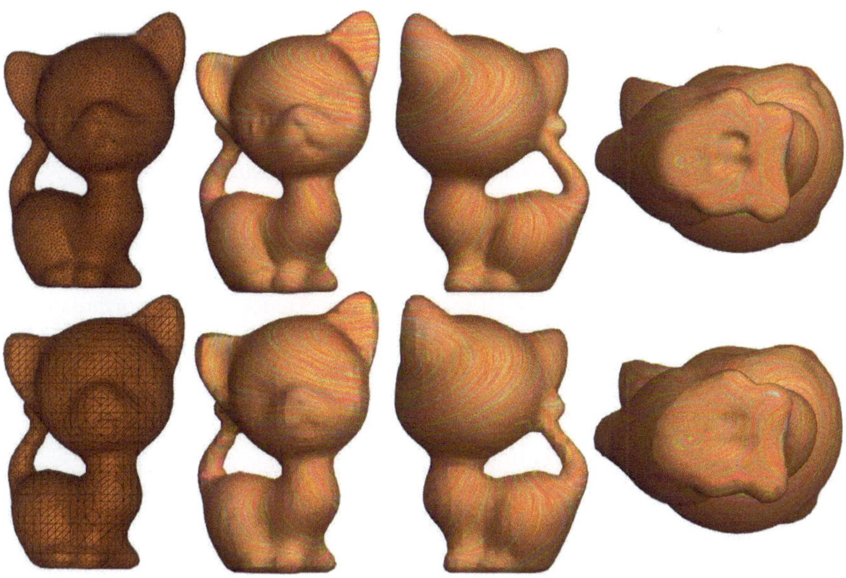

FIG. 9 The extrinsic extraction of the commutator vector field yields similar results on a uniformly tessellated mesh (*top*) and a highly nonuniform tessellation of the same model (*bottom*).

4.4 The Lie bracket as a linear transformation on vector fields

In some applications, instead of computing $D^{\mathcal{F}}_{[v,w]}$ for two given vector fields, we wish to compute a vector field w whose bracket with v fulfils some property, e.g., its norm is minimal. We therefore define the following operator.
Definition 11. Given a vector field v, the operator $\mathrm{ad}_v : \mathbb{R}^{3m} \to \mathbb{R}^{n\times 3}$ is given by:

$$\mathrm{ad}_v(w) = D^{\mathcal{F}}_{[v,w]} X. \tag{62}$$

Note that ad_v is linear and can be written in the following concise form.
Lemma 17. *The operator* ad_v *can be written as:*

$$\mathrm{ad}_v = \begin{pmatrix} \widetilde{D}^{\mathcal{F}}_v & 0 & 0 \\ 0 & \widetilde{D}^{\mathcal{F}}_v & 0 \\ 0 & 0 & \widetilde{D}^{\mathcal{F}}_v \end{pmatrix} - \begin{pmatrix} \overline{\widetilde{D}^{\mathcal{F}}_{\mathrm{v}_1}} \\ \overline{\widetilde{D}^{\mathcal{F}}_{\mathrm{v}_2}} \\ \overline{\widetilde{D}^{\mathcal{F}}_{\mathrm{v}_3}} \end{pmatrix} \in \mathbb{R}^{3n\times 3m} \quad , \tag{63}$$

where $\mathrm{v}_i \in \mathbb{R}^m$ *are the coordinate functions of* v.

Proof. From Lemma 9, we have that $D^{\mathcal{F}}_v x_i = \mathrm{v}_i$, $i \in 1,\ldots,3$. Hence, from Definitions 9 and 10 we have

$$D^{\mathcal{F}}_{[v,w]} x_i = \widetilde{D}^{\mathcal{F}}_v \mathrm{w}_i - \widetilde{D}^{\mathcal{F}}_w \mathrm{v}_i = \widetilde{D}^{\mathcal{F}}_v \mathrm{w}_i - \overline{\widetilde{D}^{\mathcal{F}}_{\mathrm{v}_i}} w. \tag{64}$$

The result follows, by considering all three components, and taking $\mathrm{w} = \begin{pmatrix} \mathrm{w}_1 \\ \mathrm{w}_2 \\ \mathrm{w}_2 \end{pmatrix}$. □

Note that one could potentially apply interpolation and projection on ad_v to obtain a vector field on the faces.

4.4.1 Application: Commutation-guided rescaling

In parameterization applications it is often required that the parameterization aligns with two direction fields, which are often computed from the curvature directions of the surface. However, given two arbitrary vector fields, their flows generate a parameterization only if they commute. Hence, we can formulate the following optimization problem for the smooth case. Given two vector fields v, u find a smooth scalar function f such that $\| [v, fu] \|_{L^2}$ is minimal:

$$\min_{f} \left\{ \frac{1}{2} \int_{\mathcal{M}} \| [v, fu] \|^2 \mathrm{d}a + \frac{\alpha}{2} \int_{\mathcal{M}} |\nabla f|^2 \mathrm{d}a \right\} \tag{65}$$

$$\text{subject to} \int_{\mathcal{M}} |f|^2 \mathrm{d}a = 1. \tag{66}$$

Our discrete framework allows for an efficient evaluation of the above expression by computing the eigenvectors of the matrix:

$$E = F(v,u)^T F(v,u) + \alpha L,$$

where $F(v, u)f$ discretizes the smooth action on functions $[v, fu]$, and is given by $F(v, u) = \mathrm{ad}_v[u]_\bullet I^{\mathcal{V}}_{\mathcal{F}}$, and L is the standard Laplace–Beltrami operator, given by $L = -\mathrm{divgrad}$.

Fig. 10 illustrates the result this procedure. Here, v is a smooth vector field with unit norm, and u is its rotation by $\pi/2$. The two fields have nonzero bracket, as the colour coding of the norm of $[v, u]$ shows, where the bracket's norm is concentrated around the singularity locations. Using the optimization problem mentioned previously, we find a function φ which minimizes $\| [v, \varphi u] \|^2_{L^2}$. The resulting norm is smaller and more evenly distributed, as can be seen in the rightmost image (see, e.g., the tips of the fingers).

4.4.2 Application: As-commuting-as-possible vector fields

Instead of optimizing for a scalar when two directions are given as above, we are also interested in finding a set of vector fields which commute with respect to another vector field v. Since the linear operator ad_v encodes the bracket operation on vector fields, we can easily extract commuting fields by computing its kernel. Notice that v itself lies in the discrete kernel, by construction. Unfortunately, this is also true for any vector field which is parallel to v. As the set of discrete commuting fields contains many nonsmooth vector fields (e.g., any subset of vectors in v can be rotated by π), we add a regularizer term to the energy expression. We formulate the following optimization problem for the smooth case:

$$\min_{u} \left\{ \frac{1}{2} \int_{\mathcal{M}} \| \mathrm{ad}_v u \|^2 \mathrm{d}a + \frac{\lambda}{2} \int_{\mathcal{M}} \| \nabla u \|^2 \mathrm{d}a \right\}$$
$$\text{subject to} \int_{\mathcal{M}} \| u \|^2 \mathrm{d}a = 1,$$

where ∇ denotes here the covariant derivative of vector fields. In the discrete setup, the above equation is expressed as an eigenvalue problem of the following sparse matrix:

$$E = \mathrm{ad}_v^T \mathrm{ad}_v + \lambda L,$$

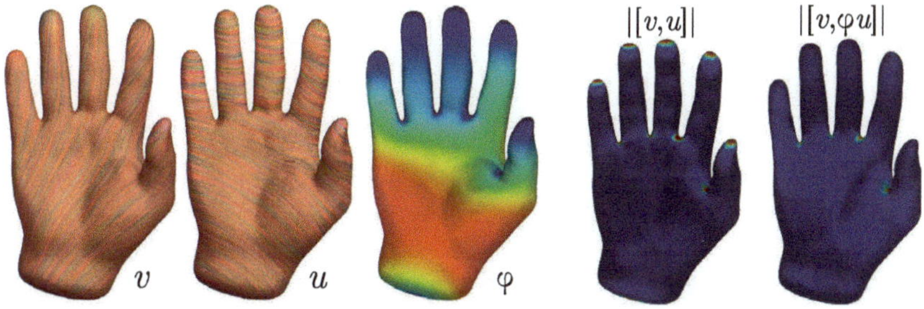

FIG. 10 Two smooth vector fields with nonzero bracket $[v, u]$ and a smooth rescaling function φ which minimizes $\| [v, \varphi u] \|^2_{L^2}$. Note that after rescaling the bracket's norm is smaller and more evenly distributed on the surface (see, e.g., the tips of the fingers).

where L is the vector Laplacian. We used the discrete connection Laplacian with a 1-ring neighbourhood of faces given by Singer and Wu (2012), however, any other vector Laplacian can be used instead, e.g., the one suggested by Knöppel et al. (2013). Since we are interested in tangent vector fields, we further modify the matrix E which is of size $3m \times 3m$ by $E_B = B^T E B$, where $B \in \mathbb{R}^{3m\times 2m}$ encodes an arbitrary orthonormal basis per face. Fig. 11 shows a collection of vector fields which are approximately commuting with a given one, computed using this approach.

4.5 Integrating the Lie bracket operator

The functional flow map defines a discrete flow map of a vector field v, by using the matrix exponential of $D_v^{\mathcal{V}}$. Applying the functional flow map allows us to transport *functions* on the flow of v. Similarly, we use the matrix exponential of ad_v, to transport *vector fields* on the flow of v. To that end, we define the following operator:

$$\mathrm{Ad}_{\Phi_{v,t}} = \exp\left(t[I_{\mathcal{F}}^{\mathcal{V}}]\mathrm{ad}_v\right) \in \mathbb{R}^{3m\times 3m} \quad , \tag{67}$$

where $[\cdot]$, replicates the input matrix 3 times along the diagonal.

The discrete approximation given in Eq. (67) is potentially useful in the context of fluid simulation. Indeed, while ideal fluids on surfaces can be simulated using function transport as was shown in Azencot et al. (2014a), the situation for ideal fluids in volume is more involved as it includes vectorial transport. For instance, Elcott et al. (2007) simulated fluids on tetrahedral meshes by advecting the vorticity field over the flow lines of the velocity field. Since our operators can be easily adapted to tetrahedral meshes, an operator related to $\mathrm{Ad}_{\Phi_{v,t}}$ can be similarly used to integrate the volumetric Euler equations. This is an interesting direction for future work.

4.5.1 Application: Interpolation of vector fields

The flow transportation machinery allows us to interpolate between two vector fields in a more intuitive way than pointwise interpolation. Specifically,

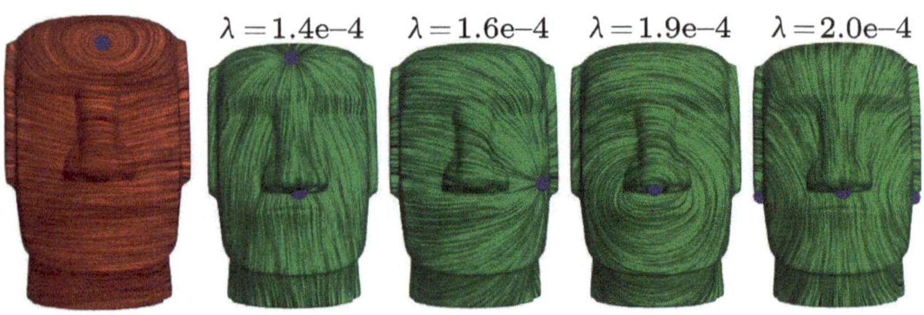

FIG. 11 As-commuting-as-possible vector fields, with respect to the vector field on the *left*, the *blue points* are the singularities.

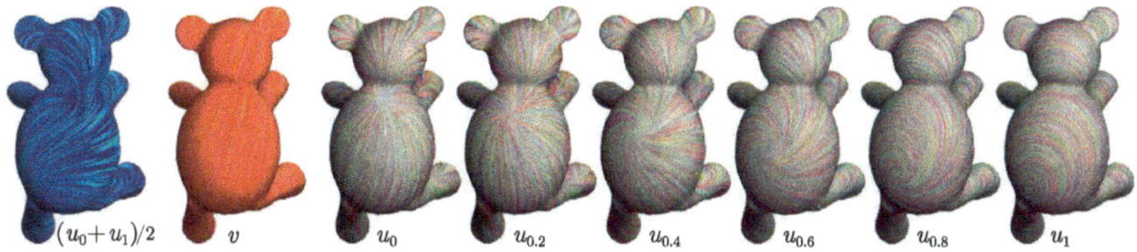

FIG. 12 Interpolating between the vector fields u_0 and u_1 by transporting both on the flow of v. See the text for details.

consider the vector fields u_0 and u_1 shown in Fig. 12. They both have two singularities: one visible on the back, and another one on the left leg. If we simply interpolate the vector fields pointwise, we would get two singularities on the back (Fig. 12, left).

However, if there exists another vector field v, with a corresponding flow $\Phi_{v,t}$, such that the singularity of u_0 is mapped to the singularity of u_1 under the map $\Phi_{v,1}$, then we can transport both vector fields on the flow of v and combine the result to get the interpolated vector field u_t. Specifically, for $t \in [0, 1]$ we use

$$u_t = (1-t)\mathrm{Ad}_{\Phi_{v,t}} u_0 + t\mathrm{Ad}_{\Phi_{v,t-1}} u_1. \tag{68}$$

A few frames from the result are shown in Fig. 12. Note that for $t=0$ and $t=1$ we get u_0 and u_1, respectively, as expected, and for intermediate t values we get a smooth interpolation of both the location of the singularity and the behaviour of the vector field.

It is worth noting that given two arbitrary vector fields, it is in general a difficult problem to find a corresponding vector field v which will transport u_0 to u_1 in a meaningful way. However, we believe that such a problem could perhaps be treated using optimal transportation machinery, as has previously been done for function flow in Solomon et al. (2014). Hence, our vector field transport could potentially be a key ingredient in extending such methods to vector fields.

4.6 Outlook

The mixed operators we have defined could potentially find multiple applications in geometry processing. Commuting flows can be used for local parameterization, and vector field transport could be used for numerically simulating partial differential equations, or as a building block in more complex setups for computing optimal transportation flows. Furthermore, the ad operator can be used to discretize the directional derivative of the matrix exponential, potentially allowing to compute a time-varying vector field whose flow corresponds to a given map.

5 Conclusion and future work

The algebraic approach presented in this chapter, in which tangent vector fields are represented globally instead of locally, as linear operators instead of pointwise directions, has proven to be beneficial in many applications, and provides ample opportunity for future work.

From the theoretical perspective, the most interesting open question is whether the space of tangent vector fields on a triangle mesh, and the corresponding flows, can be discretized such that the properties of the smooth case carry over to the discrete setting. Specifically, in the smooth case, the flows and tangent vector fields form a Lie group and a Lie algebra, respectively. In the discrete case, it would be most natural to describe these using a matrix Lie group and algebra. The answer to this question hinges on finding a definition of directional derivative operators, whose commutator is again a tangent vector field, as in the smooth case. The mixed Lie bracket that we have presented in Definition 10 has some of the required properties, e.g., the 1-ring sparsity structure, but not all. Finding such a definition, or proving it cannot exist, remains an open question, to the best of our knowledge.

From the practical perspective, the operators described in this chapter are ripe for use in geometry processing applications. The most obvious use case is parameterization using commuting tangent vector fields. Other potential applications include vector field tracing, vector field visualization, and numerical simulation of PDEs.

References

Absil, P.A., Mahony, R., Sepulchre, R., 2009. Optimization Algorithms on Matrix Manifolds. Princeton University Press.

Al-Mohy, A.H., Higham, N.J., 2011. Computing the action of the matrix exponential, with an application to exponential integrators. SIAM J. Sci. Comput. 33 (2), 488–511.

Arnold, V.I., 1989. Mathematical Methods of Classical Mechanics, vol. 60. Springer Science & Business Media.

Azencot, O., Ben-Chen, M., Chazal, F., Ovsjanikov, M., 2013. An operator approach to tangent vector field processing. Comput. Graph. Forum 32 (5), 73–82.

Azencot, O., Weißmann, S., Ovsjanikov, M., Wardetzky, M., Ben-Chen, M., Wiley Online Library, 2014a. Functional fluids on surfaces. Comput. Graph. Forum 33 (5), 237–246.

Azencot, O., Weißmann, S., Ovsjanikov, M., Wardetzky, M., Ben-Chen, M., 2014b. Functional fluids on surfaces. Comput. Graph. Forum 33 (5), 237–246.

Azencot, O., Ovsjanikov, M., Chazal, F., Ben-Chen, M., 2015a. Discrete derivatives of vector fields on surfaces—an operator approach. ACM Trans. Graph. 34 (3), 29:1–29:13.

Azencot, O., Vantzos, O., Wardetzky, M., Rumpf, M., Ben-Chen, M., 2015b. Functional thin films on surfaces. In: Proceedings of the Symposium on Computer Animation.

Azencot, O., Vantzos, O., Ben-Chen, M., 2016. Advection-Based Function Matching on Surfaces. Comput. Graph. Forum 35, 55–64.

Azencot, O., Corman, E., Ben-Chen, M., Ovsjanikov, M., 2017. Consistent functional cross field design for mesh quadrangulation. ACM Trans. Graph. 36 (4), 92:1–92:13.

Azencot, O., Vantzos, O., Ben-Chen, M., 2018. An explicit structure-preserving numerical scheme for EPDiff. Comput. Graph. Forum 37 (5), 107–119.

Bommes, D., Zimmer, H., Kobbelt, L., 2009. Mixed-integer quadrangulation. ACM Trans. Graph. (TOG) 28 (3), 77.

Botsch, M., Kobbelt, L., Pauly, M., Alliez, P., Lévy, B., 2010. Polygon Mesh Processing. CRC Press.

Crane, K., Desbrun, M., Schröder, P., 2010. Trivial connections on discrete surfaces. Comput. Graph. Forum (SGP) 29 (5), 1525–1533.

Desbrun, M., Hirani, A.N., Leok, M., Marsden, J.E., 2005. Discrete exterior calculus. Preprint, arXiv:math.DG/0508341.

Elcott, S., Tong, Y., Kanso, E., Schröder, P., Desbrun, M., 2007. Stable, circulation-preserving, simplicial fluids. ACM Trans. Graph. (TOG) 26 (1), 4.

Godunov, S.K., 1997. Ordinary Differential Equations With Constant Coefficient. vol. 169. American Mathematical Soc.

Hochbruck, M., Ostermann, A., 2010. Exponential integrators. Acta Numer. 19, 209–286.

Knöppel, F., Crane, K., Pinkall, U., Schröder, P., 2013. Globally optimal direction fields. ACM Trans. Graph. (TOG) 32 (4), 59.

Knöppel, F., Crane, K., Pinkall, U., Schröder, P., 2013. Globally optimal direction fields. ACM Trans. Graph. 32 (4), 59:1–59:10.

Lee, J., 2012. Introduction to Smooth Manifolds (Graduate Texts in Mathematics, vol. 218). Springer. ISBN: 9781441999818.

Moler, C., Van Loan, C., 2003. Nineteen dubious ways to compute the exponential of a matrix, twenty-five years later. SIAM Rev. 45 (1), 3–49.

Morita, S., 2001. Geometry of Differential Forms, Translations of Mathematical Monographs. vol. 201. American Mathematical Society, Providence, RI.

Ovsjanikov, M., Ben-Chen, M., Solomon, J., Butscher, A., Guibas, L., 2012. Functional maps: a flexible representation of maps between shapes. ACM Trans. Graph. (TOG) 31 (4), 30.

Petersen, P., Axler, S., Ribet, K.A., 2006. Riemannian Geometry, vol. 171. Springer.

Polthier, K., Preuss, E., 2003. Identifying vector field singularities using a discrete Hodge decomposition. In: Hege, H.C., Polthier, K. (Eds.), Visualization and Mathematics III. Springer, pp. 113–134.

Ray, N., Sokolov, D., 2014. Robust polylines tracing for n-symmetry direction field on triangulated surfaces. ACM Trans. Graph. (TOG) 33 (3), 30.

Singer, A., Wu, H.-T., 2012. Vector diffusion maps and the connection Laplacian. Commun. Pure Appl. Math. 65 (8), 1067–1144.

Solomon, J., Rustamov, R., Guibas, L., Butscher, A., 2014. Earth mover's distances on discrete surfaces. ACM Trans. Graph. (TOG) 33 (4), 67.

Zhang, E., Mischaikow, K., Turk, G., 2006. Vector field design on surfaces. ACM Trans. Graph. 25 (4), 1294–1326.

Chapter 4

Active contour methods on arbitrary graphs based on partial differential equations

Christos Sakaridis[a], Nikos Kolotouros[b], Kimon Drakopoulos[c] and Petros Maragos[d,*]
[a]*Department of Information Technology and Electrical Engineering, ETH Zürich, Zurich, Switzerland*
[b]*Department of Computer and Information Science, University of Pennsylvania, Philadelphia, PA, United States*
[c]*Marshall School of Business, University of Southern California, Los Angeles, CA, United States*
[d]*School of Electrical and Computer Engineering, National Technical University of Athens, Athens, Greece*
[*]*Corresponding author: e-mail: maragos@cs.ntua.gr*

Chapter Outline

Abstract

This chapter formulates and compares two approaches that have been developed to discretize over arbitrary graphs the partial differential equations (PDEs) of classical active contour models with level sets, which have been widely used in image analysis and computer vision. The first approach takes a finite difference path and proposes geometric approximations of the fundamental continuous differential operators that are involved in active contour PDEs, namely gradient and curvature, on graphs with

Handbook of Numerical Analysis, Vol. 20. https://doi.org/10.1016/bs.hna.2019.07.002
© 2019 Elsevier B.V. All rights reserved.

arbitrary vertex and edge configuration. The second approach leverages finite elements to approximate the solution of these PDEs on graphs with arbitrary vertex configuration that constitute a triangulation. We present numerical algorithms and compare both approaches while using them to successfully apply the popular models of geodesic active contours (GACs) and active contours without edges (ACWE) to arbitrary 2D graphs for graph cluster detection and image segmentation.

Keywords: Active contours on graphs, Graph segmentation, Discretized partial differential equations

AMS Classification Codes: 35Q68, 35R02, 60D05, 65D25, 65M60, 68U10

1 Introduction

Evolution of curves via active contour models has been applied extensively in computer vision for image segmentation. In the standard image setting which involves a regular grid of pixels, the discretization of partial differential equations (PDEs) governing the motion of active contours is well-established (Osher and Sethian, 1988) and ensures proper convergence of the contour to object boundaries. Recently, active contour models based on level set PDE formulations have been extended (Drakopoulos and Maragos, 2012; Kolotouros and Maragos, 2017; Sakaridis et al., 2017) to handle more general inputs in the form of graphs whose vertices are arbitrarily distributed in a two-dimensional Euclidean space. In particular, the input in this case consists of:

1. an undirected graph $\mathcal{G}=(\mathcal{V},\mathcal{E})$, where $\mathcal{V}=\{\mathbf{v}_i\in\mathbb{R}^2 : i=1,\ldots,n\}$ and $\mathcal{E}\subseteq\mathcal{V}\times\mathcal{V}$, and
2. a real-valued function $I:\mathcal{V}\rightarrow\mathbb{R}$ defined on the vertices of the graph, which resembles the image function in the standard grid setting. For the rest of the paper, we will use the term image function to refer to I.

Applications of segmentation of such graphs span not only image processing but also geographical information systems and generally any field where data can assume the form of a set of pointwise samples of a real-valued function. The arbitrary spatial structure of these graphs poses a significant challenge to the discretization of active contour PDEs.

This chapter presents two fundamentally distinct approaches that have been developed to accomplish the aforementioned discretization. The first approach, which is introduced in Drakopoulos and Maragos (2012) and Sakaridis et al. (2017), takes a *finite difference* path and proposes geometric approximations of the continuous differential operators that are involved in active contour PDEs, namely *gradient* and *curvature*, on graphs with arbitrary vertex and edge configuration. The second approach, which is proposed in Kolotouros and Maragos (2017), leverages *finite elements* to approximate the solution of these PDEs on graphs with arbitrary vertex configuration that constitute a triangulation. Both approaches have been used in Sakaridis et al. (2017) and Kolotouros and Maragos (2017) to successfully apply the widely used models of geodesic active contours (GACs) (Caselles et al., 1997) and active contours without edges (ACWE) (Chan and Vese, 2001) to arbitrary 2D graphs for graph and image segmentation.

The chapter is structured as follows. Section 2 reviews related work on active contours, graph-based morphology and segmentation, and PDE-based methods on graphs, and provides the necessary background on the employed active contour models. Section 3 outlines the finite difference approach of Sakaridis et al. (2017) for active contours on graphs. It is composed of Section 3.1 on the geometric gradient approximation on arbitrary graphs and its asymptotic consistency and accuracy in the limit of infinite vertices for the class of random geometric graphs (Penrose, 2003), Section 3.2 on the geometric curvature approximation and its convergence in probability, and Section 3.3 on improved Gaussian smoothing on graphs for initialization of the GAC model through normalization. Section 4 outlines the finite element approach of Kolotouros and Maragos (2017). More specifically, Section 4.1 discusses the problem formulation and analyzes the key aspects of the finite element approximation for active contour models. Section 4.2 presents an extension of the previous framework for locally constrained active contour models that can additionally be used for speeding up contour evolution. Last, Section 5 presents experimental results of the two approaches on segmentation from the translation of GACs (Caselles et al., 1997) and ACWE (Chan and Vese, 2001) to arbitrary graphs created from regular images or containing geographical data and Section 6 recapitulates the main components of the chapter and provides a discussion on future research directions.

2 Background and related work

Active contour models for curve evolution towards image edges originate from "snakes" (Kass et al., 1988). These early approaches could not in general handle topological changes of the contour, for instance splitting into two disjoint parts to detect the boundaries of two distinct objects. The level set method (Osher and Sethian, 1988) and its related discretized PDEs solved this problem and enabled improved alternative approaches such as the geometric active contour model (Caselles et al., 1993) which was initially introduced and subsequently complemented to establish the GAC model (Caselles et al., 1997). The former model involves two forces that govern curve motion: a balloon force that expands or shrinks it, and a curvature-dependent force that maintains its smoothness. The latter model adds an extra spring force that attracts the contour towards salient image edges. Both methods embed the active contour as a *level set* of a function $u(x, y, t)$, which is termed the embedding function and is the unknown in the PDE that models curve evolution, allowing the use of numerical schemes of the type proposed in Osher and Sethian (1988). The formulation of the PDE of the GAC model is the following:

$$\frac{\partial u}{\partial t} = \underbrace{\kappa \|\nabla u\| g(I)}_{\text{curvature}} - \underbrace{c \|\nabla u\| g(I)}_{\text{balloon}} + \underbrace{\nabla u \cdot \nabla g(I)}_{\text{spring}}, \tag{1}$$

where κ is the curvature of the level sets of u at time t, $g(I)$ is an image-dependent function that stops the contour at edges, commonly referred to as the edge stopping function, and c is a constant.

The curvature-dependent force is also used for regularization in ACWE (Chan and Vese, 2001), which abandon the aforementioned edge-driven paradigm and leverage the Mumford–Shah functional (Mumford and Shah, 1989) to formulate a piecewise constant segmentation model, with automatic detection of interior contours and reduced sensitivity to initialization. The formulation of the ACWE model is the following:

$$\begin{aligned} \frac{\partial u}{\partial t} &= \delta_\epsilon(u)\left(\mu\kappa - \nu - \lambda_1(I - c_1(u))^2 + \lambda_2(I - c_2(u))^2\right), \\ c_1(u) &= \text{average}(I)\text{ in }\{u \geq 0\},\ c_2(u) = \text{average}(I)\text{ in }\{u < 0\}, \end{aligned} \tag{2}$$

where κ is the curvature of the level sets of u at time t, δ_ϵ is a regularized version of the Dirac δ function, and μ, ν, λ_1, and λ_2 are positive constants. This model is further studied in Chan et al. (2006), where the authors prove that the global minimizers of the original nonconvex problem for given values of c_1 and c_2 can be recovered via solving a convex reformulation of it.

Graphs have long been connected to image processing, in part through their study in terms of mathematical morphology. The application of morphological transforms on neighbourhood graphs was established in Vincent (1989), while a wide variety of graph structures, algorithms for their construction and early applications in computer vision were surveyed in Jaromczyk and Toussaint (1992). The notion of structuring element in classical morphology was extended to graphs in Heijmans et al. (1992), where the proposed structuring graph enables a generalization of neighbourhood functions on a graph beyond the one induced by its set of edges. Morphological operators on graphs have been studied further in Cousty et al. (2009b), where the lattice of the subgraphs of a graph is considered in order to define filters that treat the graph as a whole.

Recently, several works, including Gilboa and Osher (2007, 2008), Drakopoulos and Maragos (2012), Sakaridis et al. (2017), Bertozzi and Flenner (2012), Merkurjev et al. (2013, 2015), van Gennip et al. (2014), Ta et al. (2011), Lézoray et al. (2012), Lozes et al. (2014), Desquesnes et al. (2013) and Elmoataz et al. (2015, 2017), have focused on the construction of PDE-based rather than algebraically defined morphological operators on graphs, which are then used as one of the basic ingredients to define active contour models as well as other PDE-based optimization schemes on graphs. All these works are based on the definition of a gradient operator on graphs. However, Gilboa and Osher (2007, 2008), Bertozzi and Flenner (2012), Merkurjev et al. (2013, 2015), van Gennip et al. (2014), Ta et al. (2011), Lézoray et al. (2012), Lozes et al. (2014), Desquesnes et al. (2013) and Elmoataz et al. (2015, 2017) work on weighted graphs and they all start from defining a *discrete* gradient at a vertex as a vector whose dimensionality is the same as the cardinality of the vertex's neighbourhood. This type of gradient is a difference operator, which *replaces* the original continuous gradient operator and enables the adaptation of continuous PDE schemes on graphs by defining analogous partial difference equations. Earlier works that used partial difference equations on graphs that mimic the variational formulation of boundary value problems include Chung and Berenstein (2005) and Bensoussan and Menaldi (2005). This class of approaches invariably sacrifices consistency of the resulting scheme with the original, continuous PDE formulation, as it removes the link between the discrete graph domain and the underlying continuum; even the initial continuous definitions of the operators in Gilboa and Osher (2008) are eventually substituted by discrete counterparts in order to apply the proposed optimization framework to graphs. The foundation of these approaches can be based on the theory of *discrete calculus* (Grady and Polimeni, 2010; Hirani, 2003), which provides a framework for reformulating continuous energies on graphs such that the corresponding solutions share similar properties (Bougleux et al., 2009; Couprie et al., 2011b, 2013; Elmoataz et al., 2008; Grady and Alvino, 2009; Lézoray and Grady, 2012). On the other hand, Drakopoulos and Maragos (2012) and Sakaridis et al. (2017) consider unweighted graphs and *approximate* the original, continuous gradient at each vertex. Their approach effectively constitutes a generalized numerical scheme for solving the original, continuous active contour PDE on graphs that have arbitrary spatial vertex configuration and hence greater complexity than the standard 2D image grid, as was described in Section 1. Thus, their approach contributes towards a *discretized calculus on arbitrary graphs*, in the spirit of Courant et al. (1967) who discretized classical PDEs of mathematical physics on regular grid graphs. They establish approximations of the gradient and curvature operators on graphs which are *asymptotically consistent* in the limit of large graphs, in order to ensure a stable *discretization* of level set-based active contour models on arbitrary 2D graphs, as has been previously achieved for the 2D image grid. The asymptotic upper bound that is established in Sakaridis et al. (2017) for the error of their geometric gradient approximation on random geometric graphs is the first of its kind.

A different class of approaches to graph segmentation which has gained a lot of interest in the image processing community is based on graph cuts. These approaches, in contrast to Sakaridis et al. (2017) and Kolotouros and Maragos (2017), usually operate on a regular image grid and define weighted edges between image pixels based on certain cues like spatial or appearance proximity, in order to find a cut of minimal cost for the resulting weighted graph. The cost of a cut is normalized in Shi and Malik (2000) so that balanced partitions are preferred. Approximate solutions to multilabel problems are proposed in Boykov et al. (2001), guaranteeing constant-factor optimality. A link between geodesic active contours and graph cuts is established in

Boykov and Kolmogorov (2003), where the graph is constructed so that the cost of the cut corresponds to the contour's length under the induced anisotropic metric, and this link is extended to the arbitrary graph setting in Drakopoulos and Maragos (2012). The random walk algorithm of Grady (2006) assigns unlabelled pixels to user-defined seeds interactively. Efficient algorithms for watershed-like segmentation that are formulated as graph cuts are introduced in Cousty et al. (2009a, 2010). The power watershed framework of Couprie et al. (2011a) unites and generalizes several graph-based optimization methods for image segmentation by expressing their energies in a common, parametric form.

The analysis of consistency of algorithms operating on point clouds or graphs which are constructed from point clouds has received a lot of interest, especially for machine learning tasks such as clustering. The focus of this line of research is on proving that optimizers of functionals which are defined on the discrete input converge to optimizers of the limiting functional which is defined on the underlying continuum as the number of points or graph vertices goes to infinity. The consistency of k-means is studied in Pollard (1981). More recently, von Luxburg et al. (2008), Belkin and Niyogi (2006) and Ting et al. (2010) examine spectral clustering and graph Laplacians in terms of consistency by analyzing the convergence of the respective eigenvalues and eigenvectors. The works of Arias-Castro et al. (2012), Maier et al. (2013) and García Trillos and Slepčev (2016) on the consistency of various graph-cut methods are more closely related to the analysis in Sakaridis et al. (2017): they define a spatial scale parameter, which controls the connectivity of the graph through the edge weights and depends on the number of vertices, and derive certain conditions on this parameter in order for the respective graph-cut algorithm to be consistent. This analysis is performed in the setting of pointwise convergence in Arias-Castro et al. (2012) and Maier et al. (2013), whereas García Trillos and Slepčev (2016) obtain results on Γ-convergence. In comparison, the analysis of consistency of the approximations in Sakaridis et al. (2017) also involves a spatial scale parameter, the radius of the graph, which is presented in Definition 1, with similar function as in the above works. In a slightly different setting, the work in van Gennip and Bertozzi (2012) proves Γ-convergence of the graph-based Ginzburg–Landau functional that was developed for binary clustering in Bertozzi and Flenner (2012) to the discrete anisotropic total variation that models the min-cut cost; in this case, the spatial scale parameter pertains to the scale of the diffuse interface that is induced by the Ginzburg–Landau functional and it also depends on the number of vertices. An important distinction between these works and Sakaridis et al. (2017) is that the latter proves consistency at the level of operators that are used in active contour models on graphs, not at the level of active contour algorithms themselves. Given that the aforementioned approaches do not cover the case of input which is presented in Section 1 and includes an image function I defined on the vertices of the graph, the study of the consistency of level set-based active contour algorithms on graphs constitutes an interesting topic for future research; theoretical results in Sakaridis et al. (2017) are a promising first step in this direction.

Finally, the approach in Sakaridis et al. (2017) bears some resemblance to unsupervised clustering of data that are represented as graphs embedded in Euclidean domains, where techniques based on nonnegative matrix factorization (Lee and Seung, 2001; Yang et al., 2012; Yuan and Oja, 2005) have been applied successfully. Nonetheless, the framework of Sakaridis et al. (2017) is not directly applicable to this setting, since the input is limited to the set of vertex locations $\mathcal{V}$ and it does not include an image function I defined on the vertices (cf. Section 1), which is essential for formulating active contour models. In the same sense, Sakaridis et al. (2017) cannot be extended to stochastic block models (Snijders and Nowicki, 1997) for graph segmentation, given that these models do not associate each vertex with a scalar value (which would model the image function) either.

3 Active contours on graphs via geometric approximations of gradient and curvature

3.1 Geometric gradient approximation on graphs

The first term of level set-based active contour evolution models to be approximated is the gradient of the bivariate embedding function. A general method is thus developed in Sakaridis et al. (2017) for calculating the gradient of a real-valued, bivariate function that is implicitly defined on a continuous domain, although its values are known only at a finite set of 2D points, which constitute the vertices of the graph. The context is the same as in the preceding work of Drakopoulos and Maragos (2012).

3.1.1 Formulation

Compared to the gradient approximations proposed in Drakopoulos and Maragos (2012), the geometric gradient approximation of Sakaridis et al. (2017) leverages the local spatial configuration of vertices in the formulation of the approximation. More specifically, it introduces the concept of the *neighbour angle*, i.e., the angle around a vertex which is "occupied" by each of its neighbours. The motivation for this approach comes from the following lemma in bivariate calculus.

Lemma 1. *The gradient of a differentiable function* $u : \mathbb{R}^2 \to \mathbb{R}$ *at point* $\mathbf{x}$ *is*

$$\nabla u(\boldsymbol{x}) = \frac{\displaystyle\int_0^{2\pi} D_\phi u(\boldsymbol{x})\, \boldsymbol{e}_\phi\, d\phi}{\pi}, \tag{3}$$

where $\mathbf{e}_\phi$ *is the unit vector in direction* ϕ *and* $D_\phi u(\mathbf{x})$ *is the directional derivative of* u *at* $\mathbf{x}$ *in this direction, defined by*

$$D_\phi u(\boldsymbol{x}) = \lim_{h\to 0} \frac{u(\boldsymbol{x}+h\boldsymbol{e}_\phi) - u(\boldsymbol{x})}{h}.$$

Based on Lemma 1, the goal is to approximate the gradient at a vertex of the graph by substituting the integral

$$\mathcal{I} = \int_0^{2\pi} D_\phi u(\mathbf{x})\, \mathbf{e}_\phi \, d\phi \tag{4}$$

with a sum over all the neighbours of the vertex. To this end, let us first introduce several key concepts.

The Euclidean distance between vertices v and w of a graph $\mathcal{G}$ is denoted by $d(v, w)$ and the unit vector in the direction of the edge vw starting at v is denoted by $\mathbf{e}_{vw}$. We define $\phi(w) \in [0, 2\pi)$ as the angle between the vector $\mathbf{e}_{vw}$ and the horizontal axis, as in Fig. 1. A vertex v will be alternatively denoted by $\mathbf{v}$ to declare its position vector. Moreover, we denote by $\mathcal{N}(v)$ the set of neighbours of v in $\mathcal{G}$, with cardinality $N(v)$. For the sake of brevity in notation, this cardinality will be written simply as N. We write $\mathcal{N}(v) = \{w_1, w_2, \ldots, w_N\}$ so that the angles $\phi(w_i)$ are in ascending order. Based on this ordering, we define the angle around v "occupied" by w_i, which is called *neighbour angle*, as

$$\Delta\phi(w_i) = \begin{cases} \dfrac{\phi(w_{i+1}) - (\phi(w_N) - 2\pi)}{2} & \text{if } i = 1, \\ \dfrac{\phi(w_1) + 2\pi - \phi(w_{i-1})}{2} & \text{if } i = N, \\ \dfrac{\phi(w_{i+1}) - \phi(w_{i-1})}{2} & \text{otherwise.} \end{cases} \tag{5}$$

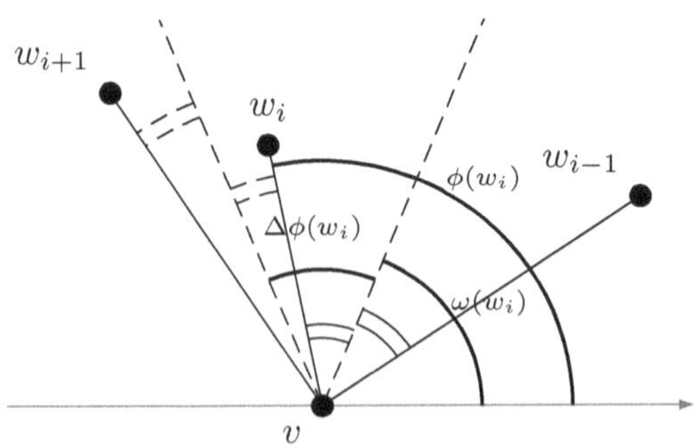

FIG. 1 Angles $\phi(w_i)$, $\Delta\phi(w_i)$ and $\omega(w_i)$. *From Sakaridis, C., Drakopoulos, K., Maragos, P. 2017. Theoretical analysis of active contours on graphs. SIAM J. Imaging Sci. 10 (3), 1475–1510.*

In a similar fashion, we define the angle corresponding to the bisector between two consecutive neighbours as

$$\omega(w_i) = \begin{cases} \dfrac{\phi(w_i) + \phi(w_N) - 2\pi}{2} & \text{if } i = 1, \\ \dfrac{\phi(w_i) + \phi(w_{i-1})}{2} & \text{otherwise.} \end{cases} \tag{6}$$

A visual representation of the neighbour angle is provided in Fig. 1. Using the above notation, the geometric gradient approximation at v is given by the following formula:

$$\nabla u(v) \approx \frac{\sum_{i=1}^{N} \dfrac{u(w_i) - u(v)}{d(v, w_i)} \mathbf{e}_{vw_i} \, \Delta\phi(w_i)}{\pi}. \tag{7}$$

The directional derivative term in (3) is approximated by the difference quotient of the function along each edge. On the other hand, the angle differential is handled through the neighbour angles, which effectively constitute a Voronoi tessellation of the circle around v, created from its neighbours. The reasoning behind this approach is to use information about the change of u along each particular direction that comes from the neighbour which is *closest* to this direction.

If the neighbour angles were not taken into account, we would place equal importance on all neighbours of v and return to an approximation similar to the weighted sum that was introduced in Drakopoulos and Maragos (2012):

$$\nabla u(v) \approx \frac{\sum_{i=1}^{N} \dfrac{u(w_i) - u(v)}{d(v, w_i)} \mathbf{e}_{vw_i}}{N}. \tag{8}$$

3.1.2 Convergence for random geometric graphs

In the remaining theoretical analysis of this section, we mainly focus on a certain type of graphs, namely *random geometric graphs* (Penrose, 2003).

Definition 1. A random geometric graph (RGG) $\mathcal{G}(n,\rho(n))$ is comprised of a set $\mathcal{V}$ of vertices and a set $\mathcal{E}$ of edges. The set $\mathcal{V}$ consists of n points distributed uniformly at random and independently in a bounded region $D \subset \mathbb{R}^2$. The set $\mathcal{E}$ of edges is defined through the radius $\rho(n)$ of the graph: an edge connects two vertices v and w if and only if their distance is at most $\rho(n)$, i.e., $d(v, w) \leq \rho(n)$.

An instance of an RGG is given in Fig. 2. For RGGs, the approximation of (7) converges in probability to the true value of the gradient as the number of vertices increases, under some conditions on the radius which constrain

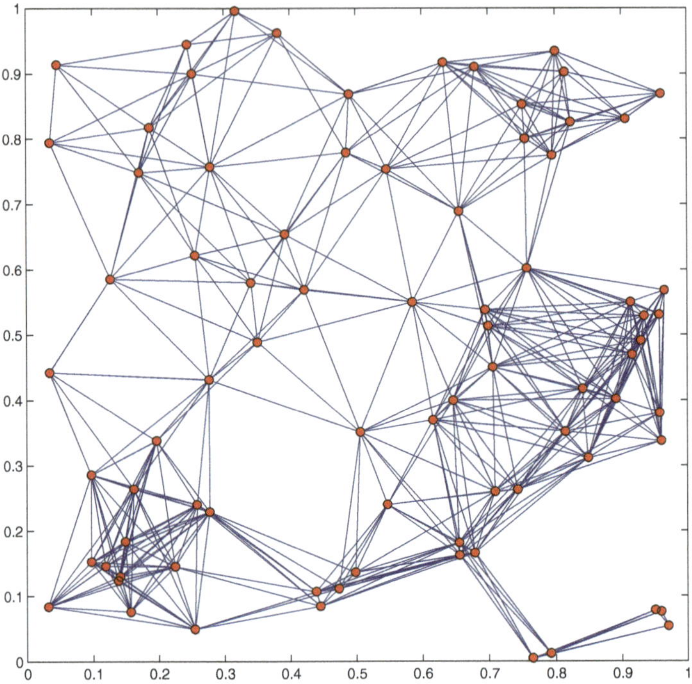

FIG. 2 A random geometric graph embedded in $D = [0, 1]^2$, with $n = 80$ vertices and radius $\rho = 0.25$. *From Sakaridis, C., Drakopoulos, K., Maragos, P. 2017. Theoretical analysis of active contours on graphs. SIAM J. Imaging Sci. 10 (3), 1475–1510.*

the density of the graph. Before stating the related theorem, we remind the reader of some definitions for the asymptotic notations which are used in the analysis.

Definition 2. Let f and g be two nonnegative functions. Then,

$$\begin{aligned}
f(n) \in O(g(n)) &\Leftrightarrow \exists k > 0\ \exists n_0\ \forall n \geq n_0 : f(n) \leq kg(n), \\
f(n) \in \Theta(g(n)) &\Leftrightarrow f(n) \in O(g(n)) \wedge g(n) \in O(f(n)), \\
f(n) \in o(g(n)) &\Leftrightarrow \forall k > 0\ \exists n_0\ \forall n \geq n_0 : f(n) < kg(n), \\
f(n) \in \omega(g(n)) &\Leftrightarrow \forall k > 0\ \exists n_0\ \forall n \geq n_0 : f(n) > kg(n).
\end{aligned}$$

Theorem 1. *Let $u : \mathbb{R}^2 \to \mathbb{R}$ be a differentiable function and $\mathcal{G}(n,\rho(n))$ an RGG embedded in $D = [0, 1]^2$, with $\rho(n) \in \omega\left(n^{-1/2}\right) \cap o(1)$. For every vertex v of $\mathcal{G}$, the gradient approximation of (7) converges in probability to $\nabla u(v)$.*

The proof of Theorem 1 is provided in Sakaridis et al. (2017).

3.1.3 Asymptotic analysis of approximation error

Note that Drakopoulos and Maragos (2012) also proved that two of their gradient approximations converge in probability to the true value of the gradient. Going one step further, Sakaridis et al. (2017) obtain an asymptotic bound on the *rate* of convergence to the true gradient as the size of the RGG grows large. Since the framework in RGGs is stochastic, their result involves the *expectation* of the approximation error.

Let us denote the error in approximating $\mathcal{I}$ with

$$\mathcal{S} = \sum_{i=1}^{N} \frac{u(w_i) - u(v)}{d(v, w_i)} \mathbf{e}_{vw_i} \Delta\phi(w_i) \tag{9}$$

by $\mathcal{E} = \mathcal{S} - \mathcal{I}$. Comparing the two expressions, we deduce that the approximation with $\mathcal{S}$ is threefold:

1. Directional derivatives along edges are approximated with difference quotients.
2. The approximate value for the directional derivative along each edge is used as a constant estimate for all the directions "falling into" the respective neighbour angle.
3. The unit vector in the direction of each edge is also used for all the directions corresponding to the respective neighbour angle.

The calculation of the error that is performed in the proof of the following theorem involves construction of intermediate expressions between $\mathcal{S}$ and $\mathcal{I}$, bounding the magnitudes of the resulting differences and combining these individual bounds using the triangle inequality.

Theorem 2. *Let $u: \mathbb{R}^2 \to \mathbb{R}$ be a differentiable function and $\mathcal{G}(n, \rho(n))$ an RGG embedded in $D = [0, 1]^2$, with $\rho(n) \in \omega(n^{-1/2}) \cap o(1)$. For the gradient approximation error at every vertex v of $\mathcal{G}$, it holds that*

$$\mathrm{E}[\|\mathcal{E}\|] \in O\left(\rho(n) + \frac{1}{n\rho^2(n)}\right). \tag{10}$$

The full proof of Theorem 2 is given in Sakaridis et al. (2017). The radius of the graph is effectively the factor that determines the strictness of the asymptotic bound. To provide better intuition, we study the case when $\rho(n) \in \Theta(n^{-a})$, $a \in (0, 1/2)$. Substituting in (10), we obtain

$$\mathrm{E}[\|\mathcal{E}\|] \in O(n^b),\ b = \begin{cases} -a & \text{if } a \in \left(0, \frac{1}{3}\right], \\ -1 + 2a & \text{if } a \in \left(\frac{1}{3}, \frac{1}{2}\right). \end{cases} \tag{11}$$

The strictest upper bound is $O(n^{-1/3})$, it is achieved for $a = 1/3$ and it constitutes a trade-off between minimizing the first error term, which calls for small radii, and the other two terms, which require more neighbours and consequently larger radii. On the other hand, when $a \notin (0, 1/2)$, the conditions of Theorems 1 and 2 for $\rho(n)$ are not met and convergence to the true value of the gradient is not guaranteed in general. For instance, for $a = 0$, we get $\rho(n) \in \Theta(1)$, which means that the radius does not approach 0 in the limit. In turn, this implies that the difference quotients are not guaranteed to converge to the respective directional derivatives, since the distance $d(v, w) \leq \rho(n)$ does not go to 0 in the limit. In addition, for $a = 1/2$, it holds that $\rho(n) \in \Theta(n^{-1/2})$ and hence $\rho^2(n) \in \Theta(1/n)$. As a result, the expected number of neighbours is $\mathrm{E}[N] = (n - 1)\pi\rho^2(n) \in \Theta(1)$. In other words, the sum $\mathcal{S}$ is finite in the limit, which forbids convergence to the integral $\mathcal{I}$.

In Theorems 1 and 2, the domain D of the RGG is assumed to be the unit square. It is, however, straightforward to generalize the results of both theorems to arbitrary rectangular regions, since their proofs in Sakaridis et al. (2017) only use this assumption for calculating the area of D. This generalization involves applying a uniform scaling to both coordinates by $1/\sqrt{|D|}$, so that the transformed region has unitary area. The values of u are also scaled by the same factor. These steps ensure that all terms in (7) remain unaffected by the transformation. The two theorems are then applicable to the transformed input. In order to transfer the results back to the original input, one just needs to scale the radius ρ by the constant factor $\sqrt{|D|}$, which leaves the asymptotic bounds in both theorems unaffected.

3.1.4 Practical application

Despite convergence of the geometric gradient approximation to the true value of the gradient in the case of RGGs, in practice there is a nonnegligible error for graphs with finite number of vertices. This error is propagated to the embedding function of the active contour after each update and may be accumulated after several iterations. To mitigate this, Sakaridis et al. (2017) apply smoothing filtering on the approximate gradient values at a local, neighbourhood level as an empirical means to eliminate potential outliers by taking into account the values at neighbouring vertices. This smoothing is also applicable to curvature, as we discuss in Section 3.2. The smoothing filter can be either an average or a median filter, receiving as input the set of function values at the vertex itself and all its neighbours. In the case of curvature this is straightforward, while for gradient, each of the two vector components are filtered separately. Application of smoothing filtering is statistically motivated for the case of smooth (differentiable) functions by showing that using the neighbours of a vertex to form an ensemble of estimators of the approximated quantity at that vertex reduces the variance of the estimation compared to the basic approximation while not changing the bias.

Experimental validation of the gradient and curvature approximations as well as the aforementioned smoothing filtering is performed by using closed-form functions defined on RGGs. In the experiments that follow in the rest of Sections 3 and 5, the radius of an RGG is chosen as $\rho(n) = 0.6n^{-1/3}$ unless otherwise mentioned, so as to achieve the strictest asymptotic bound for gradient approximation error according to the results of Section 3.1.3. Besides, in Sections 3.1.4 and 3.2.2, all RGGs are embedded in $[0, 1]^2$. For each graph, the function's gradient and the curvature of its level sets are approximated at each vertex and afterwards the results are filtered with an average or median filter. The analytical expressions of the function's gradient and curvature are then compared to the estimates. Performance is measured using a global, graph-level error metric which is called *relative error* and denoted by e_r. The error at each individual vertex of the graph is defined as the difference between the approximate value and the true analytical value, and the relative error is simply the ratio of the energy of the error signal to the energy of the true signal on the entire graph:

$$e_r = \frac{E_{\text{error}}}{E_{\text{analytical}}}. \tag{12}$$

The relative error of the geometric gradient approximation is evaluated for an isotropic Gaussian on RGGs whose size ranges from 1000 to 10,000 vertices. The analytical form of the Gaussian is

$$\exp\left\{-\left[(x-x_0)^2+(y-y_0)^2\right]/2\sigma^2\right\}, \tag{13}$$

with $\sigma = 0.25$ and $x_0 = y_0 = 0.5$. Fig. 3A shows average values of e_r over 10 different graphs for each size to reduce variance in the reported performance. Using either an average or a median filter reduces the relative error substantially irrespective of size. Based on this result, smoothing filtering is generally applied for gradient in practice when performing active contour evolution.

Another practical consideration about the geometric gradient approximation (7) is that its absolute error increases as the magnitude of the true gradient grows large, i.e., when the function exhibits abrupt variations. This does not pose a problem for the calculation of gradient direction (which is relevant as input for approximating the curvature), since the latter does not depend on the range of the function's variation around the examined vertex. Utilizing all incident edges in the weighted sum of (7) ensures that all available information in the neighbourhood of the vertex is used to estimate which direction the gradient points to, as emphasized in Drakopoulos and Maragos (2012). However, the estimated gradient *magnitude* with the geometric gradient approximation (7) is prone to greater error, as it depends on the range of the function's variation. The use of difference quotients in (7) accentuates this effect for dense graphs, where distances between neighbouring vertices that appear in the denominator of the quotients approach zero. Thus, the preferable approximation for gradient magnitude in practice is the maximum absolute

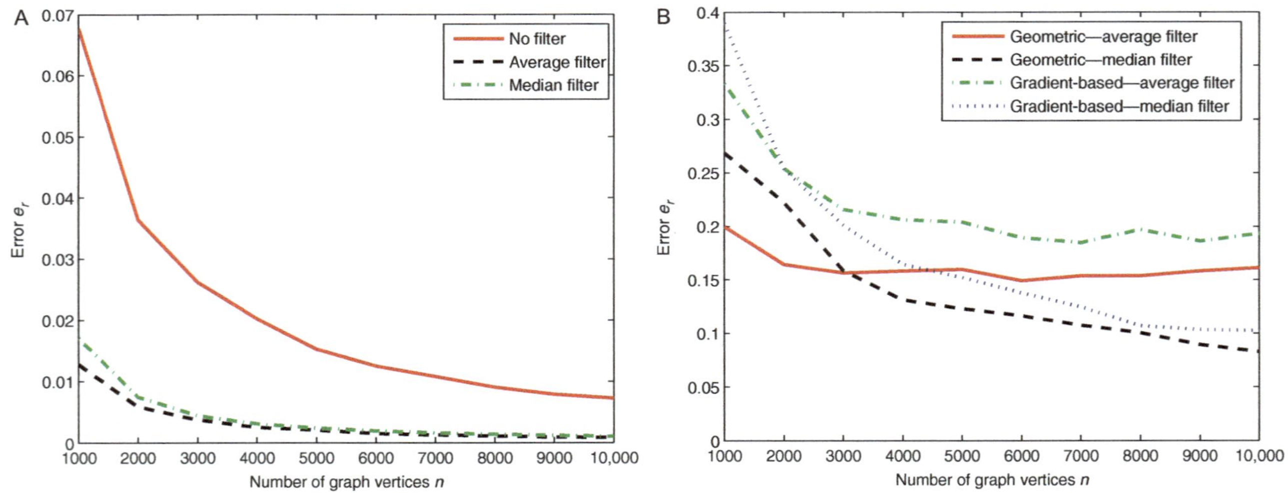

FIG. 3 (A) Relative error of gradient approximations for a Gaussian function defined on RGGs of increasing size. The geometric approximation with no filtering is compared to its filtered versions with an average or median filter. (B) Relative error of curvature approximations for a conic function defined on RGGs of increasing size. All four combinations of type of approximation (geometric or gradient-based) and smoothing filter (average or median) are compared. *Modified from Sakaridis, C., Drakopoulos, K., Maragos, P. 2017. Theoretical analysis of active contours on graphs. SIAM J. Imaging Sci. 10 (3), 1475–1510.*

difference of values of the function along edges that are incident on v, introduced in Drakopoulos and Maragos (2012):

$$\|\nabla u(v)\| \approx \max_{w \in \mathcal{N}(v)} \{|u(w) - u(v)|\}. \tag{14}$$

3.2 Geometric curvature approximation on graphs

After having devised an approximation scheme for the gradient of an embedding function, the next step is to build upon this scheme in order to estimate the curvature of the level sets of this function. The difference from the gradient case is that the input gradient values for curvature approximation are already approximate themselves, i.e., a *cascaded* approximation is attempted. Therefore, the error in curvature approximation on a graph is expected to accumulate compared to gradient approximation error on the same graph, since the estimated curvature at a vertex inherits the error of the estimated gradients at its neighbouring vertices.

3.2.1 Formulation

The geometric curvature approximation that is proposed in Sakaridis et al. (2017) is based on the expression of curvature as the divergence of the unit gradient field $\mathbf{F} = \nabla u / \|\nabla u\|$ of the embedding function u:

$$\kappa(v) = \text{div } \mathbf{F}(v), \; \nabla u(v) \neq \mathbf{0}. \tag{15}$$

The integral definition of divergence as

$$\text{div } \mathbf{F}(v) = \lim_{S \to \{\mathbf{v}\}} \frac{\oint\limits_{\Gamma(S)} \mathbf{F} \cdot \mathbf{n} \, d\ell}{|S|} \tag{16}$$

can then be used as a basis for approximating curvature, where S is a region with area $|S|$ and boundary $\Gamma(S)$ and $\mathbf{n}$ is the outward unit normal to this boundary. In Drakopoulos and Maragos (2012), the integral in (16) is approximated using a polygonal region to form a finite sum over the neighbours of the vertex (as shown in Drakopoulos and Maragos (2012, p. 8)). However, certain arrangements of the neighbours of the examined vertex can lead to regions with ill-defined area, boundary and normals when using the approach of Drakopoulos and Maragos (2012), which prevents the presentation of theoretical guarantees for the convergence of this approach.

To tackle these problems, the geometric curvature approximation of Sakaridis et al. (2017) employs the *neighbour angles* that were introduced in Section 3.1, in order to define the region S in (16) in a more compact and principled fashion. Let $S(v, \rho, \theta_0, \theta)$ be a circular sector centred at v, with radius ρ, occupying an angle $\theta > 0$ and whose rightmost radius is in the direction θ_0. For vertex v, $S(v)$ is formed as a union of circular sectors, each of

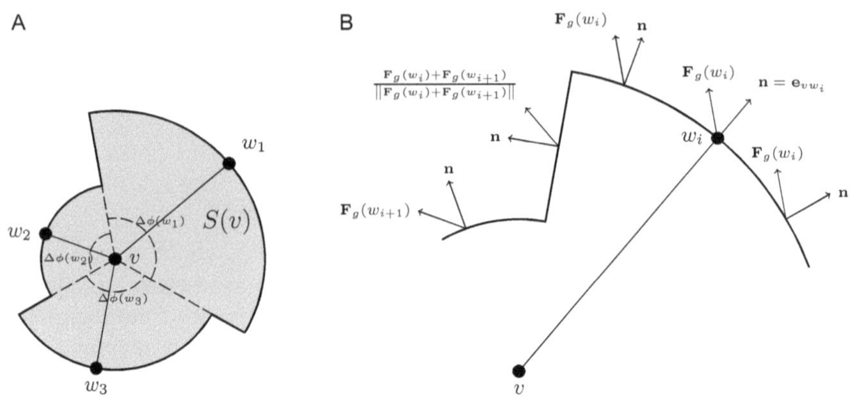

FIG. 4 (A) Definition of region $S(v)$ through the neighbour angles. (B) The profile of values of $\mathbf{F}_g$ at a part of the boundary of $S(v)$ which corresponds to neighbour w_i of v. *Modified from Sakaridis, C., Drakopoulos, K., Maragos, P. 2017. Theoretical analysis of active contours on graphs. SIAM J. Imaging Sci. 10 (3), 1475–1510.*

them corresponding to a neighbour of v, as we show in Fig. 4A. More formally, for each neighbour w of v, the respective circular sector is $S(v, d(v, w), \omega(w), \Delta\phi(w))$. The area of $S(v)$ can then be expressed as

$$|S(v)| = \sum_{i=1}^{N} \frac{\Delta\phi(w_i)}{2} d^2(v, w_i). \tag{17}$$

The challenge imposed by this construction of $S(v)$ is the choice of suitable values for $\mathbf{F}$ along the boundary of this region, given only its values at the locations of neighbours of v. The resulting boundary consists of arcs, each of which contains a neighbour of v, and line segments which connect these arcs. The proposed approximation fixes the value of $\mathbf{F}$ along each arc at the geometric gradient approximation computed for the corresponding neighbour w using (7), $\mathbf{F}_g(w)$. Moreover, for every line segment, the normalized mean of the approximate values of $\mathbf{F}$ along the two neighbouring arcs is used. The idea is again to use information from the closest vertex, which should be more reliable. We visualize the described configuration in Fig. 4B.

Using the above approximations, the line integral in (16) is substituted with a sum of simple line integrals over single arcs and line segments, which have closed analytical forms. If we denote the integral over the arc $C_a(w_i)$ containing neighbour w_i by $I_a(w_i)$ and the integral over the line segment $C_l(w_i)$ that connects the arcs $C_a(w_i)$ and $C_a(w_{i+1})$ by $I_l(w_i)$, we obtain

$$I_a(w_i) = d(v, w_i)\, \mathbf{F}_g(w_i) \cdot (\sin(\omega(w_{i+1})) - \sin(\omega(w_i)), \cos(\omega(w_i)) - \cos(\omega(w_{i+1}))) \tag{18}$$

and

$$I_l(w_i) = (d(v,w_{i+1}) - d(v,w_i))\,\frac{\mathbf{F}_g(w_i) + \mathbf{F}_g(w_{i+1})}{\|\mathbf{F}_g(w_i) + \mathbf{F}_g(w_{i+1})\|} \cdot (\sin(\omega(w_{i+1})),\ -\cos(\omega(w_{i+1}))). \tag{19}$$

The geometric approximation of curvature proposed in Sakaridis et al. (2017) is given by

$$\kappa(v) \approx \frac{\sum_{i=1}^{N} I_a(w_i) + I_l(w_i)}{|S(v)|}. \tag{20}$$

While there was no proof of convergence for the curvature approximation of Drakopoulos and Maragos (2012), the geometric curvature approximation of Sakaridis et al. (2017) is proved to converge in probability for RGGs, as in the geometric gradient approximation case, although the conditions are stronger for curvature. The full proof of the following Theorem 3 is given in the Supplementary Material of Sakaridis et al. (2017).

Theorem 3. *Let $\mathcal{G}(n,\rho(n))$ be an RGG embedded in $D = [0,\,1]^2$, with $\rho(n) \in \omega(n^{-1/2}) \cap o(1)$ and v a vertex of $\mathcal{G}$. If $u : \mathbb{R}^2 \to \mathbb{R}$ is continuously differentiable and $\nabla u(v) \neq \mathbf{0}$, then the curvature approximation of (20) converges in probability to $\kappa(v)$.*

An alternative, gradient-based curvature approximation is additionally presented in Sakaridis et al. (2017) by leveraging the differential definition of divergence instead of the integral definition of (16). This gradient-based curvature approximation is also proved to converge in probability for RGGs, under slightly stricter conditions than the geometric curvature approximation.

3.2.2 Practical application

The neighbourhood-based smoothing filtering that has been introduced in Section 3.1.4 is also applicable to the output of curvature approximations. In this case, the same type of filter is applied both to the gradient estimates before using them for computing curvature and to the final curvature estimates. In Fig. 3B, we present the results of an experiment similar to that in Fig. 3A, this time focusing on the curvature of a function that corresponds to an elliptical cone and comparing average versus median filtering as well as the geometric versus the gradient-based curvature approximation. The form of this function is

$$\sqrt{\frac{(x - x_0)^2}{\alpha^2} + \frac{(y - y_0)^2}{\beta^2}}, \tag{21}$$

where $\alpha = 0.4$, $\beta = 0.3$, $x_0 = -0.25$ and $y_0 = 0.5$. Median filtering is superior: the median-filtered approximations exhibit lower error than the corresponding average-filtered ones over almost the entire range of graph sizes (except for the smallest sizes). More importantly, the relative error of median-filtered curvature is strictly decreasing for increasing graph size both with the geometric and the gradient-based approximation, in contrast to the average-filtered cases, where the error stops decreasing around 4000 vertices. Due to these facts, median filtering is preferred for smoothing gradient and curvature. In addition, the geometric approximation induces a consistently smaller error than the gradient-based approximation.

3.3 Gaussian smoothing on graphs

Apart from the approximations of gradient and curvature, an additional ingredient which is required in particular for the GAC model (1) to operate on graphs is the edge-dependent stopping function g which helps attract the active contour to salient boundaries. The function $g(\|\nabla I_\sigma\|):\mathbb{R}_+ \to [0,1]$ is typically a decreasing function of the gradient magnitude of a *smoothed* version I_σ of the original image function I, where smoothing acts as a regularization to limit the effect of small local variations on the evolution of the contour. For this smoothing, Drakopoulos and Maragos (2012) propose a simple graph-based isotropic Gaussian filter, which is improved in Sakaridis et al. (2017) with two alternative formulations that both include normalization to account for nonuniform spatial vertex configurations in the arbitrary graph setting.

The isotropic 2D Gaussian filter with standard deviation σ is defined as

$$G_\sigma(\mathbf{x}) = \frac{1}{2\pi\sigma^2}\exp\left(-\frac{\|\mathbf{x}\|^2}{2\sigma^2}\right). \tag{22}$$

Smoothing is performed in Drakopoulos and Maragos (2012) via a simple graph-based convolution of (22) with the image function:

$$I_\sigma(\mathbf{v}) = \sum_{\mathbf{w}\in\mathcal{V}} I(\mathbf{w})G_\sigma(\mathbf{v}-\mathbf{w}). \tag{23}$$

The stopping function g is then calculated through the following formula:

$$g(\|\nabla I_\sigma\|) = \frac{1}{1+\frac{\|\nabla I_\sigma\|^2}{\lambda^2}}. \tag{24}$$

However, the arbitrary graph setting introduces nonuniformities: in some parts of the graph, the vertices might be distributed more densely than in other parts. This implies that (23) will operate counter-intuitively, introducing variations to the smoothed image in regions of the graph where the original image function is constant. To demonstrate this behaviour, we use a simple binary image of a disk, shown in Fig. 5A. The result of applying (23) on this image is shown in Fig. 5B. Not only has the range of image values changed,

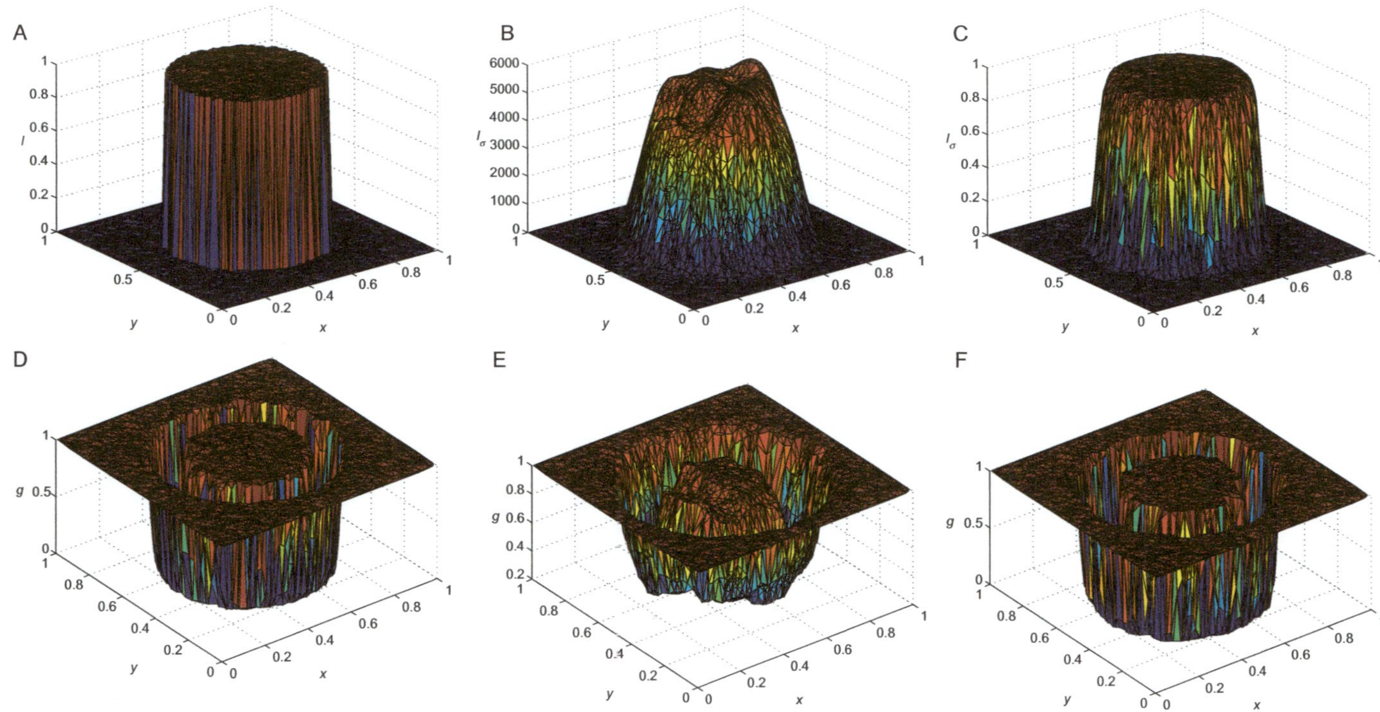

FIG. 5 Comparison of methods for Gaussian smoothing and computation of stopping function g. The original image on the graph (a disk) is shown in (A). (B) and (C) contain smoothed versions I_σ of the original image using simple Gaussian filtering (Drakopoulos and Maragos 2012) and normalized Gaussian filtering (Sakaridis et al. 2017) respectively. (D), (E) and (F) contain results for g using the three compared methods, i.e., Gaussian derivative filtering with separate normalization (Sakaridis et al. 2017), simple Gaussian filtering (Drakopoulos and Maragos 2012), and normalized Gaussian filtering (Sakaridis et al. 2017) respectively. The results in (C), (D) and (F) are obtained using $\sigma = 0.02$ and $\lambda = 0.05$, while those in (B) and (E) are obtained using $\sigma = 0.05$ and $\lambda = 1000$. The approximation of (14) is used to compute the gradient magnitude for simple Gaussian filtering and normalized Gaussian filtering. For all three methods, $\|\nabla I_\sigma\|$ is filtered with a median filter before feeding it to (24) for computing g. *Modified from Sakaridis, C., Drakopoulos, K., Maragos, P. 2017. Theoretical analysis of active contours on graphs. SIAM J. Imaging Sci. 10 (3), 1475–1510.*

but the interior of the original disk also exhibits significant variations in image values. This shortcoming is propagated to $\|\nabla I_\sigma\|$ and g, as shown in Fig. 5E. There is a deviation of g from the ideal value of 1 in the interior of the disk and a variation in its values as well, which means that the gradient of g is not equal to $\mathbf{0}$ in the interior of the disk as it should.

This issue is tackled in Sakaridis et al. (2017) by including a normalization term in (23) to address nonuniformities:

$$I_\sigma(\mathbf{v}) = \frac{\sum_{\mathbf{w}\in\mathcal{V}} I(\mathbf{w}) G_\sigma(\mathbf{v}-\mathbf{w})}{\sum_{\mathbf{w}\in\mathcal{V}} G_\sigma(\mathbf{v}-\mathbf{w})}. \tag{25}$$

This method is termed normalized Gaussian filtering and its result for the examined disk image is shown in Fig. 5C. The smoothed image is now similar to the respective output of simple Gaussian filtering in the usual image processing setting with a regular grid. As a result, the profile of the corresponding g function (shown in Fig. 5F) meets expectations.

An alternative formulation of Gaussian smoothing proposed in Sakaridis et al. (2017) is based on the fact that the input of the stopping function is the gradient of the smoothed image rather than the smoothed image itself. Since the derivatives of the Gaussian filter have closed analytical forms, it is possible to exchange the convolution with the gradient operator and convolve the image directly with Gaussian derivatives in order to obtain the gradient of I_σ without performing numerical approximations. In this case, normalization is not straightforward as in normalized Gaussian filtering: Gaussian derivatives assume both positive and negative values. This difficulty is circumvented by splitting the vertices into two sets, according to the sign of the Gaussian derivative with respect to the processed vertex, and performing separate normalization for each of these sets. This separation can be easily expressed in terms of the vertices' coordinates. If we denote $\mathbf{v} = (v_1, v_2)$, then Gaussian derivative filtering with separate normalization is defined as

$$\nabla I_\sigma(\mathbf{v}) = \left(\frac{\sum\limits_{\substack{\mathbf{w}\in\mathcal{V}:\\ w_1\geq v_1}} I(\mathbf{w}) \frac{\partial G_\sigma(\mathbf{v}-\mathbf{w})}{\partial x}}{\sum\limits_{\substack{\mathbf{w}\in\mathcal{V}:\\ w_1\geq v_1}} \frac{\partial G_\sigma(\mathbf{v}-\mathbf{w})}{\partial x}} + \frac{\sum\limits_{\substack{\mathbf{w}\in\mathcal{V}:\\ w_1< v_1}} I(\mathbf{w}) \frac{\partial G_\sigma(\mathbf{v}-\mathbf{w})}{\partial x}}{-\sum\limits_{\substack{\mathbf{w}\in\mathcal{V}:\\ w_1< v_1}} \frac{\partial G_\sigma(\mathbf{v}-\mathbf{w})}{\partial x}}, \frac{\sum\limits_{\substack{\mathbf{w}\in\mathcal{V}:\\ w_2\geq v_2}} I(\mathbf{w}) \frac{\partial G_\sigma(\mathbf{v}-\mathbf{w})}{\partial y}}{\sum\limits_{\substack{\mathbf{w}\in\mathcal{V}:\\ w_2\geq v_2}} \frac{\partial G_\sigma(\mathbf{v}-\mathbf{w})}{\partial y}} + \frac{\sum\limits_{\substack{\mathbf{w}\in\mathcal{V}:\\ w_2< v_2}} I(\mathbf{w}) \frac{\partial G_\sigma(\mathbf{v}-\mathbf{w})}{\partial y}}{-\sum\limits_{\substack{\mathbf{w}\in\mathcal{V}:\\ w_2< v_2}} \frac{\partial G_\sigma(\mathbf{v}-\mathbf{w})}{\partial y}} \right). \tag{26}$$

The result of Gaussian derivative filtering with separate normalization on the examined disk image in Fig. 5D is at least as satisfactory as in the normalized Gaussian filtering case of Fig. 5F and clearly superior to simple Gaussian filtering of Drakopoulos and Maragos (2012).

4 Active contours on graphs using a finite element framework

4.1 Problem formulation and numerical approximation

In this section we will discuss a method for solving active contour evolution equations on graphs using the finite element method as presented by Kolotouros and Maragos (2017). Finite element analysis is a powerful framework that enables us to solve complex PDEs in a simple and elegant manner. The main difference between finite difference and finite element methods is that the former try to approximate the differential operators by discretizing them whereas the latter approximate the solution with functions belonging to finite dimensional function spaces. This can be useful, especially in the case of small graphs where the accurate approximation of differential operators is a challenging problem. Additionally, compared to finite differences, the finite element method is more suitable for functions defined on irregular domains or for modelling discontinuities. There is a number of works that attempt to apply the finite element method for solving level set equations. The work in Mourad et al. (2005) presented a finite element solution to the level set equation by enforcing that the level set function remains a signed distance function throughout the evolution. A solution to the problem that exploits the partition-of-unity property of the finite element shape functions was proposed in Valance et al. (2008). A generic finite element algorithm approach for solving the Hamilton–Jacobi and front propagation equation was developed in Barth and Sethian (1998).

In this section we will consider general active contour models that can be modelled with level set equations of the form

$$\begin{aligned} F(u)\frac{\partial u}{\partial t} &= \operatorname{div}(G(u)\nabla u) + H(u) \\ \nabla u \cdot \mathbf{n} &= 0 \text{ on the boundary } \partial\Omega \\ u(x,y,0) &= \text{dist}^*(x,y), \end{aligned} \tag{27}$$

where F, G and H are functionals of u and dist*(x, y) is the signed distance function from the initial curve. Popular active contour models that can be expressed in the above form are

- *Erosion/Dilation*:

$$F(u) = \frac{1}{\|\nabla u\|}, \quad G(u) = 0, \quad H(u) = c, \tag{28}$$

- *Geometric Active Contours*:

$$F(u) = \frac{1}{g\|\nabla u\|}, \quad G(u) = \frac{1}{\|\nabla u\|}, \quad H(u) = 0, \tag{29}$$

- *Geodesic Active Contours*:

$$F(u) = \frac{1}{\|\nabla u\|}, \quad G(u) = \frac{g}{\|\nabla u\|}, \quad H(u) = \beta, \tag{30}$$

where $g(x, y)$ is the edge stopping function defined in (24).

The core of the Finite Element method is that it converts PDEs to integral equations that can then be solved using function approximations. It can be verified that using some simple manipulations, (27) can be converted to an equivalent integral equation

$$\iint_{\Omega} F(u)\frac{\partial u}{\partial t}\phi \, dxdy = -\iint_{\Omega} G(u)\nabla u \cdot \nabla \phi \, dxdy + \iint_{\Omega} H(u)\phi \, dxdy \tag{31}$$

The above form is known as the weak form and it must hold for all functions $\phi \in H^1(\Omega)$. $H^1(\Omega)$ is the Sobolev space consisting of all functions defined in Ω whose first order derivatives—in the distributional sense—belong to $L^2(\Omega)$.

The previous steps did not involve any numerical approximation. The original problem was converted to an integral form that is expected to be equivalent to the original problem. Next, the solution will be approximated using the Galerkin method. Another main idea of the finite element analysis is that it does not discretize continuous differential operators using finite differences but instead approximates the solution of the equations with functions belonging to finite dimensional subspaces of $H^1(\Omega)$. Let V_n be a n-dimensional subspace of $H^1(\Omega)$ and $\{\phi_i\}_{i=1}^n$ a basis of V_n. The approximate solution $\bar{u}$ can be written as a linear combination of the basis functions and since the problem is time-dependent, we allow the coefficients of the linear combination to be functions of time. Thus we have

$$\bar{u}(x,y,t) = \sum_{i=1}^{n} c_i(t)\phi_i(x,y). \tag{32}$$

We demand that (31) holds at least for all the functions that belong to the subspace V_n. Since (31) is a linear functional in ϕ and $\{\phi_i\}_{i=1}^n$ form a basis of V_n, this is equivalent to demanding that it holds for each of the basis functions ϕ_i. If we use the fact that

$$\frac{\partial \bar{u}}{\partial t}(x,y,t) = \sum_{i=1}^{n} \dot{c}_i(t)\phi_i(x,y), \tag{33}$$

and substitute in (31) we get

$$\sum_{i=1}^{n} \dot{c}_i \iint_{\Omega} F(\bar{u})\phi_i\phi_j \, dxdy = -\sum_{i=1}^{n} \iint_{\Omega} G(\bar{u})\nabla\phi_i \cdot \nabla\phi_j \, dxdy + \iint_{\Omega} H(\bar{u})\phi_j, \; j = 1,\ldots n. \tag{34}$$

This is a nonlinear system of ordinary differential equations (ODEs) that can be written in the form

$$\mathbf{A}(\mathbf{c})\dot{\mathbf{c}} = \mathbf{b}(\mathbf{c}), \tag{35}$$

where $\mathbf{A} = \{A_{ij}\}$is a $n \times n$ matrix that depends on $\mathbf{c}$ and $\mathbf{b} = \{b_i\}_1^n$, $\mathbf{c} = \{c_i\}_1^n$ are n-dimensional vectors with

$$\mathrm{A}_{ij}(\mathbf{c}) = \mathrm{A}_{ji}(\mathbf{c}) = \iint_{\Omega} F\left(\sum_{k=1}^{n} c_k \phi_k\right) \phi_i \phi_j \, dxdy, \tag{36}$$

and

$$b_i(\mathbf{c}) = -\sum_{j=1}^{n} \iint_{\Omega} G\left(\sum_{k=1}^{n} c_k \phi_k\right) \nabla\phi_i \cdot \nabla\phi_j \, dxdy + \iint_{\Omega} H\left(\sum_{k=1}^{n} c_k \phi_k\right) \phi_i \, dxdy. \tag{37}$$

The initial condition for the above system of differential equations can be obtained from the projection $\bar{u}(x,y,0)$ of $u(x, y, 0)$ onto the subspace V_n, i.e.,

$$\bar{u}(x,y,0) = \sum_{i=1}^{n} \langle u(x,y,0), \phi_i \rangle \phi_i, \tag{38}$$

and thus

$$c_i(0) = \langle u(x,y,0), \phi_i \rangle, \tag{39}$$

where $\langle \cdot,\cdot \rangle$is the usual inner product defined on $H^1(\Omega)$ with

$$\langle f,g \rangle = \iint_{\Omega} fg \, dxdy + \iint_{\Omega} \nabla f \cdot \nabla g \, dxdy. \tag{40}$$

Consider a graph $\mathcal{G}(\mathcal{V},\mathcal{E})$. We assume that the graph has a total of n vertices and each vertex v_i is a point in Ω and can be described by its respective planar coordinates (x_i, y_i). Note that every finite set of points in the plane can be mapped into the unit square by applying a translation followed by a scaling. Thus if we apply an appropriate geometrical transformation we can map the set of vertices to a subset of Ω. For each vertex $v_i \in \mathcal{V}$ we choose a function ϕ_i that belongs to $H^1(\Omega)$ with the following properties:

1. $\{\phi_i\}_{i=1}^{n}$ are linearly independent, and
2. for any nonadjacent vertices v_i, v_j, $\mathrm{supp}(\phi_i) \cap \mathrm{supp}(\phi_j) = \emptyset$.

We can see that the functions $\{\phi_i\}_{i=1}^{n}$ form a basis of some subspace S of $H^1(\Omega)$. Also, for reasonably smooth functions ϕ_i, $\mathrm{supp}(\nabla\phi_i) \backslash \mathrm{supp}(\phi_i)$ is a null set and thus $\mathrm{supp}(\nabla\phi_i) \cap \mathrm{supp}(\nabla\phi_j)$ is also a null set. Thus we can conclude that if v_i and v_j are two nonadjacent vertices of $\mathcal{G}$, then $\langle \phi_i, \phi_v \rangle = 0$ and thus ϕ_i, ϕ_j are orthogonal.

Proposition 1. *For any graph $\mathcal{G}(\mathcal{V},\mathcal{E})$ with $v_i = (x_i, y_i) \in \Omega$ there is at least one set of functions $\{\phi_i\}_{i=1}^{n}$ with the properties 1 and 2.*

For a proof of this fact we refer the reader to Kolotouros and Maragos (2017).

Next we will describe how we can obtain the solution of (35) on a graph $\mathcal{G}(\mathcal{V},\mathcal{E})$ equipped with the functions $\{\phi_i\}_{i=1}^{n}$. First, from (36) we can see that $\mathrm{A}_{ij}(\mathbf{c}) = 0$ if v_i is not a neighbour of v_j. If the number of edges for each vertex is small compared to the total number of vertices then $\mathbf{A}(\mathbf{c})$ is sparse. Additionally the summation in (37) reduces to a summation in the neighbourhood of v_i and thus

$$b_i(\mathbf{c}) = -\sum_{v_j \in N_i \cup \{v_i\}} \iint_{\Omega} G\left(\sum_{k=1}^{n} c_k \phi_k\right) \nabla \phi_i \cdot \nabla \phi_j \, dxdy + \iint_{\Omega} H\left(\sum_{k=1}^{n} c_k \phi_k\right) \phi_i \, dxdy, \tag{41}$$

where N_i is the neighbourhood of vertex v_i.

For the rest of the analysis the class of graphs will be restricted to the family of Delaunay graphs and more specifically the case of Delaunay graphs that are constructed from the Delaunay triangulation of a finite set of points in the unit square. Delaunay triangulation is a method of dividing the convex hull of a set points into triangles that tends to avoid creating sharp triangles, a property that ensures good convergence results for the solution of PDEs. For more details about the Delaunay triangulation and the algorithms used to produce it, we refer the reader to Persson and Strang (2004). Images can be thought of as a special case of Delaunay graphs with the vertices being the pixels of the image. In the case of graphs, the image function I is defined as in Section 1, i.e., only on the vertices of the graph. For the case of the image gradient ∇I, however, since we will be using numerical integration, it is more convenient to define it as a 2D function defined on Ω. This function is piecewise constant in each triangle of the triangulation and its value is the gradient of the plane that is formed by the values of I in the three vertices of the triangle. Similarly, $g = g(\|\nabla I\|)$ will also be constant in each triangle of the triangulation.

Let $\mathcal{G}(\mathcal{V}, \mathcal{E})$ be a Delaunay graph and $\{T_i\}_{i=1}^{m}$ the triangles of the Delaunay triangulation. Kolotouros and Maragos (2017) approximate the solution with a function belonging to the space of continuous functions on Ω which are linear in each triangle of the triangulation. We will denote this space as S_{lin}. More formally

$$S_{lin} = \{f \in C(\Omega) : f|_{T_i} = a_i x + b_i y + c_i, \text{ for } i = 1, \ldots, m\}. \tag{42}$$

This choice of subspace provides good convergence properties and also simplifies calculations as we shall see later on. After the selection of the subspace we have to choose a basis of S_{lin}. A possible choice satisfying the aforementioned properties are the pyramid functions. For each vertex v_i of the graph we define the function ϕ_i as

- $\phi_i(v_i) = 1$.
- $\phi_i(v_j) = 0$, for $j \neq i$.
- ϕ_i is linear in each triangle.

It is easy to verify that $\text{supp}(\phi_i) \cap \text{supp}(\phi_j) = \emptyset$ if v_i is not a neighbour of v_j and $\phi_i \in S_{lin}$. It can also be proven that $\{\phi_i\}_{i=1}^{n}$ are linearly independent

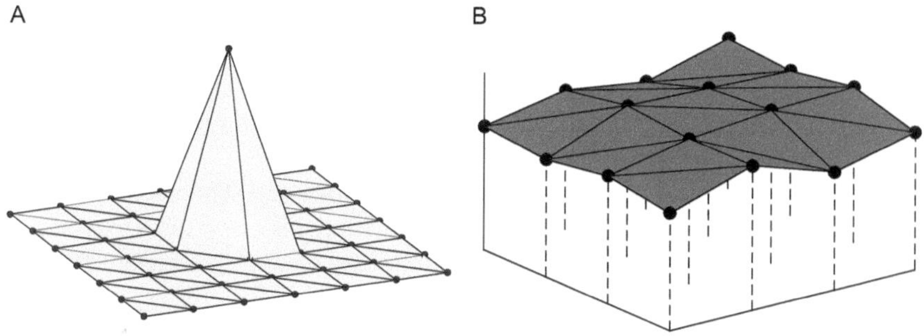

FIG. 6 (A) Pyramid function ϕ_i centred in node v_i for a rectangular grid. (B) Interpolation of discrete data in a 4 × 4 rectangular grid using the ϕ_i functions. With *solid circles* we represent the original discrete data.

and thus form a basis of S_{lin}. In Fig. 6A we can see the form of the ϕ_i functions for the case of a rectangular grid. With this choice we can see that ϕ_i overlaps only with the six other ϕ_j that correspond to the vertices v_j that are adjacent to v_i. Thus each row of **A** contains at most seven nonzero elements and at the same time only seven different coefficients c_i appear in these expressions. We can think of ϕ_i as functions that interpolate discrete data in a rectangular grid. Assume for example that we have samples of a 2D function f at n points (x_i, y_i) on the plane. By creating the Delaunay graph that corresponds to the above set of points we can construct a continuous function $\bar{f}$ that approximates f as

$$\bar{f}(x,y) = \sum_{i=1}^{N} f(x_i,y_i)\phi_i(x,y). \tag{43}$$

From the above construction and the properties of ϕ_i, it is easy to verify that $\bar{f}(x_i,y_i) = f(x_i,y_i)$ and in each triangle of the triangulation it essentially performs a form of linear interpolation. In Fig. 6B we can see an example of an interpolation of discrete data in a rectangular grid.

The final remaining step is the discretization in time. For a system of ODEs we have several options for the approximation of derivatives, each one with its advantages and disadvantages and the choice of a particular method depends on the specific properties of each problem. We want to calculate the solution in a subset of the time interval $[0, +\infty)$ starting from the initial condition until we reach convergence. Let t_k, $k \geq 0$ be the sequence of time points at which we calculate the solution with $t_0 = 0$ and $t_k = k\Delta t$. Here we assume that we use a fixed time step Δt. Also with $\mathbf{c}_k$ we will denote the approximation of $\mathbf{c}(t_k)$. Kolotouros and Maragos (2017) used the explicit Euler method which approximates the time derivative with the forward difference

$$\dot{\mathbf{c}}(t_k) = \frac{\mathbf{c}_{k+1} - \mathbf{c}_k}{\Delta t}. \tag{44}$$

If we substitute this into (35) we obtain

$$\mathbf{A}(\mathbf{c}_k)\mathbf{c}_{k+1} = \mathbf{A}(\mathbf{c}_k)\mathbf{c}_k + \Delta t \cdot (\mathbf{c}_k). \tag{45}$$

This reduces to solving a linear system for each time step. This method is easy to implement but puts limits on the choice of time step Δt. It is necessary to choose a very small time step—often in the order of $1/n^2$—to ensure that the curve evolution is stable.

Kolotouros and Maragos (2017) also provide an analysis of the complexity of the presented algorithm. For the sake of simplicity let us consider the case of the Delaunay graphs that correspond to images, with the graph structure depicted in Fig. 6A. We assume that we have a $h \times w$ image. The resulting graph will have a total of $n = hw$ vertices. Since each row of $\mathbf{A}$ contains at most seven nonzero elements, $\mathbf{A}(\mathbf{c}_k)$ can be computed in linear time in each time step. Also due to its sparsity, matrix multiplication can be performed in linear time. Observing that $\mathbf{b}(\mathbf{c}_k)$ can also be computed in linear time, the right-hand side needs $O(n)$ time in each step. With an appropriate node labelling, $\mathbf{A}(\mathbf{c}_k)$ takes the form of a band matrix with a band length of $\min(h,w) = O(\sqrt{n})$. For a $n \times n$ band matrix with band length d the solution of a linear system needs $O(nd^2)$ operations. So $O(n^2)$ operations are required in each time step, which is prohibitive for large images. As far as the time evolution is concerned, the total number of steps until convergence cannot be specified a priori, since it depends on the shape and position of the initial curve. If the initial curve is close to the object boundaries, only a few time steps are needed until convergence.

4.2 Locally constrained contour evolution

In this part we present an extension of the previous framework to solve curve evolution models of the form

$$\begin{aligned} F(u)\frac{\partial u}{\partial t} &= \delta_\epsilon(u)(\operatorname{div}(G(u)\nabla u) + H(u)) \\ \nabla u \cdot \mathbf{n} &= 0 \text{ on the boundary } \partial\Omega \\ u(x,y,0) &= \text{dist}^*(x,y). \end{aligned} \tag{46}$$

where $\delta_\epsilon(x)$ is an approximation of the Dirac δ function, with $\delta_\epsilon \to \delta$, as $\epsilon \to 0$. A typical choice for δ_ϵ is the piecewise constant approximation

$$\delta_\epsilon(x) = \begin{cases} 1/\epsilon, & |x| \le \epsilon, \\ 0, & |x| > \epsilon \end{cases}. \tag{47}$$

The term δ_ϵ constrains the curve evolution in a small area near the current position of the curve, i.e., the 0-level set. One popular active contour model that can be represented using (46) is the ACWE model of Chan and Vese (2001).

The first part of the discussion will involve showing how (46) can be approximated on graphs. Define the sets

$$\mathcal{U}_t^+ = \{(x,y) \in \mathcal{V} : u(x,y,t) > 0\}, \tag{48}$$

$$\mathcal{U}_t^- = \{(x,y) \in \mathcal{V} : u(x,y,t) < 0\}, \tag{49}$$

$$\mathcal{U}_t^0 = \{(x,y) \in \mathcal{V} : u(x,y,t) = 0\}. \tag{50}$$

$\mathcal{U}_t^+, \mathcal{U}_t^-$ and $\mathcal{U}_t^0$ are the set of points that are outside, inside and on the curve at time t. Also we define the set of *active points* at time t as

$$\mathcal{U}_t^* = \partial\mathcal{U}_t^+ \cup \partial\mathcal{U}_t^- \cup \mathcal{U}_t^0, \tag{51}$$

where ∂S denotes the boundary of the set $S \subseteq \mathcal{V}$ and is defined by

$$\partial S = \{v \in S : \exists v' \in \mathcal{V} \backslash S \text{ s.t. } v' \sim v\}. \tag{52}$$

The set of active points represent the points that are within "unit" distance from the current position of the active contour. Fig. 7 depicts visually how the active points are computed.

Then—up to a positive scaling factor—$\delta_\epsilon(u)$ can be approximated on $\mathcal{G}$ at time t with

$$\delta_t^{\mathcal{G}}(v) \sim \begin{cases} 1, & \text{if } v \in \mathcal{U}_t^* \\ 0, & \text{elsewhere.} \end{cases} \tag{53}$$

Using the above definition, the evolution equation for (46) can be approximated as

$$\tilde{\mathbf{A}}(\mathbf{c})\dot{\mathbf{c}} = \tilde{\mathbf{b}}(\mathbf{c}), \tag{54}$$

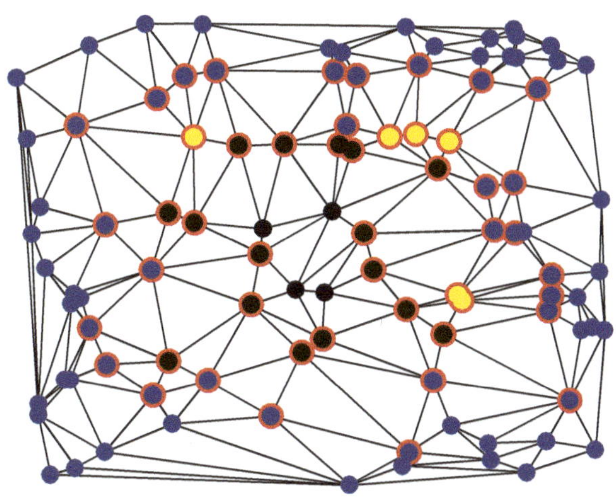

FIG. 7 Illustration of active points for a Delaunay graph with 100 vertices. *Blue nodes*: Vertices outside the contour. *Black nodes*: Vertices inside the contour. *Yellow nodes*: Vertices on the contour. Nodes with *red boundary*: Active points.

where $\tilde{\mathbf{A}}(\mathbf{c})$ is a $n \times n$ matrix with

$$\tilde{\mathrm{A}}_{ij}(\mathbf{c}) = \begin{cases} \mathrm{A}_{ij}(\mathbf{c}), & \text{if } v_i \in \mathcal{U}_t^* \\ 1, & \text{if } v_i \notin \mathcal{U}_t^* \text{ and } j = i, \\ 0, & \text{elsewhere} \end{cases}$$

and $\tilde{\mathbf{b}}(\mathbf{c})$ a n-dimensional vector with

$$\tilde{b}_i(\mathbf{c}) = \begin{cases} b_i(\mathbf{c}), & \text{if } v_i \in \mathcal{U}_t^* \\ 0, & \text{elsewhere} \end{cases}. \tag{55}$$

It is easy to observe that for all nonactive vertices v_i, $\dot{c}_i = 0$ and thus the curve evolution is indeed constrained at the subset $\mathcal{U}_t^*$.

The above formulation will help us reduce significantly the computational complexity of each time step. Consider the explicit Euler approximation of (54)

$$\tilde{\mathbf{A}}(\mathbf{c}_k)(\mathbf{c}_{k+1} - \mathbf{c}_k) = \Delta t_k \tilde{\mathbf{b}}(\mathbf{c}_k). \tag{56}$$

For all vertices v_i that are not in the set of active points $(c_i)_{k+1} - (c_i)_k = 0$. So, for all $v_i \in \mathcal{U}_t^*$, $(c_i)_{k+1}$ does not depend on the values $(c_j)_k$, where $v_j \notin \mathcal{U}_{t_k}^*$. Thus the corresponding values $\mathrm{A}_{ij}(\mathbf{c}_k)$ have no influence on the solution of the linear system, so they can be set to 0. Let $\mathbf{A}_k^*$ be a $|\mathcal{U}_{t_k}^*| \times |\mathcal{U}_{t_k}^*|$ submatrix of $\tilde{\mathbf{A}}(\mathbf{c_k})$ that is constructed by keeping only the elements A_{ij} with v_i and $v_j \in \mathcal{U}_{t_k}^*$. Similarly $\mathbf{b}_k^*$ is a $|\mathcal{U}_t^*|$-dimensional vector that is derived from $\mathbf{b}(\mathbf{c}_k)$ by discarding all the elements v_i with $v_i \notin \mathcal{U}_t^*$. If we set $\Delta\mathbf{c}_k = \mathbf{c}_{k+1} - \mathbf{c}_k$ and $\Delta\mathbf{c}_k^* = \mathbf{I}_k^* \Delta\mathbf{c}_k$, where $\mathbf{I}_k$ is the $|\mathcal{U}_{t_k}^*| \times n$ projection matrix with $\mathrm{I}_{ii} = 1$ and $\mathrm{I}_{ij} = 0$ if $i \neq j$ then we get the equivalent linear system

$$\mathbf{A}_k^* \Delta\mathbf{c}_k^* = \Delta t_k \mathbf{b}_k^*. \tag{57}$$

After computing $\Delta\mathbf{c}_k$ from (57) the update rule for the coefficients $\mathbf{c}$ in matrix notation becomes

$$\mathbf{c}_{k+1} = \mathbf{c}_k + \mathbf{I}_k^{\mathrm{T}} \Delta\mathbf{c}_k^*. \tag{58}$$

Most level set evolution models require reinitialization of the embedding function to ensure correct results and there have been several approaches to alleviate this need, as in Li et al. (2005). In this framework we avoid the need for reinitialization in an indirect way; to ensure the stability of the numerical method and make it possible to use large time steps, we normalize the embedding function after each time step. Specifically, $\bar{u}(x,y,t)$ and consequently the vector of coefficients $\mathbf{c}_k$ is saturated outside the interval $[-r,\ r]$. A typical choice of r is $r = \max|u_0(x,y)|$or equivalently $r = \max|\mathbf{c}_0|$. With the use of normalization we can choose a time step $\Delta t_k \sim 1/n$, which is significantly larger than the time step used for the solution of (27). A similar approach cannot be adopted for the solution of the full evolution equation because in each time step the evolution domain covers the whole graph and although the embedding function will be bounded, large time steps will produce noisy artifacts.

As in the previous section, Kolotouros and Maragos (2017) provide an estimate for the computational complexity of the proposed algorithm for the case of Delaunay graphs that correspond to images. For a $h \times w$ image with n pixels, each of $\mathcal{U}_{t_k}^+, \mathcal{U}_{t_k}^-$ and $\mathcal{U}_{t_k}^0$ is the union of a number of 1D curves. Generally, typical 1D curves in an image grid with n pixels contain $O(\sqrt{n})$ pixels. Thus, in almost all practical cases $\mathcal{U}_{t_k}^*$ will contain $O(\sqrt{n})$ pixels. Since the number of edges in the planar Delaunay graph is bounded by $3n - 6$, a naive calculation of $\mathcal{U}_{t_k}^*$ using the definitions of its components requires $O(n)$ operations. Additionally, it is easy to verify that given the set of active points, the elements of $\mathbf{A}_k{}^*$ and $\mathbf{b}_k{}^*$ can be computed in $O(n)$ time. Since $\mathbf{A(c_k)}$ is a sparse band matrix, $\mathbf{A_k}^*$ will also be a band matrix, but its band length can often be $O(|\mathcal{U}_{t_k}^*|)$. However, using the Reverse Cuthill–McKee algorithm (George and Liu, 1981) which can be implemented in $O(|\mathcal{U}_{t_k}^*|)$ time we can obtain a permuted matrix with a band length of $O(\sqrt{|\mathcal{U}_{t_k}^*|})$ on average. Thus the solution of the constrained linear system is expected to require $O(n)$ operations. Consequently, we have shown that on average, each time step of the algorithm requires $O(n)$ operations. If we compare this result with the number of operations required for the full curve evolution, we can see that the constrained curve evolution algorithm is faster by an order of magnitude. Its linear complexity with respect to the number of graph vertices makes it feasible to be used in practical applications.

These results point towards a fast implementation of general active contour models that can take the form of (27). The level set-based approach enables us to handle topological changes in the curve in a solid and efficient manner, but introduces an extra computational burden, because we have to evolve a 2D function instead of a curve. The answer to this problem is to try to limit our focus in a small area of interest near the curve—often referred as band—and evolve the embedding function in this subset of the image. Similar methods, called narrow-band methods, were described for the case of images in Adalsteinsson and Sethian (1995), Whitaker (1998) and Peng et al. (1999). Instead of evaluating the curve evolution on the whole graph, the curve evolution is constrained in a small band near the curve. This approximation is based on the assumption that the embedding function evolves in such a way that points outside or inside the curve will not change status, at least until the moment that the active contour reaches them. The above assumption is valid for the majority of active contour methods, such as simple Erosion/Dilation as well as the Geometric and Geodesic Active Contours.

5 Experimental results

Using the graph-based approximations of the main differential terms of active contour models with level sets, Sakaridis et al. (2017) apply finite difference-based algorithms to solve the respective PDEs that define these models on arbitrary graphs. The input comprises a graph $\mathcal{G}=(\mathcal{V},\mathcal{E})$ and an image function I, as specified in Section 1. The two representative active contour models which are used to develop these algorithms are the GAC model (Caselles et al., 1997) presented in (1) and the ACWE model (Chan and Vese, 2001) presented in (2). We first present the GAC algorithm and corresponding results and then proceed to ACWE.

The finite difference algorithm of Sakaridis et al. (2017) for GACs on graphs includes the following steps:

1. Compute $g(\|\nabla I_\sigma\|)$, using either (25) and (14) or (26) to compute $\|\nabla I_\sigma\|$. In both cases, median filtering is applied to $\|\nabla I_\sigma\|$ before plugging it into (24) for computing g. Then, compute the magnitude of g's gradient using (14) and its direction using (7).
2. Choose a subset X of $\mathcal{V}$ which *contains* the objects to be segmented and initialize the embedding function with the signed distance function from the boundary of X, denoted by u_0. By convention, u_0 is positive inside X.
3. Iterate for $r\in\mathbb{N}$

$$u_r = u_{r-1} + \Delta t((\kappa - c)\|\nabla u_{r-1}\|g + \nabla g\cdot\nabla u_{r-1}) \tag{59}$$

until convergence. In the difference equation (59), Δt and c are positive constants and κ is the estimated curvature of the level sets of u_{r-1} using (20). The direction of ∇u_{r-1} is computed with (7) and its magnitude with (14). Median filtering is applied both to κ and ∇u_{r-1} before using them in (59).

In practice, after each iteration of step 3 of the algorithm, u_r is smoothed with a median filter before proceeding to the next iteration. The parameters involved in the algorithm are the time step Δt of the difference equation, the balloon force constant c, the scale σ of the Gaussian smoothing filter and parameter λ of the stopping function g (24). Tuning their values depending on the input at hand is pivotal in obtaining satisfactory segmentation results. In the following experiments of Section 3, $\Delta t = 0.005$ and $c = 2$ unless otherwise specified.

A simple analysis of the computational complexity of each iteration (59) of the algorithm by examining each term on the right-hand side of (59) and aggregating the individual contributions of all terms shows that the total complexity of each iteration is $\Theta(n + m)$, which is linear in the number of vertices and in the number of edges.

We first discuss and experimentally explore the practical aspect of graph construction when applying such graph-based algorithms on arbitrarily spaced data as well as regular pixel grids. In the case of arbitrary graphs, the raw input often consists only of $\mathcal{V}$ and I, without any information about the edges. This setting leaves us free to choose the type of edge structure of the graph. Sakaridis et al. (2017) experiment with two types: RGGs and Delaunay triangulations (DTs) (Jaromczyk and Toussaint, 1992). In the case of regular images on a grid, graph-based active contour methods are still relevant, since the input image can be sampled at a number of arbitrary locations that will serve as the vertices. This brings us to the previous setting where edges can be defined freely. A baseline for such sampling is to place vertices uniformly at random on the image domain. A more sophisticated strategy is to extract vertex locations via watershed transformation. In particular, watershed transformation is applied directly to the gradient of the image and the vertices are placed at the centroids (ultimate erosions) of the resulting superpixels. For the grayscale image with four distinct coins in Fig. 8A, we compare the results of the GAC algorithm of Sakaridis et al. (2017) for all four possible combinations of the aforementioned vertex placement strategies (random, watershed) and choices of edge structure (RGG, DT) in Fig. 8C–F. To ensure a fair comparison, the number of randomly placed vertices is approximately the same as in the watershed case. The most accurate segmentation is achieved with DT and watershed-placed vertices (Fig. 8F). Moreover, DT consistently outperforms RGG and watershed-based vertex placement outperforms random placement. Consequently, the DT edge structure is preferable to the random geometric one and usage of watershed transformation to place vertices when a full image is available is preferable to random placement, as it captures image particularities into the spatial structure of the graph. In the rest of the experiments of Sakaridis et al. (2017) that are presented in this section, construction of the graphs follows these choices.

Fig. 9 presents results of the algorithm of Sakaridis et al. (2017) for GACs on graphs on two natural colour images coming from the Berkeley Segmentation Dataset BSDS500 (Martin et al., 2001). The images were converted to grayscale for applying the algorithm. In general, the algorithm segments the dominant objects in the images successfully, even though background clutter and thin protrusions or concavities of the objects' boundaries sometimes cause minor inaccuracies.

An interesting application of the examined graph segmentation framework is related to geographical data, where the two spatial coordinates are longitude and latitude and the image function can encode information about any type of real-valued signal defined at the vertices of the graph. Such a signal is the signal strength of a cellular network, such as that shown in Fig. 10A. The coordinates and the values of signal strength are normalized and a very small time step $\Delta t = 10^{-4}$ is used to guarantee convergence, which requires more iterations until termination than in the previous experiments. Other parameters are set as $\sigma = 0.03$, $\lambda = 0.02$ and $c = 20$. The segmented set of vertices in Fig. 10D using the GAC algorithm of Sakaridis et al. (2017) comprises two

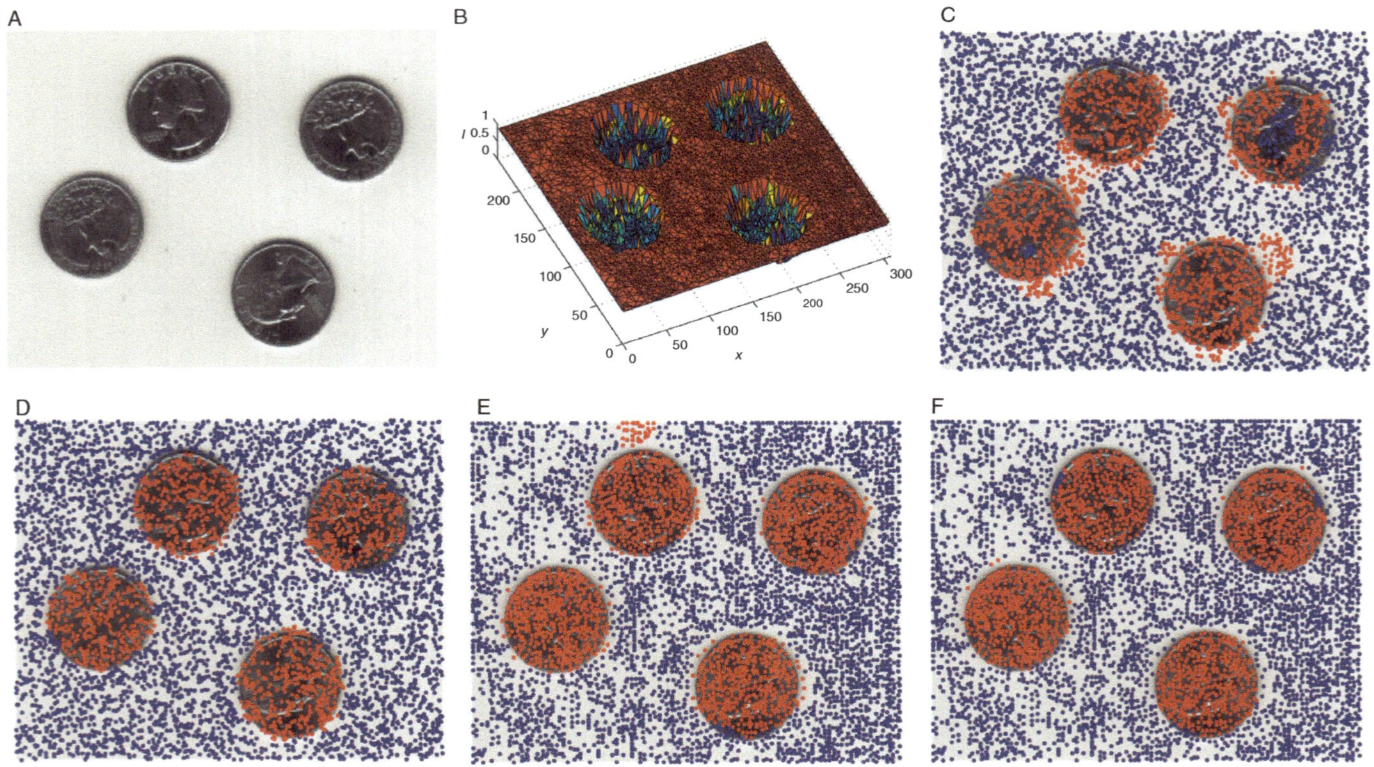

FIG. 8 Segmentation of grayscale objects on graphs with the GAC algorithm of Sakaridis et al. (2017). (A) Full grayscale image with four coins (B) image function on watershed-based DT (C)–(F) final segmentation results overlaid on original image, with segmented objects shown in *red* and background in *blue*. We use (C) randomly placed vertices with random geometric structure, $\sigma = 0.02$ and $\lambda = 0.03$, (D) randomly placed vertices with DT structure, $\sigma = 0.02$ and $\lambda = 0.03$, (E) watershed-placed vertices with random geometric structure, $\sigma = 0.008$ and $\lambda = 0.07$, and (F) watershed-placed vertices with DT structure, $\sigma = 0.01$ and $\lambda = 0.07$. The total number of iterations to obtain the final segmentation result is (C) 2200, (D) 4000, (E) 3000 and (F) 4000. *Modified from Sakaridis, C., Drakopoulos, K., Maragos, P. 2017. Theoretical analysis of active contours on graphs. SIAM J. Imaging Sci. 10 (3), 1475–1510.*

FIG. 9 Segmentation of images from BSDS500 dataset with the GAC algorithm of Sakaridis et al. (2017). The detected boundary, marked white, is defined as the boundary of the union of the watershed superpixels for which the corresponding graph vertex belongs to the final contour's interior. We use $\sigma = 0.005$ and $\lambda = 0.03$ in (A), and $\sigma = 0.01$ and $\lambda = 0.02$ in (B). The total number of iterations to obtain the final segmentation result is (A) 3400 and (B) 5600. *Modified from Sakaridis, C., Drakopoulos, K., Maragos, P. 2017. Theoretical analysis of active contours on graphs. SIAM J. Imaging Sci. 10 (3), 1475–1510.*

regions, the southern of which is characterized by an increased signal strength compared to the rest of the graph. The graph segmentation framework with active contours is tailored for geographical data with arbitrary spatial configuration such as the above case, which grants greater flexibility when collecting measurements.

At this point, we present a short theoretical and experimental comparison between the finite difference-based GAC algorithm of Sakaridis et al. (2017) that was presented previously and the finite element method (FEM) for GACs on graphs that is developed by Kolotouros and Maragos (2017) and is detailed in Section 4. Even though the algorithm of Kolotouros and Maragos (2017) inherits the advantages of the well-established framework of finite elements, the finite difference algorithm of Sakaridis et al. (2017) features greater generality and lower computational complexity per iteration. On the other hand, the FEM algorithm of Kolotouros and Maragos (2017) achieves slightly better segmentation results in practice.

In particular, the finite difference framework of Sakaridis et al. (2017) is applicable for any type of connectivity pattern defined through the set of edges $\mathcal{E}$ of the input graph. For instance, when $\mathcal{E}$ is given as part of the input and models connections of vertices that follow an arbitrary, nonlocal pattern which is known a priori, the framework of Sakaridis et al. (2017) is still

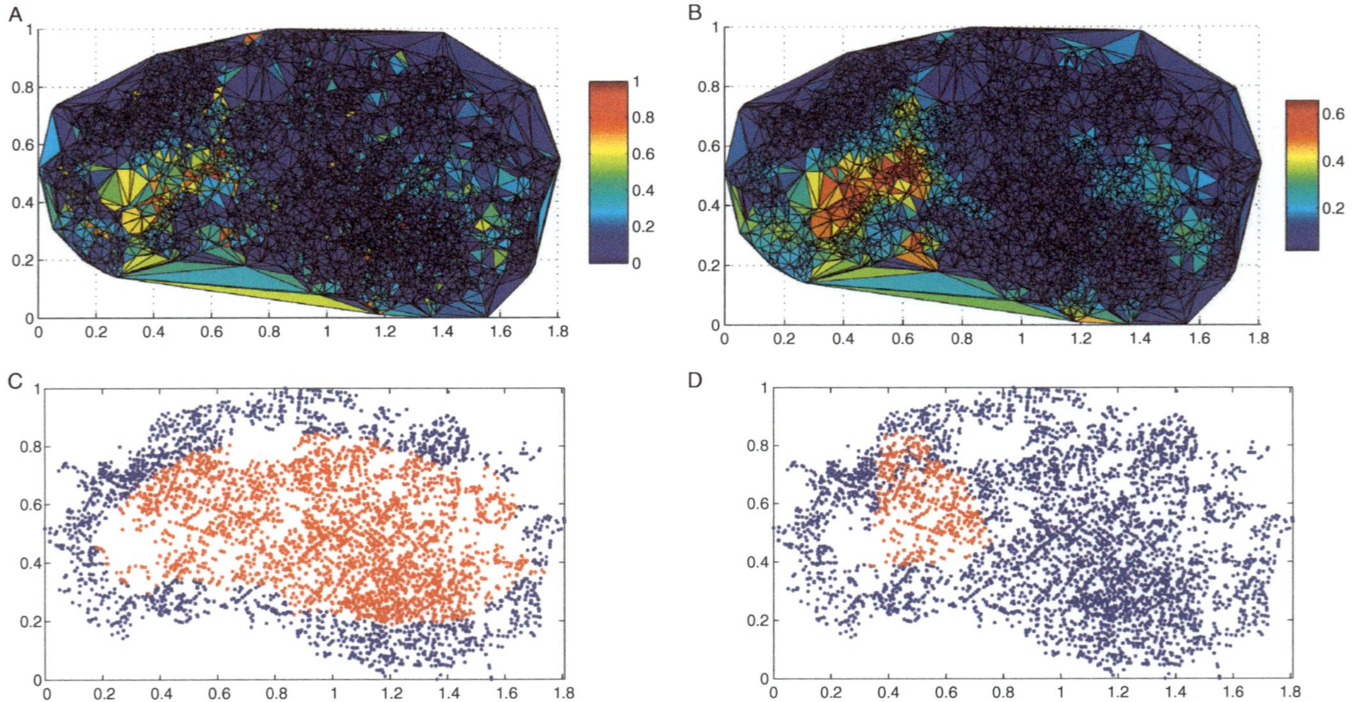

FIG. 10 Segmentation of signal strength data of a cellular network on a graph with the GAC algorithm of Sakaridis et al. (2017). (A) Normalized data on graph. (B) Smoothed signal strength. (C) Initial contour. (D) Final segmentation result after 40,000 iterations, with vertices in the contour's interior shown in *red* and the rest in *blue*. *From Sakaridis, C., Drakopoulos, K., Maragos, P. 2017. Theoretical analysis of active contours on graphs. SIAM J. Imaging Sci. 10 (3), 1475–1510.*

relevant. On the contrary, the FEM framework of Kolotouros and Maragos (2017) only applies to triangulations; notably, it cannot handle the aforementioned case with arbitrary edge structure, or other commonly used types of graphs such as k nearest neighbour graphs or RGGs. Furthermore, in the case of triangulations, the complexity of each iteration of the finite difference-based GAC algorithm is $O(n)$, i.e., *linear* in the number of vertices n (which follows from the general $\Theta(n + m)$ complexity and planarity of triangulations), whereas the respective complexity of each iteration of the FEM-based GAC algorithm is $O(n^2)$, i.e., *quadratic* in n.

In Fig. 11 we compare the two approaches on the coins image and an image from BSDS500 (Martin et al., 2001). The graphs are formed as DTs with watershed-placed vertices. Whenever possible, we use the same or at least similar parameter values for the two approaches, so that a fair comparison is ensured. The results of the finite difference GAC algorithm are presented in Fig. 11A and C and the respective results of the FEM GAC algorithm are presented in Fig. 11B and D. The two methods demonstrate similar overall performance on both images, with the FEM algorithm producing slightly more accurate boundaries in challenging cases such as the bird's tail.

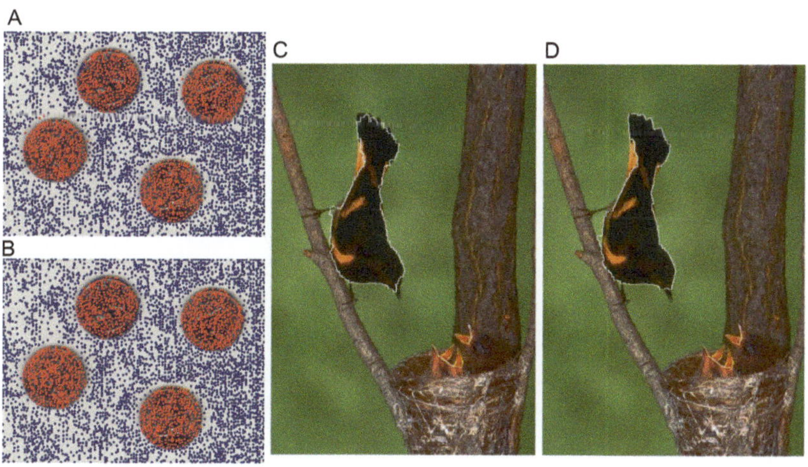

FIG. 11 Comparison between the finite difference GAC algorithm of Sakaridis et al. (2017) and the FEM GAC algorithm of Kolotouros and Maragos (2017). For the coins image, the foreground is shown in *red* and the background in *blue* and we use $\sigma = 0.01$ and $\lambda = 0.07$ for both methods. (A) The result of Sakaridis et al. (2017) after 4000 iterations. (B) The result of Kolotouros and Maragos (2017) after 301 iterations, using $c = 10$ and $\Delta t = 0.001$. For the bird image, the detected boundary is marked *white* and we use $\sigma = 0.005$ for both methods. (C) The result of Sakaridis et al. (2017) after 4400 iterations, using $\lambda = 0.05$. (D) The result of Kolotouros and Maragos (2017) after 4001 iterations, using $\lambda = 0.07$, $c = 10$ and $\Delta t = 0.001$. *From Sakaridis, C., Drakopoulos, K., Maragos, P. 2017. Theoretical analysis of active contours on graphs. SIAM J. Imaging Sci. 10 (3), 1475–1510.*

We now proceed to the presentation of the ACWE algorithm of Sakaridis et al. (2017) and related results. The finite difference algorithm of Sakaridis et al. (2017) for ACWE on graphs includes the following steps:

1. Choose a subset X of $\mathcal{V}$ and initialize the embedding function with the signed distance function from the boundary of X, denoted by u_0. By convention, u_0 is positive inside X.
2. Iterate for $r \in \mathbb{N}$:

$$c_1 = \text{average}\,\{I(v) : u_{r-1}(v) \geq 0\},\; c_2 = \text{average}\,\{I(v) : u_{r-1}(v) < 0\}, \quad (60)$$

$$u_r = u_{r-1} + \Delta t\left(\delta_\epsilon(u_{r-1})\left(\mu\kappa - \nu - \lambda_1(I - c_1)^2 + \lambda_2(I - c_2)^2\right)\right) \quad (61)$$

until convergence. In the difference equation (61), Δt, ϵ, μ, ν, λ_1 and λ_2 are positive constants and κ is the estimated curvature of the level sets of u_{r-1} using (20). Moreover, δ_ϵ is the derivative of a regularized version of the Heaviside function which was introduced in Chan and Vese (2001) as:

$$\delta_\epsilon(x) = H'_\epsilon(x) = \left(\frac{1}{2}\left(1 + \frac{2}{\pi}\arctan\left(\frac{x}{\epsilon}\right)\right)\right)' = \frac{\epsilon}{\pi(x^2 + \epsilon^2)}. \quad (62)$$

The standard ACWE model is modified in Sakaridis et al. (2017) for colour images, so that it uses all three colour channels instead of reducing the image to a simpler grayscale version. In order for the averages and Euclidean distances to be meaningful, the original RGB values of the vector-valued image function $\mathbf{I}$ at graph vertices are transformed to CIELAB values, as the latter colour space is more perceptually uniform. The update equations are modified to operate on multiple colour channels:

$$\mathbf{c}_1 = \text{average}\,\{\mathbf{I}(v) : u_{r-1}(v) \geq 0\},\; \mathbf{c}_2 = \text{average}\,\{\mathbf{I}(v) : u_{r-1}(v) < 0\} \quad (63)$$

and

$$u_r = u_{r-1} + \Delta t\left(\delta_\epsilon(u_{r-1})\left(\mu\kappa - \nu - \lambda_1\|\mathbf{I} - \mathbf{c}_1\|^2 + \lambda_2\|\mathbf{I} - \mathbf{c}_2\|^2\right)\right). \quad (64)$$

In Fig. 12, we present results that are analogous to Figs. 8F and 9A and B by replacing the GAC algorithm with the ACWE algorithm and demonstrate the utility of the geometric approximations proposed in Sakaridis et al. (2017) for successfully translating various active contour models to graphs. For the colour images from BSDS500 in Fig. 12A and C, the modification of ACWE in (63) and (64) is used. The segmentation results are of high quality, e.g., capturing the thin feathers at the eagle's wing very precisely.

Last we present some additional qualitative results when applying the FEM framework for solving more complicated active contour models. Kolotouros and Maragos (2017) presented Region-based Geodesic Active Contours (RBGAC), an interactive segmentation algorithm where the users have to specify some initial markers for the foreground and background and

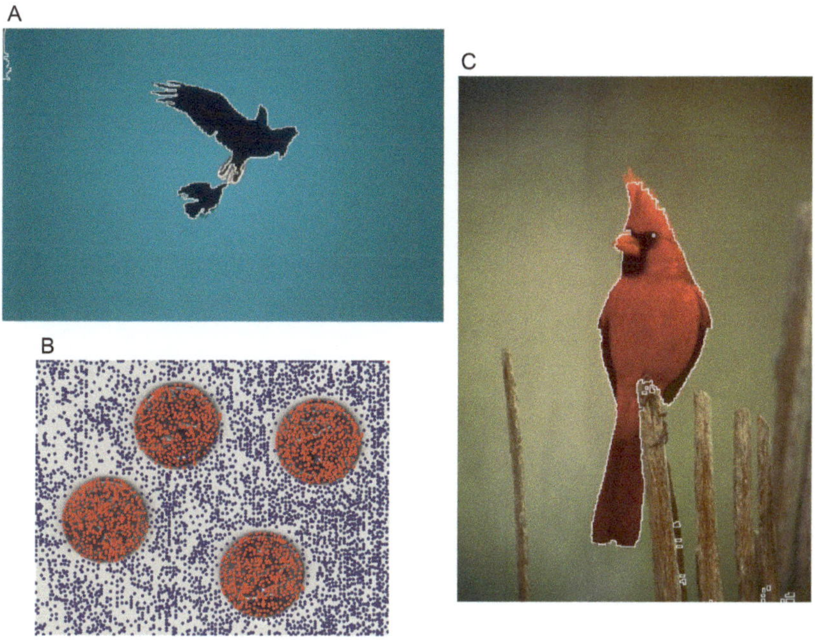

FIG. 12 Segmentation of images with the ACWE algorithm of Sakaridis et al. (2017). (A) and (C) Images from BSDS500 with detected boundaries marked *white*. (B) Coins image with detected foreground shown in *red* and background in *blue*. We use $\Delta t = 0.05$, $\nu = 0$, $\lambda_1 = \lambda_2 = 1$ and $\epsilon = \text{median}_{vw \in \mathcal{E}}\{d(v,w)\}$ in all cases, $\mu = 0.1$ in (A) and $\mu = 0.5$ in (B) and (C). The total number of iterations to obtain the final segmentation result is (A) 1000, (B) 7000 and (C) 2000. *Modified from Sakaridis, C., Drakopoulos, K., Maragos, P. 2017. Theoretical analysis of active contours on graphs. SIAM J. Imaging Sci. 10 (3), 1475–1510.*

FIG. 13 Illustration of RBGAC algorithm. *Left column*: Images and seeds. The *red curves* are the initial contours and the *blue polygons* mark the regions of interest. *Right column*: Segmentation results.

then the algorithm iterates between expanding and shrinking contours to compute the segmentation. An example use of this method can be seen in Fig. 13. Also in Fig. 14 we depict a quantitative comparison between RBGAC and several other popular interactive segmentation algorithms.

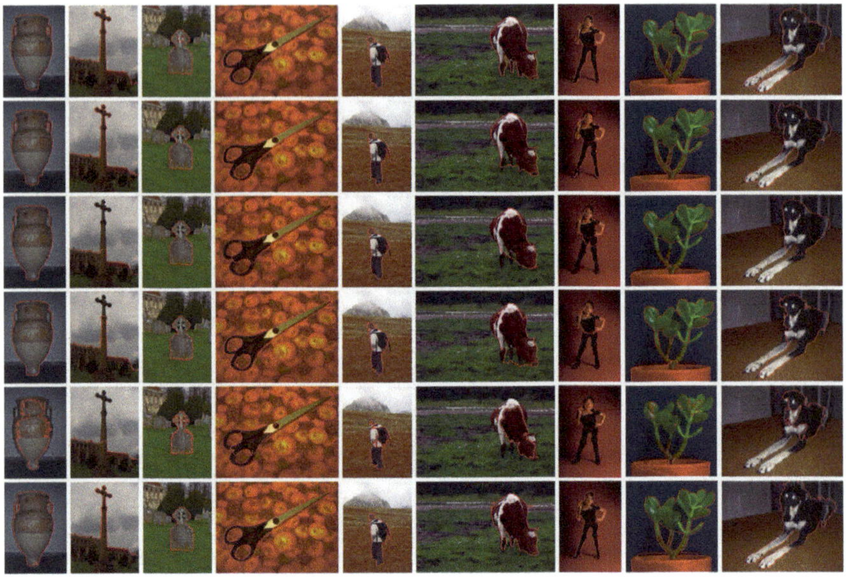

FIG. 14 Segmentation results. Columns 1–5: GrabCut Dataset. Columns 6–9: PASCAL Dataset. From *top* to *bottom*: Ground Truth, segmentation results from GrabCut, Laplacian Coordinates, Power Watershed, ACWE and RBGAC. *Data from Kolotouros, N., Maragos P. 2017. A finite element computational framework for active contours on graphs. ArXiv e-prints 1710.04346.*

6 Conclusion

In this chapter, we have delineated two approaches to translate level set-based active contour models to the general case of arbitrary graphs, using either finite differences (Drakopoulos and Maragos, 2012; Sakaridis et al., 2017) or the finite element method (Kolotouros and Maragos, 2017) for the required discretization of the employed PDEs. We have overviewed the approximations introduced in Sakaridis et al. (2017) for the gradient and curvature terms in active contour models for 2D graphs with arbitrary vertex and edge configuration and presented the main theoretical results of Sakaridis et al. (2017) regarding asymptotic consistency and accuracy of these approximations for the class of random geometric graphs. In addition, a detailed presentation of the formulation of the finite element method that is proposed in Kolotouros and Maragos (2017) for the approximate solution of active contour PDEs through discretization on arbitrary 2D Delaunay triangulations has been provided. The effectiveness of the two examined methods has been demonstrated experimentally by using them to apply the popular GAC and ACWE models on regular as well as arbitrary graphs and successfully segment regular images as well as unorganized 2D geographical data.

Although the framework of Sakaridis et al. (2017) applies to any type of graph, the presented theoretical analysis considers random geometric graphs, whose definition simplifies convergence proofs for the proposed approximations. However, judging from segmentation quality, more regular graph structures, such as Delaunay triangulations, lead to more accurate results. Therefore, an interesting direction in this line is the theoretical study of the approximations of Sakaridis et al. (2017) on graphs formed as Delaunay triangulations, ideally establishing similar asymptotic consistency results to those shown in Sakaridis et al. (2017).

Last but not least, both methods (Kolotouros and Maragos, 2017; Sakaridis et al., 2017) are formulated for the case of 2D graphs. While this type of input is sufficient to model signals such as regular images and geographical profiles, there are several areas of application that involve signals defined in higher-dimensional—often 3D—Euclidean spaces, including medical image processing and geology. Thus, it is worth generalizing the above methods to more dimensions. For the approach of Sakaridis et al. (2017) in particular, this requires generalizing the approximation formulae for gradient and curvature. With regard to gradient, the main challenge is the calculation of neighbour angles (cf. (5)) in higher dimensions. For each vertex v, this would involve the calculation of solid angles which are induced by the Voronoi tessellation of the ambient d-dimensional space that is defined by the set of neighbours $\mathcal{N}(v)$, analogously to Fig. 1. The general formula of Ribando (2006) for measuring simplicial solid angles in Euclidean spaces of arbitrary dimensionality would likely have to be utilized to this end. As far as curvature is concerned, the quantity of interest is the mean curvature and we foresee that the geometric type of approximation we highlighted in Section 3.2 will require meticulous calculations in order to be extended to more dimensions, whereas the gradient-based approximation we briefly mentioned, which uses the differential definition of divergence, is easier to extend.

References

Adalsteinsson, D., Sethian, J.A., 1995. A fast level set method for propagating interfaces. J. Comput. Phys. 118 (2), 269–277. ISSN 0021-9991. https://doi.org/10.1006/jcph.1995.1098.

Arias-Castro, E., Pelletier, B., Pudlo, P., 2012. The normalized graph cut and Cheeger constant: from discrete to continuous. Adv. Appl. Probab. 44 (4), 907–937.

Barth, T.J., Sethian, J.A., 1998. Numerical schemes for the Hamilton-Jacobi and level set equations on triangulated domains. J. Comput. Phys. 145 (1), 1–40.

Belkin, M., Niyogi, P., 2006. Convergence of Laplacian eigenmaps. In: Advances in Neural Information Processing Systems (NIPS).

Bensoussan, A., Menaldi, J.-L., 2005. Difference equations on weighted graphs. J. Convex Anal. 12 (1), 13–44.

Bertozzi, A.L., Flenner, A., 2012. Diffuse interface models on graphs for classification of high dimensional data. Multiscale Model. Simul. 10 (3), 1090–1118.

Bougleux, S., Elmoataz, A., Melkemi, M., 2009. Local and nonlocal discrete regularization on weighted graphs for image and mesh processing. Int. J. Comput. Vis. 84, 220–236.

Boykov, Y., Kolmogorov, V., 2003. Computing geodesics and minimal surfaces via graph cuts. In: Proc. Int. Conf. Comput. Vis.

Boykov, Y., Veksler, O., Zabih, R., 2001. Fast approximate energy minimization via graph cuts. IEEE Trans. Pattern Anal. Mach. Intell. 23 (11), 1222–1239.

Caselles, V., Catté, F., Coll, T., Dibos, F., 1993. A geometric model for active contours in image processing. Numer. Math. 66, 1–31.

Caselles, V., Kimmel, R., Sapiro, G., 1997. Geodesic active contours. Int. J. Comput. Vis. 22 (1), 61–79.

Chan, T.F., Vese, L.A., 2001. Active contours without edges. IEEE Trans. Image Process. 10 (2), 266–277.

Chan, T.F., Esedoglu, S., Nikolova, M., 2006. Algorithms for finding global minimizers of image segmentation and denoising models. SIAM J. Appl. Math. 66 (5), 1632–1648.

Chung, S.-Y., Berenstein, C.A., 2005. ω-Harmonic functions and inverse conductivity problems on networks. SIAM J. Appl. Math. 65 (4), 1200–1226.

Couprie, C., Grady, L., Najman, L., Talbot, H., 2011a. Power watershed: a unifying graph-based optimization framework. IEEE Trans. Pattern Anal. Mach. Intell. 33 (7), 1384–1399.

Couprie, C., Grady, L., Talbot, H., Najman, L., 2011b. Combinatorial continuous maximum flow. SIAM J. Imaging Sci. 4 (3), 905–930.

Couprie, C., Grady, L., Najman, L., Pesquet, J.-C., Talbot, H., 2013. Dual constrained TV-based regularization on graphs. SIAM J. Imaging Sci. 6 (3), 1246–1273.

Courant, R., Friedrichs, K., Lewy, H., 1967. On the partial differential equations of mathematical physics. IBM J. Res. Dev. 11 (2), 215–234.

Cousty, J., Bertrand, G., Najman, L., Couprie, M., 2009a. Watershed cuts: minimum spanning forests and the drop of water principle. IEEE Trans. Pattern Anal. Mach. Intell. 31 (8), 1362–1374.

Cousty, J., Najman, L., Serra, J., 2009b. Some morphological operators in graph spaces. In: Proc. International Symposium on Mathematical Morphology.

Cousty, J., Bertrand, G., Najman, L., Couprie, M., 2010. Watershed cuts: thinnings, shortest path forests, and topological watersheds. IEEE Trans. Pattern Anal. Mach. Intell. 32 (5), 925–939.

Desquesnes, X., Elmoataz, A., Lézoray, O., 2013. Eikonal equation adaptation on weighted graphs: fast geometric diffusion process for local and non-local image and data processing. J. Math. Imaging Vision 46 (2), 238–257.

Drakopoulos, K., Maragos, P., 2012. Active contours on graphs: multiscale morphology and graphcuts. IEEE J. Sel. Topics Signal Process. 6 (7), 780–794.

Elmoataz, A., Lézoray, O., Bougleux, S., 2008. Nonlocal discrete regularization on weighted graphs: a framework for image and manifold processing. IEEE Trans. Image Process. 17 (7), 1047–1060.

Elmoataz, A., Toutain, M., Tenbrinck, D., 2015. On the p-Laplacian and ∞-Laplacian on graphs with applications in image and data processing. SIAM J. Imaging Sci. 8 (4), 2412–2451.

Elmoataz, A., Lozes, F., Toutain, M., 2017. Nonlocal PDEs on graphs: from tug-of-war games to unified interpolation on images and point clouds. J. Math. Imaging Vision 57 (3), 381–401.

García Trillos, N., Slepčev, D., 2016. Continuum limit of total variation on point clouds. Arch. Ration. Mech. Anal. 220 (1), 193–241.

George, A., Liu, J.W.H., 1981. Computer Solution of Large Sparse Positive Definite Systems. Prentice Hall.

Gilboa, G., Osher, S., 2007. Nonlocal linear image regularization and supervised segmentation. Multiscale Model. Simul. 6 (2), 595–630.

Gilboa, G., Osher, S., 2008. Nonlocal operators with applications to image processing. Multiscale Model. Simul. 7 (3), 1005–1028.

Grady, L., 2006. Random walks for image segmentation. IEEE Trans. Pattern Anal. Mach. Intell. 28 (11), 1768–1783.

Grady, L., Alvino, C., 2009. The piecewise smooth Mumford-Shah functional on an arbitrary graph. IEEE Trans. Image Process. 18 (11), 2547–2561.

Grady, L., Polimeni, J., 2010. Discrete Calculus: Applied Analysis on Graphs for Computational Science. Springer Science & Business Media.

Heijmans, H.J.A.M., Nacken, P., Toet, A., Vincent, L., 1992. Graph morphology. J. Vis. Commun. Image Represent. 3 (1), 24–38.

Hirani, A. N., 2003. Discrete Exterior Calculus (Ph.D. dissertation). Cal. Tech.

Jaromczyk, J.W., Toussaint, G.T., 1992. Relative neighborhood graphs and their relatives. Proc. IEEE 80 (9), 1502–1517.

Kass, M., Witkin, A., Terzopoulos, D., 1988. Snakes: active contour models. Int. J. Comput. Vis. 1 (4), 321–331.

Kolotouros, N., Maragos, P., 2017. A finite element computational framework for active contours on graphs. ArXiv e-prints 1710.04346.

Lee, D.D., Seung, H.S., 2001. Algorithms for non-negative matrix factorization. In: Advances in Neural Information Processing Systems (NIPS).

Lézoray, O., Grady, L., 2012. Image Processing and Analysis with Graphs: Theory and Practice, Digital Imaging and Computer Vision, CRC Press.

Lézoray, O., Elmoataz, A., Ta, V.-T., 2012. Nonlocal PdEs on graphs for active contours models with applications to image segmentation and data clustering. In: ICASSP, pp. 873–876.

Li, C., Xu, C., Gui, C., Fox, M.D., 2005. Level set evolution without re-initialization: a new variational formulation. In: Proceedings of the IEEE Conference on Computer Vision and Pattern Recognition (CVPR'05).

Lozes, F., Elmoataz, A., Lézoray, O., 2014. Partial difference operators on weighted graphs for image processing on surfaces and point clouds. IEEE Trans. Image Process. 23 (9), 3896–3909.

Maier, M., von Luxburg, U., Hein, M., 2013. How the result of graph clustering methods depends on the construction of the graph. ESAIM: Probab. Stat. 17, 370–418.

Martin, D., Fowlkes, C., Tal, D., Malik, J., 2001. A database of human segmented natural images and its application to evaluating segmentation algorithms and measuring ecological statistics. In: Proc. Int. Conf. Comput. Vis., vol. 2, pp. 416–423.

Merkurjev, E., Kostic, T., Bertozzi, A.L., 2013. An MBO scheme on graphs for classification and image processing. SIAM J. Imaging Sci. 6 (4), 1903–1930.

Merkurjev, E., Bae, E., Bertozzi, A.L., Tai, X.-C., 2015. Global binary optimization on graphs for classification of high-dimensional data. J. Math. Imaging Vision 52 (3), 414–435.

Mourad, H.M., Dolbow, J., Garikipati, K., 2005. An assumed-gradient finite element method for the level set equation. Int. J. Numer. Methods Eng. 64 (8), 1009–1032.

Mumford, D., Shah, J., 1989. Optimal approximations by piecewise smooth functions and associated variational problems. Commun. Pure Appl. Math. 42 (5), 577–685.

Osher, S., Sethian, J.A., 1988. Fronts propagating with curvature dependent speed: algorithms based on Hamilton-Jacobi formulations. J. Comput. Phys. 79, 12–49.

Peng, D., Merriman, B., Osher, S., Zhao, H., Kang, M., 1999. A PDE-based fast local level set method. J. Comput. Phys. 155 (2), 410–438. ISSN 0021-9991. https://doi.org/10.1006/jcph.1999.6345.

Penrose, M., 2003. Random Geometric Graphs. Oxford University Press.

Persson, P.-O., Strang, G., 2004. A simple mesh generator in MATLAB. SIAM Rev. 46 (2), 329–345.

Pollard, D., 1981. Strong consistency of k-means clustering. Ann. Statist. 9 (1), 135–140.

Ribando, J.M., 2006. Measuring solid angles beyond dimension three. Discret. Comput. Geom. 36 (3), 479–487.

Sakaridis, C., Drakopoulos, K., Maragos, P., 2017. Theoretical analysis of active contours on graphs. SIAM J. Imaging Sci. 10 (3), 1475–1510.

Shi, J., Malik, J., 2000. Normalized cuts and image segmentation. IEEE Trans. Pattern Anal. Mach. Intell. 22 (8), 888–905.

Snijders, T., Nowicki, K., 1997. Estimation and prediction for stochastic blockmodels for graphs with latent block structure. J. Classif. 14 (1), 75–100.

Ta, V.-T., Elmoataz, A., Lézoray, O., 2011. Nonlocal PDEs-based morphology on weighted graphs for image and data processing. IEEE Trans. Image Process. 20 (6), 1504–1516.

Ting, D., Huang, L., Jordan, M.I., 2010. An analysis of the convergence of graph Laplacians. In: Proc. International Conference on Machine Learning.

Valance, S., De Borst, R., Réthoré, J., Coret, M., 2008. A partition-of-unity-based finite element method for level sets. Int. J. Numer. Methods Eng. 76 (10), 1513–1527.

van Gennip, Y., Bertozzi, A.L., 2012. Γ-convergence of graph Ginzburg-Landau functionals. Adv. Differ. Equ. 17 (11/12), 1115–1180.

van Gennip, Y., Guillen, N., Osting, B., Bertozzi, A.L., 2014. Mean curvature, threshold dynamics, and phase field theory on finite graphs. Milan J. Math. 82 (1), 3–65.

Vincent, L., 1989. Graphs and mathematical morphology. Signal Process. 16, 365–388.

von Luxburg, U., Belkin, M., Bousquet, O., 2008. Consistency of spectral clustering. Ann. Statist. 36 (2), 555–586.

Whitaker, R.T., Hingham, M.A., USA, 1998. A level-set approach to 3D reconstructionfrom range data. Int. J. Comput. Vision 29 (3), 203–231. ISSN 0920-5691. https://doi.org/10.1023/A:1008036829907.

Yang, Z., Hao, T., Dikmen, O., Chen, X., Oja, E., 2012. Clustering by nonnegative matrix factorization using graph random walk. In: Advances in Neural Information Processing Systems (NIPS).

Yuan, Z., Oja, E., 2005. Projective nonnegative matrix factorization for image compression and feature extraction. In: Proc. Scandinavian Conf. Image Anal.

Chapter 5

Fast operator-splitting algorithms for variational imaging models: Some recent developments

Roland Glowinski[a,b], Shousheng Luo[c,d] and Xue-Cheng Tai[b,e,*]
[a]*Department of Mathematics, University of Houston, Houston, TX, United States*
[b]*Department of Mathematics, Hong Kong Baptist University, Kowloon Tong, Hong Kong*
[c]*Beijing Computational Science Research Center, Beijing, PR China*
[d]*School of Mathematics and Statistics, Data Analysis Technology Lab, Institute of Applied Mathematics, Henan University, Kaifeng, Henan Province, PR China*
[e]*University of Bergen, Department of Mathematics, Bergen, Norway*
[*]*Corresponding author: e-mail: tai@math.uib.no*

Chapter Outline

Handbook of Numerical Analysis, Vol. 20. https://doi.org/10.1016/bs.hna.2019.08.002
© 2019 Elsevier B.V. All rights reserved.

Abstract
We present in this chapter fast operator-splitting-based algorithms for the solutions of variational problems from image processing. The models we consider use geometrical information and rely on the minimization of appropriate energy functionals. These energy functionals are nonquadratic, possibly nonconvex and nonsmooth, making their minimization a nontrivial endeavour. We show in this chapter that operator-splitting methods based on the Lie scheme, and variants of it, can provide algorithms that are efficient, robust and nearly parameter free. These new algorithms are simpler to use, and often more efficient, than those relying on alternating direction methods of multipliers (ADMM). We want to emphasize that one can use the methods discussed in this chapter for a wide range of applications; actually it has already been done for several of them.

Keywords: Image processing, Variational method, Geometric information, Total variation, Euler's elastica, Gaussian/mean curvature, Willmore energy, Operator splitting, Lie scheme

AMS Classification Codes: 68U10, 94A08, 65C20, 49M27, 49M99, 65K10

1 Introduction

One is currently witnessing the increasing popularity of geometrical information-based variational methods for the formulation and solutions of *image processing problems*. This approach can produce superior results for difficult situations and has good robustness properties. However, the numerical solutions of the related mathematical problems may be very difficult. During the past decade, various algorithms have been proposed for the solution of these image processing related variational problems, among them those relying on the *alternating direction method of multipliers* (ADMM), a primal-dual method introduced in Glowinski and Marroco (1975) (to the best of our knowledge). Evidences of the popularity of ADMM related algorithms in image processing can be found in Glowinski et al. (2017) (see also Glowinski et al. (2016b) and the references therein). However, ADMM has some drawbacks when the variational models become complicated, implying that one has to split the related minimization problems into several subproblems (typically three or more), making, among other possible difficulties, the *adjustment of the augmentation parameters* a nontrivial task.

It is known (since Chan and Glowinski, 1978 in particular) that ADMM type algorithms are strongly related to operator-splitting methods such as *Peaceman–Rachford's* and *Douglas–Rachford's* (see, e.g., Glowinski, 2014, 2015 for historical details). The goal of this chapter is to investigate how to keep the advantages of ADMM type algorithms for image processing applications, while reducing at the same time the number of parameters ("parameter free" being the ultimate goal). In order to reach these objectives when solving complicated image processing problems, we are going to employ appropriate operator-splitting methods, namely the *Lie scheme* and its *Marchuk–Yanenko* variant, both scheme being very popular in Computational Mechanics and Physics (see Marchuk (1990), Chan and Glowinski (1978) and Glowinski et al. (2017) for details). Properly applied, these schemes lead to subproblems which are easy to solve and which are nearly parameter free. In addition, due to their *implicit* nature these schemes have good stability properties, leading to robust and (relatively) cheap image processing methods.

We structured as follows the remaining part of this chapter: several image processing variational models are introduced in Section 2. In Section 3, one discusses the *Lie scheme* and its *Marchuk–Yanenko* variant. Both schemes are applied in Sections 4–7 to the computational treatment of a variety of image processing variational models. Finally, in Section 8 we draw conclusions and discuss related future work.

2 Regularizers and associated variational models for image restoration

2.1 Generalities

Image processing is about manipulating an image to extract useful information. Mathematically, one can represent an image by a two-variable real-valued function, whose variations (possibly discontinuous) contain most of the important image information. Thus, being able to reproduce accurately these variations is one of the central issues of image processing. Accordingly, one of the image processing critical issues is to identify the right functionals to measure the image and its variations and possible discontinuities. Below, we will present some classical such functionals and a new one (Willmore bending energy), using the following geometrical approach.

Let $f:\Omega\mapsto\mathbb{R}$ be an observed image defined over a domain $\Omega\subset\mathbb{R}^2$ (*colour images* are vector-valued functions; they will not be discussed in this chapter). Typically, the variational approach in image restoration involves the minimization of an energy functional of the following form

$$E(v)=F(v;f)+\theta R(v) \tag{1}$$

over an appropriate functional space $\mathcal{V}$. In (1), θ (>0) is a parameter whose role is to balance the two terms in the right-hand side of (1). The first term

F is usually a data fitting term with respect to image f (the imaging data); usually, F is known in the image processing community as the *fidelity functional*. In the second term, functional R measures the regularity (or lack of regularity) of v; R is known as the *regularization functional*. There are many possibilities concerning the choice of F and R. To explain the basic ideas we consider the simple case where the fidelity functional F is defined by

$$F(v;f) = \frac{1}{2}\int_{\Omega}(v-f)^2 dx, \tag{2}$$

where $dx = dx_1 dx_2$. Concerning the regularization functionals, we will use below geometrical approaches to present three classical functionals R and also a new one (the Willmore bending functional).

For a given image, the most visual information is contained in the *level set curves*. We recall that for a two-variable function (image) v, the level set associated with $c \in \mathbb{R}$ is the set defined by

$$\Gamma_c(v) = \{x = (x_1, x_2) \in \Omega \mid v(x) = c\},$$

which is normally a contour in $\mathbb{R}^2$. The geometric information contained in Γ_c (such as length and curvature) will be used to define the regularization functionals.

2.2 Total variation regularization

Summing the lengths of all the level set curves together, and using the *coarea* formula (Evans, 2018), we obtain

$$R_1(v) = \int_{-\infty}^{\infty} |\Gamma_c(v)| dc = \int_{\Omega} |\nabla v| dx, \tag{3}$$

where $|\Gamma_c(v)|$ denotes the length of curve $\Gamma_c(v)$. Functional $R_1(v)$ is clearly the well-known *total variation regularizer* proposed by *Rudin*, *Osher* and *Fatemi* (ROF model) in Rudin et al. (1992). The mathematical and denoising properties of the ROF model have been discussed extensively, together with the iterative treatment of its discrete analogues obtained by finite difference and finite element approximations (see Acar and Vogel, 1994; Bellettini et al., 2002; Chambolle and Lions, 1997; Chambolle and Pock, 2019; Kärkkäinen and Majava, 2000a,b; Kärkkäinen et al., 2001; Shi et al., 2012; Vese, 2001). It is worth mentioning that the total variation (TV) functional in (3) plays also an important role in the modelling of *viscoplasticity* phenomena as shown in, e.g., Glowinski and Le Tallec (1989), Glowinski and Wachs (2011), Dean et al. (2007), Fortin and Glowinski (1983) and Marchuk (1990) (see also the references therein).

Since its occurrence in 1992, ROF has become the most popular image restoration model. Indeed, ROF yields distinct results for image denoising, while preserving the edges and contours of objects. However, since the regularization part of the ROF model relies only on the length of the level curves (as shown by (3)), it possesses some significant drawbacks under some circumstances, such as the loss of image contrast, the smearing of corners and the staircase effect (see Chambolle and Lions, 1997; Chan et al., 2006; Meyer, 2001; Strong and Chan, 2003; Zhu et al., 2013a). In order to remedy these unfavourable properties, models using higher order derivatives have been proposed (Ambrosio and Masnou, 2003a; Bredies et al., 2010; Brito-Loeza and Chen, 2010b; Chan et al., 2000, 2002; Chen et al., 2006; Lysaker and Tai, 2006; Lysaker et al., 2003; Mumford, 1994; You and Kaveh, 2000) and Hintermüller and Papafitsoros (2019) (a survey), some of them discussed below.

2.3 The Euler elastica regularization

Beside the length, the *curvature* of the level curves is another important intrinsic attribute. Combining in an appropriate fashion the level curve and curvature information, one obtains a regularization functional known as the *Euler elastica*. The Euler elastica energy functional for image processing was introduced and further discussed in Mumford (1994), Chan et al. (2002), Ambrosio and Masnou (2003a), Ambrosio and Masnou (2003b), Masnou and Morel (1998) and Sochen et al. (1998). It is widely applied to various image processing problems, such as *denoising* (Ambrosio and Masnou, 2003a,b; Masnou, 2002; Masnou and Morel, 1998; Tai et al., 2011; Zhang et al., 2017), *inpainting* (Brito-Loeza and Chen, 2010a; Chan et al., 2002; Esedoglu and Shen, 2002; Yashtini and Kang, 2016), *illusory contour* (Kang et al., 2014; Masnou and Morel, 2006), *image segmentation* (Bae et al., 2017; Cai et al., 2013; Esedoglu and March, 2003; Nitzberg et al., 1993; Tan et al., 2019; Zhang and Chen, 2016; Zhu et al., 2006, 2013b) and (Kang et al., 2019) (a survey chapter in this volume of the Handbook) and *image interpolation* (zooming) (Tai et al., 2011). Mathematical properties of the Euler elastica model are discussed in Chan et al. (2002). It is observed in Bertalmío and Levine (2014) that the curvature of the level curves is less corrupted than the image itself, justifying thus the curvature based denoising approach.

A brief derivation of the elastica model is given just below:

For a general curve $\Gamma \subset \mathbb{R}^2$ with curvature κ, the *Euler elastica energy* is defined as:

$$\int_\Gamma (a + b|\kappa|^2)ds,$$

where s refers to the arc-length, and a, b (both >0) are weighting parameters. For a given function (or image) v, it follows from Carmo (1976) that the curvature of the level curve Γ_c is

$$\kappa = \nabla \cdot \left(\frac{\nabla v}{|\nabla v|}\right).$$

Thus the Euler elastica energy for the level curve Γ_c is

$$l(c) = \int_{\Gamma_c} \left(a + b \left[\nabla \cdot \left(\frac{\nabla v}{|\nabla v|} \right) \right]^2 \right) ds.$$

Summing up the Euler elastica energies for all the curves Γ_c, $c \in (-\infty, +\infty)$, and using again the coarea formula (Evans, 2018), one obtains the total Euler elastica energy for the given function (image) v, namely:

$$\begin{aligned}\int_{-\infty}^{+\infty} l(c) dc &= \int_{-\infty}^{+\infty} \int_{\Gamma_c} \left(a + b \left[\nabla \cdot \left(\frac{\nabla v}{|\nabla v|} \right) \right]^2 \right) dsdc \\ &= \int_{\Omega} \left(a + b \left[\nabla \cdot \left(\frac{\nabla v}{|\nabla v|} \right) \right]^2 \right) |\nabla v| dx.\end{aligned}$$

Replacing κ^2 by $|\kappa|^p$ for $p \in [1, +\infty)$, one obtains the so-called *generalized Euler elastica energy*

$$R_2(v) = \int_{\Omega} \left(a + b \left[\nabla \cdot \left(\frac{\nabla v}{|\nabla v|} \right) \right]^p \right) |\nabla v| dx. \tag{4}$$

Most applications rely on either $p = 1$ (Bae et al., 2017; Tan et al., 2019) or $p = 2$ (Tai et al., 2011). In this chapter we will consider only $p = 2$.

The regularization functionals we have been considering so far, namely R_1 and R_2 have been derived from the geometrical properties of the level set curves. The two functionals we are going to consider in the following paragraphs originate from the *curvature properties* of the graph of function v.

2.4 L^1-Mean curvature and L^1-Gaussian curvature energies

Let v be a real-valued function defined over Ω. Its graph $\Gamma(v)(= \{(x_1, x_2, x_3) \mid (x_1, x_2) \in \Omega, x_3 = v(x_1, x_2)\})$ is a surface of $\mathbb{R}^3$ (Kimmel et al., 2000; Sochen et al., 1998; Zhu et al., 2013a, a), whose properties can be used to define regularization functionals (Brito-Loeza et al., 2016; Elsey and Esedog-lu, 2009; Gong and Sbalzarini, 2013; Myllykoski et al., 2015; Zhu et al., 2013a). The *mean* and *Gaussian curvatures* are surface properties that have been used to define image regularization functionals (the mean curvature in Zhu et al. (2013a), Zhu and Chan (2012) and Myllykoski et al. (2015), the Gaussian curvature in Brito-Loeza et al. (2016) and Gong and Sbalzarini (2013)). Gaussian curvature (Elsey and Esedog-lu, 2009) and mean curvature (Brito-Loeza and Chen, 2013) are also popular in *surface fairing*.

It follows from, e.g., Carmo, 1976; Goldman, 2005, that the mean curvature H and the Gaussian curvature K of $\Gamma(v)$ are given by

$$H = \nabla \cdot \left(\frac{\nabla v}{\sqrt{1 + |\nabla v|^2}} \right), \quad K = \frac{\det \mathbf{D}^2 v}{(1 + |\nabla v|^2)^2}, \tag{5}$$

where $\mathbf{D}^2 v$ is the *Hessian* of function v, that is $\mathbf{D}^2 v = \left(\frac{\partial^2 v}{\partial x_i \partial x_j}\right)_{1 \leq i, j \leq 2}$, implying

$$\det \mathbf{D}^2 v = \frac{\partial^2 v \partial^2 v}{\partial x_1^2 \partial x_2^2} - \left|\frac{\partial^2 v}{\partial x_1 \partial x_2}\right|^2.$$

A classical regularization functional for image restoration associated with the *mean curvature* is the L^1-mean curvature functional R_3 defined (Myllykoski et al., 2015; Zhu and Chan, 2012; Zhu et al., 2013a) by

$$R_3(v) = \int_\Omega \left| \nabla \cdot \left(\frac{\nabla v}{\sqrt{1 + |\nabla v|^2}} \right) \right| dx. \tag{6}$$

In Zhu et al. (2013a), one can find a discussion of the properties of R_3 when applied to *image segmentation* and *noise removal*. On the basis of numerical experiments comparing with various methods, it appears that the variational model based on R_3 is superior to ROF (see Rudin et al., 1992) and to Euler elastica (see Tai et al., 2011) for noise removal, while preserving object corners and image contrasts (see Zhu and Chan, 2012 for details). In Zhu et al. (2013a) and Myllykoski et al. (2015), one applies ADMM to the numerical solution of image processing problems relying on the regularization functional R_3.

An alternative to the regularization functional R_3 is provided by the following L^1-Gaussian curvature functional

$$\int_\Omega \frac{|\det \mathbf{D}^2 v)|}{(1 + |\nabla v|^2)^2} dx. \tag{7}$$

The rationality for the use of functional (7) and applications to image denoising are discussed in Brito-Loeza et al. (2016).

2.5 Willmore bending energy

For a two-dimensional surface S, the *Willmore bending energy* (Willmore, 1996) is defined (with obvious notation) by:

$$\int_S (H^2 - K) dS, \tag{8}$$

The Willmore energy combines (nonlinearly) both the mean and Gaussian curvatures. Suppose that S is the graph of a function; we observe that in (8) the integration is done over S, not over the function domain of definition. Using the functional in (8) for regularization purposes leads to very challenging minimization problems. It is a very challenging task to minimize (8), so various approximation methods are proposed, see Bretin et al. (2015) and the references therein.

If we assume the surface function is periodic on the image domain, the integral of Gaussian curvature in (8) is a constant. This leads to the regularization functional R_4 defined by

$$R_4(v) = \int_{\Gamma(v)} H^2 dS = \int_\Omega \left[\nabla \cdot \left(\frac{\nabla v}{\sqrt{1+|\nabla v|^2}} \right) \right]^2 \sqrt{1+|\nabla v|^2} dx. \quad (9)$$

The full Willmore bending energy (as in (8)) will be treated in some forthcoming work. To avoid confusion with the L^1-mean curvature energy (defined by (6)), we still use the term Willmore energy (or functional) for R_4 defined by (9).

Replacing H^2 by $|H|^p$, for $p \in [1, +\infty)$, and introducing two nonnegative numbers a and b, one obtains the so-called *generalized Willmore bending energy*, that we still denote by R_4 and define by

$$R_4(v) = \int_\Omega \left[a + b \left| \nabla \cdot \left(\frac{\nabla v}{\sqrt{1+|\nabla v|^2}} \right) \right|^p \right] \sqrt{1+|\nabla v|^2} dx. \quad (10)$$

In this chapter, the only case to be considered corresponds to $a = 0$, $b = 1$ and $p = 2$. Generalization to other values of a, b and p is easy.

2.6 Summary

In Sections 2.2–2.5 we identified several *regularization functionals*. Combining them with the *fidelity functional* defined by (2), one obtains the following family of variational models for image processing

$$\arg \min_{v \in \mathcal{V}_i} \left\{ \theta R_i(v) + \frac{1}{2} \int_\Omega (f - v)^2 dx \right\}, \quad i = 1, 2, 3, 4, \quad (11)$$

where $\mathcal{V}_i$ is a well-chosen functional space, and the regularization functionals R_i are those discussed in Sections 2.2–2.5.

With the exception of the *ROF model* (the one corresponding to $i = 1$), to the best of our knowledge the "good" functional spaces $\mathcal{V}_i$ have yet to be identified if $i = 2$, 3 and 4, implying that the mathematical treatment of the corresponding variational models is largely *formal*. For $i = 1$ (the ROF model) it is well known that the right space $\mathcal{V}_1$ is the space $BV(\Omega)$ of the functions with *bounded variation* over Ω (see, e.g., Tartar (2007) for the definition and basic properties of $BV(\Omega)$).

3 Basic results, notations and an introduction to operator-splitting methods

3.1 Basic results and notations

In Sections 4–6 of this chapter we will assume that $\Omega = (0, L_1) \times (0, L_2)$ with $L_1, L_2 > 0$. We will assume also that the functions we are dealing with are periodic over Ω in the Ox_1 and Ox_2 directions, a very common assumption in image processing.

In Deng et al. (2019), we proved (in a semiformal way) the following

Proposition 1. *Suppose that u is solution to problem* (11) *for* $i = 2$ (*Euler elastica model*). *We have then*

$$\int_\Omega u dx = \int_\Omega f dx. \tag{12}$$

One can extend easily the proof of Proposition 1 to the three other models in (11) ($i = 1, 3, 4$). From relation (12), it makes sense to introduce

$$\mathcal{V}_f = \{v \in \mathcal{V} \mid \int_\Omega (f - v) dx = 0\}.$$

Later, we will make use of the set S_1 defined by

$$S_1 = \{\mathbf{q} \in (\mathcal{L}^2(\Omega))^2 \mid \exists\, v \in \mathcal{V}_f, \text{ such that } \mathbf{q} = \nabla v\}. \tag{13}$$

Let us define the space $H_p^1(\Omega)$ (a space of *doubly periodic functions*) by

$$H_p^1(\Omega) = \{\varphi \in H^1(\Omega) \mid \varphi(0, x_2) = \varphi(L_1, x_2), \quad x_2 \in (0, L_2), \\ \varphi(x_1, 0) = \varphi(x_1, L_2), \quad x_1 \in (0, L_1)\},$$

and let us consider $\mathbf{q}(= (q_1, q_2)) \in \left(\mathcal{L}^2(\Omega)\right)^2$. It follows from, e.g., the Chapter 1 of Glowinski (2015) that the linear variational problem

$$\begin{cases} v_\mathbf{q} \in H_p^1(\Omega), \\ \displaystyle\int_\Omega \nabla v_\mathbf{q} \cdot \nabla w dx = \int_\Omega \mathbf{q} \cdot \nabla w dx, \forall w \in H_p^1(\Omega), \\ \displaystyle\int_\Omega v_\mathbf{q} dx = \int_\Omega f dx, \end{cases} \tag{14}$$

has a unique solution. This solution is the unique function in $H^1(\Omega)$ verifying

$$\begin{cases} \nabla^2 v_\mathbf{q} = \nabla \cdot \mathbf{q} \text{ (in the sense of distributions)}, \\ v_\mathbf{q}(0, x_2) = v_\mathbf{q}(L_1, x_2), \text{ a.e. on } (0, L_2), v_\mathbf{q}(x_1, 0) = v_\mathbf{q}(x_1, L_2), \text{ a.e. on } (0, L_1), \\ \left(\dfrac{\partial v_\mathbf{q}}{\partial x_1} - q_1\right)(0, x_2) = \left(\dfrac{\partial v_\mathbf{q}}{\partial x_1} - q_1\right)(L_1, x_2), \text{ a.e. on } (0, L_2), \\ \left(\dfrac{\partial v_\mathbf{q}}{\partial x_2} - q_2\right)(x_1, 0) = \left(\dfrac{\partial v_\mathbf{q}}{\partial x_2} - q_2\right)(x_1, L_2), \text{ a.e. on } (0, L_1), \\ \displaystyle\int_\Omega (f - v_\mathbf{q}) dx = 0. \end{cases}$$

We will denote $v_\mathbf{q}$ by $\nabla_f^{-1}\mathbf{q}$. It is easy to see that

$$\nabla_f^{-1}\mathbf{q} = v \text{ if } \mathbf{q} = \nabla v \text{ and } v \in \mathcal{V}_f.$$

For any $\mathbf{q} \in (\mathcal{L}^2(\Omega))^2$, we define the functional G by

$$G(\mathbf{q}) = \frac{1}{2}\int_\Omega (\nabla_f^{-1}\mathbf{q} - f)^2 dx, \ \ \mathbf{q} \in (\mathcal{L}^2(\Omega))^2$$

implying that the fidelity functional F, in (1) and (2), verifies

$$F(v; f) = G\left(\nabla_f^{-1}\mathbf{q}\right), \ \ \text{if } \mathbf{q} = \nabla v. \tag{15}$$

Let us consider the sets X and Y with $Y \subset X$ strictly. The (proper) functional $I_Y : X \to \mathbb{R}\bigcup\{+\infty\}$ defined by

$$I_Y(\mathbf{x}) = \begin{cases} 0 & \text{if } \ \mathbf{x} \in Y, \\ +\infty & \text{if } \ \mathbf{x} \in X \backslash Y, \end{cases} \tag{16}$$

is called the *indicator functional* of set Y with respect to X. Indicator functionals will be quite useful in the following parts of this chapter.

We recall that a functional $J : X \to \mathbb{R}\bigcup\{+\infty\}$ is proper if there exists $\mathbf{x}_0 \in X$ such that $J(\mathbf{x}_0) \in \mathbb{R}$ and $J(\mathbf{x}) > -\infty$, $\forall\, \mathbf{x} \in X$. All the functionals encountered in this chapter are proper.

3.2 The Lie and Marchuk–Yanenko operator-splitting schemes for the time discretization of initial value problems

Since we are going to use these two schemes extensively in the sequel, the *Lie* and *Marchuk–Yanenko* schemes deserve a (brief) introduction, in order to make this chapter reasonably self-contained.

3.2.1 Generalities

Let us consider the following (autonomous) initial value problem

$$\begin{cases} \dfrac{d\phi}{dt} + A(\phi) = 0 \ \text{ on } \ (0, T) \ (\text{with } 0 < T \leq +\infty), \\ \phi(0) = \phi_0, \end{cases} \tag{17}$$

where operator A maps a vector space Φ into itself (modulo a linear duality map if necessary); we suppose that $\phi_0 \in \Phi$. We suppose also that operator A has a nontrivial decomposition, such as

$$A = \sum_{j=1}^{J} A_j, \tag{18}$$

with $J \geq 2$ (by nontrivial we mean that the operators A_j are individually simpler than A).

A question arising naturally is clearly:

Can one take advantage of decomposition (18) to solve problem (17)?

It is common knowledge (in the Computational Physics and Mechanics community, at least) that the answer to the above question is definitely *yes*.

Many schemes have been designed to take advantage of decomposition (18) when solving (17). Below, we will briefly discuss the simplest one, namely the *Lie scheme*.

3.3 Time discretization of the initial value problem (17) by the Lie scheme

Let τ (>0) be a *time-discretization step* (we suppose τ uniform, for simplicity): we denote $n\tau$ by t^n. With ϕ^n denoting an approximation of $\phi(t^n)$, the Lie scheme reads as follows (see, e.g., Chorin et al. (1978) and Chapter 6 of Glowinski (2003)):

$$\phi^0 = \phi_0; \tag{19}$$

then, for $n \geq 0$, $\phi^n \rightarrow \phi^{n+1}$ as follows:

For $j = 1, \ldots, J$, solve

$$\begin{cases} \dfrac{d\phi_j}{dt} + A_j(\phi_j) = 0 \ \text{ on } \ (t^n, t^{n+1}), \\ \phi_j(t^n) = \phi^{n+(j-1)/J}, \end{cases} \tag{20}$$

and set

$$\phi^{n+j/J} = \phi_j(t^{n+1}). \tag{21}$$

If the space Φ where problem (17) is taking place is finite dimensional, and if the operators, A_j, $j = 1, 2, \ldots, J$, are smooth enough, then $\|\phi^n - \phi(t^n)\| = O(\tau)$, function ϕ being the solution of (17).

Remark 1. The above scheme applies also for *multivalued* operators (such as the subdifferential of proper, lower semicontinuous, convex functionals) but first-order accuracy is not guaranteed anymore for such a case. We will encounter this type of situations in Sections 4–7 of this chapter.

Remark 2. Scheme (19) and (21) is semiconstructive in the sense that one still has to solve the subinitial value problems in (20) for every j (unless some of them are sufficiently simple to have a closed form solution). Suppose that one discretizes these subproblems using just *one step of the backward Euler scheme*. The resulting scheme reads as follows:

$$\phi^0 = \phi_0; \tag{22}$$

then, for $n \geq 0$, assuming that $\phi^{n+(j-1)/J}$ is known, compute $\phi^{n+j/J}$ via the solution of

$$\frac{\phi^{n+j/J} - \phi^{n+(j-1)/J}}{\tau} + A_j(\phi^{n+j/J}) = 0, \tag{23}$$

for $j = 1, \ldots, J$. If operator A_j is *multivalued*, one should replace $(23)_j$ by

$$\phi^{n+j/J} + \tau A_j(\phi^{n+j/J}) \ni \phi^{n+(j-1)/J}. \tag{24}$$

Scheme (22) and (23) is known as the *Marchuk–Yanenko scheme* (see, e.g., Refs. Marchuk (1990) and Glowinski (2003) for more details, and applications (in Computational Fluid Dynamics in particular)). As the Lie scheme, the Marchuk–Yanenko scheme is *generically first-order accurate at best.* However, this low order accuracy is compensated by this scheme simplicity, flexibility and robustness, making it a method of choice for those situations where J is large and some of the operators A_j are poorly differentiable and even nonsmooth. The numerical simulation of time dependent *viscoplastic flows* provides such situations as shown in, e.g., Dean et al. (2007), as does also *image processing* as shown in Sections 4–7.

Remark 3. Suppose that in (23) (resp., (24)), operator A_j is the differential (resp., the generalized differential) of a (proper) functional J_j. Quite often, from the context of the problem under consideration, it makes sense to consider (23) or (24) as the *optimality condition* associated with the following optimization problem:

$$\phi^{n+j/J} = \underset{\phi \in \Phi}{\operatorname{argmin}} \left\{ \frac{1}{2} \| \phi - \phi^{n+(j-1)/J} \|^2 + \tau J_j(\phi) \right\}, \tag{25}$$

giving to scheme (22) and (23) the flavour of a *descent proximal algorithm.* Relation (25) assumes that Φ is a Hilbert space.

3.4 Asymptotic properties of the Lie and Marchuk–Yanenko schemes

Consider now the *steady-state problem* associated with (17), namely

$$A(\phi) = 0. \tag{26}$$

Assume that problem (17) has steady-state solutions. These solutions are necessarily solutions of problem (26). A classical method to solve (26) is then to integrate (17) on the time interval $(0, +\infty)$, with (if possible) ϕ_0 properly chosen. Suppose that decomposition (18) holds. It is then tempting to use either the Lie or Marchuk–Yanenko schemes to capture these steady-state solutions. It is where one encounters the *main drawback* of the Lie and Marchuk–Yanenko schemes, namely their *asymptotic inconsistency*. Indeed, one observes the following phenomena, generically:

1. If converging for $j = 1, \ldots, J$, the sequences $(\phi^{n+j/J})_{n \geq 0}$ converge to different limits, the distance between these limits being $O(\tau)$ at best.
2. None of the above limits is a steady-state solution, but their distance to a steady-state solution converges to 0 as $\tau \to 0$ (if a steady-state solution does exist).

This behaviour has been proved in, e.g., Chapter 6 in Glowinski (2003), for simple situations where the space Φ is finite dimensional and the operators A_j affine.

There are several ways to overcome asymptotic inconsistency: among them, *averaging* and *symmetrization*. Actually, the simplest way to reduce the so-called *splitting error* is to take a small time step τ. That's the approach followed by the authors of this chapter.

Remark 4. The *Douglas–Rachford* and *Peaceman–Rachford* alternating direction schemes are asymptotically consistent, but they have drawbacks of their own, their (relatively) slow convergence to steady-state solutions in particular. They have also difficulties at handling large values of J, particularly if some of the operators A_j are nonsmooth or multivalued. We suggest to the interested readers to have a look at Glowinski et al. (2016a). The article Glowinski et al. (2016a) is dedicated to the numerical solution of

$$|\nabla\psi| - \frac{|1-\mathbf{V}\cdot\nabla\psi|}{c} = 0, \tag{27}$$

an *Eikonal type equation* from *Acoustics*, modelling *sound propagation in moving media*, where c (>0) is the medium at rest local speed of sound, and $\mathbf{V}$ is the medium motion velocity). The article Glowinski et al. (2016a) contains a thorough comparison of operator-splitting methods applied to the solution of problem (27). Among these splitting methods, let us mention the Lie and Marchuk–Yanenko splitting schemes, and the Peaceman–Rachford and Douglas–Rachford alternating direction methods.

4 Operator splitting method for Euler elastica energy functional

4.1 Formulation of the problem and operator-splitting solution methods

Following Deng et al. (2019), we are going to discuss in this section the numerical solution of problem (11) when $i = 2$, that is of the minimization problem (11) associated with the Euler elastica functional. For $p = 2$ in (4), the Euler elastica minimization problem for image restoration reads as

$$\min_{v\in\mathcal{V}}\left\{\int_\Omega\left[a+b\left(\nabla\cdot\frac{\nabla v}{|\nabla v|}\right)^2\right]|\nabla v|dx+\frac{\theta^{-1}}{2}\int_\Omega(v-f)^2dx\right\}, \tag{28}$$

where the nonnegative coefficients a and b are chosen according to the application at hand.

The *nonconvexity* and the *nonsmoothness* of the functional in (28) made the above minimization problem a very challenging one. Indeed, to the best of our knowledge, there are not many efficient numerical methods to solve problem (28). Among the existing solution methods, let us mention graph cut methods (Bae et al., 2011; El-Zehiry and Grady, 2010), a linear programming method (Schoenemann et al., 2009), augmented Lagrangian methods (ALM) (Tai et al., 2011; Yashtini and Kang, 2016). Moreover, a linearization method was used in Zhang et al. (2017), to simplify the ALM algorithm discussed in Tai et al. (2011). In Yashtini and Kang (2016), one discussed a split-Bregman method applied to the linearized Euler elastica model from Zhang et al. (2017). In Chambolle and Pock (2019) and Bredies et al. (2013), one tried to convexify problem (28), in order to make its solution easier. Recently, a method based on the Lie and Marchuk–Yanenko operator-splitting schemes was proposed in Deng et al. (2019), for the solution of problem (28). One can see more details on the algorithms for Euler's elastica in Kang et al. (2019), especially for image segmentation. This is the method we are going to describe now.

An important issue related to $\frac{\nabla v}{|\nabla v|}$ is that this vector-valued function makes no sense on this part of Ω where $\nabla v = \mathbf{0}$. An obvious (and once popular) way to overcome this difficulty is to replace $|\nabla v|$ by $\sqrt{|\nabla v|^2 + \epsilon^2}$, where ϵ is a small parameter. A more sophisticated way to proceed (borrowed from *visco-plasticity* (Dean et al., 2007; Glowinski and Wachs, 2011)) is to replace $\frac{\nabla v}{|\nabla v|}$ by a vector-valued function $\boldsymbol{\mu}$ $(= (\mu_1, \mu_2))$ verifying

$$\boldsymbol{\mu} \cdot \nabla v = |\nabla v|, \, |\boldsymbol{\mu}| \leq 1 \tag{29}$$

with $|\boldsymbol{\mu}| = \sqrt{\mu_1^2 + \mu_2^2}$. Then, problem (28) can be rewritten as

$$\min_{(v, \boldsymbol{\mu}) \in W} \left\{ \int_\Omega \left(a + b|\nabla \cdot \boldsymbol{\mu}|^2 \right) |\nabla v| dx + \frac{\theta^{-1}}{2} \int_\Omega (f - v)^2 dx \right\}, \tag{30}$$

where (formally)

$$W = \left\{ (v, \boldsymbol{\mu}) \in H_p^1(\Omega) \times H_p(\Omega, \text{div}) \,|\, \boldsymbol{\mu} \cdot \nabla v = |\nabla v|, |\boldsymbol{\mu}| \leq 1 \right\}, \tag{31}$$

and

$$H_p(\Omega, \text{div}) = \{\boldsymbol{\mu} = (\mu_1, \mu_2) \in H(\Omega, \text{div}) \,|\, \mu_1(0, \cdot) = \mu_1(L_1, \cdot), \mu_2(\cdot, 0) = \mu_2(\cdot, L_2)\},$$

that is, the real-valued function μ_1 (resp., μ_2) is L_1 (resp., L_2)-periodic in the Ox_1 (resp., Ox_2) direction. We defined the space $H_p^1(\Omega)$ in Section 3.1, and recall that space $H(\Omega,\ \text{div})$ is defined by $H(\Omega, \text{div}) = \left\{ \boldsymbol{\mu} \in \left(\mathcal{L}^2(\Omega)\right)^2 \,|\, \nabla \cdot \boldsymbol{\mu} \in \mathcal{L}^2(\Omega) \right\}$, $\nabla \cdot \boldsymbol{\mu}$ being in the sense of distributions.

Let us define the functional set S_2 by

$$S_2 = \left\{ (\mathbf{q}, \boldsymbol{\mu}) \in \left(\mathcal{L}^2(\Omega)\right)^2 \times \left(\mathcal{L}^2(\Omega)\right)^2 \,|\, \mathbf{q} \cdot \boldsymbol{\mu} = |\mathbf{q}|, |\boldsymbol{\mu}| \leq 1 \right\}. \tag{32}$$

Then, one can reformulate problem (28) or (30) as:

$$\min_{(\mathbf{q},\boldsymbol{\mu})\in\widehat{W}}\left\{\int_{\Omega}\left(a+b|\nabla\cdot\boldsymbol{\mu}|^2\right)|\mathbf{q}|dx+\frac{\theta^{-1}}{2}\int_{\Omega}\left(\nabla_f^{-1}\mathbf{q}-f\right)^2+I_{S_1}(\mathbf{q})+I_{S_2}(\mathbf{q},\boldsymbol{\mu})\right\},\tag{33}$$

with $\widehat{W}=(\mathcal{L}^2(\Omega))^2\times H_p(\Omega,\text{div})$, where $\nabla_f^{-1}\mathbf{q}$ has been defined in Section 3.1, and where I_{S_1} (resp., I_{S_2}) is the indicator functional of S_1 (resp., S_2). We will denote by T_1 the functional $(\mathbf{q},\boldsymbol{\mu})\rightarrow\int_{\Omega}\left(a+b|\nabla\cdot\boldsymbol{\mu}|^2\right)|\mathbf{q}|dx$. Suppose now that the pair $(\mathbf{p}, \boldsymbol{\lambda})$ is a solution of the minimization problem (33). Then $u=\nabla_f^{-1}\mathbf{p}$ is a solution to problem (28) and the following optimality conditions hold:

$$\begin{cases}\partial_{\mathbf{q}}T_1(\mathbf{p},\boldsymbol{\lambda})+\theta^{-1}D_{\mathbf{q}}G(\mathbf{p})+\partial_{\mathbf{q}}I_{S_1}(\mathbf{p})+\partial_{\mathbf{q}}I_{S_2}(\mathbf{p},\boldsymbol{\lambda})\ni\mathbf{0},\\ D_{\mu}T_1(\mathbf{p},\boldsymbol{\lambda})+\partial_{\mu}I_{S_2}(\mathbf{p},\boldsymbol{\lambda})\ni\mathbf{0},\end{cases}\tag{34}$$

where D (resp., ∂) denotes classical (resp., generalized) differentials. We associate with (34) the following initial value problem (dynamical flow):

$$\begin{cases}\begin{cases}\dfrac{\partial\mathbf{p}}{\partial t}+\partial_{\mathbf{q}}T_1(\mathbf{p},\boldsymbol{\lambda})+\theta^{-1}D_{\mathbf{q}}G(\mathbf{p})+\partial_{\mathbf{q}}I_{S_1}(\mathbf{p})+\partial_{\mathbf{q}}I_{S_2}(\mathbf{p},\boldsymbol{\lambda})\ni\mathbf{0},\\ \gamma\dfrac{\partial\boldsymbol{\lambda}}{\partial t}+D_{\mu}T_1(\mathbf{p},\boldsymbol{\lambda})+\partial_{\mu}I_{S_2}(\mathbf{p},\boldsymbol{\lambda})\ni\mathbf{0},\end{cases} & \text{in }\Omega\times(0,+\infty),\\ (\mathbf{p}(0),\boldsymbol{\lambda}(0))=(\mathbf{p}_0,\boldsymbol{\lambda}_0).\end{cases}\tag{35}$$

In (35): (i) $(\mathbf{p}_0, \boldsymbol{\lambda}_0)\in S_2$ (ii) γ is a strictly positive real-valued function whose selection will be (briefly) discussed later, and (iii) $\phi(t)$ denotes the function $x\rightarrow\phi(x, t)$ (a very popular notation in the *evolution equation* community).

Let

$$A_1=\begin{pmatrix}\partial_{\mathbf{q}}T_1\\ D_{\mu}T_1\end{pmatrix},\quad A_2=\begin{pmatrix}\partial_{\mathbf{q}}I_{S_2}\\ \partial_{\mu}I_{S_2}\end{pmatrix},\quad A_3=\begin{pmatrix}\theta^{-1}D_{\mathbf{q}}G+\partial_{\mathbf{q}}I_{S_1}\\ 0\end{pmatrix}.$$

Then problem (35) is clearly a particular case of problem (17) and (18), with $T=+\infty$ and $J=3$. From this observation, the *Lie scheme*, (19) and (20), is a natural candidate for the solution of the initial value problem (35). Applying therefore scheme (17) and (18) to the solution of problem (35), one obtains

$$(\mathbf{p}^0,\boldsymbol{\lambda}^0)=(\mathbf{p}_0,\boldsymbol{\lambda}_0)\tag{36}$$

For $n\geq 0$, $(\mathbf{p}^n,\boldsymbol{\lambda}^n)\rightarrow\left(\mathbf{p}^{n+1/3},\boldsymbol{\lambda}^{n+1/3}\right)\rightarrow\left(\mathbf{p}^{n+2/3},\boldsymbol{\lambda}^{n+2/3}\right)\rightarrow\left(\mathbf{p}^{n+1},\boldsymbol{\lambda}^{n+1}\right)$.

1st Fractional step: Solve

$$\begin{cases} \begin{cases} \dfrac{\partial \mathbf{p}}{\partial t} + \partial_{\mathbf{q}} T_1(\mathbf{p}, \boldsymbol{\lambda}) \ni \mathbf{0}, \\ \gamma \dfrac{\partial \boldsymbol{\lambda}}{\partial t} + D_{\mu} T_1(\mathbf{p}, \boldsymbol{\lambda}) = \mathbf{0}, \end{cases} \quad \text{in } \Omega \times (t^n, t^{n+1}), \\ (\mathbf{p}(t^n), \boldsymbol{\lambda}(t^n)) = (\mathbf{p}^n, \boldsymbol{\lambda}^n), \end{cases} \tag{37}$$

and set

$$(\mathbf{p}^{n+1/3}, \boldsymbol{\lambda}^{n+1/3}) = (\mathbf{p}(t^{n+1}), \ \ \boldsymbol{\lambda}(t^{n+1})). \tag{38}$$

2nd Fractional step: Solve

$$\begin{cases} \begin{cases} \dfrac{\partial \mathbf{p}}{\partial t} + \partial_{\mathbf{q}} I_{S_2}(\mathbf{p}, \boldsymbol{\lambda}) \ni \mathbf{0}, \\ \gamma \dfrac{\partial \boldsymbol{\lambda}}{\partial t} + \partial_{\mu} I_{S_2}(\mathbf{p}, \boldsymbol{\lambda}) \ni \mathbf{0}, \end{cases} \quad \text{in } \Omega \times (t^n, t^{n+1}), \\ (\mathbf{p}(t^n), \boldsymbol{\lambda}(t^n)) = (\mathbf{p}^{n+1/3}, \boldsymbol{\lambda}^{n+1/3}), \end{cases} \tag{39}$$

and set

$$(\mathbf{p}^{n+2/3}, \boldsymbol{\lambda}^{n+2/3}) = (\mathbf{p}(t^{n+1}), \ \ \boldsymbol{\lambda}(t^{n+1})). \tag{40}$$

3rd Fractional step: Solve

$$\begin{cases} \begin{cases} \dfrac{\partial \mathbf{p}}{\partial t} + \theta^{-1} D_{\mathbf{q}} G(\mathbf{p}) + \partial_{\mathbf{q}} I_{S_1}(\mathbf{p}) \ni \mathbf{0}, \\ \gamma \dfrac{\partial \boldsymbol{\lambda}}{\partial t} = \mathbf{0}, \end{cases} \quad \text{in } \Omega \times (t^n, t^{n+1}), \\ (\mathbf{p}(t^n), \boldsymbol{\lambda}(t^n)) = (\mathbf{p}^{n+2/3}, \boldsymbol{\lambda}^{n+2/3}), \end{cases} \tag{41}$$

and set

$$(\mathbf{p}^{n+1}, \boldsymbol{\lambda}^{n+1}) = (\mathbf{p}(t^{n+1}), \ \ \boldsymbol{\lambda}^{n+2/3}). \tag{42}$$

In practice, we will employ the *Marchuk–Yanenko* variant of the Lie scheme, that is (from (23) to (24)), and assume that γ does not depend of n):

$$(\mathbf{p}^0, \boldsymbol{\lambda}^0) = (\mathbf{p}_0, \lambda_0). \tag{43}$$

For $n \geq 0$, $(\mathbf{p}^n, \boldsymbol{\lambda}^n) \to (\mathbf{p}^{n+1/3}, \boldsymbol{\lambda}^{n+1/3}) \to (\mathbf{p}^{n+2/3}, \boldsymbol{\lambda}^{n+2/3}) \to (\mathbf{p}^{n+1}, \boldsymbol{\lambda}^{n+1})$ as follows:

$$\begin{cases} \dfrac{\mathbf{p}^{n+1/3} - \mathbf{p}^n}{\tau} + \partial_{\mathbf{q}} T_1(\mathbf{p}^{n+1/3}, \boldsymbol{\lambda}^n) \ni \mathbf{0}, \\ \gamma \dfrac{\boldsymbol{\lambda}^{n+1/3} - \boldsymbol{\lambda}^n}{\tau} + D_{\mu} T_1(\mathbf{p}^{n+1/3}, \boldsymbol{\lambda}^{n+1/3}) = \mathbf{0}, \end{cases} \quad \text{in } \Omega \Rightarrow (\mathbf{p}^{n+1/3}, \boldsymbol{\lambda}^{n+1/3}), \tag{44}$$

$$\begin{cases} \dfrac{\mathbf{p}^{n+2/3}-\mathbf{p}^{n+1/3}}{\tau}+\partial_{\mathbf{q}} I_{S_2}(\mathbf{p}^{n+2/3},\boldsymbol{\lambda}^{n+2/3}) \ni \mathbf{0}, \\ \gamma\dfrac{\boldsymbol{\lambda}^{n+2/3}-\boldsymbol{\lambda}^{n+1/3}}{\tau}+\partial_{\mu} I_{S_2}(\mathbf{p}^{n+2/3},\boldsymbol{\lambda}^{n+2/3}) \ni \mathbf{0}, \end{cases} \text{in } \Omega \ \Rightarrow \ (\mathbf{p}^{n+2/3},\boldsymbol{\lambda}^{n+2/3}), \tag{45}$$

$$\begin{cases} \dfrac{\mathbf{p}^{n+1}-\mathbf{p}^{n+2/3}}{\tau}+D_{\mathbf{q}}G(\mathbf{p}^{n+1})+\partial_{\mathbf{q}} I_{S_1}(\mathbf{p}^{n+1}) \ni \mathbf{0}, \\ \gamma\dfrac{\boldsymbol{\lambda}^{n+1}-\boldsymbol{\lambda}^{n+2/3}}{\tau}=\mathbf{0}, \end{cases} \text{in } \Omega \ \Rightarrow \ (\mathbf{p}^{n+1},\boldsymbol{\lambda}^{n+1}). \tag{46}$$

4.2 On the solution of problem (44)

It follows from (44) that the vector-valued function $\mathbf{p}^{n+1/3}$ is nothing but the solution of the following minimization problem:

$$\mathbf{p}^{n+1/3} = \arg \min_{\mathbf{q}\in(\mathcal{L}^2(\Omega))^2} \left\{ \frac{1}{2}\int_\Omega |\mathbf{q}-\mathbf{p}^n|^2 dx + \tau\int_\Omega \left(a+b|\nabla\cdot\boldsymbol{\lambda}^n|^2\right)|\mathbf{q}|dx \right\}. \tag{47}$$

Problem (47) has a closed form solution (obtained by *shrinking* (Donoho, 1995; Fortin and Glowinski, 1983; Glowinski and Le Tallec, 1989; Tai et al., 2011; Wang et al., 2008)), namely

$$\mathbf{p}^{n+1/3} = \sup\left\{0, 1-\frac{c}{|\mathbf{p}^n|}\right\}\mathbf{p}^n, \tag{48}$$

where $c = \tau a + \tau b|\nabla\cdot\boldsymbol{\lambda}^n|^2$. Once $\mathbf{p}^{n+1/3}$ is known, one observes that the second equation in (44) shows that the vector-valued function $\boldsymbol{\lambda}^{n+1/3} = (\lambda_1^{n+1/3}, \lambda_2^{n+1/3})$ is the solution (at least formally) of the following *elliptic problem* with *periodic boundary conditions*

$$\begin{cases} \gamma\boldsymbol{\lambda}^{n+1/3} - 2b\tau\nabla\left(|\mathbf{p}^{n+1/3}|\nabla\cdot\boldsymbol{\lambda}^{n+1/3}\right) = \gamma\boldsymbol{\lambda}^n \text{ in } \Omega, \\ \lambda_1^{n+1/3}(0,\cdot) = \lambda_1^{n+1/3}(L_1,\cdot), \lambda_2^{n+1/3}(\cdot,0) = \lambda_2^{n+1/3}(\cdot,L_2), \\ \left(|\mathbf{p}^{n+1/3}|\nabla\cdot\boldsymbol{\lambda}^{n+1/3}\right)(0,\cdot) = \left(|\mathbf{p}^{n+1/3}|\nabla\cdot\boldsymbol{\lambda}^{n+1/3}\right)(L_1,\cdot), \\ \left(|\mathbf{p}^{n+1/3}|\nabla\cdot\boldsymbol{\lambda}^{n+1/3}\right)(\cdot,0) = \left(|\mathbf{p}^{n+1/3}|\nabla\cdot\boldsymbol{\lambda}^{n+1/3}\right)(\cdot,L_2). \end{cases} \tag{49}$$

It follows from, e.g., Tai et al. (2011), that the periodic conditions verified by $\boldsymbol{\lambda}^{n+1/3}$ imply that one can use *FFT* to solve problem (49) (see Tai et al. (2011) for details).

4.3 On the solution of problem (45)

Quite obviously, (39) is the Euler–Lagrange system associated with the following optimization problem

$$\min_{(\mathbf{q},\boldsymbol{\mu})\in S_2}\left\{\int_\Omega |\mathbf{q}-\mathbf{p}^{n+1/3}|^2 dx + \int_\Omega \gamma|\boldsymbol{\mu}-\boldsymbol{\lambda}^{n+1/3}|^2 dx\right\}. \tag{50}$$

Actually, one can solve problem (50) *pointwise*, since it is equivalent to

$$\begin{cases}(\mathbf{p}^{n+2/3}(x),\boldsymbol{\lambda}^{n+2/3}(x)) = \operatorname{argmin}_{(\mathbf{q},\boldsymbol{\mu})\in\sigma} h_{n+1/3}(\mathbf{q},\boldsymbol{\mu};x),\\ \text{a.e. } x\in\Omega,\end{cases} \tag{51}$$

where

$$\sigma = \{(\mathbf{q},\boldsymbol{\mu})\in\mathbb{R}^2\times\mathbb{R}^2 \mid \mathbf{q}\cdot\boldsymbol{\mu} = |\mathbf{q}|,\ |\boldsymbol{\mu}|\le 1\}$$

and

$$h_{n+1/3}(\mathbf{q},\boldsymbol{\mu};x) = \left|\mathbf{q}-\mathbf{p}^{n+1/3}(x)\right|^2 + \gamma\left|\boldsymbol{\mu}-\boldsymbol{\lambda}^{n+1/3}(x)\right|^2.$$

The key idea to solve the local problem (51) is to observe that $\sigma=\sigma_0\bigcup\sigma_1$, where

$$\sigma_0 = \{(\mathbf{q},\boldsymbol{\mu})\in\mathbb{R}^2\times\mathbb{R}^2 \mid \mathbf{q}=\mathbf{0},\ |\boldsymbol{\mu}|\le 1\},$$

$$\sigma_1 = \{(\mathbf{q},\boldsymbol{\mu})\in\mathbb{R}^2\times\mathbb{R}^2 \mid \mathbf{q}\neq\mathbf{0},\ \mathbf{q}\cdot\boldsymbol{\mu} = |\mathbf{q}|,\ |\boldsymbol{\mu}| = 1\}.$$

It follows clearly from the above relations that the solution to problem (51) verifies

$$\begin{aligned}(\mathbf{p}^{n+2/3}(x),\boldsymbol{\lambda}^{n+2/3}(x)) = \operatorname{argmin}\Big\{& h_{n+1/3}(\mathbf{p}_0^{n+2/3}(x),\boldsymbol{\lambda}_0^{n+2/3}(x);x),\\ & h_{n+1/3}(\mathbf{p}_1^{n+2/3}(x),\boldsymbol{\lambda}_1^{n+2/3}(x);x)\Big\}, \text{ a.e. on } \Omega,\end{aligned} \tag{52}$$

where $(\mathbf{p}_0^{n+2/3}(x),\boldsymbol{\lambda}_0^{n+2/3}(x))$ and $(\mathbf{p}_1^{n+2/3}(x),\boldsymbol{\lambda}_1^{n+2/3}(x))$ are the minimizers of $h_{n+1/3}(\cdot,\cdot\,;x)$ over σ_0 and σ_1, respectively.

Computing $(\mathbf{p}_0^{n+2/3}(x),\boldsymbol{\lambda}_0^{n+2/3}(x))$ is simple since $\mathbf{p}_0^{n+2/3}(x)=\mathbf{0}$, implying

$$\begin{cases}\boldsymbol{\lambda}_0^{n+2/3}(x) = \arg\min_{\boldsymbol{\mu}\in\mathbb{R}^2,\,|\boldsymbol{\mu}|\le 1}\left|\boldsymbol{\mu}-\boldsymbol{\lambda}^{n+1/3}(x)\right| = \dfrac{\boldsymbol{\lambda}^{n+1/3}(x)}{\max\{1,|\boldsymbol{\lambda}^{n+1/3}(x)|\}},\\ \text{a.e. on } \Omega.\end{cases} \tag{53}$$

As expected, computing $\left(\mathbf{p}_1^{n+2/3}(x),\boldsymbol{\lambda}_1^{n+2/3}(x)\right)$ is more complicated. We have

$$\mathbf{q} = \beta\boldsymbol{\mu} \tag{54}$$

with $\beta = |\mathbf{q}| > 0$, $|\boldsymbol{\mu}| = 1$ and $\mathbf{q} \cdot \boldsymbol{\mu} = |\mathbf{q}|$. Relation $|\boldsymbol{\mu}| = 1$ implies

$$h_{n+1/3}(\mathbf{q}, \boldsymbol{\mu}; x) = |\mathbf{q}|^2 - 2\mathbf{q} \cdot \mathbf{p}^{n+1/3} - 2\gamma\boldsymbol{\mu} \cdot \boldsymbol{\lambda}^{n+1/3} + const,$$

which implies in turn that minimizing functional $(\mathbf{q}, \boldsymbol{\mu}) \rightarrow h_{n+1/3}(\mathbf{q}, \boldsymbol{\mu}; x)$ is equivalent to solving

$$\min_{\beta>0} \min_{\boldsymbol{\mu}\in\mathbb{R}^2,\, |\boldsymbol{\mu}|=1} \left\{\frac{1}{2}\beta^2 - \beta\boldsymbol{\mu} \cdot \mathbf{p}^{n+1/3} - \gamma\boldsymbol{\mu} \cdot \boldsymbol{\lambda}^{n+1/3}\right\}. \tag{55}$$

For β given, the solution of the inner problem in (55) (the $\boldsymbol{\mu}$-problem) is

$$\boldsymbol{\mu}(\beta; x) = \frac{\beta\mathbf{p}^{n+1/3} + \gamma\boldsymbol{\lambda}^{n+1/3}}{|\beta\mathbf{p}^{n+1/3} + \gamma\boldsymbol{\lambda}^{n+1/3}|}, \tag{56}$$

implying that problem (55) reduces to

$$\min_{\beta>0} \left\{\frac{1}{2}\beta^2 - |\beta\mathbf{p}^{n+1/3} + \gamma\boldsymbol{\lambda}^{n+1/3}|\right\}. \tag{57}$$

There are many methods to solve problem (57) (Newton's method, bisection method, etc.). In Deng et al. (2019) we used the simple *fixed point method defined* by

$$\beta^{(k+1)} = \max\left\{0, \frac{\mathbf{p}^{n+1/3} \cdot (\beta^{(k)}\mathbf{p}^{n+1/3} + \gamma\boldsymbol{\lambda}^{n+1/3})}{|\beta^{(k)}\mathbf{p}^{n+1/3} + \gamma\boldsymbol{\lambda}^{n+1/3}|}\right\}. \tag{58}$$

Suppose that $\beta^* = \lim_{k\rightarrow+\infty}\beta^k$. We define $(\mathbf{p}_1^{n+2/3}, \boldsymbol{\lambda}_1^{n+2/3})$ by

$$\begin{cases} \boldsymbol{\lambda}_1^{n+2/3}(x) = \dfrac{\beta^*(x)\mathbf{p}^{n+1/3}(x) + \gamma(x)\boldsymbol{\lambda}^{n+1/3}(x)}{|\beta^*(x)\mathbf{p}^{n+1/3}(x) + \gamma(x)\boldsymbol{\lambda}^{n+1/3}(x)|}, \\ \mathbf{p}_1^{n+2/3}(x) = \beta^*\boldsymbol{\lambda}_1^{n+2/3}(x). \end{cases} \tag{59}$$

Remark 5. As one can guess, choosing γ properly is of some importance, particularly for problem (45) where a strong coupling takes place between the unknowns $\mathbf{p}$ and $\boldsymbol{\lambda}$. Following the analysis reported in Section 3.5 of Deng et al. (2019), the authors suggested taking in (44)–(46)$\gamma(= \gamma^n)$ defined by

$$\gamma^n(x) = \max\left\{|\mathbf{p}^{n+1/3}(x)|^2, \hat{\alpha}\right\},$$

where $\hat{\alpha}$ is a small positive number we introduced to avoid troubles on those parts of Ω where $\mathbf{p}^{n+1/3} = \mathbf{0}$. The computations reported in Deng et al. (2019) make use of $\hat{\alpha} = \sqrt{\tau}$.

4.4 On the solution of problem (46)

It is obvious that

$$\lambda^{n+1} = \lambda^{n+2/3}. \tag{60}$$

On the other hand, $\mathbf{p}^{n+1}$ is nothing but the solution to the following minimization problem

$$\mathbf{p}^{n+1} = \arg\min_{\mathbf{q}\in S_1} \left\{ \int_\Omega |\mathbf{q} - \mathbf{p}^{n+2/3}|^2 dx + \tau\theta^{-1} \int_\Omega |\nabla_f^{-1}\mathbf{q} - f|^2 dx \right\}. \tag{61}$$

From the definition of S_1, problem (61) is equivalent to

$$\begin{cases} \mathbf{p}^{n+1} = \nabla u^{n+1}, \\ u^{n+1} = \arg\min_{v\in H_p^1(\Omega)} \left\{ \frac{1}{2}\int_\Omega |\nabla v|^2 dx + \frac{\tau\theta^{-1}}{2}\int_\Omega |v-f|^2 dx - \int_\Omega \mathbf{p}^{n+2/3}\cdot\nabla v dx \right\}. \end{cases} \tag{62}$$

In PDE form, the Euler–Lagrange equation associated with problem (62), reads as an elliptic equation with periodic boundary conditions, namely:

$$\begin{cases} -\nabla^2 u^{n+1} + \tau\theta^{-1}u^{n+1} = \tau\theta^{-1}f - \nabla\cdot\mathbf{p}^{n+2/3}, \text{ in } \Omega, \\ u^{n+1}(0,\cdot) = u^{n+1}(L_1,\cdot),\ u^{n+1}(\cdot,0) = u^{n+1}(\cdot,L_2), \\ \left(\frac{\partial u^{n+1}}{\partial x_1} - p_1^{n+2/3}\right)(0,\cdot) = \left(\frac{\partial u^{n+1}}{\partial x_1} - p_1^{n+2/3}\right)(L_1,\cdot), \\ \left(\frac{\partial u^{n+1}}{\partial x_2} - p_2^{n+2/3}\right)(\cdot,0) = \left(\frac{\partial u^{n+1}}{\partial x_2} - p_2^{n+2/3}\right)(\cdot,L_2). \end{cases} \tag{63}$$

The efficient solution of problem (63) can be achieved by *FFT*.

The numerical results reported in Deng et al. (2019) validate the methodology discussed in Section 4. They rely on a space approximation by finite differences, an approach well suited to rectangular shaped domains Ω.

5 An operator-splitting method for the Willmore energy-based variational model

5.1 Formulation of the problem and operator-splitting solution methods

We will discuss in this section an operator-splitting method for the numerical solution of an *image restoration* problem relying on a *Willmore regularization*. The variational problem reads as follows:

$$u = \arg\min_{v\in\mathcal{V}} \left\{ \int_\Omega \left(\nabla\cdot\frac{\nabla v}{\sqrt{1+|\nabla v|^2}} \right)^2 \sqrt{1+|\nabla v|^2}\, dx + \frac{\theta^{-1}}{2}\int_\Omega (v-f)^2 dx \right\}. \tag{64}$$

Following the approach taken in Section 4 for the Euler elastica model, and in Zhu et al. (2013a) and Myllykoski et al. (2015), we introduce $\mathbf{q} = \nabla v$ and $\boldsymbol{\mu} = \frac{\mathbf{q}}{\sqrt{1+|\mathbf{q}|^2}}$, and (assuming that *periodicity* still holds) define

$$S_3(\mathbf{q}, \boldsymbol{\mu}) = \left\{ (\mathbf{q}, \boldsymbol{\mu}) | \mathbf{q} \in (\mathcal{L}^2(\Omega))^2, \boldsymbol{\mu} = \frac{\mathbf{q}}{\sqrt{1+|\mathbf{q}|^2}} \right\}. \tag{65}$$

One can reformulate (formally) problem (64) as the following close variant of problem (33)

$$\min_{(\mathbf{q},\boldsymbol{\mu}) \in \widehat{W}} \left\{ \int_\Omega (\nabla \cdot \boldsymbol{\mu})^2 \sqrt{1+|\mathbf{q}|^2} dx + \frac{\theta^{-1}}{2} \int_\Omega (\nabla_f^{-1} \mathbf{q} - f)^2 dx + I_{S_1}(\mathbf{q}) + I_{S_3}(\mathbf{q}, \boldsymbol{\mu}) \right\}, \tag{66}$$

where: (i) the space $\widehat{W}$ is as in (33), (ii) the function $\nabla_f^{-1}\mathbf{q}$ is as in Section 3.1, and (iii) the functional I_{S_3} is the indicator functional of the set S_3. In the following, we will denote by T_2 the functional $(\mathbf{q}, \boldsymbol{\mu}) \to \int_\Omega |\nabla \cdot \boldsymbol{\mu}|^2 \sqrt{1+|\mathbf{q}|^2} dx$.

Suppose now that the pair $(\mathbf{p}, \boldsymbol{\lambda})$ is a solution to the minimization problem (66). We have then (formally)

$$\begin{cases} D_\mathbf{q} T_2(\mathbf{p}, \boldsymbol{\lambda}) + \theta^{-1} D_\mathbf{q} G(\mathbf{p}) + \partial_\mathbf{q} I_{S_3}(\mathbf{p}, \boldsymbol{\lambda}) + \partial_\mathbf{q} I_{S_1}(\mathbf{p}) \ni \mathbf{0}, \\ D_\mu T_2(\mathbf{p}, \boldsymbol{\lambda}) + \partial_\mu I_{S_3}(\mathbf{p}, \boldsymbol{\lambda}) \ni \mathbf{0}, \end{cases} \tag{67}$$

where G is defined as in Section 3.1. Proceeding as in Section 4 for the Euler elastica model, we introduce the following initial value problem (a variant of system (35)):

$$\begin{cases} \begin{cases} \frac{\partial \mathbf{p}}{\partial t} + D_\mathbf{q} T_2(\mathbf{p}, \boldsymbol{\lambda}) + \theta^{-1} D_\mathbf{q} G(\mathbf{p}) + \partial_\mathbf{q} I_{S_3}(\mathbf{p}, \boldsymbol{\lambda}) + \partial_\mathbf{q} I_{S_1}(\mathbf{p}) \ni \mathbf{0}, \\ \gamma \frac{\partial \boldsymbol{\lambda}}{\partial t} + D_\mu T_2(\mathbf{p}, \boldsymbol{\lambda}) + \partial_\mu I_{S_3}(\mathbf{p}, \boldsymbol{\lambda}) \ni \mathbf{0}, \end{cases} & \text{in } \Omega \times (0, +\infty), \\ (\mathbf{p}(0), \boldsymbol{\lambda}(0)) = (\mathbf{p}_0, \boldsymbol{\lambda}_0). \end{cases} \tag{68}$$

Similarly, let

$$A_1 = \begin{pmatrix} \partial_\mathbf{q} T_2 \\ D_\mu T_2 \end{pmatrix}, \quad A_2 = \begin{pmatrix} \partial_\mathbf{q} I_{S_3} \\ \partial_\mu I_{S_3} \end{pmatrix}, \quad A_3 = \begin{pmatrix} \theta^{-1} D_\mathbf{q} G + \partial_\mathbf{q} I_{S_1} \\ 0 \end{pmatrix}.$$

Applying the Lie scheme with $J = 3$ to the solution of problem (68), one obtains

$$(\mathbf{p}^0, \boldsymbol{\lambda}^0) = (\mathbf{p}_0, \boldsymbol{\lambda}_0). \tag{69}$$

For $n \geq 0$, $(\mathbf{p}^n, \lambda^n) \to \left(\mathbf{p}^{n+1/3}, \lambda^{n+1/3}\right) \to \left(\mathbf{p}^{n+2/3}, \lambda^{n+2/3}\right) \to \left(\mathbf{p}^{n+1}, \lambda^{n+1}\right)$ as follows:

1st Fractional step: Solve

$$\begin{cases} \begin{cases} \dfrac{\partial \mathbf{p}}{\partial t} + D_{\mathbf{q}} T_2(\mathbf{p}, \lambda) = \mathbf{0}, \\ \gamma \dfrac{\partial \lambda}{\partial t} + D_\mu T_2(\mathbf{p}, \lambda) = \mathbf{0}, \end{cases} \text{in } \Omega \times (t^n, t^{n+1}), \\ (\mathbf{p}(t^n), \lambda(t^n)) = (\mathbf{p}^n, \lambda^n), \end{cases} \tag{70}$$

and set

$$(\mathbf{p}^{n+1/3}, \lambda^{n+1/3}) = (\mathbf{p}(t^{n+1}), \ \lambda(t^{n+1})). \tag{71}$$

2nd Fractional step: Solve

$$\begin{cases} \begin{cases} \dfrac{\partial \mathbf{p}}{\partial t} + \partial_{\mathbf{q}} I_{S_3}(\mathbf{p}, \lambda) \ni \mathbf{0} \\ \gamma \dfrac{\partial \lambda}{\partial t} + \partial_\mu I_{S_3}(\mathbf{p}, \lambda) \ni \mathbf{0} \end{cases} \text{in } \Omega \times (t^n, t^{n+1}), \\ (\mathbf{p}(t^n), \lambda(t^n)) = (\mathbf{p}^{n+1/3}, \lambda^{n+1/3}), \end{cases} \tag{72}$$

and set

$$(\mathbf{p}^{n+2/3}, \lambda^{n+2/3}) = (\mathbf{p}(t^{n+1}), \ \lambda(t^{n+1})). \tag{73}$$

3rd Fractional step: Solve

$$\begin{cases} \begin{cases} \dfrac{\partial \mathbf{p}}{\partial t} + \theta^{-1} D_{\mathbf{q}} G(\mathbf{p}) + \partial_{\mathbf{q}} I_{S_1}(\mathbf{p}) \ni \mathbf{0} \\ \gamma \dfrac{\partial \lambda}{\partial t} = \mathbf{0} \end{cases} \text{in } \Omega \times (t^n, t^{n+1}), \\ (\mathbf{p}(t^n), \lambda(t^n)) = (\mathbf{p}^{n+2/3}, \lambda^{n+2/3}), \end{cases} \tag{74}$$

and set

$$(\mathbf{p}^{n+1}, \lambda^{n+1}) = (\mathbf{p}(t^{n+1}), \ \lambda^{n+2/3}). \tag{75}$$

Following Section 4, we will employ in practice the *Marchuk–Yanenko* variant of the Lie scheme, that is (from (22), (23), and assuming, for simplicity, that γ does not depend on n):

$$(\mathbf{p}^0, \lambda^0) = (\mathbf{p}_0, \lambda_0). \tag{76}$$

For $n \geq 0$, $(\mathbf{p}^n, \boldsymbol{\lambda}^n) \to \left(\mathbf{p}^{n+1/3}, \boldsymbol{\lambda}^{n+1/3}\right) \to \left(\mathbf{p}^{n+2/3}, \boldsymbol{\lambda}^{n+2/3}\right) \to \left(\mathbf{p}^{n+1}, \boldsymbol{\lambda}^{n+1}\right)$ as follows:

$$\begin{cases} \dfrac{\mathbf{p}^{n+1/3} - \mathbf{p}^n}{\tau} + D_{\mathbf{q}} T_2(\mathbf{p}^{n+1/3}, \boldsymbol{\lambda}^n) = \mathbf{0}, \\ \gamma \dfrac{\boldsymbol{\lambda}^{n+1/3} - \boldsymbol{\lambda}^n}{\tau} + D_{\mu} T_2(\mathbf{p}^{n+1/3}, \boldsymbol{\lambda}^{n+1/3}) = \mathbf{0}, \end{cases} \quad \text{in } \Omega \Rightarrow (\mathbf{p}^{n+1/3}, \boldsymbol{\lambda}^{n+1/3}), \tag{77}$$

$$\begin{cases} \dfrac{\mathbf{p}^{n+2/3} - \mathbf{p}^{n+1/3}}{\tau} + \partial_{\mathbf{q}} I_{S_3}(\mathbf{p}^{n+2/3}, \boldsymbol{\lambda}^{n+2/3}) \ni \mathbf{0}, \\ \gamma \dfrac{\boldsymbol{\lambda}^{n+2/3} - \boldsymbol{\lambda}^{n+1/3}}{\tau} + \partial_{\mu} I_{S_3}(\mathbf{p}^{n+2/3}, \boldsymbol{\lambda}^{n+2/3}) \ni \mathbf{0}, \end{cases} \quad \text{in } \Omega \Rightarrow (\mathbf{p}^{n+2/3}, \boldsymbol{\lambda}^{n+2/3}), \tag{78}$$

$$\begin{cases} \dfrac{\mathbf{p}^{n+1} - \mathbf{p}^{n+2/3}}{\tau} + \theta^{-1} D_{\mathbf{q}} G(\mathbf{p}^{n+1}) + \partial_{\mathbf{q}} I_{S_1}(\mathbf{p}^{n+1}) \ni \mathbf{0}, \\ \gamma \dfrac{\boldsymbol{\lambda}^{n+1} - \boldsymbol{\lambda}^{n+2/3}}{\tau} = \mathbf{0}, \end{cases} \quad \text{in } \Omega \Rightarrow (\mathbf{p}^{n+1}, \boldsymbol{\lambda}^{n+1}). \tag{79}$$

Since problem (79) is identical to problem (46), the solution method discussed in Section 4.4 still applies here. On the other hand, we still have to discuss the solution of problems (77) and (78). We will do it in Sections 5.2 and 5.3.

5.2 On the solution of problem (77)

It is clear from (77) that (formally)

$$\mathbf{p}^{n+1/3} = \arg \min_{\mathbf{q} \in (\mathcal{L}^2(\Omega))^2} \left\{ \int_\Omega |\mathbf{q} - \mathbf{p}^n|^2 dx + \tau \int_\Omega (\nabla \cdot \boldsymbol{\lambda}^n)^2 \sqrt{1 + |\mathbf{q}|^2} dx \right\}. \tag{80}$$

From the strict convexity and coercivity of the functional in (80), the related minimization problem has a unique solution. By differentiation, we have the following Euler–Lagrange equation:

$$(\mathbf{p}^{n+1/3} - \mathbf{p}^n) + \tau (\nabla \cdot \boldsymbol{\lambda}^n)^2 \frac{\mathbf{p}^{n+1/3}}{\sqrt{1 + |\mathbf{p}^{n+1/3}|^2}} = \mathbf{0},$$

which implies in turn that

$$\left(1 + \tau \frac{(\nabla \cdot \boldsymbol{\mu}^n)^2}{\sqrt{1 + |\mathbf{p}^{n+1/3}|^2}}\right) \mathbf{p}^{n+1/3} = \mathbf{p}^n. \tag{81}$$

We observe that the vector-valued functions $\mathbf{p}^n$ and $\mathbf{p}^{n+1/3}$ are collinear and also that $\mathbf{p}^n(x) = \mathbf{0}$ (resp., $\mathbf{p}^n(x) \neq \mathbf{0}$) implies $\mathbf{p}^{n+1/3} = \mathbf{0}$ (resp., $\mathbf{p}^{n+1/3} \neq \mathbf{0}$). We have more since the nonnegative real-valued function $|\mathbf{p}^{n+1/3}|$ is the unique solution of the nonlinear functional algebraic equation

$$\left(1+\tau\frac{|\nabla\cdot\boldsymbol{\mu}^n|^2}{\sqrt{1+\rho^2}}\right)\rho = |\mathbf{p}^n|. \tag{82}$$

One can solve Eq. (82) pointwise: Indeed, let us denote $\rho(x)$ by z and $|\mathbf{p}^n(x)|$ by r. We have then to solve a.e. on Ω the single variable equation

$$z+\tau|(\nabla\cdot\boldsymbol{\mu}^n)(x)|^2\frac{z}{\sqrt{1+|z|^2}} = r. \tag{83}$$

Eq. (83) has a unique solution. The fact that one has to pick a small τ, in order to reduce the splitting error, suggests solving Eq. (83) using a Newton-like algorithm initialized by r. The algorithm we advocate reads as follows:

$$\begin{cases} z^0 = r. \\ \text{For } k \geq 0,\ z^k \to z^{k+1} \text{ via} \\ z^{k+1} = \max\left(0, \dfrac{g(z^k)}{g'(z^k)}\right), \end{cases} \tag{84}$$

where function g is defined by $g(z) = z+\tau|(\nabla\cdot\boldsymbol{\mu}^n)(x)|^2\frac{z}{\sqrt{1+|z|^2}} - r$ (other initializations are possible; one can take for example $z^0 = |\mathbf{p}^{(n-1)+1/3}(x)|$). Once function $|\mathbf{p}^{n+1/3}|$ is known, relation (81) provides $\mathbf{p}^{n+1/3}$.

The second equation in (77) implies that $\boldsymbol{\lambda}^{n+1/3}$ is solution of the following linear elliptic system with periodic boundary conditions:

$$\begin{cases} \gamma\boldsymbol{\lambda}^{n+1/3} - 2\tau\nabla\left(\sqrt{1+|\mathbf{p}^{n+1/3}|^2}\nabla\cdot\boldsymbol{\lambda}^{n+1/3}\right) = \gamma\boldsymbol{\lambda}^n, \quad \text{in } \Omega, \\ \lambda_1^{n+1/3}(0,\cdot) = \lambda_1^{n+1/3}(L_1,\cdot),\ \lambda_2^{n+1/3}(\cdot,0) = \lambda_2^{n+1/3}(\cdot,L_2), \\ \left(\sqrt{1+|\mathbf{p}^{n+1/3}|^2}\nabla\cdot\boldsymbol{\lambda}^{n+1/3}\right)(0,\cdot) = \sqrt{1+|\mathbf{p}^{n+1/3}|^2}\nabla\cdot\boldsymbol{\lambda}^{n+1/3}(L_1,\cdot), \\ \left(\sqrt{1+|\mathbf{p}^{n+1/3}|^2}\nabla\cdot\boldsymbol{\lambda}^{n+1/3}\right)(\cdot,0) = \sqrt{1+|\mathbf{p}^{n+1/3}|^2}\nabla\cdot\boldsymbol{\lambda}^{n+1/3}(0,L_2). \end{cases} \tag{85}$$

Problem (85) is a simple variant of problem (49) from Section 4.2. It follows from, e.g., Tai et al. (2011) that the periodic conditions verified by $\boldsymbol{\lambda}^{n+1/3}$ imply that one can use *FFT* to solve problem (85) (see again Deng et al. (2019) for details).

5.3 On the solution of problem (78)

We consider (78) as an optimal system for the following minimization problem

$$(\mathbf{p}^{n+2/3}, \boldsymbol{\lambda}^{n+2/3}) = \arg\min_{(\mathbf{q},\boldsymbol{\mu})\in S_3} \left\{ \int_\Omega |\mathbf{q} - \mathbf{p}^{n+1/3}|^2 dx + \int_\Omega \gamma |\boldsymbol{\mu} - \boldsymbol{\lambda}^{n+1/3}|^2 dx \right\}. \tag{86}$$

Problem (86) can be solved pointwise since we have, a.e. on Ω,

$$(\mathbf{p}^{n+2/3}(x), \boldsymbol{\lambda}^{n+2/3}(x)) = \underset{(\mathbf{q},\boldsymbol{\mu})\in\sigma_3}{\operatorname{argmin}} \left\{ |\mathbf{q} - \mathbf{p}^{n+1/3}(x)|^2 + \gamma(x)|\boldsymbol{\mu} - \boldsymbol{\lambda}^{n+1/3}(x)|^2 \right\}, \tag{87}$$

where $\sigma_3 = \left\{ (\mathbf{q},\boldsymbol{\mu}) \in \mathbb{R}^2 \times \mathbb{R}^2 \mid \boldsymbol{\mu} = \frac{\mathbf{q}}{\sqrt{1+|\mathbf{q}|^2}} \right\}$.

By elimination of $\boldsymbol{\mu}$, problem (87) becomes

$$\begin{cases} \mathbf{p}^{n+2/3}(x) = \arg\min_{\mathbf{q}\in\mathbb{R}^2} \left\{ \frac{1}{2}|\mathbf{q} - \mathbf{p}^{n+1/3}|^2 + \frac{\gamma}{2}\left| \frac{\mathbf{q}}{\sqrt{1+|\mathbf{q}|^2}} - \boldsymbol{\lambda}^{n+1/3} \right|^2 \right\} \\ \qquad = \arg\min_{\mathbf{q}\in\mathbb{R}^2} \left\{ \frac{|\mathbf{q}|^2}{2}\left(1 + \frac{\gamma}{1+|\mathbf{q}|^2}\right) - \mathbf{q}\cdot\left(\mathbf{p}^{n+1/3} + \frac{\gamma}{\sqrt{1+|\mathbf{q}|^2}}\boldsymbol{\lambda}^{n+1/3}\right) \right\}. \end{cases} \tag{88}$$

Vector $\mathbf{p}^{n+2/3}(x)$ is solution of a 2×2 nonlinear system obtained by differentiation of the functional in (88). This system can be solved by the Newton's method initialized by $\mathbf{p}^{n+1/3}$ (or $\mathbf{p}^{(n-1)+2/3}$). Actually, one encounters in Myllykoski et al. (2015) a problem very close to (88), when using ADMM to solve L^1- mean curvature related denoising problems. Following the approach in Myllykoski et al. (2015), we observe that the Schwarz inequality in $\mathbb{R}^2$ implies that the vectors $\mathbf{p}^{n+2/3}(x)$ and $\mathbf{p}^{n+1/3}(x) + \frac{\gamma(x)}{\sqrt{1+|\mathbf{p}^{n+2/3}(x)|^2}}\boldsymbol{\lambda}^{n+1/3}(x)$ are positively co-linear, that is, there exists $\alpha(=\alpha(x)) \geq 0$ such that:

$$\mathbf{p}^{n+2/3}(x) = \alpha\left(\mathbf{p}^{n+1/3}(x) + \frac{\gamma(x)}{\sqrt{1+|\mathbf{p}^{n+2/3}(x)|^2}}\boldsymbol{\lambda}^{n+1/3}(x) \right). \tag{89}$$

Let us define $A^n(x)(\subset \mathbb{R}^2)$ by

$$A^n(x) = \left\{ (\beta, r) \in \mathbb{R}_+ \times \mathbb{R}_+ \mid r = \beta\left| \mathbf{p}^{n+1/3}(x) + \frac{\gamma(x)}{\sqrt{1+r^2}}\boldsymbol{\lambda}^{n+1/3}(x) \right| \right\}.$$

It follows then from (88) to (89) that $\mathbf{p}^{n+2/3}(x)=\alpha(\mathbf{p}^{n+1/3}(x)+\frac{\gamma(x)}{\sqrt{1+\rho^2}}\boldsymbol{\lambda}^{n+1/3}(x))$, where the pair (α,ρ) is solution to

$$(\rho,\alpha)=\underset{(\beta,r)\in A^n(x)}{\operatorname{argmin}}\left\{\frac{\beta^2}{2}\left|\mathbf{p}^{n+1/3}+\frac{\gamma\boldsymbol{\lambda}^{n+1/3}}{\sqrt{1+r^2}}\right|^2\left(1+\frac{\gamma}{1+r^2}\right)-\beta\left|\mathbf{p}^{n+1/3}+\frac{\gamma\boldsymbol{\lambda}^{n+1/3}}{\sqrt{1+r^2}}\right|^2\right\}. \tag{90}$$

Problem (90) simplifies since from the definition of the set $A^n(x)$ it reduces to

$$\alpha=\frac{\rho}{\left|\mathbf{p}^{n+1/3}+\frac{\gamma\boldsymbol{\lambda}^{n+1/3}}{\sqrt{1+\rho^2}}\right|} \tag{91}$$

with

$$\rho=\arg\min_{r\in\mathbb{R}^+}\left\{\frac{r^2}{2}\left(1+\frac{\gamma}{1+r^2}\right)-r\left|\mathbf{p}^{n+1/3}+\frac{\gamma\boldsymbol{\lambda}^{n+1/3}}{\sqrt{1+r^2}}\right|\right\}. \tag{92}$$

The single variable optimization problem (92) can be solved by many methods, including Newton's initialized by $\rho^0=|\mathbf{p}^{n+1/3}|$ or $|\mathbf{p}^{(n-1)+2/3}|$.

5.4 Estimating γ

By differentiation, it follows from relation $\boldsymbol{\lambda}(t)=\frac{\mathbf{p}(t)}{\sqrt{1+|\mathbf{p}(t)|^2}}$ that

$$\dot{\boldsymbol{\lambda}}=\frac{\left(1+|\mathbf{p}|^2\right)\dot{\mathbf{p}}-(\mathbf{p}\cdot\dot{\mathbf{p}})\mathbf{p}}{\left(1+|\mathbf{p}|^2\right)^{3/2}}. \tag{93}$$

For simplicity, we use notation $\dot{\phi}=\partial\phi/\partial t$ above and later. Relation (93) implies

$$|\dot{\boldsymbol{\lambda}}|^2+\frac{2+|\mathbf{p}|^2}{\left(1+|\mathbf{p}|^2\right)^3}(\mathbf{p}\cdot\dot{\mathbf{p}})^2=\frac{|\dot{\mathbf{p}}|^2}{1+|\mathbf{p}|^2},$$

which implies in turn

$$|\dot{\boldsymbol{\lambda}}(t)|\le\frac{1}{\sqrt{1+|\mathbf{p}(t)|^2}}|\dot{\mathbf{p}}(t)|. \tag{94}$$

For small number τ, relation

$$\left|\mathbf{p}^{n+2/3}-\mathbf{p}^{n+1/3}\right|^2+\gamma\left|\boldsymbol{\lambda}^{n+2/3}-\boldsymbol{\lambda}^{n+1/3}\right|^2\approx\tau^2\left(\left|\frac{\partial\mathbf{p}}{\partial t}(t^n)\right|^2+\gamma\left|\frac{\partial\boldsymbol{\lambda}}{\partial t}(t^n)\right|^2\right)$$

and (94) suggest taking

$$\gamma(=\gamma^n) = 1 + |\mathbf{p}^{n+1/3}|^2 \tag{95}$$

to balance the two quadratic terms in (87).

6 An operator-splitting method for the L^1-mean curvature variational model

In this section we are going to discuss (rather briefly) the solution of the variational problem (11) associated with the regularizing functional R_3 defined by (6), implying that the problem under consideration reads as

$$u = \arg\min_{v \in \mathcal{V}} \left\{ \int_\Omega \left| \nabla \cdot \frac{\nabla v}{\sqrt{1+|\nabla v|^2}} \right| dx + \frac{\theta^{-1}}{2} \int_\Omega (v-f)^2 dx \right\}. \tag{96}$$

In Zhu and Chan (2012), Zhu et al. (2013a) and Myllykoski et al. (2015) one has proposed methods for the numerical solution of problem (96). The method in Zhu and Chan (2012) is of the time-marching type (giving it the flavour of a gradient descent algorithm), while the methods in Zhu et al. (2013a) and Myllykoski et al. (2015) are of the ADMM type. The method in Zhu et al. (2013a) (resp., Myllykoski et al., 2015) makes use of $\mathbf{q} = (\nabla v, -1)$ and $\boldsymbol{\mu} = \frac{\mathbf{q}}{|\mathbf{q}|}$ (resp., $\mathbf{q} = \nabla v$ and $\boldsymbol{\mu} = \frac{\mathbf{q}}{\sqrt{1+|\mathbf{q}|^2}}$). A closely related augmented ADMM algorithm is discussed in Sun and Chen (2014); it makes use of the vector-valued function $\boldsymbol{\mu} = \frac{\nabla v}{\sqrt{1+|\nabla v|^2}}$ and of the constraint $\nabla v - \boldsymbol{\mu}\sqrt{1+|\nabla v|^2} = \mathbf{0}$. The operator-splitting method to be discussed in this section has many commonalities with the methods we described in Sections 4 and 5. We will take advantage of these commonalities to shorten our presentation. The method we are going to describe relies on the following additional unknown functions $\mathbf{q} = \nabla v$, $\boldsymbol{\mu} = \frac{\mathbf{q}}{\sqrt{1+|\mathbf{q}|^2}}$ and $\eta = \nabla \cdot \boldsymbol{\mu}$. The sets S_1 and S_3 being as in Sections 3 and 5, we introduce S_4 defined by

$$S_4 = \{(\boldsymbol{\mu}, \eta) \in H_p(\Omega; \mathrm{div}) \times \mathcal{L}^2(\Omega) \mid \eta = \nabla \cdot \boldsymbol{\mu}\}. \tag{97}$$

We can then reformulate (formally) the minimization problem in (96) as:

$$\min_{\mathbf{q}, \boldsymbol{\mu}, \eta} \left\{ \int_\Omega |\eta| dx + \theta^{-1} \int_\Omega (\nabla_f^{-1}\mathbf{q} - f)^2 dx + I_{S_1}(\mathbf{q}) + I_{S_3}(\mathbf{q}, \boldsymbol{\mu}) + I_{S_4}(\boldsymbol{\mu}, \eta) \right\} \tag{98}$$

with $(\mathbf{q}, \boldsymbol{\mu}, \eta) \in \left(\mathcal{L}^2(\Omega)\right)^2 \times H_p(\Omega; \mathrm{div}) \times \mathcal{L}^2(\Omega)$, and operator ∇_f^{-1} as in Section 3. We will denote by T_3 the functional $\eta \to \int_\Omega |\eta| dx$.

Suppose ($\mathbf{p}$, $\boldsymbol{\lambda}$, ξ) is a minimizer of the (proper) functional in (98); we have then

$$\mathbf{p} = \nabla u, \boldsymbol{\lambda} = \frac{\mathbf{p}}{\sqrt{1+|\mathbf{p}|^2}}, \xi = \nabla \cdot \boldsymbol{\lambda} \tag{99}$$

and the following optimality necessary conditions

$$\begin{cases} \theta^{-1} D_{\mathbf{q}} G(\mathbf{p}) + \partial_{\mathbf{q}} I_{S_1}(\mathbf{p}) + \partial_{\mathbf{q}} I_{S_3}(\mathbf{p}, \boldsymbol{\lambda}) \ni \mathbf{0}, \\ \partial_{\mu} I_{S_3}(\mathbf{p}, \boldsymbol{\lambda}) + \partial_{\mu} I_{S_4}(\boldsymbol{\lambda}, \xi) \ni \mathbf{0}, \\ \partial_{\eta} T_3(\xi) + \partial_{\eta} I_{S_4}(\boldsymbol{\lambda}, \xi) \ni \mathbf{0}, \end{cases} \tag{100}$$

with functional G as in Section 3.

Proceeding as in Sections 4 and 5, we associate with the optimality system (100) the following initial value problem (gradient flow):

$$\begin{cases} \begin{cases} \dfrac{\partial \mathbf{p}}{\partial t} + \theta^{-1} D_{\mathbf{q}} G(\mathbf{p}) + \partial_{\mathbf{q}} I_{S_1}(\mathbf{p}) + \partial_{\mathbf{q}} I_{S_3}(\mathbf{p}, \boldsymbol{\lambda}) \ni \mathbf{0}, \\ \gamma_1 \dfrac{\partial \boldsymbol{\lambda}}{\partial t} + \partial_{\mu} I_{S_3}(\mathbf{p}, \boldsymbol{\lambda}) + \partial_{\mu} I_{S_4}(\boldsymbol{\lambda}, \xi) \ni \mathbf{0}, & \text{in } \Omega \times (0, +\infty), \\ \gamma_2 \dfrac{\partial \xi}{\partial t} + \partial_{\eta} T_3(\xi) + \partial_{\eta} I_{S_4}(\boldsymbol{\lambda}, \xi) \ni \mathbf{0}, \end{cases} \\ (\mathbf{p}(0), \boldsymbol{\lambda}(0), \xi(0)) = (\mathbf{p}_0, \boldsymbol{\lambda}_0, \xi_0), \end{cases} \tag{101}$$

where γ_1 and γ_2 are two strictly positively real-valued functions (their choice will be discussed later), and where $I_{S_1}(\mathbf{p}_0) = 0, I_{S_3}(\mathbf{p}_0, \boldsymbol{\lambda}_0) = 0, I_{S_4}(\boldsymbol{\lambda}_0, \xi_0) = 0$.

Let

$$A_1 = \begin{pmatrix} 0 \\ 0 \\ \partial_{\eta} T_3 \end{pmatrix}, \quad A_2 = \begin{pmatrix} 0 \\ \partial_{\mu} I_{S_4} \\ \partial_{\eta} I_{S_4} \end{pmatrix}, \quad A_3 = \begin{pmatrix} \partial_{\mathbf{q}} I_{S_3} \\ \partial_{\mu} I_{S_3} \\ 0 \end{pmatrix}, \quad A_4 = \begin{pmatrix} \theta^{-1} D_{\mathbf{q}} G + \partial_{\mathbf{q}} I_{S_1} \\ 0 \\ 0 \end{pmatrix}.$$

Applying the Lie scheme with $J = 4$ to the solution of problem (101), one obtains

$$(\mathbf{p}^0, \boldsymbol{\lambda}^0, \xi^0) = (\mathbf{p}_0, \boldsymbol{\lambda}_0, \xi_0). \tag{102}$$

For $n \geq 0$, $(\mathbf{p}^n, \boldsymbol{\lambda}^n, \xi^n) \to \left(\mathbf{p}^{n+1/4}, \boldsymbol{\lambda}^{n+1/4}, \xi^{n+1/4}\right) \to \left(\mathbf{p}^{n+2/4}, \boldsymbol{\lambda}^{n+2/4}, \xi^{n+2/4}\right) \to \left(\mathbf{p}^{n+3/4}, \boldsymbol{\lambda}^{n+3/4}, \xi^{n+3/4}\right) \to (\mathbf{p}^{n+1}, \boldsymbol{\lambda}^{n+1}, \xi^{n+1})$ as follows:

1st Fractional step: Solve

$$\begin{cases} \gamma_2 \dfrac{\partial \xi}{\partial t} + \partial_{\eta} T_3(\xi) \ni \mathbf{0}, \quad \text{in } \Omega \times (t^n, t^{n+1}), \\ \xi(t^n) = \xi^n, \end{cases} \tag{103}$$

and set

$$\mathbf{p}^{n+1/4} = \mathbf{p}^n, \quad \boldsymbol{\lambda}^{n+1/4} = \boldsymbol{\lambda}^n, \quad \xi^{n+1/4} = \xi(t^{n+1}). \tag{104}$$

2nd Fractional step: Solve

$$\begin{cases} \begin{cases} \gamma_1 \dfrac{\partial \boldsymbol{\lambda}}{\partial t} + \partial_\mu I_{S_4}(\boldsymbol{\lambda}, \xi) \ni \mathbf{0}, \\ \gamma_2 \dfrac{\partial \xi}{\partial t} + \partial_\eta I_{S_4}(\boldsymbol{\lambda}, \xi) \ni \mathbf{0}, \end{cases} \quad \text{in } \Omega \times (t^n, t^{n+1}), \\ (\boldsymbol{\lambda}(t^n), \xi(t^n)) = (\boldsymbol{\lambda}^{n+1/4}, \xi^{n+1/4}), \end{cases} \tag{105}$$

and set

$$\mathbf{p}^{n+2/4} = \mathbf{p}^{n+1/4}, \quad \boldsymbol{\lambda}^{n+2/4} = \boldsymbol{\lambda}(t^{n+1}), \quad \xi^{n+2/4} = \xi(t^{n+1}). \tag{106}$$

3rd Fractional step: Solve

$$\begin{cases} \begin{cases} \dfrac{\partial \mathbf{p}}{\partial t} + \partial_{\mathbf{q}} I_{S_3}(\mathbf{p}, \boldsymbol{\lambda}) \ni \mathbf{0}, \\ \gamma_1 \dfrac{\partial \boldsymbol{\lambda}}{\partial t} + \partial_\mu I_{S_3}(\mathbf{p}, \boldsymbol{\lambda}) \ni \mathbf{0}, \end{cases} \quad \text{in } \Omega \times (t^n, t^{n+1}), \\ (\mathbf{p}(t^n), \boldsymbol{\lambda}(t^n)) = (\mathbf{p}^{n+2/4}, \boldsymbol{\lambda}^{n+2/4}), \end{cases} \tag{107}$$

and set

$$\mathbf{p}^{n+3/4} = \mathbf{p}(t^{n+1}), \quad \boldsymbol{\lambda}^{n+3/4} = \boldsymbol{\lambda}(t^{n+1}), \quad \xi^{n+3/4} = \xi^{n+2/4}. \tag{108}$$

4th Fractional step: Solve

$$\begin{cases} \dfrac{\partial \mathbf{p}}{\partial t} + D_{\mathbf{q}} G(\mathbf{p}) + \partial_{\mathbf{q}} I_{S_1}(\mathbf{p}) \ni \mathbf{0} \text{ in } \Omega \times (t^n, t^{n+1}), \\ \mathbf{p}(t^n) = \mathbf{p}^{n+3/4}, \end{cases} \tag{109}$$

and set

$$\mathbf{p}^{n+1} = \mathbf{p}(t^{n+1}), \quad u^{n+1} = \nabla_f^{-1} \mathbf{p}^{n+1}, \quad \boldsymbol{\lambda}^{n+1} = \boldsymbol{\lambda}^{n+3/4}, \quad \xi^{n+1} = \xi^{n+3/4}. \tag{110}$$

The Marchuk–Yanenko scheme associated with scheme (102)–(110) reads as:

$$(\mathbf{p}^0, \boldsymbol{\lambda}^0, \xi^0) = (\mathbf{p}_0, \boldsymbol{\lambda}_0, \xi_0). \tag{111}$$

For $n \geq 0 (\mathbf{p}^n, \boldsymbol{\lambda}^n, \xi^n) \to \left(\mathbf{p}^{n+1/4}, \boldsymbol{\lambda}^{n+1/4}, \xi^{n+1/4}\right) \to \left(\mathbf{p}^{n+2/4}, \boldsymbol{\lambda}^{n+2/4}, \xi^{n+2/4}\right)$ $\to \left(\mathbf{p}^{n+3/4}, \boldsymbol{\lambda}^{n+3/4}, \xi^{n+3/4}\right) \to (\mathbf{p}^{n+1}, \boldsymbol{\lambda}^{n+1}, \xi^{n+1})$ as follows:

$$\begin{cases} \mathbf{p}^{n+1/4} = \mathbf{p}^n, \\ \boldsymbol{\lambda}^{n+1/4} = \boldsymbol{\lambda}^n, \\ \gamma_2 \dfrac{\xi^{n+1/4} - \xi^n}{\tau} + \partial_\eta T_3(\xi^{n+1/4}) \ni \mathbf{0} \text{ in } \Omega, \end{cases} \tag{112}$$

$$\begin{cases} \mathbf{p}^{n+2/4} = \mathbf{p}^{n+1/4}, \\ \gamma_1 \dfrac{\boldsymbol{\lambda}^{n+2/4} - \boldsymbol{\lambda}^{n+1/4}}{\tau} + \partial_\mu I_{S_4}(\boldsymbol{\lambda}^{n+2/4}, \xi^{n+2/4}) \ni \mathbf{0} \text{ in } \Omega, \\ \gamma_2 \dfrac{\xi^{n+2/4} - \xi^{n+1/4}}{\tau} + \partial_\eta I_{S_4}(\boldsymbol{\lambda}^{n+2/4}, \xi^{n+2/4}) \ni \mathbf{0} \text{ in } \Omega, \end{cases} \tag{113}$$

$$\begin{cases} \dfrac{\mathbf{p}^{n+3/4}-\mathbf{p}^{n+2/4}}{\tau}+\partial_{\mathbf{q}} I_{S_3}(\mathbf{p}^{n+3/4},\boldsymbol{\lambda}^{n+3/4}) \ni \mathbf{0}, \quad \text{in } \Omega, \\ \gamma_1 \dfrac{\boldsymbol{\lambda}^{n+3/4}-\boldsymbol{\lambda}^{n+2/4}}{\tau}+\partial_{\mu} I_{S_3}(\mathbf{p}^{n+3/4},\boldsymbol{\lambda}^{n+3/4}) \ni \mathbf{0}, \quad \text{in } \Omega, \\ \xi^{n+3/4}=\xi^{n+2/4}, \end{cases} \tag{114}$$

$$\begin{cases} \dfrac{\mathbf{p}^{n+1}-\mathbf{p}^{n+3/4}}{\tau}+D_{\mathbf{q}}\, G(\mathbf{p})+\partial_{\mathbf{q}} I_{S_1}(\mathbf{p}^{n+1}) \ni \mathbf{0} \quad \text{in } \Omega, \\ \boldsymbol{\lambda}^{n+1}=\boldsymbol{\lambda}^{n+3/4}, \\ \xi^{n+1}=\xi^{n+3/4}. \end{cases} \tag{115}$$

Since we discussed in Sections 4 and 5 the solution of close variants of problems (114) and (115), we just need to focus on the solution of problems (112) and (113):

The third equation in (112) is the (formal) Euler–Lagrange "equation" of the following minimization problem:

$$\xi^{n+1/4}=\arg \min_{\eta \in \mathcal{L}^2(\Omega)} \left\{ \int_\Omega \frac{\gamma_2}{2}|\eta-\xi^n|^2 dx+\tau \int_\Omega |\eta| dx \right\}. \tag{116}$$

Problem (116) can be solved pointwise, leading to the closed form solution below

$$\xi^{n+1/4}(x)=\max\left\{0, 1-\frac{\tau}{\gamma_2(x)|\xi^n(x)|}\right\}\xi^n(x), \text{ a.e. in } \Omega. \tag{117}$$

We still have to discuss the solution of problem (113): The second and third "equations" (inclusions) form a system which is (formally) the optimality system for

$$(\boldsymbol{\lambda}^{n+2/4},\xi^{n+2/4})=\arg \min_{(\boldsymbol{\mu},\eta)\in S_4} \left\{ \int_\Omega \gamma_1 \left|\boldsymbol{\mu}-\boldsymbol{\lambda}^{n+1/4}\right|^2 dx+\int_\Omega \gamma_2|\eta-\xi^{n+1/4}|^2 dx \right\}. \tag{118}$$

From the definition of S_4 (see (97)), problem (118) reduces to

$$\begin{cases} \xi^{n+2/4}=\nabla\cdot\boldsymbol{\lambda}^{n+2/4}, \\ \boldsymbol{\lambda}^{n+2/4}=\arg \min\limits_{\boldsymbol{\mu}\in H_p(\Omega;\text{div})} \left\{ \int_\Omega \gamma_1 \left|\boldsymbol{\mu}-\boldsymbol{\lambda}^{n+1/4}\right|^2 dx+\int_\Omega \gamma_2|\nabla\cdot\boldsymbol{\mu}-\xi^{n+1/4}|^2 dx \right\}. \end{cases} \tag{119}$$

Written in variational form, the Euler–Lagrange equation of the minimization problem in (119) reads as

$$\begin{cases} \boldsymbol{\lambda}^{n+2/4} \in H_p(\Omega;\, \mathrm{div}), \\ \displaystyle\int_\Omega \gamma_1 \boldsymbol{\lambda}^{n+2/4} \cdot \boldsymbol{\mu} dx + \int_\Omega \gamma_2 \left(\nabla \cdot \boldsymbol{\lambda}^{n+2/4}\right)(\nabla \cdot \boldsymbol{\mu}) dx = \\ \qquad \displaystyle\int_\Omega \gamma_1 \boldsymbol{\lambda}^{n+1/4} \cdot \boldsymbol{\mu} dx + \int_\Omega \gamma_2 \xi^{n+1/4} \nabla \cdot \boldsymbol{\mu} dx, \quad \forall \boldsymbol{\mu} \in H_p(\Omega; \mathrm{div}). \end{cases} \tag{120}$$

It follows from (120) that the vector-valued function $\boldsymbol{\lambda}^{n+2/4}$ is also solution to the following elliptic linear boundary value problem:

$$\begin{cases} \gamma_1 \boldsymbol{\lambda}^{n+2/4} - \nabla\left(\gamma_2 \nabla \cdot \boldsymbol{\lambda}^{n+2/4}\right) = \gamma_1 \boldsymbol{\lambda}^{n+1/4} - \nabla\left(\gamma_2 \xi^{n+1/4}\right), \quad \text{in } \Omega, \\ \gamma_2\left(\nabla \cdot \boldsymbol{\lambda}^{n+2/4} - \xi^{n+1/4}\right)(0,\,\cdot) = \gamma_2\left(\nabla \cdot \boldsymbol{\lambda}^{n+2/4} - \xi^{n+1/4}\right)(L_1,\,\cdot), \\ \gamma_2\left(\nabla \cdot \boldsymbol{\lambda}^{n+2/4} - \xi^{n+1/4}\right)(\cdot\,,0) = \gamma_2\left(\nabla \cdot \boldsymbol{\lambda}^{n+2/4} - \xi^{n+1/4}\right)(\cdot\,,L_2). \end{cases} \tag{121}$$

Problem (120) and (121) is a closed variant of problem (49) from Section 4.2. After appropriate space approximation, one can use *FFT* to solve the discrete analogues of problem (120) and (121) (see Deng et al., 2019; Tai et al., 2011).

Concerning the choice of γ_1 and γ_2, by analogy with Section 5, we suggest taking $\gamma_1 (= \gamma_1^n) = 1 + |\mathbf{p}^{n+2/4}|^2$. Choosing γ_2 is more complicated: The default solution is to take $\gamma_2 = \gamma_1$, but as shown in, e.g., Section 5.1 of Myllykoski et al. (2015) for a related problem (originating also from the L^1-mean curvature model), other possibilities exist (after space discretization one can take, for example, $\gamma_2 = \gamma_1 h^2$, where h is the space discretization step size).

7 Operator-splitting methods for the ROF model

7.1 Generalities: synopsis

It follows from Rudin et al. (1992) that the *ROF* model for image restoration reads as

$$u = \arg \min_{v \in BV(\Omega)} \left\{ \int_\Omega |\nabla v| dx + \frac{\theta^{-1}}{2} \int_\Omega (v - f)^2 dx \right\}. \tag{122}$$

As one of the most popular models in image processing (if not the most popular), the numerical solution of problem (122) has motivated (and still does) a most abundant literature. Beside the time-marching method discussed in Rudin et al. (1992), let us mention among many others: primal-dual methods (Carter, 2001; Chambolle, 2004; Chambolle and Pock, 2011; Chan et al., 1999; Zhu and Chan, 2008; Zhu et al., 2010), augmented Lagrangian and other splitting methods (Huang et al., 2008; Wang et al., 2007, 2008; Yang et al., 2009), split-Bregman methods (Goldstein and Osher, 2009; Tai and Wu, 2009), a domain decomposition method (Duan and Tai, 2011) and graph cut methods (Darbon and Sigelle, 2006a,b).

In this section we are going to discuss two methods for the solution of the convex problem (122), both of them relying on the Lie and Marchuk–Yanenko schemes. The first method is a simplified variant of the method we discussed in Section 4 for the solution of the Euler elastica problem (28); it is obtained by taking $a = 1$ and $b = 0$ in (28). The second method is novel (to the best of our knowledge); it relies on a (kind of) *double deconvexification* of problem (122).

7.2 A first operator-splitting method

For clarification we are going to address the solution of problem (122) with some references to the elastica problem we discussed in Section 4.

The starting point of our discussion is the (formal) equivalence between problem (122) and

$$(u, \mathbf{p}) = \arg \min_{(v, \mathbf{q}) \in W} \left\{ \int_\Omega |\mathbf{q}| dx + \frac{\theta^{-1}}{2} \int_\Omega |v - f|^2 dx \right\}, \tag{123}$$

with

$$W = \{(v, \mathbf{q}) \in H^1(\Omega) \times (\mathcal{L}^2(\Omega))^2 | \nabla v - \mathbf{q} = \mathbf{0}\}.$$

Let us denote by $\widetilde{S}_1$ the set defined by

$$\widetilde{S}_1 = \{\mathbf{q} \in (L^2(\Omega))^2 \mid \exists\, v \in H^1(\Omega), \text{ such that } \mathbf{q} = \nabla v \text{ and } \int_\Omega (f - v) dx = 0\}.$$

Let us denote ∇u by $\mathbf{p}$. There is then equivalence between (123) and

$$\mathbf{p} = \arg \min_{\mathbf{q} \in \tilde{S}_1} \left\{ \int_\Omega |\mathbf{q}| dx + \frac{\theta^{-1}}{2} \int_\Omega (\widetilde{\nabla}_f^{-1} \mathbf{q} - f)^2 dx \right\}, \tag{124}$$

where $\widetilde{\nabla}_f^{-1}\mathbf{q}$ is the unique solution in $H^1(\Omega)$ of

$$\begin{cases} \nabla^2 v = \nabla \cdot \mathbf{q}, \text{ in } \Omega, \\ (\nabla v - \mathbf{q}) \cdot \mathbf{n} = 0 \text{ on } \partial\Omega, \\ \int_\Omega (v - f) dx = 0, \end{cases}$$

where $\mathbf{n}$ is the unit outward normal vector at the boundary $\partial\Omega$ of Ω. Let us denote by $I_{\tilde{S}_1}$ the indicator function of $\widetilde{S}_1$ in $(\mathcal{L}^2(\Omega))^2$, that is

$$I_{\tilde{S}_1}(\mathbf{q}) = \begin{cases} 0 & \text{if } \mathbf{q} \in \widetilde{S}_1, \\ +\infty & \text{if } \mathbf{q} \in (\mathcal{L}^2(\Omega))^2 \backslash \widetilde{S}_1. \end{cases}$$

There is equivalence, clearly, between problem (124) and

$$\mathbf{p} = \arg \min_{\mathbf{q} \in \left(\mathcal{L}^2(\Omega)\right)^2} \left\{ \int_\Omega |\mathbf{q}| dx + \frac{\theta^{-1}}{2} \int_\Omega \left| \widetilde{\nabla}_f^{-1} \mathbf{q} - f \right|^2 dx + I_{\tilde{S}_1}(\mathbf{q}) \right\}. \tag{125}$$

Let us denote by T_4 (resp., G) the functional defined by $T_4 = \int_\Omega |\mathbf{q}| dx$ (resp., $G(\mathbf{q}) = \frac{1}{2} \int_\Omega \left| \widetilde{\nabla}_f^{-1} \mathbf{q} - f \right|^2 dx$). The vector-valued function $\mathbf{p}$ is characterized by the following inclusion

$$\partial_{\mathbf{q}} T_4(\mathbf{p}) + \theta^{-1} D_{\mathbf{q}} G(\mathbf{p}) + \partial_{\mathbf{q}} I_{\tilde{S}_1}(\mathbf{p}) \ni \mathbf{0}. \tag{126}$$

We associate with (126) the following initial value problem:

$$\begin{cases} \dfrac{\partial \mathbf{p}}{\partial t} + \partial_{\mathbf{q}} T_4(\mathbf{p}) + \theta^{-1} D_{\mathbf{q}} G(\mathbf{p}) + \partial_{\mathbf{q}} I_{\tilde{S}_1}(\mathbf{p}) \ni \mathbf{0} \text{ in } \Omega \times (0, +\infty), \\ \mathbf{p}(0) = \mathbf{p}_0. \end{cases} \tag{127}$$

A reasonable (and natural) choice for $\mathbf{p}_0$ is to take $\mathbf{p}_0 = \nabla u_0$, function u_0 being defined by

$$\begin{cases} u_0 - \varepsilon \nabla^2 u_0 = f \text{ in } \Omega, \\ \dfrac{\partial u_0}{\partial \mathbf{n}} = 0 \text{ on } \partial\Omega, \end{cases}$$

(with $\epsilon \approx h^2$ where h is the space discretization step size, in practice). Let $A_1 = \partial_{\mathbf{q}} T_4$ and $A_2 = D_{\mathbf{q}} G + \partial_{\mathbf{q}} I_{\tilde{S}}$. Applying the Lie scheme to the solution of (127), we obtain

$$\mathbf{p}^0 = \mathbf{p}_0. \tag{128}$$

For $n \geq 0$, $\mathbf{p}^n \rightarrow \mathbf{p}^{n+1/2} \rightarrow \mathbf{p}^{n+1}$ as follows:

1st Fractional step: Solve

$$\begin{cases} \dfrac{\partial \mathbf{p}}{\partial t} + \partial_{\mathbf{q}} T_4(\mathbf{p}) \ni \mathbf{0} \text{ in } \Omega \times (t^n, t^{n+1}), \\ \mathbf{p}(t^n) = \mathbf{p}^n, \end{cases} \tag{129}$$

and set

$$\mathbf{p}^{n+1/2} = \mathbf{p}(t^{n+1}). \tag{130}$$

2nd Fractional step: Solve

$$\begin{cases} \dfrac{\partial \mathbf{p}}{\partial t} + \theta^{-1} D_{\mathbf{q}} G(\mathbf{p}) + \partial_{\mathbf{q}} I_{\tilde{S}_1}(\mathbf{p}) \ni \mathbf{0} \text{ in } \Omega \times (t^n, t^{n+1}), \\ \mathbf{p}(t^n) = \mathbf{p}^{n+1/2}, \end{cases} \tag{131}$$

and set

$$\mathbf{p}^{n+1} = \mathbf{p}(t^{n+1}). \tag{132}$$

The Marchuk–Yanenko approximation of the Lie scheme (128)–(132) reads as

$$\mathbf{p}^0 = \mathbf{p}_0. \tag{133}$$

For $n \geq 0$, $\mathbf{p}^n \rightarrow \mathbf{p}^{n+1/2} \rightarrow \mathbf{p}^{n+1}$ via the solution of

$$\frac{\mathbf{p}^{n+1/2} - \mathbf{p}^n}{\tau} + \partial_{\mathbf{q}} T_4\left(\mathbf{p}^{n+1/2}\right) \ni \mathbf{0}, \tag{134}$$

$$\frac{\mathbf{p}^{n+1} - \mathbf{p}^{n+1/2}}{\tau} + \theta^{-1} D_{\mathbf{q}} G\left(\mathbf{p}^{n+1}\right) + \partial_{\mathbf{q}} I_{\tilde{S}_1}\left(\mathbf{p}^{n+1}\right) \ni \mathbf{0}. \tag{135}$$

Relation (134) characterizes $\mathbf{p}^{n+1/2}$ as the unique solution of

$$\mathbf{p}^{n+1/2} = \arg \min_{\mathbf{q} \in (\mathcal{L}^2(\Omega))^2} \left\{ \frac{1}{2} \int_\Omega |\mathbf{q} - \mathbf{p}^n|^2 dx + \tau \int_\Omega |\mathbf{q}| dx \right\}. \tag{136}$$

Problem (136) can be solved pointwisely leading to the following closed form solution

$$\mathbf{p}^{n+1/2}(x) = \max\left\{0, 1 - \frac{\tau}{|\mathbf{p}^n(x)|}\right\} \mathbf{p}^n(x), \text{ a.e. in } \Omega. \tag{137}$$

Finally, it follows from (135) that

$$\begin{cases} \mathbf{p}^{n+1} = \nabla u^{n+1}, \\ \begin{cases} -\nabla^2 u^{n+1} + \theta^{-1} \tau u^{n+1} = \theta^{-1} \tau f - \nabla \cdot \mathbf{p}^{n+1/2} \text{ in } \Omega, \\ \left(\nabla u^{n+1} - \mathbf{p}^{n+1/2}\right) \cdot \mathbf{n} = 0 \text{ on } \partial\Omega. \end{cases} \end{cases} \tag{138}$$

Solving Neumann's problems like the one in (138) is routine nowadays (on rectangular domains in particular). We observe that algorithm (136)–(138) (almost) coincides with the algorithm one obtains by taking $a = 1$ and $b = 0$ in algorithm (43)–(46); the two algorithms differ by the boundary conditions.

7.3 A second operator-splitting method

In order to derive the second operator splitting based method for the solution of problem (122), we observe the (formal) equivalence between the minimization problem (122) and

$$(u, \nabla u, \boldsymbol{\lambda}) = \arg \min_{(v, \mathbf{q}, \boldsymbol{\mu}) \in \widetilde{W}} \left\{ \int_\Omega \mathbf{q} \cdot \boldsymbol{\mu} dx + \frac{\theta^{-1}}{2} \int_\Omega |f - v|^2 dx \right\} \tag{139}$$

with

$$\widetilde{W} = \left\{ (v, \mathbf{q}, \boldsymbol{\mu}) \in H^1(\Omega) \times \left(\mathcal{L}^2(\Omega)\right)^2 \times \left(\mathcal{L}^2(\Omega)\right)^2, \mathbf{q} = \nabla v, \boldsymbol{\mu} \cdot \mathbf{q} = |\mathbf{q}|, |\boldsymbol{\mu}| \leq 1 \right\}.$$

Formally, $\boldsymbol{\lambda} = \nabla u/|\nabla u|$, a relation which makes no sense on those parts of Ω where function ∇u vanishes. Let us denote ∇u by $\mathbf{p}$ and define set S_2 by (32). We observe also that there is equivalence between problem (139) and

$$(\mathbf{p},\boldsymbol{\lambda}) = \underset{(\mathbf{q},\boldsymbol{\mu})\in(\mathcal{L}^2(\Omega))^4}{\operatorname{argmin}} \left\{ \int_\Omega \mathbf{q}\cdot\boldsymbol{\mu}dx + \frac{\theta^{-1}}{2}\int_\Omega \left|\widetilde{\nabla}_f^{-1}\mathbf{q} - f\right|^2 dx + I_{S_2}(\mathbf{q},\boldsymbol{\mu}) + I_{\tilde{S}_1}(\mathbf{q}) \right\} \tag{140}$$

with $\mathcal{L}^2(\Omega))^4 = \left(\mathcal{L}^2(\Omega)\right)^2 \times \left(\mathcal{L}^2(\Omega)\right)^2$, where function $\widetilde{\nabla}_f^{-1}\mathbf{q}$ and set $\widetilde{S}_1$ are as in Section 7.2, and I_{S_2} is the indicator function of S_2 with S_2 as in Section 4.1, that is

$$I_{S_2}(\mathbf{q},\boldsymbol{\mu}) = \begin{cases} 0 & \text{if } (\mathbf{q},\boldsymbol{\mu}) \in S_2, \\ +\infty & \text{if } (\mathbf{q},\boldsymbol{\mu}) \in \left(\mathcal{L}^2(\Omega)\right)^2 \times \left(\mathcal{L}^2(\Omega)\right)^2 \backslash S_2. \end{cases}$$

With functional G as in Section 7.2, an optimality system associated with problem (140) reads as:

$$\begin{cases} \boldsymbol{\lambda} + \partial_{\mathbf{q}} I_{S_2}(\mathbf{p},\boldsymbol{\lambda}) + \theta^{-1} D_{\mathbf{q}} G(\mathbf{p}) + \partial_{\mathbf{q}} I_{\tilde{S}_1}(\mathbf{p}) \ni \mathbf{0}, \\ \mathbf{p} + \partial_{\mu} I_{S_2}(\mathbf{p},\boldsymbol{\lambda}) \ni \mathbf{0}. \end{cases} \tag{141}$$

Next, we associate with (141) the following initial value problem

$$\begin{cases} \begin{cases} \dfrac{\partial \mathbf{p}}{\partial t} + \boldsymbol{\lambda} + \partial_{\mathbf{q}} I_{S_2}(\mathbf{p},\boldsymbol{\lambda}) + \theta^{-1} D_{\mathbf{q}} G(\mathbf{p}) + \partial_{\mathbf{q}} I_{\tilde{S}_1}(\mathbf{p}) \ni \mathbf{0}, \\ \gamma \dfrac{\partial \boldsymbol{\lambda}}{\partial t} + \mathbf{p} + \partial_{\mu} I_{S_2}(\mathbf{p},\boldsymbol{\lambda}) \ni \mathbf{0}, \end{cases} & \text{in } \Omega \times (0, +\infty), \\ (\mathbf{p}(0),\boldsymbol{\lambda}(0)) = (\mathbf{p}_0,\boldsymbol{\lambda}_0), \end{cases} \tag{142}$$

where γ is a strictly positive function, and $(\mathbf{p}_0, \boldsymbol{\lambda}_0) \in S_2$.

Let A_1 be the identity operator, $A_2 = (\partial_{\mathbf{q}} I_{S_2}, \partial_{\mu} I_{S_2})^T$ and $A_3 = (\partial_{\mathbf{q}} I_{\tilde{S}_2} + \theta^{-1} D_{\mathbf{q}} G, 0)^T$. Applying the Machuk–Yanenko scheme to the solution of problem (142), we obtain

$$(\mathbf{p}^0, \boldsymbol{\lambda}^0) = (\mathbf{p}_0, \boldsymbol{\lambda}_0). \tag{143}$$

For $n \geq 0$, $(\mathbf{p}^n, \boldsymbol{\lambda}^n) \to \left(\mathbf{p}^{n+1/3}, \boldsymbol{\lambda}^{n+1/3}\right) \to \left(\mathbf{p}^{n+2/3}, \boldsymbol{\lambda}^{n+2/3}\right) \to \left(\mathbf{p}^{n+1}, \boldsymbol{\lambda}^{n+1}\right)$ as follows: Solve

$$\begin{cases} \dfrac{\mathbf{p}^{n+1/3} - \mathbf{p}^n}{\tau} + \boldsymbol{\lambda}^{n+1/3} = \mathbf{0}, \\ \gamma \dfrac{\boldsymbol{\lambda}^{n+1/3} - \boldsymbol{\lambda}^n}{\tau} + \mathbf{p}^{n+1/3} = \mathbf{0}, \end{cases} \tag{144}$$

$$\begin{cases} \dfrac{\mathbf{p}^{n+2/3} - \mathbf{p}^{n+1/3}}{\tau} + \partial_{\mathbf{q}} I_{S_2}\left(\mathbf{p}^{n+2/3}, \boldsymbol{\lambda}^{n+2/3}\right) \ni \mathbf{0}, \\ \gamma \dfrac{\boldsymbol{\lambda}^{n+2/3} - \boldsymbol{\lambda}^{n+1/3}}{\tau} + \partial_{\mu} I_{S_2}\left(\mathbf{p}^{n+2/3}, \boldsymbol{\lambda}^{n+2/3}\right) \ni \mathbf{0}, \end{cases} \tag{145}$$

and

$$\begin{cases} \dfrac{\mathbf{p}^{n+1} - \mathbf{p}^{n+2/3}}{\tau} + \partial_{\mathbf{q}} I_{\tilde{S}_1}\left(\mathbf{p}^{n+1}\right) + \theta^{-1} D_{\mathbf{q}} G\left(\mathbf{p}^{n+1}\right) \ni \mathbf{0}, \\ \gamma \dfrac{\boldsymbol{\lambda}^{n+1} - \boldsymbol{\lambda}^{n+2/3}}{\tau} = \mathbf{0}. \end{cases} \tag{146}$$

Solving Problem (144): Problem (144) can be rewritten (pointwise) as

$$\begin{pmatrix} 1 & \tau \\ \tau & \gamma(x) \end{pmatrix} \begin{pmatrix} \mathbf{p}^{n+1/3}(x) \\ \boldsymbol{\lambda}^{n+1/3}(x) \end{pmatrix} = \begin{pmatrix} \mathbf{p}^{n}(x) \\ \gamma(x)\boldsymbol{\lambda}^{n}(x) \end{pmatrix}. \tag{147}$$

Assuming (from Remark 6) that $\gamma(x)(=\gamma^n(x))) > \sqrt{\tau}$, we have

$$\det \begin{pmatrix} 1 & \tau \\ \tau & \gamma(x) \end{pmatrix} = \gamma(x) - \tau^2 > 0$$

for $0 < \tau < 1$, implying that the linear system (147) is well-posed, its solution being given by

$$\begin{cases} \mathbf{p}^{n+1/3}(x) = \gamma(x) \dfrac{\mathbf{p}^{n}(x) - \tau \boldsymbol{\lambda}^{n}(x)}{\gamma(x) - \tau^2}, \\ \boldsymbol{\lambda}^{n+1/3}(x) = \dfrac{\gamma(x)\boldsymbol{\lambda}^{n}(x) - \tau \mathbf{p}^{n}(x)}{\gamma(x) - \tau^2}, \end{cases} \quad \text{a.e. in } \Omega. \tag{148}$$

Solving Problem (145): Problem (145) being identical to problem (45), the solution method discussed in Section 4.3 still applies here.

Solving Problem (146): It follows from (146) that

$$\begin{cases} \left(\mathbf{p}^{n+1}, \boldsymbol{\lambda}^{n+1}\right) = \left(\nabla u^{n+1}, \boldsymbol{\lambda}^{n+2/3}\right), \\ \text{with function } u^{n+1} \text{ being solution of,} \\ \begin{cases} -\nabla^2 u^{n+1} + \theta^{-1} \tau u^{n+1} = \theta^{-1} \tau f - \nabla \cdot \mathbf{p}^{n+2/3} \text{ in } \Omega, \\ \left(\nabla u^{n+1} - \mathbf{p}^{n+2/3}\right) \cdot \mathbf{n} = 0 \text{ on } \partial\Omega. \end{cases} \end{cases} \tag{149}$$

Remark 6. Concerning the choice of γ, Remark 5 from Section 4.3 still applies here. We recall that in Remark 5 we mentioned that in Deng et al. (2019) (an Euler elastica related article) one advocates that $\gamma(=\gamma^n)$ verifies $\gamma \geq \sqrt{\tau}$.

Remark 7. Schemes (144)–(146) "enjoy" several variants. We will mention only one, namely the one where one replaces (148) (and therefore (144)) by:

$$\begin{cases} \mathbf{p}^{n+1/3}(x) = \gamma(x)\dfrac{\mathbf{p}^n(x) - \tau\boldsymbol{\lambda}^n(x)}{\gamma(x) - \tau^2}, \\ \boldsymbol{\lambda}^{n+1/3}(x) = P_{\overline{D}}\left(\dfrac{\gamma(x)\boldsymbol{\lambda}^n(x) - \tau\mathbf{p}^n(x)}{\gamma(x) - \tau^2}\right), \end{cases} \quad \text{a.e. in } \Omega, \tag{150}$$

where, in (150), $\overline{D}$ is the closed unit disk of $\mathbb{R}^2$ centred at (0, 0), and $P_{\overline{D}}$ is the orthogonal projection operator from $\mathbb{R}^2$ onto $\overline{D}$, that is

$$P_{\overline{D}}(\boldsymbol{\mu}) = \frac{\boldsymbol{\mu}}{\max\left\{1, \sqrt{\mu_1^2 + \mu_2^2}\right\}}, \ \forall \boldsymbol{\mu} = (\mu_1, \mu_2) \in \mathbb{R}^2. \tag{151}$$

8 Conclusion

Via ADMM and split-Bregman methods, operator-splitting-based algorithms have been associated with image processing for many years already (see, e.g., Glowinski et al., 2017 and the references therein). However, none of these methods has the flexibility and robustness of the *Lie* and *Marchuk–Yanenko schemes* discussed in this chapter. These new methods (in the context of imaging science) allow for an easy solution of those complicated variational problems associated with sophisticated restoration models such as Euler elastica, L^1-mean curvature. The algorithms we presented in Sections 4–7 are (almost) parameter free, the main parameter to adjust being the time discretization step.

However, the algorithms we presented are largely heuristic implying that a mathematical analysis of their convergence properties would be welcome, including the existence of solutions for some of the subproblems associated with the fractional steps. The continuous model discretization brings many open problems worth studying, the convergence as $h \to 0$ being one of them. Another direction for investigations is of course applications to imaging variational problems beyond the ones we discussed in this chapter.

Acknowledgements

R.G. was supported by the Hong Kong Baptist University and by the Kennedy Wong Foundation. S.L. was supported by the Programs for Science and Technology Development of Henan Province (192102310181). X.C.T. was supported by the startup grant at Hong Kong Baptist University, grants RG(R)-RC/17-18/02-MATH and FRG2/17-18/033.

References

Acar, R., Vogel, C.R., 1994. Analysis of bounded variation penalty methods for ill-posed problems. Inverse Prob. 10 (6), 1217–1229.

Ambrosio, L., Masnou, S., 2003a. A direct variational approach to a problem arising in image reconstruction. Interfaces Free Bound. 5 (1), 63–82.

Ambrosio, L., Masnou, S., 2003b. On a variational problem arising in image reconstruction. In: Colli, P., Verdi, C., Visintin, A. (Eds.), Free Boundary Problems. Springer, pp. 17–26.

Bae, E., Shi, J., Tai, X.-C., 2011. Graph cuts for curvature based image denoising. IEEE Trans. Image Process. 20 (5), 1199–1210.

Bae, E., Tai, X.-C., Zhu, W., 2017. Augmented Lagrangian method for an Euler's elastica based segmentation model that promotes convex contours. Inverse Prob. Imaging 11 (1), 1–23.

Bellettini, G., Caselles, V., Novaga, M., 2002. The total variation flow in R^N. J. Differ. Eq. 184 (2), 475–525.

Bertalmío, M., Levine, S., 2014. Denoising an image by denoising its curvature image. SIAM J. Imaging Sci. 7 (1), 187–211.

Bredies, K., Kunisch, K., Pock, T., 2010. Total generalized variation. SIAM J. Imaging Sci. 3 (3), 492–526.

Bredies, K., Pock, T., Wirth, B., 2013. A convex, lower semicontinuous approximation of Euler's elastica energy. SIAM J. Math. Anal. 51 (6), 3232–3258. https://doi.org/10.1137/090745854.

Bretin, E., Masnou, S., Oudet, E., 2015. Phase-field approximations of the Willmore functional and flow. Numer. Math. 131 (1), 115–171.

Brito-Loeza, C., Chen, K., 2010a. Fast numerical algorithms for Euler's elastica inpainting model. Int. J. Modern Math. 5 (2), 157–182.

Brito-Loeza, C., Chen, K., 2010b. On high-order denoising models and fast algorithms for vector-valued images. IEEE Trans. Image Process. 19 (6), 1518–1527.

Brito-Loeza, C., Chen, K., 2013. Fast iterative algorithms for solving the minimization of curvature-related functionals in surface fairing. Int. J. Comput. Math. 90 (1), 92–108.

Brito-Loeza, C., Ke, C., Uc-Cetina, V., 2016. Image denoising using the Gaussian curvature of the image surface. Numer. Methods Partial Differ. Equ. 32 (3), 1066–1089.

Cai, X., Chan, R., Zeng, T., 2013. A two-stage image segmentation method using a convex variant of the Mumford-Shah model and thresholding. SIAM J. Imaging Sci. 6 (1), 368–390.

Do Carmo, M. P., 1976. Differential Geometry of Curves and Surfaces, Translated From the Portuguese. Prentice-Hall, Inc., Englewood Cliffs, NJ.

Carter, J. L., 2001. Dual Methods for Total Variation-based Image Restoration (Ph.D. thesis), UCLA Los Angeles, CA, USA.

Chambolle, A., 2004. An algorithm for total variation minimization and applications. J. Math. Imaging Vis. 20 (1–2), 89–97.

Chambolle, A., Lions, P.-L., 1997. Image recovery via total variation minimization and related problems. Numer. Math. 76 (2), 167–188. ISSN 0945-3245. https://doi.org/10.1007/s002110050258.

Chambolle, A., Pock, T., 2011. A first-order primal-dual algorithm for convex problems with applications to imaging. J. Math. Imaging Vis. 40 (1), 120–145.

Chambolle, A., Pock, T., 2019. Total roto-translational variation. Numer. Math. 142 (3), 611–666.

Chan, T.F., Glowinski, R., 1978. Finite-element approximation and iterative solution of a class of mildly nonlinear elliptic equations. Technical report, Stanford report STAN-CS-78-674, Computer Science Department, Stanford University, Palo Alto, CA.

Chan, T.F., Golub, G.H., Mulet, P., 1999. A nonlinear primal-dual method for total variation-based image restoration. SIAM J. Sci. Comput. 20 (6), 1964–1977.

Chan, T.F., Marquina, A., Mulet, P., 2000. High-order total variation-based image restoration. SIAM J. Sci. Comput. 22 (2), 503–516.

Chan, T.F., Kang, S.H., Shen, J., 2002. Euler's elastica and curvature-based inpainting. SIAM J. Appl. Math. 564–592.

Chan, T.F., Esedoglu, S., Park, F., Yip, A., 2006. Total variation image restoration: Overview and recent developments. In: Paragios, N., Chen, Y., Faugeras, O. (Eds.), Handbook of Mathematical Models in Computer Vision. Springer, pp. 17–31.

Chen, Y., Levine, S., Rao, M., 2006. Variable exponent, linear growth functionals in image restoration. SIAM J. Appl. Math. 66 (4), 1383–1406.

Chorin, A.J., Hughes, T.J.R., McCracken, M.F., Marsden, J.E., 1978. Product formulas and numerical algorithms. Commun. Pure Appl. Math. 31 (2), 205–256.

Darbon, J., Sigelle, M., 2006a. Image restoration with constrained total variation. Part II: Levelable functions, convex priors and non-convex cases. J. Math. Imaging Vis. 26 (3), 277–292.

Darbon, J., Sigelle, M., 2006b. Image restoration with discrete constrained total variation. Part I: Fast and exact optimization. J. Math. Imaging Vis. 26 (3), 261–276.

Dean, E.J., Glowinski, R., Guidoboni, G., 2007. On the numerical simulation of Bingham viscoplastic flow: old and new results. J. Non-Newtonian Fluid Mech. 142 (1), 36–62.

Deng, L., Glowinski, R., Tai, X.-C., 2019. A new operator splitting method for Euler's elastica model. SIAM J. Imaging Sci. 12 (2), 1190–1230.

Donoho, D.L., 1995. De-noising by soft-thresholding. IEEE Trans. Inform. Theory 41 (3), 613–627.

Duan, Y., Tai, X.-C., 2011. Domain decomposition methods with graph cuts algorithms for total variation minimization. Adv. Comput. Math. 1–25. ISSN 1019-7168. https://doi.org/10.1007/s10444-011-9213-4.

El-Zehiry, N.Y., Grady, L., 2010. Fast global optimization of curvature. In: IEEE Computer Society Conference on Computer Vision and Pattern Recognition, pp. 3257–3264.

Elsey, M., Esedog-lu, S., 2009. Analogue of the total variation denoising model in the context of geometry processing. Multiscale Model. Simul. 7 (4), 1549–1573.

Esedoglu, S., March, R., 2003. Segmentation with depth but without detecting junctions. J. Math. Imaging Vis. 18 (1), 7–15.

Esedoglu, S., Shen, J., 2002. Digital inpainting based on the Mumford-Shah-Euler image model. Eur. J. Appl. Math. 13 (4), 353–370.

Evans, L.C., 2018. Measure Theory and Fine Properties of Functions. Routledge.

Fortin, M., Glowinski, R., 1983. Augmented Lagrangian methods in quadratic programming. In: Fortin, M., Glowinski, R. (Eds.), Studies in Mathematics and its Applications, vol. 15. Elsevier, pp. 1–46.

Glowinski, R., 2003. Finite element methods for incompressible viscous flow. In: Ciarlet, P.G., Lions, J.L. (Eds.), Handbook of Numerical Analysis, vol. 4. Elsevier, pp. 3–1176.

Glowinski, R., 2014. On alternating direction methods of multipliers: a historical perspective. In: Fitzgibbon, W., Kuznetsov, Y.A., Neittaanmäki, P., Pironneau, O. (Eds.), Modeling, Simulation and Optimization for Science and Technology, vol. 34. Springer, Dordrecht, pp. 59–82.

Glowinski, R., 2015. Variational Methods for the Numerical Solution of Nonlinear Elliptic Problems. SIAM, Philadelphia.

Glowinski, R., Le Tallec, P., 1989. Augmented Lagrangian and Operator-Splitting Methods in Nonlinear Mechanics. vol. 9. SIAM, Philadelphia.

Glowinski, R., Marroco, A., 1975. Sur l'approximation, par éléments finis d'ordre un, et la résolution, par pénalisation-dualité d'une classe de problèmes de dirichlet non linéaires. ESAIM: Math. Model. Numer. Anal. 9 (R2), 41–76.

Glowinski, R., Wachs, A., 2011. On the numerical simulation of viscoplastic fluid flow. In: Glowinski, R., Xu, J. (Eds.), Handbook of Numerical Analysis, vol. 16. Elsevier, pp. 483–717.

Glowinski, R., Leung, S., Qian, J., 2016a. Operator-splitting based fast sweeping methods for isotropic wave propagation in a moving fluid. SIAM J. Sci. Comput. 38 (2), A1195–A1223.

Glowinski, R., Pan, T.-W., Tai, X.-C., 2016b. Some facts about operator-splitting and alternating direction methods. In: Glowinski, R., Osher, S., Yin, W. (Eds.), Splitting Methods in Communication, Imaging, Science, and Engineering. Springer, Switzerland, pp. 19–94.

Glowinski, R., Osher, S., Yin, W., 2017. Splitting Methods in Communication, Imaging, Science, and Engineering. Springer, Switzerland.

Goldman, R., 2005. Curvature formulas for implicit curves and surfaces. Comput. Aided Geom. Des. 22 (7), 632–658.

Goldstein, T., Osher, S., 2009. The split Bregman method for L^1 regularized problems. SIAM J. Imaging Sci. 2 (2), 323–343.

Gong, Y., Sbalzarini, I.F., 2013. Local weighted Gaussian curvature for image processing. In: IEEE International Conference on Image Processing, pp. 534–538.

Hintermüller, M., Papafitsoros, K., 2019. Generating structured nonsmooth priors and associated primal-dual methods. In: Handbook of Numerical Analysis, vol. 20. Elsevier, pp. 437–502.

Huang, Y., Ng, M.K., Wen, Y.-W., 2008. A fast total variation minimization method for image restoration. Multiscale Model. Simul. 7 (2), 774–795.

Kang, S.H., Tai, X.-C., Zhu, W., 2019. Survey of fast algorithms for Euler's elastica-based image segmentation. In: Handbook of Numerical Analysis, vol. 20. Elsevier, pp. 533–552.

Kang, S.H., Zhu, W., Jianhong, J., 2014. Illusory shapes via corner fusion. SIAM J. Imaging Sci. 7 (4), 1907–1936.

Kärkkäinen, T., Majava, K., 2000a. Nonmonotone and monotone active-set methods for image restoration, Part 1: Convergence analysis. J. Optim. Theory Appl. 106 (1), 61–80.

Kärkkäinen, T., Majava, K., 2000b. Nonmonotone and monotone active-set methods for image restoration, Part 2: Numerical results. J. Optim. Theory Appl. 106 (1), 81–105.

Kärkkäinen, T., Majava, K., Mäkelä, M.M., 2001. Comparison of formulations and solution methods for image restoration problems. Inverse Prob. 17 (6), 1977–1995.

Kimmel, R., Malladi, R., Sochen, N., 2000. Images as embedded maps and minimal surfaces: movies, color, texture, and volumetric medical images. Int. J. Comput. Vis. 39 (2), 111–129.

Lysaker, M., Tai, X.-C., 2006. Iterative image restoration combining total variation minimization and a second-order functional. Int. J. Comput. Vis. 66 (1), 5–18.

Lysaker, M., Lundervold, A., Tai, X.-C., 2003. Noise removal using fourth-order partial differential equation with applications to medical magnetic resonance images in space and time. IEEE Trans. Image Process. 12 (12), 1579–1590.

Marchuk, G.I., 1990. Splitting and alternating direction methods. In: Handbook of Numerical Analysis, vol. 1. Elsevier, pp. 197–462.

Masnou, S., 2002. Disocclusion: a variational approach using level lines. IEEE Trans. Image Process. 11 (2), 68–76.

Masnou, S., Morel, J.-M., 1998. Level lines based disocclusion. In: IEEE International Conference on Image Processing, pp. 259–263.

Masnou, S., Morel, J.-M., 2006. On a variational theory of image amodal completion. Rend. Sem. Mat. Univ. Padova 116, 211–252.

Meyer, Y., 2001. Oscillating Patterns in Image Processing and Nonlinear Evolution Equations: The Fifteenth Dean Jacqueline B. Lewis Memorial Lectures. vol. 22. American Mathematical Society.

Mumford, D., 1994. Elastica and computer vision. In: Bajaj, C.L. (Ed.), Algebraic Geometry and Its Applications. Springer, pp. 491–506.

Myllykoski, M., Glowinski, R., Kärkkäinen, T., Rossi, T., 2015. A new augmented Lagrangian approach for L^1-mean curvature image denoising. SIAM J. Imaging Sci. 8 (1), 95–125.

Nitzberg, M., Mumford, D., Shiota, T., 1993. Filtering, Segmentation, and Depth, vol. 662. Springer-Verlag, Berlin.

Rudin, L., Osher, S., Fatemi, E., 1992. Nonlinear total variation based noise removal algorithms. Phys. D 60, 259–268.

Schoenemann, T., Kahl, F., Cremers, D., 2009. Curvature regularity for region-based image segmentation and inpainting: a linear programming relaxation. In: IEEE International Conference on Computer Vision, pp. 17–23.

Shi, Y., Wang, L.-L., Tai, X.-C., 2012. Geometry of total variation regularized L^p-model. J. Comput. Appl. Math. 236 (8), 2223–2234. ISSN 03770427. https://doi.org/10.1016/j.cam.2011.09.043.

Sochen, N., Kimmel, R., Malladi, R., 1998. A general framework for low level vision. IEEE Trans. Image Process. 7 (3), 310–318.

Strong, D., Chan, T.F., 2003. Edge-preserving and scale-dependent properties of total variation regularization. Inverse Probl. 19 (6), S165–S187.

Sun, L., Chen, K., 2014. A new iterative algorithm for mean curvature-based variational image denoising. BIT Numer. Math. 54 (2), 523–553.

Tai, X.-C., Wu, C., 2009. Augmented Lagrangian method, dual methods and split Bregman iteration for ROF model. In: Scale Space and Variational Methods in Computer Vision, pp. 502–513.

Tai, X.-C., Hahn, J., Chung, G.J., 2011. A fast algorithm for Euler's elastica model using augmented Lagrangian method. SIAM J. Imaging Sci. 4 (1), 313–344.

Tan, L., Li, L., Liu, W., Sun, J., Zhang, M., 2019. A novel Euler's elastica based segmentation approach for noisy images via using the progressive hedging algorithm. arXiv preprint arXiv:1902.07402.

Tartar, L., 2007. An Introduction to Sobolev Spaces and Interpolation Spaces. Springer Science & Business Media, Berlin.

Vese, L., 2001. A study in the BV space of a denoising-deblurring variational problem. Appl. Math. Optim. 44 (2), 131–161.

Wang, Y., Yin, W., Zhang, Y., 2007. A fast algorithm for image deblurring with total variation regularization. Technical report.

Wang, Y., Yang, J., Yin, W., Zhang, Y., 2008. A new alternating minimization algorithm for total variation image reconstruction. SIAM J. Imaging Sci. 1 (3), 248–272.

Willmore, T.J., 1996. Riemannian Geometry. Oxford University Press.

Yang, J., Yin, W., Zhang, Y., Wang, Y., 2009. A fast algorithm for edge-preserving variational multichannel image restoration. SIAM J. Imaging Sci. 2 (2), 569–592.

Yashtini, M., Kang, S.H., 2016. A fast relaxed normal two split method and an effective weighted TV approach for Euler's elastica image inpainting. SIAM J. Imaging Sci. 9 (4), 1552–1581.

You, Y.-L., Kaveh, M., 2000. Fourth-order partial differential equations for noise removal. IEEE Trans. Image Process. 9 (10), 1723–1730.

Zhang, J., Chen, K., 2016. A new augmented Lagrangian primal dual algorithm for elastica regularization. J. Algorithms Comput. Technol. 10 (4), 325–338.

Zhang, J., Chen, R., Deng, C., Wang, S., 2017. Fast linearized augmented Lagrangian method for Euler's elastica model. Numer. Math.: Theory Methods Appl. 10 (01), 98–115. ISSN 1004-8979. https://doi.org/10.4208/nmtma.2017.m1611.

Zhu, M., Chan, T.F., 2008. An efficient primal-dual hybrid gradient algorithm for total variation image restoration. UCLA CAM Report 34.

Zhu, W., Chan, T.F., 2012. Image denoising using mean curvature of image surface. SIAM J. Imaging Sci. 5 (1), 1–32.

Zhu, W., Chan, T.F., Esedoglu, S., 2006. Segmentation with depth: a level set approach. SIAM J. Sci. Comput. 28 (5), 1957–1973.

Zhu, M., Wright, S.J., Chan, T.F., 2010. Duality-based algorithms for total-variation-regularized image restoration. Comput. Optim. Appl. 47 (3), 377–400.

Zhu, W., Tai, X.-C., Chan, T.F., 2013a. Augmented Lagrangian method for a mean curvature based image denoising model. Inverse Prob. Imaging 7 (4), 1409–1432.

Zhu, W., Tai, X.-C., Chan, T.F., 2013b. Image segmentation using Euler's elastica as the regularization. J. Sci. Comput. 57 (2), 414–438.

Chapter 6

From active contours to minimal geodesic paths: New solutions to active contours problems by Eikonal equations

Da Chen[a,b] and Laurent D. Cohen[a,*]

[a]*University Paris Dauphine, PSL Research University, CEREMADE, CNRS, UMR 7534, Paris, France*

[b]*Centre Hospitalier National d'Ophtalmologie des Quinze-Vingts, Paris, France*

[*]*Corresponding author: e-mail: cohen@ceremade.dauphine.fr*

Chapter Outline

Handbook of Numerical Analysis, Vol. 20. https://doi.org/10.1016/bs.hna.2019.07.009
© 2019 Elsevier B.V. All rights reserved.

Abstract

In this chapter, we give an overview of part of our previous work based on the minimal geodesic path framework and the Eikonal partial differential equation (PDE). We show that by designing adequate Riemannian and Randers geodesic metrics the minimal paths can be utilized to search for solutions to almost all of the active contour problems and to the Euler–Mumford elastica problem, which allows to blend the advantages from minimal geodesic paths and those original approaches, i.e., the active contours and elastica curves. The proposed minimal path-based models can be applied to deal with a broad variety of image analysis tasks such as boundary detection, image segmentation and tubular structure extraction. The numerical implementations for the computation of minimal paths are known to be quite efficient thanks to the Eikonal solvers such as the Finsler variant of the fast marching method introduced in Mirebeau (2014b).

Keywords: Minimal path, Eikonal partial differential equation, Geodesic distance, Finsler metric, Riemannian metric, Randers metric, Euler–Mumford elastica curve, Region-based active contour model

AMS Classification Codes: 68U10, 62H35

1 Introduction

The original active contour model proposed by Kass et al. (1988) is a variational approach for the applications of boundary detection, image segmentation and object tracking. It is capable of locally minimizing an energy functional for the purpose of detecting an optimal and continuous curve, either open or closed, to depict the image features of interest such as image edges. The energy functional of this model can be established using the image grey levels or gradients. A broad variety of region-based active contour approaches have been intensively studied in order to develop the original active contour model (Kass et al., 1988). They take into account a type of more robust regional homogeneity penalization way for image segmentation. As a consequence, these region-based variant approaches are able to search for suitable solutions in more complicated situations. In the other hand, the minimal geodesic path model was first introduced by Cohen and Kimmel (1997) in order to search for the global minimum of a simplified version of the functional used in the original active contour model (Kass et al., 1988). In essence, this simplified active contour energy functional can be regarded as a weighted curve length and it can be solved through the Eikonal partial differential equation (PDE). The objective of this chapter is to show how the terms that are commonly used in the active contour models can be handled using different kinds of metrics for geodesic paths, in particular Randers metrics.

In its basic formulation, the minimal geodesic path model extracts image features by a globally optimal curve between two prescribed points, usually provided by the user. A minimal path can be tracked through the solution to the Eikonal PDE. A large number of well-established Eikonal solvers have been exploited such as the fast marching methods (Mirebeau, 2014a, b, 2018; Sethian, 1999). The global optimality and the efficient solutions lead to a series of successful minimal path-based applications as reviewed in Peyré et al. (2010). However, the original minimal geodesic path model (Cohen and Kimmel, 1997) depends only on the first-order derivative term of a curve. Thus this model cannot impose curvature to define the regularization term which corresponds to the curve rigidity property. Finally, in the minimal path-based image segmentation scenario, almost all of the existing geodesic metrics (Appia and Yezzi, 2011; Benmansour and Cohen, 2009; Mille et al., 2015) are derived from the image boundary features, which may limit their performance on complicated image segmentation tasks. In the remaining of this chapter, we proposed different geodesic metrics to overcome the aforementioned drawbacks suffered by the existing minimal path models.

In Kimmel (2003) and Kimmel and Bruckstein (2003), the authors proposed a listing of the active contour energy terms for image segmentation. In our chapter, we propose to revisit Kimmel's chapter (Kimmel, 2003), named *fast edge integration*, with minimal path methods to address the active contour problems.

1.1 Outline

In this chapter, we illustrate different types of geodesic metrics and their respective applications in image analysis based on the minimal path framework and the Eikonal PDE. The structure of this chapter is organized as follows:

- In Section 2, we present the preliminary background on the active contour models involving the approaches, respectively, driven by edge- and region-based terms.
- In Section 3, we review the existing minimal path models associated with both of the Riemannian and Randers metrics, the corresponding Eikonal PDEs for the computation of the geodesic distance as well as the gradient descent ordinary differential equations (ODEs) for tracking geodesic curves.
- In Section 4, we present two minimal path models for the minimization of the geometric active contour models with alignment terms (Kimmel, 2003; Kimmel and Bruckstein, 2003). We, respectively, induce a Randers metric and an anisotropic Riemannian metric from the variants of the asymmetric and symmetric alignment terms. In this case, the active contour problems with different alignment terms are transferred to the corresponding minimal path problems.
- In Section 5, we present a Finsler elastica minimal path model (Chen et al., 2017) to solve the Mumford–Euler elastica curve problem (Mumford, 1994). This is done by using a strongly anisotropic and asymmetric Randers geodesic metric that is established over an orientation-lifted space.

The orientation dimension is used to represent the curve directions. The introduced Finsler elastica minimal paths can be naturally applied for the applications of retinal imaging and boundary detection.

- In Section 6, we present a new method that applies the Randers minimal path model to address the region-based active contour problems (Chen et al., 2016, 2019). The used Randers metric is derived from the region-based homogeneity term as well as the curve length-based regularization term, relying on the solution to divergence equation-constrained minimization problem. In this case, we can make use of the Randers Eikonal PDE to solve a broad variety of region-based active contour problems.

2 Active contour models

The active contour models originated from the work proposed in Kass et al. (1988) have obtained successful applications in the fields of boundary detection, image segmentation, object tracking and shape modelling. Basically, the goal of an active contour model can be summarized as seeking a regular parametric curve for the delineation of the image feature of interest, such as the object boundaries, by minimizing an energy functional. In general, the minimization of an active contour energy functional can be carried out by the curve evolution scheme. The crucial point for this scheme is the estimation of the gradient descent flow with adequate boundary conditions. Basically, this scheme amounts to searching for a parametric curve $\mathcal{C}_\tau : (u,\tau) \in [0,1] \times \mathbb{R}_0^+ \mapsto \mathcal{C}_\tau(u) \in \Omega$, where $u \in [0, 1]$ is the curve parameter and $\tau \in \mathbb{R}_0^+$ represents the (artificial) time and $\Omega \subset \mathbb{R}^2$ denotes the image domain. The curve $\mathcal{C}_\tau$ is supposed to converge to the target boundary when $\tau \to \infty$.

2.1 Edge-based active contour model

2.1.1 The computation of the image gradient features

Let $I : \Omega \to \mathbb{R}$ be a grey level image. The standard Euclidean gradient of a (smoothed) image I can be defined in conjunction with a Gaussian kernel G_σ with variance σ:

$$\nabla I_\sigma = \left(\partial_x G_\sigma * I,\ \ \partial_y G_\sigma * I\right)^T, \tag{1}$$

where $*$ denotes the convolution operator, and $\partial_x G_\sigma$, $\partial_y G_\sigma$ are the first-order partially differentials of the Gaussian kernel G_σ, respectively, along the axes x and y. The magnitude $g : \Omega \to \mathbb{R}_0^+$ of the image gradient ∇I_σ is a scalar-valued function that can be formulated by

$$g = \| \nabla I_\sigma \| . \tag{2}$$

In the context of boundary detection and image segmentation, the magnitude g often serves as the edge appearance feature map.

For a vector-valued image $\mathbf{I} = (I_1, I_2, I_3)$ in a RGB colour space, we first consider a Jacobi matrix $\mathbf{W}_\sigma(\mathbf{x})$ of size 2×3 to construct the associated image boundary features for a (smoothed) colour image $\mathbf{I}$, i.e.

$$\mathbf{W}_\sigma(\mathbf{x}) = \begin{pmatrix} \partial_x G_\sigma * I_1, & \partial_x G_\sigma * I_2, & \partial_x G_\sigma * I_3 \\ \partial_y G_\sigma * I_1, & \partial_y G_\sigma * I_2, & \partial_y G_\sigma * I_3 \end{pmatrix}(\mathbf{x}). \tag{3}$$

The edge appearance feature map g for a colour image can be estimated using the Frobenius norm of the matrix $\mathbf{W}_\sigma(\cdot)$. In other words, one can build the magnitude g as follows:

$$g = \sum_{i=1}^{3} \sqrt{(\partial_x G_\sigma * I_i)^2 + (\partial_y G_\sigma * I_i)^2}. \tag{4}$$

2.1.2 The original active contour model

The original edge-based active contour model or the snakes model (Kass et al., 1988) aims to minimize the following energy functional

$$E_{\text{snake}}(\mathcal{C}) = \int_0^1 \left(\alpha_1 \| \mathcal{C}'(u)\|^2 + \alpha_2 \| \mathcal{C}''(u)\|^2 + P(\mathcal{C}(u))\right) du, \tag{5}$$

where $\mathcal{C} \in H^2([0,1],\Omega)$ is a regular curve with nonvanishing velocity and the parameters α_1 and α_2 are real positive constants controlling the importance for each component. The functions $\mathcal{C}'$ and $\mathcal{C}''$ of the forms

$$\mathcal{C}'(u) := \frac{d}{du}\mathcal{C}(u), \quad \mathcal{C}''(u) := \frac{d}{du}\mathcal{C}'(u), \tag{6}$$

are, respectively, the first- and second-order derivatives of the curve $\mathcal{C}$. Note that the squared norm $\|\mathcal{C}''(\cdot)\|^2$ is relevant to the squared curvature of $\mathcal{C}$. The function $P : \Omega \to \mathbb{R}_0^+$, which is referred to as *potential*, is an image data-driven function. It is defined to have low values around the image structures of interest and high values otherwise. For instances, the potential P can be set as the decreasing function of the image appearance feature map g or can be derived from the image grey levels (Kass et al., 1988).

The curve evolution equation can be derived from the Euler–Lagrange equation of the energy functional E_{snake}, which reads

$$\frac{\partial \mathcal{C}_\tau}{\partial \tau} = \underbrace{\alpha_1 \mathcal{C}''_\tau - \alpha_2 \mathcal{C}''''_\tau}_{\text{Internal Forces}} - \underbrace{\nabla P(\mathcal{C}_\tau)}_{\text{External Force}}, \tag{7}$$

where $\mathcal{C}''''_\tau(u)$ is the fourth-order derivative of the curve $\mathcal{C}_\tau$ with respect to u. The first two terms in the evolution equation (7) obtained from the curve itself is referred to as the internal forces which impose the regularity of the curve, while the last term is the external force used to attract the curve to the features (e.g., object edges) of the target.

2.1.3 Balloon force

Cohen (1991) introduced an external force, named the balloon force, to relax the demanding initialization requirement of the original active contours model (Kass et al., 1988). According to Cohen (1991), the balloon force $\mathbf{F}_{\text{balloon}}$ can be expressed by

$$\mathbf{F}_{\text{balloon}} = \mathcal{N}_{\mathcal{C}}, \tag{8}$$

where $\mathcal{N}_{\mathcal{C}}$ is the *outward* unit normal to the curve $\mathcal{C}$. The balloon force $\mathbf{F}_{\text{balloon}}$ is able to drive the curve expanding from the interior region of $\mathcal{C}$ to the desired features such as the object boundaries. Thus the initial guess should be located inside the target and allowed to be far from the boundaries.

In conjunction with the balloon force $\mathbf{F}_{\text{balloon}}$, the corresponding curve evolution equation eventually turns out to be

$$\frac{\partial \mathcal{C}_\tau}{\partial \tau} = \alpha_1 \mathcal{C}''_\tau - \alpha_2 \mathcal{C}''''_\tau + \alpha_3 \mathbf{F}_{\text{balloon}}(\mathcal{C}_\tau) - \frac{\nabla P(\mathcal{C}_\tau)}{\|\nabla P(\mathcal{C}_\tau)\|}, \tag{9}$$

where $\alpha_3 \in \mathbb{R}^+$ is a weighting parameter that controls the importance of the balloon force during the curve evolution.

2.1.4 Geodesic active contour model

Let Lip([0, 1], Ω) be the set of all the Lipschitz continuous curves $\mathcal{C}: [0,1] \to \Omega$. The geodesic active contour model proposed in Caselles et al. (1997) and Yezzi et al. (1997) takes into account a potential function P to build a weighted curve length

$$\mathcal{L}_{\text{geo}}(\mathcal{C}) = \int_0^1 P(\mathcal{C}(u)) \, \| \mathcal{C}'(u) \| \, du. \tag{10}$$

The weighted curve length $\mathcal{L}_{\text{geo}}$ is intrinsic since it is independent of the parameterization of the curve $\mathcal{C}$. Comparing to the energy functional (5) considered in the original snakes model (Kass et al., 1988), the term which relies on the second-order derivative of the curve has been removed.

An extremum curve that minimizes the weighted curve length $\mathcal{L}_{\text{geo}}$ is a geodesic path. In Caselles et al. (1997) and Yezzi et al. (1997), a locally minimizing curve can be generated by the curve evolution scheme. Denoting by $\kappa_\tau : [0,1] \to \mathbb{R}$ the curvature of $\mathcal{C}_\tau$, the curve evolution equation obtained through the Euler–Lagrange equation of the weighted curve length $\mathcal{L}_{\text{geo}}$ reads

$$\frac{\partial \mathcal{C}_\tau}{\partial \tau} = P(\mathcal{C}_\tau)\kappa_\tau \mathcal{N}_\tau - \langle \nabla P(\mathcal{C}_\tau), \mathcal{N}_\tau \rangle \mathcal{N}_\tau, \tag{11}$$

where the operator $\langle \cdot, \cdot \rangle$ stands for the Euclidean scalar product, $\mathcal{N}_\tau$ is the *inward* unit normal to the curve $\mathcal{C}_\tau$, and ∇P is the Euclidean gradient of the function P with respect to positions. In the homogeneous regions of an image, the gradients of the potential P tend to vanish, leading to the following Euclidean curve shortening flow

$$\frac{\partial \mathcal{C}_\tau}{\partial \tau} \propto \kappa_\tau \mathcal{N}_\tau,$$

where $\propto$ is the positively proportional operator. When the curve $\mathcal{C}_\tau$ is near an edge at some time τ, the values of $P(\mathcal{C}_\tau(\cdot))$ are low and the curve evolution will be dominated by the second term of Eq. (11). Therefore, it is a good choice to initialize the curve $\mathcal{C}_\tau|_{\tau=0}$ surrounding the target. The drawback of the solution to the original active contour energy (5) is the high possibility of being stuck in a local minimum. The level set method introduced by Osher and Sethian (1988) is a very popular way for solving the curve evolution problem (Caselles et al., 1997; Malladi et al., 1995; Yezzi et al., 1997), but it still can be stopped by unexpected local minima. This is why the authors in Cohen and Kimmel (1997) introduced the use of the minimal geodesic path to find the global minimum of a simplified active contour energy functional, i.e., the weighted curve length (10). The global minimum is attained by solving a nonlinear PDE, called the Eikonal PDE, which will be briefly introduced in Section 3.1.

2.1.5 Geometric active contour models with alignment terms

The aforementioned geodesic active contour model (Caselles et al., 1997; Yezzi et al., 1997) as well as the fast implementation approach (Goldenberg et al., 2001) utilizes a direction-independent potential function P to measure the length of a curve. Kimmel and Bruckstein (2003) introduced an anisotropic variant of the geodesic active contour model based on an image data-driven vector field $V_{\text{align}} : \Omega \to \mathbb{R}^2$, leading to a new weighted curve length formula

$$\mathcal{L}_\varrho(\mathcal{C}) = \int_0^1 \varrho(\langle V_{\text{align}}(\mathcal{C}), \mathcal{N}\rangle)\|\mathcal{C}'\|\, du, \tag{12}$$

where ϱ is a scalar-value function and $\langle V_{\text{align}}(\mathcal{C}), \mathcal{N}\rangle$ is an alignment measure term. For the applications of boundary detection and image segmentation, the vector field V_{align} can be naturally built by the image gradient vector field ∇I_σ.

Kimmel and Bruckstein (2003) suggested two effective types of ϱ for the alignment measure term. The first one is to choose $\varrho(\langle \nabla I_\sigma(\mathcal{C}), \mathcal{N}\rangle) = \langle \nabla I_\sigma(\mathcal{C}), \mathcal{N}\rangle$ which yields an asymmetric functional

$$\mathcal{L}_{\text{align}}(\mathcal{C}) = \int_0^1 \langle \nabla I(\mathcal{C}), \mathcal{N}\rangle \|\mathcal{C}'\|\, du. \tag{13}$$

The curve evolution equation for the evolving curve $\mathcal{C}_\tau$ with respect to the functional (13) can be expressed by

$$\frac{\partial \mathcal{C}_\tau}{\partial \tau} = \Delta I_\sigma(\mathcal{C}_\tau)\mathcal{N}_\tau, \tag{14}$$

where Δ is the Laplacian operator such that $\Delta I_\sigma = \text{div}(\nabla I_\sigma)$. The extremum curves $\mathcal{C}_\tau$ as $\tau \to \infty$ corresponds to a maximizer of the functional $\mathscr{L}_{\text{align}}$.

The second choice introduced by Kimmel and Bruckstein (2003) for the function ϱ is to set

$$\varrho(\langle \nabla I(\mathcal{C}), \mathcal{N} \rangle) = |\langle \nabla I(\mathcal{C}), \mathcal{N} \rangle|,$$

which yields a robust and symmetric variant of the asymmetric case (13) as follows:

$$\mathscr{L}^+_{\text{align}}(\mathcal{C}) = \int_0^1 |\langle \mathcal{N}, \nabla I_\sigma(\mathcal{C}) \rangle| \, \|\mathcal{C}'\| \, du. \tag{15}$$

The curve evolution equation in this case has a form

$$\frac{\partial \mathcal{C}_\tau}{\partial \tau} = \text{sign}(\langle \nabla I_\sigma, \mathcal{N}_\tau \rangle) \Delta I_\sigma(\mathcal{C}_\tau) \mathcal{N}_\tau. \tag{16}$$

The maximization of the alignment-based functionals $\mathscr{L}_{\text{align}}$ and $\mathscr{L}^+_{\text{align}}$ can be, respectively, solved through the curve evolution equations (14) and (16) in conjunction with the level set framework (Osher and Sethian, 1988). An optimal curve generated by means of the evolution equation (14) (resp., the evolution equation (16)) corresponds to the local maximum of the alignment-based functional $\mathscr{L}_{\text{align}}$ (resp., the functional $\mathscr{L}^+_{\text{align}}$). In Sections 4.1 and 4.2, we will introduce a Randers metric and an anisotropic Riemannian metric, respectively, for the global minimization of the slight variants of the alignment-based functionals $\mathscr{L}_{\text{align}}$ and $\mathscr{L}^+_{\text{align}}$.

2.2 The piecewise smooth Mumford–Shah model and the piecewise constant reduction model

The edge-based active contour approaches aforementioned, which take the image gradients as the edge appearance features for image boundary extraction, are quite efficient and effective models. However, the image gradients are very sensitive to the image noise and spurious edges, which also likely yield strong edge appearance features. In order to address this issue, the region-based active contour approaches exploit more complicated homogeneity criteria to establish the energy functionals. Significant examples of these region-based approaches involve the piecewise smooth Mumford–Shah model (Mumford and Shah, 1989) as well as the associated variant approaches such as Zhu and Yuille (1996), Cohen (1997), Chan and Vese (2001a) and Brox and Cremers (2009).

2.2.1 *The piecewise smooth Mumford–Shah model*

Let K be a closed subset of the image domain $\Omega \subset \mathbb{R}^2$, which is made up of a finite set of smooth curves. The goal for the original piecewise smooth Mumford–Shah model is to seek the set K and a piecewise smooth function $f: \Omega \to \mathbb{R}$ by minimizing the energy functional

$$E_{\mathrm{MS}}(f,K)=\alpha\int_{\Omega}(I-f)^2d\mathbf{x}+\hat{\alpha}\int_{\Omega\backslash K}\|\nabla f\|^2d\mathbf{x}+\mathrm{Per}(K), \tag{17}$$

where α and $\hat{\alpha}$ are two positive constants which control the relative importance between different terms, Per(K) is the perimeter of the set K, and f is referred to as an image data fitting function. The first integration term defines a similarity measure by estimating the approximation errors between the image grey levels I and the data fitting function f. The second integration is a smoothness penalty term that imposes a prior to the function f.

Since the original work in Mumford and Shah (1989), many efforts have been devoted to seeking practical solutions to the minimization of the piecewise smooth Mumford–Shah problem (17). Among them, the explicit curve evolution scheme is a widely considered tool for minimizing the energy functional (17), which can be implemented either by the level set method (Chan and Vese, 2001b; Osher and Sethian, 1988; Tsai et al., 2001; Vese and Chan, 2002) or by the convex relaxation framework (Bresson et al., 2007; Chan et al., 2006; Pock et al., 2009).

2.2.2 Piecewise constant reduction of the full Mumford–Shah model

The piecewise constant active contour model (Chan and Vese, 2001a; Chan et al., 2000; Cohen, 1997) is a practical variant of the full Mumford–Shah model. Instead of using a piecewise smooth function f to approximate the image data, the piecewise constant model or the active contours without edges model (ACWE) assumes that the image grey levels within each region can be approximated by the mean intensity value estimated in the corresponding region.

Let $\mathcal{C}$ be a simple and closed curve. We denote by $A_{\mathcal{C}}\subset\Omega$ the open and bounded subset enclosed by the curve $\mathcal{C}$. In the case of binary-valued image segmentation scenario, the piecewise constant active contour model (Chan and Vese, 2001a; Cohen, 1997) considers an energy functional

$$E_{\mathrm{ACWE}}(\mathcal{C},\mu_{\mathrm{in}},\mu_{\mathrm{out}})=\alpha\int_{A_{\mathcal{C}}}(I-\mu_{\mathrm{in}})^2d\mathbf{x}+\alpha\int_{\Omega\backslash A_{\mathcal{C}}}(I-\mu_{\mathrm{out}})^2d\mathbf{x}+\mathscr{L}_{\mathrm{Euclid}}(\mathcal{C}), \tag{18}$$

where μ_{in} and μ_{out} are, respectively, the mean grey levels inside and outside the closed curve $\mathcal{C}$ and the term $\mathscr{L}_{\mathrm{Euclid}}(\mathcal{C})$ stands for the Euclidean curve length of $\mathcal{C}$ which serves as a regularization term for the active contour model. Note that in the original Chan–Vese piecewise constant model (Chan and Vese, 2001a), there is a term related to the area of the region $A_{\mathcal{C}}$, which has been ignored here.

2.2.3 Minimization of the piecewise constant active contour model

Chan and Vese (2001a) proposed a solution for the minimization of the active contour energy functional E_{ACWE} based on the level set method (Osher and Sethian, 1988; Zhao et al., 1996). Basically, the level set formulation relies on a Lipschitz function $\phi:\Omega\rightarrow\mathbb{R}$ such that the open and bounded subset $A_{\mathcal{C}}$ can be represented by $A_{\mathcal{C}}:=\{\mathbf{x}\in\Omega;\ \phi(\mathbf{x})>0\}$ and its boundary $\mathcal{C}$ can be identified from the zero level set of ϕ, i.e., $\mathcal{C}=\{\mathbf{x}\in\Omega;\ \phi(\mathbf{x})=0\}$. In numerical implementation, the level set function ϕ is often set to be a signed Euclidean distance function.

Let H be the Heaviside function which can be formulated by

$$H(a)=\begin{cases}1, & a\geq 0,\\ 0, & a<0.\end{cases}$$

The characteristic function of the subset $A_{\mathcal{C}}$ can be defined by the Heaviside function of the corresponding level set function. Nevertheless, one can reformulate the functional (18) using the Heaviside function H and the level set function ϕ as follows:

$$\begin{aligned}\mathcal{E}_{\text{ACWE}}(\phi,\mu_{\text{in}},\mu_{\text{out}})=&\ \alpha\int_{\Omega}(I-\mu_{\text{in}})^2H(\phi)d\mathbf{x}\\ &+\alpha\int_{\Omega}(I-\mu_{\text{out}})^2(1-H(\phi))d\mathbf{x}\\ &+\int_{\Omega}\delta(\phi)\ \|\nabla\phi\|\ d\mathbf{x},\end{aligned}\tag{19}$$

where $\delta=H'$ is the one-dimensional Dirac measure (Chan and Vese, 2001a).

As introduced in Chan and Vese (2001a), the minimization of the functional $\mathcal{E}_{\text{ACWE}}$ formulated in Eq. (19) can be addressed by iteratively evaluating the following two steps:

(1). Keeping the level set function ϕ fixed and minimizing the functional E_{ACWE} with respect to the scalar values μ_{in} and μ_{out} will yield

$$\mu_{\text{in}}^*=\frac{\int_{\Omega}IH(\phi)d\mathbf{x}}{\int_{\Omega}H(\phi)d\mathbf{x}},\quad \mu_{\text{out}}^*=\frac{\int_{\Omega}I(1-H(\phi))d\mathbf{x}}{\int_{\Omega}(1-H(\phi))d\mathbf{x}}.\tag{20}$$

(2). Based on the scalar values μ_{in}^*, μ_{out}^* derived from Eq. (20), one solves

$$\min_{\phi}\ \mathcal{E}_{\text{ACWE}}(\phi,\mu_{\text{in}}^*,\mu_{\text{out}}^*),\tag{21}$$

with respect to the level set function ϕ.

According to Chan and Vese (2001a), finding the numerical solution to the minimization problem (21) requires a regularized function H_ϵ which approximates the Heaviside function H as $\epsilon\rightarrow 0$. By computing the Euler–Lagrange equation for the level set function ϕ, one can obtain the following level set evolution equation

$$\frac{\partial\phi}{\partial\tau}=\delta_\epsilon(\phi)\left[\alpha(I-\mu_{\text{out}}^*)^2-\alpha(I-\mu_{\text{in}}^*)^2-\text{div}\left(\frac{\nabla\phi}{\|\nabla\phi\|}\right)\right]\tag{22}$$

where $\delta_\epsilon:=H'_\epsilon$ is the regularized version of δ.

3 Minimal paths for edge-based active contours problems

3.1 Cohen-Kimmel minimal path model

The minimal path model introduced by Cohen and Kimmel (1997) is a powerful tool in the fields of image analysis and medical imaging (Cohen, 2006; Peyré et al., 2010). It is designed to search for the global minimum of the weighted curve length energy (Caselles et al., 1997; Yezzi et al., 1997), measured along a curve $\mathcal{C}\in \text{Lip}([0,1],\Omega)$ as follows:

$$\mathscr{L}_{\text{IR}}(\mathcal{C}) = \int_0^1 (c_1 + P(\mathcal{C}(u)))\, \|\mathcal{C}'(u)\|\, du \tag{23}$$

where c_1 is a positive constant used for regularization and the function $P: \Omega \to \mathbb{R}^+$ is a scalar-valued potential. In the context of boundary detection, the potential P is usually set as a decreasing function of the image gradient magnitude (Cohen and Kimmel, 1997).

Given a source point $\mathbf{s}$ and a target point $\mathbf{x}$ in the domain Ω, a geodesic path or a minimal path $\mathcal{G}_{\mathbf{s},\mathbf{x}} \in \text{Lip}([0,1],\Omega)$ with $\mathcal{G}_{\mathbf{s},\mathbf{x}}(0)=\mathbf{s}$ and $\mathcal{G}_{\mathbf{s},\mathbf{x}}(1)=\mathbf{x}$ is a curve that globally minimizes the length $\mathscr{L}_{\text{IR}}$ among all the paths in the set Lip([0, 1], Ω). In other words, a geodesic path $\mathcal{G}_{\mathbf{s},\mathbf{x}}$ can be determined by

$$\mathcal{G}_{\mathbf{s},\mathbf{x}} = \underset{\mathcal{C}\in\text{Lip}([0,1],\Omega)}{\arg\min}\ \mathscr{L}_{\text{IR}}(\mathcal{C}), \quad s.t. \begin{cases} \mathcal{C}(0)=\mathbf{s}, \\ \mathcal{C}(1)=\mathbf{x}. \end{cases} \tag{24}$$

Tracking such a geodesic path from the source point $\mathbf{s}$ to a target point $\mathbf{x}$ relies on a geodesic distance map $\mathscr{D}_{\mathbf{s}}: \Omega \to \mathbb{R}_0^+$. At each point $\mathbf{x}$, the value of the geodesic distance $\mathscr{D}_{\mathbf{s}}(\mathbf{x})$ is equal to the minimal curve length from $\mathbf{s}$ to $\mathbf{x}$

$$\begin{aligned} \mathscr{D}_{\mathbf{s}}(\mathbf{x}) &= \inf_{\mathcal{C}\in\text{Lip}([0,1],\Omega)} \{\mathscr{L}_{\text{IR}}(\mathcal{C}); \mathcal{C}(0)=\mathbf{s}, \mathcal{C}(1)=\mathbf{x}\} \\ &= \mathscr{L}_{\text{IR}}(\mathcal{G}_{\mathbf{s},\mathbf{x}}). \end{aligned} \tag{25}$$

It is known that the geodesic distance map $\mathscr{D}_{\mathbf{s}}$ is the unique viscosity solution to the Hamiltonian–Jacobi–Bellman equation or the Eikonal equation

$$\begin{cases} \|\nabla\mathscr{D}_{\mathbf{s}}(\mathbf{x})\| = c_1 + P(\mathbf{x}), & \forall \mathbf{x}\in\Omega\backslash\{\mathbf{s}\}, \\ \mathscr{D}_{\mathbf{s}}(\mathbf{s}) = 0, \end{cases} \tag{26}$$

where $\nabla\mathscr{D}_{\mathbf{s}}$ is the standard Euclidean gradient of the geodesic distance map $\mathscr{D}_{\mathbf{s}}$.

Solving a gradient descent ODE on the distance map $\mathscr{D}_{\mathbf{s}}$ can generate a geodesic path, or a minimal path $\hat{\mathcal{G}}_{\mathbf{x},\mathbf{s}}$, which connects from $\mathbf{x}$ to $\mathbf{s}$. It corresponds to a back-propagation processing from the target point $\mathbf{x}$ to the source point $\mathbf{s}$. The gradient descent ODE can be expressed by

$$\frac{d\hat{\mathcal{G}}_{\mathbf{x},\mathbf{s}}(u)}{du} = -\frac{\nabla\mathscr{D}_{\mathbf{s}}(\hat{\mathcal{G}}_{\mathbf{x},\mathbf{s}}(u))}{\|\nabla\mathscr{D}_{\mathbf{s}}(\hat{\mathcal{G}}_{\mathbf{x},\mathbf{s}}(u))\|}, \tag{27}$$

with boundary condition $\hat{\mathcal{G}}_{\mathbf{x},\mathbf{s}}(0)=\mathbf{x}$. Note that the geodesic curve $\hat{\mathcal{G}}_{\mathbf{x},\mathbf{s}}$ is parameterized by its arc-length. The gradient descent ODE (27) can be numerically solved by Heun's or Runge–Kutta's methods.

In Fig. 1, we show an example for the computation of a geodesic path between a given pair of prescribed points on an image from the BSDS500 dataset (Arbelaez et al., 2011). In Fig. 1A, the red and yellow dots indicate the source point and the end point, respectively. The corresponding geodesic distance map and the target geodesic path are shown in Fig. 1B and C, respectively. The potential P is computed by in terms of image gradient features

$$P(\mathbf{x}) = \exp(-\beta g(\mathbf{x})),$$

where β is a positive constant and g is the magnitude of the colour image gradients, see Eq. (4). We apply the isotropic fast marching method (Sethian, 1999) for the estimation of the geodesic distance map.

In Fig. 2 we show an example of applying a set of minimal geodesic paths to extract a curvilinear tree structure. In this example, in order to find

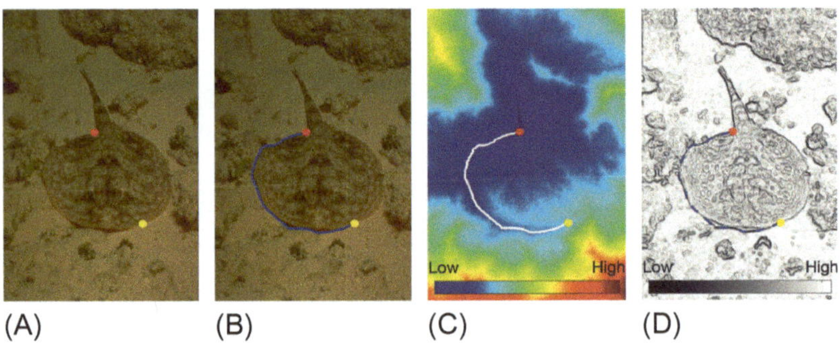

FIG. 1 An example for the minimal path extraction using the Cohen-Kimmel model (Cohen and Kimmel, 1997). (A) The source point and the end point which are, respectively, indicated by *red* and *yellow dots*. (B) The geodesic path indicated by the *blue line*. This path follows the target boundary between the given points. (C) Geodesic distance map. (D) The potential P with the geodesic path represented by *blue dash line*.

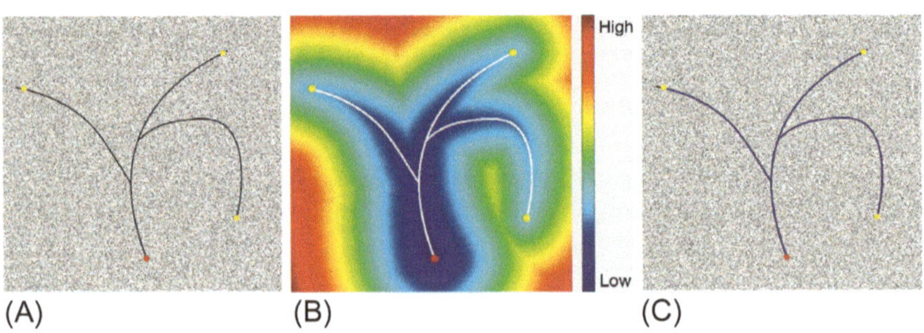

FIG. 2 An example for the extraction of a curvilinear tree structure using minimal paths. (A) A synthetic image blurred by noise. The *red* and *yellow dots* denote the source point and the end point, respectively. (B) The geodesic distance map. (C) The extracted minimal paths which are indicated by *blue lines*.

the whole tree structure, we give three end points which are located at the end of each bifurcation. These end points are indicated by yellow dots as illustrated in Fig. 2A. The geodesic distance map and the associated minimal geodesic paths (represented by blue lines) are shown in Fig. 2B and C, respectively. In this experiment, we build the potential P using the image grey levels as follows:

$$P(\mathbf{x}) = c_2 + I(\mathbf{x}),$$

where the scalar value $c_2 \in \mathbb{R}^+$ is a regularization factor.

3.2 Finsler and Randers minimal paths

A Finsler metric is a continuous map $\mathfrak{F}_\mathrm{F} : \Omega \times \mathbb{R}^2 \to [0, +\infty]$ over the space $\Omega \times \mathbb{R}^2$. For each point $\mathbf{x} \in \Omega$, the Finsler metric $\mathfrak{F}_\mathrm{F}(\mathbf{x}, \vec{\mathbf{u}})$ can be defined through an asymmetric norm, which is a convex and 1-homogeneous function on its second argument. Basically, the asymmetry property of a Finsler metric allows to take advantages of path directions during the computation of the geodesic distance and of the geodesic paths, as explored in Melonakos et al. (2008) and Chen and Cohen (2018).

The weighted length of a curve $\mathcal{C} \in \mathrm{Lip}([0,1],\Omega)$ can be measured with respect to a Finsler metric $\mathfrak{F}_\mathrm{F}$ (see Eq. (32)) by

$$\mathscr{L}_\mathrm{F}(\mathcal{C}) = \int_0^1 \mathfrak{F}_\mathrm{F}(\mathcal{C}(u), \mathcal{C}'(u))du. \tag{28}$$

Similar to the isotropic case, the geodesic distance map $\mathscr{D}_\mathbf{s} : \Omega \to \mathbb{R}_0^+$ associated to the weighted curve length $\mathscr{L}_\mathrm{F}$ can be formulated by

$$\mathscr{D}_\mathbf{s}(\mathbf{x}) = \inf_{\mathcal{C} \in \mathrm{Lip}([0,1],\Omega)} \{\mathscr{L}_\mathrm{F}(\mathcal{C}); \mathcal{C}(0) = \mathbf{s}, \mathcal{C}(1) = \mathbf{x}\}. \tag{29}$$

In order to estimate the geodesic distance map $\mathscr{D}_\mathbf{s}$, one can solve the following Finsler Eikonal equation:

$$\begin{cases} \sup\limits_{\vec{\mathbf{v}} \neq \mathbf{0}} \dfrac{\langle \nabla \mathscr{D}_\mathbf{s}(\mathbf{x}), \vec{\mathbf{v}} \rangle}{\mathfrak{F}_\mathrm{F}(\mathbf{x}, \vec{\mathbf{v}})} = 1, & \forall \mathbf{x} \in \Omega \backslash \{\mathbf{s}\} \\ \mathscr{D}_\mathbf{s}(\mathbf{s}) = 0, & \end{cases} \tag{30}$$

The corresponding gradient descent ODE on the geodesic distance map $\mathscr{D}_\mathbf{s}$ with respect to the Finsler metric $\mathfrak{F}_\mathrm{F}$ can be written as

$$\begin{cases} \dfrac{d\hat{\mathcal{G}}_{\mathbf{x},\mathbf{s}}(u)}{du} = \arg\min\limits_{\|\vec{\mathbf{v}}\|=1} \dfrac{\langle \nabla \mathscr{D}_\mathbf{s}(\hat{\mathcal{G}}_{\mathbf{x},\mathbf{s}}(u)), \vec{\mathbf{v}} \rangle}{\mathfrak{F}_\mathrm{F}(\hat{\mathcal{G}}_{\mathbf{x},\mathbf{s}}(u), \vec{\mathbf{v}})}, \\ \hat{\mathcal{G}}_{\mathbf{x},\mathbf{s}}(0) = \mathbf{x}. \end{cases} \tag{31}$$

The target geodesic path $\mathcal{G}_{\mathbf{s},\mathbf{x}}$ can be computed by reparameterizing the geodesic path $\hat{\mathcal{G}}_{\mathbf{x},\mathbf{s}}$.

In the following, we focus on the Randers metric (denoted by $\mathfrak{F}_{\text{Randers}}$) and the Riemannian metric (denoted by $\mathfrak{R}$), which are two particular cases of Finsler metric. The Eikonal PDEs with respect to these two types of geodesic metrics $\mathfrak{F}_{\text{Randers}}$ and $\mathfrak{R}$ are presented in Sections 3.2.1 and 3.2.2, respectively.

3.2.1 Randers minimal paths

Let $\mathbb{S}_2^+$ be the set of all the positive definite symmetric matrices of size 2×2. A Randers metric (Randers, 1941) is comprised of a positive definite symmetric tensor field $\mathcal{M}:\Omega \to \mathbb{S}_2^+$ and a vector field $\omega:\Omega \to \mathbb{R}^2$

$$\mathfrak{F}_{\text{Randers}}(\mathbf{x}, \vec{\mathbf{u}}) = \sqrt{\langle \vec{\mathbf{u}}, \mathcal{M}(\mathbf{x})\, \vec{\mathbf{u}} \rangle} + \langle \omega(\mathbf{x}), \vec{\mathbf{u}} \rangle, \tag{32}$$

where $\langle \cdot, \cdot \rangle$ stands for the standard Euclidean scalar product on $\mathbb{R}^2$. A Randers metric $\mathfrak{F}_{\text{Randers}}$ of the form (32) should obey the positive definiteness condition over the domain $\Omega \times \mathbb{R}^2$. In other words, one has $\mathfrak{F}_{\text{Randers}}(\mathbf{x}, \vec{\mathbf{u}}) > 0$ for any point $\mathbf{x} \in \Omega$ and any vector $\vec{\mathbf{u}}$, if and only if

$$\langle \omega(\mathbf{x}), \mathcal{M}^{-1}(\mathbf{x})\omega(\mathbf{x}) \rangle < 1. \tag{33}$$

For the applications of image analysis, the tensor field $\mathcal{M}$ and the vector field ω can be constructed dependently on the tasks, as formulated in Sections 5 and 6.

The Eikonal equation (30) with respect to a Randers metric $\mathfrak{F}_{\text{Randers}}$ is equivalent to the following nonlinear PDE (Chen and Cohen, 2018; Mirebeau, 2017)

$$\|\nabla \mathscr{D}_{\mathbf{s}}(\mathbf{x}) - \omega(\mathbf{x})\|_{\mathcal{M}^{-1}(\mathbf{x})} = 1, \quad \forall \mathbf{x} \in \Omega \backslash \{\mathbf{s}\}, \tag{34}$$

with boundary condition $\mathscr{D}_{\mathbf{s}}(\mathbf{s}) = 0$.

3.2.2 Riemannian minimal paths

The Randers metric $\mathfrak{F}_{\text{Randers}}$ of the form (32) gets to be a Riemannian metric $\mathfrak{R}$ of a general form, if the vector field ω obeys that $\omega \equiv \mathbf{0}$, i.e.,

$$\mathfrak{R}(\mathbf{x}, \vec{\mathbf{u}}) = \sqrt{\langle \vec{\mathbf{u}}, \mathcal{M}(\mathbf{x})\, \vec{\mathbf{u}} \rangle}. \tag{35}$$

Then the Eikonal equation associated to the Riemannian metric $\mathfrak{R}$ defined in Eq. (35) can be formulated by

$$\begin{cases} \|\nabla \mathscr{D}_{\mathbf{s}}(\mathbf{x})\|_{\mathcal{M}^{-1}(\mathbf{x})} = 1, & \forall \mathbf{x} \in \Omega \backslash \{\mathbf{s}\}, \\ \mathscr{D}_{\mathbf{s}}(\mathbf{s}) = 0. \end{cases} \tag{36}$$

In this case, the gradient descent ODE defined in (31) gets to

$$\begin{cases} \dfrac{d\hat{\mathcal{G}}_{\mathbf{x},\mathbf{s}}(u)}{du} = -\dfrac{\mathcal{M}^{-1}(\hat{\mathcal{G}}_{\mathbf{x},\mathbf{s}}(u))\nabla \mathscr{D}_{\mathbf{s}}(\hat{\mathcal{G}}_{\mathbf{x},\mathbf{s}}(u))}{\|\mathcal{M}^{-1}(\hat{\mathcal{G}}_{\mathbf{x},\mathbf{s}}(u))\nabla \mathscr{D}_{\mathbf{s}}(\hat{\mathcal{G}}_{\mathbf{x},\mathbf{s}}(u))\|}, \\ \hat{\mathcal{G}}_{\mathbf{x},\mathbf{s}}(0) = \mathbf{x}. \end{cases} \tag{37}$$

This ODE can be solved by Mirebeau's ODE solver (Mirebeau, 2014a).

4 Minimal paths for alignment active contours

Inspired by the geometric active contour models with edge-based alignment terms (Kimmel, 2003; Kimmel and Bruckstein, 2003), we introduce a Randers metric and an anisotropic Riemannian metric based on the image gradients for the application of boundary detection. These metrics are induced using the variant forms of the asymmetric and symmetric edge-based alignment terms.

4.1 Randers alignment minimal paths

We introduce a method, named the Randers alignment minimal path model, for boundary detection. The basic idea is to integrate the asymmetric alignment term $\mathscr{L}_{\text{align}}$ formulated in Eq. (13), and a Euclidean curve length-based regularization term to establish a new edge-based energy functional

$$\begin{aligned}\mathcal{L}_{\text{align}}(\mathcal{C}) &= \mathscr{L}_{\text{Euclid}}(\mathcal{C}) - \beta\mathscr{L}_{\text{align}}(\mathcal{C})\\ &= \int_0^1 \|\mathcal{C}'\|\,du - \beta\int_0^1 \langle\mathcal{N},\nabla I_\sigma(\mathcal{C})\rangle\ \|\mathcal{C}'\|\,du\\ &= \int_0^1 (1-\beta\langle\mathcal{N},\nabla I_\sigma(\mathcal{C})\rangle)\ \|\mathcal{C}'\|\,du,\end{aligned} \tag{38}$$

where $\beta \in \mathbb{R}^+$ is a parameter that controls the relative importance between the regularization term $\mathcal{L}_{\text{Euclid}}$ and the edge-based asymmetric align term $\mathscr{L}_{\text{align}}$. Note that ∇I_σ represents the gradient vector field of the smoothed image I, see Eq. (1).

The objective in this section is to globally minimize $\mathcal{L}_{\text{align}}$ by searching for a geodesic path between two given points. Let $\mathbf{M}$ be a rotation matrix with angle $\pi/2$ and let ξ be the rotated image gradient vector field

$$\xi(\mathbf{x}) = \mathbf{M}\,\nabla I_\sigma(\mathbf{x}). \tag{39}$$

In this case, a vector $\xi(\mathbf{x})$ indicates the direction of a boundary should have at an edge point $\mathbf{x}$.

We reformulate the energy functional $\mathcal{L}_{\text{align}}$ by

$$\begin{aligned}\mathcal{L}_{\text{align}}(\mathcal{C}) &= \int_0^1 (1-\beta\langle\mathbf{M}\mathcal{N},\mathbf{M}\,\nabla I_\sigma(\mathcal{C})\rangle)\ \|\mathcal{C}'\|\,du,\\ &= \int_0^1 (\|\mathcal{C}'\| - \beta\langle\mathcal{C}',\xi(\mathcal{C})\rangle)du,\\ &= \int_0^1 \mathfrak{F}_{\text{align}}(\mathcal{C}(u),\mathcal{C}'(u))du,\end{aligned}$$

where $\mathfrak{F}_{\text{align}}$ can be expressed by

$$\mathfrak{F}_{\text{align}}(\mathbf{x},\vec{\mathbf{u}}) = \|\vec{\mathbf{u}}\| - \langle\vec{\mathbf{u}},\beta\xi(\mathbf{x})\rangle. \tag{40}$$

Now we can see that $\mathcal{L}_{\text{align}}$ has been formulated as a weighted curve length associated to a Randers metric. The function $\mathfrak{F}_{\text{align}}$ is required to obey the positive definiteness condition (33) in order to globally minimize $\mathcal{L}_{\text{align}}$ through the Eikonal PDE framework. One simple solution is to limit the range of the parameter β as follows

$$\beta < \left(\inf_{\mathbf{x} \in \Omega} \|\xi(\mathbf{x})\| \right)^{-1}.$$

However, a small value of β may reduce the importance of the image data, thus increasing the risk of shortcuts.

As a second choice, we make use of a new vector field $\tilde{\xi} : \Omega \to \mathbb{R}^2$ to replace the term $\beta\xi$ used in the Randers metric $\mathfrak{F}_{\text{align}}$. The considered vector field $\tilde{\xi}$ can be expressed by

$$\tilde{\xi}(\mathbf{x}) = \begin{cases} \varphi(\beta \|\xi(\mathbf{x})\|) \dfrac{\xi(\mathbf{x})}{\|\xi(\mathbf{x})\|}, & \text{if } \|\xi(\mathbf{x})\| \neq 0, \\ \mathbf{0}, & \text{otherwise.} \end{cases} \tag{41}$$

where $\varphi : \mathbb{R}^+ \to \mathbb{R}^+$ is a scalar function

$$\varphi(a) = 1 - \exp(-a), \quad \forall a > 0. \tag{42}$$

By the vector field $\tilde{\xi}$, we obtain a new Randers metric $\mathcal{F}_{\text{align}}$

$$\mathcal{F}_{\text{align}}(\mathbf{x}, \vec{\mathbf{u}}) = \|\vec{\mathbf{u}}\| - \langle \vec{\mathbf{u}}, \tilde{\xi}(\mathbf{x}) \rangle. \tag{43}$$

One can see that the positive definiteness condition characterized by (33) will always hold for the Randers metric $\mathcal{F}_{\text{align}}$.

Notice that through the vector field $\tilde{\xi}$, we actually minimize a new weighted curve length

$$\begin{aligned} \tilde{\mathcal{L}}_{\text{align}}(\mathcal{C}) &= \int_0^1 \mathcal{F}_{\text{align}}(\mathcal{C}'(u), \mathcal{C}(u))\, du \\ &= \int_0^1 \|\mathcal{C}'\|\, du - \int_0^1 \langle \mathcal{C}', \tilde{\xi}(\mathcal{C}) \rangle\, du, \end{aligned} \tag{44}$$

where the second term in the second line of Eq. (44) can be treated as a variant of the asymmetric alignment term $\beta\mathscr{L}_{\text{align}}$.

The asymmetric edge feature at each point $\mathbf{x}$ still can be characterized by the vector $\tilde{\xi}(\mathbf{x})$. Specifically, a large value of $\|\tilde{\xi}(\mathbf{x})\|$ indicates a high possibility that the point $\mathbf{x}$ belongs to an image edge. Moreover, when $\mathbf{x}$ is located at an edge, the direction $\tilde{\xi}(\mathbf{x})/\|\tilde{\xi}(\mathbf{x})\| = \xi(\mathbf{x})/\|\xi(\mathbf{x})\|$ is proportional to the tangent of the image edge at point $\mathbf{x}$.

In Fig. 3B, we visualize the vector field $\tilde{\xi}$ through the colour coding scheme (Baker et al., 2011). In this figure, we only illustrate the vectors

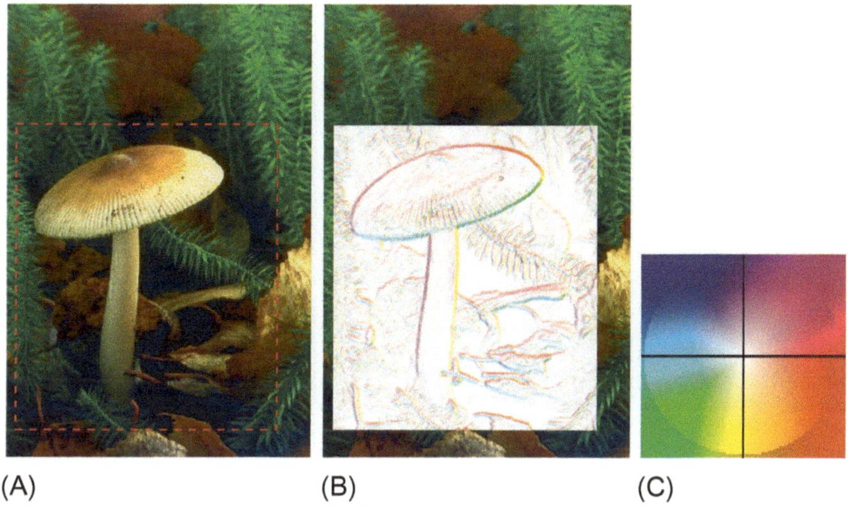

FIG. 3 Visualization of the vector field $\widetilde{\xi}$. (A) The original image. (B) The visualization for the vector field $\widetilde{\xi}$ using colours. We only show the vectors within the rectangle (indicated by *red dash line*) in (A). (C) colour coding.

$\widetilde{\xi}(\cdot)$ within a rectangle as shown in Fig. 3A. In Fig. 3C, we show the colour coding, where the direction of a vector is coded by hue while the norm is coded by saturation (Baker et al., 2011). The colour image used in this experiment is obtained from the BSDS500 dataset (Arbelaez et al., 2011).

Notice that when dealing with a vector-valued image $\mathbf{I} := (I_1, I_2, I_3)$, we build the vector field ξ through the following equation:

$$\xi = \mathbf{M}\left(\sum_{i=1}^{3}\partial_x G_\sigma * I_i, \sum_{i=1}^{3}\partial_y G_\sigma * I_i\right)^T.$$

The asymmetry property of the Randers metrics $\mathfrak{F}_{\text{align}}$ and $\mathcal{F}_{\text{align}}$ allows to exploit the asymmetric edge features for boundary detection and image segmentation. In Fig. 4, we show the boundary detection results using the Randers metric $\mathcal{F}_{\text{align}}$ on a colour image. In Fig. 4A, the red dot and the green dot, respectively, denote the given source and end points. In this experiment, the objective is to extract a geodesic path to depict the boundary of the object. The geodesic distance map superimposed in the original image is shown in Fig. 4C. Note that we adopt the Finsler variant of the fast marching method introduced by Mirebeau (2014b) as the Eikonal solver. The computation of the geodesic distance map is terminated once the end point, denoted by the yellow dot, is reached by the geodesic distance front (Deschamps and Cohen, 2001).

In Fig. 5, we illustrate the asymmetry property of the Randers alignment minimal paths. In Fig. 5A, the geodesic path is obtained by taking the red

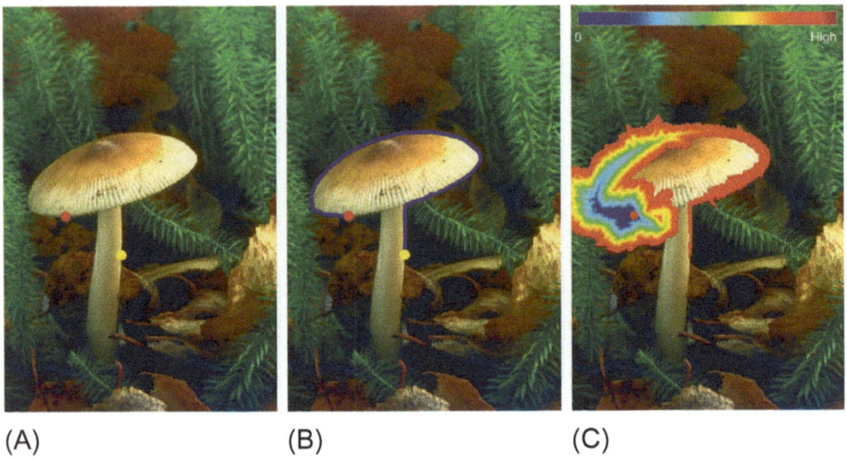

FIG. 4 An example for boundary detection through an asymmetric Randers alignment minimal path. (A) The original image and the seed and end points which are indicated by *red* and *yellow dots*, respectively. (B) The obtained minimal path (indicated by *blue line*) associated to the Randers metric $\mathcal{F}_{\text{align}}$. (C) The geodesic distance map superimposed on the original image.

FIG. 5 The asymmetry property of the Randers alignment minimal paths. (A) A Randers alignment minimal path generated by, respectively, taking the *red* and *yellow dots* as the source and end points. (B) Another Randers alignment minimal path generated by exchanging the source and end points used in (A). (C) Combining the minimal paths shown in (A) and (B) will generate a closed contour which passes through the target boundary.

dot as the source point and the blue dot as the end point. While in Fig. 5B, we exchange the source and end points used in Fig. 5A. One can obtain a different minimal path to the one shown in Fig. 5A. Then both minimal paths can be used to form a closed curve which depicts the complete object boundary, as shown in Fig. 5C.

FIG. 6 The Randers alignment minimal paths associated to different values of β. (A)–(D) The Randers alignment minimal path extraction results as the values of β increase. In particular, we set $\beta = 0$ in (A) which yields a straight line segment. In each figure, the *red* and *yellow dots* indicate the source point and the end point, respectively.

In Fig. 6, we show the Randers alignment minimal path extraction results with respect to different values of the parameter β. Specifically, we increase the values of β from Fig. 6A–D. In particular, we use $\beta = 0$ in Fig. 6A which generates a straight line segment. One can see that a small value of β which yields a weakly asymmetric Randers metric may generate a geodesic path with unexpected shortcuts, as shown in Fig. 6B and C. For a suitable value of β, the obtained minimal path can successfully depict the object boundary, as shown in Fig. 6D.

4.2 Riemannian alignment minimal paths

In this section, we induce an anisotropic Riemannian metric from a variant form of the symmetric alignment term $\mathscr{L}^+_{\text{align}}$ which is defined in Eq. (15). By integrating a regularization term and the symmetric alignment term, we first introduce an energy functional with a sufficiently small β

$$
\begin{aligned}
\mathcal{L}^+_{\text{align}}(\mathcal{C}) &= \int_0^1 \sqrt{1-\beta^2|\langle \mathcal{N}, \nabla I_\sigma(\mathcal{C})\rangle|^2}\; \|\mathcal{C}'\|\, du \\
&= \int_0^1 \sqrt{1-\beta^2|\langle \mathbf{M}\,\mathcal{N}, \mathbf{M}\,\nabla I_\sigma(\mathcal{C})\rangle|^2}\; \|\mathcal{C}'\|\, du \\
&= \int_0^1 \sqrt{\|\mathcal{C}'\|^2 - \beta^2|\langle \mathcal{C}', \xi(\mathcal{C})\rangle|^2}\, du \\
&= \int_0^1 \sqrt{\langle \mathcal{C}'(u), \mathcal{M}_{\text{align}}(\mathcal{C}(u))\, \mathcal{C}'(u)\rangle}\, du.
\end{aligned} \tag{45}
$$

The tensor field $\mathcal{M}_{\text{align}} : \Omega \to \mathbb{S}_2^+$ can be formulated by

$$\mathcal{M}_{\text{align}}(\mathbf{x}) = \mathrm{I_d} - \beta^2 \xi(\mathbf{x})\xi(\mathbf{x})^T, \tag{46}$$

providing that β is sufficiently small, where $\xi = \mathbf{M}\ \nabla I_\sigma$ is the rotated image gradient vector field with angle $\pi/2$. In order to ensure the matrix $\mathcal{M}_{\text{align}}(\mathbf{x})$ to be positive symmetric definite for any point $\mathbf{x}$, we replace the term $\beta^2\xi(\cdot)\xi(\cdot)^T$ in Eq. (46) by $\tilde{\xi}(\cdot)\tilde{\xi}(\cdot)^T$ defined in Eq. (41).

Hence we can obtain a new tensor field

$$\tilde{\mathcal{M}}_{\text{align}}(\mathbf{x}) = \mathrm{I_d} - \tilde{\xi}(\mathbf{x})\tilde{\xi}(\mathbf{x})^T,$$

The matrix $\tilde{\mathcal{M}}_{\text{align}}(\mathbf{x})$ will always be positive definite symmetric for each point $\mathbf{x}$ since the inequality $\|\tilde{\xi}(\mathbf{x})\| < 1$ always holds. Finally, we can define a new Riemannian metric

$$\mathcal{R}_{\text{align}}(\mathbf{x}, \vec{\mathbf{u}}) = \|\vec{\mathbf{u}}\|_{\tilde{\mathcal{M}}_{\text{align}}(\mathbf{x})}. \tag{47}$$

Following (Di Zenzo, 1986; Sapiro, 1997), we take into account the matrix $\mathbf{W}_\sigma(\mathbf{x})$ (defined in Eq. (3)) to estimate the vector $\xi(\mathbf{x})$ from a colour image $\mathbf{I}$. More precisely, we set

$$\xi(\mathbf{x}) = g(\mathbf{x})\upsilon(\mathbf{x}), \quad \forall \mathbf{x} \in \Omega$$

where the vector $\upsilon(\mathbf{x})$ is the eigenvector corresponding to the smaller eigenvalue of the matrix $\mathbf{W}_\sigma(\mathbf{x})\mathbf{W}_\sigma(\mathbf{x})^T$ and g is the Frobenius norm of $\mathbf{W}_\sigma(\mathbf{x})$, *see* Eq. (4).

In Fig. 7B, we show the geodesic path associated to the Riemannian metric $\mathcal{R}_{\text{align}}$ on a colour image from the Grabcut dataset (Rother et al., 2004). The red and yellow dots, respectively, denote the given source and end points. In Fig. 7C, we show the corresponding geodesic distance map superimposed on the original image.

In Fig. 8, we illustrate the Riemannian alignment minimal paths with respect to different values of the parameter β. In the first row, from columns 1 to 3, the minimal paths indicated by the blue lines are obtained by setting $\beta = 4/g_{\max}$, $8/g_{\max}$ and $10/g_{\max}$, respectively, where $g_{\max}$ is a scalar value defined by

$$g_{\max} := \max_{\mathbf{x}\in\Omega} \|\ \xi(\mathbf{x})\ \|.$$

(A) (B) (C)

FIG. 7 Alignment minimal path derived from the Riemannian metric $\mathcal{R}_{\text{align}}$. (A) The original image with the *red* and *green dots*, respectively, indicating the source and end points. (B) The obtained minimal path between the given points. (C) The geodesic distance map superimposed on the original image.

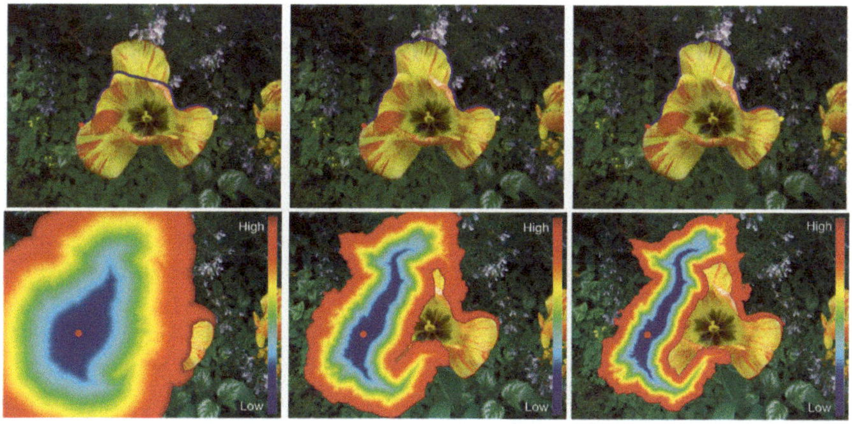

FIG. 8 The Riemannian alignment minimal paths associated to different values of β. From columns 1 to 3, we show the results corresponding to increasing values of β. The *first row* shows the minimal paths between two points and the *second row* illustrates the geodesic distance maps superimposed on the original images.

In the first row, we illustrate the respective geodesic distance maps associated to different values of β. For a small value of β, we find a shortcut as shown in row 1 and column 1. When increasing the values of β, one can obtain desired minimal paths which pass through the boundaries of the target, as illustrated in columns 2–3 of Fig. 8.

5 Orientation-lifted Randers minimal paths for Euler–Mumford elastica problem

In this section, we denote by $\mathbb{S}^1 = [0, 2\pi)$ an orientation space with periodic boundary condition, based on which we can define an orientation-lifted domain $\mathbb{D} = \Omega \times \mathbb{S}^1$. In this case, each point $\hat{\mathbf{x}} = (\mathbf{x}, \theta) \in \mathbb{D}$ is an ordered pair including a position $\mathbf{x}$ in the domain Ω and an orientation θ in the orientation space $\mathbb{S}^1$.

The original Euler–Mumford elastica bending energy functional (Mumford, 1994; Nitzberg and Mumford, 1990) assigns to each regular curve $\mathcal{C} \in H^2([0,1], \Omega)$ a curvature-dependent length which can be expressed by

$$\mathscr{L}_{\mathrm{Elastica}}(\mathcal{C}) = \int_0^1 (1 + \alpha \kappa^2(u)) \, \|\mathcal{C}'(u)\| \, du, \tag{48}$$

where $\alpha \in \mathbb{R}^+$ is a scalar-valued parameter which controls the importance of the curvature $\kappa : [0,1] \to \mathbb{R}$ in the functional $\mathscr{L}_{\mathrm{Elastica}}$.

In order to extract the image features such as object boundaries or tubular structure centrelines, we take into account the following data-driven elastica bending energy (Chen et al., 2017):

$$\mathcal{L}_{\mathrm{Elastica}}(\mathcal{C}) = \int_0^1 \Phi(\mathcal{C}(u), \mathcal{C}'(u)) \big(1 + \alpha \kappa^2(u)\big) \, \|\mathcal{C}'(u)\| \, du, \tag{49}$$

The function $\Phi:\Omega\times\mathbb{R}^2\to\mathbb{R}^+$ is a data-driven function relying on both positions and directions. Around the image features of interest and along the proper directions, the orientation-dependent function Φ is supposed to have small values, and large values, otherwise. The elastica bending energy functional $\mathcal{L}_{\text{Elastica}}$ is a geometric variant of the original active contour energy (5) which also takes the curvature-dependent term for regularization. The goal of both the Euler–Mumford elastica model (Mumford, 1994) and the original active contour model (Kass et al., 1988) is relevant to one another, i.e., finding an optimal curve of low curvature integration value, which simultaneously tends to pass through the image features of interest.

For the sake of simplicity, we will set the data-driven function $\Phi\equiv 1$ in Sections 5.1 and 5.2 and present a Randers minimal path solution to the Euler–Mumford Elastica problem (48). The minimization of the bending energy functional $\mathcal{L}_{\text{Elastica}}$ including the data-driven function will be introduced in Section 5.3.

5.1 Euler–Mumford elastica problem and its Finsler metric interpretation

The goal in this section is to search for an optimal curve which minimizes the elastica bending energy functional

$$\mathcal{L}_{\text{Elastica}}(\mathcal{C})=\int_0^1(1+\alpha\,\kappa(u)^2)\,\|\mathcal{C}'(u)\|\,du,\quad \mathcal{C}\in H^2([0,1],\Omega),\tag{50}$$

by the Eikonal equation-based minimal path framework. The basic idea is to establish a Finsler geodesic metric to interpret $\mathcal{L}_{\text{Elastica}}$ as a weighted curve length. Since the Eikonal equation is a first-order PDE, we should transform the squared curvature term to a first-order term via an auxiliary parametric function of orientations.

We denote by $\varsigma:[0,1]\to\mathbb{S}^1$ a parametric function over the interval [0, 1], where ς is defined being such that

$$(\cos(\varsigma(u)),\sin(\varsigma(u)))^T=\frac{\mathcal{C}'(u)}{\|\mathcal{C}'(u)\|}.\tag{51}$$

By a short computation, one can obtain the following expression

$$\frac{d}{du}\left(\frac{\mathcal{C}'(u)}{\|\mathcal{C}'(u)\|}\right)=\kappa(u)\,\|\mathcal{C}'(u)\|\left(\frac{\mathcal{C}'(u)}{\|\mathcal{C}'(u)\|}\right)^{\perp},\tag{52}$$

where $^{\perp}$ is a perpendicular operator, i.e., $\vec{\mathbf{u}}^{\perp}$ is a vector perpendicular to $\vec{\mathbf{u}}$.

On the other hand, we have the following equations

$$\begin{aligned}&\frac{d}{du}(\cos(\varsigma(u)),\sin(\varsigma(u)))^T\\&=\varsigma'(u)(-\sin(\varsigma(u)),\cos(\varsigma(u)))^T\\&=\varsigma'(u)\left(\frac{\mathcal{C}'(u)}{\|\mathcal{C}'(u)\|}\right)^{\perp}.\end{aligned}\tag{53}$$

In terms of Eqs. (52) and (53), we can express the curvature $\kappa(u)$ by

$$\kappa(u) = \frac{\varsigma'(u)}{\| \mathcal{C}'(u) \|}, \quad \forall u \in [0,1]. \tag{54}$$

Now we have reformulated the curvature $\kappa(\cdot)$ as a ratio of two first-order terms $\varsigma'(\cdot)$ and $\| \mathcal{C}'(\cdot) \|$. By incorporating the formula in Eq. (54) to the elastica bending length $\mathscr{L}_{\text{Elastica}}$, one obtains that

$$\mathscr{L}_{\text{Elastica}}(\mathcal{C}) = \int_0^1 \left(\| \mathcal{C}'(u) \| + \frac{\alpha |\varsigma'(u)|^2}{\| \mathcal{C}'(u) \|} \right) du, \tag{55}$$

following the constraint in Eq. (51).

Let $\Gamma = (\mathcal{C}, \varsigma) \in \text{Lip}([0,1], \mathbb{D})$ be a canonical orientation-lifted curve and one has $\Gamma'(u) = (\mathcal{C}'(u), \varsigma'(u))$, $\forall u \in [0,1]$. Now the elastica bending energy functional $\mathscr{L}_{\text{elastica}}$ of the form (55) can be rewritten by

$$\mathscr{L}_{\text{Eastica}}(\mathcal{C}) = \int_0^1 \mathfrak{F}_\infty(\Gamma(u), \Gamma'(u))\, du, \tag{56}$$

where $\mathfrak{F}_\infty : \mathbb{D} \times \mathbb{R}^3 \to [0, +\infty]$ is an orientation-lifted Finsler metric. It is defined for any orientation-lifted point $\hat{\mathbf{x}} = (\mathbf{x}, \theta) \in \mathbb{D}$ and any vector $\tilde{\mathbf{u}} = (\vec{\mathbf{u}}, \nu) \in \mathbb{R}^3$ by

$$\mathfrak{F}_\infty(\hat{\mathbf{x}}, \tilde{\mathbf{u}}) = \begin{cases} \|\vec{\mathbf{u}}\| + \dfrac{\alpha |\nu|^2}{\|\vec{\mathbf{u}}\|}, & \text{if } \vec{\mathbf{u}} \propto \vec{\mathbf{p}}_\theta \\ +\infty, & \text{otherwise,} \end{cases} \tag{57}$$

where $\vec{\mathbf{p}}_\theta = (\cos\theta, \sin\theta)^T$ is a unit vector associated to an orientation $\theta \in \mathbb{S}^1$ and where $\vec{\mathbf{u}} \propto \vec{\mathbf{p}}_\theta$ means that the vector $\vec{\mathbf{u}}$ is positively proportional to the vector $\vec{\mathbf{p}}_\theta$, i.e., $\vec{\mathbf{u}} = \|\vec{\mathbf{u}}\| \, \vec{\mathbf{p}}_\theta$.

5.2 Finsler elastica geodesic path for approximating the elastica curve

The orientation-lifted Finsler metric $\mathfrak{F}_\infty$ defined in Eq. (57) involves the infinite value $+\infty$, which will introduce difficulties to the numerical computation of the geodesic distance maps and the associated minimal geodesic paths. In order to address this problem, here we consider a relaxed orientation-lifted Randers metric $\mathfrak{F}_\lambda$ with a scalar factor $\lambda \gg 1$, which can be expressed by

$$\mathfrak{F}_\lambda(\hat{\mathbf{x}}, \tilde{\mathbf{u}}) = \sqrt{\lambda^2 \, \|\vec{\mathbf{u}}\|^2 + 2\alpha\lambda |\nu|^2} - (\lambda - 1)\langle \vec{\mathbf{u}}, \vec{\mathbf{p}}_\theta \rangle, \tag{58}$$

where $\alpha \in \mathbb{R}^+$ is a positive weighing parameter as used in Eq. (57). In the following, we refer to the Randers metric $\mathfrak{F}_\lambda$ as the *Finsler elastica metric*.

The metric $\mathfrak{F}_\lambda$ can also be reformulated in the form (32) through a tensor field $\mathcal{M}_\lambda : \mathbb{D} \to \mathbb{S}_3^+$ and a vector field $\omega_\lambda : \mathbb{D} \to \mathbb{R}^3$, which can be, respectively, formulated for any point $\hat{\mathbf{x}} = (\mathbf{x}, \theta)$ as

$$\mathcal{M}_\lambda(\hat{\mathbf{x}}) = \mathrm{diag}(\lambda^2, \lambda^2, 2\alpha\lambda), \quad \omega_\lambda(\hat{\mathbf{x}}) = (\lambda - 1)(\cos\theta, \sin\theta, 0)^T. \tag{59}$$

One can claim that the positive definiteness condition formulated in Eq. (33) can be satisfied:

$$\langle \omega_\lambda(\hat{\mathbf{x}}), \mathcal{M}_\lambda^{-1}(\hat{\mathbf{x}})\omega_\lambda(\hat{\mathbf{x}})\rangle = (1 - \lambda^{-1})^2 < 1.$$

Now we give the analysis for the relation between the orientation-lifted Finsler metric $\mathfrak{F}_\infty$ and the Finsler elastica metric $\mathfrak{F}_\lambda$. As $\lambda \to \infty$, we can express the metric $\mathfrak{F}_\lambda$ as follows:

$$\begin{aligned}
\mathfrak{F}_\lambda(\hat{\mathbf{x}}, \tilde{\mathbf{u}}) &= \sqrt{\lambda^2 \, \|\vec{\mathbf{u}}\|^2 + 2\alpha\lambda|\nu|^2} - (\lambda - 1)\langle \vec{\mathbf{u}}, \vec{\mathbf{p}}_\theta\rangle \\
&= \lambda \, \|\vec{\mathbf{u}}\| \left(\sqrt{1 + \frac{\alpha|\nu|^2}{\lambda \, \|\vec{\mathbf{u}}\|}} + \mathcal{O}(\lambda^{-2}) \right) - (\lambda - 1)\langle \vec{\mathbf{u}}, \vec{\mathbf{p}}_\theta\rangle \\
&= \|\vec{\mathbf{u}}\| + \frac{\alpha|\nu|^2}{\|\vec{\mathbf{u}}\|} + (\lambda - 1)(\|\vec{\mathbf{u}}\| - \langle \vec{\mathbf{u}}, \vec{\mathbf{p}}_\theta\rangle) + \mathcal{O}(\lambda^{-1})
\end{aligned}$$

The value of the term $(\lambda - 1)(\|\vec{\mathbf{u}}\| - \langle \vec{\mathbf{u}}, \vec{\mathbf{p}}_\theta\rangle)$ will get to 0 when the directions $\vec{\mathbf{u}}$ and $\vec{\mathbf{p}}_\theta$ are positively proportional to each other. Thus, one can see that $\mathfrak{F}_\lambda \to \mathfrak{F}_\infty$ pointwisely, as $\lambda \to \infty$.

The control set or the unit ball is a basic tool for visualizing the geometry distortion of a metric. For the metrics $\mathfrak{F}_\infty$ and $\mathfrak{F}_\lambda$, the sets of the unit balls can be, respectively, defined as

$$B_\infty(\hat{\mathbf{x}}) = \{\tilde{\mathbf{u}} = (\vec{\mathbf{u}}, \nu) \in \mathbb{R}^3; \mathfrak{F}_\infty(\hat{\mathbf{x}}, \tilde{\mathbf{u}}) \le 1\}, \tag{60}$$

and

$$B_\lambda(\hat{\mathbf{x}}) = \{\tilde{\mathbf{u}} = (\vec{\mathbf{u}}, \nu) \in \mathbb{R}^3; \mathfrak{F}_\lambda(\hat{\mathbf{x}}, \tilde{\mathbf{u}}) \le 1\}. \tag{61}$$

In order to characterize the sets B_∞ and B_λ, we first define two scalar values μ_1 and μ_2 as follows (Chen et al., 2017)

$$\mu_1 = \langle \vec{\mathbf{u}}, \vec{\mathbf{p}}_\theta\rangle, \quad \mu_2 = \langle \vec{\mathbf{u}}, \vec{\mathbf{p}}_\theta^{\perp}\rangle.$$

Hence any vector $\tilde{\mathbf{u}} = (\vec{\mathbf{u}}, \nu) \in B_\infty(\hat{\mathbf{x}})$ can be characterized by the following inequalities

$$\mu_1 > 0, \quad \mu_2 = 0, \quad \text{and} \quad \mu_1 + \alpha \frac{|\nu|^2}{\mu_1} \le 1. \tag{62}$$

Taking into account the Eq. (62), one obtains

$$\left(\mu_1 - \frac{1}{2}\right)^2 + \alpha \, |\nu|^2 \le \frac{1}{4}. \tag{63}$$

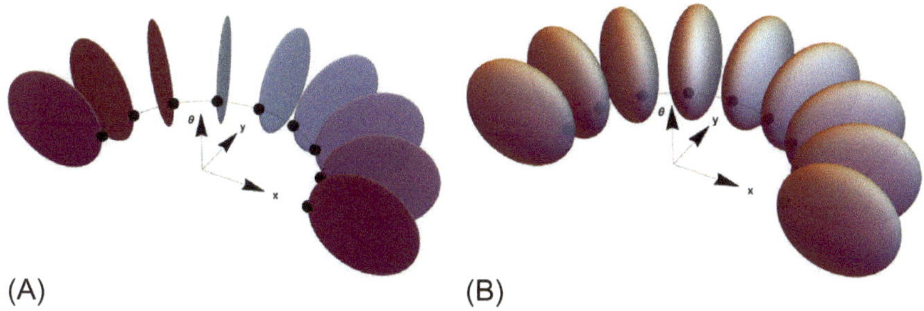

FIG. 9 Control sets derived from the metrics $\mathfrak{F}_\infty$ and $\mathfrak{F}_\lambda$. (A) Control sets of the metrics $\mathfrak{F}_\infty$ associated to different orientations θ are flat 2D disks embedded in 3D space, which aligned with the direction $\vec{\mathbf{p}}_\theta$. (B) Control sets of the Finsler elastica metrics $\mathfrak{F}_\lambda$ are ellipsoids. We set $\alpha = 1$ for the construction of the metrics $\mathfrak{F}_\infty$ and $\mathfrak{F}_\lambda$. *From Chen, D., Mirebeau, J.-M., Cohen, L. D., 2017. Global minimum for a Finsler elastica minimal path approach. Int. J. Comput. Vis. 122 (3), 458–483.*

Thus the ball $B_\infty(\hat{\mathbf{x}})$ can be characterized by a flat 2D ellipse which is embedded in the 3D tangent space with the origin being on its boundary. In particular, if one sets $\alpha = 1$, the unit ball $B_\infty(\hat{\mathbf{x}})$ gets to be a flat 2D disk of radius 1/2 as shown in Fig. 9A.

When the factor $\lambda < \infty$, one can point out that any vector $\tilde{\mathbf{u}} = (\vec{\mathbf{u}}, \nu) \in B_\lambda(\hat{\mathbf{x}})$ can be characterized by the following inequality

$$\frac{\lambda}{2}\mu_2^2 + a_\lambda\left(\mu_1 - \frac{b_\lambda}{2}\right)^2 + \alpha\,|\nu|^2 \le \frac{c_\lambda}{4}, \tag{64}$$

where the scalar values a_λ, b_λ, c_λ are all $1 + \mathcal{O}(1/\lambda)$. Hence the ball $B_\lambda(\hat{\mathbf{x}})$ is an ellipsoid and is almost flat in the direction $\vec{\mathbf{p}}_\theta^{\perp}$ due to the large value of $\lambda/2$, as shown in Fig. 9B for which we set $\alpha = 1$. One can see that the unit ball $B_\lambda(\hat{\mathbf{x}})$ will converge to $B_\infty(\hat{\mathbf{x}})$ in the sense of the Hausdorff distance, as $\lambda \to \infty$.

In Fig. 10B, we show the minimal paths derived from the metric $\mathfrak{F}_\lambda$. We denote by the red and black dots the source and end positions, respectively. In Fig. 10A, the directions at each position are indicated by arrows.

5.3 Data-driven Finsler elastica metric

In order to apply the Finsler elastica minimal paths for image segmentation and tubular structure centreline extraction, we should take into account the image data carried out by the orientation score $\psi : \mathbb{D} \to \mathbb{R}_0^+$ which is defined over the orientation-lifted domain $\mathbb{D}$.

The data-driven Finsler elastica metric $\mathcal{F}_\lambda$ can be expressed by

$$\mathcal{F}_\lambda(\hat{\mathbf{x}}, \tilde{\mathbf{u}}) = \exp\left(-\beta\frac{\psi(\hat{\mathbf{x}})}{\|\psi\|_\infty}\right)\mathfrak{F}_\lambda(\hat{\mathbf{x}}, \tilde{\mathbf{u}}), \tag{65}$$

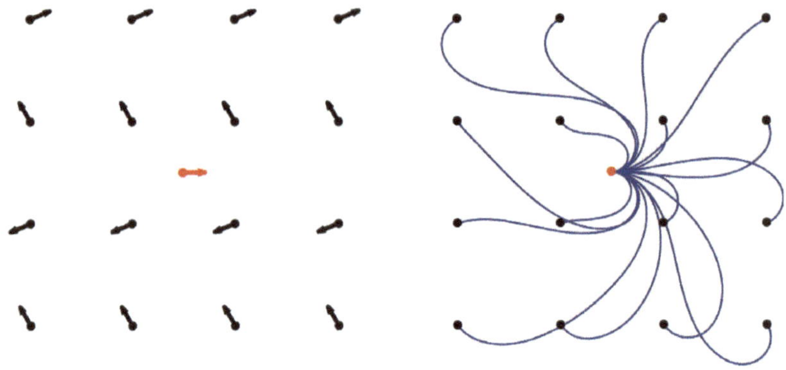

FIG. 10 Finsler elastica curves associated with the metric $\mathfrak{F}_\lambda$. (A) The source position and the end positions are denoted by the *red* and *black dots*, respectively. Each of the positions is assigned a direction indicated by a *arrow*. (B) Minimal paths (*blue lines*) associated to the initializations shown in (A).

or

$$\mathcal{F}_\lambda(\hat{\mathbf{x}},\tilde{\mathbf{u}}) = \left(1+\beta\frac{\psi(\hat{\mathbf{x}})}{\|\psi\|_\infty}\right)\mathfrak{F}_\lambda(\hat{\mathbf{x}},\tilde{\mathbf{u}}), \tag{66}$$

where β is a positive constant. From Eqs. (65) and (66), we can see that the data-driven function Φ (see Eq. (49)) is replaced by the orientation score-based functions $\exp(-\beta\psi(\hat{\mathbf{x}})/\|\psi\|_\infty)$ or $1+\beta\psi(\hat{\mathbf{x}})/\|\psi\|_\infty$. Note that in the experiments of this section, we make use of the metric defined in (65) for the data-driven Finsler elastica minimal path computation.

The computation of the orientation score ψ should be task-dependent. We, respectively, denote by ψ_{edge} and ψ_{tube} the orientation score ψ for image segmentation and for tubular structure centreline extraction, each of which can be estimated by steerable filters (Freeman and Adelson, 1991).

Let $\mathbf{I}=(I_1,I_2,I_3):\Omega\to\mathbb{R}^3$ be a RGB colour image with three channels. For boundary detection and image segmentation applications, we use the m-order Canny-like steerable filter (Jacob and Unser, 2004) with m an odd integer to build the orientation score ϕ_{edge}. The m-order canny-like steerable kernel $\mathbf{G}_m^\theta$ can be expressed by

$$\mathbf{G}_m^\theta(\mathbf{x}) = \sum_{a=1}^{m}\sum_{b=1}^{a} f_{a,b}(\theta)\frac{\partial^{a-b}}{\partial x^{a-b}}\frac{\partial^b}{\partial y^b}G_\sigma(\mathbf{x}), \tag{67}$$

where the function G_σ is a Gaussian kernel with variance σ. The coefficients $f_{a,b}$ are independent of the positions $\mathbf{x}$ but relying on the orientation θ and we refer to Jacob and Unser (2004) for the computation of $f_{a,b}$. Based on the kernels $\mathbf{G}_m^\theta$, the orientation score ψ_{edge} can be constructed by

$$\psi_{\text{edge}}(\mathbf{x},\theta)=\sum_{i=1}^{3}\left|\mathbf{G}_m^\theta * I_i(\mathbf{x})\right|. \tag{68}$$

The orientation score ψ_{tube} can be computed through a multiscale tubular structure enhancing filter such as the classical vesselness filter (Frangi et al., 1998) or the optimally oriented flux (OOF) filter (Law and Chung, 2008). In this section, we use the OOF filter to derive the orientation score ψ_{tube}, where the kernel $\mathbf{F}_r$ at a scale r can be expressed as

$$\mathbf{F}_r(\mathbf{x})=\left(\mathbf{1}_r*\begin{pmatrix}\partial_{xx}G_\sigma & \partial_{xy}G_\sigma\\ \partial_{yx}G_\sigma & \partial_{yy}G_\sigma\end{pmatrix}\right)(\mathbf{x}). \tag{69}$$

where $\mathbf{1}_r$ is the indicator function of a disk structure with radius r. In other words, the function $\mathbf{1}_r$ can be defined as a step function relying on a radius r such that

$$\mathbf{1}_r(\mathbf{x})=\begin{cases}1, & \text{if } \|\mathbf{x}\|<r,\\ 0, & \text{otherwise.}\end{cases}$$

Then we can define a multiscale response function at position $\mathbf{x}$ and scale r for a given image $I:\Omega\to\mathbb{R}$ through the kernel $\mathbf{F}_r$

$$\mathbf{Q}(\mathbf{x},r)=(\mathbf{F}_r*I)(\mathbf{x}).$$

Note that the response $\mathbf{Q}(\mathbf{x}, r)$ is a symmetric matrix of size 2×2. Let us denote by $\eta_1(\mathbf{x}, r)$, $\eta_2(\mathbf{x}, r)$ the two eigenvalues of the matrix $\mathbf{Q}(\mathbf{x}, r)$ obeying that $\eta_2(\cdot)\geq\eta_1(\cdot)$. We suppose that the grey levels inside the tubular structure are lower than background. In this case, for each point $\mathbf{x}$ we can estimate an optimal scale map $R_*:\Omega\to\mathbb{R}^+$ by

$$R_*(\mathbf{x})=\arg\max_r\ \{\eta_2(\mathbf{x},r)\}.$$

Then for a point $\mathbf{x}$ inside the vessel region, the eigenvalues of the matrix $\mathbf{Q}(\mathbf{x}, R_*(\mathbf{x}))$ satisfy that $\eta_2(\mathbf{x}, R_*(\mathbf{x}))\gg\eta_1(\mathbf{x}, R_*(\mathbf{x}))\approx 0$.

The orientation score ψ_{tube} can be computed by

$$\psi_{\text{tube}}(\mathbf{x},\theta)=\max\{\langle\vec{\mathbf{n}}_\theta,\mathbf{Q}(\mathbf{x},R_*(\mathbf{x}))\,\vec{\mathbf{n}}_\theta\rangle,0\}, \tag{70}$$

where $\vec{\mathbf{n}}_\theta=(-\sin\theta,\cos\theta)^T$ is a unit vector associated to an orientation $\theta\in\mathbb{S}^1$.

The minimal geodesic paths derived from the data-driven Finsler elastica metric should be as smooth as possible and simultaneously follow the desired image features. In Fig. 11, we illustrate an example for boundary detection using the data-driven Finsler elastica metric. In Fig. 11A, the arrows indicate the directions at the given positions (denoted by red and yellow dots). In Fig. 11B, the obtained minimal geodesic path indicated by blue line can detect an image edge of high Euclidean length, which satisfies the properties of the Finsler elastica minimal paths as discussed earlier.

In Fig. 12, we show the geodesic paths derived from the data-driven Finsler elastica metric on a synthetic image. The arrows in Fig. 12A and

(A) (B)

FIG. 11 An example for the minimal path derived from the data-driven Finsler elastica metric $\mathcal{F}_\lambda$ based on the orientation score ψ_{edge}. (A) The original image with source position (*red dot*) and end position (*yellow dot*). The *arrows* indicate the directions assigned to each position. (B) The obtained minimal path which are indicated by the *blue line*.

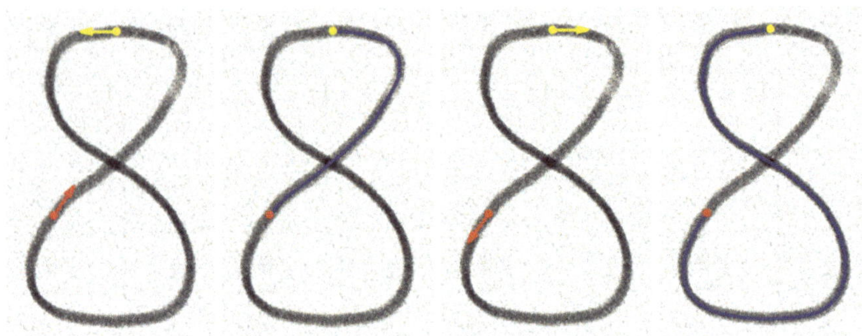

(A) (B) (C) (D)

FIG. 12 Centreline extraction for a tubular structure based on data-driven Finsler elastica metric using the orientation score ϕ_{tube}. (A) and (C) Prescribed positions with *arrows* denoting the respective tangents. The *red dots* are the source positions and the *yellow dots* are the end positions. (B) and (D) The minimal paths corresponding to the initializations shown in (A) and (C).

C denote the corresponding directions assigned to the given positions indicated by the red and yellow dots. The geodesic paths shown in Fig. 12B and D are computed using the same source and end positions but opposite directions. One can see that these geodesic paths tend to pass through the centreline of the target tubular structure, and simultaneously try to avoid sharp turnings as much as possible.

In Fig. 13, we apply the Finsler elastica minimal path model for retinal vessel centreline tracking. In this experiment, only the positions are provided. Each position will be assigned two directions corresponding to the orientations maximizing the orientation score ψ_{tube} over the space $[0, 2\pi)$. In this case, the source position (resp., end position) will generate two orientation-lifted source (resp., end) points. Hence we will obtain four pairs of orientation-lifted source and end points, which correspond to four geodesic paths. Among them, the geodesic

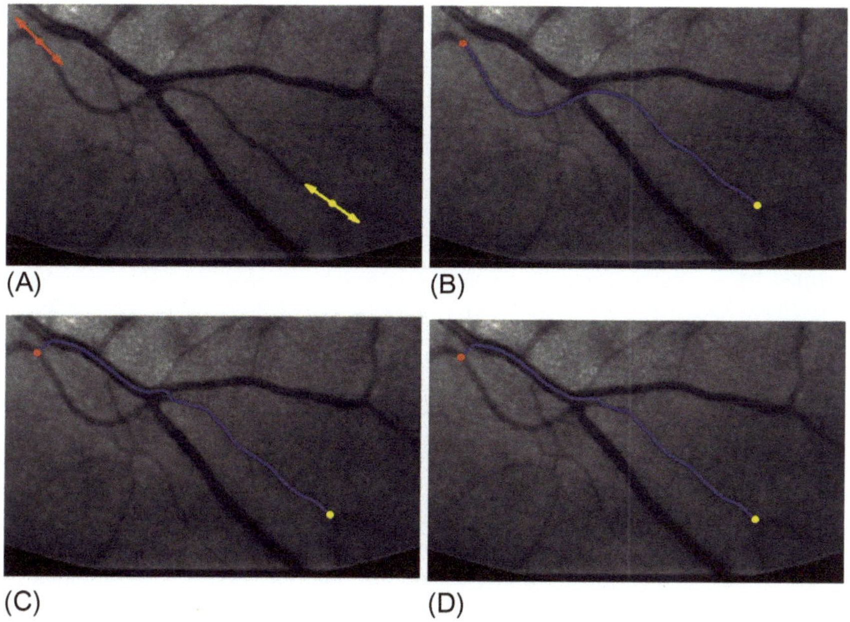

FIG. 13 Centreline tracking for a retinal vessel. (A) A retinal image patch. The *red* and *yellow dots* are the given source and end positions. The *arrows* denote the directions at the corresponding positions. (B) The extracted minimal path associated to the data-driven Finsler elastica metric using the orientation score ϕ_{tube}. (C) and (D) The extracted minimal paths associated to isotropic metric and anisotropic Riemannian metric, respectively.

path with the smallest geodesic distance value will be chosen as the desired path, see Fig. 13B. For the purpose of comparison, we show the minimal paths, respectively, derived from the isotropic Riemannian metric (Li and Yezzi, 2007) and the anisotropic Riemannian metric (Benmansour and Cohen, 2011), as shown in Fig. 13C and D. One can point out that these paths from the Riemannian metrics travel the segments belonging to different vessels, while the path from the Finsler elastica metric can track the correct vessel. Note that both of the Riemannian metrics used in this experiment are constructed by the OOF filter (Law and Chung, 2008).

6 Randers minimal paths for region-based active contours

In this section, we introduce a new minimal path model (Chen et al., 2016, 2019) for solving the active contour problem involving a region-based homogeneity penalization term, using the minimal path and Eikonal PDE framework.

6.1 Hybrid active contour model

A hybrid active contour model (Cohen, 1997; Kimmel and Bruckstein, 2003; Paragios and Deriche, 2002; Sagiv et al., 2006) invokes an energy functional that is made up of a region-based homogeneity penalization functional $\mathscr{H}_{\text{region}}$ and an edge-based weighted curve length $\mathscr{L}_{\text{IR}}$ (see Eq. (23)), i.e.

$$E_{\text{hybrid}}(\mathcal{C}) = \alpha \mathscr{H}_{\text{region}}(\mathbf{1}_{A_\mathcal{C}}) + \mathscr{L}_{\text{IR}}(\mathcal{C}), \tag{71}$$

where $\alpha \in \mathbb{R}^+$ is a parameter that controls the relative importance of the two terms $\mathscr{H}_{\text{region}}$ and $\mathscr{L}$, $\mathcal{C} \in \text{Lip}([0,1],\Omega)$ is a simple and closed curve and Ω represents the image domain. The function $\mathbf{1}_{A_\mathcal{C}} : \Omega \to \{0,1\}$ is the characteristic function of the open and bounded subset $A_\mathcal{C}$ enclosed by $\mathcal{C}$, which can be defined by

$$\mathbf{1}_{A_\mathcal{C}}(\mathbf{x}) = \begin{cases} 1, & \forall \mathbf{x} \in A_\mathcal{C}, \\ 0, & \forall \mathbf{x} \in \Omega \backslash A_\mathcal{C}. \end{cases}$$

The weighted curve length $\mathscr{L}_{\text{IR}}$ plays the role for regularization, which can be measured using an edge-based potential $P : \Omega \to \mathbb{R}^+$. In this section, we make use of the following method to construct the potential

$$P(\mathbf{x}) = \exp(\beta(\| g \|_\infty - g(\mathbf{x}))), \tag{72}$$

where $\beta \in \mathbb{R}^+$ is a contrast parameter and the function g represents the magnitude of the image gradient vector field, as defined in Eqs. (2) and (4). One can see that by Eq. (72) the potential P obeys that $\inf_{\mathbf{x}\in\Omega}\{P(\mathbf{x})\} = 1$ and appears to have low values around the image edges.

Let $\tilde{\mathcal{C}} \in \text{Lip}([0,1],\Omega)$ be a fixed simple and closed curve that is parameterized in a clockwise order. Supposing that the region $A_\mathcal{C}$ is close to $A_{\tilde{\mathcal{C}}}$ and by the differentiability of the region-based functional $\mathscr{H}_{\text{region}}$, we can obtain the following approximation (Chen et al., 2019)

$$\mathscr{H}_{\text{region}}(\mathbf{1}_{A_\mathcal{C}}) \approx c + \int_\Omega \rho_{\tilde{\mathcal{C}}}(\mathbf{x}) \mathbf{1}_{A_\mathcal{C}}(\mathbf{x}) d\mathbf{x}, \tag{73}$$

where $\rho_{\tilde{\mathcal{C}}}$ is the gradient of the functional $\mathscr{H}_{\text{region}}$ at $\mathbf{1}_{A_{\tilde{\mathcal{C}}}}$ and $c \in \mathbb{R}$ is a scalar value that is independent to the characteristic function $\mathbf{1}_{A_\mathcal{C}}$.

One can point out that only the second term in the right-hand side of Eq. (73) relies on the characteristic function $\mathbf{1}_{A_\mathcal{C}}$ of the region $A_\mathcal{C}$. For convenience, let us denote this term by

$$\begin{aligned} \mathcal{J}_{\tilde{\mathcal{C}}}(\mathbf{1}_{A_\mathcal{C}}) &= \int_\Omega \rho_{\tilde{\mathcal{C}}}(\mathbf{x}) \mathbf{1}_{A_\mathcal{C}}(\mathbf{x}) d\mathbf{x} \\ &= \int_{A_\mathcal{C}} \rho_{\tilde{\mathcal{C}}}(\mathbf{x})\, d\mathbf{x}. \end{aligned} \tag{74}$$

In our model, solving the hybrid active contour problem through the curve evolution scheme amounts to iteratively searching for a family of successive simple and closed curves $\mathcal{C}_k$ (indexed by k) such that the final curve $\mathcal{C}_\infty$ can be used to depict the target boundary.

In the kth iteration ($k \geq 1$), the input is a curve $\mathcal{C}_k$ obtained from the last iteration and the output is a new simple and closed curve $\mathcal{C}_{k+1}$. By setting

$\tilde{\mathcal{C}} := \mathcal{C}_k$, we can obtain the gradient $\rho_{\mathcal{C}_k}$ of the functional $\mathscr{H}_{\text{region}}$ at $\mathbf{1}_{\mathcal{C}_k}$. Then the curve $\mathcal{C}_{k+1}$ can be generated by solving

$$\min_{\mathcal{C}\in\Xi_{\mathcal{C}_k}} \{\alpha \mathcal{J}_{\mathcal{C}_k}(\mathbf{1}_{A_\mathcal{C}}) + \mathscr{L}_{\text{IR}}(\mathcal{C})\}, \tag{75}$$

where $\Xi_{\mathcal{C}_k}$ is a set of simple and closed curves. The construction of $\Xi_{\mathcal{C}_k}$ relies on $\mathcal{C}_k$ and will be introduced in the next section. In order to solve the problem (75) by the minimal path model and the Eikonal PDE, we transform the region-based functional $\mathcal{J}_{\mathcal{C}_k}(\mathbf{1}_{A_\mathcal{C}})$ into a weighted curve length associated to a Randers metric using the divergence theorem, providing that the curve $\mathcal{C}_{k+1}$ is chosen from the set $\Xi_{\mathcal{C}_k}$.

6.2 A Randers metric interpretation to the hybrid energy

In the kth iteration, let $U_{\mathcal{C}_k} = \{\mathbf{x}\in\Omega; d(\mathbf{x},\mathcal{C}_k) < r\}$ be a tubular neighbourhood of the simple and closed curve $\mathcal{C}_k$ with radius r, where $d(\mathbf{x},\mathcal{C}) = \min_u \| \mathbf{x} - \mathcal{C}(u) \|$ denotes the Euclidean distance between $\mathbf{x}$ and $\mathcal{C}$. Then the set $\Xi_{\mathcal{C}_k}$, which is made up of simple and closed curves, can be defined by

$$\Xi_{\mathcal{C}_k} = \{\mathcal{C}\in \text{Lip}([0,1], U_{\mathcal{C}_k}); A_{\mathcal{C}_k}\backslash U_{\mathcal{C}_k} \subset A_\mathcal{C}\}.$$

For any simple and closed curve $\mathcal{C}\in\Xi_{\mathcal{C}_k}$, one has $A_\mathcal{C}\backslash U_{\mathcal{C}_k} = A_{\mathcal{C}_k}\backslash U_{\mathcal{C}_k}$. This means that the region $A_\mathcal{C}$ can be decomposed as

$$\begin{aligned} A_\mathcal{C} &= (A_\mathcal{C}\backslash U_{\mathcal{C}_k}) \cup (A_\mathcal{C}\cap U_{\mathcal{C}_k}) \\ &= (A_{\mathcal{C}_k}\backslash U_{\mathcal{C}_k}) \cup (A_\mathcal{C}\cap U_{\mathcal{C}_k}). \end{aligned}$$

In this case, we can reformulate the region-based functional $\mathcal{J}_{\mathcal{C}_k}$ as follows:

$$\begin{aligned} \mathcal{J}_{\mathcal{C}_k}(\mathbf{1}_{A_\mathcal{C}}) &= \int_{A_\mathcal{C}\cap U_{\mathcal{C}_k}} \rho_{\mathcal{C}_k}(\mathbf{x}) \mathbf{1}_{U_{\mathcal{C}_k}}(\mathbf{x}) d\mathbf{x} + \int_{A_{\mathcal{C}_k}\backslash U_{\mathcal{C}_k}} \rho_{\mathcal{C}_k}(\mathbf{x}) d\mathbf{x} \\ &= \int_0^1 \langle \vartheta_{\mathcal{C}_k}(\mathcal{C}(u)), \mathcal{N}(u)\rangle \| \mathcal{C}'(u) \| du + \int_{A_{\mathcal{C}_k}\backslash U_{\mathcal{C}_k}} \rho_{\mathcal{C}_k}(\mathbf{x}) d\mathbf{x}, \end{aligned} \tag{76}$$

where $\mathbf{1}_{U_{\mathcal{C}_k}}$ is defined as the characteristic function of the tubular neighbourhood $U_{\mathcal{C}_k}$, $\mathcal{N}$ is the outward normal to the curve $\mathcal{C}$, and $\vartheta_{\mathcal{C}_k} : \mathbb{R}^2 \to \mathbb{R}^2$ is a vector field which satisfies the following divergence equation over the domain $\mathbb{R}^2$

$$\text{div}\ \vartheta_{\mathcal{C}_k} = \rho_{\mathcal{C}_k} \mathbf{1}_{U_{\mathcal{C}_k}}. \tag{77}$$

One can always find such a vector field $\vartheta_{\mathcal{C}_k}$ due to the existence of the solution to the divergence equation (77).

In Eq. (76), the second term is independent to the curve $\mathcal{C}$ and thus we only take into account the first term in the following calculation. Let $\mathbf{M}$ be the rotation matrix of angle $\pi/2$. Integrating the weighted curve length $\mathscr{L}_{\text{IR}}$ and the first term in Eq. (76), we obtain that

$$\begin{aligned}
&\int_0^1 \alpha\langle \vartheta_{\mathcal{C}_k}(\mathcal{C}(u)),\mathcal{N}(u)\rangle \| \mathcal{C}'(u) \| du + \int_0^1 P(\mathcal{C}(u)) \| \mathcal{C}'(u) \| du \\
&= \int_0^1 \alpha\langle \mathbf{M}\vartheta_{\mathcal{C}_k}(\mathcal{C}(u)),\mathcal{C}'(u)\rangle du + \int_0^1 P(\mathcal{C}(u)) \| \mathcal{C}'(u) \| du \\
&= \int_0^1 \mathfrak{F}_{\mathcal{C}_k}(\mathcal{C}(u),\mathcal{C}'(u))du,
\end{aligned} \tag{78}$$

where the metric $\mathfrak{F}_{\mathcal{C}_k}$ has a form of

$$\mathfrak{F}_{\mathcal{C}_k}(\mathbf{x}, \vec{\mathbf{u}}) = P(\mathbf{x}) \, \|\vec{\mathbf{u}}\| + \langle \alpha \, \mathbf{M}\vartheta_{\mathcal{C}_k}(\mathbf{x}), \, \vec{\mathbf{u}}\rangle. \tag{79}$$

The positive definiteness condition (33) for $\mathfrak{F}_{\mathcal{C}_k}$ requires that

$$\alpha \, \| \vartheta_{\mathcal{C}_k}(\mathbf{x}) \| < P(\mathbf{x}), \quad \forall \mathbf{x} \in U_{\mathcal{C}_k}. \tag{80}$$

However, the requirement (80) is difficult to satisfy. Hence we consider a new inequality formulated by

$$\alpha \, \| \vartheta_{\mathcal{C}_k}(\mathbf{x}) \| < \inf_{\mathbf{y}\in\Omega} \| P(\mathbf{y}) \| = 1. \tag{81}$$

Now we have induced a Randers metric formulated in Eq. (79), which embeds the region-based homogeneity information as well as the image gradient features. This gives us the possibility of applying the minimal path framework to solve the region-based image segmentation problem. In next section, we will introduce a new vector field to approximate $\vartheta_{\mathcal{C}_k}$, such that the inequality (81) will always hold.

6.3 Practical implementations

In this section, we first introduce a method for computing the vector field $\vartheta_{\mathcal{C}_k}$ used in the kth iteration. In order to satisfy (81), the vector field $\vartheta_{\mathcal{C}_k}$ is expected to be as small as possible. Since the next evolving curve lies within the tubular neighbourhood $U_{\mathcal{C}_k}$, we consider to find a vector field $\vartheta_{\mathcal{C}_k} : U_{\mathcal{C}_k} \to \mathbb{R}^2$ by solving the following minimization problem on the domain $U_{\mathcal{C}_k}$

$$\text{minimize} \int_{U_{\mathcal{C}_k}} \| \vartheta_{\mathcal{C}_k}(\mathbf{x}) \|^2 d\mathbf{x}, \quad s.t. \quad \text{div} \, \vartheta_{\mathcal{C}_k} = \rho_{\mathcal{C}_k} \quad \text{on} \quad U_{\mathcal{C}_k}. \tag{82}$$

As discussed in Chen et al. (2016, 2019), the value of $\sup_{\mathbf{x}} \| \vartheta_{\mathcal{C}_k}(\mathbf{x}) \|$ is bounded by the area of the tubular neighbourhood $U_{\mathcal{C}_k}$. Since the tubular neighbourhood $U_{\mathcal{C}_k}$ acts as the search space for the evolving curves, reducing the area of $U_{\mathcal{C}_k}$ may lead the proposed model being stuck in unexpected local minima. In order to address this issue, we make use of a new vector field $\varpi_{\mathcal{C}_k} : U_{\mathcal{C}_k} \to \mathbb{R}^2$ derived by the nonlinear mapping φ defined in Eq. (42) such that

$$\varpi_{\mathcal{C}_k}(\mathbf{x}) = \begin{cases} \frac{\varphi(\tilde{\alpha} \, \| \vartheta_{\mathcal{C}_k}(\mathbf{x}) \|)}{\| \vartheta_{\mathcal{C}_k}(\mathbf{x}) \|} \vartheta_{\mathcal{C}_k}(\mathbf{x}), & \text{if } \| \vartheta_{\mathcal{C}_k}(\mathbf{x}) \| \neq 0, \\ \mathbf{0}, & \text{otherwise}, \end{cases} \tag{83}$$

where $\tilde{\alpha} \in \mathbb{R}^+$ is a constant.

Replacing the vector field $\alpha\vartheta_{\mathcal{C}_k}$ (see Eq. (79)) by $\varpi_{\mathcal{C}_k}$, one can obtain a new Randers metric:

$$\mathcal{F}_{\mathcal{C}_k}(\mathbf{x}, \vec{\mathbf{u}}) = P(\mathbf{x})\,\|\vec{\mathbf{u}}\| + \langle \mathbf{M}\varpi_{\mathcal{C}_k}(\mathbf{x}), \vec{\mathbf{u}}\rangle. \tag{84}$$

In each iteration k, the weighted curve length measured along a curve $\mathcal{C} \in \Xi_{\mathcal{C}_k}$ can be formulated by

$$\text{Length}_{\mathcal{F}_{\mathcal{C}_k}}(\mathcal{C}) = \int_0^1 \mathcal{F}_{\mathcal{C}_k}(\mathcal{C}(u), \mathcal{C}'(u))du. \tag{85}$$

For any point $\mathbf{x} \in U_{\mathcal{C}_k}$ the vector $\vartheta_{\mathcal{C}_k}(\mathbf{x})$ is positively proportional to $\varpi_{\mathcal{C}_k}(\mathbf{x})$ in the sense of the magnitude and of the direction. This gives the relevance between the original region-based active contour problem and the minimal path computation associated to the geodesic metric $\mathcal{F}_{\mathcal{C}_k}$. More analysis for the reasonability of the use of the Randers metric $\mathcal{F}_{\mathcal{C}_k}$ can be found in Chen et al. (2019).

6.4 Application to image segmentation

We make use of the piecewise constant homogeneity term (Chan and Vese, 2001a) to build the Randers metric $\mathcal{F}_{\mathcal{C}_k}$ in Eq. (84) and illustrate how to apply $\mathcal{F}_{\mathcal{C}_k}$ for image segmentation. The objective is to search for a family of successive $\mathcal{C}_k$, each of which is simple and closed, to depict the object boundary, as $k \to \infty$.

The region-based homogeneity term in the piecewise constant variant of the Mumford–Shah model can be expressed as follows

$$\mathscr{H}_{\text{region}}(\mathbf{1}_{A_\mathcal{C}}, \mu_{\text{in}}, \mu_{\text{out}}) = \int_\Omega (I - \mu_{\text{in}})^2 \mathbf{1}_{A_\mathcal{C}} d\mathbf{x} + \int_\Omega (I - \mu_{\text{out}})^2 (1 - \mathbf{1}_{A_\mathcal{C}}) d\mathbf{x}.$$

Recall that in the kth iteration of the curve evolution scheme, the input is the curve $\mathcal{C}_k$ ($k \geq 1$) and the output is $\mathcal{C}_{k+1}$. Note that when $k = 1$, $\mathcal{C}_1$ is the initial curve. We can estimate the gradient $\rho_{\mathcal{C}_k}$ of the region-based functional $\mathscr{H}_{\text{region}}$, which can be formulated by

$$\rho_{\mathcal{C}_k} = (I(\mathbf{x}) - \mu_{\text{in}}[\mathcal{C}_k])^2 - (I - \mu_{\text{out}}[\mathcal{C}_k])^2, \tag{86}$$

where $\mu_{\text{in}}[\mathcal{C}_k]$ and $\mu_{\text{out}}[\mathcal{C}_k]$ are, respectively, the mean grey levels inside and outside the curve $\mathcal{C}_k$.

Based on the gradient $\rho_{\mathcal{C}_k}$ and the tubular neighbourhood $U_{\mathcal{C}_k}$, one can solve the PDE-constrained minimization problem (82) to obtain the vector field $\vartheta_{\mathcal{C}_k}$ and also to obtain $\varpi_{\mathcal{C}_k}$. Finally, the Randers metric $\mathcal{F}_{\mathcal{C}_k}$ can be constructed using Eq. (84).

Interactive image segmentation method. By solving the gradient descent ODE, one can obtain an open geodesic path between two points within a given domain. However, for image segmentation, the objective is to seek simple and closed curves. Thus we need to solve an issue: how to derive a closed and simple curve $\mathcal{C}_{k+1}$ in each iteration k, providing that the Randers metric $\mathcal{F}_{\mathcal{C}_k}$ is given. In this section, we consider an interactive segmentation scheme to obtain the evolving curve $\mathcal{C}_{k+1}$, which depends on a set $\{\mathbf{q}_i\}_{1\leq i\leq m}$ of m ($m \geq 3$) prescribed vertices distributed on the target boundary in a clockwise order. These vertices will be fixed in the course of curve evolution. Each evolving curve $\mathbf{q}_i$ is supposed to pass through each of these vertices in a clockwise order. The initial curve $\mathcal{C}_1$ is a polygon which is generated using the vertices $\mathbf{q}_i$. The simple and closed curve $\mathcal{C}_{k+1}$ can be generated by the end-to-end concatenation of a set of geodesic paths obtained using $\mathcal{F}_{\mathcal{C}_k}$. Each of these geodesic paths, denoted by $\mathcal{G}_{i,k+1}$ with $1 \leq i \leq m$, connects a vertex $\mathbf{q}_i$ to its successive one.

Now we present the method for computing the geodesic paths $\mathcal{G}_{i,k+1}$ to obtain the curve $\mathcal{C}_{k+1}$. Recall that the input curve $\mathcal{C}_k$ passes through all of the fixed vertices $\{\mathbf{q}_i\}_{1\leq i\leq m}$. In this case, we are able to decompose $\mathcal{C}_k$ into a set $\{\mathcal{G}_{i,k}\}_i$ of m subpaths. Each path $\mathcal{G}_{i,k}$, defined over the integral [0, 1], links a vertex $\mathbf{q}_i$ to its successive one $\mathbf{q}_{i+1}$ for $i < m$ or to $\mathbf{q}_1$ for $i = m$. Following that the tubular neighbourhood $U_{\mathcal{C}_k}$ can be decomposed into a family of disjoint subdomains $\mathcal{U}_i \subset U_{\mathcal{C}_k}$, each of which can be regarded as a narrowband region (with radius r) involving the open curve $\mathcal{G}_{i,k}$. In essence, these regions $\mathcal{U}_i$ ($i = 1, 2, \ldots, m$) can be identified by using the Voronoi index map over the domain $U_{\mathcal{C}_k}$ associated to the open curves $\mathcal{G}_{i,k}$ ($1 \leq i \leq m$) as introduced in Chen et al. (2019). In other words, a point $\mathbf{x} \in \mathcal{U}_i$ implies that $d(\mathbf{x}, \mathcal{G}_{i,k}) < d(\mathbf{x}, \mathcal{G}_{j,k}), \forall j \neq i$.

With these definitions in hands, we can track a geodesic path $\mathcal{G}_{i,k+1}$ in the corresponding region $\mathcal{U}_i \cup \{\mathbf{q}_i, \mathbf{q}_{i+1}\}$ if $i < m$, or $\mathcal{U}_m \cup \{\mathbf{q}_m, \mathbf{q}_1\}$ if $i = m$. The use of these regions $\mathcal{U}_i \cup \{\mathbf{q}_i, \mathbf{q}_{i+1}\}$ for geodesic path computation is to avoid the curve self-crossing issue and also to reduce the computation time. The decomposition of the tubular neighbourhood $U_{\mathcal{C}_k}$ can be efficiently implemented by the fast marching method with a Voronoi index map estimation procedure. Once all of the geodesic paths $\mathcal{G}_{i,k+1}$ ($1 \leq i \leq m$) are generated, we can construct a simple and closed curve $\mathcal{C}_{k+1}$ by the concatenation of the paths $\mathcal{G}_{i,k+1}$ in an end-to-end manner. Finally, the curve $\mathcal{C}_{k+1}$ is regarded as the output of the kth iteration.

Note that for a pair of successive vertices ($\mathbf{q}_i$, $\mathbf{q}_{i+1}$), the extracted geodesic path $\mathcal{G}_{i,k+1}$ is the globally optimal curve for the length $\text{Length}_{\mathcal{F}_{\mathcal{C}_k}}$ (see Eq. (85)) within the domain $\mathcal{U}_i \cup \{\mathbf{q}_i, \mathbf{q}_{i+1}\}$.

In Fig. 14, we show an example for this interactive image segmentation procedure providing that a set of three ordered vertices are given. These vertices are illustrated by red dots, see Fig. 14A. In Fig. 14B, the blue straight segment lines connecting each pair of red dots represent the initial curve. In Fig. 14C–E, we show the intermediate curve evolution results, where the blue contour is generated by the concatenation of a set of the Randers minimal paths. The final curve is shown in Fig. 14F. Again, the vertices denoted by red dots are fixed during the curve evolution.

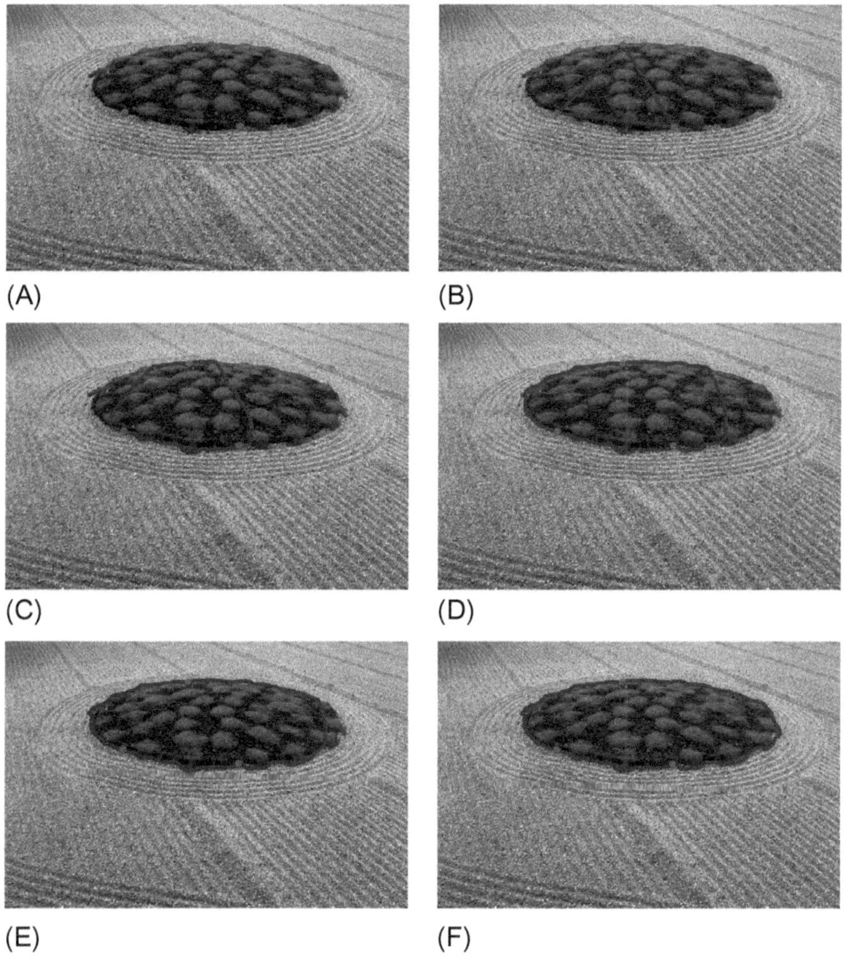

FIG. 14 Randers minimal paths for region-based image segmentation. (A) A set of prescribed vertices indicated by *red dots*. (B) The initial contour indicated by the *blue line*. (C)–(E) Intermediate steps of the curve evolution processing. (F) The final curve generated by the concatenation of a set of Randers geodesic paths. One can see that all the closed curves indicated by *blue lines* pass through the *red dots*.

Remark 1. The method presented in this section assumes that the vertices are provided by user and are fixed during the curve evolution. As a matter of fact, in each iteration k, these vertices can be sampled from the input curve $\mathcal{C}_k$. In this case, one can initialize the proposed model by providing a closed curve and the vertices will evolve in the course of the curve evolution processing (Chen et al., 2016).

Remark 2. Note that a variant of the edge-based balloon active contour model can also be addressed by the framework proposed in this section. The balloon force $\mathbf{F}_{\text{balloon}}$ in Eq. (8) can be obtained by minimizing the following term

$$\int_{A_{\mathcal{C}}} -1d\mathbf{x}, \tag{87}$$

yielding a curve evolution flow

$$\frac{\partial \mathcal{C}_\tau}{\partial \tau} = \mathcal{N}_\tau. \tag{88}$$

Recall that $\mathcal{N}_\tau$ is the outward normal to $\mathcal{C}_\tau$.

Now we can establish a variant form of the isotropic geodesic active contour functional (Caselles et al., 1997; Yezzi et al., 1997) with the balloon force term

$$E_{\text{balloon}}(\mathcal{C}) = \alpha \int_{A_{\mathcal{C}}} -1d\mathbf{x} + \mathcal{L}_{\text{IR}}(\mathcal{C}). \tag{89}$$

In this case, for a given simple and closed curve $\tilde{\mathcal{C}}$, the gradient of E_{balloon} thus turns out to be a constant function, i.e., $\rho_{\tilde{\mathcal{C}}} \equiv 1$. This edge-based active contour model thus can be solved by the Randers metric-based model proposed in this section.

Furthermore, an alternative choice for the regularization term in Eqs. (71) and (89) can be constructed using an anisotropic Riemannian metric $\mathfrak{R}$ (see Eq. (35)), which allows to take into account the path directions for image segmentation (Chen and Cohen, 2017; Chen et al., 2019).

7 Conclusion

In this chapter, we review the methods for the construction of the Riemannian and Randers metrics in terms of various active contour terms. We show that by these metrics the edge- and region-based active contour problems and the Euler–Mumford elastica problem can be efficiently solved by the minimal path framework based on the Eikonal PDE. Moreover, we also exploit the relevant applications of minimal paths associated to the proposed metrics in image analysis such as image segmentation, boundary detection as well as tubular structure centreline tracking, which are able to blend the benefits from the minimal path framework and the active contour models. The well-established fast marching methods, which are the Eikonal solvers, allow efficient and practical implementations for these applications.

Acknowledgements

The authors thank Dr. Jean-Marie Mirebeau from Université Paris-Saclay for his fruitful discussion and regular collaboration. This research has been partially funded by Roche pharma (project AMD_short) and by a grant from the French Agence Nationale de la Recherche ANR-16-RHUS-0004 (RHU TRT_cSVD). Fig. 9 is reprinted by permission from Springer Nature Customer Service Centre GmbH: Springer Nature, International Journal of Computer Vision, Global minimum for a Finsler elastica approach, *Da Chen, Jean-Marie Mirebeau and Laurent D. Cohen* (Chen et al., 2017).

References

Appia, V., Yezzi, A., 2011. Active geodesics: region-based active contour segmentation with a global edge-based constraint. In: Proc. ICCV, 1975–1980.

Arbelaez, P., Maire, M., Fowlkes, C., Malik, J., 2011. Contour detection and hierarchical image segmentation. IEEE Trans. Pattern Anal. Mach. Intell. 33 (5), 898–916.

Baker, S., Scharstein, D., Lewis, J., Roth, S., Black, M.J., Szeliski, R., 2011. A database and evaluation methodology for optical flow. Int. J. Comput. Vis. 92 (1), 1–31.

Benmansour, F., Cohen, L.D., 2009. Fast object segmentation by growing minimal paths from a single point on 2D or 3D images. J. Math. Imaging Vis. 33 (2), 209–221.

Benmansour, F., Cohen, L.D., 2011. Tubular structure segmentation based on minimal path method and anisotropic enhancement. Int. J. Comput. Vis. 92 (2), 192–210.

Bresson, X., Esedog-lu, S., Vandergheynst, P., Thiran, J., Osher, S., 2007. Fast global minimization of the active contour/snake model. J. Math. Imaging Vis. 28 (2), 151–167.

Brox, T., Cremers, D., 2009. On local region models and a statistical interpretation of the piecewise smooth Mumford-Shah functional. Int. J. Comput. Vis. 84 (2), 184–193.

Caselles, V., Kimmel, R., Sapiro, G., 1997. Geodesic active contours. Int. J. Comput. Vis. 22 (1), 61–79.

Chan, T.F., Vese, L.A., 2001. Active contours without edges. IEEE Trans. Image Process. 10 (2), 266–277.

Chan, T.F., Vese, L.A., 2001. A level set algorithm for minimizing the Mumford-Shah functional in image processing. In: Proc. IEEE Workshop on Variational and Level Set Methods in Computer Vision, 161–168.

Chan, T.F., Sandberg, B.Y., Vese, L.A., 2000. Active contours without edges for vector-valued images. J. Vis. Commun. Image Represent. 11 (2), 130–141.

Chan, T.F., Esedoglu, S., Nikolova, M., 2006. Algorithms for finding global minimizers of image segmentation and denoising models. SIAM J. Appl. Math. 66 (5), 1632–1648.

Chen, D., Cohen, L.D., 2017. Anisotropic edge-based Balloon Eikonal active contours. In: Proc. GSI, pp. 782–790.

Chen, D., Cohen, L.D., 2018. Fast asymmetric fronts propagation for image segmentation. J. Math. Imaging Vis. 60 (6), 766–783.

Chen, D., Mirebeau, J.-M., Cohen, L.D., 2016. Finsler geodesics evolution model for region based active contours. In: Proc. BMVC.

Chen, D., Mirebeau, J.-M., Cohen, L.D., 2017. Global minimum for a Finsler elastica minimal path approach. Int. J. Comput. Vis. 122 (3), 458–483.

Chen, D., Mirebeau, J.-M., Cohen, L.D., 2019. Eikonal region-based active contours for image segmentation. Int. J. Comput. Vis. (submitted for publication).

Cohen, L.D., 1991. On active contour models and balloons. CVGIP: Image Understand 53 (2), 211–218.

Cohen, L.D., 1997. Avoiding local minima for deformable curves in image analysis. In: Curves and Surfaces with Applications in CAGD, pp. 77–84.

Cohen, L., 2006. Minimal paths and fast marching methods for image analysis. In: Handbook of Mathematical Models in Computer Vision, Springer, pp. 97–111.

Cohen, L.D., Kimmel, R., 1997. Global minimum for active contour models: a minimal path approach. Int. J. Comput. Vis. 24 (1), 57–78.

Deschamps, T., Cohen, L.D., 2001. Fast extraction of minimal paths in 3D images and applications to virtual endoscopy. Med. Image Anal. 5 (4), 281–299.

Di Zenzo, S., 1986. A note on the gradient of a multi-image. Comput. Vis. Graph. Image Process. 33 (1), 116–125.

Frangi, A.F., Niessen, W.J., Vincken, K.L., Viergever, M.A., 1998. Multiscale vessel enhancement filtering. In: Proc. MICCAI, pp. 130–137.

Freeman, W.T., Adelson, E.H., 1991. The design and use of steerable filters. IEEE Trans. Pattern Anal. Mach. Intell. 13 (9), 891–906.

Goldenberg, R., Kimmel, R., Rivlin, E., Rudzsky, M., 2001. Fast geodesic active contours. IEEE Trans. Image Process. 10 (10), 1467–1475.

Jacob, M., Unser, M., 2004. Design of steerable filters for feature detection using canny-like criteria. IEEE Trans. Pattern Anal. Mach. Intell. 26 (8), 1007–1019.

Kass, M., Witkin, A., Terzopoulos, D., 1988. Snakes: active contour models. Int. J. Comput. Vis. 1 (4), 321–331.

Kimmel, R., 2003. Fast edge integration. In: Geometric Level Set Methods in Imaging, Vision, and Graphics, Springer, pp. 59–77.

Kimmel, R., Bruckstein, A.M., 2003. Regularized Laplacian zero crossings as optimal edge integrators. Int. J. Comput. Vis. 53 (3), 225–243.

Law, M.W., Chung, A.C., 2008. Three dimensional curvilinear structure detection using optimally oriented flux. In: Proc. ECCV, pp. 368–382.

Li, H., Yezzi, A., 2007. Vessels as 4-D curves: global minimal 4-D paths to extract 3-D tubular surfaces and centerlines. IEEE Trans. Med. Imaging 26 (9), 1213–1223.

Malladi, R., Sethian, J., Vemuri, B.C., 1995. Shape modeling with front propagation: a level set approach. IEEE Trans. Pattern Anal. Mach. Intell. 17 (2), 158–175.

Melonakos, J., Pichon, E., Angenent, S., Tannenbaum, A., 2008. Finsler active contours. IEEE Trans. Pattern Anal. Mach. Intell. 30 (3), 412–423.

Mille, J., Bougleux, S., Cohen, L.D., 2015. Combination of piecewise-geodesic paths for interactive segmentation. Int. J. Comput. Vis. 112 (1), 1–22.

Mirebeau, J.-M., 2014. Anisotropic fast-marching on Cartesian grids using lattice basis reduction. SIAM J. Numer. Anal. 52 (4), 1573–1599.

Mirebeau, J.-M., 2014. Efficient fast marching with Finsler metrics. Numer. Math. 126 (3), 515–557.

Mirebeau, J.-M., 2017. Anisotropic fast-marching on Cartesian grids using Voronoi's first reduction of quadratic forms. Preprint https://hal-uvsq.archives-ouvertes.fr/UNIV-PARIS-SACLAY/hal-01507334v2.

Mirebeau, J.-M., 2018. Fast-marching methods for curvature penalized shortest paths. J. Math. Imaging Vis. 60 (6), 784–815.

Mumford, D., 1994. Elastica and computer vision. In: Algebraic Geometry and its Applications, Springer, pp. 491–506.

Mumford, D., Shah, J., 1989. Optimal approximations by piecewise smooth functions and associated variational problems. Commun. Pure Appl. Math. 42 (5), 577–685.
Nitzberg, M., Mumford, D., 1990. The 2.1-D sketch. In: Proc. ICCV, pp. 138–144.
Osher, S., Sethian, J.A., 1988. Fronts propagating with curvature-dependent speed: algorithms based on Hamilton-Jacobi formulations. J. Comput. Phys. 79 (1), 12–49.
Paragios, N., Deriche, R., 2002. Geodesic active regions: a new framework to deal with frame partition problems in computer vision. J. Vis. Commun. Image Represent. 13 (1), 249–268.
Peyré, G., Péchaud, M., Keriven, R., Cohen, L.D., 2010. Geodesic methods in computer vision and graphics. Found. Trends. Comput. Graph. Vis. 5 (3–4), 197–397.
Pock, T., Cremers, D., Bischof, H., Chambolle, A., 2009. An algorithm for minimizing the Mumford-Shah functional. In: Proc. ICCV, pp. 1133–1140.
Randers, G., 1941. On an asymmetrical metric in the four-space of general relativity. Phys. Rev. 59 (2), 195.
Rother, C., Kolmogorov, V., Blake, A., 2004. Grabcut: interactive foreground extraction using iterated graph cuts. ACM Trans. Graph. 23 (3), 309–314.
Sagiv, C., Sochen, N.A., Zeevi, Y.Y., 2006. Integrated active contours for texture segmentation. IEEE Trans. Image Process. 15 (6), 1633–1646.
Sapiro, G., 1997. Color snakes. Comput. Vis. Image Underst. 68 (2), 247–253.
Sethian, J.A., 1999. Fast marching methods. SIAM Rev. 41 (2), 199–235.
Tsai, A., Yezzi, A., Willsky, A.S., 2001. Curve evolution implementation of the Mumford-Shah functional for image segmentation, denoising, interpolation, and magnification. IEEE Trans. Image Process. 10 (8), 1169–1186.
Vese, L.A., Chan, T.F., 2002. A multiphase level set framework for image segmentation using the Mumford and Shah model. Int. J. Comput. Vis. 50 (3), 271–293.
Yezzi, A., Kichenassamy, S., Kumar, A., Olver, P., Tannenbaum, A., 1997. A geometric snake model for segmentation of medical imagery. IEEE Trans. Med. Imaging 16 (2), 199–209.
Zhao, H., Chan, T.F., Merriman, B., Osher, S., 1996. A variational level set approach to multiphase motion. J. Comput. Phys. 127 (1), 179–195.
Zhu, S., Yuille, A., 1996. Region competition: unifying snakes, region growing, and Bayes/MDL for multiband image segmentation. IEEE Trans. Pattern Anal. Mach. Intell. 18 (9), 884–900.

Chapter 7

Computable invariants for curves and surfaces☆

Oshri Halimi[a], Dan Raviv[b,*], Yonathan Aflalo[c] and Ron Kimmel[d]
[a]*Electrical Engineering Department, Technion - Israel Institute of Technology, Haifa, Israel*
[b]*School of Electrical Engineering, Faculty of Engineering, Tel Aviv University, Tel Aviv-Yafo, Israel*
[c]*Alibaba Research, Haifa, Israel*
[d]*Computer Science Department, Technion - Israel Institute of Technology, Haifa, Israel*
[*]*Corresponding author: e-mail: darav@tauex.tau.ac.il*

Chapter Outline

Abstract
During the last decade the trend in image analysis has been to shift from axiomatically derived measures into ones that are extracted empirically from data samples. The problem is that accurate geometric data are often unavailable and thus, for proper data augmentation, the research community has resorted yet again to axiomatic construction of invariant measures for curves and surfaces.

☆Scale and affine invariants of curves and surfaces for shape analysis and recognition.

Handbook of Numerical Analysis, Vol. 20. https://doi.org/10.1016/bs.hna.2019.07.004
© 2019 Elsevier B.V. All rights reserved.

For some geometric problems, such as shape classification and matching, it is appealing to adopt learning approaches due to their potential accuracy and computational efficiency. Nevertheless, even within the deep learning arena, geometric invariants offer a natural criterion for the learning process. Using geometric invariants one can overcome the need for annotated data and replace it by a purely geometric measure, leading to an unsupervised or a semisupervised learning schemes. Here, we review such constructions that are useful for the geometric analysis of visual information. The measures we explore include the construction of a scale or similarity invariant arc-length for curves and surfaces, an affine invariant one, resulting spectral geometries, and potential signatures that reflect the result of the discrepancies between the corresponding metric spaces. As an example we study novel signatures for surfaces known as the self functional map, which allow us to translate the problem of shape matching into that of computing distances between matrices.

Keywords: Scale invariant, Affine invariant, Differential geometry, Laplace–Beltrami operator, Shape analysis, Gaussian curvature

AMS Classification Codes: 53B21, 58D17

1 Introduction

Invariants play an important and critical role in the design of computational learning systems. At the core of a successful method one can find the definition of an appropriate and relevant measure of similarity. Shapes or images may look different with respect to a one measure, yet, provided an alternative metric, significant relationships between the observed objects can be revealed. The study of invariants in computer vision has a long history. Differential geometry of planar curves that are subject to projective transformations (Bruckstein et al., 1992, 1993), scale and affine deformations (Pauwels et al., 1995), and local scale (Cohignac et al., 1994) have been explored and developed through the short history of computer vision. Other, approaches applied concepts like scale-space (Bruckstein et al., 1997) and semidifferential signatures (Moons et al., 1995) laid down the foundations to more recent algorithms. Images, at the other end, required a less puristic approach. The input was not only geometric in flavour, and in the search for invariants researchers had to deal with parameters that incorporated changes in colour, scale, noise, and blur, to name just a few. Lowe's milestone paper on scale invariant feature transform (SIFT) (Lowe, 2004), introduced a shift in paradigm in the way automated systems extracted meaningful information for images. Successful follow-ups were presented like the SURF (Bay et al., 2006), affine-SIFT (ASIFT) (Morel and Yu, 2009), MSER (Donoser and Bischof, 2006), and its modification in Kimmel et al. (2011). Axioms and fundamental equations were nicely defined and summarized in Alvarez et al. (1993), and invariant detection of edges, diffusion, and active contours were presented in Olver et al. (1996). Once depth came into play, images started being interpreted also through their geometric structures, and a new field came to life based on the explicit marriage between geometry and photometry. An example for such models are faces for which the geometric structure was shown to be as important as the photometric texture or surface albedo, for example, for recognition and antispoofing tasks. The face geometric structure is somewhat flexible and stretchable, and a variety of models were suggested over the years to deal with recognition, synthesis, and inference problems for such data (Bronstein et al., 2005b). Another interesting thread in that research direction was the synthesis of a geometric structure of a face from an image (Masi et al., 2016; Richardson et al., 2016, 2017; Sela et al., 2017). Faces are just one example for nonrigid structures that we like to explore, and indeed many methods were suggested to try to compare and find dissimilarities between nonrigid surfaces.

One approach for measuring shape invariance is to embed the given surfaces into a new space. Using specific embeddings, the hard to compute intrinsic isometry is replaced by much simpler parametrizable transformations—like global rotation, translation, and reflection (Euclidean isometry). Shapes invariance can be treated with conformal geometry tools, locally preserving the angles on the original surface. For example, in Wang et al. (2007) quasi-conformal maps for matching surfaces were explored. Shape descriptors based on conformal geometry were proposed in Ben-Chen and Gotsman (2008), using the conformal factor to quantify dissimilarity between shapes. Mapping to conformal disks or spheres was suggested in Lipman and Funkhouser (2009). Comparing shapes using conformal Wasserstein distance was suggested in Lipman and Daubechies (2011), measuring the optimal mass transportation distance between conformal factor profiles obtained by mapping surfaces to a unit disc. For an introduction to computational conformal geometry we refer to Gu and Yau (2008), see also Aflalo et al. (2013b). A different type of embedding tries to preserve the metric properties of the original curved surface. Embedding of the surface into a low-dimensional Euclidean space that translates the geodesic distances between surface points into distances in the Euclidean space was suggested in Elad and Kimmel (2003), creating isometry invariant signatures in the embedding space, referred to as *canonical forms*. Transforming from shapes to their canonical forms, the intrinsic metric is replaced by an extrinsic Euclidean metric and the intrinsic isometry between the nonrigid shapes is replaced by a rigid transformation between the canonical forms. Canonical forms were used for face recognition in Bronstein et al. (2005b). Diffusion maps (Coifman and Lafon, 2006) can be regarded as yet another type of embedding into a Euclidean space, preserving the diffusion distances defined on the surface. Trying to avoid unbounded distortion, Bérard et al. (1994) suggested to embed the Riemannian manifold in the heat kernel space. This approach was later used to evaluate intrinsic distances and features (Ovsjanikov et al., 2009, 2010; Sun et al., 2009). One useful tool emerged from this research provided intrinsic distances, known as diffusion distances, that are based on the heat flow defined on the given geometry (Coifman et al., 2005). Those average distances successfully replaced geodesic ones in some applications gaining some stability at the obvious expense of accuracy.

In Shamai and Kimmel (2017), the geodesic distance descriptor (GDD) embedding was proposed, mapping the original surface to the complex Euclidean space thereby avoiding embedding errors that were apparent in the canonical forms. Spherical spaces embedding were suggested in Elad et al. (2005) and Bronstein et al. (2005a). Finally, graph embedding was proposed in Hamza and Krim (2006), Wang et al. (2008), and Qiu and Hancock (2007).

A related school of thought promoted treating each surface as a metric space and using the Gromov–Hausdorff distance between such geometric structures, as a measure of shape similarity. The Gromov–Hausdorff framework for the comparison of nonrigid shapes was first suggested in Mémoli and Sapiro (2005). The same definition of the Gromov–Hausdorff distance (Burago et al., 2001) as the distortion of embedding one metric space into another, lead to the introduction of the generalized MDS framework (GMDS) (Bronstein et al., 2006), utilizing metric invariance for nonrigid shape matching. In a follow up paper (Bronstein et al., 2010b), the geodesic distances used in the Gromov–Hausdorff distance were replaced with diffusion distances. While in Mémoli (2011) the Gromov-Wasserstein distance was defined as a reformulation of the distance measure between the shapes. In Aflalo and Kimmel (2013) and Aflalo et al. (2016b), respectively, the MDS and GMDS models were translated to the spectral domain. Recently, a GMDS-based deep neural network architecture (Halimi et al., 2019) introduced the first completely unsupervised correspondence learning method and provided an evidence that a purely geometric criterion can replace annotated data and result in a performance on par with supervised schemes. Other recent attempt towards that direction has been made by Roufosse and Ovsjanikov (2018) that utilized global structural properties of the maps, such as bijectivity or approximate isometry, for unsupervised correspondence learning.

At the core of this review stands the extension of dealing with bendable geometric structures, to richer families of transformations. In nature, excluding some degenerate cases, every bending must be accompanied with some stretching. Can we redefine the basic length element such that our measures would be robust to such stretching? Traditionally, we consider a Riemannian manifold to be charted by local mapping to a Euclidean space. Such an assumption provides an isometric invariant property, better known as bending invariant. In many practical cases, considering only local bending is enough, as stretching is insignificant. Yet, the matching results could deteriorate when these assumptions are violated. One approach for handling those cases can be done by defining local length measurements (metrics) which are invariant to different group of deformations such as scale, equi-affine and affine. The scale invariant metric (Aflalo et al., 2013a) copes with isotropic stretching using the curvature as a way to normalize the regular length. The equi-affine metric, first introduced in Blaschke (1923) and further explored in Su (1983), was the main motivation of Raviv et al. (2014) who used those constructions to compute new geodesics and novel spectral distances. Probably the most recent result along this line of thought, a composition of two metric construction methodologies (Raviv and Kimmel, 2014) gave birth to the fully affine invariant metric. This novel metric is not restricted by the equi-affine volume preserving limitation.

This chapter is organized as follows: The scale invariant metric is introduced in Section 2, where we explore the one- and two-dimensional cases. We then discuss the case of implicit level set representation and the conformal property in two dimensions. Section 3 presents the equi-affine metric in one and two dimensions and its corresponding numerical implementations. Next, Section 4 shows how scale and equi-affine metrics can be integrated into the construction of an affine invariant metric. Finally, Section 5 reviews a few applications that utilize metric invariants to construct invariant metric spaces that help to define shape signatures.

2 Scale invariant metric

2.1 Scale invariant arc-length for planar curves

We begin our exploration of invariants with a simple scenario of curves in $\mathbb{R}^2$. Let the planar parametric curve C be given by a continuous regular function $C(p)$, such that,

$$\begin{aligned} C &: [0,1] \to \mathbb{R}^2 \\ C(p) &= (x(p), y(p)), \end{aligned} \tag{1}$$

and let its derivative $\frac{\partial C}{\partial p} = (\dot{x}, \dot{y})$ be defined for all p. That is, we consider a curve $C(p) = (x(p), y(p))$ as a mapping of $p \in [0, 1]$ to the plane. For each $p \in [0, 1]$, the Euclidean arch length is defined by

$$s(p) = \int_0^p \sqrt{\dot{x}^2 + \dot{y}^2} d\hat{p}, \tag{2}$$

and as a differential form $ds = |\dot{C}|dp$, while the Euclidean arc-length of the scaled curve $\tilde{C} = \alpha C$, for some positive scalar α, becomes

$$\tilde{s}(p) = \int_0^p \sqrt{\alpha^2 \dot{x}^2 + \alpha^2 \dot{y}^2} d\hat{p} \ = \ \alpha s(p), \tag{3}$$

since $\frac{\partial \alpha x}{\partial p} = \alpha \frac{\partial x}{\partial p}$. Thus, as a differential form, $d\tilde{s} = \alpha ds$.

Choosing a length element that would be invariant to scaling can be done by considering higher derivatives of the curve. Specifically, we use the curvature as a local normalization factor. The scale invariant arc-length would then obviously be different from the Euclidean one. Let the curvature $\kappa(p)$ be given by

$$\kappa(p) = \frac{|\dot{x}\ddot{y} - \dot{y}\ddot{x}|}{(\dot{x}^2 + \dot{y}^2)^{\frac{3}{2}}}, \tag{4}$$

and its scaled version

$$\tilde{\kappa}(p) = \frac{\alpha^2}{\alpha^3}\kappa = \frac{1}{\alpha}\kappa. \tag{5}$$

Theorem 1. *$d\theta = |\kappa| ds$ is a scale invariant differential arc-length.*

Proof. From (3) and (5) it follows that $|\tilde{\kappa}| d\tilde{s} = |\kappa| ds$. That is, $|\kappa| ds$ is indeed scale invariant. □

Note that once the curvature vanishes, that is, $\kappa = 0$, the metric is not defined. One cannot use the curvature to normalize for scale if there is no meaningful geometric structure that supports this construction. We thus refer to such a metric as a *pseudo-metric*.

2.2 Scale invariant metric for implicitly defined planar curves

Let us reexamine the local invariant metric using a level set representation of planar curves, following Raviv and Raskar (2015). A curve can be given as the zero (or any other constant) level set of an implicit function $F(x, y)$, specifically define $\{C = (x, y) : \forall x, y \text{ s.t. } F(x, y) = 0\}$. Recall the classical notations for the gradient and the Hessian,

$$\begin{aligned} \nabla F &= \left(\frac{\partial F}{\partial x}, \frac{\partial F}{\partial y}\right)^T = (F_x, F_y)^T \\ H(F) &= \begin{pmatrix} F_{xx} & F_{xy} \\ F_{yx} & F_{yy} \end{pmatrix}. \end{aligned} \tag{6}$$

Since the gradient of F is perpendicular to the tangent of its level set curves, it must be parallel to the normal of the level set contours. This leads to the known curvature formula (Goldman, 2005; Spivak, 1970)

$$\kappa = \frac{(\overline{\nabla} F)^T H(F) \overline{\nabla} F}{|\nabla F|^3}, \tag{7}$$

where $\overline{\nabla} F = (-F_y, F_x)^T$ is the tangent to the planar curve. Note that the sign of κ depends on the sign of F. Changing the sign from F to $\hat{F} = -F$, would yield the same level sets for which $\hat{\kappa} = -\kappa$, assuming the normal direction is defined by the direction of the gradient of the function.

Denoting by $\tilde{x} = \alpha x$ and $\tilde{y} = \alpha y$, we consider the scaled level set $\tilde{F}$

$$\tilde{F} = F(\tilde{x}, \tilde{y}) = F(\alpha x, \alpha y). \tag{8}$$

Since $\frac{\partial \tilde{F}}{\partial \tilde{x}} = \frac{\partial F}{\partial x}\frac{\partial x}{\partial \tilde{x}} = \frac{1}{\alpha}\frac{\partial F}{\partial x}$, we readily have that

$$\nabla\tilde{F} = (\tilde{F}_{\tilde{x}},\tilde{F}_{\tilde{y}})^T = \alpha^{-1}(F_x, F_y)^T = \alpha^{-1}\nabla F,$$
$$H(\tilde{F}) = \begin{pmatrix} \tilde{F}_{\tilde{x}\tilde{x}} & \tilde{F}_{\tilde{x}\tilde{y}} \\ \tilde{F}_{\tilde{y}\tilde{x}} & \tilde{F}_{\tilde{y}\tilde{y}} \end{pmatrix} = \alpha^{-2}H(F). \tag{9}$$

From (9) we see that

$$\tilde{\kappa} = \frac{\alpha^{-1}\alpha^{-2}\alpha^{-1}}{((\alpha^{-2})^{0.5})^3}\kappa = \frac{\alpha^{-4}}{\alpha^{-3}}\kappa = \frac{1}{\alpha}\kappa. \tag{10}$$

From which we conclude, yet again, that

$$\tilde{\kappa} = \frac{(\overline{\nabla}\tilde{F})^T H(\tilde{F})\overline{\nabla}\tilde{F}}{||\nabla\tilde{F}||^3} = \frac{1}{\alpha}\kappa. \tag{11}$$

Theorem 1 indeed holds for the level set representation.

2.3 Scale invariant metric for surfaces

Consider a compact two-dimensional Riemannian manifold $\{\mathcal{S},g\}$ given by the surface $\mathcal{S}$ with a metric tensor g. Let us further assume that $\mathcal{S}$ is embedded in $\mathbb{R}^3$ by a regular map $S:\Omega\subset\mathbb{R}^2\to\mathbb{R}^3$. The mapping S from the parametrization plane Ω to $\mathbb{R}^3$ is referred to as a parametric surface. Similar to the one-dimensional case shown above, we next search for a renormalization of the first fundamental form that would be invariant to local isotropic stretching or local scaling of the Euclidean space. A scale invariant metric for surfaces was introduced in Aflalo et al. (2013a), where the authors showed that multiplying the coefficients of the metric tensor (also known as the first fundamental form) by the absolute value of the Gaussian curvature produces a scale invariant pseudo-metric.

The Gaussian curvature K can be defined by the ratio of the determinants of the second (b_{ij}) and the first (g_{ij}) fundamental forms, namely,

$$K \equiv \frac{b}{g} = \frac{\det(b_{ij})}{\det(g_{ij})} = \frac{b_{11}b_{22} - b_{12}^2}{g_{11}g_{22} - g_{12}^2}, \tag{12}$$

where the second fundamental form is given by the inner product $b_{ij} = \langle S_{ij}, n\rangle$, and the surface normal $n = \frac{S_1\times S_2}{\|S_1\times S_2\|}$, and the metric tensor for the Euclidean case is given by $g_{ij} = \langle S_i, S_j\rangle$. In this chapter, lower indices for the fundamental forms are associated with their corresponding derivatives, for example, $S_1 \equiv \partial S(u, v)/\partial u$ for $(u, v)\in\Omega$, and $S_2 = \partial S/\partial v$, $S_{11} = \partial^2 S/\partial u^2$, etc. The determinants of the fundamental forms are denoted by $g\equiv\det(g_{ij})$ and $b = \det(b_{ij})$, respectively.

Scaling by α the coordinates of the surface S, scales the coefficients of the first fundamental form by α^2 as

$$\tilde{g}_{ij} = \langle\alpha S_i, \alpha S_j\rangle = \alpha^2\langle S_i, S_j\rangle = \alpha^2 g_{ij}. \tag{13}$$

It scales the coefficients of the second fundamental form by α, since

$$\tilde{b}_{ij} = \langle \alpha S_{ij}, n \rangle = \alpha \langle S_{ij}, n \rangle = \alpha b_{ij}. \tag{14}$$

These relations lead to the construction of an invariant metric first introduced in Aflalo et al. (2013a).

Theorem 2. *$|K| g_{ij}$ is a scale invariant pseudo-metric for surfaces.*

Proof. We plugin equations (13) and (14) into the Gaussian curvature of the scaled surface $\tilde{K}$,

$$|\tilde{K}| \tilde{g}_{ij} = \left| \frac{b}{g} \right| \tilde{g}_{ij} = \frac{\alpha^2}{\alpha^4} |K| \alpha^2 g_{ij} = |K| g_{ij}, \tag{15}$$

that concludes the proof that $|K| g_{ij}$ is indeed a scale invariant pseudo-metric. □

2.4 Approximating the scale invariant Laplace–Beltrami operator for surfaces

Modelling the surface as a two-dimensional Riemannian manifold $\{\mathcal{S}, g\}$, possibly with boundary, the Laplace–Beltrami operator (LBO) generalizes the classical Laplacian operator defined on flat domains to Riemannian manifolds. In local coordinates, the LBO can be expressed as

$$\Delta_g \equiv \frac{1}{\sqrt{g}} \partial_i \sqrt{g} g^{ij} \partial_j, \tag{16}$$

where we used Einstein's summation convention, by which (g_{ij}) is the metric matrix, and $g \equiv \det(g_{ij})$. One index as subscript followed by the same index as superscript, or vice-versa, means summation over the relevant indices. Here, g^{ij} are the elements of the inverse of the metric tensor (g_{ij}), that is, $(g^{ij}) = (g_{ij})^{-1}$.

The LBO admits an eigendecomposition

$$\begin{aligned} -\Delta_g \phi_i(s) &= \lambda_i \phi_i(s) \qquad s \in S \backslash \partial S \\ \langle \nabla_g \phi_i(s), \hat{n}(s) \rangle &= 0 \qquad s \in \partial S, \end{aligned} \tag{17}$$

with Neumann boundary condition (17) for a surface with boundary ∂S, where $\hat{n}(s)$ is the normal to the surface along the boundary, and ∇_g is the intrinsic gradient defined on the Riemannian manifold. The eigendecomposition yields a discrete set of eigenfunctions that are invariant to isometries. The eigenfunctions of the LBO on a regularly parametrized torus define the Fourier transform, while its decomposition on general compact Riemannian manifolds produces eigenfunctions that, when ordered by their corresponding eigenvalues or frequencies, have been shown to be optimal and unique in representing smooth functions on the manifold (Aflalo et al., 2015, 2016a; Brezis and Gómez-Castro, 2017). The invariance of the operator with respect to the isometry it was constructed by, makes it particularly useful for shape analysis. For example, pose variations can be modelled as isometries. Thus, articulated shapes would share similar spectral properties with respect to Δ_g.

The scale invariant metric can be used to define a scale invariant LBO from which a scale invariant spectral geometry could be constructed. Here, we show that such an invariant metric, by which surfaces are treated as a conformal pseudo-Riemannian manifolds, defines a Laplacian that involves up to second-order derivatives. As with conformal geometry, this stability property happens only for two-dimensional manifolds. Specifically, we show that for triangulated surfaces, the LBO can be approximated by an equation that involves only the angles of the triangles. This construction allows stable numerical approximation of quantities like its spectral decomposition which was found to be useful in shape analysis. In its continuous general setting the LBO is defined in Eq. (16). For two-dimensional manifolds we have

$$(g^{ij}) \equiv (g_{ij})^{-1} = \begin{pmatrix} g_{11} & g_{12} \\ g_{12} & g_{22} \end{pmatrix}^{-1} = \frac{1}{g}\begin{pmatrix} g_{22} & -g_{12} \\ -g_{12} & g_{11} \end{pmatrix}. \tag{18}$$

Given the scale invariant pseudo-metric $\tilde{g}_{ij} = |K| g_{ij}$, we have

$$\tilde{g} = K^2 g, \tag{19}$$

and

$$\begin{aligned}
(\tilde{g}^{ij}) &= \begin{pmatrix} |K|g_{11} & |K|g_{12} \\ |K|g_{12} & |K|g_{22} \end{pmatrix}^{-1} \\
&= \frac{1}{K^2 g}\begin{pmatrix} |K|g_{22} & -|K|g_{12} \\ -|K|g_{12} & |K|g_{11} \end{pmatrix} \\
&= \frac{1}{|K|}\frac{1}{g}\begin{pmatrix} g_{22} & -g_{12} \\ -g_{12} & g_{11} \end{pmatrix} \\
&= |K|^{-1}(g^{ij}).
\end{aligned} \tag{20}$$

Thus,

$$\begin{aligned}
\Delta_{\tilde{g}} &\equiv \frac{1}{\sqrt{\tilde{g}}}\partial_i \sqrt{\tilde{g}}\tilde{g}^{ij}\partial_j \\
&= \frac{1}{\sqrt{K^2 g}}\partial_i \sqrt{K^2 g}|K|^{-1} g^{ij}\partial_j \\
&= |K|^{-1}\frac{1}{\sqrt{g}}\partial_i |K|\sqrt{g}|K|^{-1} g^{ij}\partial_j \\
&= |K|^{-1}\Delta_g.
\end{aligned} \tag{21}$$

Note that the resulting expression involves only second-order derivatives that are required for the definition of the LBO Δ_g and the Gaussian curvature K. In other words, although we are now considering a richer group of transformations that involves an additional parameter beyond translation and rotation, the order of derivatives required for computing the Laplacian remains the same.

Given a triangulated surfaces with N vertices $\{v_1, \ldots, v_i, \ldots, v_N\}$, for which the one-ring neighbourhood $\mathcal{N}_i$ is defined by the vertices connected by a single edge to the vertex v_i. Define the angular deficiency about the ith vertex v_i as

$$Q_i = \left| 2\pi - \sum_{v_j \in \mathcal{N}_i} \theta_j^i \right|, \tag{22}$$

where θ_j^i is the angle of the triangle with the edge (v_i, v_j) for $v_j \in \mathcal{N}_i$, as shown in Fig. 1. Without loss of generality, angles are defined by the edge of a triangle visited in a counter-clockwise fashion about the vertex v_i. The absolute value of the Gaussian curvature at v_i of our triangulated surface can be approximated by the Gauss–Bonnet discrete approximation (Banchoff, 1970) as

$$K_i = \frac{Q_i}{A_i}, \tag{23}$$

where A_i is the area about the vertex i. Note, that in our discussion, the specific way by which A_i is approximated is insignificant, see Pinkall and Polthier (1993) for an example of computing A_i. As an operator, we can define

$$K = \begin{pmatrix} K_1 & & & & \\ & \ddots & & 0 & \\ & & K_i & & \\ & 0 & & \ddots & \\ & & & & K_N \end{pmatrix}$$

and similarly for Q and A. Then, $K \equiv A^{-1}Q$.

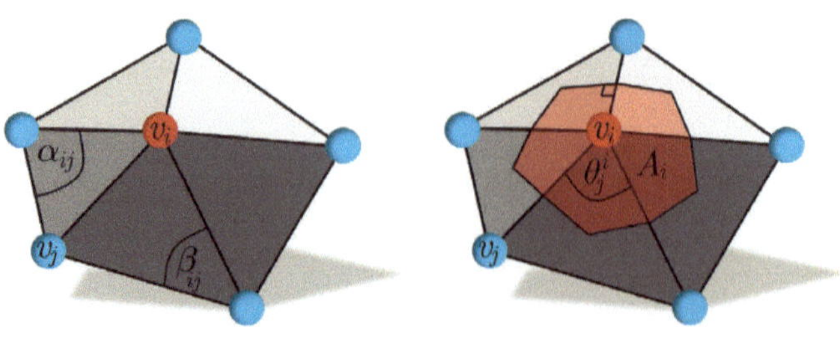

FIG. 1 The measures used to define the Laplacian and the Gaussian curvature.

Let the cotangent weights matrix W be defined as

$$W_{ij} = \begin{cases} \sum_{v_j \in \mathcal{N}_i} \omega_{ij} & \text{if } i = j \\ -\omega_{ij} & \text{if } i \neq j, v_j \in \mathcal{N}_i \\ 0 & \text{otherwise,} \end{cases} \tag{24}$$

for $\omega_{ij} \equiv \cot\alpha_{ij} + \cot\beta_{ij}$, where α_{ij} and β_{ij} are the angles opposite to the edge (v_i, v_j), see Fig. 1.

According to Pinkall and Polthier (1993), the LBO can be approximated by the cotangent weights equation given by

$$L = A^{-1}W, \tag{25}$$

where W are the cotangent weights matrix. The scale invariant LBO for two-dimensional triangulated surfaces can then be approximated by

$$\begin{aligned} \tilde{L} &= K^{-1}A^{-1}W \\ &= (A^{-1}Q)^{-1}A^{-1}W \\ &= (Q^{-1}A)A^{-1}W \\ &= Q^{-1}AA^{-1}W \\ &= Q^{-1}W. \end{aligned} \tag{26}$$

Explicitly, the elements of $\tilde{L}$ read

$$\tilde{L}_{ij} = \begin{cases} \left|2\pi - \sum_{v_j \in \mathcal{N}_i} \theta_j^i\right|^{-1} \sum_{v_j \in \mathcal{N}_i} (\cot\alpha_{ij} + \cot\beta_{ij}) & \text{if } i = j \\ -\left|2\pi - \sum_{v_j \in \mathcal{N}_i} \theta_j^i\right|^{-1} (\cot\alpha_{ij} + \cot\beta_{ij}) & \text{if } i \neq j, v_j \in \mathcal{N}_i \\ 0 & \text{otherwise.} \end{cases} \tag{27}$$

Computing the decomposition of the above possibly singular operator, we resort to the generalized eigenvalue problem

$$W\psi_i = \lambda_i Q\psi_i. \tag{28}$$

Note that the operator involves only the angles of the given triangulated mesh.

In the following numeric example, the heat kernel signature (HKS) (Sun et al., 2009) is derived from the regular metric and the scale invariant one, and the corresponding signatures are compared. The heat kernel signature is defined using the eigenfunctions of LBO

The HKS is a linear combination of the squared LBO eigenfunctions given by

$$\text{HKS}(s,t) = \sum_i e^{-\lambda_i t}\phi_i^2(s), \quad s \in S. \tag{29}$$

Fig. 2 presents the invariance of the HKS signature at several times t_i, textured mapped onto a centaur and its locally scaled version. The same experiment

FIG. 2 The heat kernel signature at different times textured mapped onto the surface for the regular metric (*left frame*) and the invariant one (*right frame*). The four shapes at each row, *left* to *right*, capture the HKS values at $t = 10$, 50, 100, and 500, respectively.

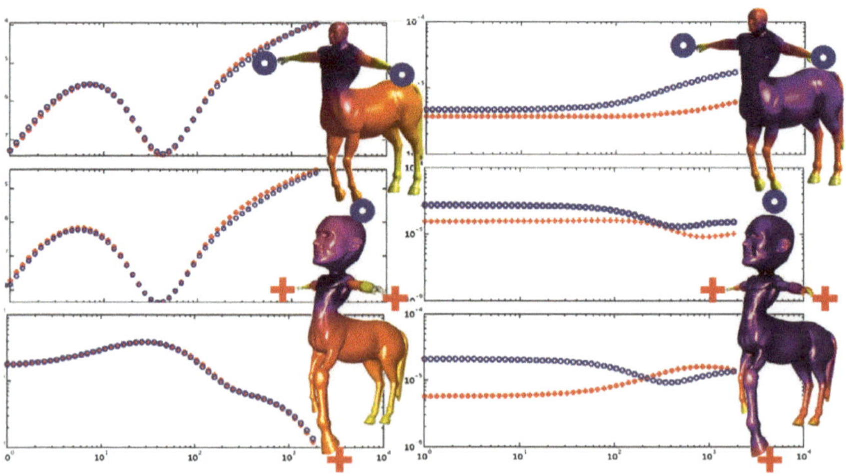

FIG. 3 Scaled heat kernel signatures for the regular metric (*right*), and the invariant version (*left*). The *blue circles* represent the signatures for three points on the original surface, while the *red plus signs* are computed from the deformed version. Using a log–log axes we plot the scaled-HKS as a function of t.

applies also for a normal horse and one with enlarged head that looks like a mule. Fig. 3 displays the inconsistency of corresponding signatures when using the regular metric (right) and the consistency achieved with the invariant one (left). The signatures are extracted at three points as indicated in the figure:

two finger tips; one on the right and one on the left hand of the centaur, and the horseshoe of the front left leg. In the graphs, the signature value at each time t_i is scaled with respect to the integral of the signatures at that time over the (invariant) surface, i.e., $\mathrm{HKS}(s,t)/\int_S \mathrm{HKS}(s,t)da(s)$ as done in Sun et al. (2009) for presentation prepossess. As can be observed, the proposed metric produces invariant nontrivial signatures.

3 Equi-affine invariant metric

3.1 Equi-affine invariant arc-length

The origin of the equi-affine structure goes back to the 19th century, presented by Blaschke and captured the imagination of researchers as it demonstrated yet again the existence of interesting geometries beyond the Euclidean one. An affine transformation of the plane can be expressed by six parameters as

$$\mathbf{Ax}+\mathbf{b}=\begin{pmatrix} a & b \\ c & d \end{pmatrix}\begin{pmatrix} x \\ y \end{pmatrix}+\begin{pmatrix} c \\ d \end{pmatrix}, \tag{30}$$

where $\det(\mathbf{A}) > 0$. The area preserving affine group, also called *special-affine* or *equi-affine*, is defined by restricting $\det(\mathbf{A}) = 1$, which leaves us with five degrees of freedom.

Dealing with planar curves, one way to construct an equi-affine arc-length p is by detecting a quantity which is preserved by the equi-affine transformation. While rotation transformation preserves the angles between vectors, a general equi-affine transformation changes the relative angles, turning squares into parallelograms. However, equi-affine transformations preserve the area of the shapes. Specifically, for an equi-affine invariant arc-length ν, the area of the parallelogram formed by the velocity and acceleration vectors of $C(\nu)$ should be preserved under the equi-affine transformation. Without loss of generality we can write

$$|C_\nu \times C_{\nu\nu}| = |\det(C_\nu, C_{\nu\nu})| = 1. \tag{31}$$

We construct the arc-length as $d\mu = f(\det(C_p, C_{pp}))dp$, where the function f is determined by demanding reparametrization invariance of the geometric structure that defines the arc-length. For that goal, let ξ be a reparametrization that maps p into q, such that $p = \xi(q)$. We require that

$$f(\det(C_p, C_{pp}))d\nu = f(\det(C_q, C_{qq}))dq. \tag{32}$$

Using the chain rule,

$$\begin{aligned} C_q &= C_p\dot{\xi} \\ C_{qq} &= C_{pp}\dot{\xi}^2 + C_p\ddot{\xi}. \end{aligned} \tag{33}$$

Hence,

$$\begin{aligned}\det(C_q,C_{qq}) &= \det\left(\begin{pmatrix}\dot{\xi} & \\ \ddot{\xi} & \dot{\xi}^2\end{pmatrix}(C_p,C_{pp})\right)\\ &= \det\begin{pmatrix}\dot{\xi} & \\ \ddot{\xi} & \dot{\xi}^2\end{pmatrix}\cdot\det(C_p,C_{pp})\\ &= \dot{\xi}^3\det(C_p,C_{pp}).\end{aligned} \tag{34}$$

Combining (32) with $dp=\dot{\xi}dq$ yields

$$f(\det(C_p,C_{pp}))dp \;= f(\dot{\xi}^3\det(C_p,C_{pp}))\frac{dp}{\dot{\xi}}, \tag{35}$$

by which the equi-affine arc-length is defined.

Theorem 3. *$d\upsilon = |\det(C_p,C_{pp})|^{\frac{1}{3}}dp$ is an equi-affine invariant arc-length.*

Proof. Follows immediately from (35). □

3.2 Equi-affine metric for surfaces

Moving from curves to surfaces opened the door to many applications in modern shape analysis and image processing. The equi-affine metric for surfaces (Blaschke, 1923; Su, 1983) was presented as a theoretical framework and was recently applied to shape analysis and further explored using spectral geometry (Raviv et al., 2011a), and for geodesic measurements (Raviv et al., 2011b) in different applications.

The Euclidean metric tensor can be obtained from the reparameterization invariant arc-length ds of a parametrized curve $C(s)$ on $\mathcal{S}$. As the simplest Euclidean invariant is length, we search for the parameterization s that would satisfy $|C_s| = 1$, or explicitly,

$$\begin{aligned}1 &= \langle C_s,C_s\rangle = \langle S_s,S_s\rangle\\ &= \left\langle\frac{\partial S}{\partial u}\frac{du}{ds}+\frac{\partial S}{\partial v}\frac{dv}{ds},\frac{\partial S}{\partial u}\frac{du}{ds}+\frac{\partial S}{\partial v}\frac{dv}{ds}\right\rangle\\ &= ds^{-2}\left(g_{11}du^2+2g_{12}dudv+g_{22}dv^2\right),\end{aligned} \tag{36}$$

where, as before,

$$g_{ij} = \langle S_i, S_j\rangle, \tag{37}$$

An infinitesimal displacement ds on the surface $S(u, v)$ is thereby given by

$$ds^2 = g_{11}du^2 + 2g_{12}dudv + g_{22}dv^2. \tag{38}$$

The metric coefficients g_{ij} translate the surface parametrization coordinates u and v into a Euclidean invariant distance measure on the surface. This distance should not change under Euclidean transformations of the surface $\mathbf{R}S + \mathbf{b}$, where $\mathbf{R}$ is a rotation matrix in $\mathbb{R}^3$. It would also be preserved w.r.t. isometric (length preserving) transformations.

The equi-affine transformation, defined by the linear operator $\mathbf{A}S + \mathbf{b}$, where $\det(\mathbf{A}) = 1$, requires a different treatment, see Blaschke (1923) and Su (1983). Consider the curve $C \in \mathcal{S}$, parametrized by w. The equi-affine transformation is volume preserving, and thus, its invariant metric is constructed by restricting the volume defined by S_u, S_v, and C_{ww} to one. That is,

$$\begin{aligned} 1 &= \det(S_u, S_v, C_{ww}) \\ &= \det(S_u, S_v, S_{ww}) \\ &= \det\left(S_u, S_v, S_{uu}\frac{du^2}{dw^2} + 2S_{uv}\frac{du\,dv}{dw\,dw} + S_{vv}\frac{dv^2}{dw^2} + S_u\frac{d^2u}{dw^2} + S_v\frac{d^2v}{dw^2}\right) \\ &= dw^{-2}\det\left(S_u, S_v, S_{uu}du^2 + 2S_{uv}dudv + S_{vv}dv^2\right) \\ &= dw^{-2}\left(r_{11}du^2 + 2r_{12}dudv + r_{22}dv^2\right), \end{aligned} \tag{39}$$

resulting in an equi-affine invariant arc-length on the surface

$$dw^2 = r_{11}du^2 + 2r_{12}dudv + r_{22}dv^2 \tag{40}$$

where now, the metric elements are given by

$$r_{ij} = \det\left(S_1, S_2, S_{ij}\right), \tag{41}$$

and we extended the shorthand notation to second-order derivatives by which $S_{11} = \frac{\partial^2 S}{\partial u^2}$, $S_{22} = \frac{\partial^2 S}{\partial v^2}$, and $S_{12} = \frac{\partial^2 S}{\partial u \partial v}$. Note that the second fundamental form in the Euclidean case is given by $b_{ij} = \sqrt{g}r_{ij}$, where $g = \det(g_{ij}) = g_{11}g_{22} - g_{12}^2$. The equi-affine reparametrization invariant metric (Blaschke, 1923; Su, 1983) reads

$$q_{ij} = |r|^{-\frac{1}{4}}r_{ij}, \tag{42}$$

where $r = \det(r_{ij}) = r_{11}r_{22} - r_{12}^2$.

Theorem 4. *$q_{ij} = |r|^{-\frac{1}{4}}r_{ij}$ is an equi-affine invariant metric.*

It is easy to see from the construction that the quadratic form is equi-affine invariant and it can be shown that renormalization is needed for the reparameterization invariant property. We refer to Raviv et al. (2011a) and Raviv et al. (2011b) for two different proofs. Intuitively, the normalization factor $|r|^{-1/4}$ reflects the number of derivatives involved in the construction of r_{ij}. Note that this is a quasi-metric as it is not defined everywhere.

4 Affine metric

4.1 Affine invariant arc-length

The equi-affine metric played an important role in many successful solutions to computer vision and image processing problems over the years, see, for example, Alvarez et al. (1993) and Raviv and Kimmel (2014). The volume preservation limitation of the equi-affine invariant metric was a limiting factor in many applications. In Raviv and Kimmel (2014) it was demonstrated that scale invariant and equi-affine invariant metric can be combined into an affine invariant metric without the volume preserving constraint. We summarize this construction in the next section. The same logic can also be applied to planar curves as we will demonstrate in the next section.

In the affine invariant case we try to construct a scale invariant normalization factor which is also invariant to the equi-affine transformations. Specifically, we showed earlier that by multiplying the Euclidean arc-length by the curvature we obtain a similarity invariant metric. Following this line of thought, we will show that for planar curves the square root of the equi-affine curvature can be used to normalize the equi-affine metric generating an affine arc-length.

The equi-affine curvature μ was defined in Blaschke (1923) and Su (1983) by

$$\mu(v) = \det(C_{vv}, C_{vvv}), \tag{43}$$

and for a planar curve with arbitrary parameterization $C(p) = (x(p), y(p))$, we readily have

$$\mu(p) = \frac{4\det(C_{pp}, C_{ppp}) + \det(C_p, C_{pppp})}{3\left(\det(C_p, C_{pp})\right)^{\frac{5}{3}}} - \frac{5}{9}\frac{\left(\det(C_p, C_{ppp})\right)^2}{\left(\det(C_p, C_{pp})\right)^{\frac{8}{3}}}, \tag{44}$$

Using the above invariant measures we are ready to define an affine invariant arc-length for planar curves,

Theorem 5. *The differential arc-length*

$$dv = (\mu(p))^{\frac{1}{2}}\left(\det(C_p, C_{pp})\right)^{\frac{1}{3}} dp, \tag{45}$$

is affine invariant.

Proof. First, by construction, dv is equi-affine invariant, as both the metric and the curvature are invariant to equi-affine transformations. By the same argument, it is also geometric or reparameterization invariant. Next, let us show that (45) is also scale invariant.

Note that scaling the curve C by α, we have $\hat{C} = \alpha C$, and

$$\hat{C}_p = \alpha C_p. \tag{46}$$

The same relation holds also for C_{pp}, C_{ppp} and C_{pppp}, etc. It follows that,

$$\begin{aligned} d\hat{\upsilon} &= \left(\det\left(\hat{C}_p, \hat{C}_{pp}\right)\right)^{\frac{1}{3}} dp, \\ &= \left(\det\left(\alpha C_p, \alpha C_{pp}\right)\right)^{\frac{1}{3}} dp \\ &= \alpha^{\frac{2}{3}} d\upsilon. \end{aligned} \tag{47}$$

At the other end, plugging the derivatives of $\hat{C}$ into Eq. (44), we have

$$\hat{\mu} = \alpha^{-\frac{4}{3}} \mu. \tag{48}$$

From (47) and (48) we conclude that

$$d\hat{v} = \sqrt{\hat{\mu}}\, d\hat{\upsilon} = \sqrt{\mu}\, d\upsilon = dv, \tag{49}$$

is indeed an affine invariant arc-length. □

4.2 Affine metric for surfaces

Similar to the curves, we can combine the scale invariant and the equi-affine invariant metrics to construct a full affine invariant metric for surfaces as first shown in Raviv and Kimmel (2014). There, it was shown that the volume preserving restriction can be eliminated. The idea was to use the Gaussian curvature, computed with respect to the equi-affine invariant metric, in order to construct a new metric by multiplying the equi-affine metric elements by the equi-affine Gaussian curvature. This way, the resulting metric would been both intrinsic and affine invariant. Specifically, consider the surface $(\mathcal{S}, q)$, where q_{ij} is the equi-affine invariant pseudo-metric, and compute the equi-affine Gaussian curvature $\mathcal{K}$ of $(\mathcal{S}, q)$ at each point. The affine invariant pseudo-metric can then be defined by Raviv and Kimmel (2014)

$$h_{ij} = |\mathcal{K}| q_{ij}. \tag{50}$$

Using Brioschi formula (Gray et al., 2006) we can evaluate the Gaussian curvature directly from the metric and its first and second derivatives. Specifically, given the metric tensor q_{ij}, we have

$$\mathcal{K} \equiv \frac{\beta - \gamma}{\det^2(q)}, \tag{51}$$

where

$$\beta=\det\begin{pmatrix} -\frac{1}{2}q_{11,vv}+q_{12,vu}-\frac{1}{2}q_{22,uu} & \frac{1}{2}q_{11,u} & q_{12,u}-\frac{1}{2}q_{11,v} \\ q_{12,v}-\frac{1}{2}q_{22,u} & q_{11} & q_{12} \\ \frac{1}{2}q_{22,v} & q_{12} & q_{22} \end{pmatrix} \tag{52}$$

$$\gamma=\det\begin{pmatrix} 0 & \frac{1}{2}q_{11,v} & \frac{1}{2}q_{22,u} \\ \frac{1}{2}q_{11,v} & q_{11} & q_{12} \\ \frac{1}{2}q_{22,u} & q_{12} & q_{22} \end{pmatrix},$$

here $q_{ij,u}$ denotes the derivation of q_{ij} with respect to u, and in a similar manner $q_{ij,uv}$ is the second derivative w.r.t. u and v. Same notations follow for $q_{ij,v}$, $q_{ij,vv}$, and $q_{ij,uu}$.

Note that $\mathcal{K}$ is an equi-affine invariant Gaussian curvature. That is an immediate outcome from the Brioschi formula, as the curvature is evaluated directly from the metric and its derivatives which are invariant to equi-affine deformations.

Theorem 6. *h_{ij} is an affine invariant pseudo-metric.*

Proof. Let $\mathcal{K}$ be the equi-affine invariant Gaussian curvature, then, we will show that the metric defined by $h_{ij}=|\mathcal{K}|q_{ij}$ is scale invariant.

Scaling the surface S by α, the corresponding equi-affine invariant components are

$$\begin{aligned} \tilde{r}_{ij}&=\det(\alpha S_1,\alpha S_2,\alpha S_{ij})=\alpha^3 r_{ij} \\ \tilde{r}_{ij,u}&=\alpha^3 r_{ij,u} \\ \tilde{r}_{ij,uv}&=\alpha^3 r_{ij,uv} \\ \det(\tilde{r})&=\alpha^6\det(r), \end{aligned} \tag{53}$$

that leads to

$$\tilde{q}_{ij}=\frac{\tilde{r}_{ij}}{(\det(\tilde{r}))^{\frac{1}{4}}}=\frac{\alpha^3 r_{ij}}{(\alpha^6\det(r))^{\frac{1}{4}}}=\alpha^{\frac{3}{2}}q_{ij}, \tag{54}$$

that yields $\det(\tilde{q})=\left(\alpha^{\frac{3}{2}}\right)^2\det(q)=\alpha^3\det(q)$. Denote the equi-affine Gaussian curvature of the scaled surface $\tilde{S}=\alpha S$ by $\tilde{\mathcal{K}}$. We have that

$$\tilde{\beta} = \left(\alpha^{\frac{3}{2}}\right)^3 \beta,$$
$$\tilde{\gamma} = \left(\alpha^{\frac{3}{2}}\right)^3 \gamma, \tag{55}$$
$$\tilde{\mathcal{K}} = \frac{\tilde{\beta} - \tilde{\gamma}}{\det^2(\tilde{q})} = \frac{\left(\alpha^{\frac{3}{2}}\right)^3 (\beta - \gamma)}{(\alpha^3)^2 \det^2(q)} = \alpha^{-\frac{3}{2}}\mathcal{K}.$$

It immediately follows that

$$|\tilde{\mathcal{K}}|\tilde{q}_{ij} = |\tilde{\mathcal{K}}|\alpha^{\frac{3}{2}} q_{ij} = |\alpha^{-\frac{3}{2}}\mathcal{K}|\alpha^{\frac{3}{2}} q_{ij} = |\mathcal{K}| q_{ij} \tag{56}$$

which concludes the proof. □

4.2.1 *Voronoi diagram*

An interesting application of the affine invariant metric can be seen in a Voronoi diagram depicted in Fig. 4. Distances are measured using diffusion distances based on Euclidean and affine invariant metrics. We can see that the affine invariant metric is indeed invariant with respect to the affine stretching.

Another interesting experiment, first presented in Raviv and Kimmel (2014), demonstrates the invariance properties of eigenfunctions of the LBO defined by various metrics and different deformations. Fig. 5 presents the 9th eigenfunction textured mapped onto the surface using a Euclidean metric, the scale invariant, equi-affine, and the affine invariant pseudo metric. At each row a different deformation of the surface is shown. Local scaling was applied to the shapes at the second row, while surfaces on the third row underwent volume preserving equi-affine stretching. The bottom row shows a full affine transformation that was applied to the surface including local scaling. The accumulated histogram values of the corresponding 9th eigenfunction are plotted at the top of each column. Blue is used for the original shape, red for the locally scaled one, green for the equi-affine transformed, and black for the affine transformed version.

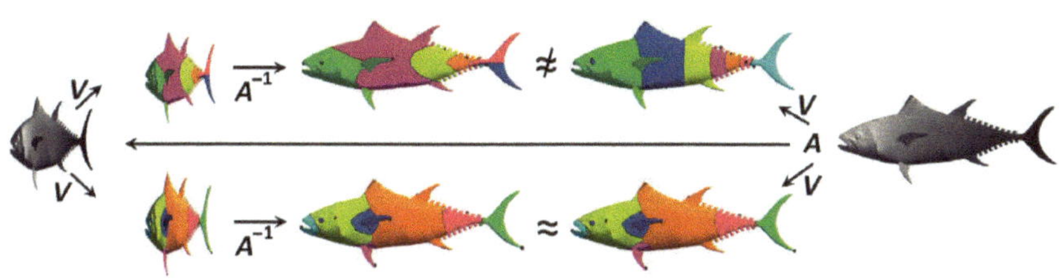

FIG. 4 Voronoi diagrams of 10 points selected by farthest point sampling. Diffusion distances were used based on Euclidean metric (*top*) and an affine invariant one (*bottom*).

FIG. 5 The 9th LBO eigenfunction textured mapped on the surface using four different metrics, from *left* to *right*: Euclidean, scale-invariant, equi-affine, and affine. Deformations from *top* to *bottom*: None, local scale, equi-affine, and affine. At the top, the accumulated (histogram) values of the eigenfunction per invariant metric are displayed. The *blue curve* depicts the original shape, *red* the locally scaled one, *green* the equi-affine, and *black* the affine transformation.

4.3 Approximating the Gaussian curvature

The Gaussian curvature for smooth surfaces can be defined using the global Gauss–Bonnet theorem, see do Carmo (1976). This property is often used to approximate the Gaussian curvature of triangulated surfaces, see Polthier and Schmies (1998). Given a vertex in a triangulated mesh that is shared by p triangles such that the angle of each triangle at that vertex is given by θ_i, where $i \in 1, \ldots p$. The Gaussian curvature K at that vertex can be approximated by

$$K \cong \frac{1}{\frac{1}{3}\sum_{i=1}^{p} \mathcal{A}_i}\left(2\pi - \sum_{i=1}^{p} \theta_i\right), \tag{57}$$

where $\mathcal{A}_i$ is the area of the ith triangle, and θ_i is the corresponding angle of the ith triangle at the vertex for which K is being approximated.

So, as an alternative for computing the Gaussian curvature with respect to any given metric using the Brioschi formula, we could use the Gauss–Bonnet deficiency angles approximation given in Eq. (57), where angles and areas are computed using the given elements of the metric tensor. Consider the triangle $ABC = \{S(u_0, v_0), S(u_0 + du, v_0), S(u_0, v_0 + dv)\}$, infinitesimally defined on the surface $S(u, v)$ with metric (q_{ij}). Denote by θ_A, θ_B, θ_C the angles measured at the vertices A, B, C, respectively, and by l_A, l_B, l_C the edges opposite to the vertices A, B, C, respectively. For simplicity of notations let $du = dv = 1$. The lengths of the edges would then be

$$\begin{aligned} l_A^2 &= (-du\, dv)(q_{ij})(-du\, dv)^T = (-1\ 1)(q_{ij})(-1\ 1)^T = q_{11} - 2q_{12} + q_{22} \\ l_B^2 &= (0 dv)(q_{ij})(0\, dv)^T = (0\ 1)(q_{ij})(0\ 1)^T = q_{22} \\ l_C^2 &= (du\, 0)(q_{ij})(du\, 0)^T = (1\ 0)(q_{ij})(1\ 0)^T = q_{11}. \end{aligned} \tag{58}$$

From the law of cosines we readily have that

$$\cos\theta_A = \frac{q_{11} + q_{22} - (q_{11} - 2q_{12} + q_{22})}{2\sqrt{q_{11}q_{22}}} = \frac{q_{12}}{\sqrt{q_{11}q_{22}}}. \tag{59}$$

Next, the area of the above triangle can be expressed by the metric coefficients as $\mathcal{A}_{ABC} = \frac{1}{2}\sqrt{\det(q_{ij})}$. We readily have that

$$\sin^2\theta_A = 1 - \left(\frac{q_{12}}{\sqrt{q_{11}q_{22}}}\right)^2 = \frac{q_{11}q_{22} - q_{12}^2}{q_{11}q_{22}}. \tag{60}$$

The area of the triangle is half the length of its base multiplied by its height,

$$\mathcal{A}_{ABC} = \frac{1}{2}\sqrt{q_{22}}\left(\sqrt{q_{11}}\sqrt{\frac{q_{11}q_{22} - q_{12}^2}{q_{11}q_{22}}}\right) = \frac{1}{2}\sqrt{\det(q_{ij})}. \tag{61}$$

Consider the surface S given by its three coordinate functions $x(u, v)$, $y(u, v)$ and $z(u, v)$, where, $x, y, z : \Omega \subset \mathbb{R}^2 \to \mathbb{R}$, and $S : \Omega \in \mathbb{R}^2 \to \mathbb{R}^3$

$$S(u,v) = \begin{pmatrix} x(u,v) \\ y(u,v) \\ z(u,v) \end{pmatrix}. \tag{62}$$

In order to evaluate the equi-affine metric, the surface is first locally approximated by a quadratic form. For each triangle we approximate

$$S(u,v) = \begin{pmatrix} x(u,v) \\ y(u,v) \\ z(u,v) \end{pmatrix} \approx B \begin{pmatrix} 1 \\ u \\ v \\ uv \\ u^2 \\ v^2 \end{pmatrix}. \tag{63}$$

where the matrix $B_{3\times 6}$ contains 18 parameters. The local coefficients matrix B is evaluated from six surface points (vertices). Three vertices belong to the triangle for which the metric is evaluated, and three to its nearest neighbouring triangles. The equi-affine metric coefficients for each triangle are then evaluated from our local quadratic approximation of $S(u, v)$ and its corresponding derivatives (see Raviv and Kimmel (2014) for more details). Finally, for each vertex, the Gaussian curvature with respect to the equi-affine metric is approximated using Eq. (57). The corresponding angles and areas in Eq. (57) are defined by the metric, according to Eq. (59) and (61), respectively. In order to construct the full affine invariant metric we need to scale the equi-affine metric by the equi-affine Gaussian curvature. To that end, the curvature is linearly interpolated, from its values at the vertices, at the centre of each triangle.

5 Applications

5.1 Self functional maps: A song of shapes and operators

5.1.1 A question of representation

Considering the boundary surface that identifies a shape in 3D, we could think of a few possible representations for this surface. One is an explicit representation as a function $S : \Omega \subset \mathbb{R}^2 \to \mathbb{R}^3$ from the Euclidean plane to the 3D Euclidean space. For a general surface, the existence of a direct representation is neither guaranteed nor unique. Another possibility is an implicit representation, as a level set of a function defined in the 3D Euclidean space $\mathcal{S}(x,y,z) = c$, where $\mathcal{S} : \mathbb{R}^3 \to \mathbb{R}$ and c is a constant. One choice for the function $\mathcal{S}$ can be the *signed distance function* from a given closed, orientable surface, where $\mathcal{S}$ admits positive values for points in the exterior of the shape and negative values at the interior. Choosing $c = 0$, the zero level set of $\mathcal{S}$ provides an implicit representation. The implicit representation is obviously not unique. While these representations (when they exist) encode completely the surface, they are incapable of providing an immediate answer to the fundamental geometrical question: *Are two surfaces isometric?* or more generally *Are two surfaces related by a special transformation?*

In light of the last statement, we may ask, is it possible to find a representation for a surface that would provide a natural answer to the previous questions. In the sequel, we show that self functional maps (Halimi and Kimmel, 2018) is such a representation. Using self functional maps the question *Are two surfaces isometric?* is translated to the problem of measuring the algebraic distances between matrices.

5.1.2 Hearing shapes using invariant operators

Some operators have the special property that they commute with isometries. Such operators can be shown to be *isometry invariant*, that is, when computed for two isometric shapes the operator admit the same spectrum. Additionally, the set of eigenfunctions on the second shape is the result of mapping the set of eigenfunctions on the first shape, using the isometric transformation. A classic example is the LBO. Assume we have two surfaces: $\mathcal{X}$, and $\mathcal{Y}$, and an isometric transformation between them $T:\mathcal{X}\rightarrow\mathcal{Y}$. Denoting the space of scalar functions defined upon $\mathcal{X}$ by $\mathcal{F}(\mathcal{X},\mathbb{R})$, by definition of the commutation relation, for every $f\in\mathcal{F}(\mathcal{X},\mathbb{R})$,

$$(\Delta_{g\mathcal{X}}f)\circ T^{-1}=\Delta_{g\mathcal{Y}}(f\circ T^{-1}). \tag{64}$$

Namely, applying $\Delta_{g\mathcal{X}}$ to f defined on the source shape $\mathcal{X}$ and then mapping the values of the resulting function to the target shape $\mathcal{Y}$, using the transformation T, is equivalent to mapping f to the shape $\mathcal{Y}$ using the transformation T, and then applying $\Delta_{g\mathcal{Y}}$ to the mapped function. The commutation property dictates that $\Delta_{g\mathcal{X}}$ and $\Delta_{g\mathcal{Y}}$ have exactly the *same spectrum*. To convince ourselves that it is true, assume that the eigenvalues are a nondecreasing sequence $0=\lambda_0\leq\lambda_1\leq\lambda_2\leq\ldots$. Assume that λ_i is an eigenvalue of $\Delta_{g\mathcal{X}}$, with the respective eigenfunction $\phi_i^{\mathcal{X}}$,

$$\Delta_{g\mathcal{X}}\phi_i^{\mathcal{X}}=\lambda_i\phi_i^{\mathcal{X}}. \tag{65}$$

When the eigenspace is of dimension $d>1$, i.e., $\lambda_{m-1}\neq\lambda_m=\lambda_{m+1}=\cdots=\lambda_{m+d-1}\neq\lambda_{m+d}$, for some m, we assume that $\{\phi_m^{\mathcal{X}},\phi_{m+1}^{\mathcal{X}},\ldots,\phi_{m+d-1}^{\mathcal{X}}\}$ is an orthogonal set, defined up to a unitary transformation in the eigenspace, with respect to the standard inner product between scalar functions on the surface $\mathcal{S}$

$$\langle f,g\rangle\equiv\int_S f(s)g(s)da(s)\ \ f,g:S\rightarrow\mathbb{R} \tag{66}$$

Then, substituting Eq. (65) into Eq. (64) we get

$$\lambda_i(\phi_i^{\mathcal{X}}\circ T^{-1})=(\lambda_i\phi_i^{\mathcal{X}})\circ T^{-1}=(\Delta_{g\mathcal{X}}\phi_i^{\mathcal{X}})\circ T^{-1}=\Delta_{g\mathcal{Y}}(\phi_i^{\mathcal{X}}\circ T^{-1}), \tag{67}$$

implying that λ_i is an eigenvalue of $\Delta_{g\mathcal{Y}}$ as well, with the respective eigenfunction $\phi_i^{\mathcal{Y}}=\phi_i^{\mathcal{X}}\circ T^{-1}$. The other direction relation can be shown similarly, by using the equivalent commutation relation $(\Delta_{g\mathcal{Y}}g)\circ T=\Delta_{g\mathcal{X}}(g\circ T)$ for every $g\in\mathcal{F}(\mathcal{Y},\mathbb{R})$. Note that we could perform the same general analysis for a different operator that commutes with a transformation rather than isometry (a rigid transformation, for example). Doing so, we would have reached the following conclusion: *if an operator commutes with a given transformation, its spectrum and its eigenfunctions are invariant under the transformation.* This observation motivates the approach of using invariant operators to represent geometric shapes, due to the guarantee that an operator that commutes with a specific transformation will have the same spectral properties on any pair of shapes that are related by this transformation. Naturally, we would like to have a guarantee also for the opposite direction, namely, that for a pair of shapes that are not related by this transformation the spectrum will be unique. The famous question "Can we hear the shape of a drum?" was formulated by Mark Kac in 1966. In the context of 2D-Reimannian manifolds, the question is whether the *spectrum* of LBO determines the shape of the surface up to isometries. In other words, is the spectrum distinct between nonisometric shapes or can we find an example of two *isospectral* but *nonisometric* shapes. In 1992 a counter example was given by a pair of different isospectral planar concave polygons (Gordon et al., 1992). Numerous similar examples of planar isospectral domains were constructed later (Buser et al., 2010). On the other hand, it was proved that with the restriction to certain convex planar regions with analytic boundary, the answer to Kac's question is positive (Zelditch, 2000). To the best of our knowledge, for nonconvex analytic planar domains and for nonplanar surface domains the question is still open. Specifically, the question remained open for the domains that are of interest to the computer vision community, namely nonplanar 2-manifolds. Reuter et al. (2006) suggested to use the spectrum of LBO as a signature for isometric surfaces, called Shape-DNA. Recent work (Cosmo et al., 2018) suggests a numerical procedure called *isospectralization*, that is, a deformation of a shape until its Laplacian spectrum matches that of another.

5.1.3 Shape from eigenfunctions

A question with a flavour similar to the classical hearing shapes of drums is whether the *eigenfunctions* of the LBO determine the shape up to isometries. It might be that the eigenfunctions carry additional information beyond the spectrum that makes it possible to determine the shape. Unlike the spectrum that can be compared directly as two sequences of numbers, the eigenfunctions are continuous functions defined on different domains. The eigenfunctions of $\Delta_{g_\mathcal{X}}$ are defined on $\mathcal{X}$, and similarly, the eigenfunctions of $\Delta_{g_\mathcal{Y}}$ are defined on $\mathcal{Y}$. Therefore, these functions cannot be compared directly. When the shapes are related by an isometric transformation $T:\mathcal{X}\to\mathcal{Y}$, the mapping dictates the relation

$$\phi_i^{\mathcal{Y}} = \phi_i^{\mathcal{X}} \circ T^{-1}. \tag{68}$$

Still, generally we do not assume to have access to this correspondence, when we compare two shapes.

Self functional maps suggest an indirect way to compare the eigenfunctions $\phi_i^{\mathcal{X}}, \phi_i^{\mathcal{Y}}$ of the shapes $\mathcal{X}, \mathcal{Y}$. Assume we have a second *isometry invariant* operator with the respective eigenfunctions $\{\tilde{\phi}_j^{\mathcal{X}}\},\{\tilde{\phi}_j^{\mathcal{Y}}\}$. We can use $\{\tilde{\phi}_j^{\mathcal{X}}\}$ as a basis for scalar functions on the shape $\mathcal{X}$ and similarly use the basis $\{\tilde{\phi}_j^{\mathcal{Y}}\}$ as a basis defined on shape $\mathcal{Y}$. Using the invariant basis functions we can ask whether the representation of $\phi_i^{\mathcal{X}}$ in the basis $\{\tilde{\phi}_j^{\mathcal{X}}\}$ is the same as the representation of $\phi_i^{\mathcal{Y}}$ in the basis $\{\tilde{\phi}_j^{\mathcal{Y}}\}$. It can be shown that if the eigenfunctions $\{\phi_i^{\mathcal{X}}\}$ and $\{\phi_i^{\mathcal{Y}}\}$ are isometry invariant, that is, related under the isometric mapping T, such that

$$\phi_i^{\mathcal{Y}} = \phi_i^{\mathcal{X}} \circ T^{-1}, \quad \forall i \in \mathbb{N}, \tag{69}$$

then, the representation coefficients in the basis functions $\{\tilde{\phi}_j^{\mathcal{X}}\}$ and $\{\tilde{\phi}_j^{\mathcal{Y}}\}$ will be the same. Thus, self functional maps suggest a method to compare the eigenfunctions of the LBO and to reformulate the question "can we hear the shape of a drum?" in terms of the eigenfunctions. The key idea is that we can measure the eigenfunctions of one invariant operator, using the eigenfunctions of a second invariant operator. Before we continue, we would like to show that in the case of isometric shapes the representation coefficients are the same.

Theorem 7. *Let $\mathcal{X}$ and $\mathcal{Y}$ be two isometric shapes, for which there exists an isometric transformation $T:\mathcal{X}\rightarrow\mathcal{Y}$.*

Additionally, let $\{\tilde{\phi}_j^{\mathcal{X}}\}$ and $\{\tilde{\phi}_j^{\mathcal{Y}}\}$ be the orthonormal eigenfunctions of an isometry invariant operator, such that,

$$\tilde{\phi}_j^{\mathcal{Y}} = \tilde{\phi}_j^{\mathcal{X}} \circ T^{-1}, \quad \forall j \in \mathbb{N}. \tag{70}$$

Then, for every i, it holds that $\{a_{ji}\}$ and $\{b_{ji}\}$ are identical series, where $\{a_{ji}\}$ and $\{b_{ji}\}$ are the representation coefficients of the ith LBO eigenfunction, in the basis functions $\{\tilde{\phi}_j^{\mathcal{X}}\}$ and $\{\tilde{\phi}_j^{\mathcal{Y}}\}$, respectively.

$$\phi_i^{\mathcal{X}} = \sum_j a_{ji}\tilde{\phi}_j^{\mathcal{X}}, \quad \forall i \in \mathbb{N} \tag{71}$$

$$\phi_i^{\mathcal{Y}} = \phi_i^{\mathcal{X}} \circ T^{-1} = \sum_j b_{ji}\tilde{\phi}_j^{\mathcal{Y}}, \quad \forall i \in \mathbb{N}. \tag{72}$$

Proof. Substituting Eq. (71) in (72) we get

$$\phi_i^{\mathcal{Y}} = \phi_i^{\mathcal{X}} \circ T^{-1} = \left(\sum_j a_{ji}\tilde{\phi}_j^{\mathcal{X}}\right) \circ T^{-1} = \sum_j a_{ji}\left(\tilde{\phi}_j^{\mathcal{X}} \circ T^{-1}\right), \quad \forall i \in \mathbb{N}, \tag{73}$$

where the last equality is due to linearity. Using the relation (70), the last term can be written as $\sum_j a_{ji}\tilde{\phi}_j^{\mathcal{Y}}$. Therefore,

$$\phi_i^{\mathcal{Y}} = \sum a_{ji} \tilde{\phi}_j^{\mathcal{Y}}, \quad \forall i \in \mathbb{N}. \tag{74}$$

At the other end, according to Eq. (72),

$$\phi_i^{\mathcal{Y}} = \sum_j b_{ji} \tilde{\phi}_j^{\mathcal{Y}}, \quad \forall i \in \mathbb{N}. \tag{75}$$

The orthonormality of the basis $\{\tilde{\phi}_j^{\mathcal{Y}}\}$ requires that

$$a_{ij} = b_{ij}, \quad \forall i, j \in \mathbb{N}^2. \tag{76}$$

□

Self functional maps encode these representation coefficients in a matrix form, for each shape. Thus, for two isometric shapes we expect the self functional maps to be identical, due to the equality we showed above that holds for isometric shapes.

5.1.4 A tale of two metrics

In the previous section we saw that for two isometric shapes, the representation coefficients of LBO eigenfunctions in the basis formed by the eigenfunctions of a second isometric invariant operator, are identical. A natural question that comes in mind is to what extent is the representation discriminative. Similar to the question about the LBO spectrum, we would like to ask whether identical representation guarantees isometry? Taking the case where the second operator is also the LBO, as an extreme example. Then, for each shape $\mathcal{X}$ and for every i it holds that $\tilde{\phi}_i^{\mathcal{X}} = \phi_i^{\mathcal{X}}$. In this case the coefficients are given by $a_{ij} = \delta_{ij}$ (where δ_{ij} is Kronecker delta). Obviously, these coefficients cannot serve as a unique signature. At the other end, if the second operator is significantly different from the LBO, we expect a wider discrete support than a Kronecker delta, namely we expect that in the following representation

$$\phi_i^{\mathcal{X}} = \sum_j a_{ji} \tilde{\phi}_j^{\mathcal{X}}, \quad \forall i \in \mathbb{N} \tag{77}$$

more coefficients will be dominant. Representing a_{ji} in a matrix form, we would expect a rich off-diagonal pattern, if the operators cannot be simultaneously diagonalized. If the pattern has high variation across shapes, we would expect that the only case where two shapes will exhibit the same pattern will be when they are isometric. In other words, we would expect that adversarial examples, if they exist, would be negligible in practice. In Section 5.1.7 we show some empirical evidence for this hypothesis. For now, let us show how different metric definitions can be utilized to construct isometric invariant operators that are significantly different from each other. Taking, for example, the standard Euclidean metric (g_{ij}) and the scale invariant metric $(\tilde{g}_{ij})$ that was introduced in Section 2.3.

Assigning the standard Euclidean metric tensor (g_{ij}) to the surface S, it can be treated as a Riemannian manifold denoted as $\{S, g\}$. For the same surface, we can assign also the scale invariant metric tensor ($\tilde{g}_{ij}$), resulting in a different Riemannian manifold denoted by $\{S, \tilde{g}\}$. Given the Riemannian metric tensor, geometric quantities such as lengths of curves, distances on the manifold, angles and curvatures can be defined on the surface, in terms of the metric tensor. In this example, the relation between the metrics is conformal. As a consequence, for each $p \in S$, the angle between each pair of vectors in the tangent plane T_p, will be the same, no matter if we use the standard metric or the scale invariant metric tensor to measure this angle. Nevertheless, distances on the surface are measured differently for each Riemannian manifold. Each of these two metrics is sensitive to different types of deformations that can be applied to a given surface. Two given shapes can be said to be isometric from a point of view of one metric and nonisometric from a point of view of a different one. To exemplify this statement, consider a sphere of radius one, the pairwise distances on the surface overlap in both geometries, as the geodesic arc-length and the scale invariant arc-length coincide. Scaling the sphere by a uniform factor results in nonisometric surfaces from the regular perspective and isometric surfaces from the scale invariant one. Returning to the original sphere with radius one, assuming the sphere is cut symmetrically and the two halves are glued with a thin stripe to form a spherocylinder with $R = 1, h \to 0$, as shown in Fig. 6. From the Euclidean perspective, when $h \ll 1$, the deformed sphere is almost isometric to the original one as the introduction of the infinitesimal cylinder connecting the two hemispheres has almost no influence on the geodesic distances. In contrast, from the scale invariant point of view, this deformation has a dramatic effect, since on the cylinder part the Gaussian curvature vanishes. The introduction of the infinitesimal cylinder leads to trajectories of zero-length between every pair of points along the equator, effectively shrinking all the points along the equator to a single point. Next, assume the spherocylinder height h is gradually extended.

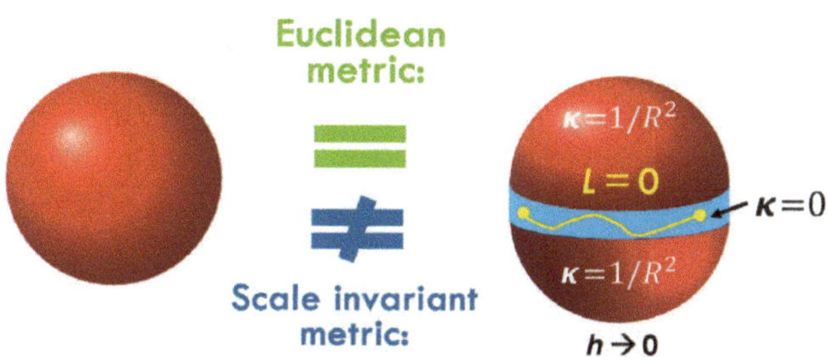

FIG. 6 The introduction of the cylinder between the hemispheres creates a trajectory of zero-length between any pair of points that were originally on the equator, separated apart with a finite distance. From the scale invariant perspective all the points on the equator coincide at a single point, after the transformation of the sphere to a spherocylinder, while from the Euclidean perspective both shapes are isometric when $h \to 0$.

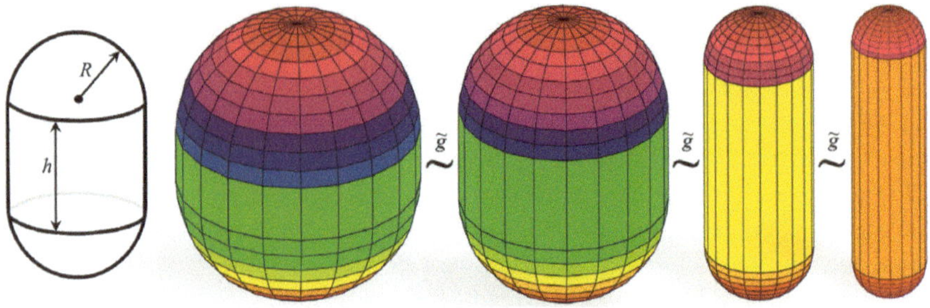

FIG. 7 Spherocylinder. All the following spherocylinders are isometric with respect to the scale invariant metric.

As a consequence, the geodesic distances between pairs of points on different hemispheres would increase when the Euclidean metric is used, resulting in nonisometric family of surfaces. While from the scale invariant point of view, all resulting spherocylinders with varying h are still isometric to one another. In fact, any composition of uniform scaling and cylinder height elongation would result in an isometric surface, from the perspective of the scale-invariant metric and nonisometric from the Euclidean perspective, see Fig. 7.

These examples emphasizes the fact that similarity is influenced by the method used to measure distances on the surface, that is, by the definition of a metric tensor. Considering two different metric tensors can be thought of as two different view points of the same surface, where each is sensitive to different class of deformation.

5.1.5 Introduction to functional maps

Let us put aside for a moment our discussion on self functional maps, which is basically a matrix representation for a surface, and focus on a completely different problem, the problem of nonrigid shape correspondence. In this section we explain the idea of functional maps and show how functional maps were introduced as a framework that addresses the problem of shape correspondence. In the next section we would define properly self functional maps and show how they can be considered as a special case of functional maps.

Functional map (Ovsjanikov et al., 2012) is a tool for transferring functions between surfaces without establishing an explicit point-to-point correspondence between the surfaces. Instead, the functional mapping is constructed using linear constrains, derived from partial knowledge about the mapping. Let $\mathcal{X}$, and $\mathcal{Y}$ be two shapes, relating to each other by the transformation, $T:\mathcal{X}\rightarrow\mathcal{Y}$. Let $\mathcal{F}(\mathcal{X},\mathbb{R})$ be the real function space defined on $\mathcal{X}$ and $\mathcal{F}(\mathcal{Y},\mathbb{R})$ be the real function space defined on $\mathcal{Y}$. $T_F:\mathcal{F}(\mathcal{X},\mathbb{R})\rightarrow\mathcal{F}(\mathcal{Y},\mathbb{R})$ maps every scalar function f defined on $\mathcal{X}$ to its corresponding function $g=f\circ T^{-1}$ defined on $\mathcal{Y}$. T_F is called *the functional representation of the mapping* T. It has been shown that T and T_F can be recovered from each other.

Clearly, if T_F is known, the vertex-correspondence can be calculated by mapping delta functions concentrated at each vertex from one shape to the other. Specifically, it has been shown that if the transformation T is known, at least for some descriptor functions, the functional mapping can be constructed. Assume the orthonormal bases $\{\phi_i^{\mathcal{X}}\}$ and $\{\phi_i^{\mathcal{Y}}\}$ defined on each surface, respectively, the functional maps formalism does not impose a specific choice of basis functions on the shapes $\mathcal{X}$ and $\mathcal{Y}$, though the LBO eigenfunctions are usually used. Suppose a function f is defined on $\mathcal{X}$, with the basis expansion

$$f(x)=\sum_i \hat{f}_i\phi_i^{\mathcal{X}}(x) \quad x\in\mathcal{X}. \tag{78}$$

The mapped function $g = T_F(f) = f \circ T^{-1}$ on $\mathcal{Y}$ can be expanded in the basis $\{\phi_i^{\mathcal{Y}}\}$, as

$$g(y)=\sum_j \hat{g}_j\phi_j^{\mathcal{Y}}(y) \quad y\in\mathcal{Y}, \tag{79}$$

Let us explore the relation between the expansion coefficients $\{\hat{f}_i\}$ and $\{\hat{g}_i\}$. Using the definition of g as the mapping of f and Eq. (78), we can write

$$g=f\circ T^{-1}=\left(\sum_i \hat{f}_i\phi_i^{\mathcal{X}}\right)\circ T^{-1}=\sum_i \hat{f}_i(\phi_i^{\mathcal{X}}\circ T^{-1}), \tag{80}$$

where the last equality is due to linearity. Next, for every i we can expand the function $\phi_i^{\mathcal{X}}\circ T^{-1}$, as a function defined on $\mathcal{Y}$, in the basis $\{\phi_i^{\mathcal{Y}}\}$,

$$\phi_i^{\mathcal{X}}\circ T^{-1}\equiv T_F(\phi_i^{\mathcal{X}})=\sum_j \langle T_F(\phi_i^{\mathcal{X}}),\phi_j^{\mathcal{Y}}\rangle\phi_j^{\mathcal{Y}}, \quad \forall i\in\mathbb{N} \tag{81}$$

substituting Eq. (81) into Eq. (80), we get

$$g=\sum_i \hat{f}_i\left(\sum_j \langle T_F(\phi_i^{\mathcal{X}}),\phi_j^{\mathcal{Y}}\rangle\phi_j^{\mathcal{Y}}\right)=\sum_j\left(\sum_i \langle T_F(\phi_i^{\mathcal{X}}),\phi_j^{\mathcal{Y}}\rangle\hat{f}_i\right)\phi_j^{\mathcal{Y}}. \tag{82}$$

Comparing Eq. (82) and Eq. (79), due to the uniqueness of the representation in an orthonormal basis, we obtain that

$$\hat{g}_j=\sum_i \langle T_F(\phi_i^{\mathcal{X}}),\phi_j^{\mathcal{Y}}\rangle\hat{f}_i=\sum_i C_{ij}\hat{f}_i, \tag{83}$$

where the matrix C is the functional map matrix, defined by

$$C_{ij}\equiv\langle T_F(\phi_i^{\mathcal{X}}),\phi_j^{\mathcal{Y}}\rangle. \tag{84}$$

We observe that the relation between the coefficients $\{\hat{f}_i\}$ and $\{\hat{g}_j\}$ is *linear*. This is a useful result! It means that if we have a function f defined on shape

$\mathcal{X}$ and we want to map it to the shape $\mathcal{Y}$ we should use the following procedure: First we have to decompose f in the basis $\{\phi_i^{\mathcal{X}}\}$, denoting by $\hat{f}$ the vector of coefficients. Then, in order to obtain $\hat{g}$, the vector of coefficients of g in the basis $\{\phi_i^{\mathcal{Y}}\}$, all we have to do is to multiply $\hat{f}$ by the functional map matrix

$$\hat{g} = C^T \hat{f}. \tag{85}$$

The only question that remains is how to find the matrix C. To solve for the matrix C, one needs partial knowledge of specific corresponding functions on the two surfaces. Corresponding functions are functions that preserve their values under the mapping T. The corresponding functions can be provided, for examples, by means of corresponding landmarks, corresponding segments or invariant dense descriptors, such as HKS (Sun et al., 2009), when T is isometry. Each pair of matching functions is translated to a linear constraint. Suppose we have a set of matching functions, defined on both shapes, $\mathcal{X}$ and $\mathcal{Y}$, denoted by $\{f_i : \mathcal{X} \to \mathbb{R}\}$ and $\{g_i : \mathcal{Y} \to \mathbb{R}\}$, stacked as the columns in the matrices $F \in \mathbb{R}^{n_{\mathcal{X}} \times q}, G \in \mathbb{R}^{n_{\mathcal{Y}} \times q}$, respectively, where $n_{\mathcal{X}}, n_{\mathcal{Y}}$ are the number of vertices in each shape and q is the number of corresponding pairs of functions. Each set of functions is projected on the respective basis functions $\{\phi_i^{\mathcal{X}}\}$ and $\{\phi_i^{\mathcal{Y}}\}$

$$\hat{F} = \Phi_{\mathcal{X}}^T F \tag{86}$$

$$\hat{G} = \Phi_{\mathcal{Y}}^T G \tag{87}$$

where $\Phi_{\mathcal{X}} \in \mathbb{R}^{n_{\mathcal{X}} \times k}, \Phi_{\mathcal{Y}} \in \mathbb{R}^{n_{\mathcal{Y}} \times k}$ contains the respective basis functions $\{\phi_i^{\mathcal{X}}\}$ and $\{\phi_i^{\mathcal{Y}}\}$ in their columns, and k is the number of basis functions. Usually, when the LBO eigenfunctions are used as a basis, k has the interpretation of a cut-off frequency. The functional map matrix C is the solution to the following optimization problem, that is designed to maintain all the linear constraints that emerge from each pair of corresponding functions,

$$C = \underset{C}{\arg\min} \| C^T \hat{F} - \hat{G} \|, \tag{88}$$

where the optimization problem can be solved using numeric linear solvers. We need $q \geq k$ independent constraints, so the system will not be under-determined, while due to noisy matching functions its better to have $q > k$. Thus, the requirement for specific knowledge of the point-to-point correspondence is replaced by the relaxed requirement of knowledge about correspondences of functions. Finally, once C is computed we can map any given function from $\mathcal{X}$ to $\mathcal{Y}$.

5.1.6 Self functional maps

Self functional map is a functional map between the surface to itself. Using two different metric tensors g and $\tilde{g}$, the same surface manifold S can be regarded as two different Riemannian manifolds $\{S, g\}$ and $\{S, \tilde{g}\}$, respectively. Next, we can harness the functional maps formulation to define the *self functional map* as the functional map between the two different Riemannian manifolds $\{S, g\}$ and $\{S, \tilde{g}\}$. Dealing with the same surface $\mathcal{X} = \mathcal{Y} = S$ and the transformation T is defined as the trivial map $T : S \to S$ from the surface S to itself, while the basis functions $\{\phi_i\}, \{\tilde{\phi}_i\}$ are defined as the eigenfunctions of the corresponding LBOs calculated with respect to the Riemannian metrics

$$\begin{aligned} -\Delta_g \phi_i &= \lambda_i \phi_i \\ -\Delta_{\tilde{g}} \tilde{\phi}_i &= \tilde{\lambda}_i \tilde{\phi}_i. \end{aligned} \tag{89}$$

The self functional map of the surface S is then defined as

$$C_{ij}^{S,\tilde{S}} \equiv \langle \phi_i^S, \tilde{\phi}_j^S \rangle, \tag{90}$$

where the inner product can be calculated with respect to g or $\tilde{g}$, depending on the direction of the transformation. The self functional map is a fundamental representation of the shape, reflecting the interaction between the eigenfunctions of the LBO calculated on different Riemannian manifolds obtained from the surface that defines the shape. These metrics should be different yet intrinsic in the sense that distances can be computed without relating to the embedding space. One possible way to construct the self functional map is to use the regular metric tensor g_{ij} and the scale invariant metric tensor $\tilde{g}_{ij} = |K| g_{ij}$.

As was shown in previous section, self functional maps are isometry invariant by construction since the eigenfunctions and the inner product are isometry invariant. Therefore, for two isometric surfaces S and Q, it holds that

$$C_{ij}^{Q,\tilde{Q}} \equiv \langle \phi_i^Q, \tilde{\phi}_j^Q \rangle = \langle \phi_i^S, \tilde{\phi}_j^S \rangle \equiv C_{ij}^{S,\tilde{S}}. \tag{91}$$

5.1.7 Self functional maps as surface signatures

In this section we would like to show how self functional maps can be used for the task of surface classification. Assume we have a collection of 3D shapes, we can treat each shape as a vector in a high-dimensional space, by stacking the columns of the self functional maps into a single vector. Defining distance between the vectors, the resulting pairwise distance matrix can be provided as an input to a classic classification algorithm, such as k-medoids. With a proper definition of the distance, the result will be clustering of the shapes, where each cluster corresponds to a set of isometric shapes. To define the distance properly, we should address the computational issues that emerges when the LBO eigenfunctions are calculated independently in each shape. The self functional maps are susceptible to sign ambiguities as the eigenfunctions of the Laplacian operators are defined up to a sign. Thus, the calculated C is determined up to Hadamard multiplication by a specific sign matrix. Additionally, order switching of consecutive eigenfunctions is a common situation when dealing with m, n of low order, while more complicated permutations could appear in higher m, n. Thus, we confine our model to switching between close neighbours. We define *local permutation* as order switching between *adjacent* eigenfunctions, that is, adjacent rows or columns. In order to model these ambiguities we would like to formulate the equivalence relation,

$$C \sim C' \text{ iff } C = M \circ (P_r C' P_c) \tag{92}$$

where $M = ab^T$, $a \in \{-1, 1\}^m$, $b \in \{-1, 1\}^n$ and $P_r \in \Pi_r$, $P_c \in \Pi_c$. $a_{m\times 1}$ represents the sign inversion of the rows, and $b_{n\times 1}$ the sign inversion of the columns. Π_r, Π_c denote, respectively, the set of local permutation matrices applied to the rows and columns of the self functional map C,

To resolve the sign and order ambiguities of the eigenfunctions, the self functional map should be updated by multiplication with rows and columns local permutation matrices Π_r and Π_c, respectively, and by Hadamard multiplication with a sign matrix $M \in \{-1, 1\}^{m\times n}$, namely,

$$\begin{aligned} C_{\text{final}} &= M \circ (P_r C P_c) \\ M &= ab^T, \quad a \in \{-1,1\}^m, \quad b \in \{-1,1\}^n, \\ P_r &\in \Pi_r, \quad P_c \in \Pi_c. \end{aligned} \tag{93}$$

where a represents sign inversion of the rows after rows permutation and b represents sign inversion of the columns after columns permutation.

Since shape association to a class is unknown, the ambiguity can be addressed in one of these approaches: One approach is a universal canonization of each self functional map that can be done for each shape. We can define an optimization problem in the equivalence class of the calculated self functional map and choose the optimal a, b, P_r, P_c for every shape separately. Alternatively, we can define the distance in a way it will vanish between maps in the same equivalence class. Finally, given a pair of shapes, we can find the relative optimal a, b, P_r, P_c between the functional maps, by solving an optimization problem that involves both self functional maps. The ambiguity of the self functional maps is a real challenge that hinders the utilization of the latent information encapsulated in the self functional map. In the paper (Halimi and Kimmel, 2018) the challenge was addressed by defining a robust distance, such that the distance between *locally* permuted maps is zero. The distance was defined as *Adjacent Entries Distance* between them as, $D_{AE}(C, C') =$

$$\left(\min \left(\sum_{i,j} \min_{\substack{i',j' \\ |i'-i|\le 1 \\ |j'-j|\le 1}} \| |C_{ij}| - |C'_{i',j'}| \|, \sum_{i,j} \min_{\substack{i',j' \\ |i'-i|\le 1 \\ |j'-j|\le 1}} \| |C'_{ij}| - |C_{i',j'}| \| \right) \right)^2$$

Though, the last formula is cumbersome, the meaning is pretty simple. First, in order to resolve the sign ambiguity C and C' are taken in absolute value. Then, to calculate a similar measure to L_1 squared distance, that will additionally be robust to local permutations, we adapt the original L_1 distance definition. Instead of comparing C_{ij} to the respective entry C'_{ij}, we take the closest entry in C' within a 3×3 block centred at C'_{ij}. Once we fix the entries in C and search the entries in C', and once we fix the entries in C' and search the entries in C. Since this operation is not symmetric, we take the minimal value. Finally we take the squared value of the resulting "robust L_1 distance." With the new definition of L_1 distance, the pairwise distance matrix D was provided to k-medoids algorithm, resulting in nice classification results for a variety of benchmarks, such as TOSCA, Faust, and SHREC14. Additionally, for *visualization* purpose, shape clustering can be performed by embedding the self functional maps into a Euclidean space $\mathbb{R}^d$ for low d. This was done by the following procedure.

1. The distance matrix between the shapes, $D \in \mathbb{R}_+^{N \times N}$, is calculated using the *Adjacent Entries Distance* between the self functional maps, where N is the number of the shapes in the benchmark

$$D_{ij} = D_{AE}(C_i, C_j) \quad i, j \in \{1 \ldots N\}. \tag{94}$$

2. The distance matrix is used to embed the shapes in the Euclidean space $\mathbb{R}^3$, given as an input to nonmetric MDS algorithm, using stress minimization.

Visualization of the results of the MDS embedding is provided in Figs. 8–10, for few benchmarks. For further details we refer to Halimi and Kimmel (2018).

5.2 Object recognition

The authors of Raviv and Kimmel (2014) experimented with the SHREC 2010 dataset (Bronstein et al., 2010a) using the shapeGoogle framework (Ovsjanikov et al., 2009), while introducing four new categories of deformations; equi-affine, isometry and equi-affine, affine, and a combination

FIG. 8 SHREC'14 synthetic dataset: MDS clustering of the *self functional maps* into $\mathbb{R}^3$, $m = n = 8$. *Colour* indicates ground-truth class. The clusters are well separated in 3D.

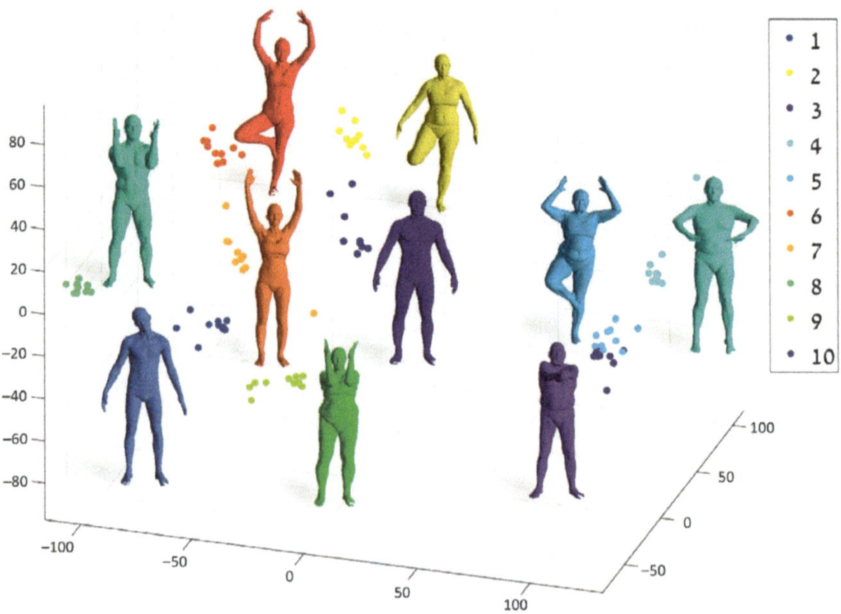

FIG. 9 FAUST: MDS clustering of the *self functional maps* into $\mathbb{R}^3$, $m = n = 8$. *Colour* indicates ground-truth class.

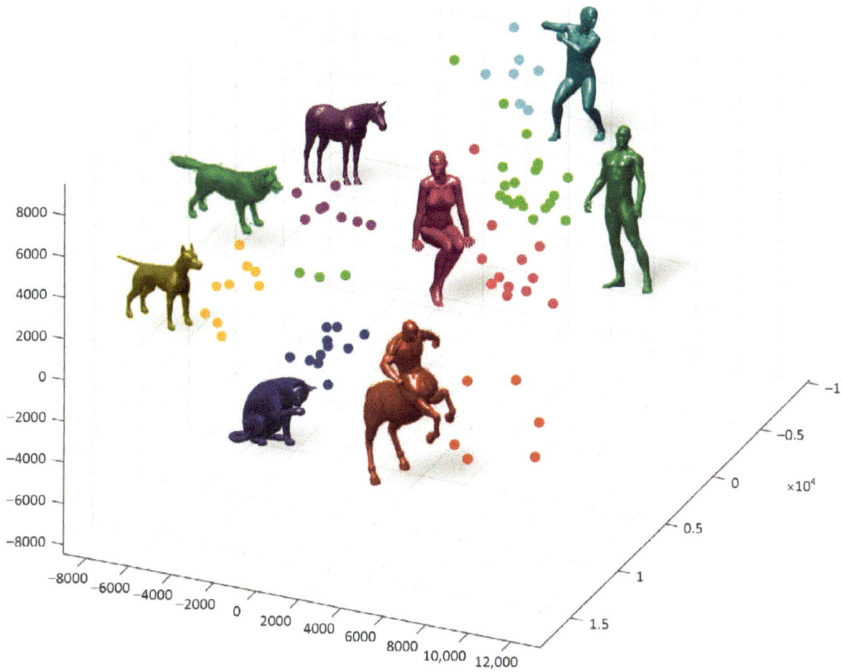

FIG. 10 MDS clustering, using *Adjacent Entries Distance*. $n = m = 100$. *Colour* indicates ground-truth class.

of isometry and affine. Table 1 shows that the new affine metric discriminative power is as good as the Euclidean one, performs well on scaling as the scale invariant metric, and similar to the equi-affine one for handling volume preserving affine transformations. Moreover, the new metric is the only one capable of dealing with full affine deformations. Note that shapes, which were considered to be locally scaled in that database, were in fact treated with an offset operation (morphological erosion) rather than scaling. This explains part of the performance degradation in the local scale examples.

Performance was evaluated using precision/recall characteristics. *Precision* $P(r)$ is defined as the percentage of relevant shapes in the first r top-ranked retrieved shapes. *Mean average precision* (mAP) is defined as $\text{mAP} = \sum_r P(r) \cdot \text{rel}(r)$, where $\text{rel}(r)$ the relevance of a given rank, was used as a single measure of performance. Intuitively, mAP is interpreted as the area below the precision-recall curve. Ideal retrieval ($\text{mAP} = 100\%$) is achieved when all queries provide the correct answer as first match. Performance results are summarized by the transformation class and strength.

TABLE 1 Performance of different metrics with shapeGoogle framework (mAP in %).

	Euclidean invariant metric					Scale invariant metric				
	Strength					*Strength*				
	1	*≤2*	*≤3*	*≤4*	*≤5*	*1*	*≤2*	*≤3*	*≤4*	*≤5*
Isometry	100.00	100.00	100.00	100.00	100.00	100.00	100.00	100.00	100.00	100.00
Micro holes	100.00	100.00	100.00	100.00	100.00	100.00	92.12	84.66	78.90	74.96
Noise	100.00	100.00	100.00	98.72	98.97	100.00	95.00	92.42	87.49	81.62
Shot noise	100.00	100.00	98.72	97.60	97.05	100.00	100.00	97.01	94.94	92.21
Scale	28.87	56.74	61.42	53.75	47.77	100.00	100.00	100.00	100.00	100.00
Local scale	100.00	100.00	98.72	91.79	87.40	100.00	97.44	94.87	90.21	82.98
Equi-affine	100.00	98.08	92.52	87.72	83.05	81.03	72.35	65.82	61.55	59.86
Iso+equi-affine	100.00	97.12	92.95	85.32	79.99	81.03	74.13	63.93	61.17	62.30
Affine	88.46	77.79	70.81	65.21	61.21	100.00	90.87	87.36	84.64	81.99
Iso+affine	88.46	77.82	73.76	68.35	64.60	100.00	93.75	89.64	84.33	83.11

	Equi-affine invariant metric					Affine invariant metric				
	Strength					*Strength*				
	1	*≤2*	*≤3*	*≤4*	*≤5*	*1*	*≤2*	*≤3*	*≤4*	*≤5*
Isometry	100.00	100.00	100.00	100.00	100.00	100.00	100.00	98.72	99.04	99.23
Micro holes	100.00	100.00	100.00	100.00	100.00	100.00	100.00	98.08	97.02	95.53
Noise	100.00	100.00	91.88	83.72	75.62	100.00	95.51	94.02	94.55	94.87
Shot noise	100.00	100.00	100.00	97.12	92.31	100.00	100.00	100.00	99.04	97.69
Scale	29.82	50.81	48.34	43.16	39.89	94.87	97.44	98.29	95.03	92.40
Local scale	100.00	100.00	95.51	87.81	81.08	100.00	100.00	98.29	96.47	92.07
Equi-affine	100.00	100.00	100.00	100.00	100.00	100.00	100.00	96.79	96.63	96.08
Iso+equi-affine	100.00	100.00	100.00	100.00	100.00	100.00	98.08	93.80	94.39	94.36
Affine	75.90	60.35	52.54	47.82	43.86	96.15	95.51	95.73	95.83	96.67
Isometry+affine	75.46	59.59	52.08	47.39	44.34	96.15	98.08	97.44	98.08	98.46

Strength is unit-less.

6 Summary

We explored invariant differential arc-length for curves and extended their construction to surfaces. It allowed us to define novel metric spaces, like the scale invariant diffusion maps, and introduce novel invariant signatures for shapes. The self functional map was suggested as one such signature. The problem of classifying shapes, that can be observed as isometrically similar appearances caused by natural poses, was reformulated as that of comparing matrices. The matrices entries correspond to the inner product of eigenfunctions produced for the same surface manifold, yet each time with a different metric. The example we favour was that of using the LBO once defined with the regular metric and once with the scale invariant one to produce two such bases. In other words, treating the same surface as two different Riemannian manifolds allowed us to translate the shape classification problem into that of classifying matrices. We followed the footsteps of René Descartes in an attempt to establish links between geometry and algebra converting geometric problems into algebraic ones.

We also utilized the fact that in order to deal with similarity one needs to multiply the two simplest invariants of a given group of transformations which does not include scaling. This way, one could extend the Euclidean arc-length for curves into a scale invariant one, and the equi-affine into an affine one. A similar construction applies to surfaces. We have also demonstrated a theoretical background and a simple numerical way to construct the Gaussian curvature for any given metric and then utilize it to handle scaling effects.

In the specific case of surfaces, the scale invariant metric established by multiplying the Euclidean one by the Gaussian curvature is intrinsic. Same property applies to the affine invariant one. In both examples, the conformal construction of the metric by which the LBO was defined enabled us to bound the number of required derivatives for the very relevant cases of two-dimensional manifolds.

With the tools and example applications provided in this chapter we hope the interested reader would be able to extend existing shape analysis methodologies and take existing supervised learning techniques into the unsupervised arena. That is, using the well-established theoretical knowledge we have about shape invariants in order to train artificial neural networks without the need for a large amount of data when such does not exist.

Acknowledgements

R.K. wishes to thank Prof. Giovanni Felder of the Institute of Theoretical Studies at ETH Zurich for his enlightening comment about the magic of conformal metrics and the Laplacian operator of two-dimensional manifolds during his visit to the institute in April 2017. O.H. is funded by Israel Ministry of Science and Technology.

References

Aflalo, Y., Kimmel, R., 2013. Spectral multidimensional scaling. Proc. Natl. Acad. Sci. 110 (45), 18052–18057.

Aflalo, Y., Kimmel, R., Raviv, D., 2013a. Scale invariant geometry for nonrigid shapes. SIAM J. Imaging Sci. 6 (3), 1579–1597.

Aflalo, Y., Kimmel, R., Zibulevsky, M., 2013b. Conformal mapping with as uniform as possible conformal factor. SIAM J. Imag. Sci. 6 (1), 78–101.

Aflalo, Y., Brezis, H., Kimmel, R., 2015. On the optimality of shape and data representation in the spectral domain. SIAM J. Imag. Sci. 8 (2), 1141–1160.

Aflalo, Y., Brezis, H., Bruckstein, A., Kimmel, R., Sochen, N., 2016a. Best bases for signal spaces. C.R. Math. 354 (12), 1155–1167.

Aflalo, Y., Dubrovina, A., Kimmel, R., 2016b. Spectral generalized multi-dimensional scaling. Int. J. Comput. Vis. 118 (3), 380–392.

Alvarez, L., Guichard, F., Lions, P.-L., Morel, J.-M., 1993. Axioms and fundamental equations of image processing. Arch. Ration. Mech. Anal. 123 (3), 199–257.

Banchoff, T.F., 1970. Critical points and curvature for embedded polyhedral surfaces. Am. Math. Mon. 77 (5), 475–485.

Bay, H., Tuytelaars, T., Van Gool, L., 2006. SURF: speeded up robust features. Lect. Notes Comput. Sci 3951, 404.

Ben-Chen, M., Gotsman, C., 2008. Characterizing shape using conformal factors. In: 3DOR, pp. 1–8.

Bérard, P., Besson, G., Gallot, S., 1994. Embedding Riemannian manifolds by their heat kernel. Geom. Funct. Anal. GAFA 4 (4), 373–398.

Blaschke, W., 1923. Vorlesungen Uber Differentialgeometrie und Geometrische Grundlagen Von Einsteins Relativitatstheorie. vol. 2 Springer.

Brezis, H., Gómez-Castro, D., 2017. Rigidity of optimal bases for signal spaces. C.R. Math. 355 (7), 780–785.

Bronstein, A.M., Bronstein, M.M., Kimmel, R., 2005a. Expression-invariant face recognition via spherical embedding. In: International Conference on Image Processing (ICIP), vol. 3, pp. 756–759.

Bronstein, A.M., Bronstein, M.M., Kimmel, R., 2005b. Three-dimensional face recognition. Int. J. Comput. Vis. 64 (1), 5–30.

Bronstein, A.M., Bronstein, M.M., Kimmel, R., 2006. Generalized multidimensional scaling: a framework for isometry-invariant partial surface matching. Proc. Natl. Acad. Sci. 103 (5), 1168–1172.

Bronstein, A.M., Bronstein, M.M., Castellani, U., Falcidieno, B., Fusiello, A., Godil, A., Guibas, L.J., Kokkinos, I., Lian, Z., Ovsjanikov, M., Patané, G., Spagnuolo, M., Toldo, R., 2010a. SHREC 2010: robust large-scale shape retrieval benchmark. In: Proc. Workshop on 3D Object Retrieval (3DOR).

Bronstein, A.M., Bronstein, M.M., Kimmel, R., Mahmoudi, M., Sapiro, G., 2010b. A Gromov-Hausdorff framework with diffusion geometry for topologically-robust non-rigid shape matching. Int. J. Comput. Vis. 89 (2–3), 266–286.

Bruckstein, A.M., Katzir, N., Lindenbaum, M., Porat, M., 1992. Similarity-invariant signatures for partially occluded planar shapes. Int. J. Comput. Vis. 7 (3), 271–285.

Bruckstein, A.M., Holt, R.J., Netravali, A.N., Richardson, T.J., 1993. Invariant signatures for planar shape recognition under partial occlusion. Comput. Vis. Image Underst. 58 (1), 49–65.

Bruckstein, A.M., Rivlin, E., Weiss, I., 1997. Scale space semi-local invariants. Image Vis. Comput. 15 (5), 335–344.

Burago, D., Burago, Y., Ivanov, S.D., Ivanov, S., 2001. A Course in Metric Geometry. vol. 33 American Mathematical Society.

Buser, P., Conway, J., Doyle, P., Semmler, K.-D., 2010. Some planar isospectral domains. arXiv preprint arXiv:1005.1839.

Cohignac, T., Lopez, C., Morel, J.M., 1994. Integral and local affine invariant parameter and application to shape recognition. In: Proceedings of 12th International Conference on Pattern Recognition.

Coifman, R.R., Lafon, S., 2006. Diffusion maps. Appl. Comput. Harmon. Anal. 21 (1), 5–30.

Coifman, R.R., Lafon, S., Lee, A.B., Maggioni, M., Nadler, B., Warner, F., Zucker, S.W., 2005. Geometric diffusions as a tool for harmonic analysis and structure definition of data: diffusion maps. Proc. Natl. Acad. Sci. U. S. A. 102 (21), 7426–7431.

Cosmo, L., Panine, M., Rampini, A., Ovsjanikov, M., Bronstein, M.M., Rodolà, E., 2018. Isospectralization, or how to hear shape, style, and correspondence. arXiv preprint arXiv: 1811.11465.

do Carmo, M.P., 1976. Differential Geometry of Curves and Surfaces. Prentice-Hall.

Donoser, M., Bischof, H., 2006. Efficient maximally stable extremal region (MSER) tracking. In: Conference on Computer Vision and Pattern Recognition (CVPR), IEEE, pp. 553–560.

Elad, A., Kimmel, R., 2003. On bending invariant signatures for surfaces. IEEE Trans. Pattern Anal. Mach. Intell. 25 (10), 1285–1295.

Elad, A., Keller, Y., Kimmel, R., 2005. Texture mapping via spherical multi-dimensional scaling. In: Proc. Scale-Space Theories in Computer Vision, pp. 443–455.

Goldman, R., 2005. Curvature formulas for implicit curves and surfaces. Comput. Aided Geom. Des. 22 (7), 632–658.

Gordon, C., Webb, D., Wolpert, S., 1992. Isospectral plane domains and surfaces via Riemannian orbifolds. Invent. Math. 110 (1), 1–22.

Gray, A., Abbena, E., Salamon, S., 2006. Modern Differential Geometry of Curves and Surfaces with Mathematica, third ed. Chapman and Hall/CRC.

Gu, X.D., Yau, S.-T., 2008. Computational Conformal Geometry. International Press of Boston.

Halimi, O., Kimmel, R., 2018. Self functional maps. In: 2018 International Conference on 3D Vision (3DV), IEEE, pp. 710–718.

Halimi, O., Litany, O., Rodola, E., Bronstein, A.M., Kimmel, R., 2019. Unsupervised learning of dense shape correspondence. In: Proceedings of the IEEE Conference on Computer Vision and Pattern Recognition, pp. 4370–4379.

Hamza, A., Krim, H., 2006. Geodesic matching of triangulated surfaces. IEEE Trans. Image Process. 15 (8), 2249–2258.

Kimmel, R., Zhang, C., Bronstein, A.M., Bronstein, M.M., 2011. AreMSER features really interesting? IEEE Trans. Pattern Anal. Mach. Intell. 33 (11), 2316–2320.

Lipman, Y., Daubechies, I., 2011. Conformal Wasserstein distances: comparing surfaces in polynomial time. Adv. Math. 227 (3), 1047–1077.

Lipman, Y., Funkhouser, T., 2009. Möbius voting for surface correspondence. In: Proc. ACM Transactions on Graphics (SIGGRAPH), vol. 28.

Lowe, D., 2004. Distinctive image features from scale-invariant keypoint. Int. J. Comput. Vis. 60 (2), 91–110.

Masi, I., Tran, A.T., Hassner, T., Leksut, J.T., Medioni, G., 2016. Do we really need to collect millions of faces for effective face recognition? In: European Conference on Computer Vision, pp. 579–596.

Mémoli, F., 2011. Gromov-Wasserstein distances and the metric approach to object matching. Found. Comput. Math. 11 (4), 417–487.

Mémoli, F., Sapiro, G., 2005. A theoretical and computational framework for isometry invariant recognition of point cloud data. Found. Comput. Math. 5 (3), 313–347.

Moons, T., Pauwels, E.J., Gool, L.J.V., Oosterlinck, A., 1995. Foundations of semi-differential invariants. Int. J. Comput. Vis. 14, 25–48.

Morel, J.M., Yu, G., 2009. ASIFT: a new framework for fully affine invariant image comparison. SIAM J. Imag. Sci. 2, 438–469.

Olver, P.J., Sapiro, G., Tannenbaum, A.R., 1996. Affine invariant detection: edges, active contours, and segments. In: 1996 Conference on Computer Vision and Pattern Recognition (CVPR'96), June 18–20, 1996, San Francisco, CA, USA, pp. 520–525.

Ovsjanikov, M., Bronstein, A.M., Bronstein, M.M., Guibas, L.J., 2009. ShapeGoogle: a computer vision approach to invariant shape retrieval. In: Proc. Non-Rigid Shape Analysis and Deformable Image Alignment (NORDIA).

Ovsjanikov, M., Mérigot, Q., Mémoli, F., Guibas, L.J., 2010. One point isometric matching with the heat kernel. In: Proc. Symposium on Geometry Processing(SGP), vol. 29, pp. 1555–1564.

Ovsjanikov, M., Ben-Chen, M., Solomon, J., Butscher, A., Guibas, L., 2012. Functional maps: a flexible representation of maps between shapes. ACM Trans. Graph.(TOG) 31 (4), 30.

Pauwels, E.J., Moons, T., Van Gool, L.J., Kempenaers, P., Oosterlinck, A., 1995. Recognition of planar shapes under affine distortion. Int. J. Comput. Vis. 14, 49–65.

Pinkall, U., Polthier, K., 1993. Computing discrete minimal surfaces and their conjugates. Exp. Math. 2 (1), 15–36.

Polthier, K., Schmies, M., 1998. Straightest geodesics on polyhedral surfaces. In: Hege, H.C., Polthier, K. (Eds.), Mathematical Visualization. Springer, Berlin, Heidelberg.

Qiu, H., Hancock, E.R., 2007. Clustering and embedding using commute times. IEEE Trans. Pattern Anal. Mach. Intell. 29 (11), 1873–1890.

Raviv, D., Kimmel, R., 2014. Affine invariant geometry for non-rigid shapes. Int. J. Comput. Vis. 111 (1), 1–11.

Raviv, D., Raskar, R., 2015. Scale invariant metrics of volumetric datasets. SIAM J. Imag. Sci. 8 (1), 403–425.

Raviv, D., Bronstein, A.M., Bronstein, M.M., Kimmel, R., Sochen, N., 2011a. Affine-invariant diffusion geometry of deformable 3D shapes. In: Proc. Computer Vision and Pattern Recognition (CVPR), pp. 2361–2367.

Raviv, D., Bronstein, A.M., Bronstein, M.M., Kimmel, R., Sochen, N., 2011b. Affine-invariant geodesic geometry of deformable 3D shapes. Comput. Graph. 35 (3), 692–697.

Raviv, D., Bronstein, A.M., Bronstein, M.M., Waisman, D., Sochen, N., Kimmel, R., 2014. Equi-affine invariant geometry for shape analysis. J. Math. Imaging Vision (JMIV) 50 (1–2), 144–163.

Reuter, M., Wolter, F.-E., Peinecke, N., 2006. Laplace-Beltrami spectra as 'shape-DNA' of surfaces and solids. Comput. Aided Des. 38 (4), 342–366.

Richardson, E., Sela, M., Kimmel, R., 2016. 3D face reconstruction by learning from synthetic data. In: 2016 Fourth International Conference on 3D Vision (3DV), IEEE, pp. 460–469.

Richardson, E., Sela, M., Or-El, R., Kimmel, R., 2017. Learning detailed face reconstruction from a single image. In: 2017 IEEE Conference on Computer Vision and Pattern Recognition (CVPR), IEEE, pp. 5553–5562.

Roufosse, J.-M., Ovsjanikov, M., 2018. Unsupervised deep learning for structured shape matching. arXiv preprint arXiv:1812.03794.

Sela, M., Richardson, E., Kimmel, R., 2017. Unrestricted facial geometry reconstruction using image-to-image translation. In: 2017 IEEE International Conference on Computer Vision (ICCV), pp. 1585–1594.

Shamai, G., Kimmel, R., 2017. Geodesic distance descriptors. In: IEEE Conference on Computer Vision and Pattern Recognition (CVPR), Hawaii, Honolulu.

Spivak, M., 1970. A Comprehensive Introduction to Differential Geometry. Publish or Perish Inc.

Su, B., 1983. Affine Differential Geometry. Science Press, Beijing, China. ISBN: 0677310609.

Sun, J., Ovsjanikov, M., Guibas, L.J., 2009. A concise and provably informative multi-scale signature based on heat diffusion. In: Proc. Symposium on Geometry Processing (SGP).

Wang, S., Wang, Y., Jin, M., Gu, X.D., Samaras, D., 2007. Conformal geometry and its applications on 3D shape matching, recognition, and stitching. IEEE Trans. Pattern Anal. Mach. Intell. 29 (7), 1209–1220.

Wang, Y., Gupta, M., Zhang, S., Wang, S., Gu, X., Samaras, D., Huang, P., 2008. High resolution tracking of non-rigid motion of densely sampled 3D data using harmonic maps. Int. J. Comput. Vis. 76 (3), 283–300.

Zelditch, S., 2000. Spectral determination of analytic bi-axisymmetric plane domains. Geom. Funct. Anal. GAFA 10 (3), 628–677.

Chapter 8

Solving PDEs on manifolds represented as point clouds and applications

Rongjie Lai[a] and Hongkai Zhao[b,*]
[a]*Department of Mathematics, Rensselaer Polytechnic Institute, Troy, NY, United States*
[b]*Department of Mathematics, University of California, Irvine, CA, United States*
[*]*Corresponding author: e-mail: zhao@math.uci.edu*

Chapter Outline

Abstract

This paper serves as a review of our series work of solving partial differential equations (PDEs) on manifolds represented as point clouds and using their solutions to conduct geometric understanding of point clouds. We first review our two systemic methods of discretizing PDEs including the moving least square method and the local mesh method. These methods of approximating differential operator on manifold-structured point clouds are based only on local approximation using nearest neighbours and achieve high order numerical convergence of the desired equations including diffusion and nondiffusion types of equations. We further discuss extensions of these methods to approximate the committor function for understanding dynamic systems and to solve PDEs on manifolds represented as incomplete inter-point distance information by

Handbook of Numerical Analysis, Vol. 20. https://doi.org/10.1016/bs.hna.2019.07.006
© 2019 Elsevier B.V. All rights reserved.

combining with low-rank matrix completion theory. As direct applications, we use solutions of PDEs on manifold-structured point clouds as bridges to link local and global information. With this strategy, we discuss a few key applications essential to geometric understanding for point clouds, including skeletons extraction from point clouds, the conformal structures construction from point clouds, and nonrigid intrinsic registration among point clouds.

Keywords: Manifold, Point cloud, Moving least square, Local mesh, Laplace–Beltrami eigenproblem, Eikonal equation, Fokker–Planck equation, Conformal mapping, Registration

AMS Classification Codes: 85D18, 65D25, 65D30, 65N25

1 Introduction

With the advance of modern technology and computing power, processing and analyzing of data in three and higher dimensions become a ubiquitous task in diverse fields in science and engineering. In 3D modelling, the representation of geometric entities, such as shapes and surfaces, is mostly collected as point clouds for raw data although the majority of them are represented by triangular meshes specifying both points and connectivity in practice. In applications as data mining or machine learning, data are usually represented in high dimensions for which laying down a mesh or grid is often impossible. Hence point cloud is the only feasible way to represent the data. For many problems in practice, data are commonly associated with certain coherent and nonlinear structure. Mathematically, this enables us to model data sets as points sampled on manifolds usually of low dimensions embedded in a possible high dimensional ambient space (Levina and Bickel, 2004; Lin and Zha, 2008; Little et al., 2009b). Different from image and signal processing which handle functions on flat domains with well-developed tools for processing and learning, manifold-structured data sets are far more challenging due to their complicated geometry. Hence, it is a crucial but challenging task to extract global information and structure directly from point clouds in many applications. Mathematically and computationally one can obtain rich intrinsic information by studying the behaviour of differential equations, such as heat equation, or eigenvalue problem for differential operators, such as Laplace–Beltrami operator, on manifolds (Belkin and Niyogi, 2002; Belkin et al., 2009; Bronstein et al., 2010; Coifman and Lafon, 2006; Lai et al., 2010; Levy, 2006; Raviv et al., 2011; Sun et al., 2009).

Classical finite difference methods (Meyer et al., 2002; Pinkall and Polthier, 1993; Taubin, 2000; Xu, 2004), finite element methods (Dziuk and Elliott, 2013), parametrization methods (Carr et al., 2006; Goldluecke and Cremers, 2009; Lui et al., 2005), and implicit methods (Bertalmio et al., 2000, 2002; Osher and Sethian, 1988) are only applicable for solving differential equations on manifolds with global triangulation or limited by the dimension of the embedded space. There has been increasing interests of solving partial differential equations (PDEs) on general d-dimensional manifolds in $\mathbb{R}^p$, where point cloud representation is the only feasible way for discretization. In the literature, a few methods have been proposed for solving differential equations directly on point clouds without a global mesh or grid. These methods are particular useful in high dimensions where a global triangulation or grid is intractable. There are three types of methods in this class. One type is kernel-based methods, where the elliptic operators (like the LB operator or the Fokker–Planck operator) on point clouds are approximated by heat diffusion in the ambient Euclidean space (Coifman and Lafon, 2006) or in the tangent space (Belkin et al., 2009) among nearby points. In other words, the metric on the manifold is approximated by Euclidean metric locally. This type of method does not need to approximate differential operators directly. The main advantage of such methods, e.g., diffusion map method (Coifman and Lafon, 2006), is their simplicity and generality. It can be applied to point clouds even not necessarily embedded in a metric space, e.-g., graph Laplacian. The resulting linear system is an M-matrix, i.e., satisfying discrete maximum principle, but not symmetric. The other method is called the point integral method (Li and Shi, 2016; Li et al., 2017). It essentially converts differential equations to integral equations using specially designed kernel functions. This method also enjoys flexibility to apply on general data sets where it may lose the convergence and geometric meaning. Both of aforementioned methods render low order approximation and are only limited to the approximation of the diffusion types of differential equations on point clouds. Recently, we propose a moving least square-based method to mimic the finite difference method for solving PDEs on point clouds (Liang and Zhao, 2013; Liang et al., 2012), where systematic methods are introduced to approximate differential operators intrinsically at each point from local approximation of the manifold and its metric by moving least squares through nearby neighbours. It also has the flexibility to design discretized differential system satisfying discrete maximum principle through point-wise constrained moving least square. In addition, we also propose a local mesh methods which relies only on local connectivity constructed through the tangent space at each point (Lai et al., 2013). This method can mimic the finite element method for solving elliptic types of equations and also solve nondiffusion types of equations such as the Eikonal equation by adapting existing fast algorithms. Both of our methods can achieve higher order accuracy than the diffusion map method and point integral method. They also enjoy the flexibility to approximate general differential operators on point clouds sampling manifolds with arbitrary dimensions and codimensions. Moreover, the computational complexity depends mainly on the true dimension of the manifold rather than the dimension of the embedded space. However, the computational cost for local reconstruction of the manifold and its metric based on moving least squares grows quickly with the dimension of the manifold.

Besides solving differential equations on manifold-structured point clouds, it is also very useful to use the tool to understand the geometry of the underlying manifold. This task of global structure analysis can be accomplished by borrowing ideas from differential geometry. In practice, it is relatively easy to have local approximation of a given point cloud but quite challenging to extract global information due to the lack of global connectivity/parametrization of the point cloud. While differential geometers have already explored possible ways to patch together local information using differential equations defined on manifolds. A lot of intrinsic and global information can be obtained by studying behaviours of differential equations, such as the Eikonal equation, the heat equation, or eigenvalue problems of the Laplace–Beltrami operator, on manifolds (Belkin and Niyogi, 2002; Belkin et al., 2009; Bronstein et al., 2010; Coifman and Lafon, 2006; Lai et al., 2010; Levy, 2006; Raviv et al., 2011; Sun et al., 2009). On the other hand, local geometry and connectivity at each point can be relatively easy to obtain using local approximation, which enables us to approximate differential operators point-wisely. This motivates us to propose geometric PDE-based methods to understand point clouds. Namely, we use PDEs to connect "dots" and piece together local information to extract global information.

This paper is served as a review for our previous work on solving PDEs on manifolds represented as point clouds and their applications (Lai and Li, 2018; Lai and Lu, 2017; Lai and Zhao, 2017; Lai et al., 2013; Liang and Zhao, 2013; Liang et al., 2012). The rest of this paper is organized as follows. In Section 2, we review our methods of solving PDEs on manifold-structured point clouds based on the moving least square method and the local mesh method. After that, we discuss extensions of these methods to handle the Fokker–Planck equation on point clouds sampled on dynamic system in Section 3, and to solve manifolds represented as incomplete inter-point distance extending recent advances of low-rank matrix completion theory in Section 4. The applications of using solution of geometric PDEs to understand and analyze point clouds are discussed in Section 5, where applications to extraction of skeleton, construction of conformal structure, and nonrigid point clouds registration are reviewed. We conclude the paper in Section 6.

2 Solving PDEs on manifolds represented as point clouds

In this section, we will review our recent work of solving PDEs on point clouds discussed in Liang et al. (2012), Liang and Zhao (2013), and Lai et al. (2013), where two systematic methods are introduced to approximate differential operators intrinsically at each point from a local approximation of the manifold and its metric by moving least squares or local mesh reconstruction through nearby neighbours. These methods achieve high order accuracy and are capable of approximating general differential operators (possible nondiffusion type) on point clouds sampling manifolds with low dimensions and possible high codimensions.

2.1 Moving least square methods

We assume that the input point cloud $\{\boldsymbol{p}_k \in \mathbb{R}^p \mid k = 1, \ldots, \ell\}$ is sampled from a d-dimensional manifold $\mathcal{M}$ in $\mathbb{R}^p$. We write the index set of the K-nearest neighbourhood(KNN) of the ith point as $\mathcal{I}(i)$ and let $\mathcal{P}(i) = \{\boldsymbol{p}_k \in \mathbb{R}^p \mid k \in \mathcal{I}(i)\}$. Note that k should be chosen to balance the accuracy and robustness of local approximation. For good accuracy, k should be small. For example, it should be enough to choose k a little larger than the degrees of freedom, i.e., the number of unknowns, in a local moving least square construction, for clean and well sampled data. On the other hand, large k will improve robustness of local construction when there is noise in the data. In our experiments, the size of $\mathcal{I}(i)$ is fixed as a constant less than 30. A more adaptive way is to choose $\mathcal{I}(i)$ according to local sampling and noise level. Based on $\mathcal{P}(i)$, its tangent space and normal space can be determined by standard principle component analysis (PCA) (Jolliffe, 2002), which is provided by the eigensystem of $Cov_i = \sum_{k \in \mathcal{I}(i)} (\boldsymbol{p}_k - \boldsymbol{c}_i)^T (\boldsymbol{p}_k - \boldsymbol{c}_i)$ with $\boldsymbol{c}_i = \frac{1}{|\mathcal{I}(i)|} \sum_{k \in \mathcal{I}(i)} \boldsymbol{p}_k$. As the intrinsic dimension of the manifold is d, if the local sampling is dense enough to resolve local features, the gap of Cov_i's eigenvalues guides the splitting $\mathbb{R}^p = \mathcal{T}_i \oplus \mathcal{N}_i$. Here $\mathcal{T}_i$ represents the tangent space spanned by $\{e_i^1, \ldots, e_i^d\}$ corresponding to the d largest eigenvectors of Cov_i, and $\mathcal{N}_i$ represents the normal space spanned by the rest of the eigenvectors of Cov_i. For a noisy data set, a technique called multiscale SVD method (Little et al., 2009a) is an effective way to estimate the intrinsic dimension. In the local coordinate system centred at $\boldsymbol{p}_i$ with orthogonal basis $\{e_i^1, \ldots, e_i^p\}$, a new *virtual coordinates* of $\boldsymbol{p}_k \in \mathcal{P}(i)$ can be obtained by $(\boldsymbol{u}_k, \boldsymbol{v}_k) = (\{\langle \boldsymbol{p}_k - \boldsymbol{p}_i, e_i^\alpha \rangle\}_{\alpha=1}^d, \{\langle \boldsymbol{p}_k - \boldsymbol{p}_i, e_i^\beta \rangle\}_{\beta=d+1}^p)$. Therefore, the manifold near $\boldsymbol{p}_i$ can be parametrized locally by the tangent plane and approximated by polynomials, $\boldsymbol{Z}_i : \mathbb{R}^d \to \mathbb{R}^{p-d}, \boldsymbol{Z}_i(\boldsymbol{u}) = (q_1(\boldsymbol{u}), \ldots, q_{p-d}(\boldsymbol{u}))$, where q_j, $j = 1, \ldots, p - d$, are polynomials of $\boldsymbol{u}$ constructed by the following moving least square formulation (Levin, 1998; Lipman, 2009):

$$\min_{q_j} \sum_{k \in \mathcal{I}(i)} w(\| \boldsymbol{p}_k - \boldsymbol{p}_i \|) \| q_j(\boldsymbol{u}_k) - v_k^j \|^2, \quad j = 1, \ldots, p - d \tag{1}$$

where $\boldsymbol{v}_k = (v_k^1, \ldots, v_k^{p-d})$. Two popular weight functions used in MLS are inverse of squared distance function $1/(x^2 + \varepsilon^2)$, and the Wendland function $w(x) = \left(1 - \frac{x}{L}\right)^4 \left(\frac{4x}{L} + 1\right)$ with $x \in [0, L]$. These two weight functions work fine for well-sampled point clouds but may be unstable for quite nonuniform ones. On the other hand, the special weight function $w(d) = \begin{cases} 1 & \text{if } d = 0 \\ 1/K & \text{if } d \neq 0 \end{cases}$ introduced in Liang et al. (2012) works better in our numerical experiments for more general data sets. We remark that the weight function in MLS does not affect the approximation order as shown in Liang and Zhao (2013) although the stability can be affected.

Similarly, a function f or a vector field $\boldsymbol{V}$ defined on the manifold $\mathcal{M}$ sampled by the point cloud $\boldsymbol{p}_k$ can also be approximated locally by polynomial(s) $f_i(\boldsymbol{u})$ or $\boldsymbol{V}_i(\boldsymbol{u})$ in the local coordinate system centred at $\boldsymbol{p}_i$. Once local polynomial approximation of the manifold and function around $\boldsymbol{p}_i$ has been constructed, the metric tensor $G = (g_{st})$ at $\boldsymbol{p}_i$ can be approximated by $g_{st} = \delta_{st} + \left\langle \frac{\partial Z_i}{\partial u_s}, \frac{\partial Z_i}{\partial u_t} \right\rangle$, which then enables us to approximate intrinsically defined differential operators at $\boldsymbol{p}_i$ including:

$$\nabla_{\mathcal{M}} f = g^{st} \frac{\partial f}{\partial u_t} \partial_{u_s}, \quad \text{div}_{\mathcal{M}} \boldsymbol{V} = \frac{1}{\sqrt{g}} \frac{\partial}{\partial u_s} \left(\sqrt{g} \ V^s \right), \quad \Delta_{\mathcal{M}} f = \frac{1}{\sqrt{g}} \frac{\partial}{\partial u_s} \left(\sqrt{g} g^{st} \frac{\partial f}{\partial u_t} \right). \tag{2}$$

where $g = det(G)$, $(g^{st}) = G^{-1}$ and the Einstein summation is used in the above formulas. Other differential operators can also be approximated similarly. Once these differential operators can be approximated, finite difference type of methods can be used to solve differential equations on point clouds.

Theoretically, if $f(\boldsymbol{u})$ is approximated by a degree m polynomial using the above moving least square formulation, it can be shown that local truncation error for approximation of the nth derivatives of f is of order $m+1-n$ in terms of the pairwise distance among local points, which leads to consistency of the above numerical schemes (Liang and Zhao, 2013; Lipman et al., 2006). If points are evenly distributed, super-convergence, i.e., of order $m + 2 - n$, can be achieved if m and n has the same parity (Liang and Zhao, 2013). In addition, using least square approximation instead of exact interpolation allows the flexibility to include certain properties of the differential operator after discretization. For examples, for an elliptic or a parabolic PDE, such as Laplace equation or heat equation, maximum principle is a very important property that should be preserved after discretization. This can be realized by recasting the moving least square problem as a constrained quadratic optimization problem where desired properties such as maximum principle preservation can be achieved (Liang and Zhao, 2013). We also would like to point out that theoretical stability analysis of these methods is still open due to the unstructured nature of the point clouds. More stable modified virtual grid difference for discretizing the Laplace-Beltrami operator on point clouds was developed in Wang et al. (2018).

2.2 Local mesh method

The local mesh method proposed in Lai et al. (2013) is another approach to approximate differential operators based on a local mesh in the tangent plane. For convenience, we take point clouds sampled on 2-dimensional manifolds as an examples. More specifically, at point $\boldsymbol{p}_i$, a local connectivity is constructed for the projected image of $\mathcal{P}(i)$ on $\mathcal{T}_i$ provided by $\hat{\boldsymbol{p}}_k = \text{Proj}_{\mathcal{T}_i}(\boldsymbol{p}_k) = \boldsymbol{p}_k - \boldsymbol{p}_i - \sum_{\alpha=d+1}^{p} \langle \boldsymbol{p}_k - \boldsymbol{p}_i, \boldsymbol{e}_i^\alpha \rangle \boldsymbol{e}_i^\alpha, \ k \in \mathcal{I}(i)$. Then, the local mesh structure near $\boldsymbol{p}_i$ can be obtained by the standard Delaunay triangulation. Denote by $\mathcal{R}(i) = \{T_i^1, \ldots, T_i^{l_i}\}$ all simplexes adjacent to $\boldsymbol{p}_i$, referred as $\boldsymbol{p}_i$'s first ring and $\mathcal{V}(i)$ all vertices in the first ring of $\boldsymbol{p}_i$. The local connectivity of $\boldsymbol{p}_i$ is provided by $\mathcal{C}_i = \{\boldsymbol{p}_i; \mathcal{V}(i), \mathcal{R}(i)\}$. Fig. 1 illustrates an example of a set of point sampled in a 2D manifold in $\mathbb{R}^3$. Within the first ring $\mathcal{R}(i)$ of $\boldsymbol{p}_i$, the differential operators at each point can be approximated by the following weighted average on the first ring of $\boldsymbol{p}_i$ in terms of triangle areas. Namely, for any function f and vector field $V^{\rightarrow}$ defined on the given point cloud $\mathcal{P}$, the gradient and divergence operators at each point are approximated as follows:

$$\nabla_{\mathcal{P}} f(p_i) = \frac{\sum_{T \in \mathcal{R}(i)} Area(T) \nabla_T f(p_i)}{\sum_{T \in \mathcal{R}(i)} Area(T)} \qquad \mathrm{div}_{\mathcal{P}} \vec{V}(p_i) = \frac{\sum_{T \in \mathcal{R}(i)} Area(T) \mathrm{div}_T \vec{V}(p_i)}{\sum_{T \in \mathcal{R}(i)} Area(T)} \tag{3}$$

In addition, we approximate the Laplace–Beltrami operator as: $\Delta_{\mathcal{P}} f(\boldsymbol{p}_i) \approx \sum_{j \in \mathcal{V}(i)} \omega_{ij}(p_i) \left(f(\boldsymbol{p}_j) - f(\boldsymbol{p}_i) \right)$, with the cotangent formula $\omega_{ij}(p_i) = \frac{cot\alpha_{ij}^1(\boldsymbol{p}_i) + cot\alpha_{ij}^2(\boldsymbol{p}_i)}{2}$, where α_{ij}^1 and α_{ij}^2 are the two angles opposite to the edge $\overline{\boldsymbol{p}_i \boldsymbol{p}_j}$.

Moreover, a finite element type of method can be applied to estimate the mass matrix and stiffness matrix. Here, we use the LB eigensystem $-\Delta_{\mathcal{M}} \phi = \lambda \phi$ to illustrate an analogue of the finite element method (FEM), on a point cloud $\mathcal{P}$ sampled on a two-dimensional manifold using its local connectivity

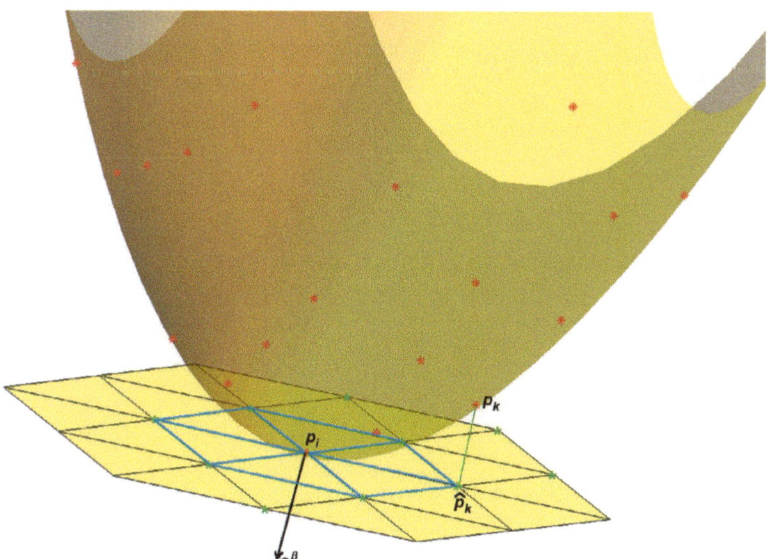

FIG. 1 *Red stars* mark the KNN of $\boldsymbol{p}_i$, *green stars* mark the projection of *red stars* on the tangent plane at $\boldsymbol{p}_i$, and *blue triangles colour-code* the connectivity of the first ring of $\boldsymbol{p}_i$.

$\mathcal{C}=\{\mathcal{C}_i=\{\boldsymbol{p}_i;\mathcal{V}(i),\mathcal{R}(i)\} \mid i=1,\ldots,N\}$. This method can be adapted to higher dimension and other types of elliptic PDEs on point clouds as well.

First we construct the mass matrix on the point cloud $\mathcal{P}$. We define the set of hat functions $\{\eta_j\}_{j=1}^N$ on $\mathcal{P}$ as the linear interpolation of $\eta_j:\mathcal{P}\rightarrow\mathbb{R}$, $\eta_j(p_i)=\delta_{ij}$ $i,j=1,\ldots,\ell$ on its first ring structure based on local connectivity. We write

$$\langle\eta_i,\eta_j\rangle=\begin{cases}\frac{1}{6}\sum_{k=1}^{l_i}Area(T_i^k), & \textit{if } i=j\\ \frac{1}{12}[Area(T_i^{i_1})+Area(T_i^{i_2})], & \textit{if } p_ip_j \text{ is the adjacent edge of } T_i^{i_1} \text{ and } T_i^{i_2}\\ 0, & \text{otherwise}\end{cases} \tag{4}$$

and define the mass matrix $\mathbb{M}=(m_{ij})$ for the point cloud $\mathcal{P}$ by:

$$m_{ij}=\int_{\mathcal{P}}\eta_i\eta_j=\frac{1}{2}(\langle\eta_i,\eta_j\rangle+\langle\eta_j,\eta_i\rangle) \tag{5}$$

Another ingredient is a numerical approximation of $\int_{\mathcal{P}}\nabla_{\mathcal{P}}e_i\cdot\nabla_{\mathcal{P}}e_j$, which forms the stiffness matrix $\mathbb{S}=(S_{ij})$ for the given point cloud $\mathcal{P}$. However, due to the lack of global connectivity, global properties such as symmetric and nonnegative definite of the stiffness matrix cannot be guaranteed due to the use of local mesh only. To be more specific, the first ring structure of p_i is not necessary compatible with the first ring structure of $\mathbf{p}_j$ although $\mathbf{p}_j$ belongs to the first ring of $\mathbf{p}_i$. To have a numerical approximation of the stiffness matrix, we first define

$$A_{ij}=\sum_{T\in\mathcal{R}(i)}\int_T\nabla_T\eta_i\cdot\nabla_T\eta_j=-\frac{1}{2}\left[\cot\alpha_{ij}^1(\mathbf{p}_i)+\cot\alpha_{ij}^2(\mathbf{p}_i)\right],\quad i\neq j \tag{6}$$

This is an approximation of $\int_{\mathcal{P}}\nabla_{\mathcal{P}}\eta_i\cdot\nabla_{\mathcal{P}}\eta_j$ based on the first ring structure at $\mathbf{p}_i$. Note that A_{ij} may not be equal to A_{ji} due to the possible incompatibility of the first ring structures of $\mathbf{p}_i$ and $\mathbf{p}_j$. One simple symmetrized definition of the stiffness matrix is the following:

$$S_{ij}=\begin{cases}\frac{1}{2}(A_{ij}+A_{ji}), & \text{if } i\neq j\\ -\sum_{k\neq i}S_{ik}, & \text{if } i=j\end{cases} \tag{7}$$

The above definition of the diagonal elements enforces the consistence condition, i.e., constant function is an eigenfunction with zero eigenvalue. In particular, if all triangles in the first ring structure are acute, off-diagonal elements are nonpositive and diagonal elements $S_{ii}=\left|\sum_{k\neq i}S_{ik}\right|$ are positive.

Hence all eigenvalues are real and nonnegative. When the distribution of points is reasonably uniform, this definition of stiffness matrix works quite well. However, when the points are distributed nonuniformly, the first ring structure of p_i is more likely incompatible with the first ring structure of p_j. Numerical tests suggest that the following definition for off-diagonals elements produces better results:

$$S_{ij} = \begin{cases} \max(A_{ij}, A_{ji}) & \text{if } A_{ij} \leq 0 \text{ and } A_{ji} \leq 0 \\ \min(A_{ij}, A_{ji}) & \text{if } A_{ij} \geq 0 \text{ and } A_{ji} \geq 0 \\ \min(A_{ij}, A_{ji}) & \text{if } A_{ij} \cdot A_{ji} < 0 \\ -\sum_{k \neq i} S_{ik} & \text{if } i = j \end{cases} \tag{8}$$

Then the solutions to the LB eigenvalue problem can be computed from the following generalized matrix eigen-problem:

$$\begin{cases} \mathbb{S}v = \lambda \mathbb{M}v, \text{ where } \quad v = (v_1, \ldots, v_N)^T \\ \phi = \sum_{i}^{N} v_i \eta_i \end{cases} \tag{9}$$

Note that the stiffness matrix $\mathbb{S}$ and the mass matrix $\mathbb{M}$ are symmetric and sparse matrices of size $N \times N$. Thus, all eigenvalues and eigenfunctions computed from the above method are real.

We would like to point out that this approach is not necessary restricted to solve LB eigensystem. Similar approach can be adapted to solve other types of PDEs, such as the Poisson equation, the heat equation, on point clouds. In particular, the proposed local mesh method can also be adapted to solve nondiffusion type of equations. Here we specifically demonstrate how to solve a special type of nonlinear hyperbolic PDE, the Eikonal equation $|\nabla_{\mathcal{M}} d(x)| = 1$, to compute distance maps and geodesics on point clouds using local mesh method. Since we have a local mesh, i.e., the first ring structure, we can directly adopt the monotone upwind discretization for triangular mesh (Kimmel and Sethian, 1998; Qian et al., 2007), which is equivalent to an approximation of the dynamic programming principle on the first ring. We adapt efficient ways to solve the Eikonal equation either using the fast marching method (Kimmel and Sethian, 1998; Sethian, 1996) or the fast sweeping method (Qian et al., 2007; Zhao, 2005), where only first ring structures of the given triangular mesh are needed. Thus, the proposed local mesh method can be directly adapted to either fast marching method or fast sweeping method to solve the Eikonal equation on point clouds. This is a more direct and more accurate approach to obtain distance maps than the previous method proposed in Mémoli and Sapiro (2005), where the Eikonal equation is solved in the tubular neighbourhood of the point cloud $\mathcal{P}$ in $\mathbb{R}^n$ to approximate the true Eikonal equation defined on $\mathcal{P}$.

Both moving least square method and local mesh method achieve high order accuracy and are capable of approximating general differential operators (possible nondiffusion type) on point clouds sampling manifolds with low dimensions and possible high codimensions. We list several numerical examples including solving the Laplace–Beltrami eigenvalue problem in Fig. 2, the heat equation in Fig. 3, the Fitzhugh–Nagumo equation in Fig. 4, and the Eikonal equation in Fig. 5 on different manifolds sampled as point clouds. We refer to our work (Lai et al., 2013; Liang and Zhao, 2013; Liang et al., 2012) for more detailed discussions and numerical experiments of the proposed methods in terms of their numerical convergence, robustness, and capability of handling high dimensional manifolds.

3 Solving Fokker–Planck equation for dynamic system

In our first application (Lai and Lu, 2017), we adapt the proposed local mesh method for solving the Fokker–Planck equation for dynamic systems, where point clouds are sampled near the transition path of the dynamic system. The understanding of transition of a stochastic system between disjoint regions in the phase space has applications in the study of chemical reactions and thermally activated process (see Metzler et al., 2014; Vanden-Eijnden, 2010; and references therein). For such physical processes, the underlying

FIG. 2 Laplace–Beltrami eigenfunctions are *colour-coded* on point clouds sampled from open and closed surfaces.

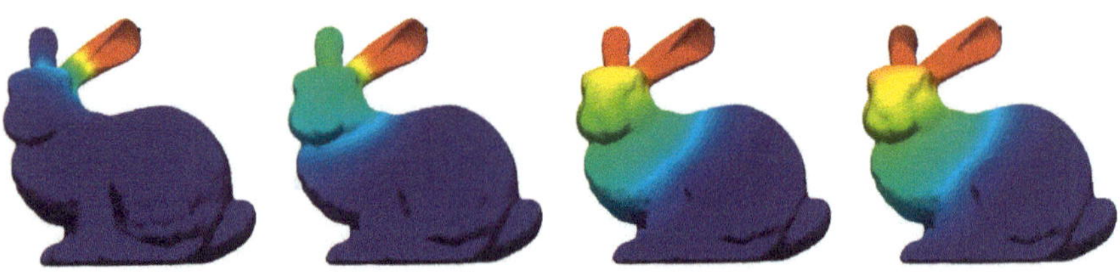

FIG. 3 Solve heat equation on Stanford bunny. The numerical solution is displayed at time $t = 0.1, 0.2, 0.4$, and 0.8. The *black dot* is the heat source. Again, we use triangular mesh for better visualization, although our solution is computed directly on point cloud.

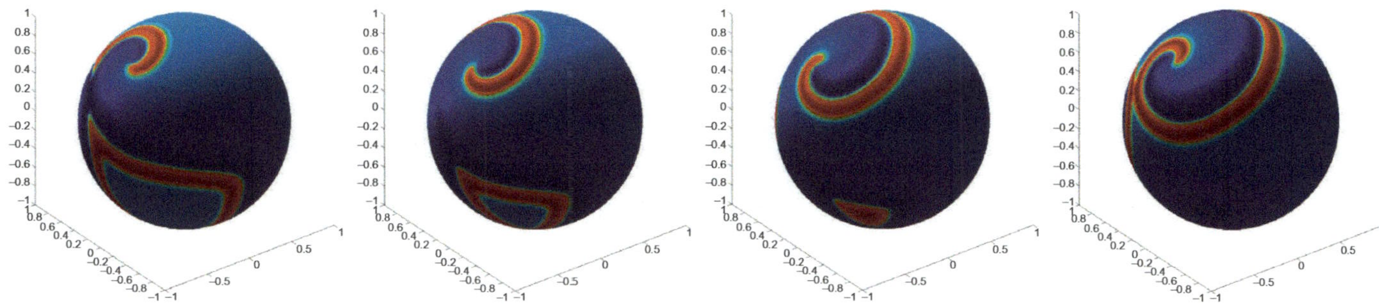

FIG. 4 Fitzhugh–Nagumo equation evolving on a sphere. The excitation variable u_s is displayed at time $t = 400, 450, 500$, and 550. We use triangular mesh for better visualization, although our solution is computed based on a point cloud.

FIG. 5 Distance maps obtained from the Eikonal equation are *colour-coded* on the point clouds, and geodesics are illustrated by *black curves*.

dynamics can be often described by a stochastic process such as the overdamped Langevin equation (also known as the Brownian dynamics):

$$\mathrm{d}X_t = -\nabla U(X_t)\,\mathrm{d}t + \sqrt{2\beta^{-1}}\,\mathrm{d}W_t, \tag{10}$$

where $X_t \in \Omega \subset \mathbb{R}^d$ denotes the current configuration of the system at time t, $U : \Omega \to \mathbb{R}$ is a potential function, $\beta = 1/(k_B T)$ is the inverse temperature (k_B being the Boltzmann constant and T the absolute temperature), and W_t is the standard d-dimensional Wiener process. Here Ω is the configurational space of the system. We are interested in understanding the transition of the dynamics (10) from a subset $A \subset \Omega$ (represents, for example, the reactant state) to a disjoint subset $B \subset \Omega$ (represents, for example, the product state of the system). Direct simulation of such transitions might be challenging since most of these systems exhibit time scale separations: It takes much longer time for the system to go from A to B compared to the intrinsic time scale of the dynamics (which limits the time step size of numerical simulation). As a result, the study of such rare transitions has been a very active area in physical chemistry and applied mathematics (Hummer, 2004; Weinan and Vanden-Eijnden, 2006; Vanden-Eijnden, 2010). In the transition path theory (Vanden-Eijnden, 2010), the central role is played by the *committor function* $q(x) = \mathbb{P}_x(\tau_B < \tau_A)$ (Hummer, 2004; Weinan and Vanden-Eijnden, 2006), which is the probability that a trajectory starting from $x \in \Omega$ first hits B rather than A. In chemical terms, the committor function can be understood as a reaction coordinate of the transition (Peters, 2016).

Given a potential function $U(x)$ defined on Ω, the proposed local mesh method can be use to solve the following PDE involving the infinitesimal generator $L = -\beta^{-1}\Delta + \nabla U \cdot \nabla$ of the process (10) on Ω represented as point clouds

$$\begin{cases} -\beta^{-1}\Delta q(x) + \nabla U(x) \cdot \nabla q(x) = 0, & x \in \Omega \backslash (A \cup B); \\ q(x) = 0, & x \in A; \\ q(x) = 1, & x \in B; \\ \nabla q(x) \cdot \vec{n}(x) = 0, & x \in \partial\Omega. \end{cases} \tag{11}$$

which has the weak formulation for any given test function $\eta(x)$ as:

$$\begin{aligned} 0 &= \int_{\Omega} (\Delta q(x) - \beta \nabla U(x) \cdot \nabla q(x)) \eta(x) e^{-\beta U(x)} \ \mathbf{dx} \\ &= \int_{\Omega} \nabla \left(\nabla q(x) \eta(x) e^{-\beta U(x)} \right) - \nabla q(x) \nabla \eta(x) e^{-\beta U(x)} \ \mathbf{dx} \\ &= - \int_{\Omega} \nabla q(x) \nabla \eta(x) e^{-\beta U(x)} \ \mathbf{dx}, \end{aligned} \tag{12}$$

Note that the Gibbs weight $e^{-\beta U(x)}$ appeared in (12) is the invariant measure $\rho(x) = Z^{-1} e^{-\beta U(x)}$ of the overdamped Eq. (10) up to a normalization constant $Z = \int_{\Omega} e^{-\beta U(x)} \ \mathbf{dx}$. Once $q(x)$ is obtained, we can immediately obtain the transition rate between A and B provided by $\nu_R = k_B T \int_{\Omega} |\nabla q(x)|^2 \rho(x) \ \mathbf{dx}$ which we will later use to quantify the approximation quality of our numerical methods. The committor function also leads the reactive current $J_R(x) = k_B T \rho(x) \nabla q(x)$ which describes the transition path.

Using the aforementioned local mesh method to compute the stiffness matrix $\mathbb{S}$ corresponding to (12). We can approximate the committor function, the solution of Eq. (11), in the following way. Let us write $C = \Omega - A \cup B$, then we obtain the following discretization of Eq. (11).

$$\begin{cases} \mathbb{S}(C, [A, B, C])(q_A, \ q_B, \ q_C)^T = 0. \\ q_A = 0 \\ q_B = 1 \end{cases} \tag{13}$$

This provides a matrix equation

$$\mathbb{S}(C, C) q_C = -\mathbb{S}(C, B) q_B \tag{14}$$

which solves the committor function we desired. We remark that $\mathbb{S}(C, [A, B, C])$, $\mathbb{S}(C, C)$, and $\mathbb{S}(C, B)$ are denoted as the restriction of the matrix $\mathbb{S}$ on the index sets of (C, [A, B, C]), (C, C), and (C, B), respectively, where we abuse the notation to interpret these sets as subsets of point cloud discretization of Ω.

3.1 Double well potential

To test the accuracy of our method, numerical experiments are conducted on 1D interval $[-1, 1]$ for the standard double well potential $U = (x^2 - 1)^2$ with $\beta = 1$, $A = [-1, -1 + 0.1]$, and $B = [1 - 0.1, 1]$. We consider points simulated by the SDE (10) using an Euler–Maruyama scheme. First, we apply both local mesh method and diffusion map method (Coifman and Lafon, 2006; Coifman et al., 2008; Rohrdanz et al., 2011; Zheng et al., 2011) to this data set, and compare to the reference solution obtained by the finite element method. As we can see from the histogram in the left image in Fig. 6,

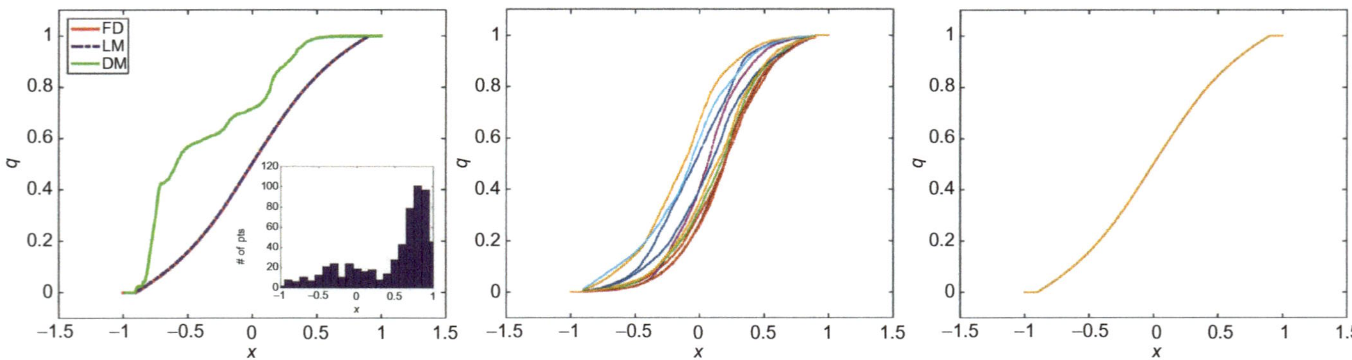

FIG. 6 *Left*: Numerical solutions using different methods with histogram of distribution of 596 points sampled on [−1, 1] showed in the *right corner*. *Middle*: 10 tests using the diffusion map method. *Right*: 10 tests using our method.

although the data do not well sample the invariant measure associated with the double well as we only choose 1000 points to discretize the SDE and keep those points in $[-1, 1]$, our method still provides almost identical result as the solution obtained from the standard finite element method. The diffusion map method on the other hand does not provide satisfactory result in this case. Furthermore, if we take 10,000 time snapshots from a long trajectory generated by the SDE (10) and repeat for 10 times with independent drawing of the point clouds. The middle picture of Fig. 6 shows results obtained from the diffusion map method, where curves with different colour-coding indicate solutions for different realization of the test. It is clear to see that the diffusion map-based method is quite sensitive to sampling quality of the point clouds, thus the approximated solutions are not quite reproducible due to the different sampling quality. As a comparison, the corresponding solutions using our method are plotted in the right picture of Fig. 6 using exactly the same samples. We found that solutions obtained from our method are stacked on top of each other as they are all essentially identical to the result obtained from the finite element method-based regular grid. This shows that our method is very robust to points sample variation, and thus yields better reproducibility, besides providing more accurate approximation of the committor function.

3.2 Rugged Mueller potential

Next we test our methods on a well-established 2*D* test problem, the rugged Mueller potential, in chemical physics. The point cloud is sampled in $\Omega = [-1.5, 1] \times [-0.5, 2]$ with the potential $U_{rm}(x,y) = \sum_{i=1}^{4} D_i e^{a_i(x-X_i)^2 + b_i(x-X_i)(y-Y_i) + c_i(y-Y_i)^2} + \gamma \sin(2k\pi x)\sin(2k\pi y)$ and the parameters chosen as $\gamma = 9$, $k = 5$, $[a_1, a_2, a_3, a_4] = [-1, -1, -6.5, 0.7]$, $[b_1, b_2, b_3, b_4] = [0, 0, 11, 0.6]$, $[c_1, c_2, c_3, c_4] = [-10, -10, -6.5, 0.7]$, $[D_1, D_2, D_3, D_4] = [-200, -100, -170, 15]$, $[X_1, X_2, X_3, X_4] = [1, 0, -0.5, -1]$, and $[Y_1, Y_2, Y_3, Y_4] = [0, 0.5, 1.5, 1]$. We also choose the reactant and product sets as $A = \{U(x,y) < -120\} \cap \{y > 0.75\}$ and $B = \{U(x,y) < -82\} \cap \{y < 0.35\}$.

We now generate point clouds by considering the number of snapshots in SDE (10) from 20,000 to 200,000 with an incremental size 10,000. We obtain a sequence of point clouds which are sampled in $[-1.5, 1] \times [-0.5, 2]$ and accumulate near the minimal energy path connecting two deep wells of U_{rm}. The local mesh method thus produces a sequence approximation of the committor function. For comparison, we use committor function q_{FE} and $\nu_R^{\text{FE}} = 0.92960$ based on the solution obtained from the finite element method on the regular grid $[-1.5, 1] \times [-0.5, 2]$ as a reference. As can be seen from Fig. 7, level contours, represented by black curve, of the approximated committor function match very well to the blue curves representing level contours of q_{FE}. Moreover, the transition rate ν_R can be approximated by evaluating $\int_\Omega |\nabla q| e^{\beta U} dx$ using the stiffness matrix constructed in (8) and approximating

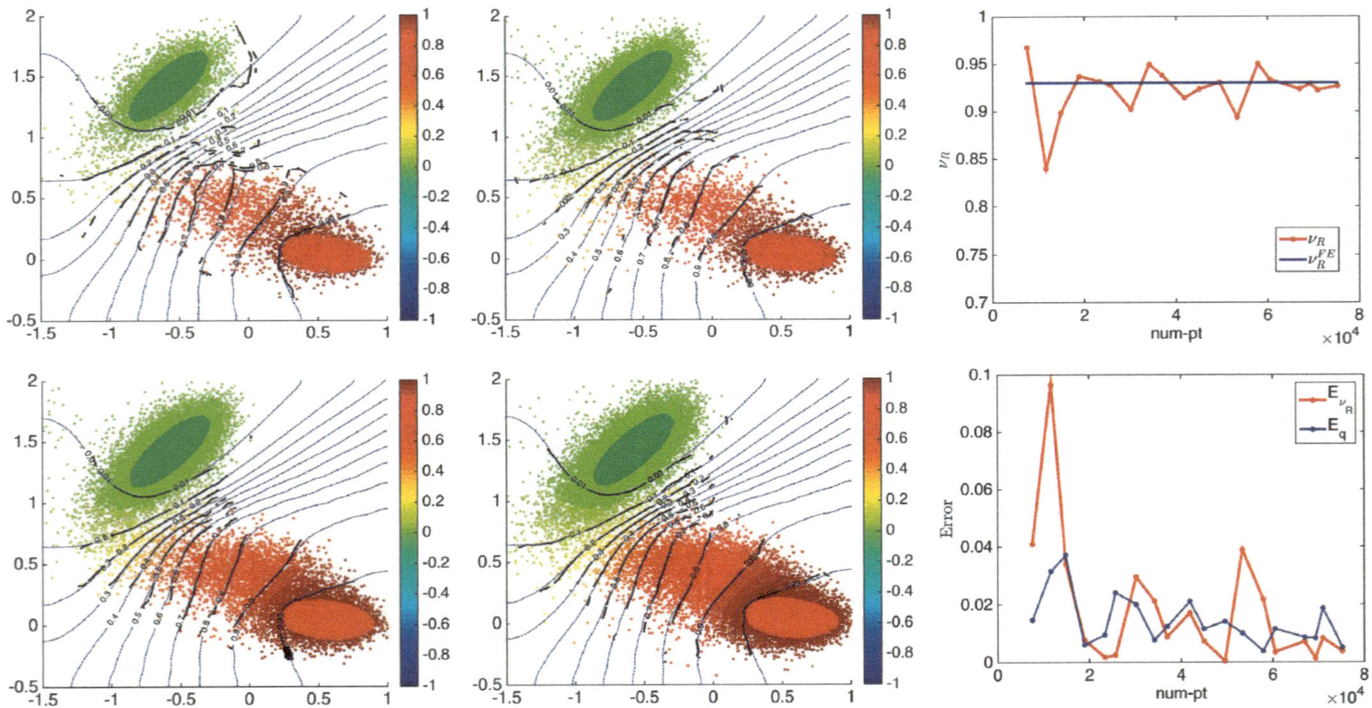

FIG. 7 *Left two columns*: Contour lines match, where *black curves* represent contours from our solutions, with different number of points generated by SDE (10) and *blue-dash curves* represent contours from solutions based on regular grid. *Top left*: 11,372 points. *Top right*: 25,711 points. *Bottom left*: 45,035 points. *Bottom right*: 69,274 points. *Right column*: Relative error of ν_R (*top*) and relative error of q (*bottom*).

$Z = \int_\Omega e^{-\beta U} \mathrm{d}x$ based on a numerical approximation of the mass matrix discussed before. After that, we measure the relative error $E_{\nu_R} = \frac{|\nu_R - \nu_R^{\mathrm{FE}}|}{\nu_R^{\mathrm{FE}}}$ between the approximated transition rate ν_R and the transition rate ν_R^{FE}. We also compute the relative error $E_q = \frac{\|q - \tilde{q}_{\mathrm{FE}}\|_2}{\|\tilde{q}_{\mathrm{FE}}\|_2}$ between q and q_{FE}. Our numerical results reported in Fig. 7 indicate that the approximation error can be controlled around 1% for moderate size of points.

As an advantage of the proposed geometry intrinsic method, the solver we designed can handle point clouds sampled from a low-dimensional manifold in a high dimension space. To illustrate the robustness of the proposed method, we next test our solver for point clouds embedded in a high dimensional ambient space with an artificial Gaussian noise. We first embed the point cloud in $\mathbb{R}^{10}$ and perturb this point cloud in $\mathbb{R}^{10}$ by adding Gaussian noise with variance $\sigma = \gamma d_{max}$ and different level γ = 10%, 20%, 50%, 100%, and also compute the reactive trajectory from the starting point by solving $\frac{dX(s)}{ds} = J_R(X(s)) = k_B T \rho(X(s)) \nabla q(X(s))$. For better visualization, Fig. 8 shows 2D projection of the Gaussian noise perturbed point clouds colour-coded with the resulting commitor function q. In addition, reactive trajectories are also plotted for different noise level with the same starting point. This figure clearly demonstrates the robustness of the proposed method.

4 Solving PDEs on incomplete distance data

Incomplete inter-point distance data are commonly used in protein structuring from NMR spectroscopy (Berger et al., 1999; Crippen and Havel, 1988; Fang and Toh, 2013; Leung and Toh, 2009) and sensor network localization (Biswas et al., 2006; Ji and Zha, 2004). In this data structure, the input of a manifold $\mathcal{M}$ has no point coordinates but only contains incomplete inter-point distance information of a unknown point set $\{\boldsymbol{p}_i\}$ sampled from $\mathcal{M}$. However, there is a lack of exploration on applying the geometric PDE methods to directly analyze this type of data. This motivates our investigation in this direction. Different from global coordinate reconstruction in the canonical distance geometry problem (Berger et al., 1999; Crippen and Havel, 1988), our recent work proposes a new strategy of understanding the incomplete inter-point distance data without global reconstruction, but via solutions of geometric PDEs on manifolds represented as incomplete inter-point distance (Lai and Li, 2018). Note that a differential operator is defined point-wisely dependent on local information of the manifold and is invariant under different choices of local coordinates. This suggests that it suffices to conduct point-wisely local coordinate reconstruction for the associated local neighbourhood, and then point-wisely approximate differential operators based on the reconstructed local coordinates. After that, numerical solvers for differential equations can be constructed and geometric understanding will be based

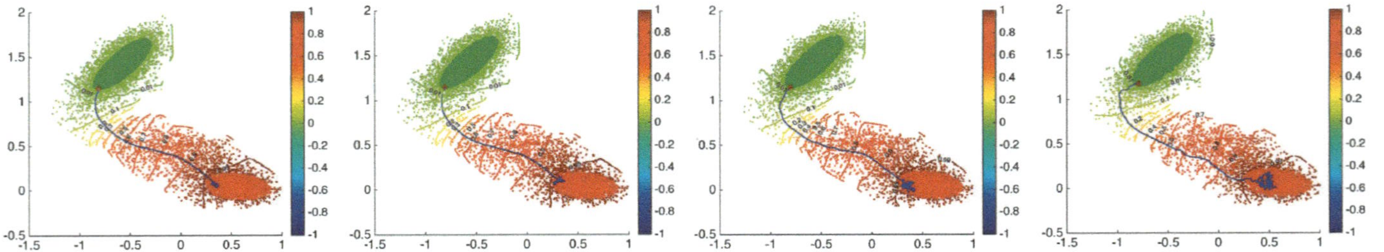

FIG. 8 The committor functions and reaction trajectories for point clouds embedded in $\mathbb{R}^{10}$ with noise levels $\gamma = 10\%$, 20%, 50%, 100% from *left* to *right*, respectively.

on solutions of PDEs. The overall framework of solving PDEs on manifold represented as incomplete inter-point distance is proposed as follows (Lai and Li, 2018):

[Step 0.] *K-nearest neighbourhood (KNN) search.* For the ith point, choose its K-nearest neighbourhood $\mathcal{I}(i)$ based on the given incomplete distance or from the prior information if it is available. For convenience, we write $\{\boldsymbol{p}_1, \boldsymbol{p}_2, \ldots, \boldsymbol{p}_K\}$ as K points for $\mathcal{I}(i)$ as the second step does not rely on a typical choice of $\mathcal{I}(i)$. This provides a local incomplete inter-point distance matrix D.

[Step 1.] *Local coordinates reconstruction via low-rank matrix completion.* If the distance matrix D has full information. The classical multidimensional scaling (MDS) method (Borg and Groenen, 2005; Kruskal and Wish, 1978) suggests that the desired coordinates can be obtained by eigen-decomposition of the associated semipositive definite (SPD) Gram matrix B provided by $B_{ii} + B_{jj} - B_{ij} - B_{ji} = D_{ij}$. One of the major challenges in our assumption is that only a proportion of the distance matrix D is available. Generally speaking, it is impossible to recover the whole matrix only from its small portion of information. Several models of semidefinite programming have been considered (Alfakih et al., 1999; Biswas et al., 2006) in this scenario by assuming D is randomly missing. Here, we consider a different model inspired by the low-rank prior information of the Gram matrix. Mathematically, let us write $\Omega \subset \{(i,j) \mid 1 \leq j < i < K\}$ as a random portion of the index set and write $\mathcal{S} = \{B = (b_{ij}) \in \mathbb{R}^{K\times K} \mid B = B^T\}$. We further denote the available part of D as a vector $D_\Omega \in \mathbb{R}^{|\Omega|}$ and define a restriction operator $\mathcal{R}_\Omega : \mathcal{S} \to \mathbb{R}^{|\Omega|}$ by $\mathcal{R}_\Omega(B)_{ij} = B_{ii} + B_{jj} - B_{ij} - B_{ji}$ with $(i, j) \in \Omega$. In addition, the input distance matrix is invariant under translation of the reconstructed coordinates $\{\boldsymbol{p}_i\}$. This translation ambiguity can be fixed by requiring $\sum_i \boldsymbol{p}_i = \mathbf{0}$, which is equivalent to row sum of B being zero. The problem becomes an optimization problem to find a low-rank SPD matrix $B \in \mathcal{S}$ satisfying $\mathcal{R}_\Omega(B) = D_\Omega$.

$$\min_{B\in\mathcal{S}} \text{Rank}(B), \quad \text{s.t. } B \succeq 0 \;\; \& \;\; \mathcal{R}_\Omega(B) = D_\Omega \tag{15}$$

Inspired by the matrix completion theory (Candès and Recht, 2008), we relax the above NP-hard problem by replacing Rank(B) as the nuclear norm $\|B\|_*$. Note that the set of symmetric matrix $\mathcal{S}$ can be viewed as $\mathbb{R}^{K(K+1)/2}$ due to a natural isomorphism $\zeta : \mathcal{S} \to \mathbb{R}^{K(K+1)/2}, B \mapsto \hat{B} = (b_{11}, \ldots, b_{K1}, b_{22}, \ldots, b_{K2}, \ldots, b_{KK})^T$. We denote $\imath = \zeta^{-1}$ and further merge the two linear constraints by defining the following linear operator:

$$\mathcal{A} : \mathbb{R}^{K(K+1)/2} \to \mathbb{R}^\Omega \times \mathbb{R}^K, \quad \hat{B} \mapsto (\mathcal{R}_\Omega \circ \imath(\hat{B}), \imath(\hat{B})\mathbf{1}) \tag{16}$$

and extend the vector D_Ω as $\tilde{D}_\Omega = (D_\Omega, \vec{0})$. Note that $\|B\|_*$ is the same as $\mathrm{Tr}(B)$ as B is positive semidefinite. Therefore, the proposed model can be written as

$$\min_{\hat{B}\in\mathbb{R}^{K(K+1)/2}} \mathrm{Tr}(\imath(\hat{B})), \quad \text{s.t.} \quad \mathcal{A}\hat{B} = \tilde{D}_\Omega, \quad \imath(\hat{B}) \succeq 0 \tag{17}$$

Once the above Gram matrix is obtained, coordinates can be reconstructed by the eigen-decomposition of B as the method used in the classical MDS. We would like to remark that the constraints and the restriction operator are different from the setting in Candès and Recht (2008) and it does not satisfy the restricted isometry property (Recht et al., 2010), therefore, theoretical analysis conducted in Candès and Recht (2008) and Recht et al. (2010) cannot be directly applied in our problem. By considering a dual basis method, we can theoretically show the exact completion can be achieved under certain coherence conditions (Tasissa and Lai, 2018).

The model (17) is a semidefinite programming (SDP) problem. It is well known that the SDP problem can be solved by interior-points method (Todd, 2001; Vandenberghe and Boyd, 1996) as a canonical choice. However, when the number of constraints in (17) is of the order $\mathcal{O}(K^2)$, the computational time can be of order $\mathcal{O}(K^6)$ (Wen et al., 2010). To save both the time and memory consumption, we use the alternating direction minimization method (ADMM) (Glowinski and Le Tallec, 1989; Wen et al., 2010) to solve the following equivalent problem with an auxiliary variable $\hat{C} = \hat{B}$.

$$\min_{\hat{B},\hat{C}\in\mathbb{R}^{K(K+1)/2}} \mathrm{Tr}(\imath(\hat{B})), \quad \text{s.t.} \quad \mathcal{A}\hat{B} = \tilde{D}_\Omega, \quad \imath(\hat{C}) \succeq 0, \quad \hat{B} = \hat{C} \tag{18}$$

This yields sequentially update $\hat{B}$ and $\hat{C}$ as

$$(\mu_1\mathcal{A}^*\mathcal{A} + \mu_2)\hat{B}^{k+1} = \mu_1\mathcal{A}^*(\tilde{D}_\Omega - H_1^k) + \mu_2(\hat{C}^k - H_2^k) - \zeta(I_n) \tag{19}$$

$$\hat{C}^{k+1} = \zeta \circ \mathcal{T}_E \circ \imath(\hat{B}^{k+1} + H_2^k) \tag{20}$$

where $\mathcal{T}_E(X) = V\mathrm{diag}(\max\{\Lambda(i,i), 0\}_{i=1}^n)V^\top$ defines an eigenvalue hard thresholding (EVHT) operator for a symmetric matrix X with eigen-decomposition $X = V\Lambda V^\top$. We summarize the above iterative method in Algorithm 1. After finding $\hat{B}$ by Algorithm 1, eigen-decomposition of $\imath(\hat{B})$ can ultimately reconstruct the coordinates of points $\{\boldsymbol{p}_1, \boldsymbol{p}_2, \ldots, \boldsymbol{p}_K\}$ from the incomplete distance D_Ω.

[Step 2.] Applying the MLS method or the local mesh method discussed in Section 2 to approximate the desired differential operators at the ith point. This provides the ith row of the discretized matrix representation of the desired differential equation.

Experimentally, considering a point cloud $\{\boldsymbol{p}_1, \ldots, \boldsymbol{p}_n\}$ uniformly sampled on a given manifold $\mathcal{M} \subset \mathbb{R}^p$, we compute the associated distance matrix D

ALGORITHM 1 Augmented Lagrangian method to solve (17)

Initialization. Set $C^0 = 0, H_1^0 = 0, H_2^0 = 0, \epsilon = 1e-6$. Set $\mu_1 = 10$, $\mu_2 = 5$.

while ($|E(k) - E(k-1)|/E(k) \geq \epsilon$) **do**

1. Solve $(\mu_1 \mathcal{A}^* \mathcal{A} + \mu_2)\hat{B}^{k+1} = \mu_1 \mathcal{A}^*(\tilde{D}_\Omega - H_1^k) + \mu_2(\hat{C}^k - H_2^k) - \zeta(I_n)$.

2. $\hat{C}^{k+1} = \zeta \circ \mathcal{T}_E \circ \iota(\hat{B}^{k+1} + H_2^k)$.

3. $H_1^{k+1} = H_1^k + \mathcal{A}\hat{B}^{k+1} - \tilde{D}_\Omega$.

4. $H_2^{k+1} = H_2^k + \hat{B}^{k+1} - \hat{C}^{k+1}$.

5. Let $E(k+1) = \mathrm{Tr}(\iota(\hat{B}^{k+1})) + \frac{\mu_1}{2} \| \mathcal{A}\hat{B}^{k+1} - \tilde{D}_\Omega \|_2^2 + \frac{\mu_2}{2} \| \hat{B}^{k+1} - \hat{C}^{k+1} \|_2^2$.

end while

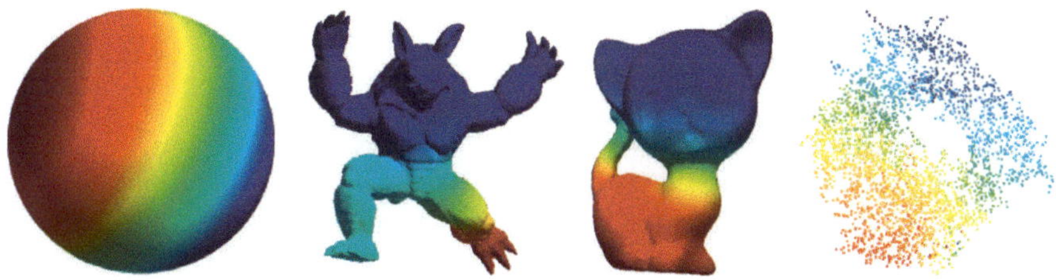

FIG. 9 The first nontrivial LB eigenfunctions are colour-coded on the manifolds based on incomplete distance. From *left to right*: The unit sphere with $D_{50\%,\ 30}$, armadillo with $D_{50\%,\ 30}$, kitten with $D_{50\%,\ 30}$ and a protein data with $D_{3\%,\ \infty}$.

provided by pairwise Euclidean distance for this point cloud. More precisely, we first uniformly randomly choose a subset $\Omega_\gamma \subset \{(i,j) \mid 1 \leq j < i \leq n\}$, such that $|\Omega_\gamma| = \lceil \gamma n(n-1)/2 \rceil$. Then incomplete distance information is provided as $D_{\gamma,K} = \{D(i,j) \mid j \in \mathcal{I}_K(i) \ \& \ (i,j) \in \Omega_\gamma,\ i = 1,\ldots,n\}$, where $\mathcal{I}_K(i)$ denotes the index set of the nearest K points to p_i. Thus, $D_{\gamma,\ K}$ roughly contains $\gamma K/n$ portion of D (a very small portion of D if $K \ll n$ as we showed in our experiments). Our numerical tests are presented to illustrate the proposed methods for solving PDEs on manifolds with incomplete inter-point distance information.

In our work (Lai and Li, 2018), we first test our method of computing the LB eigen-problem on manifolds with incomplete inter-point distance data. Our experiments illustrate numerical convergence for solving the LB eigen-problem on the unit sphere with different values of γ and K. We also test data corrupted by different level of Gaussian noise. The proposed method works for general manifolds represented with incomplete distance data. In addition, we also test numerical convergence of our method on point clouds sampled on high dimensional torus. Our quite extensive numerical experiments indicate that the proposed methods can provide good approximation based on incomplete distance information. We list a few numerical results in Fig. 9.

We would like to remark that the above local coordinate reconstruction is conducted point-wisely which can be implemented fully in parallel.

The proposed method of approximating differential operators based on local reconstruction has the advantage of computation and memory efficiency. In a different strategy, one can consider to first reconstructing coordinates, then approximate differential operators. Note that the most time-consuming part of the coordinate reconstruction based on the augmented Lagrangian method is the eigenvalue hard thresholding. Consider an incomplete distance data with n nodes, the complexity of each eigenvalue thresholding step is $\mathcal{O}(n^2m)$ for global reconstruction if only the largest m eigenvalues are computed. In the proposed local reconstruction strategy, it only needs to reconstruct coordinates of K nearest points of $\boldsymbol{p}_i$ with complexity $\mathcal{O}(nK^2m)$. When n is much larger than K, which is typically $\leq$30, the local reconstruction strategy is much more efficient (n/K^2 times faster) than global reconstruction strategy. Moreover, the local reconstruction method supports parallel computation. Additionally, since the distance matrix is not sparse, the memory need for global reconstruction is $\mathcal{O}(n^2)$. However, the local reconstruction strategy only requires $\mathcal{O}(nK^2)$ for memory. We report total computation time in Table 1 for computing the first 100 LB eigenvalues with different setting of γ and K. For large data sets, $K \ll n$, the proposed local reconstruction method can save a huge amount of time as illustrated in Table 1. For example, the local reconstruction method for solving the LB eigenvalue problem is about 150 times faster than the global reconstruction method for $n = 16{,}002$ and $K = 6$.

TABLE 1 Comparisons of time consumption (minutes) of solving the LB eigenvalue problem based on local/global reconstruction methods.

Number of points				
1002	1962	4002	7842	16002
$\gamma = 100\%$, $K = 6$				
0.26	0.51	1.01	2.03	4.05
$\gamma = 80\%$, $K = 9$				
2.28	5.60	11.17	22.28	45.02
$\gamma = 50\%$, $K = 18$				
4.03	8.09	16.14	32.44	64.71
$\gamma = 30\%$, $K = 30$				
15.13	30.19	60.42	120.95	241.63
Global reconstruction using 3% distance ($K = 6$ for MLS)				
2.09	9.86	40.13	154.40	597.06

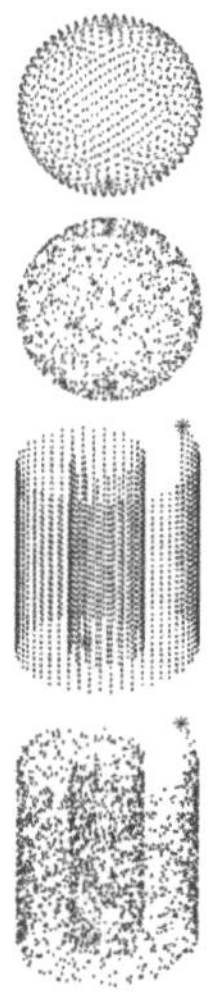

Methods \ Sample size		1002	1962	4002	7842	16,002
		Uniform sampling on S^2				
Dijkstra	E_{av}	0.0348	0.0285	0.0248	0.0232	0.0224
	E_{se}	0.008615	0.008606	0.008296	0.010642	0.011501
Our method	E_{av}	0.0113	0.0095	0.0080	0.0064	0.0022
	E_{se}	0.008100	0.005890	0.004110	0.002877	0.002158
		Nonuniform sampling on S^2				
Dijkstra	E_{av}	0.0363	0.0344	0.0319	0.0305	0.0294
	E_{se}	0.011209	0.016090	0.018380	0.016391	0.019953
Our method	E_{av}	0.0200	0.0163	0.0141	0.0124	0.0088
	E_{se}	0.012016	0.008792	0.003742	0.001736	0.002765
		Uniform sampling on a Swiss roll				
Dijkstra	E_{av}	0.0119	0.0156	0.0198	0.0200	0.0203
	E_{se}	0.013104	0.021242	0.024560	0.024311	0.026004
Our method	E_{av}	0.0065	0.0044	0.0052	0.0033	0.0022
	E_{se}	0.003127	0.001637	0.001130	0.000783	0.000620
		Nonuniform sampling on a Swiss roll				
Dijkstra	E_{av}	0.0400	0.0105	0.0114	0.0151	0.0180
	E_{se}	0.016612	0.015779	0.014573	0.016587	0.018649
Our method	E_{av}	0.0138	0.0053	0.0047	0.0029	0.0073
	E_{se}	0.004754	0.005189	0.003087	0.005171	0.007246

FIG. 10 *Top to bottom images*: Incomplete distance data $D_{60\%,20}$ for 1002 points sampled uniformly and nonuniformly on the unit sphere with fixed *north pole* (*red star*) and *south pole* (*black star*) for calculating the geodesic distance, and $D_{60\%,20}$ from 2048 points sampled uniformly and nonuniformly on the Swiss roll with a start point (*red star*) and an end point (*black star*) for calculating the geodesic distance. Table: Relative error of geodesic distances E_{av} (averaging from the start point to the other points) and E_{se} (from the start point to the end point).

In addition, we also test the problem of solving a nondiffusion type of equation, the Eiknoal equation, and illustrate the accuracy improvement of this approach compared to the canonical Dijkstra method. As illustrated in Fig. 10, we compare the relative error of geodesic distance from a start point to and end point, and from a start point to all other points using $D_{60\%,20}$ based on different samples on the unit sphere and a Swiss roll. The table in Fig. 10 shows that our approach based on the fast marching method is much more accurate than the Dijkstra's method (Dijkstra, 1959). Moreover, relative errors of our method decrease as the number of points increase. Furthermore, for the uniform sample cases, the relative error of geodesic distance from the north pole to the south pole has first-order convergence with respect to the density of points using our method, while the Dijkstra's method does not seem to converge.

5 Geometric understanding of point clouds data

Well developed mathematical theory in differential geometry and geometric analysis suggest that solutions of special type of geometric PDEs on manifolds can reflect intrinsic information of the underlying manifolds (Chavel, 1984; Jones et al., 2008; Uhlenbeck, 1976). For instance, the eigenvalue problem of the elliptic LB operator: $\Delta_{\mathcal{M}}\phi_k = -\lambda_k\phi_k,\ \ k=0,1,2,\ldots$ plays an essential role to understand intrinsic surface geometry (Chavel, 1984; Jost, 2008). The eigenfunctions of the LB operator are natural extensions of the Fourier basis functions from a Euclidean domain to a general manifold. As an intrinsic differential operator defined on a manifold, the LB operator can be used as a bridge to connect local and global geometry of the underlying manifold. Its eigenvalues and eigenfunctions can be used to study and characterize the manifold. For example, generic LB eigenfunctions of manifolds are Morse functions (Uhlenbeck, 1976), which enable LB eigenfunctions as descriptors for the topological structures of the manifold using Morse theory (Milnor, 1963). Moreover, LB eigenfunctions can be viewed as intrinsic global descriptors to analyze and compare geometric structures of manifolds in both theoretical and computational perspective (Bérard et al., 1994; Bronstein and Kokkinos, 2010; Jones et al., 2008; Lai et al., 2010, 2011; Rustamov, 2007; Sun et al., 2009). In the past few years, there have been increasing interests in using LB eigensystems for 3D surface analysis (Bronstein and Kokkinos, 2010; Bronstein et al., 2010; Lai et al., 2010; Raviv et al., 2011; Reuter, 2009; Rustamov, 2007; Shi et al., 2011, 2012; Shtern and Kimmel, 2014, 2015; Sun et al., 2009; Vallet and Levyvy, 2008), medical image analysis (Lai et al., 2009, 2010, 2011; Qiu et al., 2006; Shi et al., 2008a,b, 2009, 2010, 2011, 2012) and other types of manifold-structured data (Allard et al., 2012; Belkin et al., 2009; Coifman and Lafon, 2006; Coifman et al., 2008; Lai and Lu, 2017; Lai and Zhao, 2017; Little et al., 2009b; Rohrdanz et al., 2011; Singer, 2006; Zheng et al., 2011).

In this section, we use our discrete LB operator to solve several interesting problems related to geometric understanding of point clouds. First, we use the first nontrivial LB eigenfunction to construct global skeleton structures for point clouds. Second, based on the recent algorithm proposed in Lai et al. (2014), we adapt it to find conformal mappings from point clouds of simply connected closed surfaces onto spheres. This allows us to find the conformal structures on these surfaces, which provide the complete geometric information for point clouds. Last, we will discuss nonrigid point clouds registration using LB eigensystems proposed in Lai and Zhao (2017).

5.1 Construction of skeletons from point clouds

Based on the first nontrivial eigenfunction, a novel approach of computing surface skeletons is proposed in Shi et al. (2008c), where the construction of skeleton is realized by constructing Reeb graphs from the first nontrivial LB eigenfunction. Their idea is to define the Reeb graph $R(\phi_1)$ of ϕ_1 as the quotient space defined through the equivalent relation $x \sim y$ if $\phi_1(x) = \phi_1(y)$ for $x, y \in \mathcal{M}$. We adapt their method to construct skeletons for point clouds without connectivity information. We solve LB eigenvalue problem to obtain the first nontrivial eigenfunction ϕ_1. By connecting barycentres of level sets of ϕ_1, one can find skeleton of the point cloud.

Given the first nontrivial LB eigenfunction ϕ_1 and its local polynomial approximation Φ_1 using approach proposed in Section 2, the level contour $\phi_1 = c$ can be computed. For each p_i such that $\phi_1(p_i) \in [c - \delta, c + \delta]$, we find a point p that is the minima of $\min_p \|p - p_i\|^2 + \beta(\Phi_1(p) - c)^2$. By choosing δ small and β large, we find a point that is close to p_i and its function value is close to c. Once all level contours of ϕ_1 are computed, we take their barycentre as the skeleton point of that level set. By connecting these barycentres, the skeleton structure for the given point cloud can be obtained.

Using the first nontrivial eigenfunction computed by the MLS LB operator, we compute skeletons for several point clouds using algorithm proposed in Section 5.2. As shown in Fig. 11, the extracted skeletons agree well with our intuition of the most essential feature, or a "backbone" constituting a shape. This shows that our method is able to achieve high level understanding of the raw data of point clouds, which are only myriads of points in the 3D space as a basic representation of the geometry.

5.2 Construction of conformal mappings from point clouds

As long as the LB operator can be locally approximated for a given point cloud, we can extract global geometric structures of the given point cloud. As an example, we adapt the efficient harmonic energy minimization algorithm in Lai et al. (2014) to study global conformal structures for point clouds of genus-0 surfaces.

Given a genus-0 surface $\mathcal{M}$, its conformal structure can be obtained by a conformal map from $\mathcal{M}$ to the sphere S^2. This can be realized by computing a harmonic energy minimization problem (Gu et al., 2004). More recently, a more efficient feasible method for harmonic energy minimization problem from genus-0 surfaces to sphere is proposed in Lai et al. (2014), where authors consider the harmonic energy minimization problem from $\mathcal{M}$ to S^2 to the following optimization problem with spherical constraints:

$$\min_{\vec{F}=(f_1,f_2,f_3)} \mathcal{E}(\vec{F}) = \frac{1}{2}\int_{\mathcal{M}} \sum_{i=1,2,3} \|\nabla_{\mathcal{M}} f_i\|^2 \mathrm{d}\mathcal{M} \text{s.t.} \quad \|\vec{F}(x)\| = \sqrt{f_1^2 + f_2^2 + f_3^2} = 1, \quad \forall x \in \mathcal{M} \tag{21}$$

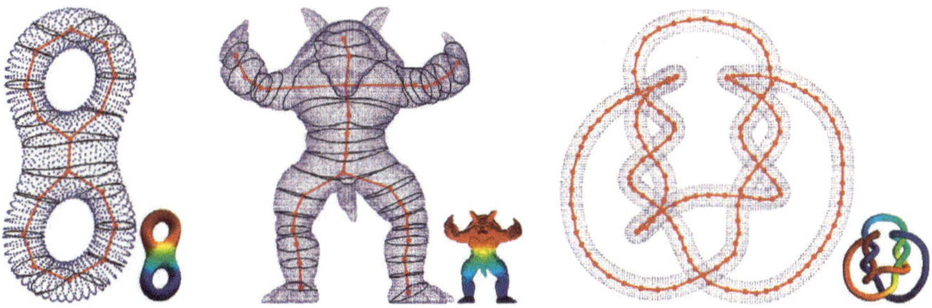

FIG. 11 Skeletons computed by our proposed algorithm. *Blue dots* are data points, *black curves* are the level contours of ϕ_1, *red dots* are centres of the corresponding level set curves and *red lines* are the resulting skeletons.

Write $H = \nabla\mathcal{E}(\vec{F}) = -(\Delta_{\mathcal{M}} f_1, \Delta_{\mathcal{M}} f_2, \Delta_{\mathcal{M}} f_3)$ as the Fréchet derivative of $\mathcal{E}(\vec{F})$ with respect to $\vec{F}$ and define a skew-symmetric matrix $A := H\vec{F}^* - \vec{F}H^*$. Similar to the algorithm for orthogonality constraint proposed in Wen and Yin (2013), an update feasible path $\vec{Y}[\tau] := \vec{F} - \frac{\tau}{2}A(\vec{F} + \vec{Y}[\tau])$ is proposed in Lai et al. (2014) to efficiently solve the above harmonic energy optimization problem. Once a conformal mapping is computed, the point cloud directly inherits its full connectivity and conformality information from the unit sphere. As shown in Fig. 12, we apply our algorithm to the point clouds of a fish and the Stanford bunny. After both connectivity and conformality are found, we apply a texture mapping on the point clouds to show their conformal structures. Locally, the checkerboard patterns are very close to squares, indicating that our algorithm successfully captures the complete geometric information on these point clouds.

5.3 Nonrigid manifolds registration using LB eigenmap

Let $(\mathcal{M}, g_{\mathcal{M}})$ and $(\mathcal{N}, g_{\mathcal{N}})$ be two isometric d-dimensional Riemannian manifolds with induced metric from $\mathbb{R}^D$. A nonrigid isometric registration between $\mathcal{M}$ and $\mathcal{N}$ is a smooth diffeomorphism $f : \mathcal{M} \to \mathcal{N}$, such that the pullback metric satisfies $f^*(g_{\mathcal{N}}) = g_{\mathcal{M}}$. Due to the nonlinearity of f, it is not straightforward to compute f in the ambient space $\mathbb{R}^D$. Instead of constructing registration directly in $\mathbb{R}^D$, $f : \mathcal{M} \to \mathcal{N}$ can be constructed by tracing the registration results in the embedding space induced from Laplace–Beltrami eigensystem. More specifically, let us write $\Phi_n = \{\lambda_k, \phi_k\}_{k=1}^n$, and $\Psi_n = \{\eta_k, \psi_k\}_{k=1}^n$ as the first n nontrivial LB eigenvalues and LB eigenfunctions of $\mathcal{M}$ and $\mathcal{N}$, respectively, which can be computed using the numerical methods discussed in Section 2. The earliest theoretical work of this idea was discussed in Bérard et al. (1994), where a heat kernel embedding is proposed for global comparison of two Riemannian manifolds, and a distance is also proposed to measure difference between two manifolds although the proposed distance is not computationally tractable. Based on global position signature, $\iota_{\mathcal{M}}^n : \mathcal{M} \to \mathbb{R}^n$, $u \mapsto p = \iota_{\mathcal{M}}^n(u) = \left(\frac{\phi_1(u)}{\lambda_1^{d/4}}, \ldots, \frac{\phi_n(u)}{\lambda_n^{d/4}} \right)$, a spectral ℓ^2 distance

FIG. 12 Point clouds of fish and bunny and the corresponding texture mappings computed by our proposed algorithm for conformal mapping.

for surface comparisons was discussed in Lai et al. (2010). However, it is also computationally challenging due to nonuniqueness of LB eigensystem from sign ambiguities for eigenfunctions or ambiguities due to the multiplicity of eigenvalues. In addition, it is not straightforward to combine LB eigensystem with other intrinsic features because they have different scale or unit.

To handle ambiguities due to arbitrary sign, multiplicity or possible order switching of LB eigensystem, we propose to align $\mathcal{P}=\iota^n_{\mathcal{M}}(\mathcal{M})\subset\mathbb{R}^n$ and $\mathcal{Q}=\iota^n_{\mathcal{N}}(\mathcal{N})\subset\mathbb{R}^n$ in $\mathbb{R}^n$ up to an orthonormal transformation. Intuitively, a one-to-one and onto map should be constructed to align $\mathcal{P}$ and $\mathcal{Q}$. In practice, however, $\mathcal{M}$ and $\mathcal{N}$ are just discrete samples of the underlying manifolds. In many cases, sampling of $\mathcal{M}$ and $\mathcal{N}$ does not necessarily have the same number of points. Thus, a one-to-one and onto map does not make sense in this case. Here, we propose to view a registration in the distribution sense and propose the registration model for $\mathcal{P}$ and $\mathcal{Q}$ based on the framework of optimal transport theory (Kantorovich, 1942; Monge, 1781; Villani, 2003, 2008).

In practice, we assume $\mathcal{M}$ and $\mathcal{N}$ are sampled by two sets of points denoted by $\{u_i\in\mathcal{M}\subset\mathbb{R}^D \mid i=1,\ldots,\ell_M\}$ with a mass distribution $\mu^{\mathcal{M}}\in\mathbb{R}^{\ell_M}_+$ and $\{v_i\in\mathcal{N}\subset\mathbb{R}^D \mid i=1,\ldots,\ell_N\}$ with a mass distribution $\mu^{\mathcal{N}}\in\mathbb{R}^{\ell_N}_+$. Hence, our task is to find a meaningful mapping or registration between two point clouds, $(\{p_i=\iota^n_{\mathcal{M}}(u_i)\}_{i=1}^{\ell_M},\mu^{\mathcal{M}})$ and $(\{q_i=\iota^n_{\mathcal{N}}(v_i)\}_{i=1}^{\ell_N},\mu^{\mathcal{N}})$, in $\mathbb{R}^n$ up to a rigid transformation. For convenience and consistency, we let $\ell_P=\ell_M,\ell_Q=\ell_N,\mu^P=\mu^{\mathcal{M}},\mu^Q=\mu^{\mathcal{N}}$ and write row vectors p_i and q_i as the ith row of matrices $P\in\mathbb{R}^{\ell_P\times n}$ and $Q\in\mathbb{R}^{\ell_Q\times n}$, respectively. In other words, we need to find an optimal transport plan $\sigma\in\mathbb{R}^{\ell_P\times\ell_Q}$ and an orthonormal matrix R such that the mass transportation from (PR, μ^P) to (Q, μ^Q) is minimal.

To avoid the high computation cost of the optimal transportation problem in high dimension, we adopt the rotation-invariant sliced Wasserstein distance (RSWD) which averages the Wasserstein distance between P and Q along each direction (Lai and Zhao, 2017). The computation efficiency of RSWD is based on the existence of analytical solution to 1D optimal transport problem using sorting algorithms. More specifically, given a direction $\theta\in S^{n-1}$, let us denote $\mu^P_{\theta,R}$ and μ^Q_θ as the measures on $\mathbb{R}$ induced by $\pi^{\theta,R}: p_i\mapsto p_iR\theta^T$ and $\pi^\theta: q_j\mapsto q_j\theta^T$ respectively. Then the discrete version of RSWD can be formulated as follows.

$$
\begin{aligned}
\mathrm{RSWD}\big((P,\mu^P),(Q,\mu^Q)\big)^2 &= \min_{R\in O(n)}\int_{S^{n-1}}\min_{\sigma\in\mathrm{ADM}(\mu^P_{\theta,R},\mu^Q_\theta)}\sum_{i,j}\sigma_{ij}\,\|p_iR\theta^T-q_j\theta^T\|^2\quad d\theta\\
&= \min_{R\in O(n)}\int_{S^{n-1}}\sum_{i,j}\sigma^{\theta,R}_{ij}\,\|p_iR\theta^T-q_j\theta^T\|^2\quad d\theta
\end{aligned}
\tag{22}
$$

where $\sigma^{\theta,R} \in \text{ADM}(\mu^P_{\theta,R}, \mu^Q_\theta) = \{\sigma \mid \sigma\vec{1} = \mu^P_{\theta,R}, \sigma^T\vec{1} = \mu^Q_\theta\}$ denotes the optimizer for the optimal transport for the two 1D point sets $\{p_i R\theta^T\}_{i=1}^{\ell_P}$ and $\{q_j\theta^T\}_{j=1}^{\ell_Q}$ solved by sorting in $\mathcal{O}(Nlog(N))$ operations. In addition, the following theorem justifies the proposed RSWD as a distance. We refer Lai and Zhao (2017) for more details about the proof.

Theorem 1. *We write* $\mathfrak{M}_n = \bigcup_{\ell=1}^{\infty}\left\{(P,\mu) = \{(p_i,\mu_i) \in \mathbb{R}^n \times \mathbb{R}_+\}_{i=1}^{\ell} \mid \sum_{i=1}^{\ell}\mu_i = 1,\right.$ $\left. p_i \neq p_j \text{ if } i \neq j\right\}$ *as the space of all possible point clouds in* $\mathbb{R}^n$ *that has* $\ell = 1, 2, \ldots$ *number of points associated with a measure* μ. *We define an equivalence relation for two point clouds* $(P,\mu^P), (Q,\mu^Q) \in \mathfrak{M}_n$, *with number of points* ℓ_P, ℓ_Q *respectively as:*

$$(P,\mu^P) \sim (Q,\mu^Q) \Leftrightarrow \begin{cases} P \text{ and } Q \text{ have the same number of points, i.e., } \ell_P = \ell_Q \\ \text{there exist } R \in O(n) \text{ and a permutation } \pi \text{ of } [1,\ldots,\ell_P] \\ \text{s.t. } p_i R = q_{\pi(i)}, \mu_i^P = \mu^Q_{\pi(i)} \end{cases}$$

Then, RSWD(·, ·) defines a distance on the space $\mathfrak{M}_n/\backsim$.

Last but not least, the LB eigenmap $\iota^n_{\mathcal{M}}$ provides a natural multiscale characterization of the manifold $\mathcal{M}$ with global information from smaller n to larger n. This observation leads to a simple and natural multiscale registration algorithm which enjoys robustness, efficiency and accuracy. First, we choose a sequence of numbers of LB eigenfunctions to construct LB eigenmap for point clouds $\{u_i \in \mathcal{M}\}_{i=1}^{\ell_{\mathcal{M}}}$ and $\{v_i \in \mathcal{N}\}_{i=1}^{\ell_{\mathcal{N}}}$. Efficient and robust registration on coarse scale can be obtained from LB eigenmap with the first few leading LB eigenvalues and corresponding eigenfunctions. Then registration at finer and finer scales can be achieved effectively by using LB eigenmap with more and more LB eigenvalues and corresponding eigenfunctions using available coarser scale registration as the initial guess.

Fig. 13 plots registration results using the first 5 nontrivial LB eigenvalues and corresponding eigenfunctions. Meanwhile, as a by-product of the algorithms, we can obtain the optimal rigid transformation matrix for registering the two point clouds after LB eigenmap, which in fact shows the ambiguity of LB eigenmap. For those surface pairs with "ground truth" map $f_T : \mathcal{M} \to \mathcal{N}$, a quantitative error analysis can be conducted by measuring the geodesic error between the numerical registration result f and f_T. The error is obtained by computing geodesic distance $d_{\mathcal{N}}(f(p), f_T(p))$ on $\mathcal{N}$ as the way used in Kim et al. (2011). In this experiment, we apply the proposed methods to two nonrigid shape matching benchmark data sets, Shape Completion and Animation of People (SCAPE) (Anguelov et al., 2005) and Tools for Surface Comparison and Analysis (TOSCA) (Bronstein et al., 2008). We further compute point-wise geodesic error $d_{\mathcal{N}}(f(p), f_T(p))$ between numerical results and ground truth correspondence included in the data set.

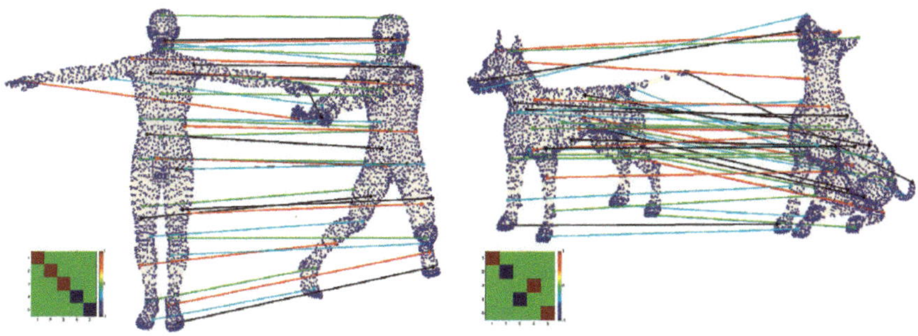

FIG. 13 Registration using the first 5 eigenvalues and corresponding eigenfunctions with $\mu^P = \mu^Q = 1/N$ and the resulting 5×5 rotation matrices.

In addition, the Princeton surface correspondence benchmark also contains geodesic errors computed from several other existing methods including blended intrinsic maps (Kim et al., 2011), Mobius voting (Lipman and Funkhouser, 2009), generalized multidimensional scaling (Bronstein et al., 2006), isometric matching using heat kernel (Ovsjanikov et al., 2010), and the deformation-driven method (Zhang et al., 2008). In Fig. 14, we conduct comparisons with these existing methods by illustrating the cumulative distribution of geodesic errors obtained in different methods. It is clear to see that our multiscale registration method outperforms those existing methods on the SCAPE data set and the TOSCA data set. We remark that two methods based on enhanced feature space were proposed recently for nonrigid shape matching (Shtern and Kimmel, 2014, 2015). Due to the enhanced feature spaces, both methods perform better registration results than purely using LB eigenmaps. As the proposed methods focus on LB eigenmaps, we do not include results from these two methods for the SCAPE and TOSCA data sets here.

6 Conclusion

In this paper, we review two recent methods developed by the authors for solving PDEs on manifolds represented as point clouds. One is called the moving least square method which mimics the finite difference method. It approximates the manifold structure and differential operators based on moving least square approximation parametrized by the tangent space locally. The other one is called the local mesh method which mimics the finite element method. By generating a local mesh for a point cloud through its tangent space, the stiffness and the mass matrix can be computed for solving elliptic and parabolic equations on the point cloud. It can also be adapted to solving other type of PDEs on point clouds. Various applications and examples are presented to demonstrate how to connect "dots" and piece together local information to extract global information by solving PDEs on point clouds.

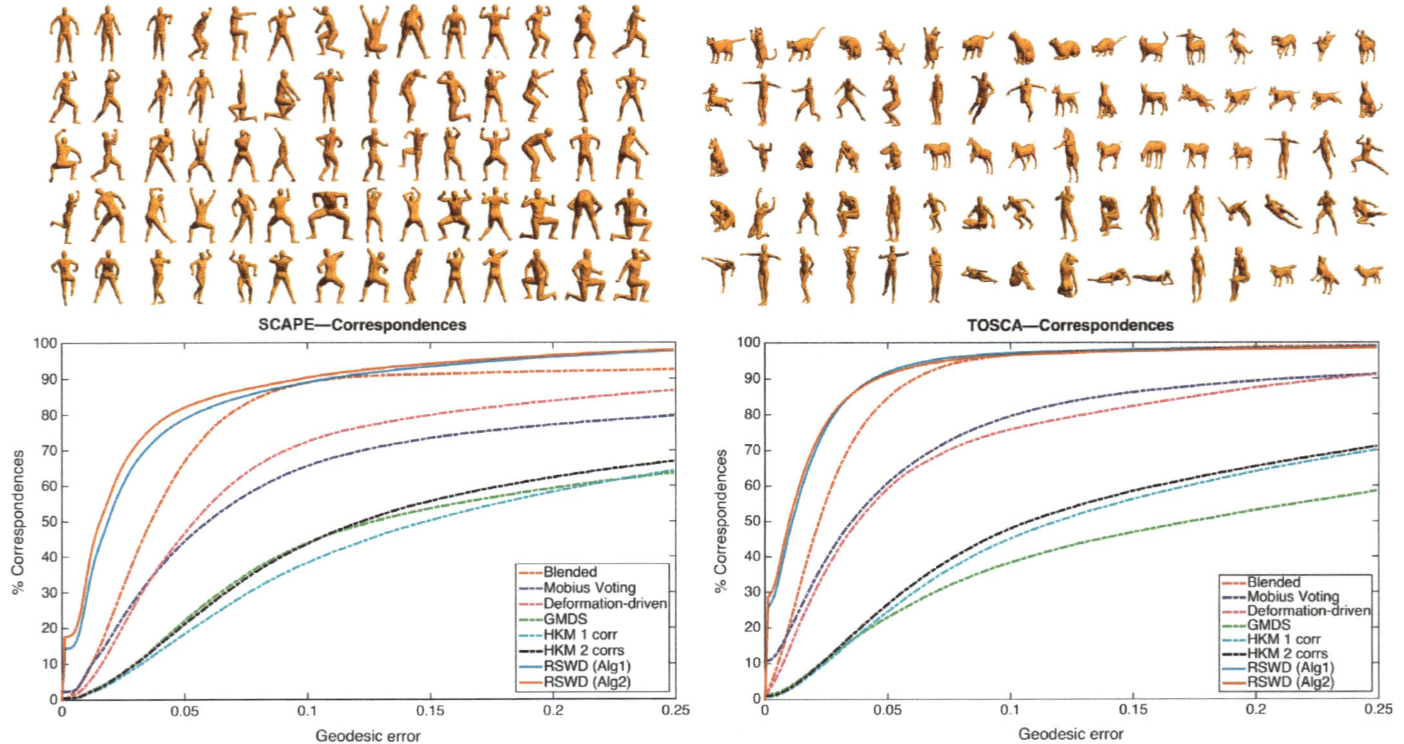

FIG. 14 Comparisons of our results with the existing methods used in Kim et al. (2011), Lipman and Funkhouser (2009), Bronstein et al. (2006), Ovsjanikov et al. (2010), and Zhang et al. (2008) on the benchmark dataset SCAPE and TOSCA.

Acknowledgements

R.L.'s work is supported in part by NSF DMS-1522645 and an NSF CAREER award, DMS-1752934. H.Z.'s work is supported in part by DMS-1622490 and DMS-1821010.

References

Alfakih, A.Y., Khandani, A., Wolkowicz, H., 1999. Solving Euclidean distance matrix completion problems via semidefinite programming. Comput. Optim. Appl. 12 (1–3), 13–30.

Allard, W.K., Chen, G., Maggioni, M., 2012. Multiscale geometric methods for data sets II: geometric multi-resolution analysis. Appl. Comput. Harmon. Anal. 32 (3), 435–462.

Anguelov, D., Srinivasan, P., Koller, D., Thrun, S., Rodgers, J., Davis, J., 2005. SCAPE: shape completion and animation of people. ACM Trans. Graphics (TOG) 24 (3), 408–416.

Belkin, M., Niyogi, P., 2002. Laplacian eigenmaps for dimensionality reduction and data representation. Neural Comput. 15, 1373–1396.

Belkin, M., Sun, J., Wang, Y., 2009. Constructing Laplace operator from point clouds in $\mathbb{R}^d$. In: Proceedings of the Twentieth Annual ACM-SIAM Symposium on Discrete Algorithms, pp. 1031–1040. http://dl.acm.org/citation.cfm?id=1496770.1496882. Philadelphia, PA, USA.

Bérard, P., Besson, G., Gallot, S., 1994. Embedding Riemannian manifolds by their heat kernel. Geom. Funct. Anal. 4 (4), 373–398.

Berger, B., Kleinberg, J., Leighton, T., 1999. Reconstructing a three-dimensional model with arbitrary errors. J. ACM (JACM) 46 (2), 212–235.

Bertalmio, M., Sapiro, G., Cheng, L.-T., Osher, S., 2000. A framework for solving surface partial differential equations for computer graphics applications. UCLA CAM Report (00-43).

Bertalmio, M., Mémoli, F., Cheng, L.-T., Sapiro, G., Osher, S., 2002. Variational problems and partial differential equations on implicit surfaces. UCLA CAM Report (02-17).

Biswas, P., Liang, T.-C., Toh, K.-C., Ye, Y., Wang, T.-C., 2006. Semidefinite programming approaches for sensor network localization with noisy distance measurements. IEEE Trans. Autom. Sci. Eng. 3 (4), 360.

Borg, I., Groenen, P. J. F., 2005. Modern Multidimensional Scaling: Theory and Applications. Springer Science & Business Media.

Bronstein, M.M., Kokkinos, I., 2010. Scale-invariant heat kernel signatures for non-rigid shape recognition. In: IEEE Conference on Computer Vision and Pattern Recognition (CVPR), pp. 1704–1711.

Bronstein, A.M., Bronstein, M.M., Kimmel, R., 2006. Generalized multidimensional scaling: a framework for isometry-invariant partial surface matching. Proc. Natl. Acad. Sci. U.S.A. 103 (5), 1168–1172.

Bronstein, A.M., Bronstein, M.M., Kimmel, R., 2008. Numerical Geometry of Non-Rigid Shapes. Springer.

Bronstein, A.M., Bronstein, M.M., Kimmel, R., Mahmoudi, M., Sapiro, G., 2010. A Gromov-Hausdorff framework with diffusion geometry for topologically-robust non-rigid shape matching. Int. J. Comput. Vis. 89 (2–3), 266–286.

Candès, E.J., Recht, B., 2008. Exact matrix completion via convex optimization. Found. Comput. Math. 9, 717–772.

Carr, N.A., Hoberock, J., Crane, K., Hart, J.C., 2006. Rectangular multi-chart geometry images. In: Symposium on Geometry Processing, pp. 181–190.

Chavel, I., 1984. Eigenvalues in Riemannian Geometry. vol. 115. Academic Press.

Coifman, R., Lafon, S., 2006. Diffusion maps. Appl. Comput. Harmon. Anal. 21, 6–30.

Coifman, R.R., Kevrekidis, I.G., Lafon, S., Maggioni, M., Nadler, B., 2008. Diffusion maps, reduction coordinates, and low dimensional representation of stochastic systems. Multiscale Model. Simul. 7, 842–864.

Crippen, G.M., Havel, T.F., 1988. Distance Geometry and Molecular Conformation. vol. 74. Research Studies Press, Taunton, UK.

Dijkstra, E.W., 1959. A note on two problems in connexion with graphs. Numer. Math. 1 (1), 269–271.

Dziuk, G., Elliott, C.M., 2013. Finite element methods for surface PDEs. Acta Numer. 22, 289–396.

Fang, X., Toh, K.-C., 2013. Using a distributed SDP approach to solve simulated protein molecular conformation problems. In: Distance Geometry, Springer, pp. 351–376.

Glowinski, R., Le Tallec, P., 1989. Augmented Lagrangian and Operator-Splitting Methods in Nonlinear Mechanics. SIAM.

Goldluecke, B., Cremers, D., 2009. Superresolution texture maps for multiview reconstruction. In: 2009 IEEE 12th International Conference on Computer Vision, IEEE, pp. 1677–1684.

Gu, X., Wang, Y., Chan, T.F., Thompson, P.M., Yau, S.-T., 2004. Genus zero surface conformal mapping and its application to brain surface mapping. IEEE Trans. Med. Imaging 23 (8), 949–958.

Hummer, G., 2004. From transition paths to transition states and rate coefficients. J. Chem. Phys. 120, 516–523.

Ji, X., Zha, H., 2004. Sensor positioning in wireless ad-hoc sensor networks using multidimensional scaling. In: INFOCOM 2004. Twenty-third Annual Joint Conference of the IEEE Computer and Communications Societies, vol. 4, pp. 2652–2661.

Jolliffe, I., 2002. Principal Component Analysis. Wiley Online Library.

Jones, P.W., Maggioni, M., Schul, R., 2008. Manifold parametrizations by eigenfunctions of the Laplacian and heat kernels. Proc. Natl. Acad. Sci U.S.A. 105 (6), 1803–1808.

Jost, J., 2008. Riemannian Geometry and Geometric Analysis. Springer Science & Business Media.

Kantorovich, L., 1942. On the translocation of masses. C.R. (Doklady) Acad. Sci. URSS (N.S.) 37, 199–201.

Kim, V.G., Lipman, Y., Funkhouser, T., 2011. Blended intrinsic maps. ACM Trans. Graph. 30 (4), 79.

Kimmel, R., Sethian, J.A., 1998. Computing geodesic paths on manifolds. Proc. Natl. Acad. Sci. 95 (15), 8431–8435.

Kruskal, J.B., Wish, M., 1978. Multidimensional Scaling. vol. 11. Sage.

Lai, R., Li, J., 2018. Solving partial differential equations on manifold from incomplete inter-point distance. SIAM J. Sci. Comput. 39 (5), 2231–2256.

Lai, R., Lu, J., 2017. Point cloud discretization of Fokker-Planck operators for committor functions. arXiv preprint arXiv:1703.09359.

Lai, R., Zhao, H., 2017. Multiscale nonrigid point cloud registration using robust sliced-Wasserstein distance via Laplace-Beltrami eigenmap. SIAM J. Imaging Sci. 10 (2), 449–483.

Lai, R., Shi, Y., Dinov, I., Chan, T.F., Toga, A.W., 2009. Laplace-Beltrami nodal counts: a new signature for 3D shape analysis. In: Proc. ISBI, pp. 694–697.

Lai, R., Shi, Y., Scheibel, K., Fears, S., Woods, R., Toga, A.W., Chan, T.F., 2010. Metric-induced optimal embedding for intrinsic 3D shape analysis. Comput. Vis. Pattern Recognit. (CVPR), 2871–2878.

Lai, R., Shi, Y., Sicotte, N., Toga, A.W., 2011. Automated corpus callosum extraction via Laplace-Beltrami nodal parcellation and intrinsic geodesic curvature flows on surfaces. In: ICCV2034–2040.

Lai, R., Liang, J., Zhao, H., 2013. A local mesh method for solving PDEs on point clouds. Inverse Prob. Imaging 7 (3), 737–755.

Lai, R., Wen, Z., Yin, W., Gu, X., Lui, L.M., 2014. Folding-free global conformal mapping for genus-0 surfaces by harmonic energy minimization. J. Sci. Comput. 58 (3), 705–725.

Leung, N.-H.Z., Toh, K.-C., 2009. An SDP-based divide-and-conquer algorithm for large-scale noisy anchor-free graph realization. SIAM J. Sci. Comput. 31 (6), 4351–4372.

Levin, D., 1998. The approximation power of moving least-squares. Math. Comput. Am. Math. Soc. 67 (224), 1517–1531.

Levina, E., Bickel, P.J., 2004. Maximum likelihood estimation of intrinsic dimension. In: Advances in Neural Information Processing Systems, pp. 777–784.

Levy, B., 2006. Laplace-Beltrami eigenfunctions: towards an algorithm that understands geometry. In: IEEE International Conference on Shape Modeling and Applications, Invited Talk.

Li, Z., Shi, Z., 2016. A convergent point integral method for isotropic elliptic equations on a point cloud. Multiscale Model. Simul. 14 (2), 874–905.

Li, Z., Shi, Z., Sun, J., 2017. Point integral method for solving Poisson-type equations on manifolds from point clouds with convergence guarantees. Commun. Comput. Phys. 22 (1), 228–258.

Liang, J., Zhao, H., 2013. Solving partial differential equations on point clouds. SIAM J. Sci. Comput. 35 (3), A1461–A1486.

Liang, J., Lai, R., Wong, T., Zhao, H., 2012. Geometric understanding of point clouds using Laplace-Beltrami operator. Comput. Vis. Pattern Recognit. (CVPR), 214–221.

Lin, T., Zha, H., 2008. Riemannian manifold learning. IEEE Trans. Pattern Anal. Mach. Intell. 30 (5), 796–809.

Lipman, Y., 2009. Stable moving least-squares. J. Approx. Theory 161 (1), 371–384.

Lipman, Y., Funkhouser, T., 2009. Möbius voting for surface correspondence. ACM Trans. Graphics (TOG) 28 (72), 1–12.

Lipman, Y., Cohen-Or, D., Levin, D., 2006. Error bounds and optimal neighborhoods for MLS approximation. In: Proceedings of the fourth Eurographics symposium on Geometry processing, pp. 71–80.

Little, A.V., Jung, Y.-M., Maggioni, M., 2009a. Multiscale estimation of intrinsic dimensionality of data sets. In: AAAI Fall Symposium: Manifold Learning and its Applications, vol. 9, p. 04.

Little, A.V., Lee, J., Jung, Y.-M., Maggioni, M., 2009b. Estimation of intrinsic dimensionality of samples from noisy low-dimensional manifolds in high dimensions with multiscale SVD. In: IEEE/SP 15th Workshop on Statistical Signal Processing, 2009. SSP'09, IEEE, pp. 85–88.

Lui, L.M., Wang, Y., Chan, T.F., 2005. Solving PDEs on manifolds with global conformal parametriazation. In: International Workshop on Variational, Geometric, and Level Set Methods in Computer VisionSpringer, pp. 307–319.

Mémoli, F., Sapiro, G., 2005. Distance functions and geodesics on submanifolds of Rd and point clouds. SIAM J. Appl. Math. 65 (4), 1227–1260.

Metzler, R., Oshanin, G., Redner, S., 2014. First-Passage Phenomena and Their Applications. World Scientific.

Meyer, M., Desbrun, M., Schröder, P., Barr, A.H., 2002. Discrete differential-geometry operators for triangulated 2-manifolds. Vis. Math. 3 (2), 52–58.

Milnor, J.W., 1963. Morse Theory. vol. 51. Princeton University Press.

Monge, G., 1781. Mémoire sur la théorie des déblais et des remblais. In: Histroire de l'Académie Royale des Sciences de Paris, avec les Mémoires de Mathématique et de Physique pour la même année, pp. 666–704.

Osher, S., Sethian, J., 1988. Fronts propagation with curvature-dependent speed: algorithms based on Hamilton-Jacobi formulations. J. Comput. Phys. 79 (1), 12–49.

Ovsjanikov, M., Mérigot, Q., Mémoli, F., Guibas, L., 2010. One point isometric matching with the heat kernel. Comput. Graphics Forum 29 (5), 1555–1564.

Peters, B., 2016. Reaction coordinates and mechanistic hypothesis tests. Ann. Rev. Phys. Chem. 67, 669–690.

Pinkall, U., Polthier, K., 1993. Computing discrete minimal surfaces and their conjugates. Exp. Math. 2 (1), 15–36.

Qian, J., Zhang, Y.-T., Zhao, H.-K., 2007. Fast sweeping methods for Eikonal equations on triangular meshes. SIAM J. Numer. Anal. 45 (1), 83–107.

Qiu, A., Bitouk, D., Miller, M.I., 2006. Smooth functional and structural maps on the neocortex via orthonormal bases of the Laplace-Beltrami operator. IEEE Trans. Med. Imag. 25 (10), 1296–1306.

Raviv, D., Bronstein, A.M., Bronstein, M.M., Kimmel, R., Sochen, N., 2011. Affine-invariant diffusion geometry for the analysis of deformable 3D shapes. CVPR, 2361–2367.

Recht, B., Fazel, M., Parrilo, P., 2010. Guaranteed minimum-rank solutions of linear matrix equations via nuclear norm minimization. SIAM Rev. 52 (3), 471–501.

Reuter, M., 2009. Hierarchical shape segmentation and registration via topological features of Laplace-Beltrami eigenfunctions. Int. J. Comput. Vis. 89 (2–3), 287–308. https://doi.org/10.1007/s11263-009-0278-1.

Rohrdanz, M.A., Zheng, W., Maggioni, M., Clementi, C., 2011. Determination of reaction coordinates via locally scaled diffusion map. J. Chem. Phys. 134, 124116.

Rustamov, R.M., 2007. Laplace-Beltrami eigenfunctions for deformation invariant shape representation. In: Eurographics Symposium on Geometry Processing.

Sethian, J.A., 1996. A fast marching level set method for monotonically advancing fronts. Proc. Natl. Acad. Sci. 93 (4), 1591–1595.

Shi, Y., Lai, R., Kern, K., Sicotte, N., Dinov, I., Toga, A.W., 2008a. Harmonic surface mapping with Laplace-Beltrami eigenmaps. In: Proc. MICCAI, pp. 147–154.

Shi, Y., Lai, R., Krishna, S., Sicotte, N., Dinov, I., Toga, A.W., 2008b. Anisotropic Laplace-Beltrami eigenmaps: bridging Reeb graphs and skeletons. In: IEEE Computer Society Conference on Computer Vision and Pattern Recognition Workshops, 2008b. CVPRW'08, IEEE, pp. 1–7.

Shi, Y., Lai, R., Krishna, S., Sicotte, N., Dinov, I., Toga, A.W., 2008c. Anisotropic Laplace-Beltrami eigenmaps: bridging Reeb graphs and skeletons. In: Computer Vision and Pattern Recognition Workshops, 1–7.

Shi, Y., Dinov, I., Toga, A.W., 2009. Cortical shape analysis in the Laplace-Beltrami feature space. In: Medical Image Computing and Computer-Assisted Intervention, Springer, pp. 208–215.

Shi, Y., Lai, R., Morra, J., Dinov, I., Thompson, P., Toga, A.W., 2010. Robust surface reconstruction via Laplace-Beltrami eigen-projection and boundary deformation. IEEE Trans. Medical Imaging 29 (12), 2009–2022.

Shi, Y., Lai, R., Gill, R., Pelletier, D., Mohr, D., Sicotte, N., Toga, A.W., 2011. Conformal metric optimization on surface (CMOS) for deformation and mapping in Laplace-Beltrami embedding space. In: International Conference on Medical Image Computing and Computer-Assisted Intervention, Springer, pp. 327–334.

Shi, Y., Lai, R., Toga, A., 2012. Unified geometry and topology correction for cortical surface reconstruction with intrinsic Reeb analysis. In: Medical Image Computing and Computer-Assisted Intervention (MICCAI), 601–608.

Shtern, A., Kimmel, R., 2014. Iterative closest spectral kernel maps. In: 2014 2nd International Conference on 3D Vision (3DV), vol. 1. IEEE, pp. 499–505.

Shtern, A., Kimmel, R., 2015. Spectral gradient fields embedding for nonrigid shape matching. Comput. Vis. Image Underst. 140, 21–29.

Singer, A., 2006. From graph to manifold Laplacian: the convergence rate. Appl. Comput. Harmon. Anal. 21 (1), 128–134.

Sun, J., Ovsjanikov, M., Guibas, L., 2009. A concise and provably informative multi-scale signature based on heat diffusion. Comput. Graphics Forum 28 (5), 1383–1392.

Tasissa, A., Lai, R., Exact reconstruction of Euclidean distance geometry problem using low-rank matrix completion. IEEE Trans. Inf. Theory 65 (5), 3124–3144.

Taubin, G., 2000. Geometric signal processing on polygonal meshes. Eurographics 4 (3), 81–96.

Todd, M.J., 2001. Semidefinite optimization. Acta Numer. 10, 515–560.

Uhlenbeck, K., 1976. Generic properties of eigenfunctions. Am. J. Math. 98 (4), 1059–1078.

Vallet, B., Levyvy, B., 2008. Spectral geometry processing with manifold harmonics. Comput. Graphics Forum (Proceedings Eurographics) 27, 251–260.

Vandenberghe, L., Boyd, S., 1996. Semidefinite programming. SIAM Rev 38, 49–95.

Vanden-Eijnden, E., 2010. Transition path theory and path-finding algorithms for the study of rare events. Ann. Rev. Phys. Chem. 61, 391–420.

Villani, C., 2003. Topics in Optimal Transportation. vol. 58. American Mathematical Soc.

Villani, C., 2008. Optimal Transport: Old and New. vol. 338. Springer.

Wang, M., Leung, S., Zhao, H., 2018. Modified virtual grid difference for discretizing the Laplace-Beltrami operator on point clouds. SIAM J. Sci. Comput. 40 (1), A1–A21.

Weinan, E., Vanden-Eijnden, E., 2006. Toward a theory of transition paths. J. Stat. Phys. 123, 503–523.

Wen, Z., Yin, W., 2013. A feasible method for optimization with orthogonality constraints. Math. Program. 142 (1–2), 397–434.

Wen, Z., Goldfarb, D., Yin, W., 2010. Alternating direction augmented Lagrangian methods for semidefinite programming. Math. Program. Comput. 2, 203–230.

Xu, G., 2004. Convergent discrete Laplace-Beltrami operators over triangular surfaces. In: Geometric Modeling and Processing, 2004. Proceedings, IEEE, pp. 195–204.

Zhang, H., Sheffer, A., Cohen Or, D., Zhou, Q., Van Kaick, O., Tagliasacchi, A., Wiley Online Library, 2008. Deformation-driven shape correspondence. Comput. Graphics Forum 27 (5), 1431–1439.

Zhao, H., 2005. A fast sweeping method for Eikonal equations. Math. Comput. 74 (250), 603–628.

Zheng, W., Rohrdanz, M.A., Maggioni, M., Clementi, C., 2011. Polymer reversal rate calculated via locally scaled diffusion map. J. Chem. Phys. 134, 144108.

Chapter 9

Tighter continuous relaxations for MAP inference in discrete MRFs: A survey

Hariprasad Kannan[a,*], Nikos Komodakis[b] and Nikos Paragios[c]
[a]*Inria Saclay, Palaiseau, France*
[b]*Ecole des Ponts ParisTech, Champs-sur-Marne, France*
[c]*CentraleSupelec, Gif-sur-Yvette, France*
[*]*Corresponding author: e-mail: hkannan@gmail.com*

Chapter Outline

Abstract

Continuous relaxations are central to MAP inference in discrete Markov random fields (MRFs). In these methods, the intractable discrete optimization problem is approximated by a continuous relaxation. The relaxation could be based on different approaches, with the linear programming (LP) relaxation being the most well studied. For continuous relaxations, the important considerations are efficiently solving the optimization formulations and improving the approximation quality of the relaxations. In this chapter, we focus on the latter topic, which is referred to as tighter continuous relaxations. We present a comprehensive survey of techniques to achieve a tighter LP relaxation, followed by a discussion of semidefinite programming (SDP) relaxation techniques. We discuss these topics by covering both theoretical motivations and algorithmic considerations. We hope this will be a ready reference for researchers working

Handbook of Numerical Analysis, Vol. 20. https://doi.org/10.1016/bs.hna.2019.06.001
© 2019 Elsevier B.V. All rights reserved.

on tighter relaxations and for practitioners who want to model their problem more accurately and efficiently. The foundation for the theoretical aspects comes from the discrete optimization community through the topics of polyhedral combinatorics and relaxation hierarchies. The structure of discrete MRFs lends a special perspective to these classic topics and we will see how several works have explored this path. Over the years, research efforts in discrete MRFs have characterized graph topologies and/or clique energies that lead to accurate MAP inference. In recent years, there have been similar efforts, to better understand the settings and their accompanying guarantees, for the algorithms that tighten the continuous relaxations. We will end this chapter by giving a taste for this emerging research topic.

Keywords: Graphical models, Discrete Markov random fields, MAP inference, Discrete and continuous optimization, Continuous relaxation, Relaxation hierarchies

Probabilistic graphical models describe a set of random variables in a compact manner by exploiting the structured nature of their interaction (Koller and Friedman, 2009; Wainwright and Jordan, 2008). In this chapter, we focus on undirected graphical models, which are referred to as Markov random fields (MRFs) (Blake et al., 2011). We have a graph $\mathcal{G}=(\mathcal{V},\mathcal{C})$, where $\mathcal{V}$ denotes the set of nodes, which are the random variables and $\mathcal{C}$ denotes the set of cliques, which represent the interactions. In this chapter, the random variables take discrete values or labels. A clique composed of two nodes is also called an edge. A clique with more than two nodes is called a hyperedge or a higher-order clique. We will refer to cliques, edges and hyperedges interchangeably, where the meaning will be clear based on the context. An important computation that we seek is finding the labelling of all the random variables, which will maximize the joint probability distribution,

$$P(\boldsymbol{X}=\boldsymbol{x})=\prod_{i\in\mathcal{V}}\Psi_i(x_i)\prod_{c\in\mathcal{C}}\Psi_c(\boldsymbol{x}_c)=\exp\left(-\sum_{i\in\mathcal{V}}\theta_i(x_i)-\sum_{c\in\mathcal{C}}\theta_c(\boldsymbol{x}_c)\right). \tag{1}$$

Here, $\boldsymbol{X}$ denotes the complete set of n discrete variables $X_1, \ldots, X_n$. The set of variables corresponding to a clique c is indicated by $\boldsymbol{X}_c$. Without loss of generality, we assume that all variables take labels from a common set of size l. A particular realization (labelling) of X_i is denoted by x_i. Thus, the labelling of a clique, $\boldsymbol{x}_c$, is a vector of node labels. Thus, a clique could take one of $l^{|c|}$ labellings. Depending on the application, we define appropriate potential functions over nodes and cliques. They take the form $\Psi_i(x_i)=\exp(-\theta_i(x_i))$ and $\Psi_c(\boldsymbol{x}_c)=\exp(-\theta_c(\boldsymbol{x}_c))$, where $\theta_i(x_i)$ and $\theta_c(\boldsymbol{x}_c)$ denote the energies (also, called potentials) corresponding to node i taking label x_i and clique c taking label $\boldsymbol{x}_c$, respectively. The joint probability distribution factorizing according to the nodes and the cliques is a fundamental property of MRFs. Based on (1), maximum a posteriori inference or MAP inference, computes the labelling $\boldsymbol{x}$ that minimizes the following energy,

$$E(\boldsymbol{x};\boldsymbol{\theta})\triangleq\sum_{i\in\mathcal{V}}\theta_i(x_i)+\sum_{c\in\mathcal{C}}\theta_c(\boldsymbol{x}_c). \tag{2}$$

However, except for some special graph topologies and/or clique potentials, this problem is NP-hard (Li et al., 2016; Shimony, 1994). Thus, there has been considerable research effort to develop approximate inference algorithms for MRFs. Several approaches exist to minimize this energy efficiently and approximately: especially methods based on graph cuts, belief propagation and continuous relaxation. Refer Szeliski et al. (2008), Blake et al. (2011) and Kappes et al. (2015) for a good survey about various methods. Among these methods, continuous relaxation-based approaches have led to state-of-the-art algorithms and also, provide a theoretical foundation to the topic of MAP inference (Komodakis et al., 2016; Wainwright et al., 2005).

First, we observe that the energy in (2) has a linear form, with respect to the energies, $\theta_i(x_i)$'s and $\theta_c(\boldsymbol{x}_c)$'s. Thus, MAP inference can be equivalently represented as an intractable integer linear program (ILP) as follows,

$$\underset{\boldsymbol{\phi}}{\text{minimize}} \quad \sum_i \sum_{x_i} \theta_i(x_i)\phi_i(x_i) + \sum_c \sum_{\boldsymbol{x}_c} \theta_c(\boldsymbol{x}_c)\phi_c(\boldsymbol{x}_c) \tag{3a}$$

$$\text{subject to} \quad \sum_{\boldsymbol{x}_{c\backslash i}} \phi_c(\boldsymbol{x}_c) = \phi_i(x_i), \quad \forall (i,c): i \in c,\ \forall x_i \tag{3b}$$

$$\begin{aligned} &\sum_{x_i} \phi_i(x_i) = 1, \quad \phi_i(x_i) \in \{0,1\} \quad \forall i \in \mathcal{V} \\ &\sum_{\boldsymbol{x}_c} \phi_c(\boldsymbol{x}_c) = 1, \quad \phi_c(\boldsymbol{x}_c) \in \{0,1\} \quad \forall c \in \mathcal{C}. \end{aligned} \tag{3c}$$

Here, for each node-label pair (i, x_i), there is an indicator variable $\phi_i(x_i)$. Similarly, for each clique-label pair $(c, \boldsymbol{x}_c)$, there is an indicator variable $\phi_c(\boldsymbol{x}_c)$. Let us define a vector $\boldsymbol{\phi}$ that is obtained by stacking all the $\phi_i(x_i)$'s and $\phi_c(\boldsymbol{x}_c)$'s together. The normalization constraints, (3c), ensure that for a given labelling of all the nodes, for each node i, only one of $\phi_i(x_i)$ will be set to 1 and for each clique c, only one of $\phi_c(\boldsymbol{x}_c)$ will be set to 1. The node-based indicator variables, $\phi_i(x_i)$'s, are enough to denote the labelling of the graphical model. However, we need the clique-based indicator variables, $\phi_c(\boldsymbol{x}_c)$'s, in order to model our optimization problem. Shortly, we will see that this approach leads to a relaxation of the integer linear program that is convex in nature. Since, nodes and cliques are having separate indicator variables, we have to make sure that the labelling of the nodes and the labelling of the cliques agree. We ensure this through the marginalization constraints given by (3b). We will see how the summation $\sum_{\boldsymbol{x}_{c\backslash i}} \phi_c(\boldsymbol{x}_c)$ is performed. For a given clique c and node $i \in c$, there will be a set of clique labellings $\boldsymbol{x}_c$'s, where the label of node i is fixed to x_i. We will denote this set as $\boldsymbol{x}_{c\backslash i}$. With a slight abuse of notation, it is a more compact way of denoting the set, $\boldsymbol{x}_c : \boldsymbol{x}_c(i) = x_i$. Since, we will come across such summations often, a shorter notation is preferable. The summation of $\phi_c(\boldsymbol{x}_c)$'s is performed over this set and the constraint says that the sum should match with $\phi_i(x_i)$.

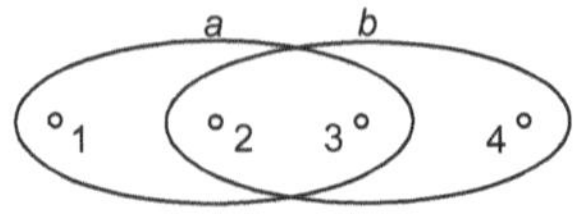

FIG. 1 A toy graphical model with nodes 1, 2, 3 and 4 and cliques a and b.

Using the toy graphical model given in Fig. 1, we will further understand these variables and constraints. In this graphical model, each clique is made of three nodes, i.e., $a = \{1, 2, 3\}$ and $b = \{2, 3, 4\}$. We will assume that each node could take one of three labels, $\{x_1, x_2, x_3\}$. Thus, both the $\phi(\boldsymbol{x}_a)$'s and the $\phi(\boldsymbol{x}_b)$'s consist of 27 elements. Consider the case where node 3 takes label x_2, i.e., $\phi_3(x_1) = 0$, $\phi_3(x_2) = 1$, $\phi_3(x_3) = 0$. We see that the normalization constraint (3c) is satisfied. Node 3 is part of both cliques a and b. There are nine elements in each set of $\phi_a(\boldsymbol{x}_a)$'s and $\phi_b(\boldsymbol{x}_b)$'s, which correspond to the case of node 3 taking a particular label. We denote these sets by $\boldsymbol{x}_{a\backslash 3}$ and $\boldsymbol{x}_{b\backslash 3}$, respectively. According to the marginalization constraint (3b), the summation of $\phi_a(\boldsymbol{x}_a)$'s (resp. $\phi_b(\boldsymbol{x}_b)$'s) over $\boldsymbol{x}_{a\backslash 3}$ (resp. $\boldsymbol{x}_{b\backslash 3}$) should equal 1.

In the quest for efficient algorithms for approximate MAP inference, one could derive continuous relaxation-based formulations for the ILP given in (3). There have been relaxations based on linear programming (LP), quadratic programming (QP), semidefinite programming (SDP) and second-order cone programming (SOCP). Kumar et al. (2009) have presented an excellent analysis of these various relaxations. A continuous relaxation is naturally going to return an approximate solution and it is worthwhile to *tighten* the relaxation in order to attain more accurate results. In this chapter, we will survey several important aspects of this topic. First, we will see various algorithms that could help to achieve a tighter LP relaxation. Second, we will see inference algorithms based on the SDP relaxation that could offer better results than the LP relaxation for some problem instances. We will conclude by indicating research efforts to characterize graph topologies and clique potentials that govern the tightness of these relaxations.

1 LP relaxation

In this section, we will see how a tractable optimization problem could be obtained by LP relaxation and present the core concepts of marginal and local polytopes. First, we observe that the ILP in (3) could be equivalently expressed as an intractable linear program. In this LP, the objective is expressed in the same way as (3a) and the decision variables, $\phi_i(x_i)$'s and $\phi_c(\boldsymbol{x}_c)$'s, take continuous values within a polytope called the *marginal* polytope, denoted as $\mathcal{M}(\mathcal{G})$. The marginal polytope is the convex hull of the feasible set of the ILP (3). Thus, the vertices of $\mathcal{M}(\mathcal{G})$ are points taking $\{0, 1\}$ integer values, satisfying the constraints, (3b) and (3c). The optimal solution of any linear program consists of one of the vertices of the polytope over which it is optimized. Thus, the solution of this LP will automatically belong to the feasible set of (3). However, one needs an exponential number of constraints to represent the facets of $\mathcal{M}(\mathcal{G})$, thus this LP is as hard the original ILP. In order to obtain a more efficient situation, we could optimize over a feasible region that is simpler than the marginal polytope. In order to characterize such tractable feasible regions, we need to understand more about the marginal polytope. The marginal polytope contains those $\boldsymbol{\phi}$'s for which the $\phi_i(x_i)$'s and $\phi_c(\boldsymbol{x}_c)$'s will be valid marginals, with respect to some joint probability distribution over all the random variables. This property can be stated as follows,

$$\begin{aligned}\mathcal{M}(\mathcal{G}) = \Big\{\boldsymbol{\phi} \mid \exists\, P \in \mathscr{P}(\{0,1,\ldots,l-1\}^{|\mathcal{V}|}) \\ \text{s.t. } \phi_i(x_i) = P(X_i = x_i)\ \forall i \in \mathcal{V},\ \phi_c(\boldsymbol{x}_c) = P(\boldsymbol{X}_c = \boldsymbol{x}_c)\ \forall c \in \mathcal{C}\Big\}\end{aligned} \tag{4}$$

where $\mathscr{P}(\{0,1,\ldots,l-1\}^{|\mathcal{V}|})$ is the set of all valid probability distributions defined over the finite set $\{0,1,\ldots,l-1\}^{|\mathcal{V}|}$. It takes an exponential number of linear constraints to ensure that a vector $\boldsymbol{\phi}$ belongs to $\mathcal{M}(\mathcal{G})$. The main idea behind LP relaxations is to enforce a tractable number of constraints, which will lead to simpler polytopes to optimize over. The most common LP relaxation takes the following form,

$$\underset{\boldsymbol{\phi}}{\text{minimize}} \quad \sum_i \sum_{x_i} \theta_i(x_i)\phi_i(x_i) + \sum_c \sum_{\boldsymbol{x}_c} \theta_c(\boldsymbol{x}_c)\phi_c(\boldsymbol{x}_c) \tag{5a}$$

$$\text{subject to} \quad \sum_{\boldsymbol{x}_{c\setminus i}} \phi_c(\boldsymbol{x}_c) = \phi_i(x_i), \quad \forall (i,c) : i \in c,\ \forall x_i \tag{5b}$$

$$\begin{aligned}&\sum_{x_i} \phi_i(x_i) = 1, \quad \phi_i(x_i) \geq 0 \quad \forall i \in \mathcal{V} \\ &\sum_{\boldsymbol{x}_c} \phi_c(\boldsymbol{x}_c) = 1, \quad \phi_c(\boldsymbol{x}_c) \geq 0 \quad \forall c \in \mathcal{C}.\end{aligned} \tag{5c}$$

The marginalization constraints (5b) and the normalization constraints (5c) should be understood similar to (3b) and (3c), respectively. In this case, we work with continuous valued variables. For this LP, the constraint polytope is called the *local* polytope because the linear constraints are marginalization constraints (5b) that are only enforced locally with respect to the cliques. This represents a substantially lesser number of constraints compared to the marginal polytope. We will denote the local polytope as $\mathcal{L}(\mathcal{G})$. The local polytope is an outer bound on the marginal polytope. More precisely, it will contain all the points that satisfy (4), since, they automatically satisfy (5b) and (5c), also. Further, $\mathcal{L}(\mathcal{G})$ contains other points that satisfy constraints (5b) and (5c) but not (4). The points contained within the local polytope are called *pseudomarginals*. This is inspired by the fact that the marginal polytope contains true marginals.

All the {0, 1} valued vertices of the marginal polytope are also vertices of the local polytope. Apart from these, there are several fractional valued vertices in the local polytope. It is difficult to visualize the marginal and local polytopes. Fig. 2 represents these polytopes in an abstract way. Suppose the LP relaxation returns a integer valued vertex, then we have the guarantee that we have recovered the exact MAP labelling of the graphical model. This is an important guarantee of the LP relaxation-based approach (Wainwright et al., 2005). However, the LP relaxation could return a fractional valued vertex. The desired MAP labelling is obtained by rounding these fractional values. Tightening the polytope over which the LP is optimized could lead to a fractional solution that could give a better integral solution after rounding or if possible an integer valued solution could be obtained. Since, the integral points belonging to the local polytope correspond to the vertices of the marginal polytope, an integral solution obtained through a tighter polytope will correspond to a vertex of the marginal polytope. We have already seen that the MAP solution will correspond to a vertex of the marginal polytope. Thus, we would have obtained the exact MAP solution, if we are lucky with our tighter relaxation.

So far, we have discussed about two linear programs: the intractable LP that is exactly equivalent to the ILP given in (3) and the LP relaxation given in (5). The formulation given in (5) looks similar to the ILP, except for one difference. The integer values constraints in (3c) have been relaxed to continuous values (5c). However, this simple difference leads to a much more tractable problem.

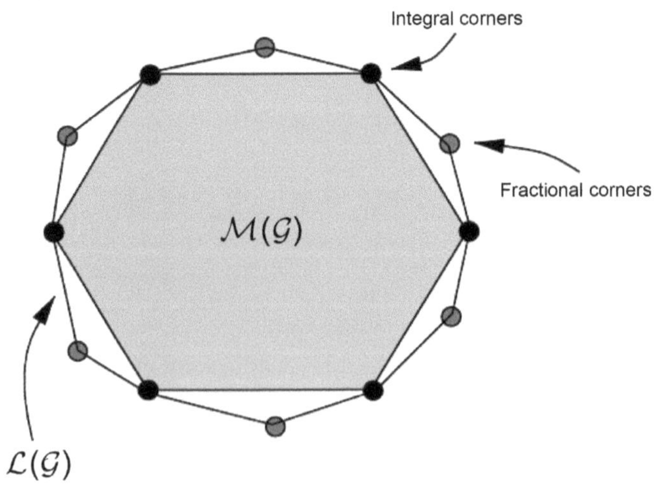

FIG. 2 A cartoonish illustration of the relationship between the marginal polytope $\mathcal{M}(\mathcal{G})$ and the local polytope $\mathcal{L}(\mathcal{G})$. All the vertices of $\mathcal{M}(\mathcal{G})$ are also vertices of $\mathcal{L}(\mathcal{G})$. $\mathcal{L}(\mathcal{G})$ also has fractional vertices. The fact that $\mathcal{L}(\mathcal{G})$ has lesser number of facets than $\mathcal{M}(\mathcal{G})$ is difficult to visualize. *Image courtesy Wainwright, M.J. Jordan, M.I., 2008. Graphical models, exponential families, and variational inference. Found. Trends Mach. Learn. 1 (1–2), 1–305.*

1.1 Tightening the polytope

Now, we will see how looser polytopes like the local polytope could be tightened to obtain more accurate results. Tighter relaxations for MAP inference are intimately related to *polyhedral combinatorics* (Schrijver, 2003). According to the combinatorial optimization community, a sequence of tighter LP relaxations called the Sherali–Adams hierarchy could approximate the ILP (3) to arbitrary precision (Laurent, 2003). In each level of the hierarchy, we obtain a polytope $\mathcal{SA}_r(\mathcal{G})$ by imposing marginalization constraints over all possible *clusters* of size r. A cluster is the most general data structure that we will come across in our discussions. In fact, any subset of nodes could be a cluster. In order to define the new set of marginalization constraints, we define a new set of variables, $\boldsymbol{\tau}_s$ for each cluster s. The marginalization constraints take the same form as (5b), with the summation computed over the cluster, while the label is held fixed for subsets (like cliques or individual nodes) of the cluster. The highest level of the hierarchy, $\mathcal{SA}_{treewidth}(\mathcal{G})$, corresponds to the marginal polytope and the size of the clusters that is required at this level is equal to the treewidth of the graph. Thus the highest level has intractable complexity. Even lower levels like $r = 2$ or 3 are computationally expensive. Since, it involves imposing constraints on all possible pairs or triplets of nodes. In practice, we consider smaller sets of clusters, like only pairs of nodes that correspond to edges of the graph could be considered. This is exactly what leads to the local polytope. In Section 6, we will see more about efforts to understand such hierarchies and their guarantees with respect to graph topology and/or clique potentials (Rowland et al., 2017; Thapper and Živný, 2016; Wainwright and Jordan, 2004b; Weller et al., 2016). For now, we will look at algorithms to tighten the relaxation, which work well in practice, while accompanied by a reasonable level of theoretical justification. We will see some of the pioneering works that algorithmically addressed the tighter relaxation problem (Komodakis and Paragios, 2008; Sontag and Jaakkola, 2007; Sontag et al., 2008; Werner, 2008). Before we go further, it is good to know that tightening the relaxation based on clusters has been pursued in the statistical physics community in the name of Kikuchi clustering (Wainwright and Jordan, 2008, section 4) and has been used within belief propagation algorithms (Yedidia et al., 2005).

When one talks about tightening a polytope, the natural algorithmic approach that comes to mind is a *cutting-plane* method. In a cutting-plane method, one progressively tightens the polytope over which the optimization is performed by adding more constraints. This is the underlying principle behind (Sontag and Jaakkola, 2007), which starts with a loose relaxation of the marginal polytope, like the local polytope and iteratively adds constraints. In the previous paragraph, we came across marginalization constraints that could be enforced with respect to clusters of larger and larger sizes in order to tighten the polytope. In the combinatorial optimization literature, other valid constraints are also possible that lead to tighter relaxations. Sontag and Jaakkola (2007) worked with *cycle inequalities* (Deza and Laurent, 2009), which could be enforced for *pairwise binary* graphs, $\mathcal{G}_{\{0,1\}}$. For any labelling of such a graph, when one traverses a cycle of edges, one will encounter a change of node labelling, only an even (including zero) number of times. This fact leads to the following constraint for a cycle of edges C and any subset $F \subseteq C$ such that $|F|$ is odd,

$$\sum_{(i,j)\in C\setminus F} (\phi_{ij}(1,0)+\phi_{ij}(0,1)) + \sum_{(i,j)\in F} (\phi_{ij}(0,0)+\phi_{ij}(1,1)) \geq 1. \tag{6}$$

Here, the $\phi_{ij}(\cdot)$ are (pseudo)marginals corresponding to the edge ij and a labelling of the edge is indicated by a pair of binary values. The ϕ_{ij}'s will be integer valued only if they belong to a vertex of the marginal polytope. Otherwise, they could take fractional values. These constraints hold for any marginal $\boldsymbol{\phi} \in \mathcal{M}(\mathcal{G}_{\{0,1\}})$ and any cycle C of the pairwise binary graph $\mathcal{G}_{\{0,1\}}$. We have seen that for any Markov random field, one of the vertices of the marginal polytope corresponds to an optimal labelling of the nodes. While optimizing over a looser polytope (like the local polytope), the cutting-plane algorithm strives to add constraints in order to reach the integer valued vertex of the marginal polytope. The pseudomarginals in an outer bounding polytope like $\mathcal{L}(\mathcal{G})$ need not satisfy the above cycle inequality constraints. Thus, enforcing these constraints will lead to a tighter polytope. The important step in any cutting-plane algorithm is about finding the constraints to add in an efficient manner, which is called the *separation* problem. In the algorithm of Sontag and Jaakkola (2007), the cycle that most violates the cycle inequality (6) could be selected. Even though the number of cycles in a general graph is exponentially large, it is possible to identify the violated cycle inequalities in polynomial time. In fact, it is of complexity $O(n^2 \log n + n|\mathcal{C}|)$ based on Dijkstra's shortest path algorithm along with a Fibonacci heap data structure.

So far, we have discussed an approach which will work for pairwise binary graphs. This will be a recurring setting in the study of tighter relaxations. Generally, extending these ideas to nodes taking multiple labels and also, to graphs with higher-order cliques lead to looser bounds, computationally more expensive algorithms and larger problem sizes due to introduction of auxiliary variables. For example, Sontag and Jaakkola (2007) have shown how to extend the cycle inequalities constraints to the multilabel scenario and also, to nonpairwise graphs. Again, they take inspiration from polyhedral combinatorics and propose an approach based on *projection* and *aggregation*. Consider a pairwise graph, with multilabel nodes. The set of labels of a node i, which we will denote as $\boldsymbol{l}_i$, can be partitioned into two sets. For ease of exposition, we will assume that all nodes take the same number of labels l. For l labels, there are $2^{(l-1)} - 1$ possible partitions and let π_i^s denote one such partition. Let $\boldsymbol{\pi}_i$ be a collection of partitions for node i. We define a binary *projection graph*, $\mathcal{G}^\pi = (\mathcal{V}^\pi, \mathcal{C}^\pi)$, where there is a node in $\mathcal{V}^\pi \forall \pi_i^s \in \boldsymbol{\pi}_i$, $\forall i \in \mathcal{V}$ and there is an edge between π_i^s and π_j^t

if $(i,j) \in \mathcal{C}$. If $\boldsymbol{\pi}_i$ consists of all possible partitions for all nodes $i \in \mathcal{V}$, then $\mathcal{G}^{\pi}$ is called the *full projection* graph. If each node's set of labels is partitioned only once, i.e., $|\boldsymbol{\pi}_i| = 1\ \forall i$, we have the *single projection* graph. In the *l-projection* graph, each $\boldsymbol{\pi}_i$ consists of one versus the rest partitions for each of the l labels. Given a marginal with respect to the original marginal polytope, $\boldsymbol{\phi} \in \mathcal{M}(\mathcal{G})$, its values could be aggregated to define marginals with respect to the marginal polytope of the binary projection graph. For this marginal polytope, it is possible to define cycle inequalities as follows,

$$\begin{aligned} &\sum_{(i,j)\in c\setminus f} \phi_{ij}^{\pi}(x_i' \neq x_j') + \sum_{(i,j)\in f} \phi_{ij}^{\pi}(x_i' = x_j') \geq 1 \\ &\text{where } \phi_{ij}^{\pi}(x_i' \neq x_j') = \sum_{\substack{x_i \in \boldsymbol{l}_i,\, x_j \in \boldsymbol{l}_j: \\ \pi_i(x_i) \neq \pi_j(x_j)}} \phi_{ij}(x_i, x_j) \\ &\phi_{ij}^{\pi}(x_i' = x_j') = \sum_{\substack{x_i \in \boldsymbol{l}_i,\, x_j \in \boldsymbol{l}_j: \\ \pi_i(x_i) = \pi_j(x_j)}} \phi_{ij}(x_i, x_j) \end{aligned} \tag{7}$$

where x_i' takes binary values $\{0, 1\}$, the projection operation $\pi_i\ (x_i)$ maps a node label, x_i, to one of these binary values, C is a cycle with respect to the original graph $\mathcal{G}$ and any $F \subseteq C$ such that $|F|$ is odd. These equations are easy to understand for single projection graphs but could be derived for the full projection and l-projection cases based on auxiliary variables. Thus, depending on whether the original graph has binary variables or multilabel variables, we obtain inequalities (6) or (7), respectively. Based on these cycle inequalities we could obtain cutting-planes to tighten the polytope. Thus, the algorithm given in Sontag and Jaakkola (2007) proceeds as follows. The first iteration involves solving an LP over a loose polytope, like the LP in (5). Then in each iteration, given the optimal pseudomarginals, cutting-planes (up to to n of them) are added and the tighter LP is solved to optimality. This is repeated until no cutting-planes are found. Even though there are specialized solvers to solve the LP relaxation in (5), specialized efficient solvers are not yet available to optimize the linear objective over a polytope which also has cycle inequalities constraints. For the binary case, it is possible to optimize using the Frank–Wolfe algorithm, using linear or semidefinite programming to optimize over the tighter polytope. For the multilabel case, a commercial solver like CPLEX is needed. Note, even though only the single projection graph leads to a surjective projection between the marginals of the multilabel graph to the marginals of the binary graph, it is preferable to work with the full projection graph. This is because we need several single projection graphs to obtain the same tightness as the full projection graph. Similarly, the full projection graph is tighter than the l-projection graph. Sontag and Jaakkola (2007) have shown experimental evidence, also, regarding better tightness of the full projection graph. Also, in order to extend this work to higher-order graphs, we need to create an intermediate pairwise graph which involves auxiliary variables corresponding to the higher-order clique potentials. Wainwright and Jordan (2008, section E.3) present one way to convert higher-order cliques to pairwise edges by adding auxiliary variables. Thus, we see how dealing with multilabel nodes and higher-order cliques lead to the addition of several auxiliary variables.

Even though the above work greatly helps to improve our understanding about tighter relaxations, it is not scalable. This is because we are not able to develop specialized algorithms for solving the LP relaxation with additional cycle inequalities based constraints. We are forced to use standard off-the-shelf LP solvers. The lack of scalability of standard LP solvers is well known in the MAP inference literature (Yanover et al., 2006). This is because they do not exploit the structure of the graphical model. Working in the dual space, leads to exploitation of the graph structure. Hence, there have been several scalable algorithms to solve the LP relaxation that are based on dual decomposition (Komodakis et al., 2011; Sontag et al., 2011). We will see how by recalling the dual presented in Komodakis et al. (2011). This is derived for the LP relaxation shown in (5) based on the decomposition of the original graph into a set of subgraphs. Based on such a decomposition, Eq. 5 can be equivalently written as,

$$\underset{\phi_i,\phi_c,\phi_{Si},\phi_{Sc}}{\text{minimize}} \sum_{S\in\mathcal{S}(\mathcal{G})} \left(\sum_{i\in S}\sum_{x_i} \theta_{Si}(x_i)\phi_{Si}(x_i) + \sum_{c\in S}\sum_{\boldsymbol{x}_c} \theta_{Sc}(\boldsymbol{x}_c)\phi_{Sc}(\boldsymbol{x}_c) \right) \tag{8a}$$

$$\left.\begin{aligned} &\text{subject to } \sum_{x_{c\setminus i}} \phi_{Sc}(\boldsymbol{x}_c) = \phi_{Si}(x_i), \quad \forall (i,c): c\in S, i\in c \\ &\sum_{x_i} \phi_{Si}(x_i) = 1, \quad \forall i\in S; \ \sum_{\boldsymbol{x}_c} \phi_{Sc}(\boldsymbol{x}_c) = 1, \quad \forall c\in S \\ &\phi_{Si}(x_i) \geq 0 \ \ \forall i\in S; \ \ \phi_c(\boldsymbol{x}_c) \geq 0 \ \ \forall c\in S \end{aligned}\right\} \forall S\in\mathcal{S}(\mathcal{G}) \tag{8b}$$

$$\phi_{Si}(x_i) = \phi_i(x_i) \quad \forall i\in\mathcal{V}, \forall x_i; \quad \phi_{Sc}(\boldsymbol{x}_c) = \phi_c(\boldsymbol{x}_c) \quad \forall c\in\mathcal{C}, \forall \boldsymbol{x}_c. \tag{8c}$$

Here $\mathcal{S}(\mathcal{G})$ is the set of subgraphs over which the original graph is decomposed. The subgraphs should be chosen such that all nodes and cliques of the original graph are covered by at least one subgraph. Also, each subgraph should share at least one node or clique with at least one other subgraph. The potentials $\theta_i^S(x_i)$ and $\theta_c^S(\boldsymbol{x}_c)$ are chosen such that $\sum_{S\in\mathcal{S}(i)} \theta_i^S(x_i) = \theta_i(x_i) \ \forall i\in\mathcal{V}$, $\sum_{S\in\mathcal{S}(c)} \theta_c^S(\boldsymbol{x}_c) = \theta_c(\boldsymbol{x}_c)$, $\forall c\in\mathcal{C}$. A simple way to achieve this is by setting the potentials as follows: $\theta_i^S(x_i) = \frac{\theta_i(x_i)}{|\mathcal{S}(i)|} \ \forall S\in\mathcal{S}(i)$ and $\theta_c^S(\boldsymbol{x}_c) = \frac{\theta_c(\boldsymbol{x}_c)}{|\mathcal{S}(c)|} \ \forall S\in\mathcal{S}(c)$. The consensus constraints (8c) ensure that this formulation is equivalent to (5). We could derive the dual by taking the consensus constraints (8c) into the objective and it takes the following form,

$$\max_{\delta}. \sum_{\mathcal{S}\in\mathcal{S}(\mathcal{G})} \min_{\phi_{\mathcal{S}i},\phi_{\mathcal{S}c}}. \left(\sum_{i\in\mathcal{S}} \sum_{x_i} (\theta_{\mathcal{S}i}(x_i) + \delta_{\mathcal{S}i}(x_i))\phi_{\mathcal{S}i}(x_i) + \sum_{c\in\mathcal{S}} \sum_{\boldsymbol{x}_c} (\theta_{\mathcal{S}c}(\boldsymbol{x}_c) + \delta_{\mathcal{S}c}(\boldsymbol{x}_c))\phi_{\mathcal{S}c}(\boldsymbol{x}_c) \right) \tag{9a}$$

$$\left.\begin{aligned} &\text{subject to } \sum_{x_{c\setminus i}} \phi_{\mathcal{S}c}(\boldsymbol{x}_c) = \phi_{\mathcal{S}i}(x_i), \quad \forall(i,c): c\in\mathcal{S}, i\in c \\ &\sum_{x_i} \phi_{\mathcal{S}i}(x_i) = 1, \quad \forall i\in\mathcal{S}; \quad \sum_{\boldsymbol{x}_c} \phi_{\mathcal{S}c}(\boldsymbol{x}_c) = 1, \quad \forall c\in\mathcal{S} \\ &\phi_{\mathcal{S}i}(x_i) \geq 0 \ \ \forall i\in\mathcal{S}; \quad \phi_c(\boldsymbol{x}_c) \geq 0 \ \ \forall c\in\mathcal{S} \end{aligned}\right\} \forall\mathcal{S}\in\mathcal{S}(\mathcal{G}) \tag{9b}$$

$$\sum_{\mathcal{S}:i\in\mathcal{S}} \delta_{\mathcal{S}i}(x_i) = 0, \quad \forall i\in\mathcal{V}, \forall x_i; \quad \sum_{\mathcal{S}:c\in\mathcal{S}} \delta_{\mathcal{S}c}(\boldsymbol{x}_c) = 0, \quad \forall c\in\mathcal{C}, \forall\boldsymbol{x}_c. \tag{9c}$$

Here, $\delta_{\mathcal{S}i}(x_i)$ are the dual variables with respect to the nodes and there is one such dual variable for each subgraph $\mathcal{S}$, each node $i\in\mathcal{S}$ and for each label x_i. $\delta_{\mathcal{S}c}(\boldsymbol{x}_c)$ are the dual variables with respect to the cliques and there is one such dual variable for each subgraph $\mathcal{S}$, each clique $c\in\mathcal{S}$ and for each possible clique labelling $\boldsymbol{x}_c$. The constraints given in (9c) ensure that we eliminate the global variables, $\phi_i(x_i)$'s and $\phi_c(\boldsymbol{x}_c)$'s, from the objective. We work only with the subgraph-based variables, $\phi_{\mathcal{S}i}(x_i)$'s and $\phi_{\mathcal{S}c}(\boldsymbol{x}_c)$'s and the objective decomposes over the subgraphs. In other words, the minimization subproblems that we observe in the objective (9a) are with respect to individual subgraphs. Usually, the subgraphs are the individual pairwise edges or cliques or *hypertrees* of the original graphical model. A hypertree is a more general way of looking at a tree. It is a tree which could be made of hyperedges, not only pairwise edges. Hence, a tree made of only pairwise edges is also a hypertree. Note that nodes, edges and cliques are instances of hypertrees, too. The convenience with hypertrees is that the above minimization could be performed exactly and efficiently. This is because the minimization subproblem in a subgraph corresponds to performing MAP inference within a graph of the same topology as the subgraph and with suitably defined potentials. A fundamental result in graphical model is that we can perform MAP inference in hypertrees, exactly and efficiently (Koller and Friedman, 2009). In general, it is possible to decompose the graph according to subproblems that are not hypertrees. We will discuss more about this aspect toward the end of the next section. At this point, a natural question comes to mind. What happens to the LP relaxation if the original graphical model is a hypertree, i.e., it has no loops? It is not only efficient to perform MAP inference efficiently in a hypertree but in fact, the local polytope for such a graphical model is tight. Further, there are other graphical models where the local polytope is tight. A very important case is when the clique potentials are *submodular* (Blake et al., 2011). Submodular clique potentials are an example of *attractive* potentials. An attractive clique potential encourages the nodes to take similar labels. Specifically, for pairwise binary graphs, submodular clique potentials satisfy the following condition,

$$\theta_{ij}(0,1) + \theta_{ij}(1,0) \geq \theta_{ij}(0,0) + \theta_{ij}(1,1) \tag{10}$$

from which we can see that similar labels are encouraged, when we try to minimize the energy (2). Another example of a graphical model for which the local polytope is tight, appears in the problem of maximum weight matching in a bipartite graph, which can be formulated as a MAP inference problem (Schrijver, 2003).

So far, we have seen three types of data structures: clusters, cycles and hypertrees. They are all subgraphs of the original graphical model. This will be a common theme to several ideas discussed in this chapter: tightening the relaxation through formulations involving subgraphs of the graphical model. These subgraphs will have more structure compared to hypertrees. At the same time, these subgraphs will have a structure that lends itself to more efficient computation. Just like hypertrees have an efficient MAP inference algorithm, there are other graph topologies and/or clique potentials that lead to efficient computation. We will see more of such special subgraphs in the coming sections and discuss insights and algorithms based on them. Even though we will encounter different graph topologies in practical problems, as a first step, we recommend the reader to imagine a two-dimensional grid graph (a planar graph, where nodes are arranged in the form of rows and columns) and visualize structures like clusters, cycles and higher-order cliques on such a graph. Grid graphs are common in computer vision problems like stereo depth estimation and image segmentation. Kappes et al. (2015) is a good starting point to explore more about various datasets for MAP inference in MRFs.

2 Cluster pursuit algorithms

We will now look further into cluster-based methods and we will describe algorithms that achieve efficiency by working in the dual space. We have already mentioned relaxation hierarchies like Sherali–Adams that involve a series of tighter problem levels. At each level, marginalization constraints are enforced based on an exhaustive set of clusters of a particular size. As mentioned in Section 1.1, a cluster could be any subset of nodes of the graph. In practice, we will consider connected subsets of nodes (formed by the union of a subset of cliques) as clusters. Let us see more clearly what it means to add cluster-based marginalization constraints. Suppose, we have a graph, with pseudomarginals $\phi_i(x_i)$'s and $\phi_c(\boldsymbol{x}_c)$'s and we add marginalization constraints based on a set of clusters, $\mathcal{S}$, then we have a new polytope defined by the following constraints,

$$\begin{aligned} &\sum_{\boldsymbol{x}_{c\setminus i}} \phi_c(\boldsymbol{x}_c) = \phi_i(x_i) \\ &\sum_{\boldsymbol{x}_{s\setminus c}} \tau_s(\boldsymbol{x}_s) = \phi_c(\boldsymbol{x}_c) \quad \forall s \in \mathcal{S} \\ &\sum_{\boldsymbol{x}_s} \tau_s(\boldsymbol{x}_s) = 1 \end{aligned} \tag{11}$$

where the summation over $\boldsymbol{x}_{s\backslash c}$ is similar to the summation over $\boldsymbol{x}_{c\backslash i}$ in (3b). Note, the $\tau_s(\boldsymbol{x}_s)$'s are auxiliary variables and do not affect the dimension of the space in which the tighter polytope resides. The polytope is in the space of the $\phi_i(x_i)$'s and $\phi_{ij}(x_i, x_j)$'s as before and is a tighter polytope compared to the local polytope. However, we cannot practically add a lot of cluster-based marginalization constraints, even if they are small ones like triplets. For many problems it is sufficient to add a smaller set of clusters, when the clusters are chosen intelligently. We could intuitively imagine that it is enough if the polytope is tightened preferentially in the neighbourhood of the optimum, instead of being tightened uniformly all-around. We need a way to judge how well a cluster tightens the relaxation in the neighbourhood of the optimum. Werner (2008) presented an approach based on the language of *constraint satisfaction problems* (CSPs), by showing that all clusters that do not satisfy a CSP subproblem could potentially tighten the relaxation. However, there can be many such clusters and we still need a way to rank them, so that a smaller subset of clusters could be added. Sontag et al. (2008) show a greedy approach to rank the clusters and shows an approach to incrementally add clusters as the algorithm proceeds. At this point of time, we would like to point out that the problem of MAP inference in MRFs is intimately related to CSPs (Thapper and Živný, 2016; Werner, 2010).

In order to develop the algorithm, it is possible to work in the primal space like Sontag and Jaakkola (2007). Working with the primal, introduces cutting-planes but no new decision variables. This is in contrast to working with the dual, which could lead to the introduction of new decision variables. However, there are advantages to working in the dual space. Working with the dual, helps in designing a principled criterion to choose the clusters. This is based on the fact that the dual, like the one shown in (9), provided a lower bound to the MAP energy in (2) (Wainwright et al., 2005). Thus, it is possible to rank the clusters based on how they may increase this lower bound. A similar criterion to rank the clusters in the primal is hard to achieve. In both the primal and the dual, we have a new optimization problem each time we add a new set of cluster-based marginalization conditions and in the case of the dual, we have new dual variables corresponding to the added clusters. Interestingly, the dual formulation is such that we can get a feasible starting point by setting the old dual variables to their final value in the previous problem and by setting the new dual variables to zero. Thus, the main advantages of working in the dual are summarized as follows:

- Principled condition to choose clusters.
- Warm starting based on values from previous iteration.
- Scalable approach that readily works with multilabel and higher-order graphs.

Sontag et al. (2008) add clusters from a predefined set of small clusters, i.e., it is based on dictionary enumeration. For ease of exposition, we will present the following equations considering a pairwise graph, $\mathcal{G}=(\mathcal{V},\mathcal{E})$. Then, we will mention how to extend these concepts to graphs with higher-order cliques. In case, the graph contains small *chordless* clusters like $\{(i,j,k)|(i,j),(j,k),(k,i)\in\mathcal{E}\}$, we will have a dictionary of such triplet clusters. In case of a grid graph, the dictionary will be made of the square faces of the graph. In order to rank the clusters exactly, one needs to define a dual optimization problem for each cluster, i.e., an optimization problem that will additionally have marginalization constraints based on that cluster. Such a dual optimization problem has to be solved to optimality to assess the benefit of adding that cluster. Higher the optimal objective, higher is the rank for the cluster. Doing this for all clusters in the dictionary is very expensive. Instead, one could exploit a property shared by various dual-based algorithms. This property is that the objective of a dual-based algorithm increases monotonically. Thus, we could run for only one iteration a dual optimization problem to which marginalization constraints w.r.t. one cluster have been added. The increase in the dual objective after one iteration is guaranteed to be a lower bound on the increase that is expected if the dual optimization problem is solved to optimality. Thus, the clusters are ranked based on this lower bound. In order to obtain a valid lower bound, we need a dual algorithm for which the objective increases monotonically. This rules out an algorithm based on supergradient ascent to optimize (9), which is not a strictly ascending algorithm. On the other hand, coordinate ascent-based approaches are strictly ascending (Meshi et al., 2014). Sontag et al. (2008) is based on an algorithm called MPLP (Globerson and Jaakkola, 2008). Note, the dual for MPLP is of a different form compared to (9) and its derivation involves the introduction of auxiliary variables, which denote duplicate copies of the edge-based pseudo-marginals, $\phi_{ij}(x_i, x_j)$.

We will denote the set of dual variables as $\boldsymbol{\lambda}$ and there are dual variables corresponding to each edge-node pair ($\lambda_{ij\to i}(x_i)$), each edge ($\lambda_{ij}(x_i, x_j)$) and each cluster-edge pair ($\lambda_{s\to ij}(x_i,x_j)$). Every iteration of the algorithm in Sontag et al. (2008) involves closed form updates of all these dual variables, similar to MPLP. We note that the size of $\boldsymbol{\lambda}$ keeps increasing as we add cluster-based constraints. Suppose, at the end of iteration t, the dual variables are $\boldsymbol{\lambda}^t$. In iteration $t+1$, we would like to compute the criterion for some cluster s. In order to run one iteration of the dual problem that involves this additional cluster, we will warm start the dual variables corresponding to the already present clusters using $\boldsymbol{\lambda}^t$ and we will initialize the dual variables corresponding to cluster s to zero. Since, we are warm starting using $\boldsymbol{\lambda}^t$, the dual increases in value only based on updates from the cluster s. This quantity has a closed form expression, which is expressed in terms of edge-based *beliefs*. For every edge $(i, j)\in s$, we could define a belief as,

$$b_{ij}(x_i,x_j)=\lambda_{ij}(x_i,x_j)+\sum_{s':e\in s'}\lambda_{s'\to ij}(x_i,x_j). \tag{12}$$

The belief, $b_{ij}(x_i, x_j)$, is directly proportional to the marginal probability of edge ij taking label (x_i, x_j). The summation in (12) is not affected by the dual variables $\lambda_{s\to ij}(x_i, x_j)$ because they are set to zero. Thus, the increase in the dual objective due to the new cluster s takes the following form,

$$d(s) = \sum_{ij\in s} \max_{x_i, x_j} b_{ij}(x_i, x_j) - \max_{\boldsymbol{x}_s}\left[\sum_{ij\in s} b_{ij}(x_i, x_j)\right]. \tag{13}$$

$d(s)$ is a lower bound on the improvement to the dual objective if cluster s is added. Thus, the clusters are ranked with respect to $d(s)$ and are added in a greedy fashion. The overall approach based on these ideas is shown in Algorithm 1.

In Algorithm 1 the integral solution could be recovered in several ways (Meshi et al., 2014; Ravikumar et al., 2010; Savchynskyy and Schmidt, 2014). Many times, a simple and efficient rounding scheme is preferable, like independently picking for each of the nodes the label with maximal pseudomarginal value. In step 4, it is sufficient to run the dual algorithm for a fixed number of iterations such that we have reached a neighbourhood of more optimal decision variables. In case, we run the dual algorithm to optimality in step 4, adding clusters is exactly equivalent to adding cutting-planes in the primal problem.

In order to extend these ideas to higher-order cliques, we first observe that we have seen three groups of nodes in our equations so far: clusters, edges and individual nodes. Also, in practice the clusters are built as unions of edges. The equations involved variables that associate a larger group of nodes with a smaller subset of that larger group. We could think about this in terms of tuples of the following form: $(A, B, \boldsymbol{x}_B)$. Here, A is a larger subset of nodes like a cluster or an edge. B is also a subset of nodes such that $B \subset A$. B could be an edge (resp. node), if A is a cluster (resp. cluster or edge). $\boldsymbol{x}_B$ is one possible labelling of the nodes in B. Such a tuple is called a *pencil* in graphical models literature (Werner, 2008). The dual variable $\lambda_{s\to ij}(x_i, x_j)$ which appeared a little earlier, is with respect to the pencil $(s, ij, (x_i, x_j))$. Thus, to use Sontag et al. (2008) for higher-order cliques, we will replace the edges by higher-order cliques.

ALGORITHM 1 Dual algorithm with iterative tightening

1: Run a dual algorithm like MPLP over the local polytope as close to the optimum as possible (say, 1000 iterations or convergence, whichever comes first).
2: Recover an integral solution and compute primal-dual gap. If the gap is less than a small threshold, then exit.
3: Add a fixed number of clusters according to the bound criterion (13).
4: Warm start the dual variables if possible and run the dual algorithm for a fixed limited number of iterations.
5: Return to step 2.

We recollect that a clique with two nodes is also called an edge and a clique with more than two nodes is called a higher-order clique or hyperedge. Thus, we will work with dual variables of the form $\lambda_{s\to c}(\boldsymbol{x}_c)$, which corresponds to the pencil $(s, c, \boldsymbol{x}_c)$. As before, s is a cluster which defines additional marginalization constraints and now, it is a union of higher-order cliques. c is one such higher-order clique and $\boldsymbol{x}_c$ is one possible labelling of c.

Another point to keep in mind is that the marginalization constraints with respect to a set of clusters $\mathcal{S}$, could be equivalently represented in another way. This equivalent representation involves a new graphical model (Werner, 2008). In this graphical model, the clusters $s \in \mathcal{S}$ are represented as cliques, apart from the original set of cliques $\mathcal{C}$. However, the clique potentials, $\theta_s(\boldsymbol{x}_s)$'s, of the clusters will all be set to zero. Thus, when we consider the LP relaxation equations (similar to (5)) for this graphical model, the objective will have the same value as the original graphical model because the additional cluster-based potentials are zero valued. When we look at the constraints, there will be additional marginalization constraints present with respect to the new clusters, $s \in \mathcal{S}$, involving new pseudomarginals $\phi_s(\boldsymbol{x}_s)$'s. We can see that these pseudomarginals are exactly like the auxiliary variables $\tau_s(\boldsymbol{x}_s)$'s that we saw in (11). Thus, these new pseudomarginals have the same effect on the original pseudomarginals, $\phi_i(x_i)$'s and $\phi_c(\boldsymbol{x}_c)$'s. They get constrained to be in a tighter polytope. The advantage of imagining such a graphical model with additional zero valued potentials is to write the LP relaxation equations in a symmetric manner using pencils. For this new graphical model, the primal form of the LP relaxation will take the following form,

$$\underset{\phi}{\text{minimize}} \sum_{i\in\mathcal{V}}\sum_{x_i}\theta_i(x_i)\phi_i(x_i) + \sum_{c\in\mathcal{C}}\sum_{\boldsymbol{x}_c}\theta_c(\boldsymbol{x}_c)\phi_c(\boldsymbol{x}_c) + \sum_{s\in\mathcal{S}}\sum_{\boldsymbol{x}_s}\theta_s(\boldsymbol{x}_s)\phi_s(\boldsymbol{x}_s) \tag{14a}$$

$$\text{subject to} \sum_{\boldsymbol{x}_{A\setminus B}}\phi_A(\boldsymbol{x}_A) = \phi_B(\boldsymbol{x}_B), \quad \forall (A, B, \boldsymbol{x}_B) \in J \tag{14b}$$

$$\begin{aligned} &\sum_{x_i}\phi_i(x_i) = 1, \quad \phi_i(x_i) \geq 0 \quad \forall i \in \mathcal{V} \\ &\sum_{\boldsymbol{x}_c}\phi_c(\boldsymbol{x}_c) = 1, \quad \phi_c(\boldsymbol{x}_c) \geq 0 \quad \forall c \in \mathcal{C} \\ &\sum_{\boldsymbol{x}_s}\phi_s(\boldsymbol{x}_s) = 1, \quad \phi_s(\boldsymbol{x}_s) \geq 0 \quad \forall s \in \mathcal{S}. \end{aligned} \tag{14c}$$

Here, the set J is made of pencils $(A, B, \boldsymbol{x}_B)$ of the following form: if A is a clique of the original graphical model, then B corresponds to the nodes of that clique; if A is a cluster added to tighten the polytope, then B corresponds to the cliques whose union is the cluster A. For example, we can see that for a pairwise graph, a cluster s could be a triplet and marginalization constraints are enforced between this triplet and the edges that make up the triplet.

Among the literature on tighter relaxations, a few of them have designed algorithms based on the dual of the primal problem given in (14) (Batra et al., 2011; Werner, 2008). That dual takes the following form,

$$\underset{\substack{h_A \ \forall A \in \mathcal{V} \cup \mathcal{C} \cup \mathcal{S} \\ y_{A \to B}(\boldsymbol{x}_B) \ \forall (A,B,\boldsymbol{x}_B) \in J}}{\text{maximize}} \quad \sum_{i \in \mathcal{V}} h_i + \sum_{c \in \mathcal{C}} h_c + \sum_{s \in \mathcal{S}} h_s \tag{15a}$$

$$\text{subject to } h_A \leq \theta_A(\boldsymbol{x}_A) + \sum_{Z|(Z,A,\boldsymbol{x}_A) \in J} y_{Z \to A}(\boldsymbol{x}_A) - \sum_{B|(A,B,\boldsymbol{x}_B) \in J} y_{A \to B}(\boldsymbol{x}_B) \tag{15b}$$
$$\forall A \in \mathcal{V} \cup \mathcal{C} \cup \mathcal{S}, \quad \forall (A,B,\boldsymbol{x}_B) \in J.$$

We observe that the added clusters influence the dual objective through the variables h_s, while they do not affect the primal objective because of the zero valued potentials. Using this dual formulation, Batra et al. (2011) have addressed a computational bottleneck in Sontag et al. (2008). The computational complexity of computing $d(s)$ given in (13) is $O(l^{|s|})$. Thus, computing this score for all the clusters enumerated in the dictionary is computationally very heavy. Batra et al. (2011) present a heuristic criterion which could be used to first select a smaller subset of the clusters from the dictionary and compute $d(s)$ only for those clusters. Before we look at this heuristic criterion, we first observe that $\widetilde{\theta}_A(\boldsymbol{x}_A) \triangleq \theta_A(\boldsymbol{x}_A) + \sum_{Z|(Z,A,\boldsymbol{x}_A) \in J} y_{Z \to A}(\boldsymbol{x}_A) - \sum_{B|(A,B,\boldsymbol{x}_B) \in J} y_{A \to B}(\boldsymbol{x}_B)$ is a *reparameterization* of $\theta_A(\boldsymbol{x}_A)$. This means that for a given labelling of all the nodes in the graphical model, we will get the same energy with either the $\theta_A(\boldsymbol{x}_A)$'s or the $\widetilde{\theta}_A(\boldsymbol{x}_A)$'s. Reparameterization is a concept that we will come across often in graphical models literature, for example, Kolmogorov (2006). For each clique $c \in \mathcal{C}$, we could define a quantity,

$$l(c) = \widetilde{\theta}_c(\boldsymbol{x}_c^p) - \min_{\boldsymbol{x}_c}. \left[\widetilde{\theta}_c(\boldsymbol{x}_c)\right] \tag{16}$$

where $\boldsymbol{x}_c^p$ is an integral labelling of the clique c that is obtained in step 2 of Algorithm 1. This heuristic is derived based on the form of the complementary slackness condition for the constraint given in (15b). Then, for a cluster s, we define a criterion based on the cliques that it is made of,

$$l(s) = \sum_{c: c \in s} l(c). \tag{17}$$

Compared to $d(s)$ in (13), this criterion could be computed with complexity $O(l^{|c|})$. Just like we did with $d(s)$, the clusters could be ranked based on the value of $l(s)$. We note that this criterion comes with poorer guarantees compared to $d(s)$ and it is best to use it as an initial preprocessing step. $l(s)$ is computed for all clusters in the dictionary, in order to select a small subset of clusters and this smaller set is ranked according to $d(s)$. We also point out that the condition $l(s) = 0$ for a cluster s is equivalent to satisfaction of the CSP subproblem in Werner (2008).

The computation of $d(s)$ (13) need not be the only computationally heavy part of an algorithm for tighter relaxation. Depending on the dual algorithm, updating the dual variables could be costly, too. For example, the MPLP-based approach of Sontag et al. (2008) updates the cluster-to-clique dual variables $(\lambda_{s\to c}(\boldsymbol{x}_c))$ with complexity that is exponential in the size of the cluster s. Sontag et al. (2009) show a way to accelerate the update of these dual variables within the same MPLP-based algorithm. This is achieved by defining a polytope, which is looser than the one defined by (11) but still tighter than $\mathcal{L}(\mathcal{G})$. This polytope is looser because the marginalization constraint between the added clusters and the cliques is over a coarser partitioning of the label space of the cluster, as follows,

$$\begin{aligned} &\sum_{\boldsymbol{x}_{c\setminus i}} \phi_c(\boldsymbol{x}_c) = \phi_i(x_i) \\ &\sum_{z_s\setminus\{z_i^s\}:i\in c} \tau_s(z_s) = \sum_{x_i\in z_i^s:i\in c} \phi_c(\boldsymbol{x}_c) \quad \forall s\in\mathcal{S} \\ &\sum_{\boldsymbol{x}_c} \phi_c(\boldsymbol{x}_c) = 1 \end{aligned} \tag{18}$$

where $\{z_s\}$ is the set of labelling of the cluster s over a coarser label space. For each node in the cluster s, the set of labels is coarsened to a smaller set $\{z_i^s\}$. Then, marginalization constraints are enforced between a cluster and the clique it contains through these coarser labels. This will reduce the complexity of updating the cluster-to-clique dual variables $(\lambda_{s\to c}(\boldsymbol{x}_c))$. A better choice of the coarser label set $\{z_i^s\}$ is important. We want to ensure that the coarse partitioning behaves as closely as possible to what $d(s)$ guarantees with respect to increase in the dual objective. Sontag et al. (2009) present a greedy approach with a threshold parameter, which leads to a trade-off between guaranteed improvement of the dual and the number of coarse states.

These cluster-based ideas could be applied to practical problems, like the ones described in Yanover et al. (2006). For example, the problems of protein side chain prediction and protein design have a graph structure that allows adding triplet-based clusters. For stereo and segmentation problems, the graph is usually grid structured and we could work with clusters consisting of the four node faces of the graphs. So far, we have been looking at how to tighten the relaxation by enforcing additional cluster-based constraints. At the same time, it is known in the graphical models literature that tighter relaxations are possible by using subgraphs which have loops (Komodakis et al., 2011, section 6.1) within a dual decomposition formulation like (9). Is this approach related to tighter formulations using cluster-based marginalization constraints?

In fact, they are exactly the same concept. When we discussed about cluster-based marginalization constraints, we presented a pair of problems: primal (14) and dual (15). The clusters had zero potentials in the primal, in order to ensure that the primal energy continues to be a function of the original node and clique potentials. However, we could define the cluster potentials in the following way: $\sum_{\mathcal{S}\in\mathcal{S}(c)}\theta_c^{\mathcal{S}}(\boldsymbol{x}_c)=\theta_c(\boldsymbol{x}_c),\ \ \forall c\in\mathcal{S}$. Then, by considering only node and cluster potentials, we ensure that the primal energy stays the same. Incidentally, this form of the cluster potentials is also a required condition for the dual in (9). Thus, having cluster-based marginalization constraints in the primal, is equivalent to having cluster-based subproblems in the dual, with suitably defined potentials. We will understand this concept more by considering cycles. We will see how tighter relaxation could be achieved either by imposing cycle-based marginalization constraints within an algorithm like Sontag et al. (2008) or by decomposing the graphical model using cycle-based subgraphs.

3 Cycles in the graph

So far, we have been looking at clusters, which in theory could be any subset of the nodes. In practice, we seek small clusters which are connected to each other. The subset of nodes could have more structure, for example, they could form a *cycle*. A cycle is a chain of node pairs with the following form: $\{(x_0,x_1),(x_1,x_2),\ldots,(x_{n-2},x_{n-1}),(x_{n-1},x_0)\}$. Cycles play an important role in understanding about tighter relaxations for MAP inference. Suppose a given LP relaxation results in fractional values as the solution. So far, we have understood this as the relaxed polytope being not tight enough around the optimal integral vertex of the marginal polytope. Such fractional solutions could be directly characterized by the presence of *frustrated cycles* in the graph. To understand what these are and what to do about them, we will first consider a pairwise graph, $\mathcal{G}=(\mathcal{V},\mathcal{E})$, for ease of exposition. Later, we will see how to extend these works to higher-order graphs. Also, we will assume that all nodes take one of l labels.

So far, we have been made aware of the possibility of various dual formulations, (9), (15) and the MPLP-based dual of Sontag et al. (2008). All these duals lower bound the MAP energy given in (2). There is yet another dual formulation that throws light on the problem of frustrated cycles. It is given in Komodakis and Paragios (2008) and is of the following form,

$$\underset{\boldsymbol{y}}{\text{maximize}}\ \sum_{i\in\mathcal{V}}\min_{x_i}.\left(\theta_i(x_i)+\sum_{j:ij\in\mathcal{E}}y_{ij}(x_i)\right) \tag{19a}$$

$$\text{subject to}\ \overline{\theta}_{ij}(x_i,x_j)-y_{ij}(x_i)-y_{ji}(x_j)\geq 0\quad \forall ij\in\mathcal{E} \tag{19b}$$

$$\sum_{ij\in\mathcal{E}_S}\overline{\theta}_{ij}(x_i,x_j)\leq\sum_{ij\in\mathcal{E}_S}\theta_{ij}(x_i,x_j). \tag{19c}$$

We will denote this dual problem as $\mathcal{D}(\mathcal{E}_S)$, where, $\mathcal{E}_S$ is a subset of edges of the graph and $\mathcal{E}$ is the set of all edges. We would like to highlight the presence of $\overline{\boldsymbol{\theta}}$ in the formulation, which is the set of all $\overline{\theta}_{ij}(x_i, x_j)$'s and we will call them *virtual* potentials. Suppose, $\mathcal{E}_S = \mathcal{E}$ then we obtain a dual problem, $\mathcal{D}(\mathcal{E})$, that is equivalent to the intractable ILP (3). If $\overline{\theta}_{ij}(x_i, x_j) = \theta_{ij}(x_i, x_j) \;\; \forall ij \in \mathcal{E}_S, \forall x_i, x_j$, then the dual, which we will call as $\mathcal{D}_{fixed}$, is equivalent to the LP relaxation (5). It comes with the same guarantees as optimizing LP relaxation over the local polytope. The optimal solutions of $\mathcal{D}_{fixed}$ are characterized by the following theorem.

Theorem 1 (Weak Optimality Condition (Werner, 2007)).
After $\mathcal{D}_{fixed}$ has been solved to optimality, there must be at least one optimal label x_i^* for node i and at least one optimal label x_j^* for node j, such that $\overline{\theta}_{ij}(x_i^*, x_j^*) = y_{ij}(x_i^*) + y_{ji}(x_j^*)$.

The above theorem is a necessary but not sufficient condition for an optimal labelling of the original MAP inference problem (2). This is expected because $\mathcal{D}_{fixed}$ is a tractable approximation to $\mathcal{D}(\mathcal{E})$, which is an NP-hard problem. Komodakis and Paragios (2008) showed how solving $\mathcal{D}(\mathcal{E}_S)$ presents a trade-off between these two extremes of tractability and accuracy. In order to see how it could be done, we need to understand the following stronger constraint that an optimal labelling for MAP inference is required to satisfy.

Theorem 2 (Strong Optimality Condition).
An optimal labelling of the graph that minimizes the energy (1) *is such that it assigns the labels (x_i^*, x_j^*) to the edge (i, j) then $\overline{\theta}_{ij}(x_i^*, x_j^*) = y_{ij}(x_i^*) + y_{ji}(x_j^*)$, for all edges.*

Solving $\mathcal{D}(\mathcal{E})$ ensures Theorem 2. Suppose, $\mathcal{D}(\mathcal{E}_S)$ has been solved by some optimization algorithm and we obtain optimal labels for all nodes. We should keep in mind that there could be multiple optimal labels for some nodes. Suppose, we start at some node i on the graph and would like to take a step to a neighbouring node j. We could do it if $\exists\, x_i^*$ and x_j^* such that $\overline{\theta}_{ij}(x_i^*, x_j^*) = y_{ij}(x_i^*) + y_{ji}(x_j^*)$. Suppose we traverse a cycle of nodes in this manner and for each node i, the step toward and away from it are based on the same optimal label x_i^*, then that cycle is not a frustrated cycle. If it is possible to traverse all possible cycles of the graph based on this criterion, then the graph does not have any frustrated cycles. An optimal solution without frustrated cycles and an optimal solution satisfying the condition in Theorem 2 are equivalent. In such a situation, $\mathcal{D}(\mathcal{E}_S)$ has managed to find the optimal MAP labelling. Suppose, a cycle in the graph could not be traversed according to the above mentioned criterion, then it is a frustrated cycle. Frustrated cycles could be of two different forms. One, where for all pairs of optimal labels for two adjacent nodes, $\overline{\theta}_{ij}(x_i^*, x_j^*) \neq y_{ij}(x_i^*) + y_{ji}(x_j^*)$. Hence, it is not possible to take a step between these two nodes. Two, where a node i is not able to satisfy $\overline{\theta}_{ij}(x_i^*, x_j^*) = y_{ij}(x_i^*) + y_{ji}(x_j^*)$ with all its neighbours based on the same optimal label x_i^*. An optimal labelling with frustrated cycles satisfy Theorem 1 only. In practice, if the clique potentials are attractive, then frustrated cycles are discouraged and the opposite is true for nonattractive potentials. In Section 1.1 we said that if the clique potentials are submodular, then the local polytope is tight. Thus, there will be no frustrated cycles in a graphical model with submodular potentials. For general clique potentials, tightening the LP relaxation leads to the elimination of such frustrated cycles. Thus, cycles play an important role in understanding and addressing the tightening of LP relaxations.

We had earlier mentioned that (19) involves the virtual potentials, $\overline{\boldsymbol{\theta}}$. If we can set $\overline{\boldsymbol{\theta}}$ properly then the optimal labelling obtained by solving (19) could have less or no frustrated cycles. Komodakis and Paragios (2008) have shown a way to iteratively modify $\overline{\boldsymbol{\theta}}$, followed by solving the dual problem (19), in order to obtain an optimal labelling with less or no frustrated cycles. They proposed to solve a series of dual problems, $\mathcal{D}(\mathcal{E}_S)$ with different $\mathcal{E}_S$. They solve these dual problems one after the other and repeatedly, until a stopping criterion is reached. Each $\mathcal{E}_S$ is picked from a dictionary of cycles. For example, in a grid graph, the dictionary could be composed of cycles that are the faces of the grid and also, the cycles going around two adjacent faces. In theory, the set of edges $\mathcal{E}_S$ in the optimization problem (19) could be any set of edges, not necessarily cycles. Since, it is frustrated cycles that we are trying to eliminate, it intuitively and algorithmically makes sense to choose $\mathcal{E}_S$ to be cycles. Thus the algorithm in Komodakis and Paragios (2008) repeatedly iterates over a dictionary of cycles.

At each iteration, we have a given set of virtual potentials, $\overline{\boldsymbol{\theta}}$ and a given cycle, $\mathcal{E}_S$. Based on the optimal dual variables $\boldsymbol{y}$ from the previous iteration, we check whether the current cycle is a frustrated cycle. Komodakis and Paragios (2008) show a principled way to "repair" such a frustrated cycle and this procedure is the heart of the paper. The approach is to modify the values of $\overline{\boldsymbol{\theta}}$ corresponding to the cycle $\mathcal{E}_S$. While making this modification, the constraints (19b) and (19c) should be maintained. We do this by picking a pair of nodes in $\mathcal{E}_S$, for which there is at least one pair of labels for which the constraint (19b) is an equality. This will always be true if the previous dual problem (19) has been solved to optimality. According to Komodakis and Paragios (2008), we modify $\overline{\boldsymbol{\theta}}$ such that for that pair of nodes, the constraint (19b) becomes a strict inequality for all pairs of labels between the two nodes. This modification has to be done while continuing to satisfy the constraint (19c). This is possible by increasing the elements in $\overline{\boldsymbol{\theta}}$ w.r.t. to the pair of nodes, while decreasing elements in $\overline{\boldsymbol{\theta}}$ w.r.t. adjacent pairs of nodes in the cycle. In a frustrated cycle, it is always possible to perform this operation, while satisfying the constraints given in (19b) and (19c). Once the cycle has been "repaired" this way, the dual problem is now solved, with the constraint (19c) enforced w.r.t. this cycle. Based on the new optimal dual variables $\boldsymbol{y}$, this cycle will not be a frustrated cycle. These steps, "repair" cycle and solve the dual, are repeated for the next cycle in the dictionary and so on.

In this process, a cycle that was repaired earlier could again become a frustrated cycle. That is why the algorithm iterates over the cycles in the dictionary repeatedly. It is possible to warm-start $\boldsymbol{y}$ for the current dual optimization problem based on the previous optimal $\boldsymbol{y}$ because the virtual potentials are the same except for the two adjacent cycles in the dictionary. This leads to considerable speed-up. Interestingly, the dual objective increases monotonically as we solve one dual problem after the other. Thus, the sequence of optimal objective values are going closer to that of $\mathcal{D}(\mathcal{E})$, the intractable MAP inference problem, where constraint (19c) has to be satisfied over the entire set of edges, $\mathcal{E}$. Based on this monotonic increase property, this algorithm could be terminated in a principled manner. For example, a good way to terminate is when there are no frustrated cycles with respect to the dictionary. Even though larger frustrated cycles could still exist, this approach is sufficient to return the MAP labelling in several practical problems. In Section 2, we have looked at tightening the relaxation using cluster-based marginalization constraints. Optimizing $\mathcal{D}(\mathcal{E}_S)$, where $\mathcal{E}_S$ is a cycle, is similar to enforcing marginalization constraints with respect to that cycle. Earlier, we worked with additional dual variables ($\lambda_{s\to c}(\boldsymbol{x}_c)$) with respect to that cluster s. In this case, we work with ($\overline{\theta}_{ij}(x_i, x_j)$) corresponding to the edges ij of the cycle $\mathcal{E}_S$, which are adjusted if the cycle is frustrated.

3.1 Searching for frustrated cycles

The main bottleneck with the approach given in Komodakis and Paragios (2008) is w.r.t. efficiency. Their approach iterates over a big dictionary of short cycles present in the graph. Also, for some graphs like sparse graphs with few short cycles, it is not clear how to enumerate the dictionary of cycles. Thus, it is desirable to choose a cycle directly from the graph, instead of choosing cycles from a predefined dictionary. Sontag et al. (2012) have presented an approach to directly assess how a cycle will improve the dual bound. Their work combines ideas presented in Sontag and Jaakkola (2007), Johnson (2008, section 3.4) and Sontag et al. (2008). Just like Sontag et al. (2008) we greedily choose a cycle that leads to best dual objective improvement based on running one iteration of a dual optimization problem. This dual problem has additional dual variables corresponding to the cycle, just like we saw in Section 2. Therefore, we have a function that serves a purpose similar to $d(s)$ in (13) for choosing a cluster. In this case, we do not choose the cycle from a dictionary of cycles. At the same time, it is possible to characterize more desirable cycles. In order to do that we first construct a graph with edge weights that are defined based on edge beliefs of the original graph. We have already come across the concept of edge beliefs in (12) and they are obtained through an optimization algorithm like the dual-based algorithm in Sontag et al. (2008). In this new graph, we seek the cycle, which among the cycles with an odd number of negative edge weights, maximizes the minimum absolute value of the edge weights present. This will lead us to cycles in the original graph that are more suited for tightening the relaxation. Johnson (2008, section 3.4) showed why this is true and also, showed an algorithm to obtain desirable cycles efficiently for graphs with binary valued nodes. For binary variables, the new graph that needs to be constructed has the same topology as the original graph. In order to obtain an approach that works for multilabel graphs, Sontag et al. (2008) observed that cycle inequalities (6 and 7) also impose constraints on an odd number of edges within a cycle. Thus, for the multilabel case, a new projection graph is constructed, which was earlier used by Sontag and Jaakkola (2007) to define cycle inequalities for graphs with multilabel nodes. Sontag et al. (2012) then applies the algorithm of Johnson (2008, section 3.4) to this projection graph.

Once a cycle has been identified, it has to be used within an algorithm like that of Sontag et al. (2008). The algorithm in Sontag et al. (2008) is based on MPLP and the per iteration complexity will be high if the cycle is long. It is possible to bias the search algorithm to return cycles with shorter length. The cycle search algorithm of Johnson (2008, section 3.4) involves the construction of a spanning tree of the newly constructed graph and choosing an initial edge based on which the cycle is constructed. By constructing a shallow spanning tree and by using an efficient least common ancestor algorithm, cycles could be found, which are of shorter length and of higher rank w.r.t. the greedy criterion. In the next section, we will see an algorithm that further improves the efficiency when cycles are present within inference algorithms like MPLP.

Empirical results show the clear benefit of this approach for graphs with few or no short cycles. There are problem instances where the frustrated cycle is a short one, i.e., it is equivalent to a small cluster. In such cases, the cluster pursuit algorithm based on $d(s)$ is able to identify this cycle very quickly compared to the above cycle-based greedy criterion. Thus, whenever possible, it is a good idea to add both cycles (found using the approach outlined above) and clusters (found using $d(s)$ in (13)) to tighten the relaxation. It should be noted that if $d(s)$ is used for finding cycles, then we need to work with a dictionary of cycles. For some graph topologies it could be difficult to construct this dictionary. Also, the approach of using $d(s)$ to a dictionary of cycles is accompanied by the following hardness results:

- Suppose, the graph has binary variables and the beliefs are obtained by running the optimization routine to optimum, then finding the cycle that maximizes $d(s)$ is equivalent to finding the most violated cycle inequality in the dual.
- Suppose, the graph has binary variables and the optimization is not run to optimum, then finding the cycle that maximizes $d(s)$ is NP-hard.
- Suppose, the graph is multilabel, then it is always NP-hard to rank cycles based on $d(s)$.

We will briefly discuss about cycles and higher-order MRFs. Many higher-order MRFs, involve both higher-order cliques and pairwise edges (e.g., the datasets in Rother et al. (2009)). In such graphs, we could consider the cycles involving the pairwise edges. In case, only higher-order cliques are present, pairwise edges with zero valued potentials could be considered and suitable cycles could be defined. This is exactly like considering zero valued cluster-based potentials in (14). We have seen that a cluster is any subset of nodes and a cycle is a special type of cluster. Thus, we could have algorithms that impose cycle-based marginalization constraints, just like cluster-based marginalization constraints that we came across in Section 2. Toward the end of that section, we mentioned the equivalence between algorithms involving cluster-based marginalization constraints and algorithms involving cluster-based subproblems within dual decomposition. The same holds for cycles too. It is very well possible to achieve tighter relaxation by considering cycles as subproblems within dual decomposition. The important works in this direction have been Yarkony et al. (2011), Wang and Koller (2013) and we will discuss them next.

3.2 Efficient MAP inference in a cycle

Suppose, one or more cycles are introduced as subgraphs in dual decomposition (9), then efficient minimization of the cycle subgraphs is a challenge. If the subgraph is a pairwise tree, with m nodes, the minimization could be performed with $O(ml^2)$ complexity using dynamic programming. This is equivalent to one pass of an algorithm called *max-product message passing*. In a tree, a node is chosen as the root and message computation starts at leaf nodes and message passing proceeds toward the root. Between two nodes, message computation takes $O(l^2)$ complexity, thus resulting in $O(ml^2)$ overall complexity. On the other hand, a naive approach for the cycle, with m nodes, would require $O(ml^3)$ complexity. This is because the complexity of inference in graphical models is exponential in the treewidth of the graph and the treewidths of a tree and a cycle are 1 and 2, respectively. Generally, inference in a cycle, involves forming a clique tree by triangulating the cycle, followed by max-product message passing on the clique tree. This involves several passes of the max-product algorithm and is of complexity $O(ml^3)$. There have been algorithms that have reduced the complexity of inference in cycles below $O(ml^3)$ (Felzenszwalb and McAuley, 2011; McAuley and Caetano, 2011). Wang and Koller (2013) build on these works and presents a more efficient algorithm. One way to triangulate a cycle is to enforce a node to be common across all the triangles, i.e., all the triangulating chords are from that node to the other nodes. This results in a clique tree in the form of a chain. Wang and Koller (2013) use the presence of the common node across all the cliques to design a more efficient algorithm. Generally, as number of passes of the max-product algorithm increases fewer labels from each node participate in the message computation. Finally, the labels narrow down to the MAP labels for each node. In the naive approach, the l^3 factor appears in the complexity because all labels from each node participate in the message computation. At the same time, identifying the active labels for a node that are enough for computing the messages is a challenging task. Wang and Koller (2013) have derived inequality-based bounds to decide which labels of a node participate in message computation. They derive these inequalities by exploiting the common node and they are able to identify the necessary labels for each node that are participating in message computation. They start with a low value for the bound that allows more labels to be selected in the initial iterations and the bound increases as the algorithm progresses. The algorithm is guaranteed to terminate and the exit condition involves the inactive labels and potentials. Thus, using their MAP algorithm for cycles within dual decomposition (9) leads to a more efficient overall algorithm.

If the optimization algorithm is like the MPLP-based algorithm of Sontag et al. (2008), a cycle is treated just like a cluster and the algorithm will enforce marginalization constraints with respect to a cycle. The update of the dual variables correspond to *min-marginal* computation for the corresponding nodes. Given a graphical model, the min-marginal of a node w.r.t. to a label, is the minimum energy (of the form (1)) for that graphical model when the label of that node is fixed. Thus, each node-label pair is associated with a min-marginal value. For example, the max-product algorithm computes the MAP labelling of a tree by computing min-marginals. However, depending on the graph, min-marginal computation is more involved than computing the MAP labelling. For example, for cycles, it is more efficient to compute the MAP labelling compared to min-marginals. Apart from the algorithm for MAP inference in cycles, Wang and Koller (2013) have shown a heuristic way to compute min-marginals in cycles and has shown good empirical performance with that approach.

3.3 Planar subproblems

So far we have been discussing about individual cycles, either we compute marginalization constraints with respect to each of the cycles or optimize each cycle as a subproblem within dual decomposition. Yarkony et al. (2011) consider the possibility of optimizing multiple cycles together as one subproblem within dual decomposition. How does such a subproblem look like? Suppose that the original problem is a planar graph (like many computer vision problems) and consider a subproblem with the same topology as the original planar graph. Within dual decomposition, optimizing such a planar subproblem is equivalent to optimizing multiple cycles at the same time. Recall that dual decomposition requires the subproblems to be optimized exactly. The natural question is how is this planar subproblem going to be optimized exactly in an efficient manner? We will have to consider specific pairwise potentials that will make this possible. For example, Schraudolph and Kamenetsky (2009) showed how exact inference could be performed efficiently for a planar Ising model with binary variables. In an Ising model, the values taken by the edge potentials depend only on whether the two nodes take same or different labels.

Yarkony et al. (2011) use planar subproblems which are of the above type. However, their original graphs are planar graphs with nodes taking multiple labels. They show that a node in the planar subproblem could be assigned binary labels, where the binary labels correspond to a binary partition of the label set of that node. They show one well motivated way to obtain a binary partition of a label set. Other approaches are also possible. In their algorithm, initially, the nonsmooth dual (9) is solved to optimality using a decomposition based on trees. Then, labels are obtained for each node by rounding the fractional relaxed variables. Then, they partition the label set of each node in a one-vs-rest manner based on this labelling. Thus a binary planar subproblem is created and added to the dual decomposition. Now, we have a dual problem, with tree-based subproblems and also, a planar subproblem. Again, this dual is solved to optimality and based on the resulting labels another binary planar subproblem is created. They fix a maximum limit on the number of binary subproblems that are added in this manner (say, 10). Their intuition is that each time a dual is solved to optimality and labels are decoded, the resulting energy is an upper bound to the actual energy (just like how the dual is a lower bound). The binary planar subproblems are encouraging the nodes to take labels corresponding to such upper bounds. Within each iteration of dual decomposition, the binary subproblem is solved to optimality based on an efficient algorithm like Schraudolph and Kamenetsky (2009) and the optimal labelling leads to a valid supergradient with respect to that subproblem (Komodakis et al., 2011, Lemma 1). As we said before this optimal binary labelling corresponds to a one-vs-rest partitioning of the original label set. The binary labels that are assigned to the nodes have to be lifted back to the original label space, in order to have a valid supergradient. If the optimal binary label of a node within this binary planar subproblem corresponds to the single label within the binary partition, then we have no problem. However, if the optimal label corresponds to the rest of the labels partition, then one of these labels is chosen at random and assigned as the optimal label according to the original label space. This could be a problem, when the total number of labels increases. Yarkony et al. (2011) have shown results with problems involving up to three labels per node. Nevertheless, this work is an innovative approach that could spark some stronger approach in the future.

Similar to clusters, cycle-based ideas could be applied to several datasets like stereo, segmentation, etc. (Kappes et al., 2015). In the case of cluster-based approaches, the algorithms could work with a dictionary of small-sized clusters, like triplets. For cycle-based approaches, it is more involved to define the set of cycles to consider. For example, a small cycle made of three nodes could as well be treated as a cluster (Sontag et al., 2012). There are certain sparse graphs like those that model a social network or the web graph, where cycle-based approaches stand out (Sontag et al., 2012).

4 Tighter subgraph decompositions

So far, we have come across a data structure called a cluster. In the broadest sense, any subgraph could be a cluster. Usually, in the context of tighter relaxation, a cluster is formed as the union of cliques. Such clusters are directly related to relaxation hierarchies like Sherali–Adams. A cycle is also a cluster, with more special structure, which could be exploited for more efficient algorithms. Also, working with subgraphs in the form of cycles through light on the concept of frustrated cycles, which one will come across in loose relaxations. There are other subgraphs with special structure, too, which will lead to tighter relaxations and more efficient algorithms.

For a pairwise graphical model, one of the most general subgraphs that could lead to a tighter relaxation is a hypertree (Wainwright and Jordan, 2008, section 4). For example, if the original graph is a pairwise grid graph, the local polytope relaxation decomposes the graph according to trees of the graph (e.g., row chains and column chains). Instead, we could treat each face of the grid as a hyperedge/clique of four nodes and decompose the graph according to hypertrees, which are row chains and column chains of the faces of the grid. This decomposition will be tighter than the local polytope for a pairwise graph. From an algorithmic point of view, say, we are using a supergradient method to optimize (9), then the challenge is to perform the minimization efficiently within each of these hypertrees. We will recall that this minimization w.r.t. a hypertree is equivalent to MAP inference w.r.t. a hypertree of the same topology, with suitably defined clique potentials. Over the years there have been algorithms for efficient MAP inference in a hypertree having cliques with special structures, like cardinality and order (Tarlow et al., 2010), linear constraint nodes (Potetz and Lee, 2008), matchings (Duchi et al., 2007) and sparse, pattern-based potentials (Komodakis and Paragios, 2009; Rother et al., 2009). Such specially structured clique potentials should be, and are often, leveraged to deal with higher-order structure in graphical models. Another interesting general approach for inference in trees is based on randomized algorithms (Noorshams and Wainwright, 2013).

Going beyond trees, Batra et al. (2010) decompose the graph according to subgraphs that are outer-planar graphs. An outer-planar graph could be drawn as a planar graph such that no node is completely surrounded by edges. Another definition is that if a graph is formed from an outer-planar graph by adding an extra node and edges from that additional node to all other nodes, then that graph will be planar. Batra et al. (2010) point out that outer-planar graphs have treewidth of 2 and could be solved by the junction tree algorithm in general. Note that treewidth of 2 means $O(ml^3)$ complexity for the junction tree algorithm, where m is the number of nodes in the subgraph. This complexity could be prohibitive. For example, we saw similar complexity for inference in cycles (Section 3.2) and discussed a more efficient approach. In the case of outer-planar graphs, Batra et al. (2010) use the efficient planar Ising model solver of Schraudolph and Kamenetsky (2009), whenever it is suitable. In this context, we would like to point out the work of Zheng et al. (2012), which also works with subgraphs of treewidth 2. They present an extended junction tree representation to solve such subgraphs.

Finally, one could decompose the graph according to subgraphs that are k-fans. In a graph with a k-fan structure, there are k inner nodes which are fully connected with each other. The rest of the nodes have edges to these k inner nodes only, not to each other. Suppose a k-fan subgraph has m nodes, the junction tree algorithm has complexity $O((m - k)l^{k+1})$. For k-fans, Bergtholdt et al. (2010) and Kappes et al. (2010) work with a search-based inference algorithm that performs faster on average, even though it has the same worst complexity. This k-fan-based model is especially suited for problems with dense graphs like deformable parts-based object localization.

4.1 Global higher-order cliques

Higher-order cliques come in different special forms. Usually, the special structure is exploited for efficiently optimizing the subproblems in dual decomposition or for efficient min-marginal computation. We came across some special clique potentials in the previous section. Apart from the nature of the clique potential, a higher-order clique is also characterized by its size. A special situation is a higher-order clique which involves all the nodes of the graph, i.e., it is a global higher-order clique. We will come across such cliques in applications like enforcing global cardinality constraint (Li and Zemel, 2014) for foreground/background segmentation, diverse M-best MAP estimation (Batra et al., 2012), average-cut in image segmentation (Mezuman and Weiss, 2012). In such problems, apart from potentials corresponding to nodes and small cliques, we have a global clique, which we will denote as $\theta_\alpha(\boldsymbol{x})$. For such a problem, the LP relaxation takes the following form,

$$\underset{\phi}{\text{minimize}} \sum_i \sum_{x_i} \theta_i(x_i)\phi_i(x_i) + \sum_c \sum_{\boldsymbol{x}_c} \theta_c(\boldsymbol{x}_c)\phi_c(\boldsymbol{x}_c) + \sum_{\boldsymbol{x}} \theta_\alpha(\boldsymbol{x})\phi_\alpha(\boldsymbol{x}) \tag{20a}$$

$$\begin{aligned} \text{subject to} \sum_{\boldsymbol{x}_{c\setminus i}} \phi_c(\boldsymbol{x}_c) &= \phi_i(x_i), \quad \forall (i,c) : i \in c \\ \sum_{\boldsymbol{x}\setminus \boldsymbol{x}_c} \phi_\alpha(\boldsymbol{x}) &= \phi_c(\boldsymbol{x}_c), \quad \forall c \in \mathcal{C} \end{aligned} \tag{20b}$$

$$\begin{aligned} &\sum_{x_i} \phi_i(x_i) = 1, \quad \phi_i(x_i) \geq 0 \quad \forall i \in \mathcal{V} \\ &\sum_{\boldsymbol{x}_c} \phi_c(\boldsymbol{x}_c) = 1, \quad \phi_c(\boldsymbol{x}_c) \geq 0 \quad \forall c \in \mathcal{C} \\ &\sum_{\boldsymbol{x}} \phi_\alpha(\boldsymbol{x}) = 1, \quad \phi_\alpha(\boldsymbol{x}) \geq 0. \end{aligned} \tag{20c}$$

The presence of the marginalization constraints allows us to write the same problem in the following simpler form,

$$\begin{aligned} \underset{\phi}{\text{minimize}} & \sum_{\boldsymbol{x}} \left(\sum_i \theta_i(x_i) + \sum_c \theta_c(\boldsymbol{x}_c) + \theta_\alpha(\boldsymbol{x}) \right) \phi_\alpha(\boldsymbol{x}) \\ \text{subject to} & \sum_{\boldsymbol{x}} \phi_\alpha(\boldsymbol{x}) = 1, \quad \phi_\alpha(\boldsymbol{x}) \geq 0. \end{aligned} \tag{21}$$

Mezuman et al. (2013) show that this problem is tight. For their proof, they use the form given in (21). In this form, the objective is linear with respect to the $\phi_\alpha(\boldsymbol{x})$'s and the constraint is a simplex constraint overall $\phi_\alpha(\boldsymbol{x})$'s. Under such a simplex constraint, the optimal solution is to assign the value 1 entirely to that global labelling $\boldsymbol{x}$ that minimizes the objective in (21). Thus, we obtain the exact global labelling and the LP relaxation is tight.

As expected, solving (20) is intractable. This is because the marginalization constraint $\sum_{\boldsymbol{x}\backslash\boldsymbol{x}_c} \phi_\alpha(\boldsymbol{x}) = \phi_c(\boldsymbol{x}_c)$ is enforced with respect to all the smaller cliques, i.e., $\forall c \in \mathcal{C}$. Mezuman et al. (2013) try for a more tractable relaxation, which is still reasonably tight, by choosing a smaller set of the smaller cliques $\mathcal{S}$ and by enforcing the marginalization constraint with respect to $\phi_\alpha(\boldsymbol{x})$ in the following way,

$$\begin{aligned} & \sum_{\boldsymbol{x}\backslash x_i} \phi_\alpha(\boldsymbol{x}) = \phi_i(x_i), \quad \forall i \in \mathcal{V} - \mathcal{V}(\mathcal{S}) \\ & \sum_{\boldsymbol{x}\backslash\boldsymbol{x}_c} \phi_\alpha(\boldsymbol{x}) = \phi_c(\boldsymbol{x}_c), \quad \forall c \in \mathcal{S}. \end{aligned} \tag{22}$$

Mezuman et al. (2013) show that if $\mathcal{S}$ forms a low treewidth subgraph, the marginalization in (22) could be performed efficiently for special types of global potentials like cardinality-based potential (Tarlow et al., 2010) and sparse, pattern-based potential (Komodakis and Paragios, 2009; Rother et al., 2009). They also show a way to iteratively build this set $\mathcal{S}$. Suppose, for a given $\mathcal{S}$, the algorithm has run to optimality and for some clique $c \notin \mathcal{S}$, the optimal labelling of those nodes w.r.t. the small clique θ_c and w.r.t. the global clique θ_α disagree, then adding c to $\mathcal{S}$ will lead to a tighter relaxation. Thus, the set $\mathcal{S}$ is grown step-by-step up to a maximum tractable size, where in each step the problem is run to optimality under marginalization constraints (22) defined by that $\mathcal{S}$.

By this stage, we have seen several algorithms that lead to a tighter LP relaxation. Some of these approaches are quite general, while some exploit special structure in the problem. We have also been exposed to fundamental concepts like the marginal and local polytopes, dual decomposition and frustrated cycles. It is time to broaden our understanding of continuous relaxations for MAP inference by discussing about semidefinite programming-based relaxations.

5 Semidefinite programming-based relaxation

Generally, a continuous relaxation for the MAP inference problem takes the form of an LP or a QP or an SDP or an SOCP. Among these relaxations, Kumar et al. (2009) showed that the LP relaxation formulation dominates a large class of QP and SOCP relaxations, from a convex formulation point of view. Thus, there has been tremendous research focus on the LP relaxation and rightly so. As far as the SDP relaxation is concerned, the comparison with respect to the LP relaxation is more involved. We have seen that the LP relaxation could be tightened based on the Sherali–Adams hierarchy. Similarly, the SDP relaxation could be tightened based on the *sum-of-squares* or the *Lasserre* hierarchy (Lasserre, 2001; Parrilo, 2003). Wainwright and Jordan (2008, section 9.3) have pointed out that at a given level of their respective hierarchies, the LP and SDP relaxations are mutually incomparable, i.e., they are neither equivalent nor one strictly dominates the other. Thus, there will be problem instances where the SDP relaxation could indeed provide a tighter relaxation than the LP relaxation. However, SDP relaxations have been less attractive because solving an SDP is computationally intensive. An SDP involves a matrix as the decision variable. Therefore, the main challenge with solving SDPs is the heavy computational cost associated with linear algebraic operations (like, singular value decomposition SVD) involving that matrix decision variable.

In order to understand these computational challenges, we will first derive the general SDP relaxation for MAP inference. This could be derived from the QP relaxation for (2) . Suppose, we have a pairwise graph $\mathcal{G}=(\mathcal{V},\mathcal{E})$, with n nodes and l labels per node. The QP relaxation takes the following form,

$$\underset{\phi}{\text{minimize}} \sum_{i\in\mathcal{V}}\sum_{x_i}\theta_i(x_i)\phi_i(x_i) + \sum_{i,j\in\mathcal{E}}\sum_{x_i,x_j}\theta_{ij}(x_i,x_j)\phi_i(x_i)\phi_j(x_j) \tag{23a}$$

$$\text{subject to} \sum_{x_i}\phi_i(x_i)=1, \quad \phi_i(x_i)\in[0,1] \quad \forall i\in\mathcal{V}. \tag{23b}$$

In the above formulation, if $\phi_i(x_i)\in\{0, 1\}$ then we obtain an ILP that is equivalent to the one given in (3). In other words, the QP relaxation is obtained from an ILP formulation of (2), which involves only node-based variables $\phi_i(x_i)$. On the other hand, the ILP in (3) involves clique-based variables $\phi_c(\boldsymbol{x}_c)$, also. The reduction in the size of the decision variables comes with a price. The QP relaxation is nonconvex. However, a remarkable property of the QP relaxation is that even though it is intractable, it is tight (Ravikumar and Lafferty, 2006). In other words, if the global optimum of the QP relaxation could be obtained, it will be the MAP labelling that we are seeking. Also, direct attempts to solve this non-convex problem don't show reliable empirical performance in terms of tighter results.

We could define a symmetric matrix variable $\Phi\in\mathcal{S}^{nl\times nl}$, such that $\Phi = \boldsymbol{\phi}\boldsymbol{\phi}^\top$. Here, $\boldsymbol{\phi}$ is obtained by stacking the variables, $\phi_i(x_i)$, across all nodes i

and labels x_i. In other words, an element of Φ is of the form, $\Phi(i;\, x_i, j\,;\, x_j) = \phi_i(x_i)\phi_j(x_j)$. We note that Φ has elements for all pairs of nodes, not just the edges of the graph. Thus, the problem in (23) could be rewritten as follows,

$$\underset{\boldsymbol{\phi},\,\Phi}{\text{minimize}} \sum_i \sum_{x_i} \theta_i(x_i)\phi_i(x_i) + \sum_{i,j} \sum_{x_i,x_j} \theta_{ij}(x_i,x_j)\Phi(i;\, x_i, j;\, x_j) \tag{24a}$$

$$\text{subject to} \sum_{x_i} \phi_i(x_i) = 1, \quad \phi_i(x_i) \in [0,1] \quad \forall i \in \mathcal{V} \tag{24b}$$

$$\Phi = \boldsymbol{\phi}\boldsymbol{\phi}^\top. \tag{24c}$$

It is still the same intractable but tight relaxation of the MAP inference problem, as (23). There is yet another way to represent this intractable problem, through a positive semidefinite constraint combined with a rank-1 constraint. It takes the following form,

$$\underset{\boldsymbol{\phi},\,\Phi}{\text{minimize}} \sum_i \sum_{x_i} \theta_i(x_i)\phi_i(x_i) + \sum_{i,j} \sum_{x_i,x_j} \theta_{ij}(x_i,x_j)\Phi(i;\, x_i, j;\, x_j) \tag{25a}$$

$$\text{subject to} \sum_{x_i} \phi_i(x_i) = 1, \quad \phi_i(x_i) \in [0,1] \quad \forall i \in \mathcal{V} \tag{25b}$$

$$\begin{bmatrix} 1 & \boldsymbol{\phi}^\top \\ \boldsymbol{\phi} & \Phi \end{bmatrix} \succeq 0 \tag{25c}$$

$$\text{rank}\left(\begin{bmatrix} 1 & \boldsymbol{\phi}^\top \\ \boldsymbol{\phi} & \Phi \end{bmatrix} \right) = 1 \tag{25d}$$

$$\Phi(i;\, x_i, i;\, x_i) = \phi_i(x_i) \quad \forall i \in \mathcal{V}. \tag{25e}$$

We repeat that the problems (23), (24) and (25) are equivalent. We will look further into the motivation behind the formulation of the constraints in (25). First, we will see how the constraint $\Phi = \boldsymbol{\phi}\boldsymbol{\phi}^\top$ in (24) is equivalently represented by (25c) and (25d). In SDP relaxation literature, the presence of the equality constraint $\Phi = \boldsymbol{\phi}\boldsymbol{\phi}^\top$ (24c) in the stricter problem, usually leads to the semidefinite constraint $\Phi - \boldsymbol{\phi}\boldsymbol{\phi}^\top \succeq \mathbf{0}$ in the relaxation Goemans and Williamson (1995). Now, the following two inequalities are equivalent,

$$\Phi - \boldsymbol{\phi}\boldsymbol{\phi}^\top \succeq \boldsymbol{0} \tag{26}$$

$$\boldsymbol{M} \triangleq \begin{bmatrix} 1 & \boldsymbol{\phi}^\top \\ \boldsymbol{\phi} & \Phi \end{bmatrix} \succeq \boldsymbol{0}. \tag{27}$$

This equivalence is based on the Schur complement formula (Boyd and Vandenberghe, 2004, section A.5.5). The semidefinite inequality (25c) along with the nonconvex rank-1 constraint (25d) lead to an equivalent representation of $\Phi = \boldsymbol{\phi}\boldsymbol{\phi}^\top$. The SDP relaxation is obtained by dropping this rank-1 constraint. Further, the constraint $\Phi(i;\, x_i,\, i;\, x_i) = \phi_i(x_i)$ is inspired by the original ILP formulation of the problem. According to (24c), $\Phi(i;\, x_i,\, i;\, x_i) = \phi_i(x_i)^2$. If $\phi_i(x_i)$ takes integral values, i.e., $\{0,\, 1\}$, then $\Phi(i;\, x_i,\, i;\, x_i) = \phi_i(x_i)^2 = \phi_i(x_i)$. This constraint tries to compensate for the removal of the rank-1 constraint by encouraging the desired structure in the output variables. Thus, the SDP relaxation for MAP inference takes the following form,

$$\underset{\boldsymbol{\phi},\, \Phi}{\text{minimize}} \sum_i \sum_{x_i} \theta_i(x_i)\phi_i(x_i) + \sum_{i,j} \sum_{x_i, x_j} \theta_{ij}(x_i, x_j)\Phi(i;\, x_i, j;\, x_j) \tag{28a}$$

$$\text{subject to} \sum_{x_i} \phi_i(x_i) = 1, \quad \phi_i(x_i) \in [0,1] \quad \forall i \in \mathcal{V} \tag{28b}$$

$$\begin{bmatrix} 1 & \boldsymbol{\phi}^\top \\ \boldsymbol{\phi} & \Phi \end{bmatrix} \succeq 0 \tag{28c}$$

$$\Phi(i;\, x_i, i;\, x_i) = \phi_i(x_i) \quad \forall i \in \mathcal{V}. \tag{28d}$$

In the SDP relaxation, the elements of Φ are equivalent to pseudomarginals of pairs of nodes, $\phi_{ij}(x_i,\, x_j)$, that we came across in the local polytope. The matrix $\boldsymbol{M}$ (27) contains pseudomarginals corresponding to nodes and pairs of nodes and it is called the *moment matrix* (Wainwright and Jordan, 2004a). $\boldsymbol{M}$ is a square matrix of side, $1 + nl$. Similar to the local polytope, $\mathcal{L}(\mathcal{G})$, the above constraints define a convex outer bound to the marginal polytope, $\mathcal{M}(\mathcal{G})$. We will denote it as $\mathcal{SDP}(\mathcal{G})$. Unlike, $\mathcal{L}(\mathcal{G})$, $\mathcal{SDP}(\mathcal{G})$ has a curved boundary. Note, in our formulation of the local polytope, we explicitly represented pseudomarginals only for pairs of nodes that correspond to the edges. However, it is possible to associate pseudomarginals with respect to all pairs of nodes and the SDP relaxation explicitly represents the pseudomarginals for all node pairs. The constraints (28d) and (28e) play a role similar to the marginalization constraints (5b) in the LP relaxation. Earlier we have seen that in the LP relaxation if the optimal variables are integer valued, we have obtained the exact MAP solution. Similarly, if the matrix $\boldsymbol{M}$ is of rank-1 in the SDP relaxation, we have obtained the exact MAP solution (Wang et al., 2016, Prop. 2).

Despite the computational challenge involving the semidefinite constraint (28d), there have been several works in recent years that have explored SDP relaxations for MAP inference. For several of these methods, the main thrust has been the efficient low-rank Burer–Monteiro approach (Burer and Monteiro, 2003). The Burer–Monteiro approach is an efficient way to solve semidefinite programs through a low-rank factorization-based approximation of the matrix variable. Suppose, $\boldsymbol{M}$ is the matrix variable, it is substituted by $\boldsymbol{YY}^\top$ everywhere in the problem formulation. We obtain an optimization problem with $\boldsymbol{Y}$, which is of low-rank, as the decision variable. One of the motivating reasons for this approach is that if an SDP satisfies certain mild conditions, the rank of the optimal decision variable is bounded by $\sqrt{2 \cdot \#constraints}$ (Pataki, 1998). Thus the optimal decision variable is guaranteed to be of considerably low-rank for many practical SDP formulations. Further, even though the Burer–Monteiro approach is nonconvex, for several problems the local minimum returned by the approach is also a global minimum or is within a bounded value from the global minimum. Over the years, there has been progress in characterizing conditions which lead to these guarantees (Boumal et al., 2016, 2018; Burer and Monteiro, 2005; Mei et al., 2017).

Given a hierarchy of LP relaxations and a hierarchy of SDP relaxations, Wainwright and Jordan (2008, section 9.3) have shown that at any given level, the corresponding LP relaxation and the SDP relaxation are mutually incomparable. As mentioned earlier, this is an important motivation to pursue SDP relaxations. $\mathcal{L}(\mathcal{G})$ and $\mathcal{SDP}(\mathcal{G})$ are the lowest levels of the LP relaxation and the SDP relaxation hierarchies, respectively. Since, these two feasible regions are incomparable, for some problems $\mathcal{L}(\mathcal{G})$ could be tighter and for some others $\mathcal{SDP}(\mathcal{G})$ could be tighter. Thus, the intersection of $\mathcal{L}(\mathcal{G})$ and $\mathcal{SDP}(\mathcal{G})$ is a better bound on $\mathcal{M}(\mathcal{G})$ than either of them separately. Peng et al. (2012) have pursued this idea and presented a dual-based algorithm. The variables $\boldsymbol{\phi}$ and Φ are pseudomarginals or beliefs and we will denote them together as one large sized variable $\boldsymbol{b}$. We could plug $\boldsymbol{b}$ within the LP relaxation objective (5a), which we will denote as $F(\boldsymbol{b})$. Then, Peng et al. (2012) solve the following problem,

$$\begin{aligned} \underset{\boldsymbol{b}^1 \in \mathcal{L}(\mathcal{G}), \boldsymbol{b}^2}{\text{minimize}} \quad & F(\boldsymbol{b}^1) + \mathbb{1}\left[\boldsymbol{b}^2 \in \mathcal{SDP}(\mathcal{G})\right] \\ \text{subject to} \quad & \boldsymbol{b}^1 = \boldsymbol{b}^2. \end{aligned} \tag{29}$$

In other words, the objective function in (5a) is minimized within the intersection of $\mathcal{L}(\mathcal{G})$ and $\mathcal{SDP}(\mathcal{G})$. The Lagrangian dual of this problem nicely decomposes according to $\boldsymbol{b}^1$ and $\boldsymbol{b}^2$, as follows,

$$\underset{\boldsymbol{\delta}}{\text{maximize}} \quad \underset{\boldsymbol{b}^1 \in \mathcal{L}(\mathcal{G})}{\text{min.}} \left(F(\boldsymbol{b}^1) + \boldsymbol{\delta}^\top \boldsymbol{b}^1\right) + \underset{\boldsymbol{b}^2 \in \mathcal{SDP}(\mathcal{G})}{\text{min.}} -\boldsymbol{\delta}^\top \boldsymbol{b}^2. \tag{30}$$

Thus, while maximizing the dual w.r.t. $\boldsymbol{\delta}$, the two subproblems could be minimized w.r.t. $\boldsymbol{b}^1$ and $\boldsymbol{b}^2$ separately. Since, the dual problem is nonsmooth, the minimizing $\boldsymbol{b}^{1*}$ of subproblem 1 and the minimizing $\boldsymbol{b}^{2*}$ of subproblem 2, give us a valid supergradient ($\boldsymbol{b}^{1*} - \boldsymbol{b}^{2*}$) to update the dual variable $\boldsymbol{\delta}$ (Komodakis et al., 2011, Lemma 1). The first subproblem is equivalent to the LP relaxation (5) with an additional linear term in the objective. Peng et al. (2012) solve it using convex belief propagation (Hazan and Shashua, 2008).

The second subproblem is a semidefinite program, a linear objective with a semidefinite constraint. It involves the decision variable $\boldsymbol{b}^2$, composed of $\boldsymbol{\phi}$ and Φ, which can be written as a moment matrix $\boldsymbol{M}$, like we saw in (27).

Taking inspiration from the Burer–Monteiro approach, Peng et al. (2012) replace $\boldsymbol{M}$ by the low-rank factorization $\boldsymbol{Y}\boldsymbol{Y}^{\top}$, where $\boldsymbol{Y} \in \mathcal{R}^{(1+nl)\times r}$. Here, r is the low valued rank of $\boldsymbol{Y}$ and the moment matrix representing $\boldsymbol{b}^2$. Expecting the moment matrix in the SDP relaxation to be low-rank makes intuitive sense because we came across a rank-1 constraint in the intractable formulation given in (25). Further, this low-rank representation ensures that the moment matrix that is implicitly computed as $\boldsymbol{Y}\boldsymbol{Y}^{\top}$ is positive semidefinite by definition. Thus, the semidefinite constraint $\boldsymbol{M} \succeq 0$ in $\mathcal{SDP}(\mathcal{G})$ need not be explicitly enforced. This simplification afforded by the Burer–Monteiro approach appears repeatedly in SDP literature. The explicitly enforced constraints are (28e). Note that the constraints (28b) are also enforced in $\mathcal{L}(\mathcal{G})$ and they need not be part of the second subproblem. Peng et al. (2012) take an augmented Lagrangian-based approach to solve this subproblem. Generally, with the low-rank representation, the question of setting the rank r needs to be addressed. In Peng et al. (2012), r was set to various values and all the results are reported. r was at most 50, in their experiments.

Let us keep in mind that the decision variables of the SDP relaxation could either be represented as a moment matrix, $\boldsymbol{M}$ or explicitly through the parts, $\boldsymbol{\phi}$ and Φ, which make up $\boldsymbol{M}$. The underlying problem is the same in both cases. The constraints (28b), (28d) and (28e) could be expressed in a suitable manner in both cases. In case the decision variables are represented by the moment matrix, the constraint $\boldsymbol{M}(0, 0) = 1$ should be explicitly enforced, also.

Intersecting $\mathcal{L}(\mathcal{G})$ with $\mathcal{SDP}(\mathcal{G})$ is one instance of a general way to get better results with the SDP relaxation. The general approach is about adding suitable linear equality and/or inequality constraints to the semidefinite program (28). This idea has been explored since the early works studying the SDP relaxation (Wainwright and Jordan, 2004a). Huang et al. (2014) present one other approach along these lines. It adds to the formulation in (28), the following nonnegativity constraints on the matrix Φ,

$$\Phi(i;\, x_i, j;\, x_j) \geq 0 \quad \forall (i,j) \in \mathcal{E}. \tag{31}$$

Empirically, these inequality constraints improve the performance of SDP relaxations, when submodular edge potentials are present. We have mentioned earlier that submodular potentials are a type of attractive potentials. Also, Huang et al. (2014) have shown that the semidefinite constraint (28d) along with the linear constraints (28e) and (28b), implicitly ensure the satisfaction of the marginalization constraints (3b) that we come across in the LP relaxation. For their formulation, Huang et al. (2014) have shown an ADMM-based approach (Boyd et al., 2011). The challenge with ADMM-based algorithms is the task of proving convergence. Wen et al. (2010) have presented an ADMM-based algorithm for solving semidefinite programs, where convergence is guaranteed only if equality constraints are present. Interestingly, the equality constraints and the inequality constraints in the SDP relaxation proposed by Huang et al. (2014) are decoupled. Specifically, the equality constraints concern the vector $\boldsymbol{\phi}$ and the diagonal blocks of Φ, while the inequality constraints concern the edges of the graph, i.e., off-diagonal blocks of Φ. Hence, Huang et al. (2014) are able to prove convergence by using the same ADMM approach as Wen et al. (2010).

A big challenge with SDPs is the memory intensive representation of the optimization variables. One needs to look for structure in the problem at hand, to reduce the number of variables. For example, Huang et al. (2014) have to deal with the primal decision variables, $\boldsymbol{\phi}$ and Φ and the dual variables corresponding to the augmented Lagrangian. These dual variables correspond to the positive semidefinite constraint, the linear equality constraints and the linear inequality constraints. Interestingly, for their problem formulation, the matrix valued dual variable corresponding to the positive semidefinite constraint need not be explicitly represented. Since, it could be expressed in terms of the other dual variables and the primal variables. Still, the representation of the matrix Φ is a challenge. Similar to Peng et al. (2012), ϕ and Φ are represented in terms of a moment matrix $\boldsymbol{M}$ (27) and Burer–Monteiro is invoked to represent $\boldsymbol{M}$ through a low-rank factorization $\boldsymbol{Y}\boldsymbol{Y}^\top$. In Peng et al. (2012), this low-rank representation led to the disappearance of the positive semidefinite constraint in the algorithm. However, in this case, one of the ADMM steps still involves projection of $\boldsymbol{Y}$ onto the positive semidefinite cone. However, $\boldsymbol{Y}$ is of rank r, only the top r eigenvalues and eigenvectors of an intermediate matrix needs to be computed using the Lanczos algorithm (Golub and Van Loan, 2013). The complexity of the Lanczos algorithm is governed by a matrix-vector multiplication operation, involving this intermediate matrix. Interestingly, the sparsity pattern of this matrix is driven by that of the graph $\mathcal{G}$, which can be advantageous for sparse graphs like grid graphs. Apart from these aspects, there is ample scope to parallelize this matrix-vector product. Similar to Peng et al. (2012), setting the rank parameter r is an open question. Huang et al. (2014) propose to solve the problem to optimality repeatedly, doubling the value of r each time. The quality of the solution could be assessed based on the primal-dual gap of the original integral problem and the relaxed dual problem. Also, as the algorithm proceeds, the label for some of the nodes are fixed according to a reliability criterion. The label of a node is fixed if $\phi_i(x_i)$ is greater than a high threshold, like 0.99 or if $\phi_i(x_i)$ is the maximum $\forall i$ and $\forall x_i$. This heuristic improves the scalability of the algorithm.

Upon closer observation, one could notice that Peng et al. (2012) and Huang et al. (2014) are essentially having the same set of constraints. Both enforce the normalization and marginalization constraints of the LP relaxation and the constraints corresponding to the SDP relaxation. However, in the case of Huang et al. (2014) the marginalization constraints of the LP relaxation are enforced implicitly through other constraints. This enabled them to approach the optimization with a different algorithmic approach. Even though Huang et al. (2014) prove the convergence of their ADMM scheme when the moment matrix $\boldsymbol{M}$ is considered explicitly, they do not prove convergence when it is represented through the low-rank factorization, $\boldsymbol{Y}\boldsymbol{Y}^\top$. However, they report good empirical performance for the latter approach for various datasets.

So far, we have been seeing a way to tighten the SDP relaxation by adding linear constraints inspired by the LP relaxation. In the context of the LP relaxation, we saw that it could be increasingly tightened by a hierarchy of problems, called the Sherali–Adams hierarchy. A similar hierarchy exists for the SDP relaxation called the *sum-of-squares* or the *Lasserre* hierarchy (Lasserre, 2001; Parrilo, 2003). Earlier we saw that even though, in theory, a given level of the Sherali–Adams hierarchy will involve all possible subsets of a given size, in practice, one could work with a smaller set of subsets like triplets or the faces of a grid. A similar idea to tighten the SDP relaxation in an efficient manner has been presented by Erdogdu et al. (2017). To understand their approach, we will look at a general way to characterize the moment matrix, $\boldsymbol{M}$. The following equations hold for the case, where the nodes are binary valued and take values 1 or -1. It is possible to extend them to the multilabel case based on larger moment matrices. In (27), the moment matrix carried the unary and pairwise pseudomarginals. In general, it could carry higher-order terms that will lead to a tighter relaxation (Wainwright and Jordan, 2004a). In more detail, given an even number d, we will consider all possible node subsets of size up to $\frac{d}{2}$, this included the null-set ($\emptyset$). We will have $\binom{n}{\leq \frac{d}{2}}$ such subsets. Thus, $\boldsymbol{M}$ is a square matrix of side $\left|\binom{n}{\leq \frac{d}{2}}\right|$. We see that the moment matrix grows exponentially in the size of d. The elements of $\boldsymbol{M}$ are indexed according to pairs of subsets of the nodes. Suppose, S and T are two such subsets, then, we index the corresponding element of $\boldsymbol{M}$ as $\boldsymbol{M}(S, T)$. The moment matrix $\boldsymbol{M}$ is desired to be a symmetric matrix. This constraints the elements of $\boldsymbol{M}$ to be a function of $S\Delta T$ only, where, Δ denotes the symmetric difference between two sets. The symmetry requirement can be expressed as follows,

$$\begin{aligned} S_1\Delta T_1 = S_2\Delta T_2 \quad &\Rightarrow \quad M(S_1,T_1) = M(S_2,T_2) \\ \forall S_1,T_1,S_2,T_2 \subseteq \mathcal{V}:\ &|S_1|,|T_1|,|S_2|,|T_2| \leq \frac{d}{2}. \end{aligned} \tag{32}$$

The moment matrix that we saw in (27) corresponds to the case of $d = 2$. In that case, we can see how the elements of $\boldsymbol{\phi}$ and Φ are indexed by (S, T), where $S,T \in \emptyset \cup \mathcal{V}$. Erdogdu et al. (2017) work with the case $d = 4$ and enforce the constraints on the elements of $\boldsymbol{M}$ dictated by (32). They achieve an efficient algorithm by invoking Burer–Monteiro and represent $\boldsymbol{M}$ through a low-rank factorization. They represent each row of the low-rank matrix as a vector $\boldsymbol{\sigma}_S \in \mathcal{R}^r$, where $S \subseteq \mathcal{V}$: $|S| \leq 2$. Thus, there will be $\left|\binom{n}{\leq 2}\right|$ such vectors. We will denote the set of node subsets as $\mathcal{S}$ and it will be made of nodes and pairs of nodes. We will see soon about the effect of the vector length r on the quality of the optimization result. Thus, we could express the constraints on the moment matrix $\boldsymbol{M}$ in terms of these vectors, as follows,

$$\|\boldsymbol{\sigma}_i\| = 1 \qquad \forall i \in \mathcal{S} \tag{33a}$$

$$\boldsymbol{\sigma}_i \cdot \boldsymbol{\sigma}_j = \boldsymbol{\sigma}_{ij} \cdot \boldsymbol{\sigma}_{\varnothing} \qquad \forall i,j \in \mathcal{S} \tag{33b}$$

$$\boldsymbol{\sigma}_i \cdot \boldsymbol{\sigma}_{ij} = \boldsymbol{\sigma}_j \cdot \boldsymbol{\sigma}_{\varnothing} \qquad \forall i,j \in \mathcal{S} \tag{33c}$$

$$\boldsymbol{\sigma}_i \cdot \boldsymbol{\sigma}_{jk} = \boldsymbol{\sigma}_k \cdot \boldsymbol{\sigma}_{ij} \qquad \forall i,j,k \in \mathcal{S} \tag{33d}$$

$$\boldsymbol{\sigma}_{ij} \cdot \boldsymbol{\sigma}_{jk} = \boldsymbol{\sigma}_{ik} \cdot \boldsymbol{\sigma}_{\varnothing} \qquad \forall i,j,k \in \mathcal{S} \tag{33e}$$

$$\boldsymbol{\sigma}_{ij} \cdot \boldsymbol{\sigma}_{kl} = \boldsymbol{\sigma}_{ik} \cdot \boldsymbol{\sigma}_{jl} \qquad \forall i,j,k,l \in \mathcal{S} \tag{33f}$$

where $.\cdot.$ is the dot product operation and $i,j,k,l \in \mathcal{V}$. The constraint (33a) is a constraint on the diagonal elements of $\boldsymbol{M}$, which are all a function of the null-set ($\varnothing$). Similar to the ILP inspired constraint (28e), the diagonal elements of $\boldsymbol{M}$ are constrained to take the value 1, since, the nodes take labels in $\{-1, 1\}$. Note that the above equations are for the case of $d = 4$, when nodes take binary labels. When the same Burer–Monteiro approach is used for $d = 2$ with binary valued nodes, we will only have the constraint (33a), just like we saw in the case of Peng et al. (2012).

We would like to point out that the sets of nodes that are associated with the moment matrix, carry a similar meaning to clusters, which we came across in the LP relaxation. For example, if we are tightening the LP relaxation, we will be working with pseudomarginals corresponding to nodes, pairs of nodes, triplets of nodes, etc. In an SDP (sum-of-squares) relaxation, we will have a moment matrix with elements corresponding to nodes, pairs of nodes, triplets of nodes, quadruplets of nodes, etc. If we consider only up to quadruplets of nodes, we will denote the relaxation as SOS(4). We can see that SOS(2) leads to a moment matrix with elements corresponding to nodes and pairs of nodes. This is what we saw in (28). Also, SOS(2) corresponds to the pioneering SDP relaxation work of Goemans and Williamson (1995). It will be a good exercise to verify the structure of the moment matrix for SOS(2) and SOS(4).

The set of vectors $\boldsymbol{\sigma}_S$ satisfying the constraints in (33) is denoted as $\boldsymbol{\Omega}(\mathcal{S})$. Thus, the SOS(4) SDP relaxation for a binary valued pairwise graph takes the following form,

$$\underset{\boldsymbol{\sigma}_S, S \in \mathcal{S}}{\text{minimize}} \quad \sum_{(i,j)\in\mathcal{E}} \theta_{ij} \boldsymbol{\sigma}_i \cdot \boldsymbol{\sigma}_j + \sum_{i \in \mathcal{V}} \theta_i \boldsymbol{\sigma}_i \cdot \boldsymbol{\sigma}_{\varnothing} \tag{34a}$$

$$\text{subject to} \quad \boldsymbol{\sigma}_S \in \boldsymbol{\Omega}(\mathcal{S}) \ \ \forall S \in \mathcal{S} : |S| \leq 2. \tag{34b}$$

The set $\mathcal{S}$ is very big. Even when $d = 4$, it is of size $\left|\binom{n}{\leq 2}\right|$, i.e., we are considering all subsets $S : |S| \leq 2$. This is because we are considering all possible pairs of nodes. Just like we did in LP relaxation, while considering clusters, we could restrict the pairs of nodes that we consider. We could divide the graph into triplets of adjacent nodes. We can always create triplets by introducing edges with zero potential and we require these triplets to cover all the nodes and edges of the original graph. We will define a $\boldsymbol{\sigma}_S$, only if S, which is size 1 or 2, is contained in one of the triplets. We will call this restricted set as $\hat{\mathcal{S}}$. Erdogdu et al. (2017) attempt to solve the following more tractable problem,

$$\underset{\boldsymbol{\sigma}_S, S \in \hat{\mathcal{S}}}{\text{minimize}} \quad \sum_{(i,j) \in \mathcal{E}} \theta_{ij} \boldsymbol{\sigma}_i \cdot \boldsymbol{\sigma}_j + \sum_{i \in \mathcal{V}} \theta_i \boldsymbol{\sigma}_i \cdot \boldsymbol{\sigma}_\varnothing \tag{35a}$$

$$\text{subject to} \quad \boldsymbol{\sigma}_S \in \boldsymbol{\Omega}(\hat{\mathcal{S}}) \quad \forall S \in \hat{\mathcal{S}} : |S| \leq 2. \tag{35b}$$

We will now see more about the effect of the low-rank parameter r, which is the length of the vectors, $\boldsymbol{\sigma}_S$. If $r = \left|\binom{n}{\leq \frac{d}{2}}\right|$, then the SDP relaxation is convex. However, we could set r to low values and get acceptable results, while improving the memory footprint. Erdogdu et al. (2017) have obtained good results in their experiments when $r = 10$. Erdogdu et al. (2017) optimize the resulting nonconvex version of (35) by constructing an augmented Lagrangian, where only constraint (33a) is explicitly retained. This augmented Lagrangian is optimized in a coordinate descent fashion, where each time the minimization is with respect to one vector $\boldsymbol{\sigma}_S$. Once, optimality is reached, some of the nodes are assigned a label of 1 or -1 based on a rounding scheme that we will describe in Section 5.1. Based on this, the $\boldsymbol{\sigma}_S$ vectors are fixed as $x_i^* \boldsymbol{\sigma}_\varnothing$ for those particular nodes and $x_i^* x_j^* \boldsymbol{\sigma}_\varnothing$ for edges between such nodes. Then, (35) is again optimized with respect to the remaining $\boldsymbol{\sigma}_S$ vectors. This process is repeated until all nodes are reliably labelled. We have seen similar shrinking of problem size in Huang et al. (2014).

There have been a few efforts to understand the Burer–Monteiro approach in the context of MAP inference. Frostig et al. (2014) have presented an analysis of the following setting: SOS(2) SDP relaxation of the binary pairwise graph. In other words, optimizing (28) over $\mathcal{SDP}(\mathcal{G}_{\{0,1\}})$. In this setting, they are able to prove that the output of the nonconvex problem saturates for a low value of the rank r. Thus, it is possible to ensure that the nonconvex Burer–Monteiro approach matches the convex SDP relaxation with a reasonably small rank r. Another work (Mei et al., 2017) analysing SOS(2), has shown that it is possible to achieve $O(\frac{1}{r})$ relative error with respect to the global optimum for specific choices of the node and edge potentials. Generally, SDP relaxations become more useful with the addition of more linear constraints, either borrowed from LP relaxation or by going up the SOS hierarchy. It will be interesting to analyse the Burer–Monteiro approximation in more involved settings.

So far, we have been looking at how Burer–Monteiro-based low-rank approximation leads to scalable SDP algorithms. We will now see another algorithmic approach to achieve a scalable SDP algorithm that is suitable for tighter MAP inference. We will look into the dual-based approach called SDCUT. It is an algorithm presented by Wang et al. (2013) to solve an SDP relaxation for *binary quadratic programs* (BQPs). A BQP is a special case of the quadratic programming problem in (24), where nodes take only binary labels. Their approach, which is called SDCUT, could be extended to multilabel nodes and thus used for MAP inference, as they show in Wang et al. (2015) and Wang et al. (2016). The advantages of this approach are its scalability and flexibility to accommodate additional linear inequality constraints to tighten the relaxation (Wang et al., 2016). SDCUT has also been useful for MAP inference in fully connected graphs (Wang et al., 2015), which we will not discuss in detail here. First, we will briefly see the derivation of SDCUT. We observe that the constraint, $\Phi(i;\, x_i,\, i;\, x_i) = \phi_i(x_i)$ (28e), along with the normalization constraint on $\phi_i(x_i)$ (28b), could be equivalently expressed as the constraint, $\mathrm{tr}(\boldsymbol{M}) = n+1$, where n is the number of nodes. Where $\boldsymbol{M}$ is the moment matrix (27). $\boldsymbol{M}$ is also constrained to be symmetric and positive semidefinite. As a result of all these constraints, the following result holds true: $\|\boldsymbol{M}\|_F \leq n + 1$ (Malick, 2007). This inequality condition could be made a part of the objective in the SDP (28a) as a regularization term, $\sigma(\|\boldsymbol{M}\|_F^2 - (n+1)^2)$ with $\sigma > 0$. Thus, we have a regularized problem with the following form,

$$\underset{\boldsymbol{M}\succeq \boldsymbol{0}}{\text{minimize}} \quad \mathrm{tr}\left(\boldsymbol{M}^\top \boldsymbol{A}\right) + \sigma \| \boldsymbol{M} \|_F^2 - \sigma(n+1)^2 \tag{36a}$$

$$\text{subject to} \quad \mathrm{tr}\left(\boldsymbol{M}^\top \boldsymbol{B}_i\right) = b_i, \quad \forall i = 1, \ldots, p \tag{36b}$$

$$\mathrm{tr}\left(\boldsymbol{M}^\top \boldsymbol{B}_j\right) \leq b_j, \quad \forall j = p+1, \ldots, m. \tag{36c}$$

Here, $\boldsymbol{A}$ is a matrix such that $\mathrm{tr}\left(\boldsymbol{M}^\top \boldsymbol{A}\right)$ is equal to the objective in (28a). The linear equality constraints (36b) and the linear inequality constraints (36c) are written in a generic manner. They represent the linear constraints that are imposed in the SDP. For MAP inference, they are necessarily composed of the SDP relaxation constraints, (28b) and (28e). They could also consist of constraints borrowed from the LP relaxation literature. The dual to this problem takes the following form,

$$\underset{\boldsymbol{u}}{\text{maximize}} \quad -\frac{1}{4\sigma} \| \boldsymbol{C}(\boldsymbol{u})_- \|_F^2 - \boldsymbol{u}^\top \boldsymbol{b} - \sigma(n+1)^2 \tag{37a}$$

$$\text{subject to} \quad u_j \geq 0, \quad \forall j = p+1, \ldots, m \tag{37b}$$

where, $\boldsymbol{C}(\boldsymbol{u}) = \sum_{i=1}^{m} u_i \boldsymbol{B}_i + \boldsymbol{A}$. Computing the gradient for this problem, involves SVD, since, $\boldsymbol{C}(\boldsymbol{u})_-$ is computed by considering only the negative eigenvalues of $\boldsymbol{C}(\boldsymbol{u})$. This is an expensive part of other algorithms based on SDP relaxation like Huang et al. (2014), also. Huang et al. (2014) ameliorated this problem by computing only the top eigenvectors and eigenvalues that is dictated by low-rank representation and also, by exploiting sparse matrix-vector product driven by the graph structure. Wang et al. (2016) resort to a parallel eigendecomposition algorithm to achieve speed-up of the SVD.

Wang et al. (2016) continue with the theme of adding more linear constraints to the basic SDP problem (28), in order to obtain a tighter relaxation. So far, we have come across various linear constraints that are part of the local polytope-based LP relaxation. Additionally, Wang et al. (2016) consider the following linear constraints that could be enforced. For the main diagonal blocks of Φ, apart from the constraints (28e), we could enforce the following constraints,

$$\Phi(i; x_i, i; x_j) = 0 \quad \forall i \in \mathcal{V},\ x_i \neq x_j. \tag{38}$$

Intuitively, the off-diagonal blocks of Φ correspond to the pseudomarginals of pairs of nodes and we could enforce marginalization constraints and nonnegativity constraints on them, like it is done in the LP relaxation. However, the main diagonal blocks of Φ are not suitable for similar constraints. The main diagonal blocks are an artefact of the semidefinite relaxation. We have already seen the constraints (28e) that could be enforced on the main diagonal. In a similar spirit, Eq. (38) enforce desired structure on Φ.

Further Wang et al. (2016) enforce cycle-based constraints that we came across in tighter LP relaxations (6, 7). These cycle-based constraints were obtained by the combinatorial optimization community, while studying the polyhedral nature of graphs with binary variables ($\mathcal{G}_{\{0,1\}}$) (Deza and Laurent, 2009; Schrijver, 2003). The cycle inequalities that we came across earlier (6) is an example of constraints that could be enforced on a graph with binary variables. Wang et al. (2016) have considered these cycle inequalities and also, *triangular* and *odd-wheel* inequalities. Consider, three nodes i, j, k and let $\delta_{i,j} \triangleq \mathbb{1}[x_i \neq x_j]$ indicate whether two nodes disagree in their labelling, then the triangle inequalities take the following form,

$$\begin{aligned} \delta_{i,j} + \delta_{j,k} + \delta_{k,i} &\leq 2 \\ \delta_{i,j} - \delta_{j,k} - \delta_{k,i} &\leq 0 \\ -\delta_{i,j} + \delta_{j,k} - \delta_{k,i} &\leq 0 \\ -\delta_{i,j} - \delta_{j,k} + \delta_{k,i} &\leq 0. \end{aligned} \tag{39}$$

Just like a cycle inequality, an odd-wheel inequality is defined on a cycle C with odd length of at least 3. These, inequalities take the following form,

$$\sum_{(i,j)\in C} \delta_{i,j} - \sum_{i\in\mathcal{V}(C)} \delta_{i,k} \leq \frac{|C|-1}{2} \tag{40}$$

where k is a node of the graph that is not in C but encircled by it. Like cycle inequalities, there are an exponential number of odd-wheel inequalities.

Also, similar, to cycle inequalities there is a polynomial time separation algorithm for finding the cycle that most violate the odd-wheel inequality (Deza and Laurent, 2009).

Thus, Wang et al. (2016) have presented a tighter MAP inference algorithm based on the dual given in (37a), along with suitable linear equality and inequality constraints. One of the issues to address is making this approach work for a graph with multilabel nodes. Wang et al. (2016) find a solution by constructing the l-projection graph (Sontag and Jaakkola, 2007) that we saw in Section 1.1 and by considering the above constraints w.r.t. that graph. However, this leads to another issue to be addressed. It is the huge problem size that happens with large graphs and nodes with multiple labels. They take a branch and bound approach to reduce the set of active labels for each node (Sun et al., 2012). Also, the huge number of constraints are handled in a cutting-planes framework. Thus, the dual problem is solved several times in this framework, where the label space is pruned each time, by updating lower and upper bounds for each node. The overall algorithm stops when these bounds are sufficiently close for all nodes.

To further improve the scalability, Wang et al. (2016) remove nodes based on persistency of a node label with respect to the optimal labelling. There have been several works regarding persistency in MRF inference (Kohli et al., 2008; Kovtun, 2003; Swoboda et al., 2016). Even though, obtaining the optimal labelling of all the nodes is an NP-hard problem, it is possible to identify whether a node has attained its optimal label or not. If it is true, we say this node has attained its persistent label. With this guarantee, it is possible to reduce the problem size and repeat the inference algorithm for the remaining nodes. Through this process, it is possible to obtain better overall labelling. Among various approaches, there is an algorithm that is referred to as QPBO (Blake et al., 2011, section 2.3). It works with binary valued graphs and assigns labels that are one of $\{0, 0.5, 1\}$. Whenever, a node receives an integral label, it is guaranteed to be its optimal label. Wang et al. (2016) have chosen QPBO based on its simplicity and efficiency.

The third important issue to be addressed, is an efficient way to handle the various linear equality and inequality constraints. This step is the most important from a tighter relaxation point of view. Some of the constraints are limited in total number and the violated ones are found by enumeration. This is applicable to nonnegativity (31), marginalization (3b) and triangle inequality (39) constraints. Cycle inequalities are added based on the separation algorithm given in Deza and Laurent (2009). If no violated cycle inequalities are present, then the algorithm searches for violated odd-wheel inequalities. The separation algorithm for the latter is also described in Deza and Laurent (2009). Each added constraint corresponds to a new dual variable and it is important to control the problem size. Thus, it is important to remove inactive constraints. The heuristic used by Wang et al. (2016) is to remove constraints which are satisfied by the current primal variables and whose dual variables are zero. We would like to point out that Huang et al. (2014) have shown how the marginalization constraints are implicitly enforced by the semidefinite constraint along with the normalization and nonnegativity constraints. Even though Wang et al. (2016) work with a perturbed problem, this condition will hold with more guarantee as the algorithm reaches the optimum. Thus, marginalization constraints need not be considered at later stages of the algorithm.

5.1 Rounding schemes for SDP relaxation

SDP relaxations have a rich array of rounding schemes to obtain integral labels for the nodes of the graph (Charikar et al., 2009; Chazelle et al., 2004; Goemans and Williamson, 1995; Raghavendra, 2011). Some SDP relaxation algorithms are designed such that we obtain pseudomarginals for the nodes (Huang et al., 2014; Peng et al., 2012; Wang et al., 2016). Then, we could assign a label to the node by sampling according to the probability distribution defined by the pseudomarginals. In fact, this is one of the possible rounding schemes for LP relaxations working with pseudomarginals (Ravikumar et al., 2010). This *naive randomized* scheme works better when the distribution of the graphical model is more concentrated.

Several rounding schemes are based on low-rank representation of the moment matrix, i.e., $\boldsymbol{M} = \boldsymbol{Y}\boldsymbol{Y}^\top$, where $\boldsymbol{Y}$ is of rank r. So far, we have come across two kinds of node labels: binary ($\{1, -1\}$) or multilabel. For binary labels only one row of Y is needed per node. For the multilabel case, one row of $\boldsymbol{Y}$ is needed for each node and each label. While discussing the work of Erdogdu et al. (2017), we used $\boldsymbol{\sigma}_{i;x_i}$ to denote the row of Y corresponding to node i and label x_i. We could design rounding schemes based on projecting $\boldsymbol{\sigma}_{i;x_i}$ onto a suitable vector. For some of the rounding schemes, we also need the vector $\boldsymbol{\sigma}_\varnothing$ that corresponds to the first row of $\boldsymbol{Y}$.

First, we will discuss some rounding schemes for binary valued nodes. The seminal work Goemans and Williamson (1995) proposed to draw a random vector $\boldsymbol{w} \in \mathcal{R}^{1\times r}$ such that $\|\boldsymbol{w}\| = 1$ and assigns to node i the label $\text{sign}(\boldsymbol{\sigma}_i{\cdot}\boldsymbol{w})$. This is called the *random hyperplane rounding* scheme. *Shifted random projection rounding* extends this idea further and has proved to be a necessary step within approximation algorithms for MAX-2SAT (Charikar et al., 2009). Peng et al. (2012) have shown results based on this scheme. A perturbation vector $\boldsymbol{p} \in \mathcal{R}^{1\times r}$ is constructed with i.i.d. normally distributed elements. Then node i is assigned the label $\text{sign}(\boldsymbol{\sigma}_i \cdot ((1-\beta)\boldsymbol{\sigma}_\varnothing + \beta\boldsymbol{p})$. We should refer Charikar et al. (2009) about setting β to a value between 0 and 1. Note that $\boldsymbol{\sigma}_\varnothing$ itself is of unit length. So, there are rounding schemes that work with $\boldsymbol{\sigma}_\varnothing$, instead of generating a random vector $\boldsymbol{w}$ of unit length.

Erdogdu et al. (2017) also work with $\boldsymbol{\sigma}_\varnothing$ and it fixes the label of a node as $\text{sign}(\boldsymbol{\sigma}_i \cdot \boldsymbol{\sigma}_\varnothing)$, if $|\boldsymbol{\sigma}_i \cdot \boldsymbol{\sigma}_\varnothing|$ exceeds a confidence threshold. We saw that Erdogdu et al. (2017) optimize (35) several times, each time with a smaller set of $\boldsymbol{\sigma}_S$ vectors. Suppose there is an outer iteration, where no node exceeds the confidence threshold, then the nodes with the highest $|\boldsymbol{\sigma}_i \cdot \boldsymbol{\sigma}_\varnothing|$ value are picked for pruning. Erdogdu et al. (2017) presented their work for nodes taking binary labels but these rounding schemes could be extended to multilabel nodes. For example, we could assign the node i the label x_i that maximizes $|\boldsymbol{\sigma}_{i;x_i} \cdot \boldsymbol{\sigma}_\varnothing|$ or that maximizes $|\boldsymbol{\sigma}_{i;x_i} \cdot ((1-\beta)\boldsymbol{\sigma}_\varnothing + \beta \boldsymbol{p})|$. Also, there are other advanced rounding schemes, like those based on Grothendieck's inequality (Braverman et al., 2013) which we will not discuss here.

5.2 Problems where SDP relaxation has helped

We will discuss problem instances where SDP relaxation-based algorithms lead to better MAP labelling. Usually, these algorithms achieve a tighter relaxation by moving up the Lasserre/SOS hierarchy or by adding constraints that tighten the LP relaxation like cycle or cluster-based constraints.

Wang et al. (2015, 2016) have shown that SDP relaxation-based methods perform better for densely connected graphs with weak unary potentials and for graphs that have a large fraction of nonsubmodular (also, called nonattractive) pairwise potentials. Wang et al. (2016) were able to perform better than DAOOPT Otten and Dechter (2012) in a problem with a fully connected graph and no unaries from the Probabilistic Inference Challenge, 2011 (PIC, 2011). DAOOPT was a top performer in PIC-2011. Huang et al. (2014) have also observed that their method performs well on dense problems from PIC-2011. They were able to do better that ficolofo (Cooper et al., 2010), a top performer with dense problems in PIC-2011.

Peng et al. (2012) have also observed that problems where edge potentials are strong are solved better by their approach. Also, when the edge potentials are not so strong, they observed that the cluster-based tightening of LP relaxation by Sontag et al. (2008) performed better. It is readily possible to define a feasible set that is the intersection of the SDP relaxation with the cluster-based tighter LP relaxation that we saw in Section 2. To achieve this one could use the optimization scheme given in Sontag et al. (2008) instead of convex belief propagation to solve the subproblem within the dual optimization scheme of Peng et al. (2012).

Scalable and efficient SDP algorithms for MAP inference are still evolving. The four main algorithms that we have discussed (Erdogdu et al., 2017; Huang et al., 2014; Peng et al., 2012; Wang et al., 2016) are good starting points to further explore this topic. At this moment, we would like to point out one particular area of research: the intersection between spectral graph theory and semidefinite programming. For the MAP labelling problems discussed so far, we could explore algorithms based on spectral methods (Cour and Shi, 2007; Olsson et al., 2007). Even though spectral methods are computationally more efficient, the optimum of the SDP relaxation can be characterized better. There have been attempts to combine the advantages of the two methods and it is an active area of research in combinatorial optimization (ICERM, 2014; Kolla, 2009; Simons Institute, 2017).

6 Characterizing tight relaxations

In Section 4, we came across particular graph topologies and clique potentials, which by themselves or in combination lead to exact MAP inference. It will be interesting to understand more general settings. Especially, settings where the relaxation tightening algorithms that we have discussed, will lead to exact MAP inference.

So far, we have been discussing about various ways to tighten LP and SDP relaxations. The Sherali–Adams hierarchy and the Lasserre hierarchy provide a guaranteed approach to tighten the LP relaxation and the SDP relaxation, respectively. There have been efforts to characterize graph topologies and/or clique potentials that could be solved exactly with schemes that are at lower levels of these hierarchies. However, these hierarchies are computationally heavy even at the lower levels and they are enforced only partially. Surprisingly, there are many problem instances which are solved exactly with such partial schemes.

In the Sherali–Adams hierarchy, a level r corresponds to enforcing the marginalization constraint (5b) over all clusters of variables of size $\leq r$ and we will denote the resulting polytope as $\mathcal{SA}_r(\mathcal{G})$. The local polytope, $\mathcal{L}(\mathcal{G})$ is related to $\mathcal{SA}_2(\mathcal{G})$. Even though, $\mathcal{L}(\mathcal{G})$ enforces marginalization constraints only on edges of the graph and not on all possible pairs of nodes, it can be shown to be exactly equivalent to $\mathcal{SA}_2(\mathcal{G})$ (Rowland et al., 2017, section 3.3). Rowland et al. (2017) have shown a similar result for $\mathcal{SA}_3(\mathcal{G})$: the polytope obtained by considering the marginalization constraints of $\mathcal{L}(\mathcal{G})$ together with the marginalization constraints of triplets obtained by triangulating the graph, is equivalent to $\mathcal{SA}_3(\mathcal{G})$.

Further, it is desirable to characterize the tightness of $\mathcal{SA}_r(\mathcal{G})$ for various values of r. Initially, there were efforts to study conditions for tightness based on the graph topology. Wainwright and Jordan (2004b) proved that a sufficient condition for $\mathcal{SA}_r(\mathcal{G})$ to be tight is to have a graph with treewidth $\leq r - 1$. We can immediately see that $\mathcal{L}(\mathcal{G})$ is tight for trees. Since, trees have treewidth 1. Chandrasekaran et al. (2008) proved that in the more difficult problem of inferring the marginal probabilities, if there are no restrictions on the clique potentials, then bounded treewidth is necessary for tractability. Similar studies focusing on clique potentials have also been pursued. Thapper and Živný (2016) showed that for a given family of potentials (finite multilabel nodes and any clique size), either $\mathcal{L}(\mathcal{G})$ is tight, i.e., all such problems are tractable irrespective of the graph topology or the problem set is NP-hard. Tractability for graphical models containing only submodular clique potentials is a result that fits in this line of work.

In recent years, settings combining restrictions on graph topology and clique potentials have been studied (Rowland et al., 2017; Weller et al., 2016). Rowland et al. (2017) and related works have considered binary pairwise models in their analysis. In this setting, every graph is associated with a *signed* graph of the same topology, where an edge is considered even if it is attractive and it is odd if it is not attractive. A *balanced* graph is one that could be partitioned into two exhaustive sets, such that all edges with both endpoints within any one set are even and edges with one endpoint in each set is odd. In other words, any cycle in a balanced graph, will have an even number of odd edges. Thus, a balanced graph does not have frustrated cycles. Many results that are satisfied by models with attractive potentials may be proven for balanced models, which form a wider class. An *almost balanced* graph contains no frustrated cycles except through one privileged node, s. Thus, removing that node makes the remaining graph balanced. Note, it is possible to efficiently test whether a graph is almost balanced. Weller et al. (2016) and Rowland et al. (2017) have shown that for balanced graphs, $\mathcal{SA}_3(\mathcal{G})$ as defined in the above more compact manner is sufficient to achieve a tight relaxation. This is useful for applications like foreground/background subtraction, which involve binary labels and the foreground and background form the two sets of the balanced graph. For an almost balanced graph, we could define a polytope $\mathcal{SA}_3^s(\mathcal{G})$: the polytope obtained by considering the marginalization constraints of $\mathcal{L}(\mathcal{G})$ together with the marginalization constraints over only triplets which contain s, the privileged node. Rowland et al. (2017) have shown that $\mathcal{SA}_3^s(\mathcal{G})$ is tight for an almost balanced graph with a privileged node s.

Similar results have been obtained for the SOS hierarchy for SDPs. While discussing the work of Erdogdu et al. (2017), we discussed about the more tractable problem given in (35). Here, $\hat{\mathcal{S}}$ is made of a set of regions such that together they cover all the nodes of the graph. If each chordless cycle of the graph is contained in at least one region in $\hat{\mathcal{S}}$, we say that the set of regions is *circular*. To present their result, Erdogdu et al. (2017) make the assumptions that the graph has zero valued unary potentials and its topology is not contractible to K_5. Given these assumptions they show that the problem in (35) defined over a circular set of covering regions is tight. They also show how to convert a graph with nonzero unary potentials to a graph with zero unary potentials but the resulting graph could be contractible to K_5. Note that the empirical results in Erdogdu et al. (2017) concern graphs which do not satisfy the assumptions, the set of covering regions do not contain all chordless cycles and they work with graphs having nonzero unary potentials for which the conversion to zero unary potentials could result in a graph that could be contractible to K_5. Their empirical results have exceeded the theoretical expectation and need further investigation.

We have given here only a taste for an active area of research. For example, Wainwright and Jordan (2004b) gave only a sufficient condition based on treewidth regarding the tightness of the Sherali–Adams hierarchy. Rowland et al. (2017) have presented analytical results which indicate that there are conditions beyond treewidth. Also, it is not necessary to study relaxation hierarchies based on exactness of the relaxation. Even if we manage to achieve low approximation error at a given level of the relaxation hierarchy, rounding schemes can be used to further improve the output (Chlamtac and Tulsiani, 2012).

7 Conclusion

Continuous relaxations for MAP inference in MRFs are an important topic of study with wide applications. Obtaining better MAP labelling in an efficient manner is the ultimate objective of this line work. The possibility of better labelling will increase the scope of applications for MRFs with various graph topologies and clique potentials. In this survey we especially better understood approaches based on LP and SDP relaxations. We discussed various algorithmic steps that could lead to tighter relaxation in both cases. In the process, we understood the theoretical underpinnings better. This is a topic where there is scope for further research both from an algorithmic point of view and from a theoretical analysis point of view. A very important perspective is provided by applications, which explore novel clique potentials and graph topologies. The effect of tighter relaxation-based inference within learning of structured prediction models is also an important avenue of research. Meshi et al. (2019) present an interesting analysis about the empirical observation of tight LP relaxations within structured prediction applications. Learning a structured prediction model is computationally heavy, which involves inference in each iteration of the process (called loss augmented inference). Efficient algorithms for tighter relaxation could further the study of this topic. At this juncture, we would like to highlight the scope for further research in SDP relaxation-based algorithms. The literature is much more evolving in that topic compared to algorithms based on LP relaxation.

References

Batra, D., Gallagher, A.C., Parikh, D., Chen, T., 2010. Beyond trees: MRF inference via outer-planar decomposition. IEEE Conference on Computer Vision and Pattern Recognition (CVPR).

Batra, D., Nowozin, S., Kohli, P., 2011. Tighter relaxations for MAP-MRF inference: a local primal-dual gap based separation algorithm. In: International Conference on Artificial Intelligence and Statistics (AISTATS).

Batra, D., Yadollahpour, P., Guzman-Rivera, A., Shakhnarovich, G., 2012. Diverse m-best solutions in Markov random fields. In: European Conference on Computer Vision (ECCV).

Bergtholdt, M., Kappes, J., Schmidt, S., Schnörr, C., 2010. A study of parts-based object class detection using complete graphs. Int. J. Comput. Vis. 87 (1–2), 93.

Blake, A., Kohli, P., Rother, C., 2011. Markov Random Fields for Vision and Image Processing. MIT Press.

Boumal, N., Voroninski, V., Bandeira, A., 2016. The non-convex Burer-Monteiro approach works on smooth semidefinite programs. In: Advances in Neural Information Processing Systems (NIPS).

Boumal, N., Voroninski, V., Bandeira, A.S., 2018. Deterministic guarantees for Burer-Monteiro factorizations of smooth semidefinite programs. arXiv preprint arXiv:1804.02008.

Boyd, S., Vandenberghe, L., 2004. Convex Optimization. Cambridge University Press.

Boyd, S., Parikh, N., Chu, E., Peleato, B., Eckstein, J., 2011. Distributed optimization and statistical learning via the alternating direction method of multipliers. Found. Trends Mach. Learn. 3 (1), 1–122.

Braverman, M., Makarychev, K., Makarychev, Y., Naor, A., 2013. The Grothendieck constant is strictly smaller than Krivine's bound. Forum Math. Pi 1, e4.

Burer, S., Monteiro, R.D.C., 2003. A nonlinear programming algorithm for solving semidefinite programs via low-rank factorization. Math. Program. 95 (2), 329–357.

Burer, S., Monteiro, R.D.C., 2005. Local minima and convergence in low-rank semidefinite programming. Math. Program. 103 (3), 427–444.

Chandrasekaran, V., Srebro, N., Harsha, P., 2008. Complexity of inference in graphical models. In: Twenty-Fourth Conference on Uncertainty in Artificial Intelligence (UAI).

Charikar, M., Makarychev, K., Makarychev, Y., 2009. Near-optimal algorithms for maximum constraint satisfaction problems. ACM Trans. Algorithms (TALG) 5 (3), 32.

Chazelle, B., Kingsford, C., Singh, M., 2004. A semidefinite programming approach to side chain positioning with new rounding strategies. INFORMS J. Comput. 16 (4), 380–392.

Chlamtac, E., Tulsiani, M., 2012. Convex relaxations and integrality gaps. Handbook on Semidefinite, Conic and Polynomial Optimization, Springer, pp. 139–169.

Cooper, M.C., De Givry, S., Sánchez, M., Schiex, T., Zytnicki, M., Werner, T., 2010. Soft arc consistency revisited. Artif. Intell. 174 (7–8), 449–478.

Cour, T., Shi, J., 2007. Solving Markov random fields with spectral relaxation. In: International Conference on Artificial Intelligence and Statistics (AISTATS).

Deza, M.M., Laurent, M., 2009. Geometry of Cuts and Metrics. vol. 15. Springer.

Duchi, J.C., Tarlow, D., Elidan, G., Koller, D., 2007. Using combinatorial optimization within max-product belief propagation. In: Advances in Neural Information Processing Systems (NIPS).

Erdogdu, M.A., Deshpande, Y., Montanari, A., 2017. Inference in graphical models via semidefinite programming hierarchies. In: Advances in Neural Information Processing Systems (NIPS).

Felzenszwalb, P.F., McAuley, J.J., 2011. Fast inference with min-sum matrix product. IEEE Trans. Pattern Anal. Mach. Intell. (PAMI) 33 (12), 2549–2554.

Frostig, R., Wang, S., Liang, P.S., Manning, C.D., 2014. Simple MAP inference via low-rank relaxations. In: Advances in Neural Information Processing Systems (NIPS).

Globerson, A., Jaakkola, T.S., 2008. Fixing max-product: convergent message passing algorithms for MAP LP-relaxations. In: Advances in Neural Information Processing Systems (NIPS).

Goemans, M.X., Williamson, D.P., 1995. Improved approximation algorithms for maximum cut and satisfiability problems using semidefinite programming. J. ACM (JACM) 42 (6), 1115–1145.

Golub, G.H., Van Loan, C.F., 2013. Matrix computations, 4th ed. JHU Press.

Hazan, T., Shashua, A., 2008. Convergent message-passing algorithms for inference over general graphs with convex free energies. In: Twenty-Fourth Conference on Uncertainty in Artificial Intelligence (UAI).

Huang, Q., Chen, Y., Guibas, L., 2014. Scalable semidefinite relaxation for maximum a posterior estimation. In: International Conference on Machine Learning (ICML).

ICERM, 2014. Semidefinite programming and graph algorithms. Accessed: 08.07.2018.

Johnson, J. K., 2008. Convex Relaxation Methods for Graphical Models: Lagrangian and Maximum Entropy Approaches (Ph.D. thesis), Massachusetts Institute of Technology, Department of Electrical Engineering and Computer Science.

Kappes, J.H., Schmidt, S., Schnörr, C., 2010. MRF inference by k-fan decomposition and tight Lagrangian relaxation. In: European Conference on Computer Vision (ECCV).

Kappes, J.H., Andres, B., Hamprecht, F.A., Schnörr, C., Nowozin, S., Batra, D., Kim, S., Kausler, B.X., Kröger, T., Lellmann, J., Komodakis, N., Savchynskyy, B., Rother, C., 2015. A comparative study of modern inference techniques for structured discrete energy minimization problems. Int. J. Comput. Vis. 115 (2), 155–184. ISSN 0920-5691. https://doi.org/10.1007/s11263-015-0809-x.

Kohli, P., Shekhovtsov, A., Rother, C., Kolmogorov, V., Torr, P., 2008. On partial optimality in multi-label MRFs. In: International Conference on Machine Learning (ICML).

Kolla, A., 2009. Merging Techniques for Combinatorial Optimization: Spectral Graph Theory and Semidefinite Programming (Ph.D. thesis), UC Berkeley.

Koller, D., Friedman, N., 2009. Probabilistic Graphical Models: Principles and Techniques. MIT Press.

Kolmogorov, V., 2006. Convergent tree-reweighted message passing for energy minimization. IEEE Trans. Pattern Anal. Mach. Intell. 28 (10), 1568–1583.

Komodakis, N., Paragios, N., 2008. Beyond loose LP-relaxations: optimizing MRFs by repairing cycles. European Conference on Computer Vision, pp. 806–820.

Komodakis, N., Paragios, N., 2009. Beyond pairwise energies: efficient optimization for higher-order MRFs. IEEE Conference on Computer Vision and Pattern Recognition (CVPR).

Komodakis, N., Paragios, N., Tziritas, G., 2011. MRF energy minimization and beyond via dual decomposition. IEEE Trans. Pattern Anal. Mach. Intell. 33 (3), 531–552.

Komodakis, N., Kumar, M.P., Paragios, N., 2016. (hyper)-graphs inference through convex relaxations and move making algorithms: contributions and applications in artificial vision. Found. Trends Comput. Graph. Vis. 10 (1), 1–102.

Kovtun, I., 2003. Partial optimal labeling search for a NP-hard subclass of (max,+) problems. In: Joint Pattern Recognition Symposium402–409.

Kumar, M.P., Kolmogorov, V., Torr, P.H.S., 2009. An analysis of convex relaxations for MAP estimation of discrete MRFs. J. Mach. Learn. Res. 10, 71–106.

Lasserre, J.B., 2001. Global optimization with polynomials and the problem of moments. SIAM J. Optim. 11 (3), 796–817.

Laurent, M., 2003. A comparison of the Sherali-Adams, Lovász-Schrijver, and Lasserre relaxations for 0-1 programming. Math. Oper. Res. 28 (3), 470–496.

Li, Y., Zemel, R., 2014. High order regularization for semi-supervised learning of structured output problems. In: International Conference on Machine Learning (ICML).

Li, M., Shekhovtsov, A., Huber, D., 2016. Complexity of discrete energy minimization problems. In: European Conference on Computer Vision (ECCV).

Malick, J., 2007. The spherical constraint in Boolean quadratic programs. J. Glob. Optim. 39 (4), 609–622.

McAuley, J.J., Caetano, T.S., 2011. Faster algorithms for max-product message-passing. J. Mach. Learn. Res. 12, 1349–1388.

Mei, S., Misiakiewicz, T., Montanari, A., Oliveira, R.I., 2017. Solving SDPs for synchronization and MaxCut problems via the Grothendieck inequality. Conference on Learning Theory, pp. 1476–1515.

Meshi, O., Jaakkola, T., Globerson, A., 2014. Smoothed coordinate descent for MAP inference. In: Advanced Structured Prediction. MITP Press, pp. 103–131.

Meshi, O., London, B., Weller, A., Sontag, D., 2019. Train and test tightness of LP relaxations in structured prediction. J. Mach. Learn. Res. 20 (13), 1–34.

Mezuman, E., Weiss, Y., 2012. Globally optimizing graph partitioning problems using message passing. In: International Conference on Artificial Intelligence and Statistics (AISTATS).

Mezuman, E., Tarlow, D., Globerson, A., Weiss, Y., 2013. Tighter linear program relaxations for high order graphical models. In: Uncertainty in Artificial Intelligence (UAI).

Noorshams, N., Wainwright, M.J., 2013. Stochastic belief propagation: a low-complexity alternative to the sum-product algorithm. IEEE Trans. Inf. Theory 59 (4), 1981–2000.

Olsson, C., Eriksson, A.P., Kahl, F., 2007. Solving large scale binary quadratic problems: spectral methods vs. semidefinite programming. In: IEEE Conference on Computer Vision and Pattern Recognition (CVPR).

Otten, L., Dechter, R., 2012. Anytime AND/OR depth-first search for combinatorial optimization. AI Commun. 25 (3), 211–227.

Parrilo, P.A., 2003. Semidefinite programming relaxations for semialgebraic problems. Math. Program. 96 (2), 293–320.

Pataki, G., 1998. On the rank of extreme matrices in semidefinite programs and the multiplicity of optimal eigenvalues. Math. Oper. Res. 23 (2), 339–358.

Peng, J., Hazan, T., Srebro, N., Xu, J., 2012. Approximate inference by intersecting semidefinite bound and local polytope. In: International Conference on Artificial Intelligence and Statistics (AISTATS).

PIC, 2011. Probabilistic inference challenge. Accessed: 08.07.2018.

Potetz, B., Lee, T.S., 2008. Efficient belief propagation for higher-order cliques using linear constraint nodes. Comput. Vis. Image Underst. 112 (1), 39–54.

Raghavendra, P., 2011. Generic techniques to round SDP relaxations. In: International Symposium on Mathematical Foundations of Computer Science, p. 34 34.

Ravikumar, P., Lafferty, J., 2006. Quadratic programming relaxations for metric labeling and Markov random field MAP estimation. In: International Conference on Machine learning (ICML).

Ravikumar, P., Agarwal, A., Wainwright, M.J., 2010. Message-passing for graph-structured linear programs: proximal methods and rounding schemes. J. Mach. Learn. Res. 11, 1043–1080.

Rother, C., Kohli, P., Feng, W., Jia, J., 2009. Minimizing sparse higher order energy functions of discrete variables. In: IEEE Conference on Computer Vision and Pattern Recognition (CVPR).

Rowland, M., Pacchiano, A., Weller, A., 2017. Conditions beyond treewidth for tightness of higher-order LP relaxations. In: International Conference on Artificial Intelligence and Statistics (AISTATS).

Savchynskyy, B., Schmidt, S., 2014. Getting feasible variable estimates from infeasible ones: MRF local polytope study. In: Nowozin Peter, S., Gehler Jeremy Jancsary, V., Lampert, C.H. (Eds.), Advanced Structured Prediction. MIT Press, pp. 133–158.

Schraudolph, N.N., Kamenetsky, D., 2009. Efficient exact inference in planar Ising models. In: Advances in Neural Information Processing Systems (NIPS).

Schrijver, A., 2003. Combinatorial Optimization: Polyhedra and Efficiency. vol. 24. Springer Science & Business Media.

Shimony, S.E., 1994. Finding MAPs for belief networks is NP-hard. Artif. Intell. 68 (2), 399–410.

Sontag, D., Jaakkola, T.S., 2007. New outer bounds on the marginal polytope. In: Advances in Neural Information Processing Systems (NIPS).

Sontag, D., Meltzer, T., Globerson, A., Jaakkola, T., Weiss, Y., 2008. Tightening LP relaxations for MAP using message passing. In: Uncertainty in Artificial Intelligence (UAI).

Sontag, D., Globerson, A., Jaakkola, T.S., 2009. Clusters and coarse partitions in LP relaxations. In: Advances in Neural Information Processing Systems (NIPS).

Sontag, D., Globerson, A., Jaakkola, T., 2011. Introduction to dual decomposition for inference. Optim. Mach. Learn. 1, 219–254.

Sontag, D., Choe, D.K., Li, Y., 2012. Efficiently searching for frustrated cycles in MAP inference. In: Uncertainty in Artificial Intelligence (UAI).

Sun, M., Telaprolu, M., Lee, H., Savarese, S., 2012. Efficient and exact MAP-MRF inference using branch and bound. In: International Conference on Artificial Intelligence and Statistics (AISTATS).

Swoboda, P., Shekhovtsov, A., Kappes, J.H., Schnorr, C., Savchynskyy, B., 2016. Partial optimality by pruning for MAP-inference with general graphical models. IEEE Trans. Pattern Anal. Mach. Intell. (PAMI) 38 (7), 1370–1382.

Szeliski, R., Zabih, R., Scharstein, D., Veksler, O., Kolmogorov, V., Agarwala, A., Tappen, M., Rother, C., 2008. A comparative study of energy minimization methods for Markov random fields with smoothness-based priors. IEEE Trans. Pattern Anal. Mach. Intell. (PAMI) 30 (6), 1068–1080.

Tarlow, D., Givoni, I., Zemel, R., 2010. HOP-MAP: efficient message passing with high order potentials. In: International Conference on Artificial Intelligence and Statistics (AISTATS).

Simons Institute, 2017. Spectral graph theory and optimization symposium. Accessed: 08.07.2018 https://simons.berkeley.edu/workshops/spectral-graph-theory-and-optimization/.

Thapper, J., Živný, S., 2016. The complexity of finite-valued CSPs. J. ACM (JACM) 63 (4), 37.

Wainwright, M.J., Jordan, M.I., 2004. Semidefinite relaxations for approximate inference on graphs with cycles. In: Advances in Neural Information Processing Systems (NIPS).

Wainwright, M.J., Jordan, M.I., 2004. Treewidth-based conditions for exactness of the Sherali-Adams and Lasserre relaxations. University of California, Berkeley. Technical Report 671.

Wainwright, M.J., Jordan, M.I., 2008. Graphical models, exponential families, and variational inference. Found. Trends Mach. Learn. 1 (1–2), 1–305.

Wainwright, M.J., Jaakkola, T.S., Willsky, A.S., 2005. Map estimation via agreement on trees: message-passing and linear programming. IEEE Trans. Inf. Theory 51 (11), 3697–3717.

Wang, H., Koller, D., 2013. A fast and exact energy minimization algorithm for cycle MRFs. In: International Conference on Machine Learning (ICML).

Wang, P., Shen, C., van den Hengel, A., 2013. A fast semidefinite approach to solving binary quadratic problems. In: IEEE Conference on Computer Vision and Pattern Recognition (CVPR).

Wang, P., Shen, C., van den Hengel, A., 2015. Efficient SDP inference for fully-connected CRFs based on low-rank decomposition. In: IEEE Conference on Computer Vision and Pattern Recognition (CVPR).

Wang, P., Shen, C., van den Hengel, A., Torr, P.H.S., 2016. Efficient semidefinite branch-and-cut for MAP-MRF inference. Int. J. Comput. Vis. 117 (3), 269–289.

Weller, A., Rowland, M., Sontag, D., 2016. Tightness of LP relaxations for almost balanced models. In: International Conference on Artificial Intelligence and Statistics (AISTATS).

Wen, Z., Goldfarb, D., Yin, W., 2010. Alternating direction augmented Lagrangian methods for semidefinite programming. Math. Program. Comput. 2 (3–4), 203–230.

Werner, T., 2007. A linear programming approach to max-sum problem: a review. IEEE Trans. Pattern Anal. Mach. Intell. 29 (7), 1165–1179.

Werner, T., 2008. High-arity interactions, polyhedral relaxations, and cutting plane algorithm for soft constraint optimisation (MAP-MRF). In: IEEE Conference on Computer Vision and Pattern Recognition (CVPR).

Werner, T., 2010. Revisiting the linear programming relaxation approach to Gibbs energy minimization and weighted constraint satisfaction. IEEE Trans. Pattern Anal. Mach. Intell. (PAMI) 32 (8), 1474–1488.

Yanover, C., Meltzer, T., Weiss, Y., 2006. Linear programming relaxations and belief propagation-an empirical study. J. Mach. Learn. Res. 7, 1887–1907.

Yarkony, J., Morshed, R., Ihler, A.T., Fowlkes, C.C., 2011. Tightening MRF relaxations with planar subproblems. In: Uncertainty in Artificial Intelligence (UAI).

Yedidia, J.S., Freeman, W.T., Weiss, Y., 2005. Constructing free-energy approximations and generalized belief propagation algorithms. IEEE Trans. Inf. Theory 51 (7), 2282–2312.

Zheng, Y., Chen, P., Cao, J.-Z., 2012. MAP-MRF inference based on extended junction tree representation. In: IEEE Conference on Computer Vision and Pattern Recognition (CVPR).

Chapter 10

Lagrangian methods for composite optimization

Shoham Sabach[a] and Marc Teboulle[b,*]
[a]*Faculty of Industrial Engineering and Managment, The Technion, Haifa, Israel*
[b]*School of Mathematical Sciences, Tel-Aviv University, Tel-Aviv-Yafo, Israel*
[*]*Corresponding author: e-mail: teboulle@post.tau.ac.il*

Chapter Outline

Abstract

Lagrangian-based methods have been on the market for over 50 years. These methods are robust and often can handle optimization problems with complex geometries through efficient computational steps. The last decade of research have generated a large volume of literature on various practical and theoretical aspects of many Lagrangian-based algorithms. This chapter reviews the basic elements of Lagrangian-based methods for composite minimization in the convex and nonconvex setting. In the convex case, the focus is on global rate of convergence results, which are derived here through a novel approach

 https://doi.org/10.1016/bs.hna.2019.04.002
© 2019 Elsevier B.V. All rights reserved.

and very simple proof technique. In the much harder nonconvex case, we survey a very recent methodology which allows to establish global pointwise convergence results for a broad class of genuine nonlinear composite semialgebraic problems.

Keywords: Lagrangian multiplier methods, Proximal multiplier algorithms, Convex and nonconvex composite minimization, Alternating minimization, Decomposition schemes, Global rate of convergence analysis, Kurdyka–Łosiajewicz property, Semialgebraic optimization, Global pointwise convergence

AMS Classification Codes: 90C25, 65K05, 90C06, 90C26

1 Introduction

Recently, we are witnessing a resurgence of Lagrangian-based methods and their related decomposition schemes. Augmented Lagrangian methods, also known as multiplier methods, are probably the first Lagrangian-based methods developed for optimization. These were introduced about half a century ago by Hestenes (1969) and Powell (1969) for solving nonlinear optimization problems with nonlinear equality constraints. Some earlier works can be traced back to the late fifties with the volume of Arrow et al. (1958) which have launched the fundamental idea of decomposition schemes (also known as splitting methods), which are intimately related to Lagrangian schemes. This renewed interest in Lagrangian methods has emerged from new applications arising in many fundamental scientific and engineering problems in image science, machine learning and other fields, which are often model led by structured and very large scale minimization problems, often nonsmooth and even nonconvex. These methods have gained popularity due to their ability to adapt to problem structures and to exploit data information, to produce scalable and tractable updating steps. A voluminous recent literature reflects this intense renewal interest on multiplier methods and their relatives decomposition algorithms. It is not our intention, neither is the scope of this review chapter, to discuss all past and more recent literature. Instead, throughout the chapter we provide some pointers from earlier foundational works, to some recent advances which have evolved over the last 50 years.

A classical reference providing the central developments and ideas underlying multiplier methods since their inception is the monograph of Bertsekas (1982), while decomposition methods can be found in the books of Lasdon (2002), and of Bertsekas and Tsitsiklis (1989) for a more modern treatment. Decomposition methods can also be developed via the abstract setting of maximal monotone operators (Lions and Mercier, 1979; Passty, 1979; Rockafellar, 1976a, b). We refer the reader to the recent monograph of Bauschke and Combettes (2017), which covers fundamental results in this direction, and also includes an extensive bibliography.

In this survey, we present the essential tools needed to build and analyze Lagrangian-based methods for a composite optimization model which is general enough to capture most instances of fundamental modern applications both in the convex and nonconvex settings. In the convex case, the focus is on global rate of convergence results, derived here through a novel approach leading to a simplified proof technique permitting unification. In the much harder nonconvex case, we review a very recent methodology for global pointwise convergence analysis to a broad class of genuine nonlinear composite semialgebraic problems. This is developed in three main sections, whose contents are now briefly outlined in some more detail.

In Section 2, we start with the main model of the chapter, which is a nonlinear composite minimization, and describe in informal ways the nature and basic steps in constructing Lagrangian-based methods through the penalty and proximal regularization concepts. This has the advantage of explaining in simple terms, and with minimal sophistication, the basic elements underlying Lagrangian-based schemes in the absence of convexity for a very general composite optimization model, and to reveal the difficulties associated with Lagrangian-based methods (which will be further elaborated more precisely later on in Section 4). We also introduce some basic results in the convex setting and the starring role of the proximal maps in Lagrangian-based schemes. Finally, we end the section with a sample of application examples, which illustrate the flexibility of the composite model, many other applications can be found in the cited literature.

In Section 3, we focus on the convex setting. When convexity is present in a given optimization model, it is well known that an augmented Lagrangian method is a *dual* method. More precisely, it is a manifestation of the proximal minimization method applied to the dual formulation of the given optimization problem. Starting with the pioneering work of Moreau (1965), followed by the proximal minimization algorithm of Martinet (1970), this decisive dual approach interpretation of multiplier methods was developed by Rockafellar, see, e.g., Rockafellar (1976a, b). Building on the more general proximal method of multipliers initially developed by Rockafellar (1976a), we present a new and simplified approach to analyze Lagrangian-based schemes for the convex composite model. This novel approach relies on a simple observation, which allows to show that all well-known Lagrangian methods and their decomposition variants can be captured and analyzed through a *single* scheme called the *perturbed* proximal method of multipliers. This allows us to derive a single unifying global rate of convergence result through an elementary proof, see Theorem 1, which as we show, can then be easily applied to obtain rate of convergence results for various fundamental decomposition schemes.

In Section 4, we move on to the much harder and challenging nonconvex setting. In sharp contrast to the convex case, where Lagrangian-based methods and their convergence properties have been extensively studied in past and current literature, in the nonconvex setting, global convergence analysis for general nonlinear composite models has remained scarce, and far from being well-understood. Only very recently some progress has been initiated in the nonconvex case, but mainly for the *linear* composite model, see, e.g., Li and Pong (2015) and references therein. The results we reviewed in this section are in fact very recent and were developed in the just published work of Bolte et al. (2018), which to the best of our knowledge, appears to be the first work capable of handling general nonconvex and nonlinear composite minimization, and producing the basis of a theoretical framework to derive global convergence analysis of Lagrangian-based methods for the broad class of semialgebraic problems.

The definitions and notations used in this chapter, in relation to convex and variational analysis, are classical and can be found in the book of Rockafellar (1970) and the monograph of Rockafellar and Wets (1998).

2 The Lagrangian framework

Consider the following general composite optimization model

$$\text{(M)} \qquad \min_{u \in \mathbb{R}^d} \{\Phi(u) := \varphi(u) + h(F(u))\}, \tag{1}$$

where $\varphi : \mathbb{R}^d \to (-\infty, +\infty]$ and $h : \mathbb{R}^m \to (-\infty, +\infty]$ are proper and lower semicontinuous functions, while $F : \mathbb{R}^d \to \mathbb{R}^m$ is a C^1 mapping.

The choice of the model (M) is for the sake of convenience to clearly delineate the role played by additional structures and properties in every part of the triplet $[\varphi, h, F]$ in the design and analysis of various Lagrangian-based algorithms that will be described and analyzed in the following parts of this chapter. In Section 2.3, we illustrate the flexibility of model (M) which can be used to describe a wide spectrum of disparate applications. In the following two sections we first describe in informal ways the underlying elements and basic mechanism of every Lagrangian-based method for tackling the general model (M), which is then followed by the fundamental proximal framework.

2.1 Lagrangian-based methods: Basic elements and mechanism

We start with an elementary and useful equivalent reformulation of the optimization model (M), which reads as

$$\text{(M)} \qquad \min_{u \in \mathbb{R}^d, v \in \mathbb{R}^m} \{\varphi(u) + h(v) : F(u) = v\}.$$

Attaching a multiplier $y \in \mathbb{R}^m$ to the equality constraint, we define the *Lagrangian* $\mathcal{L} : \mathbb{R}^d \times \mathbb{R}^m \times \mathbb{R}^m \to (-\infty, +\infty]$, which associated to problem (M), via

$$\mathcal{L}(u, v, y) \equiv \varphi(u) + h(v) + \langle y, F(u) - v \rangle, \tag{2}$$

and for any $\rho > 0$ the corresponding *augmented Lagrangian* $\mathcal{L}_\rho : \mathbb{R}^d \times \mathbb{R}^m \times \mathbb{R}^m \to (-\infty, +\infty]$ is defined by

$$\begin{aligned} \mathcal{L}_\rho(u, v, y) &:= \mathcal{L}(u, v, y) + \frac{\rho}{2}\|F(u) - v\|^2 \\ &= \varphi(u) + h(v) + \langle y, F(u) - v \rangle + \frac{\rho}{2}\|F(u) - v\|^2, \end{aligned} \tag{3}$$

namely, the *penalized* Lagrangian $\mathcal{L}(\cdot) \equiv \mathcal{L}_0(\cdot)$.

Equipped with $\mathcal{L}_\rho(\cdot)$, a basic Lagrangian scheme starts with any given triplet (u,v,y), and consists of minimizing the augmented Lagrangian $\mathcal{L}_\rho(\cdot)$ with respect to (u,v) followed by an update of the corresponding multiplier y, and reads as follows

---

Augmented Lagrangian scheme (ALS)

1. **Initialization:** Start with any $(u,v,y)\in\mathbb{R}^d\times\mathbb{R}^m\times\mathbb{R}^m$.
2. **Main step:** Update the new point (u^+,v^+,y^+) via:

$$(u^+,v^+)\ \in\operatorname{argmin}\{\mathcal{L}_\rho(u,v,y):(u,v)\in\mathbb{R}^d\times\mathbb{R}^m\}, \tag{4}$$

$$y^+=y+\rho(F(u^+)-v^+). \tag{5}$$

The rational underlying the iterative steps of ALS is quite simple and natural. Indeed, suppose we have adequate calculus rules at hand (technical details will be given later), then formally we can write the optimality conditions for ALS, which are characterized (u^+,v^+) through

$$(\mathbf{0},\mathbf{0})\in\partial_{(u,v)}\mathcal{L}_\rho(u^+,v^+,y),$$

where ∂ stands for some (sub)differential operation. The above is equivalent to the following system of two inclusions

$$\begin{aligned}&\mathbf{0}\in\partial\varphi(u^+)+\nabla F(u^+)^T(y+\rho(F(u^+)-v^+)),\\&\mathbf{0}\in\partial h(v^+)-y-\rho(F(u^+)-v^+),\end{aligned}$$

where $\nabla F\in\mathbb{R}^{m\times d}$ is the Jacobian matrix associated with F. Now, thanks to (5), the system reduces to

$$\mathbf{0}\in\partial\varphi(u^+)+\nabla F(u^+)^Ty^+\quad\text{and}\quad\mathbf{0}\in\partial h(v^+)-y^+,$$

that is, we have obtained that $\mathbf{0}\in\partial\varphi(u^+)+\nabla F(u^+)^T\partial h(v^+)$. Hence, when $F(u^+)=v^+$, the latter is exactly the optimality condition for model (M), at certain triplet $(\bar{u},\bar{v},\bar{y})$, which reads (through the usual Lagrangian $\mathcal{L}(\cdot)$) as:

$$\begin{matrix}\mathbf{0}\in\partial\varphi(\bar{u})+\nabla F(\bar{u})^T\bar{y}\\ \mathbf{0}\in\partial h(\bar{v})-\bar{y}\\ F(\bar{u})=\bar{v}\end{matrix}\quad\Longleftrightarrow\quad\mathbf{0}\in\partial\varphi(\bar{u})+\nabla F(\bar{u})^T\partial h(F(\bar{u})).$$

This simple description and resulting mechanism has the advantage of being essentially independent of any specific assumptions on the problem's data (except for satisfying some adequate calculus rules), e.g., convexity of the involved parts. We will now look at two interpretations of the ALS that will throw light on the form of the scheme. In fact, if we set $y\equiv\mathbf{0}$, the resulting ALS is nothing else but a usual *penalty scheme* applied to the problem (M)

$$\min_{u\in\mathbb{R}^d,v\in\mathbb{R}^m}\{\varphi(u)+h(v):F(u)=v\},$$

where $\rho > 0$ stands for the penalty parameter associated to the equality constraint and the term $(1/2)\|F(u)-v\|^2$ plays the role of the penalty function, which controls the "violation" of the constraint. Before we go into the second interpretation of ALS, we note that problem (M) is obviously equivalent to the following min–max reformulation

$$\text{(MM)} \qquad \min_{u\in\mathbb{R}^d, v\in\mathbb{R}^m} \max_{y\in\mathbb{R}^m} \{\mathcal{L}_0(u,v,y) \equiv \varphi(u) + h(v) + \langle y, F(u)-v\rangle\}.$$

A natural question, which emerges is why not simply use the Lagrangian $\mathcal{L}_0(\cdot)$ instead of the "complicated" augmented Lagrangian $\mathcal{L}_\rho(\cdot)$? In fact, the (MM) formulation reveals the problematic issue whereby the Lagrangian $\mathcal{L}_0(\cdot)$ precisely generates the "hard penalty" reformulation of problem (M), i.e., we have

$$\text{(MM)} \qquad \min_{u\in\mathbb{R}^d, v\in\mathbb{R}^m} \{\varphi(u) + h(v) + H(u,v)\},$$

where

$$H(u,v) = \sup_{y\in\mathbb{R}^m} \langle y, F(u)-v\rangle = \begin{cases} 0, & \text{if } F(u)=v, \\ \infty, & \text{otherwise.} \end{cases}$$

In the absence of a duality theory that would allow to reverse the order of the min–max operation (i.e., under some convexity/regularity assumptions), we are thus left with an intractable nonsmooth problem. This naturally leads to the second interpretation of the augmented Lagrangian scheme that can be viewed as a *smoothing* or *proximal regularization* of the Lagrangian $\mathcal{L}(\cdot)$, i.e., of $H(u,v)$, defined above. Indeed, let

$$H_\rho(u,v) = \sup_{\eta\in\mathbb{R}^m} \left\{ \langle \eta, F(u)-v\rangle - \frac{1}{2\rho}\|\eta - y\|^2 \right\}.$$

Then, an easy computation shows that an optimal solution, η^*, is given by

$$\eta^* = y + \rho(F(u)-v).$$

Plugging this optimal solution η^* in $H_\rho(u,v)$, we then get

$$\begin{aligned} H_\rho(u,v) &= \langle y, F(u)-v\rangle + \rho\|F(u)-v\|^2 - \frac{\rho}{2}\|F(u)-v\|^2 \\ &= \langle y, F(u)-v\rangle + \frac{\rho}{2}\|F(u)-v\|^2. \end{aligned}$$

Hence, using this observation in problem (MM), naturally leads to the following equivalent problem

$$\min_{u\in\mathbb{R}^d, v\in\mathbb{R}^m} \left\{ \varphi(u) + h(v) + \langle y, F(u)-v\rangle + \frac{\rho}{2}\|F(u)-v\|^2 \equiv \mathcal{L}_\rho(u,v,y) \right\},$$

which is exactly the minimization step of the augmented Lagrangian in ALS (see step (4)).

Much like the penalized interpretation, this second interpretation for building an augmented Lagrangian method has the advantage of being of great generality, and does not need any duality and/or convexity arguments. Furthermore, it naturally explains the regularization (smoothing) effect that occurs through the so-called proximal approach that will be described in the next subsection, and proven to be at the heart of any Lagrangian-based method. Recapitulating at this point, we end this subsection by outlining the main difficulty with Lagrangian-based methods described by ALS.

2.1.1 Main difficulties with ALS

Quite clearly, the central issue to address in ALS, is how to solve the minimization step (4) efficiently? Indeed, in all Lagrangian-based methods, the updated multiplier y^+ is given by the simple and explicit formula (5):

$$y^+ = y + \rho(F(u^+) - v^+).$$

Therefore, all the efforts are towards addressing the difficult minimization step (4). In fact, all the past, current and novel schemes in relation to Lagrangian-based methods simply rely on choosing a minimization algorithm that can address both scalability and efficiency by judiciously exploiting the specific structures and data information present in the model (M) and captured by the triplet $[\varphi, h, F]$.

In the following sections, we will demonstrate how many different variants of ALS can be built and analyzed through different choices of the minimization procedures. Moreover, one should also notice that the choice of the penalty/regularization parameter $\rho > 0$ also plays a fundamental role. This will be illustrated throughout the methods we elaborate below, both in the convex and nonconvex settings, under various structural scenarios for the triplet $[\varphi, h, F]$. In all these developments, as already alluded above, a key player is the so-called proximal mapping associated to a given function, which is introduced next.

2.2 Proximal mappings and minimization

From here, we will start with a precise description of the involved objects allowing us also to introduce a few definitions and relevant terminology from variational and convex analysis. We record some central properties of proximal mappings and their corresponding envelopes for convex functions. All the material described here is well known and relies on the seminal work of Moreau (1965). However, we present here some of the fundamental properties of proximal mappings and envelopes of convex functions within the slightly more general setting of a weighted proximal term, and for completeness we have included their simple proofs. Classical convex analysis monographs describe in far more details further properties of proximal mappings and their envelopes, see, e.g., Bauschke and Combettes (2017); Hiriart-Urruty and Lemaréchal (2001); Rockafellar and Wets (1998), and the recent books of Bertsekas (2015) and Beck (2017), which also provide more recent results, applications and references. Note that proximal mappings and related schemes can also be extended beyond the Euclidean setting, whereby the classical quadratic norm is replaced by a non-Euclidean distance-like function, see the recent survey of Teboulle (2018), which includes ample relevant references.

For convenience and flexibility towards the analysis of Lagrangian-based schemes developed in the next section, we adopt the slightly more general version of proximal mappings as follows. Let $P \in \mathbb{S}^d_{++}$, where $\mathbb{S}^d_{++}$ is the set of all $d \times d$ positive definite matrices, and $\psi : \mathbb{R}^d \to (-\infty, +\infty]$ be a proper, lower semicontinuous and convex function. Then,

$$\operatorname{prox}^{\psi}_P(z) := \operatorname{argmin}_{x \in \mathbb{R}^d} \left\{ \psi(x) + \frac{1}{2}\|x - z\|^2_P \right\}, \tag{6}$$

where, for any $x \in \mathbb{R}^d$, we define $\|x\|_P = \langle Px, x \rangle^{1/2}$. Clearly, when $P := (1/\lambda)I$ with $\lambda > 0$, we recover the usual (unweighed) definition of a proximal mapping.

The corresponding *proximal envelope* of ψ, in that case, is defined by the function

$$\psi_P(z) := \inf_{x \in \mathbb{R}^d} \left\{ \psi(x) + \frac{1}{2}\|x - z\|^2_P \right\}.$$

Below we collect some fundamental properties of $\operatorname{prox}^{\psi}_P(\cdot)$ and $\psi_P(\cdot)$.

Lemma 1 (Basic properties of proximal maps/envelopes).

Let $\psi : \mathbb{R}^d \to (-\infty, +\infty]$ be a proper, lower semicontinuous and convex function and let $P \in \mathbb{S}^d_{++}$. The following statements hold true:

(i) *The proximal mapping $\operatorname{prox}^{\psi}_P(\cdot)$ is the unique solution of the problem* (6), *characterized via*

$$P(z - \operatorname{prox}^{\psi}_P(z)) \in \partial\psi(\operatorname{prox}^{\psi}_P(z)). \tag{7}$$

(ii) *The proximal envelope $\psi_P(\cdot)$ is finite everywhere, convex and differentiable. Moreover, its gradient is given by*

$$\nabla\psi_P(z) = P(z - \operatorname{prox}^{\psi}_P(z)), \tag{8}$$

and is Lipschitz continuous with Lipschitz constant $\lambda_{\max}(P)$, *the maximal eigenvalue of the matrix P, i.e., we have*

$$\|\nabla\psi_P(z_1) - \nabla\psi_P(z_2)\| \le \lambda_{\max}(P)\|z_1 - z_2\|, \quad \forall\, z_1, z_2 \in \mathbb{R}^d.$$

(iii) *A vector $\bar{z}$ is a minimizer of ψ if and only if $\bar{z}$ is a minimizer of ψ_P and we have*

$$\inf_{x \in \mathbb{R}^d} \psi_P(x) = \inf_{x \in \mathbb{R}^d} \psi(x) = \psi(\bar{z}) = \psi(\operatorname{prox}^{\psi}_P(\bar{z})) = \psi_P(\bar{z}).$$

Proof. (i) Since $x \to \psi(x) + (1/2)\|x - z\|_P^2$ is strongly convex, it immediately follows that the minimum in (6) is uniquely attained at $p := \text{prox}_P^{\psi}(z)$ such that $\mathbf{0} \in \partial\psi(p) + P(p - z)$.

(ii) For any $\bar{x} \in \text{dom}\,\psi$ we clearly have

$$\psi_P(z) \le \psi(\bar{x}) + \frac{1}{2}\|\bar{x} - z\|_P^2 < \infty,$$

and since $\Psi(x,z) := \psi(x) + (1/2)\|x - z\|_P^2$ is jointly convex in (x,z), it follows that $\psi_P(z) = \inf_{x\in\mathbb{R}^d} \Psi(x,z)$ is convex on $\mathbb{R}^d$.

Now, to establish that $\psi_P(\cdot)$ is differentiable, first note that

$$\psi_P(z) = \psi(\text{prox}_P^{\psi}(z)) + \frac{1}{2}\|\text{prox}_P^{\psi}(z) - z\|_P^2,$$

and, for all $v \in \mathbb{R}^d$, we have

$$\psi_P(z + v) \le \psi(\text{prox}_P^{\psi}(z)) + \frac{1}{2}\|\text{prox}_P^{\psi}(z) - (z + v)\|_P^2.$$

Combining these two inequalities and some algebra yields that for any $v \in \mathbb{R}^d$

$$\gamma(v) := \psi_P(z + v) - \psi_P(z) - \langle v, P(z - \text{prox}_P^{\psi}(z))\rangle \le \frac{1}{2}\|v\|_P^2.$$

Now, since $\psi_P(\cdot)$ is convex, then so is the function $v \to \gamma(v)$. Hence, since $\gamma(\mathbf{0}) = 0$ we get that $\gamma(v) + \gamma(-v) \ge 0$. Therefore, the last inequality also implies that $\gamma(v) \ge -(1/2)\|v\|_P^2$, and hence it follows that

$$\lim_{\|v\|_P \to 0} \frac{|\gamma(v)|}{\|v\|_P} = 0,$$

showing that $P(z - \text{prox}_P^{\psi}(z)) = \nabla\psi_P(z)$. In order to prove the Lipschitz continuity of $\nabla\psi_P(\cdot)$, using (8), we have

$$\nabla\psi_P(z_1) - \nabla\psi_P(z_2) = P(z_1 - z_2) - P(\text{prox}_P^{\psi}(z_1) - \text{prox}_P^{\psi}(z_2)).$$

Since $P \in \mathbb{S}_{++}^d$ clearly P^{-1} exists and using the above we get

$$\begin{aligned}\langle P^{-1}(\nabla\psi_P(z_1) - \nabla\psi_P(z_2)), \nabla\psi_P(z_1) - \nabla\psi_P(z_2)\rangle &= \langle z_1 - z_2, \nabla\psi_P(z_1) - \nabla\psi_P(z_2)\rangle \\ &\quad - \Big\langle \text{prox}_P^{\psi}(z_1) - \text{prox}_P^{\psi}(z_2), \nabla\psi_P(z_1) \\ &\qquad - \nabla\psi_P(z_2)\rangle \\ &\le \langle z_1 - z_2, \nabla\psi_P(z_1) - \nabla\psi_P(z_2)\rangle,\end{aligned}$$

where the inequality uses the monotonicity of $\partial\psi$ (since ψ is convex), (7) and (8) proven above. From the latter inequality we immediately obtain that

$$\lambda_{\min}(P^{-1})\|\nabla\psi_P(z_1) - \nabla\psi_P(z_2)\| \le \|z_1 - z_2\|,$$

and this proves the desired result since $\lambda_{\min}(P^{-1}) = \lambda_{\max}(P)^{-1}$.

(iii) This easily follows from the very basic definition of $\psi_P(\cdot)$ and the relations established above.

□

The relations established in Lemma 1 (iii) are the basis for the so-called proximal minimization algorithm for tackling the problem $\inf_{x\in\mathbb{R}^d}\psi(x)$, which in turn is looking for a vector $z\in\mathbb{R}^d$ such that $\nabla\psi_P(z)=\mathbf{0}$ (for some $P\in\mathbb{S}^d_{++}$), i.e., thanks to (8) to a solution of the fixed point equation $z=\mathrm{prox}_P^{\psi}(z)$. Thus, given any z, the proximal minimization algorithm computes the next update via

$$z^+=\mathrm{prox}_P^{\psi}(z)=\mathrm{argmin}_{x\in\mathbb{R}^d}\left\{\psi(x)+\frac{1}{2}\|x-z\|_P^2\right\}.$$

The proximal step plays a central role and leads to the fundamental proximal inequality that will be systematically used in the sequel. Before stating this result, we also recall the following fundamental *Pythagoras three points identity*. For any $P\in\mathbb{S}^d_{++}$, we have

$$2\langle w-v,P(v-u)\rangle=\|w-u\|_P^2-\|w-v\|_P^2-\|u-v\|_P^2,\quad\forall\ u,v,w\in\mathbb{R}^d. \tag{9}$$

The next result gives the announced fundamental proximal inequality.
Lemma 2 (Proximal Inequality).
Let $\psi:\mathbb{R}^d\to(-\infty,+\infty]$ be a proper, lower semicontinuous and convex function and let $P\in\mathbb{S}^d_{++}$. Then, for $z^+=\mathrm{prox}_P^{\psi}(z)$ and for any $x\in\mathbb{R}^d$, we have

$$\psi(z^+)-\psi(x)\le\langle x-z^+,P(z^+-z)\rangle=\frac{1}{2}\left(\|x-z\|_P^2-\|x-z^+\|_P^2-\|z^+-z\|_P^2\right). \tag{10}$$

Proof. Writing the optimality condition, which characterize $z^+=\mathrm{prox}_P^{\psi}(z)$, yields

$$\mathbf{0}\in\partial\psi(z^+)+P(z^+-z).$$

Using the subgradient inequality for the convex function ψ, we have for all $x\in\mathbb{R}^d$

$$\psi(z^+)-\psi(x)\le\langle x-z^+,-\gamma\rangle,\quad\gamma\in\partial\psi(z^+),$$

and hence, from the first inclusion, it follows that

$$\psi(z^+)-\psi(x)\le\langle x-z^+,P(z^+-z)\rangle,$$

which proves the first inequality in (10). Invoking the identity (9) immediately yields the equality. □

In relation to Lagrangian-based methods, we will also need the following useful variant of the proximal inequality.
Lemma 3 (Composite proximal inequality).
Let $\psi:\mathbb{R}^d\to(-\infty,+\infty]$ be a proper, lower-semi continuous and convex function and let $S:\mathbb{R}^d\to\mathbb{R}$ be a differentiable function. Given a matrix $P\in\mathbb{S}^d_{++}$, we define:

$$z^+ := \text{argmin}_{x\in\mathbb{R}^d}\left\{\psi(x)+S(x)+\frac{1}{2}\|x-z\|_P^2\right\}. \tag{11}$$

Then, for any $x\in\mathbb{R}^d$, *we have*

$$\psi(z^+)-\psi(x)+\langle\nabla S(z^+),z^+-x\rangle\le\frac{1}{2}\left(\|x-z\|_P^2-\|x-z^+\|_P^2-\|z^+-z\|_P^2\right).$$

Proof. The proof follows through identical arguments of the proof given in Lemma 2. □

To illustrate the viability of the general optimization model (M), we end this section with some application examples.

2.3 Application examples

We briefly present few examples of challenging applications that can be cast as particular instances of the general model (M). We discuss below both convex and nonconvex models.

Example 1. *Convex Nonlinear Programming* (Bertsekas, 1999; Rockafellar, 1970) is a constrained minimization of a convex function φ under a set of equality and inequality constraints, which is explicitly given by

$$\min_{u\in\mathbb{R}^d}\left\{\varphi(u): f_j(u)\le b_j,\ (j=1,2,\ldots,p), f_j(u)=b_j,\ (j=p+1,p+2,\ldots,m)\right\},$$

where $f_j:\mathbb{R}^d\to(-\infty,+\infty]$, $j=1,2,\ldots,p$, are proper, lower semicontinuous and convex functions, while $f_j:\mathbb{R}^d\to(-\infty,+\infty]$, $j-p+1, p+2, \ldots, m$, are affine functions. Therefore, this fits into the general model (M), where

$$F(u):=(f_1(u).f_2(u),\ldots,f_m(u))^T,$$

and $h(v):=\sum_{j=1}^m h_j(v_j)$ with $h_j=\iota_{(-\infty,b_j]}(v_j)$ for $j=1,2,\ldots,p$ and $h_j=\iota_{\{b_j\}}(v_j)$ for $j=p+1,p+2,\ldots,m$ (the indicator function of a set C is defined as $\iota_C(x)=0$ for $x\in C$ and $\iota_C(x)=+\infty$ for $x\notin C$). In this case, the function h is convex and nonsmooth.

Example 2. *Regularized Image Deblurring* (Chambolle, 2004; Chan et al., 2013; Engl et al., 1996; Rudin et al., 1992; Wu and Tai, 2010) consists of finding an "image", which solves the following optimization problem

$$\min_{U\in\mathbb{R}^{p\times q}}\left\{\frac{1}{2}\|\mathcal{A}U-B\|^2+\lambda R(U)\right\},$$

where $\mathcal{A}$ is a linear mapping (usually called deblurring transformation) and $B\in\mathbb{R}^{p\times q}$. Here, we assume that both $\mathcal{A}$ and B are known in advance. The role of the regularizer $R(\cdot)$ and the regularizing parameter $\lambda>0$, is to control the obtained solution. For example, if a sparse solution is desired, then the convex regularizer $R(U)=\|U\|_1$ is very useful. Another example, which is very classic, is when one is interested in controlling the total variation of the obtained image. In this case, appropriate choice for a regularizer would be the ℓ_1-based anisotropic total variation, i.e., $R(U)=TV_1(U)$, where TV_1 stands for

$$TV_1(U)=\sum_{i=1}^{p-1}\sum_{j=1}^{q-1}\left(\left|u_{i,j}-u_{i+1,j}\right|+\left|u_{i,j}-u_{i,j+1}\right|\right)+\sum_{i=1}^{p-1}\left|u_{i,q}-u_{i+1,q}\right|+\sum_{j=1}^{q-1}\left|u_{p,j}-u_{p,j+1}\right|.$$

It is well known that $TV_1(U)=\|L^T(U)\|_1$ where $L:\mathbb{R}^{(p-1)\times q}\times\mathbb{R}^{p\times(q-1)}\to\mathbb{R}^{p\times q}$ is defined by the formula

$$L(x,y)_{i,j}=x_{i,j}+y_{i,j}-x_{i-1,j}-y_{i,j-1},\quad i=1,2,\ldots,p \text{ and } j=1,2,\ldots,q,$$

where we assume that $x_{0,j}=x_{p,j}=y_{i,0}=y_{i,q}=0$ for every $i=1,2,\ldots,p$ and $j=1,2,\ldots,q$. This means that in this case $h(\cdot)=\|\cdot\|_1$ and $F=L^T$ where the operator $L^T:\mathbb{R}^{p\times q}\to\mathbb{R}^{(p-1)\times q}\times\mathbb{R}^{p\times(q-1)}$ is the adjoint to L, which is given by

$$L^T(U)=(x,y),$$

where $x\in\mathbb{R}^{(p-1)\times q}$ and $y\in\mathbb{R}^{p\times(q-1)}$ are the matrices defined by

$$\begin{aligned} x&=u_{i,j}-u_{i+1,j},\quad i=1,2,\ldots,p-1 \text{ and } j=1,2,\ldots,q,\\ y&=u_{i,j}-u_{i,j+1},\quad i=1,2,\ldots,p \text{ and } j=1,2,\ldots,q-1. \end{aligned}$$

Example 3. *Fused Lasso* (Hilbe, 2009; Tibshirani, 1996; Tibshirani et al., 2005) aims at solving a linear system of equations, by minimizing the corresponding least squares error measure, with the additional requirements that the solution is both sparse and "smooth". More precisely, let $\{(a_i,b_i)\}_{i=1}^N$ be a set of samples, where $a_i\in\mathbb{R}^d$ and $b_i\in\mathbb{R}$, the problem can be then formulated as follows:

$$\min_{u\in\mathbb{R}^d}\left\{\frac{1}{2}\|Au-b\|^2+\lambda_1\|u\|_1+\lambda_2\|Du\|_1\right\},$$

where $A=\left[a_1^T,a_2^T,\ldots,a_N^T\right]^T\in\mathbb{R}^{N\times d}$, $b\in\mathbb{R}^N$, $D\in\mathbb{R}^{d\times(d-1)}$ defined by $(Du)_i=u_{i+1}-u_i$ and $\lambda_1,\lambda_2>0$ are regularizing parameters. This problem also fits well in our general model (M) by taking $h(v)=\|v\|_1$ and $F=[I,D]^T$.

Another closely related problem when the least squares error measure is replaced by the logistic regression measure. In this case, we obtain the following optimization problem

$$\min_{u\in\mathbb{R}^d}\left\{\frac{1}{N}\sum_{i=1}^N\log\left(1+\exp\left(b_iu^Ta_i\right)\right)+\lambda_1\|u\|_1+\lambda_2\|Du\|_1\right\},$$

where in this case $b_i\in\{-1,1\}$, $i=1,2,\ldots,N$.

In the above examples, the function h was convex. We turn to some example with a nonconvex h.

Example 4. *Orthogonality Constrained Minimization* (Absil et al., 2008; Edelman et al., 1999; Manton, 2002), also known as minimization on the Stiefel manifold, can be generally formulated as follows:

$$\min_{U \in \mathbb{R}^{p\times q}} \{\varphi(U) : U^T U = I\},$$

where the nonconvex constraint surface is called *Stiefel manifold.* In this case by choosing the model function h to be the indicator of the nonconvex set described by the Stiefel manifold, we immediately see that this challenging class of problems fits into the general model (M).

Example 5. *Blind Image Deconvolution* (Campisi and Egiazarian, 2007; Kundur and Hatzinakos, 1996; Levin et al., 2009) is one of the most challenging problems in imaging sciences. Similarly to the problem described above in Example 2, we are given a blurred image $B \in \mathbb{R}^{p\times q}$, and the goal is to recover a sharp image U and an unknown blurring transformation $\mathcal{A}$, which generated the blurred image B through the following linear deconvolution process $B = \mathcal{A}^* U + E$, where $*$ denotes the convolution operation (of the corresponding dimensions), $\mathcal{A} \in \mathbb{R}^{r\times s}$ and $U \in \mathbb{R}^{p\times q}$, while $E \in \mathbb{R}^{p\times q}$ is an additive noise. Therefore, by adopting again the least squares error measure we are focusing on the following optimization problem:

$$\min_{\mathcal{A}\in\mathbb{R}^{r\times s},\, U\in\mathbb{R}^{p\times q}} \left\{\frac{1}{2}\|\mathcal{A} * U - B\|^2 + h(U, \mathcal{A})\right\},$$

where $h : \mathbb{R}^{p\times q} \times \mathbb{R}^{r\times s} \to (-\infty, +\infty]$ is a regularizer (nonconvex and possibly nonsmooth) that controls both the recovered image U and the blurring transformation $\mathcal{A}$.

Example 6. *Regularized Structured Total Least Squares* (Beck et al., 2008; Mastronardi et al., 2004; Pruessner and O'Leary, 2003) are another interesting application in imaging sciences that seeks to recover a sharp image out of an observed blurred image. Here, the blurring transformation is also unknown and should be recovered, but assumed to have some linear structure. Given matrices A, A_1, A_2, ..., A_p in $\mathbb{R}^{n\times d}$ and a vector of measurements $b \in \mathbb{R}^n$, the resulting model consists of the following nonconvex minimization problem:

$$\min_{u\in\mathbb{R}^d,\, w\in\mathbb{R}^p} \left\{\frac{1}{2}\left\|\left(A + \sum_{i=1}^{p} w_i A_i\right)u - b\right\|^2 + \frac{1}{2}\|w\|^2 + h(u, w)\right\},$$

where $w_1, w_2, \ldots, w_p \in \mathbb{R}$ are unknown structure components, and $h : \mathbb{R}^d \times \mathbb{R}^p \to (-\infty, +\infty]$ is a regularizer that controls on one hand the recovered image u and on the other hand the structure components stored in w.

3 The convex setting

We focus on the convex setting and consider the following particular instance of model (M), which is the *linear composite* convex optimization problem defined by

$$\text{(CM)} \qquad \min_{u \in \mathbb{R}^d} \{\varphi(u) + h(Fu)\},$$

where here $\varphi : \mathbb{R}^d \to (-\infty, +\infty]$ and $h : \mathbb{R}^m \to (-\infty, +\infty]$ are proper, lower semicontinuous and convex functions, while $F : \mathbb{R}^d \to \mathbb{R}^m$ is a linear mapping.

The main goal of this section is to present a new and simple approach to analyze Lagrangian-based schemes for tackling model (CM), and derive their convergence rates. This relies on a simple observation, which allows to show that all well known (as well as other variants, cf. Remark 2) Lagrangian-based schemes can be captured and analyzed through *one perturbed* proximal method of multipliers that we introduce below. We start by recalling some basic and standard assumptions with respect to the convex problem (CM).

3.1 Preliminaries on the convex model (CM)

We first reformulate problem (CM) in the equivalent form:

$$\text{(CM)} \qquad \min_{u \in \mathbb{R}^d, v \in \mathbb{R}^m} \{\varphi(u) + h(v) : Fu = v\},$$

for which according to Section 2, the corresponding Lagrangian and augmented Lagrangian, are respectively defined by

$$\mathcal{L}(u,v,y) = \varphi(u) + h(v) + \langle y, Fu - v \rangle,$$

and, for any $\rho > 0$, by

$$\mathcal{L}_\rho(u,v,y) = \mathcal{L}(u,v,y) + \frac{\rho}{2}\|Fu - v\|^2.$$

Throughout this section the following assumption will be needed.

Assumption 1. The Lagrangian $\mathcal{L}$ has a saddle point, that is, there exists (u^*, v^*, y^*) such that

$$\mathcal{L}(u^*,v^*,y) \le \mathcal{L}(u^*,v^*,y^*) \le \mathcal{L}(u,v,y^*), \quad \forall\, u \in \mathbb{R}^d, \quad \forall\, v,y \in \mathbb{R}^m.$$

Under Assumption 1 and the convex data assumption for the triplet $[\varphi, h, F]$ in problem (CM), it follows that $p^* := \mathcal{L}(u^*,v^*,y^*) \in \mathbb{R}$. Furthermore, a saddle point (u^*,v^*,y^*) exists if and only if the following two conditions hold:

(i) (u^*,v^*) is a solution of problem (CM), i.e., $\varphi(u^*)$ and $h(v^*)$ are finite valued and $Fu^* = v^*$.

(ii) The multiplier y^* is an optimal solution of the dual problem associated to problem (CM), which is given by

$$\text{(DM)} \quad d^* = \sup_{y \in \mathbb{R}^m} \left\{ d(y) := \varphi^*\left(-F^T y\right) + h^*(y) \right\},$$

where φ^* and h^* stand for the Fenchel conjugate functions of φ and h, respectively.

Thanks to the convexity of problem (CM), recall that the existence of an optimal dual Lagrange multiplier y^* attached to the constraint $Fu = v$ is ensured under the standard constraint qualification for problem (CM) that can be formulated as follows[a]

$$\text{there exists a } \bar{u} \in \text{ri}(\text{dom}\,\varphi) \text{ and a } \bar{v} \in \text{ri}(\text{dom h}) \text{ satisfying } F\bar{u} = \bar{v},$$

and that strong duality holds, i.e., $p^* = d^* \in \mathbb{R}$. Moreover, the set of optimal dual solutions is nonempty, convex and compact.

We end these preliminaries part on the convex model (CM) with the following elementary result, which will be useful in the forthcoming analysis.

Lemma 4 (Objective and Feasibility Approximation).
Let (u^*, v^*, y^*) be a saddle point for $\mathcal{L}$. Suppose, for every $\delta \geq 0$ and $\alpha > 0$, that

$$\varphi(u) + h(v) - (\varphi(u^*) + h(v^*)) + \alpha\|Fu - v\| \leq \delta, \quad \forall\, u \in \mathbb{R}^d, \quad \forall\, v \in \mathbb{R}^m. \tag{12}$$

Then, for every couple $(u, v) \in \mathbb{R}^d \times \mathbb{R}^m$ which satisfies (12), the following assertions hold:

(i) $\varphi(u) + h(v) - (\varphi(u^*) + h(v^*)) \leq \delta$.

(ii) $\|Fu - v\| \leq 2\alpha^{-1}\delta$ *whenever* $2\|y^*\| \leq \alpha$.

Proof. Since $\alpha\|Fu - v\| \geq 0$, the first assertion follows at once from the premise (12). Now, since (u^*, v^*, y^*) is a saddle point of $\mathcal{L}$, we have that $\mathcal{L}(u^*, v^*, y^*) \leq \mathcal{L}(u, v, y^*)$, i.e.,

$$\varphi(u^*) + h(v^*) \leq \varphi(u) + h(v) + \langle y^*, Fu - v \rangle, \quad \forall\, u \in \mathbb{R}^d, \quad \forall\, v \in \mathbb{R}^m,$$

which combined with (12) yields

$$\alpha\|Fu - v\| \leq \delta + \langle y^*, Fu - v \rangle \leq \delta + \|y^*\| \cdot \|Fu - v\|,$$

and hence, with $\alpha > 0$ such that $2\|y^*\| \leq \alpha$, the desired claim (ii) immediately follows. □

---

[a]For any convex function $F : \mathbb{R}^d \to (-\infty, +\infty]$ we denote by $\text{dom}\, F := \{u : F(u) < \infty\}$ its *domain*, and for any convex set S, ri S stands for the *relative interior* of S.

3.2 Proximal method of multipliers and fundamental Lagrangian-based schemes

The proximal method of multipliers (PMM) invented by Rockafellar (1976a), emerges from the augmented Lagrangian scheme (discussed in Section 2) by adding two (weighted) quadratic proximal terms in the minimization step with respect to the two primal variables u and v. The PMM takes the following form:

Proximal method of multipliers (PMM)

1. **Input:** $M_1 \in \mathbb{S}_+^d$, $M_2 \in \mathbb{S}_+^m$ and $\mu \in (0,2)$.
2. **Initialization:** Start with any $(u^0, v^0, y^0) \in \mathbb{R}^d \times \mathbb{R}^m \times \mathbb{R}^m$.
3. **Main step:** For $k = 0, 1, \ldots$ generate the sequence $\{(u^k, v^k, y^k)\}_{k\in\mathbb{N}}$ as follows

$$(u^{k+1}, v^{k+1}) = \text{argmin}_{u,v}\left\{\mathcal{L}_\rho(u, v, y^k) + \frac{1}{2}\| u - u^k \|_{M_1}^2 + \frac{1}{2}\| v - v^k \|_{M_2}^2\right\}, \tag{13}$$

$$y^{k+1} = y^k + \mu\rho(Fu^{k+1} - v^{k+1}). \tag{14}$$

The PMM of Rockafellar (1976a) (with $M_1 = M_2 = cI$ for some $c > 0$, and $\mu = 1$) emerges from a primal-dual application of the proximal point algorithm (as opposed to a dual application, which leads to the usual ALS described above). One advantage of the PMM is that as long as $M_1, M_2 \succ \mathbf{0}$, it manufactures strongly convex minimization problems. As a consequence, as shown in Rockafellar (1976a), it can be proven that the generated sequence converges to a saddle point of the associated Lagrangian to problem (CM), under mild assumptions on the problem's data. Note that here we introduce a positive parameter $\mu \in (0, 2)$ which allows to scale the step-size in the multiplier update for a larger step-size which could enhance the overall performance of the scheme. This idea was suggested in Fortin and Glowinski (1983) with μ being the golden ratio in the context of the so-called ADMM (which we discuss below), and as shown in Fortin and Glowinski (1983), under some additional assumptions preserves the convergence of this scheme.

Clearly, the basic ALS is recovered from the PMM when $M_1 = M_2 = \mathbf{0}$. In that case, we need to assume that the minimization step is well-defined, i.e., admits a nonempty solution set. This can be warranted, for instance, under usual coercivity, or more generally through additional conditions on the problem's data that can be obtained through standard asymptotic (recession) analysis tools, see, e.g., Auslender and Teboulle (2003). In what follows, we always assume that the minimization steps are well-defined.

Remark 1. It should be noted that actually, the augmented Lagrangian $\mathcal{L}_\rho$ already includes a quadratic proximal term with respect to the variable v, and therefore we could avoid giving an additional proximal term for this variable. However, sometimes the model (CM) can be described through a more general linear constraint of the form $Fu + Gv = b$, for which it can then be useful to add a proximal term for the v variable. For simplicity of exposition, we have adopted here the simpler formulation of the form $Fu = v$, i.e., $G = -I$ and $b = \mathbf{0}$, which in general is sufficient to describe most applications of interest. All our results can be adapted for the more general linear constraint. Thus, for the sake of completeness, and potential use within the more general linear constraint, here we will keep the proximal terms for both variables u and v.

Much like the classical ALS, a main drawback of the PMM is that for convex problems like model (CM), which exhibit nice separability in the variables u and v, the presence of the *coupling quadratic term* $Q(u,v) := \| Fu - v \|^2$ destroys it. Thus, on the practical side, this renders the joint minimization a very difficult or impossible computational task. Nevertheless, as we shall see below, the PMM remains of central importance from the theoretical perspective.

To overcome the just alluded difficulty in the PMM, and beneficially exploit the structure of model (CM), we now describe three classical approaches to eliminate the quadratic coupling term which lead to the following three fundamental Lagrangian-based schemes. Recall that since the multiplier update remains the same for any Lagrangian-based scheme, we only describe the minimization steps of these methods with respect to the primal variables u and v.

- *Linearized PMM (L-PMM).* This idea was developed by Chen and Teboulle (1994) in the context of the PMM. By linearizing the coupling quadratic term Q around the current iteration $\left(u^k, v^k\right)$, the joint minimization step of PMM splits into the favourable easier steps (presuming as usual access to the prox maps of φ and h) that can be performed in *parallel*:

$$u^{k+1} = \operatorname{argmin}_u \left\{ \varphi(u) + \rho \left\langle F^T \left(Fu^k - v^k\right), u - u^k \right\rangle + \frac{1}{2} \| u - u^k \|_{M_1}^2 \right\}, \tag{15}$$

$$v^{k+1} = \operatorname{argmin}_v \left\{ h(v) + \rho \left\langle v^k - Fu^k, v - v^k \right\rangle + \frac{1}{2} \| v - v^k \|_{M_2}^2 \right\}. \tag{16}$$

This class of algorithms is particularly useful for developing parallel methods for solving nonlinear programmes with block structures.

- *Proximal alternating direction method of multipliers (PADMM).* The Proximal ADMM (with $\mu = 1$) was introduced by Eckstein (1994) (with $M_1 = c_1 I$ and $M_2 = c_2 I$ for some c_1, $c_2 > 0$). It consists of applying the alternating minimization approach to the joint minimization of PMM in order to eliminate the quadratic coupling:

$$u^{k+1} = \operatorname{argmin}_u \left\{ \mathcal{L}_\rho\left(u, v^k, y^k\right) + \frac{1}{2} \| u - u^k \|_{M_1}^2 \right\}, \tag{17}$$

$$v^{k+1} = \operatorname{argmin}_v \left\{ \mathcal{L}_\rho\left(u^{k+1}, v, y^k\right) + \frac{1}{2} \| v - v^k \|_{M_2}^2 \right\}. \tag{18}$$

The use of positive definite matrices allows for more flexibility to consider other practical variants of PADMM (see examples below). With $M_1 = M_2 = 0$, we recover the so-called well known Alternating Direction of Multipliers (ADMM) introduced by Glowinski and Marroco (1975) and Gabay and Mercier (1976). This method has received much attention in the past and recent literature, see e.g., Eckstein and Bertsekas (1992), Bertsekas and Tsitsiklis (1989), Fortin and Glowinski (1983), Gabay (1983), Glowinski and Le Tallec (1983), Tseng (1991), Yang and Zhang (2011), Boyd et al. (2011), and as shown in Eckstein and Bertsekas (1992) ADMM is in fact a special case of the Douglas–Rachford splitting method Douglas and Rachford (1956). Actually, splitting/decomposition methods take their origin through the work of Douglas and Rachford (1956) and were developed for the more general problem of finding the zero of the sum of two monotone operators via the abstract setting of maximal monotone operators (Lions and Mercier, 1979; Passty, 1979; Rockafellar, 1976a, b). We refer to the monograph of Bauschke and Combettes (2017) for an extensive and comprehensive treatment within the general operator setting, and references therein.

- *Linearized PADMM (L-PDAMM).* Here, we combine both strategies above, namely linearization and alternating minimization. For example, with a fixed v^k, by linearizing the quadratic coupling term $u \to 2^{-1}\rho \| Fu - v^k \|^2$ around the current iteration u^k, we obtain a Linearized PADMM

$$u^{k+1} = \mathrm{argmin}_u \left\{ \varphi(u) + \rho\langle F^T(Fu^k - v^k), u - u^k\rangle + \frac{1}{2}\| u - u^k \|_{M_1}^2 \right\}, \tag{19}$$

$$v^{k+1} = \mathrm{argmin}_v \left\{ \mathcal{L}_\rho(u^{k+1}, v, y^k) + \frac{1}{2}\| v - v^k \|_{M_2}^2 \right\}. \tag{20}$$

Here, the v-step is simple as long as the function h is proximal “friendly”. However, in the presence of a more general constraint (see Remark 1), we could also use a linearizing technique in this step to derive a simple step.

He and Yuan (2012) was the first work proving ergodic global rate of convergence for PADMM and L-PADMM through a variational inequality framework. Shefi and Teboulle (2014) provide a unified analysis to establish ergodic global rate of convergence for L-PMM and PADMM and L-PADMM, and also show that the latter includes the algorithm of Chambolle and Pock (2011).

We will now show that all the above methods (and other variants) can be seen through the lens of one simple scheme, which in turn leads to derive a simplified proof technique and unified analysis to establish their convergence properties.

3.3 One scheme for all: A perturbed PMM and its global rate analysis

Main observation: The just described PMM, L-PMM, PADMM and L-PADMM schemes (and other possible variants) can simply be recovered and analyzed *at once*, through what we call a *Perturbed* PMM that we introduce next.

Before proceeding, for simplicity of exposition it will be convenient to use the following compact notations.

- Let $\xi := (u, v)$, and $L := (F, -I_m)$. Then, problem (CM) reads

$$\min\{\Psi(\xi) : L\xi = 0, \xi \in \mathbb{R}^d \times \mathbb{R}^m\} \equiv \min\{\Psi(\xi) : \xi \in \mathcal{F}\},$$

with objective $\Psi(\xi) := \varphi(u) + h(v)$, and feasible set $\mathcal{F} = \{(u,v) \in \mathbb{R}^d \times \mathbb{R}^m : Fu = v\} \equiv \{\xi \in \mathbb{R}^d \times \mathbb{R}^m : L\xi = \mathbf{0}\}$.
- $M :=$ Diag(M_1, M_2), is the diagonal block matrix with diagonal blocks M_1 and M_2.
- For any sequence $\{(u^k, v^k, y^k)\}_{k\in\mathbb{N}}$ and any integer $N \geq 1$, we define the corresponding ergodic sequence $\{(\mathbf{u}^N, \mathbf{v}^N, \mathbf{y}^N)\}_{N\in\mathbb{N}}$ by

$$\mathbf{u}^N = \frac{1}{N}\sum_{k=0}^{N-1} u^{k+1}, \quad \mathbf{v}^N = \frac{1}{N}\sum_{k=0}^{N-1} v^{k+1}, \quad \mathbf{y}^N = \frac{1}{N}\sum_{k=0}^{N-1} y^{k+1}, \text{ and } \boldsymbol{\xi}^N = (\mathbf{u}^N, \mathbf{v}^N). \tag{21}$$

- For any matrix $P \in \mathbb{S}^q_+$, any $k \geq 0$ and any three vectors $a, a^k, a^{k+1} \in \mathbb{R}^q$, we define

$$\Delta_k(a, P) := \frac{1}{2}\|a - a^k\|_P^2 - \frac{1}{2}\|a - a^{k+1}\|_P^2. \tag{22}$$

When $P \equiv I_q$, the identity matrix, we simply write $\Delta_k(a, I_q) \equiv \Delta_k(a)$.

With these notations at hand, we introduce the following main scheme, which stands at the heart of our ability to unify the analysis of all the four mentioned methods.

Perturbed PMM

1. **Input:** $M = \text{Diag}(M_1, M_2)$, with $M_1 \in \mathbb{S}^d_+$, $M_2 \in \mathbb{S}^m_+$ and $\mu \in (0, 2)$.
2. **Initialization:** Start with any $(\xi^0, y^0) \equiv (u^0, v^0, y^0) \in \mathbb{R}^d \times \mathbb{R}^m \times \mathbb{R}^m$.
3. **Main step:** For $k = 0, 1, \ldots$ generate the sequence $\{(u^k, v^k, y^k)\}_{k\in\mathbb{N}}$ as follows:

   choose $d^k \in \mathbb{R}^{d\times m}$ and compute

$$\xi^{k+1} = \text{argmin}_\xi \left\{ \mathcal{L}_\rho(\xi, y^k) + \langle d^k, \xi\rangle + \frac{1}{2}\|\xi - \xi^k\|_M^2 \right\}, \tag{23}$$

$$y^{k+1} = y^k + \mu\rho L\xi^{k+1}. \tag{24}$$

The perturbation vector d^k, $k \in \mathbb{N}$, plays a central role that will be made clearer later. In fact below we will show that besides the PMM, which is obviously obtained with $d^k \equiv \mathbf{0}$, $k \in \mathbb{N}$, all fundamental Lagrangian-based schemes described above can be simply recovered and analyzed at once, through a proper choice of the vector d^k, of the matrix M and of the parameter μ.

The forthcoming simple result on the Perturbed PMM is essentially the key to establish the main convergence properties of all the Lagrangian-based algorithms just discussed above.

Lemma 5 (Fundamental Estimate for the Perturbed PMM).
Let $\{(\xi^k, y^k)\}_{k\in\mathbb{N}}$ be a sequence generated by the Perturbed PMM. For any $\xi \in \mathcal{F}$ and $y \in \mathbb{R}^m$, we have

$$\mathcal{L}(\xi^{k+1}, y) - \mathcal{L}(\xi, y) \le \Delta_k(\xi, M) + \frac{1}{\mu\rho}\Delta_k(y) + R_k(\xi, M), \tag{25}$$

where $R_k(\xi, M) := -\frac{1}{2} \| \xi^{k+1} - \xi^k \|_M^2 - \left(\frac{2-\mu}{2\mu^2\rho}\right) \| y^{k+1} - y^k \|^2 - \langle d^k, \xi^{k+1} - \xi \rangle$.

Proof. Let $S(\xi) = \langle y^k, L\xi \rangle + \frac{\rho}{2} \| L\xi \|^2 + \langle d^k, \xi \rangle$. Then, within our compact notations, the primal iteration (23) reads:

$$\xi^{k+1} = \operatorname{argmin}_\xi \left\{ \Psi(\xi) + S(\xi) + \frac{1}{2} \| \xi - \xi^k \|_M^2 \right\}.$$

Invoking Lemma 3 with the above functions Ψ and S, we obtain for any $\xi \in \mathbb{R}^d \times \mathbb{R}^m$

$$\Psi(\xi^{k+1}) - \Psi(\xi) + \langle y^k + \rho L\xi^{k+1}, L\xi^{k+1} - L\xi \rangle + \langle d^k, \xi^{k+1} - \xi \rangle \le \Delta_k(\xi, M) - \frac{1}{2} \| \xi^{k+1} - \xi^k \|_M^2 .$$

Since $\mathcal{L}(\xi, y^k) = \Psi(\xi) + \langle y^k, L\xi \rangle$, the above can be written as,

$$\mathcal{L}(\xi^{k+1}, y^k) - \mathcal{L}(\xi, y^k) - \rho\langle L\xi^{k+1}, L\xi \rangle + \langle d^k, \xi^{k+1} - \xi \rangle$$
$$\le \Delta_k(\xi, M) - \frac{1}{2} \| \xi^{k+1} - \xi^k \|_M^2 - \rho \| L\xi^{k+1} \|^2.$$

Therefore, for any $\xi \in \mathcal{F}$ (i.e., with $L\xi = 0$), using the multiplier update (see (24)), the above reduces to

$$\mathcal{L}(\xi^{k+1}, y^k) - \mathcal{L}(\xi, y^k) + \langle d^k, \xi^{k+1} - \xi \rangle \le \Delta_k(\xi, M) - \frac{1}{2} \| \xi^{k+1} - \xi^k \|_M^2 - \frac{1}{\mu^2\rho} \| y^{k+1} - y^k \|^2.$$

On the other hand, using again (24), followed by the three points identity, we have

$$\mathcal{L}(\xi^{k+1}, y) - \mathcal{L}(\xi^{k+1}, y^k) = \langle y - y^k, L\xi^{k+1} \rangle = \frac{1}{\mu\rho} \langle y - y^k, y^{k+1} - y^k \rangle$$
$$= \frac{1}{\mu\rho}\Delta_k(y) + \frac{1}{2\mu\rho} \| y^{k+1} - y^k \|^2.$$

Adding the two relations above and collecting terms, we obtain for any $\xi \in \mathcal{F}$ and any $y \in \mathbb{R}^m$,

$$\mathcal{L}(\xi^{k+1}, y) - \mathcal{L}(\xi, y^k) \le \Delta_k(\xi, M) + \frac{1}{\mu\rho}\Delta_k(y) + R_k(\xi, M),$$

where

$$R_k(\xi,M) = -\frac{1}{2}\| \xi^{k+1} - \xi^k \|_M^2 - \left(\frac{2-\mu}{2\mu^2\rho}\right) \| y^{k+1} - y^k \|^2 - \langle d^k, \xi^{k+1} - \xi \rangle.$$

Noting that $\mathcal{L}(\xi, y^k) \equiv \mathcal{L}(\xi, y)$ for any $\xi \in \mathcal{F}$, the proof is completed. □

The residual bound. As we shall see, the quantity

$$R_k(\xi,M) := -\frac{1}{2}\| \xi^{k+1} - \xi^k \|_M^2 - \left(\frac{2-\mu}{2\mu^2\rho}\right) \| y^{k+1} - y^k \|^2 - \langle d^k, \xi^{k+1} - \xi \rangle, \tag{26}$$

as derived in Lemma 5, *governs* the rate of convergence analysis of the perturbed PMM, and hence of all the four methods described above. For that purpose, we will just need to verify the following assumption, which will be shown to hold true for any of the well-known Lagrangian-based schemes, by simply choosing an adequate matrix M.

Assumption R. For any $\xi \in \mathcal{F}$, and any integer $N \geq 1$, there exists a nonnegative constant $C_\rho(\xi)$ such that:

$$\sum_{k=0}^{N-1} R_k(\xi,M) \leq \frac{C_\rho(\xi)}{2}.$$

Note that the constant $C_\rho(\cdot)$ is, of course, independent of N. It will determine the value of the constant in the bound of the efficiency estimate (rate of convergence) of each method. As an appetizer, for the PMM, we clearly have $d^k = \mathbf{0}$, $k \in \mathbb{N}$, and hence since $\mu \in (0,2)$, from (26) we obviously have that $R_k(\xi,M) \leq 0$, and hence Assumption **R** trivially holds with $C_\rho(\xi) \equiv 0$.

We are now ready to state and prove the main rate of convergence result for the perturbed PMM.

Theorem 1 (O(1/N) Rate of Convergence for the Perturbed PMM).
*Let $\{(\xi^k, y^k)\}_{k\in\mathbb{N}}$ be a sequence generated by the Perturbed PMM, and let (ξ^*, y^*) be a saddle point of problem (CM). Assume that Assumption **R** holds. Then, for any integer $N \geq 1$, and any positive $\alpha > 0$ such that $\alpha \geq 2\|y^*\|$, we have*

$$\Psi(\xi^N) - \Psi(\xi^*) \leq \frac{B_{\rho,\alpha}(\xi^*,M)}{2N}, \tag{27}$$

$$\| L\xi^N \| \leq \frac{B_{\rho,\alpha}(\xi^*,M)}{\alpha N}, \tag{28}$$

where $B_{\rho,\alpha}(\xi^*,M) := \| \xi^* - \xi^0 \|_M^2 + C_\rho(\xi^*) + \frac{1}{\mu\rho}(\| y^0 \| + \alpha)^2$.

Proof. Applying Lemma 5, and summing (25) over $k = 0, 1, \ldots, N-1$, it follows that for any $\xi \in \mathcal{F}$ and $y \in \mathbb{R}^m$,

$$\sum_{k=0}^{N-1} \mathcal{L}(\xi^{k+1}, y) - N\mathcal{L}(\xi, y) \le \frac{1}{2} \| \xi - \xi^0 \|_M^2 + \frac{1}{2\mu\rho} \| y - y^0 \|^2 + \sum_{k=0}^{N-1} R_k(\xi, M).$$

Since $\xi \to \mathcal{L}(\xi, y)$ is convex, thanks to Jensen's inequality, and Assumption **R**, it follows that

$$\mathcal{L}(\xi^N, y) - \mathcal{L}(\xi, y) \le \frac{1}{2N}\left(\| \xi - \xi^0 \|_M^2 + \frac{1}{\mu\rho} \| y - y^0 \|^2 + C_\rho(\xi) \right), \quad \forall\, \xi \in \mathcal{F}, \ \forall\, y \in \mathbb{R}^m.$$

Recalling our compact notations, and the definition of $\mathcal{L}$, applying the latter at $\xi = \xi^* \in \mathcal{F}$ we get:

$$\Psi(\xi^N) - \Psi(\xi^*) + \langle y, L\xi^N \rangle \le \frac{1}{2N}\left(\| \xi^* - \xi^0 \|_M^2 + C_\rho(\xi^*) + \frac{1}{\mu\rho} \| y - y^0 \|^2 \right), \quad \forall\, y \in \mathbb{R}^m.$$

Maximizing both sides of this inequality over $\|y\| \le \alpha$, we thus obtain,

$$\Psi(\xi^N) - \Psi(\xi^*) + \alpha \| L\xi^N \| \le \frac{1}{2N} B_{\rho,\alpha}(\xi^*, M),$$

with $B_{\rho,\alpha}(\xi^*, M) = \| \xi^* - \xi^0 \|_M^2 + C_\rho(\xi^*) + \frac{1}{\mu\rho}(\| y^0 \| + \alpha)^2$, and the first assertion (27) follows. Since, $\alpha \ge 2\|y^*\|$, applying Lemma 4 with $\delta := (2N)^{-1} B_{\rho,\alpha}(\xi^*, M)$, proves the second assertion (28). □

This generic rate of convergence result is all we need to derive rate of convergence for the four fundamental Lagrangian-based schemes mentioned above, as well as other variants, cf. Remark 2.

3.4 Special cases of the perturbed PMM: Fundamental schemes

In order to apply the result obtained above on various Lagrangian-based schemes, we need to identify the perturbation vector d^k (used in step (23)) in each method, namely to cast each method as a particular case of the perturbed PMM, and then verify for each case that Assumption **R** holds. Let us illustrate this on the PADMM.

Writing the optimality conditions of the two primal steps of the PADMM (see (17) and (18)) we get

$$\begin{cases} \mathbf{0} \in \partial\varphi(u^{k+1}) + F^T\left(y^k + \rho(Fu^{k+1} - v^k)\right) + M_1(u^{k+1} - u^k), \\ \mathbf{0} \in \partial h(v^{k+1}) - y^k - \rho(Fu^{k+1} - v^{k+1}) + M_2(v^{k+1} - v^k), \end{cases}$$

which is equivalent to

$$\begin{cases} \mathbf{0} \in \partial_u \mathcal{L}_\rho(u^{k+1}, v^{k+1}, y^k) + \rho F^T(v^{k+1} - v^k) + M_1(u^{k+1} - u^k), \\ \mathbf{0} \in \partial_v \mathcal{L}_\rho(u^{k+1}, v^{k+1}, y^k) + M_2(v^{k+1} - v^k). \end{cases}$$

The last two inclusions simply read

$$\xi^{k+1} = \operatorname{argmin}_{\xi}\left\{\mathcal{L}_\rho(\xi,y^k) + \langle d^k,\xi\rangle + \frac{1}{2}\|\xi - \xi^k\|_M^2\right\}, \text{ with } d^k \equiv \left(\rho F^T\left(v^{k+1} - v^k\right),\mathbf{0}\right),$$

thus showing that the PADMM is a particular instance of the perturbed PMM. The next result shows that Assumption **R** holds.

Lemma 6 (Residual Bound for PADMM).

Let $\{(u^k,v^k,y^k)\}_{k\in\mathbb{N}}$ *be a sequence generated by PADMM. Then, for any* M_1, $M_2 \succeq 0$, *and any* $0<\mu<2-\rho\left(\rho+\lambda_{\min}(M_2)\right)^{-1}$, *we have* $C_\rho(\xi)=\rho\|v^*-v^0\|^2$ *for every* $\xi\in\mathcal{F}$.

Proof. As just shown above, here we have that $d^k \equiv \left(\rho F^T\left(v^{k+1}-v^k\right),\mathbf{0}\right)$, and hence in this case for all $\xi\in\mathcal{F}$

$$R_k(\xi,M) = -\frac{1}{2}\|\xi^{k+1}-\xi^k\|_M^2 - \left(\frac{2-\mu}{2\mu^2\rho}\right)\|y^{k+1}-y^k\|^2 - \rho\langle v^{k+1}-v^k, Fu^{k+1}-v\rangle, \tag{29}$$

where we have used the fact that $\xi\in\mathcal{F}$, i.e., $Fu = v$. Using the multiplier update, we have

$$\begin{aligned}\rho\langle v^k - v^{k+1}, Fu^{k+1}-v\rangle &= \rho\langle v^k - v^{k+1}, v^{k+1}-v\rangle + \rho\langle v^k - v^{k+1}, Fu^{k+1}-v^{k+1}\rangle\\ &= \rho\Delta_k(v) - \frac{\rho}{2}\|v^{k+1}-v^k\|^2 + \frac{1}{\mu}\langle v^k - v^{k+1}, y^{k+1}-y^k\rangle\\ &\le \rho\Delta_k(v) - \frac{1}{2}\left(\rho-\frac{1}{s}\right)\|v^{k+1}-v^k\|^2 + \frac{s}{2\mu^2}\|y^{k+1}-y^k\|^2,\end{aligned}$$

where the last inequality follows from using the fact: $\langle a,b\rangle \le (s/2)\|a\|^2 + (1/2s)\|b\|^2$ for any $s>0$. Plugging this into (29) yields (recalling the definition of M and ξ)

$$\begin{aligned}R_k(\xi,M) \le \rho\Delta_k(v) - \frac{1}{2}\|u^{k+1}-u^k\|_{M_1}^2 - \frac{1}{2\mu^2}\left(\frac{2-\mu}{\rho} - s\right)\\ \|y^{k+1}-y^k\|^2 - \frac{1}{2}\|v^{k+1}-v^k\|_{M_2+(\rho-1/s)I}^2.\end{aligned}$$

Since $\mu\in(0,2)$, by taking $s:=(2-\mu)/\rho>0$, we get

$$M_2 + \left(\rho-\frac{1}{s}\right)I = M_2 + \left(\rho - \frac{\rho}{2-\mu}\right)I \succeq 0 \text{ whenever } 0<\mu<2-\frac{\rho}{\rho+\lambda_{\min}(M_2)}.$$

Thus, it follows from the above inequality that $R_k(\xi,M)\le\rho\Delta_k(v)$ and hence

$$\sum_{k=0}^{N-1} R_k(\xi,M) \le \frac{\rho}{2}\sum_{k=0}^{N-1}\left(\|v-v^k\|^2 - \|v-v^{k+1}\|^2\right) = \frac{\rho}{2}\left(\|v-v^0\|^2 - \|v-v^N\|^2\right) \le \frac{\rho}{2}\|v-v^0\|^2,$$

proving that $C_\rho(\xi)=\rho\|v-v^0\|^2$. □

From here, applying Theorem 1, we immediately get the rate of convergence of PADMM with its bound constant by plugging the obtained value for $C_\rho(\xi)$. Similarly, the same can be easily done for all the other decomposition schemes, which can be shown to be particular instances of the perturbed PMM. We omit the details, and just summarize below the results that can be easily verified for these methods.

- *PMM*: $d^k = (\mathbf{0}, \mathbf{0})$ with any M_1, $M_2 \succeq 0$ and $\mu \in (0,2)$. Then, we have $C_\rho(\xi^*) = 0$.
- *Linearized PMM*: $d^k = \rho\big(F^T(v^{k+1} - v^k), F(u^{k+1} - u^k)\big)$ with any

$$M_1 = P - \rho F^T F \quad \text{and} \quad M_2 = Q - \rho I,$$

where $\lambda_{\min}(P) \geq \lambda_{\max}(F^T F)$ and $\lambda_{\min}(Q) \geq \rho$. Then, for any $0 < \mu < 2 - t$ where

$$t = \max\left\{\frac{\rho}{2(\rho + \lambda_{\min}(M_2))}, \frac{\rho\lambda_{\max}(F^T F)}{2(\rho\lambda_{\max}(F^T F) + \lambda_{\min}(M_1))}\right\},$$

we have $C_\rho(\xi) = \rho \| u^* - u^0 \|^2_{F^T F} + \rho \| v^* - v^0 \|^2$

- *Proximal ADMM*: $d^k = \rho\big(F^T(v^{k+1} - v^k), \mathbf{0}\big)$ with any M_1, $M_2 \succeq 0$. Then, for any $0 < \mu < 2 - \rho\big(\rho + \lambda_{\min}(M_2)\big)^{-1}$, we have $C_\rho(\xi) = \rho \| v^* - v^0 \|^2$.
- *Linearized PADMM*: $d^k = \rho\big(F^T(v^{k+1} - v^k), \mathbf{0}\big)$ with any $M_1 = P - \rho F^T F$ (where $\lambda_{\min}(P) \geq \lambda_{\max}(F^T F)$) and $M_2 \succeq 0$. Then, for any $0 < \mu < 2 - \rho\big(\rho + \lambda_{\min}(M_2)\big)^{-1}$, we have $C_\rho(\xi) = \rho \| v^* - v^0 \|^2$.

Remark 2. Other combinations of the above scenarios can be derived. For instance, it is also easy to consider a variant of the perturbed PMM for the case where the function φ is smooth with a Lipschitz continuous gradient. In that case we can linearize the function φ, and then we can use the well-known descent lemma, to adapt the bound in the key estimate of Lemma 5, and from there derive the corresponding $O(1/N)$ rate from Theorem 1. Finally, note that the global pointwise convergence of the described algorithms can also be derived as a by-product of the above analysis using classical arguments. For the sake of brevity we omit the details.

4 The nonconvex setting

We are back to our general nonconvex and nonlinear composite model

$$\text{(M)} \qquad \min_{u \in \mathbb{R}^d} \{\Phi(u) \equiv \varphi(u) + h(F(u))\},$$

with the following problem data:

- $\varphi : \mathbb{R}^d \to \mathbb{R}$ is a continuously differentiable function.
- $h : \mathbb{R}^m \to (-\infty, +\infty]$ is a proper and lower semicontinuous function.
- $F : \mathbb{R}^d \to \mathbb{R}^m$ $(m \leq n)$ is a continuously differentiable map.

As exemplified in Section 2.3, this structure is very flexible and captures many fundamental optimization problems arising in disparate applications. As opposed to the development and analysis of Lagrangian-based schemes in the convex setting, the nonconvex model is much harder and quite challenging, and the literature remains scarce. Not only the lack of convexity already renders the problem difficult, but also the nonlinear nature of the map $F(\cdot)$, which in turn adds severe difficulties preventing us to apply the tools and approaches used in the convex setting. In this section, we briefly describe some very recent results as developed in Bolte et al. (2018), highlighting the main difficulties, and outlining a general abstract approach, which paves the way to analyze Lagrangian-based methods for the general nonconvex model (M).

4.1 The nonconvex nonlinear composite optimization—Preliminaries

As explained in Section 2, we first reformulate problem (M) in the equivalent split form:

$$(\mathrm{M}) \qquad \min_{u\in\mathbb{R}^d,\, v\in\mathbb{R}^m} \{\varphi(u)+h(v) : F(u)=v\}.$$

The classical *Lagrangian* $\mathcal{L}:\mathbb{R}^d\times\mathbb{R}^m\times\mathbb{R}^m\to(-\infty,+\infty]$ is defined, as before, by

$$\mathcal{L}(u,v,y)\equiv\varphi(u)+h(v)+\langle y,F(u)-v\rangle,$$

and the associated *augmented Lagrangian* is defined as follows

$$\mathcal{L}_\rho(u,v,y):=\varphi(u)+h(v)+\langle y,F(u)-v\rangle+\frac{\rho}{2}\|F(u)-v\|^2,$$

where $\rho>0$ is a penalty parameter.

To ensure the well-posedness of the algorithms to come, we assume for any fixed $y\in\mathbb{R}^m$, that

$$\inf_{u,v}\mathcal{L}_\rho(u,v,y)>-\infty.$$

To work with nonconvex functions we need some appropriate subdifferentials that we now record. All these results can be found in greater details in the book of Rockafellar and Wets (1998).

Definition 1 (Subdifferentials).

Let $\psi:\mathbb{R}^d\to(-\infty,+\infty]$ be a proper and lower semicontinuous function with $\operatorname{dom}\psi:=\{z\in\mathbb{R}^d:\psi(z)<\infty\}$.

(i) For a given $z\in\operatorname{dom}\psi$, the *Frechet subdifferential* of ψ at z, denoted by $\hat{\partial}\psi(z)$ is the set of all vectors $v\in\mathbb{R}^d$ satisfying

$$\liminf_{\substack{x\to z\\ x\neq z}}\frac{\psi(x)-\psi(z)-\langle v,x-z\rangle}{\|x-z\|}\geq 0.$$

(ii) The *limiting subdifferential* of ψ at z, denoted by $\partial\psi(z)$ is the set of all vectors $v \in \mathbb{R}^d$ for which there exist two sequences $\{z^k\}_{k\in\mathbb{N}}$ and $\{v^k\}_{k\in\mathbb{N}}$ such that $v^k \in \partial\psi(z^k)$ and

$$\lim_{k\to\infty}\left(z^k,\psi\left(z^k\right),v^k\right)=(z,\psi(z),v).$$

(iii) The *horizon subdifferential* of ψ at z, denoted by $\partial^\infty\psi(z)$ is the set of all vectors $v \in \mathbb{R}^d$ for which there exist a sequence of real numbers $\{t_k\}_{k\in\mathbb{N}}$ such that $\lim_{k\to\infty} t_k = 0$ and two sequences $\{z^k\}_{k\in\mathbb{N}}$ and $\{v^k\}_{k\in\mathbb{N}}$ such that $v^k \in \partial\psi(z^k)$ and

$$\lim_{k\to\infty}\left(z^k,\psi\left(z^k\right),t_k v^k\right)=(z,\psi(z),v).$$

To ensure the validity of calculus rules for problem (M), we need to assume a constraint qualification. We record below a central one that is relevant to our model (M).

Assumption 2. The following constraint qualification condition holds for problem (M),

$$[\text{CQ}] \quad \nabla F(u)^T z = \mathbf{0}, \;\; z \in \partial^\infty h(F(u)) \Longrightarrow z = \mathbf{0}.$$

In the classical nonlinear programming case, which is a special case of model (M), (cf. Example 1), the general condition [CQ] simply reduces to the classical *Mangasarian-Fromovitz* constraint qualification condition: there is no vector $z \neq \mathbf{0}$, which satisfies $\nabla F(u)^T z = \mathbf{0}$.

The condition [CQ] is also crucial to guarantee the subdifferential calculus rules and the chain rule, which are needed to write the first order optimality condition for model (M). More precisely, if u is a local minimum of problem (M) satisfying condition [CQ], then there exists $z \in \mathbb{R}^m$ such that $\nabla\varphi(u) + \nabla F(u)^T z = \mathbf{0}$ with $z \in \partial h(F(u))$. Thus, for model (M), under [CQ], the set of critical points of problem (M) reads as:

$$\operatorname{crit}\Phi = \left\{u \in \mathbb{R}^d : \mathbf{0} \in \nabla\varphi(u) + \nabla F(u)^T \partial h(F(u))\right\}.$$

Next, we introduce a fundamental and classical concept of regularity needed to handle the nonlinearity of the map $F(\cdot)$, see Milnor (1931).

Definition 2 (Uniform Regularity).

Let $\Omega \subset \mathbb{R}^d$ be an open set, $F:\Omega \to \mathbb{R}^m$ be a C^1 map, and $\emptyset \neq S \subset \Omega$. The mapping F is uniformly regular on S with constant $\gamma > 0$ if the following holds:

$$\left\|\nabla F(u)^T v\right\| \geq \gamma\|v\|, \;\; \forall u \in S, \;\; v \in \mathbb{R}^m.$$

For a given $u \in \Omega$, we can equivalently define

$$\gamma(F,u) := \min\left\{\left\|\nabla F(u)^T v\right\| : \|v\| = 1\right\},$$

and that $\gamma(F,u)$ is nonzero means that $\nabla F(u)$ is surjective or that $\nabla F(u)\nabla F(u)^T$ is positive definite.

A main issue in Lagrangian-based methods is to handle *simultaneously* the penalty parameter, the data constants and the constraint qualification condition. However, as we know, augmented Lagrangian methods are based on *relaxing* the classical Lagrangian, and hence are obviously *unfeasible*. The looser is the relaxation, the more unfeasible is the method. The looseness of the relaxation is measured through the penalty parameter ρ, which is used to penalize the constraint $F(u) = v$ in $\mathcal{L}_\rho$, namely:

$$\lim_{\rho \to +\infty} \mathcal{L}_\rho(u, v, y) = \begin{cases} \varphi(u) + h(F(u)), & \text{if } F(u) = v, \\ +\infty, & \text{otherwise.} \end{cases}$$

While in the convex setting, and with a linear map F, we can globally control the behaviour of the points generated by a Lagrangian-based scheme, in the nonconvex and nonlinear setting, the lack of feasibility creates new and severe difficulties in the sense that an unfeasible method generates points that might be out of control, inducing three major obstacles:

- Constraint qualification condition may fail.
- Problem's data input, such as global Lipschitz constants, may become unknown, or out of reach.
- The inherent min–max dynamics of Lagrangian-based methods clearly imply that the augmented Lagrangian function *alternatively* increases and decreases. Therefore, we cannot warrant/measure the crucial *descent* property needed in the convergence analysis through it.

In the forthcoming subsection, we describe the key ideas and building blocks of the theory, which allow to address these three main obstacles.

4.2 ALBUM—Adaptive Lagrangian-based multiplier method

In the remaining parts of this section, we have briefly outlined the fundamental ideas and theoretical tools that were developed in Bolte et al. (2018), and we refer the reader to that work for the specific details, proofs, more results and references.

The main underlying ideas allowing us to address the alluded obstacles rely on the following ingredients:

- Define an *information zone* to be a region for which *regularity is under control* and data constants are known.
- Build a *generic Lagrangian-based scheme with an adaptive regime* aiming at detecting the information zone, and forcing the iterates to enter and stay within this zone.
- Once the zone is found, using the adaptive regime, *detect an adequate Lyapunov function* ensuring the necessary descent properties.

We start by introducing the notion of information zone, which is in fact an enlargement of the feasible set.

Definition 3 (Information Zone—Enlargement of the Feasible Set).
An information zone is a subset $\mathcal{Z}$ of $\mathbb{R}^d$ such that there exists $\overline{d} \in (0, +\infty]$ for which

$$\mathcal{Z} \supset \left\{u \in \mathbb{R}^d : \operatorname{dist}(F(u), \operatorname{dom} h) \le \overline{d}\right\} \supset \mathcal{F},$$

where $\mathcal{F} = \left\{u \in \mathbb{R}^d : F(u) \in \operatorname{dom} h\right\}$ is the feasible set of problem (M).

Equipped with an information zone $\mathcal{Z}$, our second assumption for problem (M) is *local regularity* and *local smoothness* (with respect to the information zone) of the problem's data.
Assumption 3. (i) F is uniformly regular over $\mathcal{Z}$ with positive constant γ,
(ii) ∇F is $L(F)$ Lipschitz continuous over $\mathcal{Z}$,
(iii) $\nabla \varphi$ is $L(\varphi)$ Lipschitz continuous over $\mathcal{Z}$.

Throughout the rest of this section, we assume that Assumptions 2 and 3 hold.

Our next ingredient is to introduce an adequate Lyapunov function that will allow us to measure the descent property of the proposed scheme, which due to the min–max dynamics of Lagrangian-based methods, cannot be detected through the usual augmented Lagrangian function.
Definition 4 (A Lyapunov Function).
Fix $\beta > 0$ and $w \in \mathbb{R}^d$. The function $\mathcal{E}_\beta(\cdot)$, which is defined by

$$\mathcal{E}_\beta(u, v, y, w) := \mathcal{L}_\rho(u, v, y) + \beta\|u - w\|^2,$$

is called a *Lyapunov function* associated to problem (M).

The next result records the key fact showing that $\mathcal{E}_\beta$ can be beneficially used to identify critical points of the original problem (M).
Proposition 1. (Critical Points Relationships (Bolte et al., 2018, Proposition 1)). *The following implications hold for all β, $\rho > 0$*

$$(u, v, y, u) \in \operatorname{crit} \mathcal{E}_\beta \Longrightarrow (u, v, y) \in \operatorname{crit} \mathcal{L}_\rho \Longrightarrow u \in \operatorname{crit} \Phi.$$

Both the information zone and the Lyapunov function $\mathcal{E}_\beta$ just defined above, will come into play in a dynamical (adaptive) fashion in the forthcoming algorithm. As already seen, a typical Lagrangian-based method is given by

$$\begin{aligned} (u^+, v^+) &\in \mathcal{A}_\rho(u, v, y), \\ y^+ &= y + \rho(F(u^+) - v^+), \end{aligned}$$

where $\mathcal{A}_\rho$ stands for any minimization black-box map, while the multiplier update y^+ remains the same for any Lagrangian-based scheme. We now formalize the definition of the map $\mathcal{A}_\rho$, which captures the mechanism that will essentially govern the convergence of a generic Lagrangian-based scheme, and play an essential role in controlling the relevant parameters in our generic adaptive Lagrangian-based multiplier method (*ALBUM*).

Definition 5 (Lagrangian Algorithmic Map).

A map $\mathcal{A}_\rho : \mathbb{R}^d \times \mathbb{R}^m \times \mathbb{R}^m \to \mathbb{R}^d \times \mathbb{R}^m$ is called a *Lagrangian algorithmic map* if there exist two constants a, $b > 0$ such that

$$\text{(i)} \quad \frac{a}{2}\|u^+ - u\|^2 + \mathcal{L}_\rho(u^+, v^+, y) \le \mathcal{L}_\rho(u, v, y),$$

and

$$\text{(ii)} \quad \left\|\nabla_u \mathcal{L}_\rho(u^+, v^+, y)\right\| \le b\|u^+ - u\|.$$

Note that once we choose the map $\mathcal{A}_\rho$, it fully determines the constants a and b. These constants might depend on the problem's data input or/and other algorithmic constants. This will be illustrated later on. With the above framework, we are now ready to state our generic adaptive Lagrangian-based multiplier method.

Adaptive Lagrangian-based multiplier method (ALBUM)

1. **Input:** $\mathcal{A}_\rho$ an algorithmic map, with the corresponding parameters a, $b > 0$.
2. **Initialization:** Fix δ, $\rho_0 > 0$ and start with any (u^0, v^0, y^0) where $u^{-1} \equiv u^0$.
3. **Main step:** For $k = 0, 1, \ldots$ generate the sequence $\{(u^k, v^k, y^k)\}_{k\in\mathbb{N}}$ as follows
   - 3.1. **Primal step**

$$(u^{k+1}, v^{k+1}) \in \mathcal{A}_{\rho_k}(u^k, v^k, y^k).$$

   - 3.2. **Multiplier step**

$$y^{k+1} = y^k + \rho_k\left(F(u^{k+1}) - v^{k+1}\right).$$

   - 3.3. **Adaptive step:** choose $\tau \in (0, \frac{a}{2})$, set $\beta_k := \frac{b^2}{\rho_k \gamma}$, and compute

$$\Delta_{\beta_k} := \mathcal{E}_{\beta_k}(u^k, v^k, y^k, u^{k-1}) - \mathcal{E}_{\beta_k}(u^{k+1}, v^{k+1}, y^{k+1}, u^k).$$

If $u^{k+1} \notin \mathcal{Z}$ or $\tau\|u^{k+1} - u^k\|^2 > \Delta_{\beta_k}$, set $\rho_{k+1} = \rho_k + \delta$. Otherwise, set $\rho_{k+1} = \rho_k$.

4.3 A methodology for global analysis of Lagrangian-based methods

Basically, *ALBUM* aims at forcing u^k to enter the information zone $\mathcal{Z}$ in finitely many steps and stay there. The adaptive protocol essentially turns *ALBUM* into a descent scheme for the Lyapunov function $\mathcal{E}_\beta$: it leads to a finite identification of the zone $\mathcal{Z}$, and warrants sufficient descent of $\mathcal{E}_\beta$ through the adaptive step of *ALBUM*. This crucial adaptive step relies on detecting weak feasibility, in the sense that we ask for $u^k \in \mathcal{Z}$, and not for actual feasibility, which would be obviously impossible given the inherent nonfeasibility of a Lagrangian-based scheme! Moreover, it detects a descent property for $\mathcal{E}_\beta$ (and again clearly not possible for $\mathcal{L}_\rho$!), which implicitly tunes the algorithm to match the natural step-sizes attached to φ and F.

A main remaining ingredient to derive convergence results for *ALBUM* is then a fundamental definition characterizing the required properties of a sequence generated by *ALBUM*, which is called a Lagrangian sequence.

Definition 6 (Lagrangian Sequence).
A sequence $\{w^k\}_{k\in\mathbb{N}} := \{(u^k, v^k, y^k)\}_{k\in\mathbb{N}}$ generated by *ALBUM* is called a *Lagrangian sequence*, if it satisfies the following conditions:

C1 There exists a positive constant a such that

$$\frac{a}{2}\|u^{k+1} - u^k\|^2 + \mathcal{L}_{\rho_k}(u^{k+1}, v^{k+1}, y^k) \le \mathcal{L}_{\rho_k}(u^k, v^k, y^k), \quad \forall\, k \ge 0.$$

C2 There exists a positive constant b such that

$$\|\nabla_u \mathcal{L}_{\rho_k}(u^{k+1}, v^{k+1}, y^k)\| \le b\|u^{k+1} - u^k\|, \quad \forall\, k \ge 0.$$

C3 There exists a positive constant c and $q^{k+1} \in \partial_v \mathcal{L}_{\rho_k}(u^{k+1}, v^{k+1}, y^k)$ such that

$$\|q^{k+1}\| \le c\|u^{k+1} - u^k\|, \quad \forall\, k \ge 0.$$

C4 Let $\overline{v}$ be a limit point of a subsequence $\{v^k\}_{k\in\mathcal{K}}$ of $\{v^k\}_{k\in\mathbb{N}}$, then $\limsup_{k\in\mathcal{K}\subset\mathbb{N}} h(v^k) \le h(\overline{v})$.

Before proceeding, we make the following remarks regarding conditions **C1–C4**.

- Thanks to the definition of the Lagrangian algorithmic map $\mathcal{A}_\rho$, once the map is chosen, conditions **C1** and **C2** immediately hold for the sequence $\{w^k\}_{k\in\mathbb{N}}$.
- Condition **C1** is actually a *partial descent property* of the augmented Lagrangian $\mathcal{L}_\rho(\cdot)$ only with respect to the primal variable u:

$$\frac{a}{2}\|u^{k+1} - u^k\|^2 + \mathcal{L}_{\rho_k}(u^{k+1}, v^{k+1}, y^k) \le \mathcal{L}_{\rho_k}(u^k, v^k, y^k), \quad \forall\, k \ge 0.$$

It pertains to the primal variables (u, v), since by nature the multiplier y is an "ascent variable".

- Conditions **C2** and **C3** provide (sub)gradient bounds of the augmented Lagrangian $\mathcal{L}_\rho(\cdot)$ with respect to both primal variables (u, v).
- Condition **C4** is a simple sequential assumption on h, it holds, e.g., when $h : \operatorname{dom} h \to \mathbb{R}$ is continuous.

We are now ready to state the main convergence results for *ALBUM* proven in Bolte et al. (2018, Theorems 1 and 2) which we record in the following theorem.

Theorem 2 (Sequential and Global Convergence).
Let $\{w^k\}_{k \in \mathbb{N}}$ *be generated by ALBUM that satisfies conditions* ***C1–C4****, i.e., a Lagrangian sequence. Assume that the sequence* $\{w^k\}_{k \in \mathbb{N}}$ *is bounded. Then the following two assertions hold:*

(i) *(Subsequence convergence). Let* $(\bar{u}, \bar{v}, \bar{y})$ *be a limit point of* $\{w^k\}_{k \in \mathbb{N}}$*. Then,* $\bar{u}$ *is a critical point of problem (M).*
(ii) *(Global convergence). Assume that* φ*, F and h are semialgebraic. Then, the whole sequence* $\{w^k\}_{k \in \mathbb{N}}$ *converges to* $(\bar{u}, \bar{v}, \bar{y})$ *where* $\bar{u}$ *is a critical point of problem (M).*

Note that the second part of the theorem above establishes *global convergence*, i.e., the *whole* sequence generated by *ALBUM* converges to a critical point of problem (M) with *semialgebraic data*.[b] This relies on the fundamental nonsmooth version of the so-called *Kurdyka–Łojasiewicz property*. The KL property was introduced in the seminal works of Łojasiewicz (1963) and Kurdyka (1998) for the smooth case, and Bolte et al. (2006) and Bolte et al. (2010) for the nonsmooth case, the later includes an in depth study of KL functions and many relevant references. The KL property is a crucial ingredient towards the global convergence analysis of descent schemes (Attouch et al., 2013; Bolte et al., 2014). A general methodology to prove global convergence of descent methods is derived in the work of Bolte et al. (2014) which relies on a key uniformization of the KL property (Bolte et al., 2014, Lemma 6, p. 478). While verifying the KL property of a given function is often a very difficult task, it holds for the class of semialgebraic functions, which are ubiquitous in many applications (see, e.g., Bolte et al., 2014), thus making the KL property a powerful tool for proving convergence of descent schemes for semialgebraic problems. The central result establishing that KL holds for the broad class of semialgebraic functions was derived in the seminal work of Bolte et al. (2006).

4.4 ALBUM in action: Global convergence of Lagrangian-based schemes

Now we illustrate the application of *ALBUM* on the two classical schemes PMM and Proximal ADMM, previously studied in the convex case (see Section 3). To this end, in order to put these two methods in the general framework of *ALBUM*, we need first to identify the corresponding algorithmic maps $\mathcal{A}_\rho$ that update the primal variables u and v (recall that the multiplier update remains identical for any Lagrangian-based scheme).

---

[b]Recall that a subset S of $\mathbb{R}^d$ is real semialgebraic set if constructible by finite systems of polynomial inequalities, and that a function is semialgebraic if its graph is semialgebraic.

4.4.1 A proximal multipliers method (PMM)

$$\left(u^{k+1},v^{k+1}\right)\in \operatorname{argmin}_{(u,v)}\left\{\mathcal{L}_\rho\left(u,v,y^k\right)+\frac{1}{2}\left\|u-u^k\right\|_{M_1}^2\right\},\quad (M_1\succ 0).$$

It is clear that in this case, the map $\mathcal{A}_\rho$ stands for the joint minimization of the (partial) proximal counterpart of the augmented Lagrangian $\mathcal{L}_\rho$[c].

4.4.2 A proximal alternating direction method of multipliers (PADMM)

$$v^{k+1}\in \operatorname{argmin}_v \mathcal{L}_\rho\left(u^k,v,y^k\right),$$

$$u^{k+1}\in \operatorname{argmin}_u\left\{\mathcal{L}_\rho\left(u,v^{k+1},y^k\right)+\frac{1}{2}\left\|u-u^k\right\|_{M_1}^2\right\},\quad (M_1\succ 0).$$

In this case we obviously have that the map $\mathcal{A}_\rho$ consists of alternating minimization applied on the (partial) proximal counterpart of the augmented Lagrangian $\mathcal{L}_\rho$.

It can be shown in Bolte et al. (2018) that both algorithmic maps corresponding to the PMM and PADMM schemes are Lagrangian algorithmic maps satisfying the premises of Definition 5 with the same parameters:

$$a=\lambda_{\min}(M_1)>0 \ \text{ and } \ b=\lambda_{\max}(M_1)>0,\quad (\text{since } M_1\succ 0).$$

The *adaptive version* of both methods is obtained from *ALBUM* through their corresponding maps $\mathcal{A}_\rho$ just described above. In order to derive the global convergence of any bounded sequence generated by these two adaptive versions to critical points of problem (M), we need to ensure that both methods are producing *Lagrangian sequences*, i.e., that conditions **C1–C4** hold (cf. Definition 6). This was proven in Bolte et al. (2018, Section 6), which we record in the following result.

Proposition 2. *Let* $\{w^k\}_{k\in\mathbb{N}}$ *be a sequence generated by either the Adaptive PMM or the Adaptive PADMM. Then, conditions* ***C1–C3*** *hold. If, in addition,* $\{w^k\}_{k\in\mathbb{N}}$ *is bounded, then condition* ***C4*** *holds true, and hence* $\{w^k\}_{k\in\mathbb{N}}$ *is a Lagrangian sequence.*

Equipped with this result for these adaptive versions of PMM and PADMM, we can then apply Theorem 2 to obtain global convergence to critical points of problem (M) in the nonlinear and nonconvex setting for both methods.

---

[c]Recall that the use of the partial version on the variable u is enough, since the variable v has a built-in proximal term through the definition of the augmented Lagrangian.

We end by briefly indicating other decomposition variants (as discussed previously in the convex case) that fit into the *ALBUM* framework and include

- The classical ADMM. This is obtained by taking $M_1 = 0$ in the PADMM. Under the additional assumption that $u \to \mathcal{L}_\rho(u,v,y)$ is σ-strongly convex for any fixed $v, y \in \mathbb{R}^m$, the global convergence of ADMM to critical points of the nonlinear composite model (M) can be derived.
- The proximal linearized ADMM. Here the model (M) is nonconvex but, F is assumed to be a linear map, and the condition number of FF^T is strictly smaller than 2. The scheme consists of *one gradient iteration* of the u-step and it reads as:

$$v^{k+1} \in \operatorname{argmin}_v \mathcal{L}_\rho\left(u^k, v, y^k\right),$$

$$u^{k+1} \in \operatorname{argmin}_u \left\{ \left\langle u - u^k, \nabla_u \mathcal{L}_\rho\left(u^k, v^{k+1}, y^k\right) \right\rangle + \frac{1}{2}\|u - u^k\|_{M_1}^2 \right\}, \quad (M_1 \succ 0).$$

Clearly, in that case, the u-step reduces to an easy explicit formula. The general convergence results of *ALBUM* can then be applied to this variant, including the determination of the penalty parameter in terms of the problem's data, see Bolte et al. (2018) for details.

Acknowledgements

The research of Marc Teboulle was partially supported by the Israel Science Foundation, under ISF Grant 1844-16.

References

Absil, P.A., Mahony, R., Sepulchre, R., 2008. Optimization Algorithms on Matrix Manifolds. Princeton University Press, Princeton, NJ.

Arrow, K.J., Hurwicz, L., Uzawa, H., 1958. Studies in linear and non-linear programming. In: Stanford Mathematical Studies in the Social Sciences, vol. II. Stanford University Press, Stanford, CA.

Attouch, H., Bolte, J., Svaiter, B.F., 2013. Convergence of descent methods for semi-algebraic and tame problems: proximal algorithms, forward-backward splitting, and regularized Gauss-Seidel methods. Math. Program. 137 (1–2), 91–129 (Ser. A).

Auslender, A., Teboulle, M., 2003. Asymptotic cones and functions in optimization and variational inequalities. In: Springer Monographs in MathematicsSpringer-Verlag, New York.

Bauschke, H.H., Combettes, P.L., 2017. Convex Analysis and Monotone Operator Theory in Hilbert Spaces, second ed. CMS Books in MathematicsSpringer.

Beck, A., 2017. First-Order Methods in Optimization, volume 25 of MOS-SIAM Series on Optimization. Society for Industrial and Applied Mathematics (SIAM), Philadelphia, PA.

Beck, A., Ben-Tal, A., Kanzow, C., 2008. A fast method for finding the global solution of the regularized structured total least squares problem for image deblurring. SIAM J. Matrix Anal. Appl. 30 (1), 419–443.

Bertsekas, D.P., 1982. Constrained optimization and lagrange multiplier methods. In: Computer Science and Applied MathematicsAcademic Press Inc., New York/London.

Bertsekas, D.P., 1999. Nonlinear Programming, second ed. Athena Scientific, Belmont, MA.

Bertsekas, D.P., 2015. Convex Optimization Algorithms. Athena Scientific, Belmont, MA.

Bertsekas, D.P., Tsitsiklis, J.N., 1989. Parallel and Distributed Computation. Prentice-Hall International Editions, Englewood Cliffs, NJ.

Bolte, J., Daniilidis, A., Lewis, A., 2006. The Łojasiewicz inequality for nonsmooth subanalytic functions with applications to subgradient dynamical systems. SIAM J. Optim. 17, 1205–1223.

Bolte, J., Daniilidis, A., Ley, O., Mazet, L., 2010. Characterizations of łojasiewicz inequalities: subgradient flows, talweg, convexity. Trans. Am. Math. Soc. 362 (6), 3319–3363.

Bolte, J., Sabach, S., Teboulle, M., 2014. Proximal alternating linearized minimization for non-convex and nonsmooth problems. Math. Program. 146 (1–2), 459–494 (Ser. A).

Bolte, J., Sabach, S., Teboulle, M., 2018. Nonconvex Lagrangian-based optimization: monitoring schemes and global convergence. Math. Oper. Res. 43 (3), 1210–1232.

Boyd, S., Parikh, N., Chu, E., Peleato, B., Eckstein, J., 2011. Distributed optimization and statistical learning via the alternating direction method of multipliers. Found. Trends Mach. Learn. 3 (1), 1–122.

Campisi, P., Egiazarian, K., 2007. Blind Image Deconvolution: Theory and Applications. CRC Press, Taylor & Francis Group.

Chambolle, A., 2004. An algorithm for total variation minimization and applications. J. Math. Imaging Vision 20 (1–2), 89–97.

Chambolle, A., Pock, T., 2011. A first-order primal-dual algorithm for convex problems with applications to imaging. J. Math. Imaging Vision 40 (1), 120–145.

Chan, R.H., Tao, M., Yuan, X., 2013. Constrained total variation deblurring models and fast algorithms based on alternating direction method of multipliers. SIAM J. Imaging Sci. 6 (1), 680–697.

Chen, G., Teboulle, M., 1994. A proximal-based decomposition method for convex minimization problems. Math. Programming 64 (1), 81–101 (Ser. A).

Douglas, J., Rachford, H., 1956. On the numerical solution of heat conduction problems in two and three space variables. Trans. Amer. Math. Soc. 82 (2), 421–439.

Eckstein, J., 1994. Some saddle-function splitting methods for convex programming. Optim. Methods Softw. 4 (1), 75–83.

Eckstein, J., Bertsekas, D.P., 1992. On the Douglas-Rachford splitting method and the proximal point algorithm for maximal monotone operators. Math. Programming 55 (3), 293–318 (Ser. A).

Edelman, A., Arias, T., Smith, S.T., 1999. The geometry of algorithms with orthogonality constraints. SIAM J. Matrix Anal. Appl. 20 (2), 303–353.

Engl, H.W., Hanke, M., Neubauer, A., 1996. Regularization of Inverse Problems. Mathematics and its Applications, vol. 375Kluwer Academic Publishers Group, Dordrecht.

Fortin, M., Glowinski, R., 1983. Augmented Lagrangian Methods. Studies in Mathematics and its Applications, vol. 15North-Holland Publishing Co., Amsterdam.

Gabay, D., 1983. Applications of the method of multipliers to variational inequalities. In: Fortin, M., Glowinski, R. (Eds.), Augmented Lagrangian Methods: Applications to the Solution of Boundary Value Problems. North-Holland, Amsterdam, pp. 299–340.

Gabay, D., Mercier, B., 1976. A dual algorithm for the solution of nonlinear variational problems via finite element approximation. Comput. Math. Appl. 2 (1), 17–40.

Glowinski, R., Le Tallec, P., 1983. Augmented Lagrangian and operator-splitting methods in nonlinear mechanics. In: Studies in Applied and Numerical MathematicsSociety for Industrial Mathematics.

Glowinski, R., Marroco, A., 1975. Sur l'approximation, par éléments finis d'ordre un, et la résolution, par pénalisation-dualité d'une classe de problèmes de Dirichlet non linéaires. Rev. Française Automat. Informat. Recherche Opérationnelle Sér. Rouge Anal. Numér 9, 41–76.

He, B., Yuan, X., 2012. On the $o(1/n)$ convergence rate of the Douglas-Rachford alternating direction method. SIAM J. Numer. Anal. 50 (2), 700–709.

Hestenes, M., 1969. Multiplier and gradient methods. J. Optim. Theory Appl. 4 (5), 303–320.

Hilbe, J.M., 2009. Logistic Regression Models. CRC Press, Boca Raton, FL.

Hiriart-Urruty, J.B., Lemaréchal, C., 2001. Fundamentals of Convex Analysis. Grundlehren Text Editions. Springer-Verlag, Berlin.

Kundur, D., Hatzinakos, D., 1996. Blind image deconvolution. IEEE Signal Process. Mag. 13 (3), 43–64.

Kurdyka, K., 1998. On gradients of functions definable in o-minimal structures. Ann. Inst. Fourier (Grenoble) 48 (3), 769–783.

Lasdon, L.S., 2002. Optimization Theory for Large Systems. Dover Publications, Inc., Mineola, NY

Levin, A., Weiss, Y., Durand, F., Freeman, W.T., 2009. Understanding and evaluating blind deconvolution algorithms. In: 2009 IEEE Conference on Computer Vision and Pattern Recognition, publisher, Miami, FL, pp. 1964–1971.

Li, G., Pong, T.K., 2015. Global convergence of splitting methods for nonconvex composite optimization. SIAM J. Optim. 25 (4), 2434–2460.

Lions, P.L., Mercier, B., 1979. Splitting algorithms for the sum of two nonlinear operators. SIAM J. Numer. Anal. 16 (6), 964–979.

Łojasiewicz, S., 1963. Une propriété topologique des sous-ensembles analytiques réels. In: Les Équations aux Dérivées Partielles (Paris, 1962). Éditions du Centre National de la Recherche Scientifique, Paris, pp. 87–89.

Manton, J.H., 2002. Optimization algorithms exploiting unitary constraints. IEEE Trans. Signal Process. 50 (3), 635–650.

Martinet, B., 1970. Régularisation d'inéquations variationnelles par approximation ssuccessives. Rev. Française Informat. Recherche Opérationnelle 4, 154–158.

Mastronardi, N., Lemmerling, P., Kalsi, A., O'Leary, D.P., van Huffel, S., 2004. Implementation of the regularized structured total least squares algorithms for blind image deblurring. Linear Algebra Appl. 391, 203–221.

Milnor, J., 1931. Topology From the Differentiable Viewpoint. Princeton University Press.

Moreau, J.J., 1965. Proximitéet dualité dans un espace hilbertien. Bull. Soc. Math. France 93, 273–299.

Passty, G.B., 1979. Ergodic convergence to a zero of the sum of monotone operators in Hilbert space. J. Math. Anal. Appl. 72 (2), 383–390.

Powell, M.J.D., 1969. A method for nonlinear constraints in minimization problems. In: Fletcher, R. (Ed.), Optimization. Academic Press, New York, pp. 283–298.

Pruessner, A., O'Leary, D.P., 2003. Blind deconvolution using a regularized structured total least norm algorithm. SIAM J. Matrix Anal. Appl. 24 (4), 1018–1037.

Rockafellar, R.T., 1970. Convex Analysis. Princeton Mathematical Series, No. 28. Princeton University Press, Princeton, N.J.

Rockafellar, R.T., 1976. Augmented Lagrangians and applications of the proximal point algorithm in convex programming. Math. Oper. Res. 1 (2), 97–116.

Rockafellar, R.T., 1976. Monotone operators and the proximal point algorithm. SIAM J. Control Optim. 14 (5), 877–898.

Rockafellar, R.T., Wets, R. J. B., 1998. Variational Analysis. Fundamental Principles of Mathematical Sciences, vol. 317. Springer-Verlag, Berlin.

Rudin, L.I., Osher, S.J., Fatemi, E., 1992. Nonlinear total variation based noise removal algorithms. Physica D 60, 259–268.

Shefi, R., Teboulle, M., 2014. Rate of convergence analysis of decomposition methods based on the proximal method of multipliers for convex minimization. SIAM J. Optim. 24 (1), 269–297.

Teboulle, M., 2018. A simplified view of first order methods for optimization. Math. Program. 170 (1), 67–96 (Ser. B).

Tibshirani, R., 1996. Regression shrinkage and selection via the lasso. J. Roy. Statist. Soc. Ser. B 58 (1), 267–288.

Tibshirani, R., Saunders, M., Rosset, S., Zhu, J., Knight, K., 2005. Sparsity and smoothness via the fused lasso. J. R. Stat. Soc. Ser. B Stat. Methodol. 67 (1), 91–108.

Tseng, P., 1991. Applications of a splitting algorithm to decomposition in convex programming and variational inequalities. SIAM J. Control Optim. 29 (1), 119–138.

Wu, C., Tai, X.C., 2010. Augmented Lagrangian method, dual methods, and split Bregman iteration for ROF, vectorial TV, and high order models. SIAM J. Imaging Sci. 3 (3), 300–339.

Yang, J., Zhang, Y., 2011. Alternating direction algorithms for ℓ_1-problems in compressive sensing. SIAM J. Sci. Comput. 33 (1), 250–278.

Chapter 11

Generating structured nonsmooth priors and associated primal-dual methods

Michael Hintermüller[a,b,*] and Kostas Papafitsoros[b]
[a]*Humboldt-Universität zu Berlin, Berlin, Germany*
[b]*Weierstrass Institute for Applied Analysis and Stochastics (WIAS), Berlin, Germany*
[*]*Corresponding author: e-mail: michael.hintermueller@wias-berlin.de*

Chapter Outline

Abstract
The purpose of the present chapter is to bind together and extend some recent developments regarding data-driven nonsmooth regularization techniques in image processing through the means of a bilevel minimization scheme. The scheme, considered in function space, takes advantage of a dualization framework and it is designed to produce spatially varying regularization parameters adapted to the data for well-known regularizers, e.g., total variation and total generalized variation, leading to automated (monolithic), image reconstruction workflows. An inclusion of the theory of bilevel

Handbook of Numerical Analysis, Vol. 20. https://doi.org/10.1016/bs.hna.2019.08.001
© 2019 Elsevier B.V. All rights reserved.

optimization and the theoretical background of the dualization framework, as well as a brief review of the aforementioned regularizers and their parameterization, makes this chapter a self-contained one. Aspects of the numerical implementation of the scheme are discussed and numerical examples are provided.

Keywords: Nonsmooth priors, Image processing, Total variation, Total generalized variation, Bilevel optimization, Regularization parameter selection

AMS Classification Codes: 94A08, 68U10, 49K20, 49M37, 49M15, 26A45

1 Introduction

1.1 Context

Nonsmooth regularization functionals have played a central role in the field of mathematical imaging since the introduction of total variation (TV) to basic image reconstruction problems in the 1990s (Chambolle and Lions, 1997; Rudin et al., 1992). These functionals which, in the function space setting, involve (Radon) norms of distributional derivatives are used as priors in Tikhonov-like variational regularization problems. Their discrete analogues involve versions of ℓ_1 norms thus promoting certain sparsity properties and resulting in noise and artefact reductions as well as in edge-preserving reconstructions. Particularly, the latter property has made this family of regularization functionals a popular tool in the field of image reconstruction and inverse problems.

These inverse problems are typically of the form: Given data

$$f = \text{Noisy}(Tu_{true}), \tag{1}$$

where u_{true} represents some *ground truth* image, that we wish to obtain, find a good approximation u of u_{true}. Here, u_{true} as well as u are modelled as functions from some domain $\Omega \subset \mathbb{R}^d$ into $\mathbb{R}^m$. For static black and white images u we have that $d = 2$, i.e., Ω is a two-dimensional domain, typically open, connected with Lipschitz boundary (e.g., a rectangle), and u is real-valued, i.e., $m = 1$. In this sense $u(x)$ models the intensity of the image at the point $x \in \Omega$. Generalizations to domains of higher dimension (3D, time-dependent problems) as well as multivalued images (multicolour, multimodal) are also possible. But for the sake of ease of presentation of the essentials, we focus here on real-valued images over a 2D domain, only. We denote the function space which u_{true} belongs to by $\mathcal{X}$. As indicated above, instead of u_{true} typically only some data f are available which correspond to a degraded version of u_{true}. This degradation, comes from two sources as (1) depicts. First, the application of an operator $T: \mathcal{X} \to \mathcal{Y}$, the *forward operator* reflects the transformation that u_{true} goes through along with possible further limitations. For instance, in Magnetic Resonance Imaging (MRI) T stands for a subsampling of the coefficients associated with the Fourier transform of the magnetization of the tissue; see for instance (Dong et al., 2019; Knoll et al., 2011; Lustig et al., 2008). In tomography, for example, this transformation is given by the

Radon transform, where typically only a relatively small number of line integrals is available (Natterer, 2001). Apart from transformation and incomplete measurements, a second level of degradation comes from the presence of random noise. It is the result of measurement or transmission errors, e.g., due to the heating up of digital sensors, or other sources of inaccuracies that are too complex to be fully quantified or that involve random effects. This noise can be additive, multiplicative, mixed or occur in an even more complicated way in the measurement process. Additive noise in (1) leads to

$$f = Tu_{true} + \eta, \tag{2}$$

where η represents a highly oscillatory function in some space $\mathcal{Y}$. Typically it is assumed that the mean of η is zero; as otherwise a systematic deterioration is detected and remedied by suitable calibration. In variational image processing, one then aims to recover u_{true} or at least a *sensible* approximation thereof by solving a minimization problem of the type

$$\text{minimize} \;\; \Phi(Tu, f) + J_{\alpha}(u) \;\; \text{over} \;\; u \in \widetilde{\mathcal{X}}, \tag{3}$$

where $\widetilde{\mathcal{X}}$ is a subspace of $\mathcal{X}$ containing functions whose regularity is typically dictated by J_{α}. The term $\Phi(Tu, f)$ is a *data fitting* or *fidelity term* term which measures the distance between the data f and the possible reconstruction u after the action of the forward operator. This ensures that the potential reconstruction will be close to u_{true} in a certain sense. The proper choice for Φ is often the consequence of the statistics of the noise. For instance, for Gaussian noise $\frac{1}{2}\| Tu - f \|_{L^2(\Omega)}^2$ is a suitable choice (Chambolle and Lions, 1997; Rudin et al., 1992), for salt-and-pepper noise $\| Tu - f \|_{L^1(\Omega)}$ is preferable (Benning and Burger, 2011; Chan and Esedoglu, 2005; Duval et al., 2009; Nikolova, 2004), while for Poisson noise the Kullback–Leibler divergence $\int_{\Omega} u - f \log u \, dx$ is more appropriate (Brune et al., 2013; Sawatzky et al., 2009). Mixtures of noise can also be treated and are typically addressed by infimal convolutions of the single-noise fidelity terms; see for instance (Calatroni and Papafitsoros, 2019; Calatroni et al., 2017b) as well as the references therein. From a statistical viewpoint, the choice of the data fidelity is often motivated by maximum likelihood considerations concerning the underlying noise type. For the sake of simplicity, in the examples below we confine ourselves to an L^2 fidelity term.

Minimizing just the fidelity term will not give any sensible results as this corresponds to the direct inversion of T. Due to the presence of noise and potential deficiencies in T this is, however, typically an ill-posed problem, in the sense that small variations in the data f may have an enormous deteriorating effect on u. This adverse behaviour is taken care of by an additional term J_{α}, the *regularizer*, which makes the overall minimization problem well-posed. Typically it adds some extra regularity requirement to the problem. This leads to a robust reconstruction where ideally the noise is filtered out, as well.

The parameter $\boldsymbol{\alpha}$ is the *regularization parameter* and balances the effect of the two terms in (3) thus determining the amount of regularization or filtering in the reconstructed image. It can be one or more (positive) scalar quantities, or spatially varying functions. In the first case, the regularization effect is uniform throughout the image (global effect) while in the second, this effect varies depending on the magnitude of this function in different areas of the domain (local effect). We mention already here that while in simple regularizers where the regularization parameter is included in a multiplicative way the regularization effect scales with the magnitude of $\boldsymbol{\alpha}$. This is, however, not case for more complex regularizers that involve, e.g., infimal convolutions, where two or more scalar parameters determine the regularization effect in a more complex fashion. This makes their proper selection an even more challenging task.

Indeed, even with a proper selection of both, the fidelity and regularization functionals, the problem (3) contains one more degree of freedom, which is connected to a suitable choice of $\boldsymbol{\alpha}$. It aims at producing the best reconstruction result within the capabilities of the selected Φ and J. Establishing automatic regularization parameter selection rules with the help of rich and rigorous mathematical theories that eventually lead to *monolithic* image reconstruction methods approaches lies at the heart of this chapter.

At this point we wish to pin down several of the aforementioned concepts by considering the following classical TV minimization problem

$$\min_{u \in \mathrm{BV}(\Omega)} \frac{1}{2} \| Tu - f \|_{L^2(\Omega)}^2 + \alpha |Du|(\Omega). \tag{4}$$

Without getting into too many details, we note for the time being only that here $\mathcal{Y} = L^2(\Omega)$, $\widetilde{\mathcal{X}} = \mathrm{BV}(\Omega)$ the (nonreflexive Banach) space of *functions of bounded variation* and that $J_{\boldsymbol{\alpha}}(u) = \alpha|Du|(\Omega)$, the TV of u (also denoted by $\mathrm{TV}(u)$), multiplied by α. Here Du denotes the finite Radon measure that represents the distributional derivative of u and $|Du|$ the corresponding TV measure of Du. The regularization parameter here is a simple positive scalar, i.e., $\boldsymbol{\alpha} = \alpha > 0$. For precise definitions and properties of TV and $\mathrm{BV}(\Omega)$ we refer to Section 2.1. The problem (4) and its variants have been thoroughly used in many applications, like denoising (Blomgren and Chan, 1998; Chambolle, 2004; Getreuer, 2012a; Hintermüler and Stadler, 2006; Rudin et al., 1992), deblurring (Getreuer, 2012b; Vese, 2001), inpainting (Chan and Shen, 2005; Getreuer, 2012c), zooming (Malgouyres and Guichard), image decomposition (Chan et al., 2007; Wotao et al., 2007), MRI and PET reconstruction (Benning et al., 2014; Burger et al., 2014; Sawatzky, 2011), image decompression (Alter et al., 2005; Bredies and Holler, 2012c), dejittering (Dong and Scherzer, 2017; Dong et al., 2015) to name a few. The solution structure and its dependence on the parameter α are also well studied and understood (Acar and Vogel, 1994; Allard, 2008a,b, 2009; Caselles et al., 2007; Chambolle et al., 2017; Chan and Esedoglu, 2005; Duval et al., 2009; Meyer, 2001; Nikolova, 2002; Ring, 2000; Valkonen, 2015).

As it is observed in practice, but also proven theoretically in the works above, the strength of regularization due to the TV term is proportional to the parameter α, with large values resulting in cartoon-like images of large constant areas and small values having little regularizing effect, meaning that the solution u is such that Tu converges to the data f in an appropriate sense as $\alpha \to 0$. In Fig. 1, one can see a simple TV denoising example ($T = Id$) where this behaviour is depicted. Naturally, one is interested in those values of α that produce a (close to) the *best* result with respect to some quality measure. For the peak-signal-to-noise ratio (PSNR) and the Structural Similarity Index (SSIM) see the two plots on the right-hand side of Fig. 1.

As far as TV minimization with a scalar regularization parameter is concerned, there exists several works in the literature that focus on its automatic selection, see for instance (Babacan et al., 2008; Deledalle et al., 2014; Langer, 2017; Lin et al., 2010) and the references therein. However, these works focus mainly on a scalar α which poses restrictions on the reconstruction quality as we discuss later.

A series of works relevant to this problem as well as to the present survey was initiated in De los Reyes and Schönlieb (2013), at the discrete setting in Kunisch and Pock (2013), and further explored in Calatroni et al. (2014, 2017a). There, an optimal parameter for the TV minimization is computed by means of variants of the following *bilevel optimization* scheme

$$\begin{cases} \text{minimize} \ \dfrac{1}{2}\| u_\alpha - u_{true} \|_{L^2(\Omega)}^2 \quad \text{over}\ \alpha \geq 0, \\ \text{subject to}\ u_\alpha = \underset{u \in H^1(\Omega)}{\operatorname{argmin}}\ \dfrac{1}{2}\| u - f \|_{L^2(\Omega)}^2 + \alpha \| \nabla u\|_{L^1(\Omega)} + \dfrac{\epsilon}{2}\| \nabla u \|_{L^2(\Omega)}^2 . \end{cases} \tag{5}$$

Here it is assumed that the ground truth u_{true} is known and the scheme aims at finding the scalar α that produces the reconstruction u_α that is closest to u_{true} in the L^2 sense (optimal PSNR). The reconstruction u_α is a solution to a smoothed TV reconstruction problem, where smoothing is necessary for the numerical solution. The rationale is then to use this parameter to restore data g that were produced under the same circumstances as f, as this parameter is expected to be near optimal for g, as well.

One of the drawbacks of TV minimization is the *staircasing effect*, i.e., the promotion of piecewise constant (or blocky) structures (Chambolle et al., 2017; Ring, 2000), stemming from the sparsity of the gradient which, in the discrete setting, is due to the use of the ℓ_1 norm. Although local smoothings of the TV functional, e.g., of Huber type (Burger et al., 2016; Hintermüler and Stadler, 2006) can reduce this effect, a reduction or even total elimination is typically achieved via the incorporation of higher order derivatives in the regularization process. For instance see the contributions Chambolle and Lions (1997), Chan et al. (2001), Hinterberger and Scherzer (2006), Lefkimmiatis et al. (2012), Lysaker et al. (2003), Lysaker and Tai (2006), Papafitsoros and Schönlieb (2014), Scherzer (1998), Setzer and Steidl (2008), Setzer et al. (2011), Tai et al. (2011a), Wu and Tai (2010) and also Papafitsoros (2014) for a more general review. Arguably, one the most

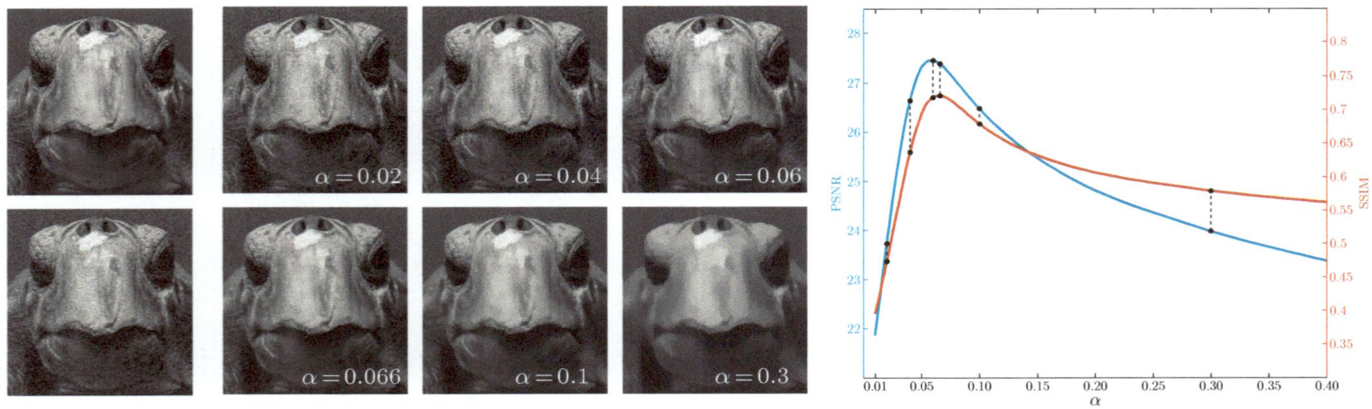

FIG. 1 TV denoising examples for different values of the scalar regularization parameter α. *Left*: Clean and noisy images. *Middle*: TV denoised images, solutions of the problem (4) with $T = Id$. The parameter α is increasing from *left* to *right* and *top* to *bottom*, corresponding to the marked points in the *right figure*. *Right*: Plots that depict the PSNR and SSIM values of the solutions vs the parameter α. The third and fourth marked points correspond to the highest PSNR and SSIM, respectively.

successful regularizers of this type is the total generalized variation (TGV) of second order (Bredies et al., 2010), defined as

$$\mathrm{TGV}^2_{\boldsymbol{\alpha}}(u) = \min_{w \in \mathrm{BD}(\Omega)} \alpha_1 |Du - w|(\Omega) + \alpha_0 |\mathcal{E}w|(\Omega), \tag{6}$$

where BD(Ω) is the space of *functions of bounded deformation* and $\mathcal{E}w$ is the distributional symmetrized gradient of w; see Section 2.2 for definitions. TGV has the ability to adapt to the regularity of the images resulting in piecewise affine reconstructions where edges are still preserved. It has already been successfully used in a variety of applications; see Bredies and Holler (2012a), Bredies and Holler (2013), Bredies and Valkonen (2011), Holler and Kunisch (2014), Knoll et al. (2011), Valkonen et al. (2013), among others. Note that here the set of regularization parameters consists of two positive scalars $\boldsymbol{\alpha} = (\alpha_0, \alpha_1)$ that balance the effect of the first and higher order terms in (6). Here the regularization strength is decided in a more complicated way than in the TV case; see for instance Fig. 2. There exist results regarding the influence of these parameters on the structure of solutions, as well as their asymptotic behaviour (Bredies et al., 2013b; Papafitsoros and Bredies, 2015; Papafitsoros and Valkonen, 2015; Pöschl and Scherzer, 2015). These studies have mostly theoretical value providing little practical guidance on how to select the parameters for (near) optimal results. We note that adaptation of the scheme (5) was done for TGV in De Los Reyes et al. (2016, 2017), producing optimal parameters with the use of pairs of ground truths and noisy images. Thus, up to now the choice of TGV parameters is mostly based on heuristics only.

So far we have mentioned regularization functionals whose parameters are scalar quantities. As we have already briefly alluded to above, this implies that the regularization strength is uniform across the image. This is unwanted when the amount of noise is nonuniform across the image. Moreover, it is also beneficial to impose weak regularization in image regions containing fine details in order for these to be better preserved and to use strong regularization in homogeneous, i.e., smooth, image areas. This can be achieved by introducing a spatially varying regularization weight, either acting on the fidelity term or the regularizer. For TV minimization, this results in the following two problems:

$$\min_{u \in \mathrm{BV}(\Omega)} \frac{1}{2} \| \sqrt{\lambda}(Tu - f) \|^2_{L^2(\Omega)} + |Du|(\Omega), \quad \lambda \in L^\infty(\Omega), \lambda \geq 0 \tag{7}$$

$$\min_{u \in \mathrm{BV}(\Omega)} \frac{1}{2} \| Tu - f \|^2_{L^2(\Omega)} + \int_\Omega \alpha \, d|Du|, \quad \alpha \in C(\bar{\Omega}), \alpha > 0. \tag{8}$$

Here the term $\int_\Omega \alpha \, d|Du|$ denotes the integral of the continuous function α with respect to the measure $|Du|$. The first instance (7) and its variants have been studied and applied to image restoration problems in several works.

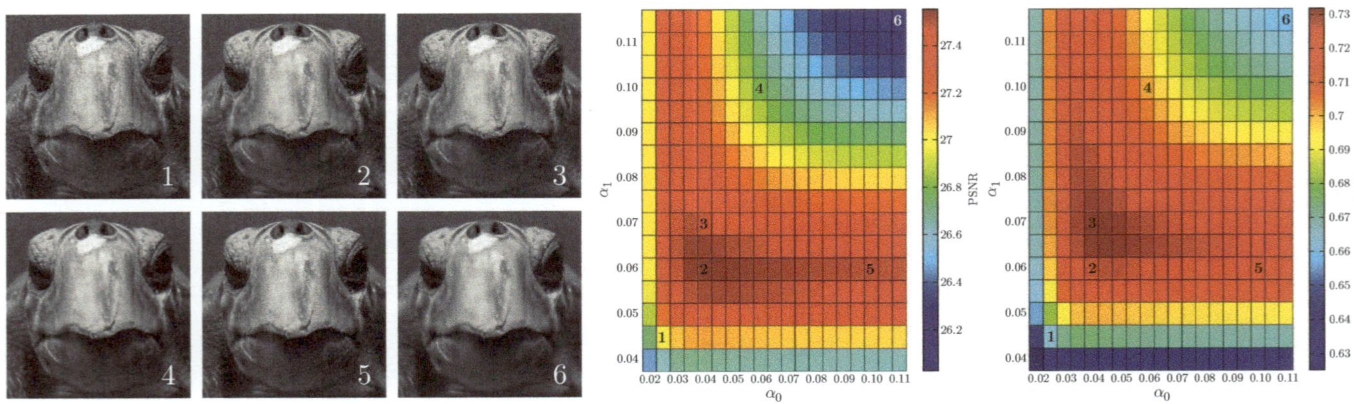

FIG. 2 *Left*: Solutions of the L^2–TGV denoising problem (using the same data as in Fig. 1) with combination of the regularization parameters (α_0, α_1) that correspond to the numbers 1–6 on the right plots. The images 2 and 3 correspond to the highest PSNR and SSIM, respectively. *Right*: PSNR and SSIM values of the denoised images with respect to the parameters α_0 and α_1.

In Dong et al. (2010a,b) grey scale and color images subject to blur and Gaussian noise are restored using that model, while in Hintermüller and Rincon-Camacho (2010) an L^1 fidelity is considered for the case of random-valued impulse noise. In finite dimensions, a technique based on a statistical multiresolution criterion can be found in Frick et al. (2012) and Hotz et al. (2012). In Almansa et al. (2008) a statistical approach with variance estimators different from the ones in Dong et al. (2010a,b) is examined. A deterministic choice rule using a presegmentation of the image and a piecewise constant fidelity weight is considered in Bertalmio et al. (2003). As this is related to the present chapter, we briefly mention the basic idea of Dong et al. (2010a). Given a normalized weight function $w \in L^\infty(\Omega \times \Omega)$ with $\int_\Omega \int_\Omega w(x, y)\, dxdy = 1$, one defines at every point a localized residual as follows

$$\mathrm{R}u(x) := \int_\Omega w(x, y)(Tu - f)^2(y)\, dy. \tag{9}$$

In Dong et al. (2010a), assuming Gaussian noise of variance σ^2, the parameter λ is updated iteratively in such a way that the localized residual at every point does not significantly exceed σ^2. The method has given satisfactory results for image denoising and deblurring. However, one disadvantage of transforming the regularization parameter into a weight function contained in the fidelity term is related to the fact that it is not immediate clear how to extend the model to data domains, like Fourier or wavelet domains. While analytically and numerically more complex, this observation supports the choice of a second model (8) which seems more appropriate in view of such data domains. While these models are equivalent for scalar regularization parameters, for spatially varying ones specific examples can be found yielding significantly different solution structures for (7) and (8), respectively; see Hintermüller et al. (2017a).

The possibility to explore the space spanned by Φ and $J_{\boldsymbol{\alpha}}$ by optimally selecting $\boldsymbol{\alpha}$, immediately leads to the question of whether one can devise a scheme such that the choice of the (best) regularization function α can be decided automatically. We note that as far as problem (8) is concerned, in order to be consistent with the functional analytic framework, the function α should be measurable with respect to the measure $|Du|$ where u is the solution. This is guaranteed when $\alpha \in C(\bar{\Omega})$ as it was ensured in Hintermüller et al. (2017) but see also Proposition 4. We note that an adaptation of the framework (5) to a spatially dependent parameter α can be found in Calatroni et al. (2017a) and Chung et al. (2017). However, this approach is mainly suitable for calibrating TV-based denoising methods for noise of spatially varying intensity and one drawback is the overfitting of the regularization parameter to the training image.

A *monolithic* approach for the automatic selection of the spatially varying function α for problems of the type (8) was extensively examined in two recent papers (Hintermüller and Rautenberg, 2017; Hintermüller et al., 2017). There, with the aim of combining the reconstruction and α-selection processes in a single mathematical framework, a bilevel optimization approach is adopted with the difference (to the aforementioned approaches) that no ground truth or any training pair is used but rather only the data f, hence the term monolithic. In the spirit of Dong et al. (2010a), one looks for solutions of the problem (8) that force the localized residuals to remain within a variance corridor, close to the variance σ^2 of the noise. This restriction of the localized residuals close to an interval $[\underline{\sigma}^2, \overline{\sigma}^2]$ is imposed by keeping the value of the following upper level objective functional small

$$F(\mathrm{R}u) = \frac{1}{2}\int_\Omega \max(\mathrm{R}u - \overline{\sigma}^2, 0)^2 dx + \frac{1}{2}\int_\Omega \min(\mathrm{R}u - \underline{\sigma}^2, 0)^2 dx. \tag{10}$$

In order to get a first understanding, the model considered in Hintermüller and Rautenberg (2017) and Hintermüller et al. (2017) aims to approximate

$$\begin{cases} \text{minimize} \quad F(\mathrm{R}u) \quad \text{over}\, u \in \mathrm{BV}(\Omega), \alpha \in C(\bar{\Omega}) \\ \text{subject to} \quad u = \underset{u \in \mathrm{BV}(\Omega)}{\operatorname{argmin}} \ \frac{1}{2} \| Tu - f \|_{L^2(\Omega)}^2 + \int_\Omega \alpha \, d|Du|. \end{cases} \tag{11}$$

Even though in principle the bilevel scheme (11) aims to achieve what was discussed above, there exist serious obstacles regarding its functional analytic treatment, well-posedness, and its suitability for numerical realization. No extra regularity is imposed on α, e.g., requiring it to be continuous or at least measurable, and lack of any source of compactness jeopardizes the well-posedness of the scheme. A possible remedy of this is to add an extra H^1 regularization term to F together with a boundedness condition requiring $\alpha \in \mathcal{A}_{ad}$ where

$$\mathcal{A}_{ad} = \{\alpha \in H^1(\Omega) : \underline{\alpha} \le \alpha \le \overline{\alpha}, \text{ a.e. in } \Omega\}, \quad 0 < \underline{\alpha} < \overline{\alpha}.$$

We further note that the first-order optimality condition of the lower level problem in (11) gives rise to a variational inequality (VI) of the second kind. It constitutes a degenerate (nonqualified) constraint (in the sense of the Karush–Kuhn–Tucker theory in Banach spaces) and imposes major challenges in analyzing the dependence $\boldsymbol{\alpha} \to u$. In order to replace the VI of second kind by a more tractable VI of the first kind (but still yielding a degenerate constraint) dualization theory may be employed. Indeed, instead of the lower level problem in (11) (primal problem) one considers its Fenchel *predual* (Ekeland and Temam, 1976; Kunisch and Hintermüller, 2004) which reads

$$\begin{aligned} &\text{minimize} \quad \frac{1}{2}\| \operatorname{div} p + T^* f \|_B^2 \quad \text{over} \quad p \in H_0(\operatorname{div}; \Omega), \\ \text{subject to} \quad &p \in K(\alpha) := \{q \in H_0(\operatorname{div}; \Omega) \cap L^\infty(\Omega, \mathbb{R}^d) : |q(x)| \le \alpha(x) \text{ for a.e. } x \in \Omega\}, \end{aligned} \tag{12}$$

where $\| u \|_B^2 = \langle w, B^{-1}u \rangle$, with $B = T^*T$ and T^* the adjoint of T, assuming invertibility; see more details in Section 2.3. Minimizers of the primal and predual problems are connected via the following optimality condition

$$Bu = \operatorname{div} p + T^* f. \tag{13}$$

We note that this dualization result was shown in Kunisch and Hintermüller (2004) and in Hintermüller and Rautenberg (2017) for α being a scalar and a continuous function, respectively. But as we show in Section 2.3 it also holds for $\alpha \in H^1(\Omega)$. Moreover, the predual problem (12) is more amenable to efficient, image resolution independent, function space-based solution algorithms, such as (inexact) *semismooth Newton* methods (Kunisch and Hintermüller, 2004), that can solve the problem up to a very high accuracy fairly quickly. Note that in view of (13) the localized residual (10) can be written in terms of the dual variable p as follows

$$\mathrm{R}u(x) = R(\operatorname{div} p)(x) = \int_\Omega w(x,y)\left(TB^{-1}\operatorname{div} p - (TB^{-1}T^* - I)f\right)^2 dy. \tag{14}$$

This leads to the following bilevel minimization problem which was the starting point in Hintermüller and Rautenberg (2017):

$$\begin{cases} \text{minimize } \; J(p,\alpha) := F(R(\operatorname{div} p)) + \dfrac{\lambda}{2}\| \alpha \|_{H^1(\Omega)}^2 \quad \text{over}\, p \in H_0(\operatorname{div};\Omega),\ \alpha \in H^1(\Omega), \\ \text{subject to } \; p \in \operatorname{argmin}\left\{ \dfrac{1}{2} \| \operatorname{div} p + T^* f \|_B^2 \colon\ p \in K(\alpha) \right\}. \end{cases} \tag{15}$$

Existence of solutions of the bilevel problem (15) and also the establishment of the dualization framework itself, which links the solution sets of the two problems via (13), are closely related to density results of associated convex intersections. In general, the latter establish equivalences of the type

$$\overline{C \cap Y}^X = C, \quad \text{for} \quad \overline{Y} = X, \tag{16}$$

where X is a Banach space and C a convex subset, typically represented by pointwise constraints, and Y a dense subset of X. As far as the bilevel problem (15) is concerned the density relation

$$\begin{aligned} &\overline{\{\phi \in C_c^\infty(\Omega, \mathbb{R}^d) \colon\ |\phi(\cdot)| \le \alpha(\cdot), \text{in}\, \Omega\}}^{H_0(\operatorname{div};\Omega)} \\ &\quad = \{p \in H_0(\operatorname{div};\Omega) \colon\ |p(\cdot)| \le \alpha(\cdot)\ \text{a.e in}\, \Omega\} \end{aligned}$$

is of particular importance. Here, we have $C = K(\alpha)$, $X = H_0(\operatorname{div};\, \Omega)$ and $Y = C_c^\infty(\Omega, \mathbb{R}^d)$. Establishing such density results can be traced back to Hintermüller and Rautenberg (2015) and to further extensions in Hintermüller et al. (2017b); see the relevant discussion in Section 2.3.

We also note that dualization can be used in order to introduce a rigorous functional analytic framework for a wide class of regularizers that are used in multimodal reconstruction problems. These typically correspond to integrands that are pointwise functions of the gradient, i.e.,

$$J(u) = \int_{\Omega} j_v(x, \nabla u(x))\, dx, \tag{17}$$

exhibit linear growth, and incorporate some a priori knowledge v. In applications, v may correspond to some prereconstructed modality. Functionals of the type (17) have been used in the discrete setting only and dualization was used in Hintermüller et al. (2018) in order to introduce a functional analytic meaning.

Further challenges exist when devising numerical solution algorithms for the bilevel problem (15). It is known that this problem falls into the class of *mathematical programmes with equilibrium constraints* (MPEC) and suffers from constraint degeneracy. For the derivation of stationarity conditions, it requires advanced nonsmooth analysis techniques other than the classical Karush–Kuhn–Tucker (KKT) theory; see Luo et al. (1996), Outrata et al. (1998), Barbu (1984), Hintermüller and Kopacka (2009) and Hintermüller et al. (2014) for the finite and the infinite dimension problem setting. An extensive discussion is given in Section 4.1.

The bilevel scheme (15) has produced promising results in applications like denoising, deblurring, wavelet inpainting and MRI reconstruction Hintermüller et al. (2017), see an illustration in Fig. 3. However, the staircasing effect, a characteristic of TV, appears to remain in a reduced form. One may extend the above framework to the weighted TGV case. For this purpose, similar dualization frameworks have to be established as well as efficient high accuracy algorithms for the solution of the corresponding dual problem need to be developed. We note that an extension of Dong et al. (2010a) to the TGV case, that is, a study of the problem

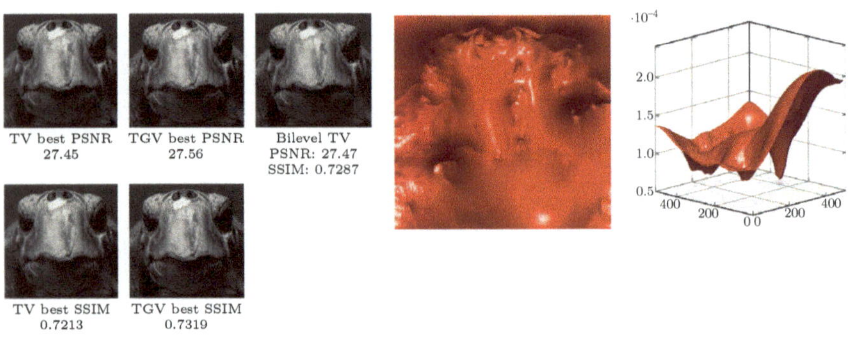

FIG. 3 *Left*: Comparison of the Bilevel weighted TV result with the best scalar TV and TGV denoising results of Figs. 1 and 2. *Right*: the weighted regularization function α which is determined automatically.

$$\min_{u\in \mathrm{BV}(\Omega)} \frac{1}{2} \| \sqrt{\lambda}(Tu-f) \|_{L^2(\Omega)}^2 + \mathrm{TGV}_{\alpha}^2(u) \tag{18}$$

was done in Bredies et al. (2013a) with good results. But the disadvantages from transforming the regularization parameter into a fidelity weight remain, as it was discussed before.

1.2 Main contributions and organization of this chapter

While in its majority, this is a review paper, it also contains some novel elements. In view of this we state clearly in the text which results are new and which are based on earlier works.

In Sections 2.1 and 2.2 we provide an overview of analytical properties of the TV and TGV functionals. We emphasize the roles of the regularization parameters in the reconstruction, like asymptotic, exact solutions as well as structural properties of the solutions of the corresponding variational problems.

In Section 2.3, after briefly reviewing the Fenchel–Rockafellar duality theory, we state the predual problems of TV-based minimizations as well as their weighted versions, and the corresponding dualization frameworks are established. We particularly emphasize the important role that density results of convex intersections play in the establishment of these frameworks.

In Section 3 we discuss algorithms for the solution of the TV variational problems focusing on function space semismooth Newton algorithms for regularized versions of predual problems.

A basic background on the theory of bilevel optimization with emphasis on different notions of stationarity is discussed in Section 4.1. An overview of the bilevel framework for the automatic determination of the regularization parameters, based on the incorporation of localized residuals on the upper objective, is given in Section 4.2. This is done for the TV case, whereas extensions to the TGV regularizer are the current focus of research. We discuss first-order optimality conditions as well as the existence of adjoint states that eventually lead to the derivation of a projected gradient scheme for the algorithmic treatment of the bilevel minimization problem.

Section 5 is dedicated to numerical examples. In particular, in Section 5.1, discretization aspects for the TV bilevel problem are discussed and in Section 5.2 numerical examples are depicted.

2 Nonsmooth priors

2.1 Total Variation

We start with some basic definitions and facts regarding the space of functions of bounded variation $\mathrm{BV}(\Omega)$. For an extensive study, we refer the reader to Ambrosio et al. (2000) but also to Attouch et al. (2014). Here $\Omega \subset \mathbb{R}^d$

denotes an open, bounded, connected, domain with Lipschitz boundary. The space $\mathrm{BV}(\Omega)$ consists of all functions in $L^1(\Omega)$ whose distributional derivative can be represented by a finite Radon measure which is denoted by $Du = (D_1 u, \ldots, D_d u)$. It can be shown that an L^1 function belongs to $\mathrm{BV}(\Omega)$ if and only if $\mathrm{TV}(u) < \infty$, i.e., it has a finite total variation, where

$$\mathrm{TV}(u) := \sup\left\{ \int_\Omega u \operatorname{div}\phi \, dx : \phi \in C_c^\infty(\Omega, \mathbb{R}^d), \ \|\phi\|_{2,\infty} \leq 1 \right\}, \tag{19}$$

It can be shown that in this case $\mathrm{TV}(u)$ is equal to the TV measure of Du evaluated in Ω, i.e., $\mathrm{TV}(u) = |Du|(\Omega)$. In the special case $u \in W^{1,1}(\Omega)$ we have $\mathrm{TV}(u) = \int_\Omega |\nabla u| \, dx$. The TV functional is convex, lower semicontinuous with respect to the strong L^1 convergence and it is a seminorm on the space $\mathrm{BV}(\Omega)$. Moreover $\mathrm{BV}(\Omega)$ is a Banach space when endowed with the norm $\| u\|_{\mathrm{BV}(\Omega)} = \| u\|_{L^1(\Omega)} + |Du|(\Omega)$.

Before we continue we would like to mention a few aspects on the isotropic and the anisotropic version of TV. In the definition (19) the finite dimensional $\|\cdot\|_{2,\infty}$ norm is defined as

$$\|\phi\|_{2,\infty} = \operatorname*{ess\,sup}_{x\in\Omega} |\phi(x)|,$$

where $|\cdot|$ is the usual Euclidean norm in $\mathbb{R}^d$. This leads to the isotropic version of TV. If in (19), the $\|\cdot\|_{\infty,\infty}$ norm is used instead, where

$$\|\phi\|_{\infty,\infty} = \operatorname*{ess\,sup}_{x\in\Omega} \max\{|\phi_i(x)| : i = 1,\ldots,d\},$$

then one is led to an anisotropic version of TV. In that case it holds that $|Du|(\Omega) = \sum_{i=1}^d |D_i u|(\Omega)$. The resulting norms are equivalent, and while the isotropic version enjoys rotational invariance when used in imaging problems, the anisotropic one is more amenable for our purposes when it comes to the algorithmic treatment of the dual problem in the bilevel framework. For this part of the paper we focus on the isotropic version, but make it explicit when there is an important difference to the anisotropic version.

The measure Du can be decomposed into its absolutely continuous and singular parts with respect to the Lebegue measure $\mathcal{L}^d$, i.e.,

$$Du = \nabla u \mathcal{L}^d + D^s u,$$

with ∇u denoting the Radon-Nikodým density of Du with respect to $\mathcal{L}^d$. Without getting into details, we mention that $D^s u$ is further decomposed into the jump part $D^j u$ and the Cantor part of the derivative $D^c u$. The jump part is concentrated on the jump set J_u of u which is the set of points where u exhibits (approximate) jump discontinuities. This is the set of points $x \in \Omega$ for which $u^+(x) > u^-(x)$ where

$$u^+(x) = \inf\left\{t \in [-\infty, \infty] : \lim_{r\to 0} \frac{\mathcal{L}^d(\{u > t\} \cap B(x, r))}{r^d} = 0\right\},$$

$$u^-(x) = \sup\left\{t \in [-\infty, \infty] : \lim_{r\to 0} \frac{\mathcal{L}^d(\{u < t\} \cap B(x, r))}{r^d} = 0\right\},$$

are the approximate upper and lower limits of u, respectively. The TV of $D^j u$ is then written as

$$|D^j u|(\Omega) = \int_{J_u} |u^+ - u^-(x)| \, d\mathcal{H}^{d-1},$$

with $\mathcal{H}^{d-1}$ denoting here the $d-1$-dimensional Hausdorff measure. The fact that BV functions are allowed to have jump discontinuities, in contrast to Sobolev objects, renders them particularly suitable for modelling image intensities as these typically exhibit discontinuities amounting to edges in images.

Besides the strong convergence in BV (norm convergence), two other types of convergence are useful. We say that $(u_n)_{n\in\mathbb{N}}$ converges weakly* in BV to u if it converges strongly in $L^1(\Omega)$ and the measures $(Du_n)_{n\in\mathbb{N}}$ converge weakly* to Du. This type of convergence is useful since it can be shown that any bounded sequence in BV has a weakly* convergent subsequence (compactness in BV). On the other hand we say that $(u_n)_{n\in\mathbb{N}}$ converges strictly to u in BV, if it converges strongly in $L^1(\Omega)$ and also $|Du_n|(\Omega) \to |Du|(\Omega)$. It can be shown that every $u \in \mathrm{BV}(\Omega)$ can be strictly approximated by $C^\infty(\bar{\Omega})$ functions. Regarding embeddings we have that $\mathrm{BV}(\Omega)$ is continuously embedded into $L^{d/d-1}(\Omega)$ if $d \geq 2$ and $L^\infty(\Omega)$ if $d = 1$.

Returning now to the corresponding TV regularization problem, which in more generality as before reads:

$$\text{minimize}\ \frac{1}{p} \| Tu - f \|_{\mathcal{Y}}^p + \alpha |Du|(\Omega) \quad \text{over}\ u \in \mathrm{BV}(\Omega). \tag{20}$$

Here $\mathcal{Y}$ is a normed space, $T : L^{d/d-1}(\Omega) \to \mathcal{Y}$ bounded linear, $\alpha > 0$ and $p \geq 1$. Existence and (potential) uniqueness of (20) have been shown in different works in the literature. Let us note that in order to show existence, in many of these works, a condition of the type $T(\mathcal{X}_\Omega) \neq 0$ is used, where $\mathcal{X}_A$ denotes the characteristic function of a set $A \subset \Omega$ of positive d-dimensional measure. This condition reflects that T does not annihilate constants. However, this property is not necessary for an existence proof, one can show that by using similar techniques as in Bredies and Holler (2014) where the result is shown for TGV instead of TV. We summarize in the following.

Theorem 1. *Let $\mathcal{Y}$ be a normed space, $T : L^{d/d-1}(\Omega) \to \mathcal{Y}$ bounded linear, $\alpha > 0$ and $p \geq 1$. Then the minimization problem (20) has at least one solution. If $p > 1$ and T is injective then the solution is unique.*

We are particularly interested in the parameterization of the problem (20), the dependence and structure of solutions with respect to such parameters. Basic asymptotic results are relatively easy to obtain and are summarized in the following:

Proposition 1. *The following asymptotic results hold:*

1. *Let* $(u_n)_{n\in\mathbb{N}} \subset \mathrm{BV}(\Omega)$ *be a sequence of solutions of the problem (20) for an increasing sequence of parameters* $(\alpha_n)_{n\in\mathbb{N}}$ *such that* $\alpha_n \to \infty$. *Then there exists a subsequence of* $(u_{n_k})_{k\in\mathbb{N}}$ *and a constant* c *such that* $u_{n_k} \to c$ *weakly* in* $\mathrm{BV}(\Omega)$ *and it also holds*
$$c \in \underset{u \text{ constant}}{\operatorname{argmin}} \ \| Tu - f\|_{\mathcal{Y}}.$$
2. *Let* $(\alpha_n)_{n\in\mathbb{N}}$ *be a decreasing parameter sequence such that* $\alpha_n \to 0$. *Then for every sequence* $(u_n)_{n\in\mathbb{N}} \subset \mathrm{BV}(\Omega)$ *of the corresponding solutions of the problem (20) there holds*
$$\| Tu_n - f\|_{\mathcal{Y}} \to 0.$$

If in addition there exists u^* *such that* $Tu^* = f$, *then there exists a sequence* $(u_n)_{n\in\mathbb{N}}$ *of solutions of the problem (20), that has a weakly* in* $\mathrm{BV}(\Omega)$ *convergent subsequence to* u^*.

In particular if T maps the set of constant functions to itself then (i) of Proposition 1 says that as the regularization parameter grows then the solutions of (20) converge up to a subsequence to a constant function which is closest to the data f, with respect to $\|\cdot\|_{\mathcal{Y}}$. If $\mathcal{Y} = L^2(\Omega)$ or $L^1(\Omega)$ for instance, then this constant will be the mean or median value of f, respectively.

The fine structure of the solutions for the TV minimization problem has also been well studied. The solutions are characterized by piecewise constant structures, the *staircasing effect*. For instance it can be shown (Bredies et al., 2013b; Ring, 2000) that in the one-dimensional denoising case, i.e., $T = Id$ and $\mathcal{Y} = L^1(\Omega), L^2(\Omega)$, the solution u will be constant in the areas where it is not equal to f (in a certain precise representative sense). In the denoising case and in higher dimensions, in Chambolle et al. (2017) the authors provide a characterization of the area that will exhibit staircasing (extended support). In Jalalzai (2015), it is shown that flat areas always occur at global extrema of the data and the extrema of the solution.

Another characteristic of the (isotropic) TV minimization is that no new discontinuities are created in the solution, up to $\mathcal{H}^{d-1}$ measure. This has only been shown so far for the denoising case.

Theorem 2. *Let* $f \in L^\infty(\Omega) \cap \mathrm{BV}(\Omega)$ *and* $\alpha > 0$. *If* u *is the solution of the problem*
$$\min_{u\in\mathrm{BV}(\Omega)} \frac{1}{2}\| u - f \|^2_{L^2(\Omega)} + \alpha|Du|(\Omega),$$
then $\mathcal{H}^{d-1}(J_u \backslash J_f) = 0$.

This result was first proven in Caselles et al. (2007) and was extended in Valkonen (2015), for a wider class of first-order regularizers including Huber-TV. We note that this result does not hold for the anisotropic TV; see Caselles et al. (2007, Remark 4) nor for the L^1 fidelity. Furthermore, there is a knowledge gap concerning the jump set inclusion when the operator T is not the identity.

2.2 Total generalized variation

TGV is a higher order extension of TV, which is highly related to functions of bounded deformation as pioneered by Suquet (1978) and then further developed by Ciarlet (1988), Temam (1985) and Temam and Strang (1980). TGV was introduced in Bredies et al. (2010) primarily to reduce the often unwanted staircasing effect. For a vector of two scalar parameters $\boldsymbol{\alpha} = (\alpha_0, \alpha_1)$, the TGV (of second order) of a function $u \in L^1(\Omega)$ is defined as follows

$$\mathrm{TGV}_{\boldsymbol{\alpha}}^2(u) = \sup\left\{ \int_\Omega u \,\mathrm{div}^2\phi\, dx : \phi \in C_c^\infty(\Omega, \mathcal{S}^{d\times d}),\ \|\phi\|_{2,\infty} \le \alpha_0,\ \|\mathrm{div}\,\phi\|_{2,\infty} \le \alpha_1 \right\}. \tag{21}$$

Here, $\mathcal{S}^{d\times d}$ denotes the space of symmetric $d \times d$ matrices. For a function $\phi \in C_c^\infty(\Omega, \mathcal{S}^{d\times d})$ the first- and second-order divergences are defined as

$$(\mathrm{div}\phi)_i = \sum_{j=1}^d \frac{\partial \phi_{ij}}{\partial x_j},\ i = 1,\ldots,d, \ \text{ and } \ \mathrm{div}^2\phi = \sum_{i=1}^d \frac{\partial^2 \phi_{ii}}{\partial x_i^2} + 2\sum_{i<j} \frac{\partial^2 \phi_{ij}}{\partial x_i \partial x_j}$$

Similarly with TV, the anisotropic version of TGV can be defined by substituting the finite dimensional norms $\|\cdot\|_{2,\infty}$ by the corresponding norm $\|\cdot\|_{\infty,\infty}$.

In Bredies and Valkonen (2011) it was shown that a function $u \in L^1(\Omega)$ has finite TGV value if and only if it belongs to BV(Ω), with the norm $\|\cdot\|_{\mathrm{BGV}} := \|\cdot\|_{L^1(\Omega)} + \mathrm{TGV}_{\boldsymbol{\alpha}}^2$ being equivalent to $\|\cdot\|_{\mathrm{BV}(\Omega)}$. Similarly to TV, TGV is a convex functional, lower semicontinuous with respect to the strong L^1 convergence. In Bredies and Holler (2014) and Bredies and Valkonen (2011) it was shown that TGV can be equivalently written as

$$\mathrm{TGV}_{\boldsymbol{\alpha}}^2(u) = \min_{w \in \mathrm{BD}(\Omega)} \alpha_1 |Du - w|(\Omega) + \alpha_0 |\mathcal{E}w|(\Omega). \tag{22}$$

Here $\mathcal{E}$ stands for the distributional symmetrized gradient and BD(Ω) is the space of functions of bounded deformation, i.e., the space of all functions $w \in L^1(\Omega, \mathbb{R}^d)$ such that $\mathcal{E}w$ is an $\mathcal{S}^{d\times d}$-valued finite Radon measure (Suquet, 1978). The formulation (22) of TGV, sometimes called the *minimum* or *primal* representation, or *differentiation cascade*, gives more insights into its regularizing mechanisms. Indeed, in areas where the data is close to an

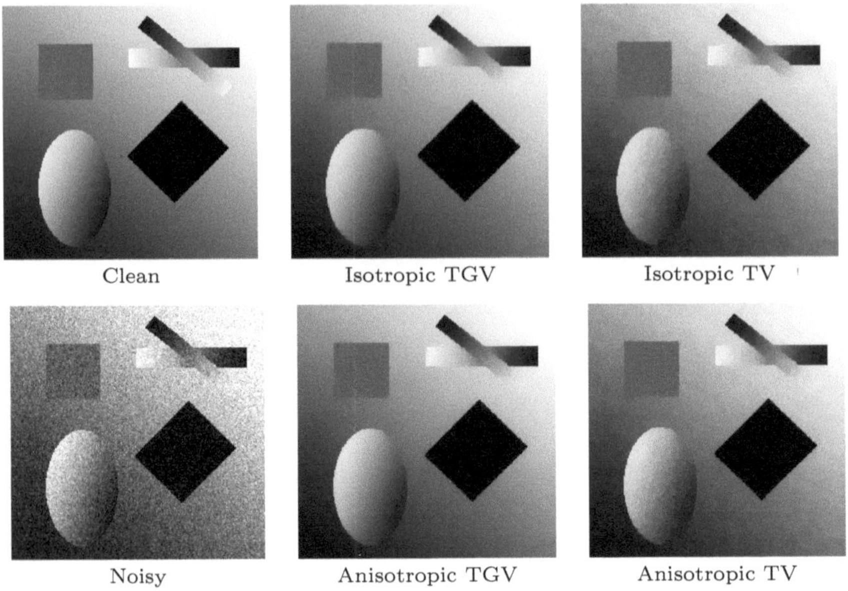

FIG. 4 Comparison between scalar TV and TGV isotropic and anisotropic denoising versions. The (scalar) regularization parameters have been manually adjusted for optimal SSIM.

affine function, locally the variable w can be equal to ∇u, resulting in a low value of the first term in (22). In this case, also the second term has also low value, since is it roughly equal to $\mathcal{E}(\nabla u) = D^2 u$ (hence the second-order derivative information). On the other hand, setting (locally) $w = 0$ in (22) gives back $\alpha_1 |Du|(\Omega)$. Thus, TGV has the possibility to behave locally both like first- and second-order TV. This means that TGV is more suitable to be used in reconstruction of piecewise smooth images. Indeed we depict a simple denoising example in Fig. 4.

The asymptotics of TGV with respect to parameters have been studied in Papafitsoros and Valkonen (2015) with some results also in Bredies and Holler (2014) and Valkonen (2017). In summary, results analogous to Proposition 1 hold for TGV as well, with the difference that only convergence of either α_0 or α_1 to zero is enough to give the results of Proposition 1 (*ii*) and the constant functions are substituted by affine ones in (*i*). The latter family of functions constitutes the kernel of the TGV functional.

Of particular interest is whether there exist combinations of parameters such that $\mathrm{TGV}_\alpha^2(u) = \alpha_1 \mathrm{TV}(u)$. This can indeed happen for certain special functions u, and in general one can show that there exist parameters such that TGV is "almost" TV. We summarize this in the following (compare also Papafitsoros and Bredies, 2015; Papafitsoros and Valkonen, 2015).

Theorem 3. *There exists a constant $C \geq 0$ depending only on the domain Ω such that if with $\boldsymbol{\alpha} = (\alpha_0, \alpha_1)$ satisfies $\alpha_0/\alpha_1 > C$, then*

$$\mathrm{TGV}_{\boldsymbol{\alpha}}^2(u) = \alpha_1 |Du - m_{\mathcal{E}}(\nabla u)|(\Omega), \quad \text{for all } u \in \mathrm{BV}(\Omega), \tag{23}$$

where for a function $g \in L^1(\Omega, \mathbb{R}^d)$ we define

$$m_{\mathcal{E}}(g) := \underset{w \in \mathrm{Ker}\mathcal{E}}{\mathrm{argmin}} \ \| g - w \|_{L^1(\Omega, \mathbb{R}^d)}. \tag{24}$$

Note here that the kernel of the symmetrized gradient Ker$\mathcal{E}$ consists exactly of all the functions of the form $r(x) = Ax + b$ where $b \in \mathbb{R}^d$ and $A \in \mathbb{R}^{d \times d}$ is a skew symmetric matrix. Theorem 3 thus states that for a large ratio α_0/α_1 TGV is (almost) equivalent to TV up to an affine correction. In practice this means that for such a combination of parameters the reconstructed images still suffer from a kind of (affine) staircasing effect; see Papafitsoros and Valkonen (2015).

As far as the structure of solutions is concerned, TGV regularized images have a piecewise affine structure, as expected. This has been studied analytically mainly in dimension one in Benning et al. (2013), Bredies et al. (2013b), Papafitsoros and Bredies (2015) and Pöschl and Scherzer (2015). Regarding to whether a result analogous to Theorem 2 holds for TGV as well is in general unknown. Under some additional regularity assumptions a corresponding result can be shown in Valkonen (2017).

In the remainder of this exhibition we primarily focus on TV as our model concept. We mention, however, that the analogous questions, e.g., concerning dualization (next section), bilevel optimization for selecting α_0, α_1 and efficient numerical realization are important for TGV, as well. Some of these problems are in the focus of currently ongoing research in mathematical image processing.

2.3 Dualization

In many circumstances, instead of considering problems of the type (20) or its TGV-version it is convenient to rather study their respective (pre)-dual problem. Dualization is here understood in the sense of Fenchel–Rockafellar. In particular, for structured classes of nonsmooth variational problems, duality theory provides a convenient framework for characterizing solutions of variational problems through their associated dual quantities. We also point out that the notion of *predualization* arises here due to the nonreflexive nature of BV(Ω) and BD(Ω), respectively. Dualizing the predual of (20) yields (20); and analogously for TGV. As we further develop this chapter we will address further advantages of this approach, both on the functional analytic as well as on the algorithmic/numerical level. We start by briefly reviewing the main duality theorem; for details see Ekeland and Temam (1976).

Consider a general variational problem of the form

$$\inf_{u\in U} F_1(u) + F_2(\Lambda u), \tag{25}$$

where U, V are Banach spaces, $\Lambda : U \to V$ is a bounded linear operator, and $F_1 : U \to \overline{\mathbb{R}}$, $F_2 : V \to \overline{\mathbb{R}}$ are proper, convex, lower semicontinuous functions. The problem (25) is referred to as the *primal* problem. The *dual* problem of (25) is given by

$$\sup_{v^*\in V^*} -F_1^*(\Lambda^* v^*) - F_2^*(-v^*). \tag{26}$$

Here, $\Lambda^* : V^* \to U^*$ denotes the adjoint operator of Λ, while $F_1^* : U^* \to \overline{\mathbb{R}}$ denotes the convex conjugate of F_1, i.e.,

$$F_1^*(u^*) = \sup_{u\in U} \langle u^*, u\rangle_{U^*,U} - F_1(u),$$

for $u^* \in U^*$, and similarly for F_2^*. We denote by $\inf \mathcal{P}_{primal}$ and by $\sup \mathcal{P}_{dual}$ the infimum of (25) and the supremum of (26), respectively. The following proposition states that provided these quantities are equal, the solutions of the two variational problems are connected by an Euler-Lagrange system. From now on ∂F denotes the subdifferential of the convex function F.

Theorem 4. *Suppose that both the primal and the dual problems* (25) *and* (26) *have solutions and that*

$$-\infty < \sup \mathcal{P}_{dual} = \inf \mathcal{P}_{primal} < +\infty \quad (\textit{zero duality gap}). \tag{27}$$

Then all solutions u, v^* of the primal and the dual problem, respectively, are related through the following first-order optimality conditions:

$$\Lambda^* v^* \in \partial F_1(u), \tag{28}$$

$$-v^* \in \partial F_2(\Lambda u), \tag{29}$$

which are also equivalent to

$$u \in \partial F_1^*(\Lambda^* v^*), \tag{30}$$

$$\Lambda u \in \partial F_2^*(-v^*). \tag{31}$$

Conversely, if $u \in U$ and $v^* \in V^*$ satisfy (28) *and* (29), *then they are solutions of* (25) *and* (26), *respectively, and* $-\infty < \sup \mathcal{P}_{dual} = \inf \mathcal{P}_{primal} < +\infty$.

There are several conditions that guarantee zero duality gap, eventually leading to the characterization (28) and (29) or (30) and (31). Below we highlight two of the most useful ones in practice.

Proposition 2. *Suppose that one of the following conditions holds:*

1. *There exists $u_0 \in U$ such that $F_1(u_0) < \infty$, $F_2(\Lambda u_0) < \infty$ and F_2 is continuous at Λu_0 (Ekeland and Temam,* 1976, *Proposition* 2.3).
2. *The set $\bigcup_{\lambda\geq 0} \lambda(dom(F_2) - \Lambda(dom(F_1)))$ is a closed subspace of Y (Attouch and Brezis,* 1986).

Then the dual problem (26) *has a solution and* (27) *holds true.*

Our target is to apply this dualization framework to a class of TV and TGV variational problems and their weighted parameter versions. For that we will first need several definitions.

Definition 1. Let $1 \leq q \leq\infty$ and $p \in L^q(\Omega, \mathbb{R}^d)$. We have $\mathrm{div} p \in L^q(\Omega)$ if there exists $w \in L^q(\Omega)$ such that for all $\phi \in C_c^\infty(\Omega)$

$$\int_\Omega \nabla\phi \cdot p\, dx = -\int_\Omega \phi w\, dx.$$

Then we define

$$W^q(\mathrm{div};\Omega) := \{p \in L^q(\Omega, \mathbb{R}^d) : \ \mathrm{div} p \in L^q(\Omega)\}$$

with the norm $\| p \|^q_{W^q(\mathrm{div};\Omega)} := \| p \|^q_{L^q(\Omega,\mathbb{R}^d)} + \| \mathrm{div} p \|^q_{L^q(\Omega)}$. Similarly we define the space $W^q(\mathrm{div}^2;\ \Omega)$ as the space of all functions $p \in L^q(\Omega, \mathcal{S}^{d\times d})$ whose first- and second-order divergence belong to L^q, equipped with the norm $\| p \|^q_{W^q(\mathrm{div}^2;\Omega)} := \| p \|^q_{L^q(\Omega)} + \| \mathrm{div} p \|^q_{L^q(\Omega,\mathbb{R}^d)} + \| \mathrm{div}^2 p \|^q_{L^q(\Omega)}$.

Using the density of C_c^∞ in L^q, it can be shown that the first- and second-order divergences are unique. Moreover both spaces above are Banach equipped with the associated norms as stated above. We refer to Bredies and Holler (2015) for a more general definition of these spaces. Note that when $q = 2$ then the standard notation is $H(\mathrm{div};\ \Omega)$ and $H(\mathrm{div}^2;\ \Omega)$; see Girault and Raviart (1986) for the former space. The spaces $W_0^q(\mathrm{div};\Omega)$ and $W_0^q(\mathrm{div}^2;\Omega)$ are defined as

$$W_0^q(\mathrm{div};\Omega) = \overline{C_c^\infty(\Omega, \mathbb{R}^d)}^{\|\cdot\|_{W^q(\mathrm{div};\Omega)}},$$
$$W_0^q(\mathrm{div}^2;\Omega) = \overline{C_c^\infty(\Omega, \mathcal{S}^{d\times d})}^{\|\cdot\|_{W^q(\mathrm{div}^2;\Omega)}}$$

Using the definitions above, the following integration by parts formulae hold true:

$$\int_\Omega \nabla\phi \cdot p\, dx = -\int_\Omega \phi\, \mathrm{div} p\, dx, \quad \text{for all } p \in W_0^q(\mathrm{div};\Omega),\ \phi \in C^\infty(\bar{\Omega}, \mathbb{R}), \tag{32}$$

$$\int_\Omega E\phi \cdot p\, dx = -\int_\Omega \phi \cdot\ \mathrm{div} p\, dx, \quad \text{for all } p \in W_0^q(\mathrm{div}^2;\Omega),\ \phi \in C^\infty(\bar{\Omega}, \mathbb{R}^d), \tag{33}$$

$$\int_\Omega \nabla\phi \cdot \mathrm{div} p\, dx = -\int_\Omega \phi\, \mathrm{div}^2 p\, dx, \quad \text{for all } p \in W_0^q(\mathrm{div}^2;\Omega),\ \phi \in C^\infty(\bar{\Omega}, \mathbb{R}), \tag{34}$$

with $E\phi$ denoting the symmetrized gradient of ϕ.

With these definitions, we can now state the following result for the scalar versions of TV and TGV.

Proposition 3. *Let* $d > 2$, $u \in L^{d/d-1}(\Omega)$ *and* α, α_0, $\alpha_1 > 0$

$$\alpha \mathrm{TV}(u) = \sup\left\{ \int_\Omega u \operatorname{div} p \, dx : p \in W_0^d(\operatorname{div};\Omega),\ \|p\|_{2,\infty} \le \alpha \right\} \tag{35}$$

$$\mathrm{TGV}_{\boldsymbol{\alpha}}^2(u) = \sup\left\{ \int_\Omega u \operatorname{div}^2 p \, dx : p \in W_0^d(\operatorname{div}^2;\Omega),\ \|p\|_{2,\infty} \le \alpha_0,\ \|\operatorname{div} p\|_{2,\infty} \le \alpha_1 \right\} \tag{36}$$

As we will see later, these equivalent characterizations of the two regularizers are important for the associated dualization framework for the corresponding regularization problems. The proof of Proposition 3 is immediate once the following two density results are shown:

$$\overline{C_\alpha}^{L^d(\Omega)} = K_\alpha, \tag{37}$$

$$\overline{C_{\boldsymbol{\alpha}}}^{L^d(\Omega)} = K_{\boldsymbol{\alpha}}, \tag{38}$$

where

$$C_\alpha := \left\{ \operatorname{div} \phi : \ \phi \in C_c^\infty(\Omega, \mathbb{R}^d),\ \|\phi\|_{2,\infty} \le \alpha \right\}, \tag{39}$$

$$K_\alpha := \left\{ \operatorname{div} p : \ p \in W_0^d(\operatorname{div};\Omega),\ \|p\|_{2,\infty} \le \alpha \right\}, \tag{40}$$

$$C_{\boldsymbol{\alpha}} := \left\{ \operatorname{div}^2 \phi : \ \phi \in C_c^\infty(\Omega, \mathcal{S}^{d\times d}),\ \|\phi\|_{2,\infty} \le \alpha_0,\ \|\operatorname{div} \phi\|_{2,\infty} \le \alpha_1 \right\}, \tag{41}$$

$$K_{\boldsymbol{\alpha}} := \left\{ \operatorname{div}^2 p : \ p \in W_0^d(\operatorname{div}^2;\Omega),\ \|p\|_{2,\infty} \le \alpha_0,\ \|\operatorname{div} p\|_{2,\infty} \le \alpha_1 \right\}. \tag{42}$$

Questions regarding density results of the above type deserve a separate discussion. They relate to a more general category of questions regarding whether for a given Banach space X, a dense subspace $\overline{Y} = X$, and a convex subset $C \subset X$, the following density result holds:

$$\overline{C \cap Y} = C. \tag{43}$$

We caution the reader that while such density results may seem available since $\overline{Y} = X$, there exist striking counterexamples for (43) to fail in general; see Hintermüller and Rautenberg (2015). However, among other results, in Hintermüller and Rautenberg (2015), the density (43) was shown for $X = H_0(\operatorname{div};\, \Omega)$ endowed with its norm, $Y = C_c^\infty(\Omega, \mathbb{R}^d)$, and

$$C = \left\{ p : \ p \in W_0^d(\operatorname{div};\Omega),\ |p(x)| \le \alpha(x), \text{ for a.e. } x \in \Omega \right\}, \quad \text{where } \alpha \in C(\bar{\Omega}),\ \alpha > 0.$$

The proof uses the theory of mollifiers, and the result is generalized to $X = W_0^d(\operatorname{div};\Omega)$. Note that this result is more general than (37) where only density for the divergences is required. In fact it is an interesting question if this result is strictly weaker, i.e., whether there are counterexamples where the density holds for the divergence (or another differential operator) but not for the full norm. We note that apart from dualization in variational imaging problems, these type of density results find applications in the study of the limiting behaviour of discretized problems, vanishing viscosity problems and others; see Hintermüller and Rautenberg (2015), Hintermüller et al. (2017b), Hintermüller and Rösel (2016) and Hintermüller and Rösel (2018).

The equalities (37) and (38), that is, the densities only for the first- and second-order divergences associated with the TV and TGV functionals were shown in Bredies and Holler (2015, Proposition 3.3) in more generality, i.e., in the context of TGV of arbitrary order; see also Bredies and Holler (2012b, Proposition 7). The associated proof exploits duality and uses scalar weights, only. It turns out that it can easily be adapted to the case where the weights are continuous functions bounded away from zero.

The case of weighted TGV, for a variety of finite dimensional norms, may also be addressed. It builds on the following result, for which $|\cdot|_r$ denotes the finite dimensional r-norm for $1 \leq r \leq \infty$, and r^* is such that $1/r + 1/r^* = 1$ with the obvious definitions for $r = 1, \infty$. Let $\boldsymbol{\alpha} = (\alpha_0, \alpha_1)$ with $\alpha_0, \alpha_1 \in C(\bar{\Omega})$ and $\alpha_0, \alpha_1 > \underline{\alpha} > 0$, $\underline{\alpha} \in \mathbb{R}$. Define the weighted TGV functional, as follows:

$$\mathrm{TGV}_{\boldsymbol{\alpha}}^2(u) = \sup\left\{ \int_\Omega u \operatorname{div}^2 \phi \, dx : \phi \in C_c^\infty(\Omega, \mathcal{S}^{d\times d}),\ |\phi(x)|_r \leq \alpha_0(x), \right.$$
$$\left. |\operatorname{div} \phi(x)|_r \leq \alpha_1(x), \text{ for all } x \in \Omega \right\} \tag{44}$$

Then this functional has also the equivalent expression

$$\mathrm{TGV}_{\boldsymbol{\alpha}}^2(u) = \min_{w \in \mathrm{BD}(\Omega)} \int_\Omega \alpha_1 \, d|Du - w|_{r^*} + \int_\Omega \alpha_0 \, d|\mathcal{E}w|_{r^*}. \tag{45}$$

Moreover, the weighted TGV functional

$$\mathrm{TGV}_{\boldsymbol{\alpha}}^2(u) = \sup\left\{ \int_\Omega u \operatorname{div}^2 \phi \, dx : \phi \in C_c^\infty(\Omega, \mathcal{S}^{d\times d}),\ |\phi(x)|_r \leq \alpha_0(x), \right.$$
$$\left. |\operatorname{div} \phi|_r \leq \alpha_1(x), \text{ for all } x \in \Omega \right\} \tag{46}$$

is also equal to

$$\sup\left\{ \int_\Omega u \operatorname{div}^2 p \, dx : p \in W_0^d(\operatorname{div}^2; \Omega),\ |p(x)|_r \leq \alpha_0(x), \right.$$
$$\left. |\operatorname{div} p(x)|_r \leq \alpha_1(x), \text{ for a.e. } x \in \Omega \right\} \tag{47}$$

for all $u \in L^{d/d-1}(\Omega)$.

Recapitulating, we have described how the following equalities can be established when $\alpha, \alpha_0, \alpha_1 \in C(\bar{\Omega})$ are bounded away from zero, where $\mathrm{TV}_\alpha(u) = \int_\Omega \alpha d|Du|$:

$$\mathrm{TV}_\alpha(u) = \sup\left\{ \int_\Omega u\, \mathrm{div} p\, dx : p \in W_0^d(\mathrm{div};\Omega),\ |p(x)|_r \le \alpha(x), \text{ for a.e. } x \in \Omega \right\}, \tag{48}$$

$$\mathrm{TGV}^2_{\boldsymbol{\alpha}}(u) = \sup\left\{ \int_\Omega u\, \mathrm{div}^2 p\, dx : p \in W_0^d(\mathrm{div}^2;\Omega),\ |p(x)|_r \le \alpha_0(x),\right.$$
$$\left. |\mathrm{div}\, p(x)|_r \le \alpha_1(x), \text{ for a.e. } x \in \Omega \right\}. \tag{49}$$

Regarding more general weight functions, at least for TV, the following proposition follows from (Hintermüller et al., 2018, Corollary 3.5, Proposition 3.8).

Proposition 4. (Hintermüller et al. (2018)).

Let $\alpha \in \mathrm{BV}(\Omega)$ with $0 \le \alpha(x) \le C$ for a.e. $x \in \Omega$. Moreover let $j(x, z) = \alpha(x) b(z)$, with $b : \mathbb{R}^d \to \mathbb{R}$ being a convex, positively 1-homogeneous, even function such that there exists $\gamma > 0$ with $b(z) \le \gamma |z|$ for all $z \in \mathbb{R}^d$. Then for every $u \in \mathrm{BV}(\Omega)$ it holds that

$$\int_\Omega j(x, \nabla u(x))\, dx + \int_\Omega j^-(x, \sigma_{D^c u})\, d|D^c u| + \int_{J_u \cap \Omega} (u^+(x) - u^-(x)) j^-(x, \sigma_{D^j u})\, d\mathcal{H}^{d-1}$$
$$= \sup\left\{ \int_\Omega u\, \mathrm{div} p\, dx : p \in W_0^d(\mathrm{div};\Omega),\ j^\circ(x, p(x)) \le 1, \textit{for a.e. } x \in \Omega \right\}, \tag{50}$$

where

$$j^\circ(x, z^*) = \sup_{z : j(x,z) \le 1} z^* \cdot z.$$

In particular, if $\alpha \in W^{1,1}(\Omega)$ with $0 \le \alpha(x) \le C$ for a.e. $x \in \Omega$ and $b(z) = |z|_{r^*}$, then

$$\int_\Omega \alpha\, d|Du|_{r^*} = \sup\left\{ \int_\Omega u\, \mathrm{div} p\, dx : p \in W_0^d(\mathrm{div};\Omega),\ |p(x)|_r \le \alpha(x), \textit{for a.e. } x \in \Omega \right\}. \tag{51}$$

The result of (50) also holds for more general integrands j that are not of the type $\alpha(x)b(x)$ but nevertheless satisfy some boundedness, convexity and 1-homogeneity assumptions. The associated proof combines some duality and lower semicontinuity results from (Amar et al., 2008). In order to derive (51) from (50) note first that if $j(x, z) = \alpha(x)b(z)$ with $b(z) = |z|_{r^*}$, then $j^\circ(x, z^*) = |z|_r/\alpha(x)$ which resolves the right-hand side. Moreover, when $\alpha \in W^{1,1}(\Omega)$, then for the set of its jump points J_α we have that $\mathcal{H}^{d-1}(J_\alpha) = 0$

which implies $|Du|(J_\alpha) = 0$ for every $u \in \mathrm{BV}(\Omega)$ (Ambrosio et al., 2000, Lemma 3.76). Then we can write α instead of α^- and $j^-(x, z) = \alpha(x)b(z)$. Thus, in this case the left-hand side of (50) equals

$$\int_\Omega \alpha(x)|\nabla u|_{r^*}\, dx + \int_\Omega \alpha(x)\left|\frac{dD^c u}{d|D^c u|}\right|_{r^*} d|D^c u| + \int_{J_u\cap\Omega} (u^+(x) - u^-(x))\alpha(x)\left|\frac{dD^j u}{d|D^j u|}\right|_{r^*} d\mathcal{H}^{d-1},$$
$$= \int_\Omega \alpha(x)|\nabla u|_{r^*}\, dx + \int_\Omega \alpha(x)\frac{d|D^c u|_{r^*}}{d|D^c u|} d|D^c u| + \int_{J_u\cap\Omega} (u^+(x) - u^-(x))\alpha(x)\frac{d|D^j u|_{r^*}}{d|D^j u|} d\mathcal{H}^{d-1},$$
$$= \int_\Omega \alpha\, d|Du|_{r^*}$$

where we have followed the properties of convex functions of measures (Demengel and Temam, 1984). We point out that the supremum on the right-hand side of (50) is finite if and only if $u \in \mathrm{BV}(\Omega)$. Also, it suffices to have $0 < \underline{\alpha} \leq \alpha(x) < C$ for a.e. $x \in \Omega$ for some strictly positive constant $\underline{\alpha}$.

2.4 Dualization of the variational regularization problems

The formulation of the TV and TGV regularizers and their weighted versions, of the type (48), (49) and (51) is the key for establishing the duals of the corresponding regularization problems. We will state a general duality result only for the weighted TV case which is connected to Proposition 4. The derivation of a corresponding result for TGV is the subject of current research.

The following proposition is a consequence of Proposition 4 and Hintermüller et al. (2018, Theorem 5.1).

Proposition 5. (Hintermüller et al., 2018)
Let $1 \leq p < \infty$, $\alpha \in W^{1,1}(\Omega)$ with $0 < \underline{\alpha} < \alpha(x) < C$ for a.e. $x \in \Omega$. Moreover let $(\mathcal{Y}, \|\cdot\|_{\mathcal{Y}})$ be a Banach space, with $f \in \mathcal{Y}$, and $T \in \mathcal{L}(L^{d/d-1}(\Omega), \mathcal{Y})$. Then, there exists a solution to the primal problem

$$\min_{u\in \mathrm{BV}(\Omega)} \frac{1}{p}\| Tu - f \|_{\mathcal{Y}}^p + \int_\Omega \alpha d|Du|_{r^*} \tag{52}$$

to its predual problem

$$\min\left\{ \left(\frac{1}{p}\| T\cdot - f \|_{\mathcal{Y}}^p\right)^* (\mathrm{div} p) : p \in W_0^d(\mathrm{div};\Omega),\ |p(x)|_r \leq \alpha(x) \text{ for a.e. } x\in\Omega \right\}, \tag{53}$$

and zero duality gap holds.

Note that the nonnegativity condition $0 < \underline{\alpha} < \alpha(x) < C$ is not only necessary for the solution of the primal problem to belong to $\mathrm{BV}(\Omega)$ but also for the establishment of the duality result itself. However, according to Hintermüller et al. (2018, Proposition) it can be omitted provided that the range of T^*, $\mathrm{Rg}(T^*)$ is closed and $\mathrm{Ker}(T)$ is finite dimensional. In that case the solution of the primal problem (52) cannot be guaranteed in $\mathrm{BV}(\Omega)$ but only to $L^{d/d-1}(\Omega)$.

In the special case where $B = T^*T$ is invertible and $p = 2$ and $\mathcal{Y}=L^2(\Omega)$, then the predual problem (53) takes a more precise form since

$$\left(\frac{1}{2}\| T\cdot -f \|_{L^2(\Omega)}^2\right)^*(u^*) = \frac{1}{2}(u^* + K^*f, B^{-1}(u^* + K^*f))$$
$$-\frac{1}{2}\|f\|_{L^2(\Omega)}^2 =: \frac{1}{2}\| u^* + K^*f \|_B^2 - \frac{1}{2}\|f\|_{L^2(\Omega)}^2 \quad (54)$$

for every $u^* \in L^d(\Omega)$; see also Kunisch and Hintermüller (2004). We note that if T^*T is not invertible then one can add a quadratic term $\frac{\gamma}{2}\| u \|_{L^2(\Omega)}^2$ for some small $\gamma > 0$ in the primal problem (52). In that case $B = \gamma I + T^*T$. Moreover the optimality conditions that correspond to (28) and (29) are

$$Bu = T^*f + \mathrm{div}p, \quad (55)$$
$$\langle(-\mathrm{div})^*u, \widetilde{p} - p)\rangle_{W_0^d(\mathrm{div};\Omega)^*, W_0^d(\mathrm{div})} \leq 0, \text{ for all } \widetilde{p} \in W_0^d(\mathrm{div}),$$
$$|p(x)|_r \leq \alpha(x) \text{ for a.e. } x \in \Omega, \quad (56)$$

where u and p solve (52) and (53), respectively; see Hintermüller and Rautenberg (2017) and Kunisch and Hintermüller (2004).

3 Numerical algorithms

In this section we review some of the most popular numerical algorithms for the solution of the nonsmooth variational problems arising in imaging, as these were discussed in the previous section. We should already mention here that the majority of the algorithms are considered in the discretized setting only, that is, in this case images are not modelled as functions on a continuous domain Ω, but rather as functions defined on a discretized grid $\Omega_h = \{1, 2, \ldots, n\} \times \{1, 2, \ldots, m\}$ where $n \times m$ is the pixel resolution of the image. Even though, this seems quite natural and in the end in practice, discretization will inevitably take place anyway, devising a solution algorithm directly in the discrete setting has its disadvantages as we shall explain below. But before we come to this, we start with a brief historical review focusing on the algorithms for the solution of the classical TV minimization problem (4).

In the initial paper (Rudin et al., 1992), the TV denoising problem was solved by using a gradient projection method where the step length was related to a (pseudo) time marching scheme with a constant step length over all time steps. This approach, however, is extremely slow even though some improved time stepping schemes have been studied, e.g., in Vogel and Oman (1996). Subsequently, several other (improved) algorithmic approaches were developed such as graph cut methods for the anisotropic TV minimization (Darbon and Sigelle, 2005), fixed point iterations (Ito and Kunisch, 2000), augmented Lagrangian-based strategies (Tai et al., 2011b; Wu and Tai, 2010; Wu et al., 2011), ADMM and Douglas–Rachford splitting (Setzer, 2011). Particular attention has been drawn to Bregman iteration methods (Getreuer, 2012b; Goldstein and Osher, 2009; Osher et al., 2005; Yin et al., 2008). A particular characteristic of this family of methods is the reduction of contrast loss (bias) which is characteristic to these derivative-based regularization methods see Brinkmann et al. (2017) as well as Benning and Burger (2018) for a review.

It has become widely accepted that the most efficient and applicable techniques to solve nonsmooth variational problems (not only the L^2–TV problem) are based on convex duality. First works can be found in Chambolle (2004), Chambolle and Lions (1997) and Chan et al. (1999). A very popular current method is the Primal-Dual hybrid gradient method (PDHGM) or otherwise called Chambolle–Pock algorithm Chambolle and Pock (2011) and its extensions; see Chambolle and Pock (2016) for a comprehensive review of the corresponding family of methods. These methods are aiming to find solutions to saddle-point problems of the following form

$$\min_{x\in X}\max_{y\in Y}\langle Kx,y\rangle + G(x) - F^*(y) \tag{57}$$

where typically X, Y are finite dimensional real vector spaces, G, F are proper, convex, lower semicontinuous functions and $K : X \to Y$ is a continuous linear operator. Using the Fenchel–Rockafellar duality theory presented in the previous section it can be shown that (x, y) solves (57) if and only if x and y solve the following primal and dual problems, respectively, i.e.,

$$\begin{aligned}&\min_{x\in X} \; F(Kx) + G(x),\\ &\max_{y\in Y} \; -G^*(-K^*y) - F^*(y).\end{aligned}$$

The Chambolle–Pock iteration then reads as follows

$$y^{k+1} = \mathrm{prox}_{\sigma F^*}(y^k + \sigma K\overline{x}^{k+1}), \tag{58}$$

$$x^{k+1} = \mathrm{prox}_{\tau G}(x^k - \tau K^* y^k), \tag{59}$$

$$\overline{x}^{k+1} = x^{k+1} + \theta(x^{k+1} - x^k). \tag{60}$$

Apparently, it consists of an alternation of a (proximal) descent in the primal variable x and an ascent in the dual variable y followed by an extrapolation step. The Chambolle–Pock algorithm is also a popular method of choice when it comes to solving inverse problems involving the TGV functional (Bredies, 2014; Bredies and Holler, 2012c; Holler and Kunisch, 2014; Knoll et al., 2011). Its popularity stems from the fact that the prox operations in (58) and (59) can typically be computed cheaply leading to an easy implementation. Unfortunately, even though one obtains good reconstructions after relatively few iterations, solving the problem to high accuracy with this and any other first-order method (given some objective stopping rule; other than closeness of successive iterates, which is not objective and may stop the iteration prematurely due to negligible progress of the chosen iterative solver) typically leads to an excessive number of iterations. Sufficiently high accuracy, however, turns out to be particularly desirable in the context of the bilevel optimization problems that we discuss in the present exposition. A second issue that is often not addressed well by first-order algorithms is (image) resolution independent convergence, which is typically connected to function space algorithms (Hintermüller et al., 2014). Resolution independence refers here to the fact that, given an objective stopping rule and an initial iterate common to all resolution levels, the number of iterations to satisfy such a stopping rule is robust under refinements of the resolution of a given image datum (over a bounded domain).

In view of the issues highlighted in the previous paragraph, we now turn our attention to Newton-type methods. In its very basic form, Newton's method aims to solve a nonlinear system

$$F(x)=0, \tag{61}$$

with $F:\mathbb{R}^n \to \mathbb{R}^n$ differentiable, by an iteration which, given some initial guess x^0, computes a root of a current linearization of (61) at every step, i.e., one finds for $k = 0, 1, 2, \ldots$ an update direction d^k with

$$\nabla F(x^k)d^k = -F(x^k), \tag{62}$$

$$x^{k+1} := x^k + d^k. \tag{63}$$

For x^0 close to x^* with the matrix $\nabla F(x^*)$ being (uniformly) invertible along the iterates, the iterates converge at a rate that depends on the order of continuity of ∇F. More specifically, the rate of convergence is (q-)quadratic when ∇F is Lipschitz, and the rate is (q-)superlinear for Hölder continuous ∇F. Relaxations on the regularity of F can still yield convergent Newton methods, with the most relevant regularity here being semismoothness (Hintermüller et al., 2002; Mifflin, 1977; Qi and Sun., 1993). Of particular interest are semismooth Newtons in Banach spaces (Chen et al., 2000; Hintermüller et al., 2002). For the rest of this section we give a basic account on such methods and how they can be used for the accurate solution of the TV (and TGV) problems.

Suppose that $F: D \subset X \to Z$, where X, Z are Banach spaces and D is an open set, and we are interested in finding $x^* \in D$ such that $F(x^*) = 0$. We assume that the map F is generalized (or *Newton*) differentiable on an open set $U \subset D$, that is, there exists a family of mappings $G: U \to \mathcal{L}(X, Z)$ such that

$$\lim_{d\to 0} \frac{1}{\|d\|_X} \|F(x+d) - F(x) - G(x+d)d\|_Z = 0, \tag{64}$$

for all $x \in U$. This condition extends a characterization of semismoothness in finite dimensions to possibly infinite dimensional Banach spaces. The operator G is referred to as the generalized (or Newton) derivative of F at x. Note that G need not be unique. However, when F is Fréchet differentiable at X, then $G(x)$ is unique and equals the Fréchet derivative of F at x. For some given $x^0 \in U$, the corresponding semismooth Newton iteration reads

$$d^k = -G(x^k)^{-1}F(x^k),\ k=0,1,2,3,\ldots \tag{65}$$

$$x^{k+1} = x^k + d^k, \tag{66}$$

where $G(x^k)$ is an arbitrary generalized derivative of F at x^k. Then the following convergence result is available (Hintermüller et al., 2002).

Proposition 6. *Let x^* be a solution of $F(x) = 0$ and F be Newton differentiable in an open neighbourhood U containing x^*. Assume furthermore that $G(x)$ is nonsingular for all $x \in U$ and that the set $\{\|G(x)^{-1}\| : x \in U\}$ is bounded. Then, provided $\|x_0 - x^*\|_X$ is sufficiently small, the semismooth Newton iteration* (65) *and* (66) *is well defined and converges superlinearly to x^*, i.e.,*

$$\lim_{\substack{x \to x^* \\ x \neq x^*}} \frac{\| x^{k+1} - x^* \|_X}{\| x^k - x^* \|_X} = 0.$$

Under suitable assumptions, it can be shown that semismooth Newton methods exhibit a discretization independent convergence (Hintermüller, 2007; Hintermüller and Ulbrich, 2004). In fact, for sufficiently small mesh sizes (for discretizing the associated Banach spaces) either the number of iterations of the discretized method is essentially constant with respect to the mesh size, or that the discretized method achieves essentially the same convergence rate independently of the mesh size. We now show how this framework can be applied for the solution of the TV-type variational problem.

We start with TV and in particular of the (anisotropic) TV predual problem with L^2 fidelity, (53) and (54):

$$\begin{aligned} &\min_{p \in W_0^d(\mathrm{div};\Omega)} \frac{1}{2} \| \mathrm{div}\, p + T^* f \|_B^2 \\ &\text{subject to } \ |p(x)|_\infty \le \alpha(x) \text{ for a.e. } x \in \Omega, \end{aligned} \tag{67}$$

where $\alpha \in W^{1,1}(\Omega)$ is a weight function. In order to address the constrained minimization problem (67) with Newton methods we substitute the box constraints by the following penalty functional (which equals as smoothed version of the Moreau–Yosida regularization with respect to the $L^2(\Omega)$-norm of the indicator of the constraint set in (67))

$$\mathcal{P}_\delta(p, \alpha) := \int_\Omega \sum_{i=1}^d (G_\delta(-(p_i + \alpha)) + G_\delta(p_i - \alpha))\, dx \tag{68}$$

where $p=(p_1, \ldots, p_d)$ and $G_\delta:\mathbb{R}\to\mathbb{R}$,

$$G_\delta(r)=\begin{cases}\frac{1}{2}r^2-\frac{\delta}{2}r+\frac{\delta^2}{6}, & \text{if } r\geq\delta,\\ \frac{r^3}{6\delta}, & \text{if } 0<r<\delta,\\ 0, & \text{if } r\leq 0,\end{cases} \tag{69}$$

for $\delta>0$. The idea is that when the quantity $\mathcal{P}_\delta(p,\alpha)$ is small then p is expected to approximately satisfy the box constraints in (68). This is achieved by multiplying this term by $1/\epsilon$ with $\epsilon>0$ sufficiently small. We note that if one is interested only in solving the reconstruction problem via a dual approach, the additional smoothing by $\delta>0$ is not needed. One may rather directly work with the aforementioned Moreau–Yosida regularization. In the context of bilevel optimization, however, the dual of the reconstruction problem involves a constraint thus asking for additional smoothing (hereby expressed by G_δ with $\delta>0$) for a subsequent derivation of stationarity conditions of the bilevel problem.

Note that $\mathcal{P}_\delta(p,\alpha)$ is well defined if both p and α enjoy L^2 regularity. As $\alpha\in W^{1,1}(\Omega)$, this would be automatically satisfied for α for $d=2$ by Sobolev embedding; otherwise L^2-regularity must be assumed in addition. As far as p is concerned, L^2 regularity results from the minimization process anyway. For uniqueness of a solution p in (67) a multiple of the L^2-norm of p may be added to the objective. The resulting modified predual problem reads

$$\min_{p\in H_0^1(\Omega)^d}\frac{\beta}{2}\|\nabla p\|^2_{L^2(\Omega)^d}+\frac{\gamma}{2}\|p\|^2_{L^2(\Omega)^d}+\frac{1}{2}\|\operatorname{div}p+T^*f\|^2_B+\frac{1}{\epsilon}\mathcal{P}_\delta(p,\alpha). \tag{70}$$

The reason for the additional H_0^1-regularization (typically with $0<\beta\ll\underline{\alpha}\leq\alpha(x)$ for $x\in\Omega$) for p is the facilitation of the convergence analysis of the Newton method in function space as well as the need for differentiability of the solution map $\alpha\mapsto p(\alpha)$, which is necessary in the context of bilevel optimization. Observe that the homogeneous Dirichlet boundary conditions on p aim to capture the boundary conditions associated with space $W_0^d(\operatorname{div};\Omega)$. It can be shown that the associated Euler–Lagrange system for (70) reads for $V=H_0^1(\Omega)^d$

$$\begin{aligned}&\langle-\beta\Delta p+\gamma p-\nabla B^{-1}\operatorname{div}p-\nabla B^{-1}T^*f,v\rangle_{V^*,V}\\&+\frac{1}{\epsilon}\langle P_\delta(p,\alpha),v\rangle_{L^2(\Omega)^d,L^2(\Omega)^d}=0,\quad\forall v\in V.\end{aligned} \tag{71}$$

Here the Laplacian $\Delta:H_0^1(\Omega)^d\to[H_0^1(\Omega)^d]^*$ is defined as $\langle\Delta p,v\rangle_{V^*,V}=\sum_{k=1}^d\int_\Omega\nabla p_k\cdot\nabla v_k\,dx$. Moreover $P_\delta:H_0^1(\Omega)^d\times L^2(\Omega)\to L^2(\Omega)^d$ is defined as

$$P_\delta(p,\alpha):=G_\delta'(p-\alpha\mathbf{1})-G_\delta'(-p-\alpha\mathbf{1}),$$

where $p-\alpha\mathbf{1}=(p_1-\alpha,\ldots,p_d-\alpha)$ and similarly for $-p-\alpha\mathbf{1}$. Furthermore,

$$G'_\delta(r) = \begin{cases} r - \dfrac{\delta}{2}, & \text{if } r \geq \delta, \\ \dfrac{r^2}{2\delta}, & \text{if } 0 < r < \delta, \\ 0, & \text{if } r \leq 0, \end{cases} \tag{72}$$

In other words Eq. (71) is of the type $F(p) = 0$ where $F : H_0^1(\Omega)^d \to [H_0^1(\Omega)^d]^*$ with

$$F(p) = -\beta\Delta p + \gamma p - \nabla B^{-1}\operatorname{div} p - \nabla B^{-1}T^*f + \frac{1}{\epsilon}P_\delta(p, \alpha). \tag{73}$$

The regularized version (70) is consistent with the original predual problem (67) in the sense that when the parameters β, γ and ϵ tend to zero then the divergences of the approximate solutions converge to the divergence of the solution of the original problem in the appropriate sense; see Hintermüller and Rautenberg (2017, Theorem 5.1). In the same work (Hintermüller and Rautenberg, 2017, Theorem 5.2), differentiability properties were established for the solution mapping $\alpha \mapsto p(\alpha)$ where $p(\alpha)$ solves (71). In particular it was shown that $p(\cdot) : L^{2+\xi}(\Omega) \to H_0^1(\Omega)^d$ is Gâteaux differentiable, for every $\xi \in (0, \overline{\xi}(d)]$, where $\overline{\xi}(d) = +\infty$ for $d = 1$, $\overline{\xi}(d) \in [0, \infty)$ for $d = 2$ and $\overline{\xi}(d) = 2d/(d-2)$ for $d > 2$. While such results are needed for deriving stationarity conditions in bilevel optimization (where α also becomes an optimization variable), we stress once again that the H_0^1-regularity (which lifts the smoothness of p) is crucial for establishing such a result.

As noted above, one may indeed choose $\delta = 0$ (Kunisch and Hintermüller, 2004). Then we get $G'_0(r) = \max(r, 0)$ which is (still) a semismooth mapping when considered from $L^r(\Omega)$ to $L^2(\Omega)$, $2 < r$; see Hintermüller et al. (2002). Even though, in this case one can still devise a convergent Newton algorithm, which is in fact equivalent to an (efficient to implement) primal-dual active strategy (Hintermüller et al., 2002; Kunisch and Hintermüller, 2004), here we prefer to keep $\delta > 0$, as we need the differentiability property of $\alpha \mapsto p(\alpha)$ for the bilevel optimization scheme.

Algorithm 1 displays the corresponding function space version of Newton's method ((65)–(66), in general terms) specified for solving Eq. (71).

Before we close this section we would like to make a few remarks regarding the term $\frac{\gamma}{2}\|p\|_{L^2}^2$ in the TV regularized predual problem (70). Similar statements apply to the TGV-context. The first remark is that in practice one can set $\gamma = 0$, which suggests that the bilateral constraints impose some type of uniqueness. Alternatively, still for $\gamma = 0$, one may add a projection onto the (nontrivial) kernel of div to the objective of the predual problem in order to guarantee uniqueness of a solution p. The other remark concerns $\gamma > 0$. In this case, this type of L^2-regularization of the predual problem, typically corresponds to a Huber regularization of the regularizing functional in

ALGORITHM 1 Function space Newton algorithm for the solution of the regularized TV dual problem (70).

Initialise: $p^0 \in H_0^1(\Omega)^d$
while stopping criterion not satisfied **do**
Find $\delta_p^k \in H_0^1(\Omega)^d$ such that

$$-\nabla B^{-1}\text{div}\delta_p^k - \beta\Delta\delta_p^k + \gamma\delta_p^k + \frac{1}{\epsilon}\left(G_\delta''(p^k - \alpha\mathbf{1}) + G_\delta''(-p^k - \alpha\mathbf{1})\right)\delta_p^k = -F(p^k), \quad \text{in } H^{-1}(\Omega)^d$$

Update p^{k+1}:

$$p^{k+1} = p^k + \delta_p^k$$

end while

the primal problem. In order to elaborate on this, consider the Huber function $\varphi : \mathbb{R} \to \mathbb{R}^+$ (Huber, 1964)

$$\varphi(t) = \begin{cases} |t| - \dfrac{\gamma}{2}, & \text{if } |t| \geq \gamma, \\ \dfrac{1}{2\gamma}t^2, & \text{if } |t| \leq \gamma \end{cases} \tag{74}$$

which constitutes a local smoothing of the absolute value function. Then one can define the (isotropic)*Huber* total variation functional as follows

$$\text{TV}_\varphi(u) = \int_\Omega \varphi(|\nabla u|)\, dx + |D^s u|(\Omega), \quad u \in \text{BV}(\Omega) \tag{75}$$

as well as the corresponding variational problem

$$\min_{u \in \text{BV}(\Omega)} \frac{1}{2} \| Tu - f \|_{L^2(\Omega)}^2 + \alpha \text{TV}_\varphi(u), \tag{76}$$

(here written only) for a scalar $\alpha > 0$. Consequently, one can show (see Hintermüler and Stadler, 2006) that the predual of (76) is (up to a constant)

$$\begin{aligned} &\min_{p \in W_0^d(\text{div};\Omega)} \frac{1}{2} \| \text{div} p + T^* f \|_B^2 + \frac{\gamma}{2\alpha} \| p \|_{L^2(\Omega)}^2 \\ &\text{such that } \ |p(x)|_\infty \leq \alpha(x) \text{ for a.e. } x \in \Omega. \end{aligned} \tag{77}$$

Also in Hintermüler and Stadler (2006) an alternative Newton approach for the solution of the TV minimization problem was proposed. Indeed, instead of using Newton to solve the first-order optimality condition of the regularized predual problem, it was used to solve the primal-dual optimality conditions of the type (28)–(29) which correspond to the primal and the predual Huber-TV problems (76), (77), respectively, with an additional small H^1-regularity term $\frac{\mu}{2}\int_\Omega |\nabla u|^2 dx$ in the primal problem. As μ, with $0 < \mu \ll \alpha$, tends to zero a solution of the Huber-TV problem is approached.

4 Bilevel optimization

In order to construct a structured nonsmooth TV- or TVG-based prior (or regularization term) we will resort to a bilevel optimization technique. Given some image data f, such an approach aims at optimizing with respect to both simultaneously, the regularization weight $\boldsymbol{\alpha}$ and the reconstruction u. For this purpose we collect several basic notions and concepts in bilevel programming. For the ease of exposition we tailor the general discussion to the problem structure which is relevant in our subsequent image processing context. For a general introduction to bilevel programming in finite dimensions we refer to Dempe (2002) and Luo et al. (1996) and to Barbu (1984) and Neittaanmaki et al. (2006) for a, respectively, limited access to infinite dimensional settings.

4.1 Background

Consider the following hierarchical optimization problem

$$\begin{aligned} &\min\ \mathcal{F}(v,\beta)\ \text{ over } (v,\beta)\\ &\text{s.t. } \beta \in B_{\text{ad}} \subset B,\\ &\qquad v \in \operatorname{argmin}\{\mathcal{G}(w;\beta) : w \in V_{\text{ad}} \subset V\}. \end{aligned} \tag{78}$$

Here, V and B denote Banach spaces, V_{ad} and B_{ad} associated nonempty, closed and convex subsets, respectively. Further, $\mathcal{F}: V \times \boldsymbol{B} \to \mathbb{R}$ and $\mathcal{G}: V \to \mathbb{R}$ are assumed continuously differentiable functions. Problem (78) constitutes a *bilevel* optimization problem for (v, β), where $\mathcal{F}$ represents the upper level objective and B_{ad} the upper level constraint set. Further, $\mathcal{G}$ is the objective of the lower level optimization problem with lower level constraints V_{ad}. Note that the optimization variable β enters the lower level problem as a parameter only, i.e., we may consider $v = v(\beta)$, assuming that the lower level problem admits a solution. Clearly, more general settings of (78) with respect to constraints and objective requirements (on both levels, respectively) are conceivable, but would go beyond the scope of this presentation.

Bilevel problems are notoriously delicate. In order to grasp some of the inherent complexities, we next discuss exemplarily some of the structural challenges. First, in case $\mathcal{G}$ is nonconvex (or V_{ad} is nonconvex), still assuming existence of a solution to the lower level problem, computing a feasible point of (78) requires to compute a global (!) solution to a nonconvex minimization problem, which resides in the constraint set of the overall problem. This, however, cannot be guaranteed by any of the current state-of-the-art solvers which makes the computation of a feasible point elusive, not even speaking of a minimizer of (78). Moreover, typically optimization solvers operate on first-order characterizations (KKT-system or Euler-Lagrange equation) of solutions. Replacing the lower level problem by KKT-conditions yields an overall minimization problem which follows a standard structure of a nonlinear programme, but this new problem would not be equivalent to the original bilevel problem. For instance, when $V_{\text{ad}} := \{w \leq b\} \subset H_0^1(\Omega)$ with $b \in H_0^1(\Omega)$ the KKT-system associated with the lower level problem reads

$$\mathcal{G}'(v;\beta) + \lambda = 0 \text{ in } H^{-1}(\Omega), \tag{79}$$

$$v \leq b, \quad \lambda \geq 0 \text{ in } H^{-1}(\Omega), \quad \langle \lambda, v - b \rangle_{H^{-1},H_0^1} = 0, \tag{80}$$

where $\langle \cdot, \cdot \rangle_{H^{-1},H_0^1}$ denotes the duality pairing between the Sobolev space $H_0^1(\Omega) =: V$ and its (topological) dual $H^{-1}(\Omega) = V^*$. In this case, using this KKT-system instead of the lower level problem would yield the following *mathematical programme with equilibrium constraints (MPEC)* (see, e.g., Hintermüller and Kopacka, 2009; Luo et al., 1996 for rather general accounts of MPECs)

$$\begin{aligned} \min \ & \mathcal{F}(v, \beta) \ \text{ over } (v, \lambda, \beta) \\ \text{s.t.} \ & \beta \in B_{\text{ad}} \subset B, \\ & \mathcal{G}'(v;\beta) + \lambda = 0 \text{ in } H^{-1}(\Omega), \\ & v \leq b, \quad \lambda \geq 0 \text{ in } H^{-1}(\Omega), \quad \langle \lambda, v - b \rangle_{H^{-1},H_0^1} = 0, \end{aligned} \tag{81}$$

which is no longer equivalent to (78), in general. Equivalence is, however, regained when $\mathcal{G}$ is convex. In this case, the KKT-system is necessary and sufficient for characterizing a solution to the lower level problem. Hence, one may study (81) analytically and also numerically. But the reader should be cautioned that (81), while appearing to be yet another nonlinear programme in Banach space as studied, e.g., in Zowe and Kurcyusz (1979), suffers from notoriously degenerate constraints. In order to illustrate this, we focus only on the *complementarity system*

$$v \leq b, \quad \lambda \geq 0 \text{ in } H^{-1}(\Omega), \quad \langle \lambda, v - b \rangle_{H^{-1},H_0^1} = 0 \tag{82}$$

in the constraint set of (81). Moreover, to ease the discussion we assume that λ admits $L^2(\Omega)$ regularity. The latter is often realistic under addition regularity requirements on the problem data. In this case we have

$$\langle \lambda, v - b \rangle_{H^{-1},H_0^1} = 0 \quad \Leftrightarrow \quad \int_\Omega \lambda\,(v - b)dx = 0 \quad \Leftrightarrow \quad \lambda\,(v - b) = 0 \text{ a.e. in } \Omega.$$

From the latter relation it is straightforward to see that the typical surjectivity requirement of linearized constraints for the existence of multipliers (see Zowe and Kurcyusz (1979); Zowe–Kurcyusz–Robinson (ZKR) constraint qualification (CQ)) fails at the presence of a *biactive set*

$$\mathcal{B} := \{x \in \Omega : \lambda(x) = 0,\ v(x) = b(x)\}$$

at a solution (v, λ, β) with $|\mathcal{B}| > 0$ where $|\cdot|$ denotes the d-dimensional Lebesgue measure. This jeopardizes the existence of (bounded) Lagrange multipliers and thus the availability of a KKT-system, which is typically the starting point for numerical solvers.

Assuming that $\mathcal{G}$ is strictly convex (over the nonempty, closed and convex set V_{ad}) yields a unique solution $v = v(\beta)$ of the lower level problem. One may now reduce (78) to a problem in β by considering $\beta \mapsto v(\beta)$. This results in

$$\begin{aligned} &\min\ \hat{\mathcal{F}}(\beta) := \mathcal{F}(v(\beta), \beta) \\ &\text{s.t. } \beta \in B_{\text{ad}} \subset B. \end{aligned} \tag{83}$$

In case B_{ad} is simple, this may at first glance appear to be a tractable minimization problem. A closer inspection, however, reveals that while $\beta \mapsto v(\beta)$ is often found to be (locally) Lipschitz continuous, this map is not Fréchet differentiable in general. Consequently, $\hat{\mathcal{F}}$ is typically not Fréchet differentiable, even if $\mathcal{F} : V \times B \to \mathbb{R}$ enjoys Fréchet differentiability. Moreover, even if all constituents in (78) are convex, the reduced objective $\hat{\mathcal{F}} : B \to \mathbb{R}$ may still be nonconvex only. Summarizing, solving (83) amounts to minimizing a nonsmooth and nonconvex objective over a constraint set B_{ad}, which is a very delicate task. Some of the involved issues relate to computability of generalized derivatives of $\hat{\mathcal{F}}$, the design of an optimization solver guaranteeing descent from one iterate to the next, and the availability of a computable stopping rule. Concerning the latter note that due to the nonsmoothness of $\hat{\mathcal{F}}$ one can only expect a set-valued generalized derivative $\partial\hat{\mathcal{F}}(\beta)$ at a solution β. Given the properties of B_{ad}, a stationarity condition of (83) then reads

$$0 \in \partial\hat{\mathcal{F}}(\beta) + N_{B_{\text{ad}}}(\beta), \tag{84}$$

where we assume that a sum rule for differentiation applies, and $N_{B_{\text{ad}}}(\beta)$ denotes the normal cone to B_{ad} at β. Note that $N_{B_{\text{ad}}}(\beta) = \partial I_{B_{\text{ad}}}(\beta)$ where $I_{B_{\text{ad}}}$ is the indicator function of B_{ad}, i.e., $I_{B_{\text{ad}}}(\beta) = 0$ if $\beta \in B_{\text{ad}}$ and $I_{B_{\text{ad}}}(\beta) = +\infty$ otherwise, and '∂' is the subdifferential of convex analysis. Now, an iterative solver of (83) typically has only one generalized gradient of $\hat{\mathcal{F}}$ at its disposal per iteration. Assuming simple constraints in B_{ad} allowing for an explicit characterization of $N_{B_{\text{ad}}}(\beta)$, this still renders checking of (84) at an iterate elusive as information on the entire set $\partial\hat{\mathcal{F}}(\beta)$ would be needed, in general. If, however, $\hat{\mathcal{F}}$ is Fréchet differentiable at β, then (84) requires to compute $\hat{\mathcal{F}}'(\beta)$ and to check whether

$$-\hat{\mathcal{F}}'(\beta) \in N_{B_{\text{ad}}}(\beta),$$

which is tractable when $N_{B_{\text{ad}}}(\beta)$ admits an explicit representation.

Let us next assume that B is reflexive, $B \ni \beta \mapsto v(\beta) \in V$ is directionally differentiable, rendering $\hat{\mathcal{F}}$ directionally differentiable as well. Starting from (84) and noting that $T_{B_{\mathrm{ad}}}(\beta) = N_{B_{\mathrm{ad}}}(\beta)^\circ := \{d \in B : \langle d, \mu \rangle_{B,B^*} \leq 0 \text{ for all } \mu \in N_{B_{\mathrm{ad}}}(\beta)\}$, we characterize a solution β to (83) by

$$\hat{\mathcal{F}}'(\beta;d) \geq 0, \quad \text{for all } d \in T_{B_{\mathrm{ad}}}(\beta). \tag{85}$$

Here, $\hat{\mathcal{F}}'(\beta;\cdot)$ denotes the directional derivative of $\hat{\mathcal{F}}$ at β and $T_{B_{\mathrm{ad}}}(\beta)$ is the tangent cone to B_{ad} at β. The latter is the polar cone of $N_{B_{\mathrm{ad}}}(\beta)$, i.e., $N_{B_{\mathrm{ad}}}(\beta)^\circ$. The characterization (85) is called *B(ouligand)*-stationarity condition. In the stated form, it is generally difficult to check numerically. With the availability of an explicit characterization of $v'(\beta;\ \cdot)$, however, this condition may be related to a specific minimization task; see, e.g., Hintermüller et al. (2014). In such cases, often also equivalence to so-called *strong stationarity* may be shown. Through the explicit characterization of $v'(\beta;\ \cdot)$, strong stationarity (or briefly S-stationarity) is a primal-dual condition (similar to the KKT-system in classical nonlinear programming, with the (Lagrange) multipliers constituting the dual variables Zowe and Kurcyusz, 1979), whereas B-stationarity is a purely primal concept. It is well known that in general it is not possible to devise a numerical solution scheme which guarantees to terminate successfully at an S-stationary point. In special instances, e.g., under certain constraint regularity, however, algorithms may discover S-stationary points; see, e.g., Anitescu et al. (2007) for a finite dimensional account and (Hintermüller et al., 2015) for a specific infinite dimensional instance.

Depending on the problem structure, besides B- and S-stationarity the notions of C(larke)- and M(ordukhovich)-stationarity are relevant. From the viewpoint of numerical computability, C-stationarity is the most attractive format; see, e.g., Hintermüller and Surowiec (2016). For more on stationarity conditions in infinite dimensions we refer to Hintermüller and Kopacka (2009) and Hintermüller et al. (2014) and the references therein, and to Scheel and Scholtes (2000) for an exposition in finite dimensions.

We end this section by returning to (78) under the assumption that $V_{\mathrm{ad}} = V$ and $\mathcal{G}$ is (twice) continuously differentiable and convex. In view of convexity of the lower level problem and (81) we may replace (78) equivalently by

$$\begin{aligned} &\min \ \ \mathcal{F}(v, \beta) \ \ \text{over}\,(v, \lambda, \beta) \\ &\text{s.t.}\ \beta \in B_{\mathrm{ad}} \subset B, \\ &\qquad \mathcal{G}'(v;\beta) = 0 \text{ in } V^*. \end{aligned} \tag{86}$$

Since $\mathcal{G}'$ is once more differentiable, one may apply classical KKT-theory in Banach space Zowe and Kurcyusz (1979) to obtain a primal-dual characterization of a solution to (86), provided $\mathcal{G}''(v, \beta)$ along with B_{ad} satisfies the ZKR-CQ. This CQ also allows to locally reduce the problem through the implicitly defined function $\beta \mapsto v(\beta)$ with $\mathcal{G}'(v(\beta), \beta) = 0$. Similar to before, this yields (83). Under our smoothness requirements and nondegeneracy assumption on $\mathcal{G}''$ at $(v(\beta), \beta)$ (through ZKR-CQ) the implicit function theorem yields (local) differentiability of $\beta \mapsto v(\beta)$ and, thus, of $\beta \mapsto \hat{\mathcal{F}}(\beta)$. In this case, (83) may be solved numerically by a projected gradient method, provided the structure of B_{ad} is sufficiently simple allowing for an efficient computation of the projection of the iterates. The latter viewpoint will be relevant in our numerical approach to computing structured nonsmooth priors in Sections 4.2 and 5 of this chapter.

4.2 Bilevel optimization—A monolithic approach

The exposition of this section is related to Hintermüller and Rautenberg (2017) and Hintermüller et al. (2017) and the references therein which address an automated statistics-based selection of the weight function $\boldsymbol{\alpha}$ in TV minimization problems. In this way, a data-motivated structuring of a nonsmooth prior will be obtained. As we mentioned in the introduction as well, the main idea of these approaches is to minimize a suitable upper level objective over both, the image u and the regularization parameter $\boldsymbol{\alpha}$ subject to u being a solution to a (regularized) variational regularization problem with parameter $\boldsymbol{\alpha}$. As a consequence of Fenchel-Rockafellar duality, equivalently the predual problem can be used instead of the primal problem in the lower level.

The upper level objective we discuss here is based on localized residuals. The latter are defined as a function $R : L^2(\Omega) \to L^\infty(\Omega)$ by

$$\mathrm{R}u(x) := \int_\Omega w(x,y)(Tu-f)^2(y)dy, \tag{87}$$

with w a localizing weight. We note that $\mathrm{R}u(x)$ can be interpreted as a local variance. Indeed, given Gaussian noise of variance σ^2 and zero mean, we have $\int_\Omega (Tu_{true}-f)^2\, dx = \int_\Omega \eta^2\, dx = \sigma^2|\Omega|$. Consequently, if a reconstructed image u is close to u_{true} then it is expected that for every $x \in \Omega$, $\mathrm{R}u(x)$ will be close to σ^2. Hence it is natural to consider an upper level objective which aims to keep Ru within certain tight bounds $\underline{\sigma}, \overline{\sigma}$ where $\underline{\sigma} < \sigma < \overline{\sigma}$. This can be achieved through the function $F : L^2(\Omega) \to \mathbb{R}$ with

$$F(v) := \frac{1}{2}\int_\Omega \max(v-\overline{\sigma}^2, 0)^2 dx + \frac{1}{2}\int_\Omega \min(v-\underline{\sigma}^2, 0)^2 dx. \tag{88}$$

Hence, by minimizing $F(\mathrm{R}u)$ one aims to find $\mathrm{R}u$ close to the variance σ^2. In order to illustrate the viability of this choice of an upper level objective we evaluate the upper objective functional $F(\mathrm{R}\cdot)$ for a series of scalar TV denoising restorations obtained from a variety of parameters α for the example image shown in the introduction (Figs. 1–3). As we see in Fig. 5, the minimum value of the functional $F(\mathrm{R}\cdot)$ is achieved for scalar parameter values α (for TV), and (α_0, α_1) (for TGV) that are not far away from those that are

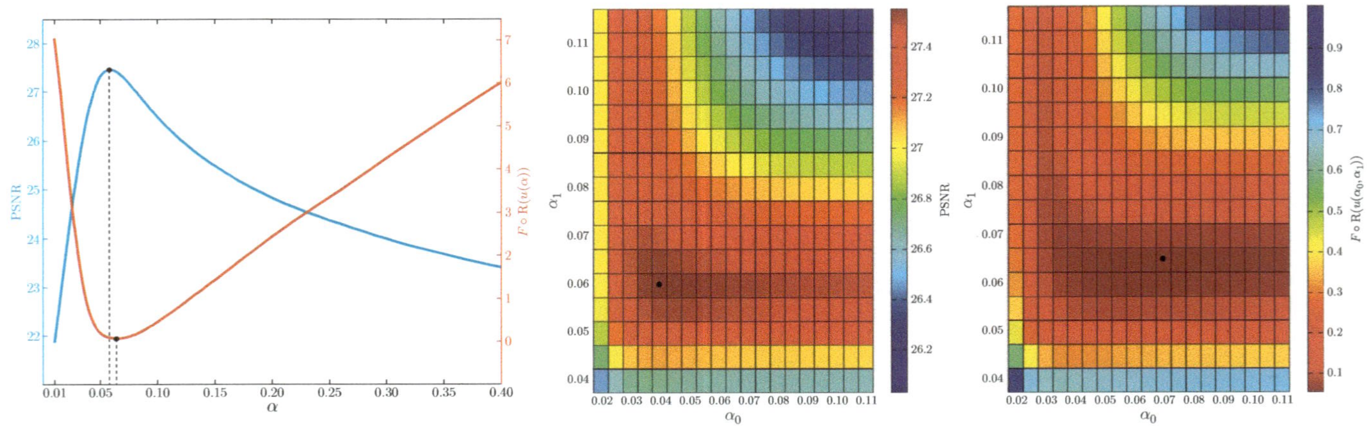

FIG. 5 Illustration for the suitability of the functional $F(\mathrm{R}\cdot)$, see (87) and (88), as an upper level objective. Evaluation of $F(\mathrm{R}u)$ where u solves the (anisotropic, unregularized) primal TV (*left*) and TGV (*right*) denoising problems for a variety of scalar regularization parameters α and (α_0, α_1), respectively. The points where this functional is minimal, respectively, is in both cases close to the points that maximize the PSNR. The minimum (for $F(\mathrm{R}\cdot)$) and maximum (for PSNR) points are denoted with a bullet in all graphs. We used the values $\underline{\sigma} = 0.00798$, $\overline{\sigma} = 0.01202$ and w a spatially invariant averaging filter of size 7×7.

optimizing the PSNR. Note moreover that in order to optimize the PSNR one needs the ground truth image u_{true}! This, however, is not the case for $F(R\cdot)$.

As we have also discussed previously we must impose also certain regularity for the regularization parameters $\boldsymbol{\alpha}$ in order for the dualization results to hold. Here we impose H^1-regularity which turns out to also facilitate existence and differentiability analysis, even though the consideration of lower regularity of α is certainly of interest. Furthermore, for reasons we discuss later, we need to constrain these parameters in box sets of the type

$$\mathcal{A}_{ad} = \{\alpha \in H^1(\Omega) : \underline{\alpha} \le \alpha \le \overline{\alpha}\}, \tag{89}$$

$$\begin{aligned} \mathcal{A}^0_{ad} &= \{\alpha_0 \in H^1(\Omega) : \underline{\alpha}_0 \le \alpha_0 \le \overline{\alpha}_0\}, \\ \mathcal{A}^1_{ad} &= \{\alpha_1 \in H^1(\Omega) : \underline{\alpha}_1 \le \alpha_1 \le \overline{\alpha}_1\}, \end{aligned} \tag{90}$$

for TV and TGV, respectively. Here $\underline{\alpha}, \overline{\alpha} \in L^2(\Omega)$ with $0 < \epsilon_0 \le \underline{\alpha}(x) < \overline{\alpha}(x) - \underline{\epsilon}_1$ in Ω for some $\underline{\epsilon}_1 > 0$ and similarly for $\underline{\alpha}_0$, $\overline{\alpha}_0$ and $\underline{\alpha}_1$, $\overline{\alpha}_1$.

Recalling that the localized residuals can be expressed in terms of the dual variables p as

$$\mathrm{R}u(x) = R(\mathrm{div}p)(x) = \int_\Omega w(x,y)\left(TB^{-1}\mathrm{div}p - (TB^{-1}T^* - I)f\right)^2 dy, \tag{91}$$

$$\mathrm{R}u(x) = R(\mathrm{div}^2 p)(x) = \int_\Omega w(x,y)\left(TB^{-1}\mathrm{div}^2 p - (TB^{-1}T^* - I)f\right)^2 dy, \tag{92}$$

for the TV and TGV problems, respectively, the associated bilevel problems (which are related to (86) with appropriate settings for the various constituents) are given by

$$\begin{cases} \min J_{\mathrm{TV}}(p,\alpha) := F(R(\mathrm{div}p)) + \dfrac{\lambda}{2}\|\alpha\|^2_{H^1(\Omega)} \quad \text{over } (p,\alpha) \in H^1_0(\Omega)^d \times \mathcal{A}_{ad} \\ \text{s.t. } p \in \underset{p \in H^1_0(\Omega)^d}{\mathrm{argmin}} \dfrac{\beta}{2}\|\nabla p\|^2_{L^2(\Omega)^d} + \dfrac{\gamma}{2}\|p\|^2_{L^2(\Omega)^d} + \dfrac{1}{2}\|\mathrm{div}p + T^* f\|^2_B + \dfrac{1}{\epsilon}\mathcal{P}_\delta(p,\alpha), \end{cases} \tag{$\mathbb{P}_{\mathrm{TV}}$}$$

and

$$\begin{cases} \min J_{\mathrm{TGV}}(p,\alpha_0,\alpha_1) := F(R(\mathrm{div}^2 p)) + \dfrac{\lambda_0}{2}\|\alpha_0\|^2_{H^1(\Omega)} + \dfrac{\lambda_1}{2}\|\alpha_1\|^2_{H^1(\Omega)} \\ \quad \text{over } (p,\alpha) \in H^2_0(\Omega,\mathcal{S}^{d\times d}) \times \mathcal{A}^0_{ad} \times \mathcal{A}^1_{ad} \\ \text{s.t. } p \in \underset{p \in H^2_0(\Omega,\mathcal{S}^{d\times d})}{\mathrm{argmin}} \dfrac{\beta}{2}\|\Delta p\|^2_{L^2(\Omega,\mathcal{S}^{d\times d})} + \dfrac{\gamma}{2}\|p\|^2_{L^2(\Omega,\mathcal{S}^{d\times d})} + \dfrac{1}{2}\|T^* f - \mathrm{div}^2 p\|^2_B \\ \qquad + \dfrac{1}{\epsilon_0}\mathcal{Q}_\delta(p,\alpha_0) + \dfrac{1}{\epsilon_1}\mathcal{P}_\delta(\mathrm{div}p,\alpha_1), \end{cases} \tag{$\mathbb{P}_{\mathrm{TGV}}$}$$

where $\mathcal{Q}_\delta$ is a smooth(ed) penalty function penalizing violations of $|p| \le \alpha_0$.

Here, $\mathcal{S}^{d\times d}$ refers to the set of real, symmetric $d \times d$ matrices.

Existence of solutions to the problems ($\mathbb{P}_{\text{TV}}$) and ($\mathbb{P}_{\text{TGV}}$) can be shown using standard arguments; see for instance (Hintermüller and Rautenberg, 2017, Theorem 6.1) for the TV case. We note that as we have shown in the dualization section, H^1-regularity of α suffices to establish the connection between the primal and predual problems in the TV case. In Hintermüller et al. (2017), α was further enforced to be a $C(\bar{\Omega})$-function which was guaranteed by a regularity result of H^1-projections onto the set $\mathcal{A}_{ad}$ which was part of the associated solution algorithm. In particular the following result was shown Hintermüller et al. (2017, Corollary 2.3).

Proposition 7. *Let $\Omega \subset \mathbb{R}^d$, $d \leq 3$, be a bounded convex set and*

$$\mathcal{A} = \{\alpha \in H^1(\Omega) :\ \underline{\alpha} \leq \alpha \leq \overline{\alpha}\},\quad \text{with}\ \underline{\alpha}, \overline{\alpha} \in H^2(\Omega), \frac{\partial \underline{\alpha}}{\partial \nu} = \frac{\partial \overline{\alpha}}{\partial \nu} = 0. \tag{93}$$

Let $P_{\mathcal{A}} : H^1(\Omega) \to \mathcal{A} \subset H^1(\Omega)$ denote the projection operator, that is

$$P_{\mathcal{A}}(\omega) := \operatorname*{argmin}_{\alpha \in \mathcal{A}} \frac{1}{2} \| \alpha - \omega \|_{H^1(\Omega)}^2, \quad \omega \in H^1(\Omega).$$

Then, if $\omega^* = P_{\mathcal{A}}(\omega)$, the following implication holds true:

$$\omega \in H^2(\Omega)\ \text{and}\ \frac{\partial \omega}{\partial \nu} = 0 \quad \Longrightarrow \quad \omega^* \in H^2(\Omega)\ \text{and}\ \frac{\partial \omega^*}{\partial \nu} = 0.$$

Furthermore,

$$\max\left(\| \omega^* \|_{H^2(\Omega)}, \| \omega^* \|_{C^{0,r}(\bar{\Omega})}\right) \leq C\left(\| L\omega \|_{L^2(\Omega)} + \| L\underline{\alpha} \|_{L^2(\Omega)} + \| L\overline{\alpha} \|_{L^2(\Omega)}\right)$$

for some $r \in (0, 1)$ and with $L = -\Delta + I$.

Hence, in order for $C(\bar{\Omega})$ regularity for the weight function α to be guaranteed by the projection it suffices that $\underline{\alpha}$, $\overline{\alpha}$ satisfy the conditions in (93). In particular this is satisfied when these are constants. Furthermore, since—as we will see later in the algorithm—this projection is performed in an iterative fashion, the initialization for α must satisfy (93).

As we have discussed in the previous section, it is not yet shown that $W^{1,1}$ regularity for α_0 and α_1 suffices to establish a dualization framework for TGV, even though one expects that it can be shown with similar arguments. Hence, for TGV we will follow the H^1-projection regularity result as described above. In fact as we will see in the (preliminary) numerics for TGV, we treat only α_1 as a spatially varying function and α_0 remains a constant.

We make another remark regarding the box constraints for the parameters (89) and (90). In De Los Reyes et al. (2016) it was shown that a

PSNR-optimizing upper level objective $\widetilde{J}(u,\alpha) = \| u(\alpha) - f \|^2_{L^2(\Omega)}$ subject to H^1 and Huber-regularized TV and TGV denoising problems produces under some mild conditions optimal scalar solutions α and (α_0, α_1) that are strictly positive. This, however, appears to require to solve a nonconvex problem to global optimality. Although, as we have already depicted in Fig. 5, the upper level objective we discuss here is not far away from optimizing the PSNR, keeping the parameters strictly positive via (89) and (90) seems indeed necessary.

Let us next address how to treat the bilevel problems $(\mathbb{P}_{\text{TV}})$ and $(\mathbb{P}_{\text{TGV}})$ algorithmically. For this purpose, let $\alpha \mapsto p(\alpha)$ and $(\alpha_0, \alpha_1) \mapsto p(\alpha_0, \alpha_1)$ denote the solution map of the lower level problems, equivalently the solutions of the optimality conditions (71) and analogously for the TGV problem. Then the problems $(\mathbb{P}_{\text{TV}})$ and $(\mathbb{P}_{\text{TGV}})$ admit the following reduced versions

$$\min \quad \hat{J}_{\text{TV}}(\alpha) := J_{\text{TV}}(p(\alpha), \alpha), \qquad \text{over}\, \alpha \in \mathcal{A}_{ad}, \tag{94}$$

$$\min \quad \hat{J}_{\text{TGV}}(\alpha_0, \alpha_1) := J_{\text{TGV}}(p(\alpha_0, \alpha_1), \alpha_0, \alpha_1), \quad \text{over}\, \alpha_0 \in \mathcal{A}^0_{ad},\ \alpha_1 \in \mathcal{A}^1_{ad}. \tag{95}$$

In view of our general discussion in the preceding subsection we have arrived at the level of (83) with corresponding settings for $\hat{\mathcal{F}}$, β, $v(\beta)$ and B_{ad} and a smooth, i.e., differentiable reduced objective. Indeed, the reduced functional $\hat{J}_{\text{TV}} : H^1(\Omega) \to \mathbb{R}$ is differentiable as a composition of differentiable functions. One can show a similar result for $\hat{J}_{\text{TGV}}$, as well. We note, however, that from now on we will consider only scalar α_0 yielding $\mathcal{A}^0_{ad} = \{\alpha_0 \in \mathbb{R} : \underline{\alpha}_0 \le \alpha_0 \le \overline{\alpha}_0\}$ where $\underline{\alpha}_0, \overline{\alpha}_0 \in \mathbb{R}$, and also we set $\lambda_0 := 0$ in $(\mathbb{P}_{\text{TGV}})$.

In order to proceed, we now recall some basic facts from optimization (Karush–Kuhn–Tucker) theory in Banach spaces (Zowe and Kurcyusz, 1979) adapted to our purposes. Let V, $\mathcal{A}$, Z be Banach spaces, $X = V \times \mathcal{A}$, $\theta : X \to \mathbb{R}$ and $g : X \to Z$ continuously Fréchet differentiable functions. Let, moreover, $\mathcal{C} \subset X$ be a nonempty, closed convex set. Consider the problem

$$\begin{cases} \min\limits_{x \in X} \theta(x) \\ \text{s.t.} \ \ x \in \mathcal{C} \text{ and } g(x) = 0. \end{cases} \tag{96}$$

and suppose that the following constraint qualification holds for a solution $\overline{x}$ of (96)

$$g'(\overline{x})[\mathcal{C}(\overline{x})] + \{\lambda g(\overline{x}) : \lambda \ge 0\} = Z, \quad \text{where}\, \mathcal{C}(\overline{x}) = \{\lambda(c - \overline{x}) : c \in \mathcal{C},\ \lambda \ge 0\}. \tag{97}$$

Then there exists an adjoint state (Lagrange multiplier) $z^* \in Z^*$ that fulfils the following condition (Zowe and Kurcyusz, 1979)

$$\theta'(\overline{x}) - z^*(g'(\overline{x})) \in C(\overline{x})^+, \tag{98}$$

where $C(\overline{x})^+ := \{x^* \in X^* : x^*(x) \geq 0, \text{ for all } x \in C(\overline{x})\}$.

Assume now that for every $\xi \in \mathcal{A}$ the equation $g(x) = 0$ has a unique solution $v(\xi) \in U$, that is $g(v(\xi), \xi) = 0$. Further assume that $g_v(v(\xi), \xi) \in \mathcal{L}(V, Z)$ is continuously differentiable. Then from the implicit function theorem (see, e.g., Hinze et al., 2009 for details), we get that $v(\xi)$ is continuously differentiable. Then one can define the reduced problem

$$\begin{cases} \min\limits_{\xi \in \mathcal{A}} \hat{\theta}(\xi) := \theta(v(\xi), \xi) \\ \text{s.t. } \ \xi \in \hat{C} := \{\xi \in \mathcal{A} : (v(\xi), \xi) \in C\}, \end{cases} \tag{99}$$

where now $\hat{\theta}$ is differentiable as well. The derivative of the reduced functional $\hat{\theta}' \in \mathcal{A}^*$ can be then computed with the help of z^* as follows

$$\hat{\theta}'(\xi) = \theta_\xi(v(\xi), \xi) + g_\xi(v(\xi), \xi)^* z^*, \tag{100}$$

where $g_\xi(v(\xi), \xi)^* \in \mathcal{L}(Z^*, \mathcal{A}^*)$ is the adjoint operator of $g_\xi(v(\xi), \xi) \in \mathcal{L}(\mathcal{A}, Z)$.

Coming back to our bilevel problems, we can easily see that the above abstract framework can be adjusted to the problems ($\mathbb{P}_{\text{TV}}$), ($\mathbb{P}_{\text{TGV}}$) and their corresponding reduced problems (94) and (95). Indeed considering ($\mathbb{P}_{\text{TV}}$) first, we set $V = H_0^1(\Omega)^d$, $\mathcal{A} = H^1(\Omega)$, $Z = V^*$, $C = V \times \mathcal{A}_{ad}$, $\theta(x) = J_{\text{TV}}(p, \alpha)$ and

$$g_{\text{TV}}(x) = -\beta \Delta p + \gamma p - \nabla B^{-1} \text{div} p - \nabla B^{-1} T^* f + \frac{1}{\epsilon} P_\delta(p, \alpha)$$

Then it can be shown that all necessary differentiability results hold; see Hintermüller and Rautenberg (2017) for details, and the corresponding condition to (98) read for the adjoint variable $q \in V$ and an optimal pair $(\overline{p}, \overline{\alpha}) \in V \times \mathcal{A}$

$$\langle \text{div}^* J_0'(\text{div} \overline{p}), p \rangle_{V^*,V} + \langle -\beta \Delta q + \gamma q - \nabla B^{-1} \text{div} q + \frac{1}{\epsilon} D_1 P_\delta(\overline{p}, \overline{\alpha}) q, p \rangle_{V^*,V} = 0, \tag{101}$$

$$\langle \lambda(-\Delta + I)\overline{\alpha} + \frac{1}{\epsilon}(D_2 P_\delta(\overline{p}, \overline{\alpha}))^* q, \alpha - \overline{\alpha} \rangle_{\mathcal{A}^*, \mathcal{A}} \geq 0, \tag{102}$$

for all $p \in V$ and for all $\alpha \in \mathcal{A}_{ad}$, where $J_0(\cdot) := F(R \cdot)$. Moreover D_1 and D_2 denote derivatives with respect to the variables p and α, respectively.

The derivative of $\hat{J}_{\text{TV}}$ can be computed as

$$\hat{J}_{\text{TV}}'(\alpha) = \lambda(-\Delta + I)\alpha + \frac{1}{\epsilon}(D_2 P_\delta(p(\alpha), \alpha))^* q(\alpha) \tag{103}$$

where $q(\alpha)$ solves (101) for $\overline{\alpha} = \alpha$ and $\overline{p} = p(\alpha)$.

Turning now to the bilevel TGV problem ($\mathbb{P}_{\text{TGV}}$), the corresponding, spaces, sets and functions will be $V = H_0^2(\Omega, \mathcal{S}^{d\times d})$, $\mathcal{A} = \mathbb{R}\times H^1(\Omega)$, $Z = V^*$, $\mathcal{C} = V\times \mathcal{A}_{ad}^0 \times \mathcal{A}_{ad}^1$, $\theta(\mathrm{x}) = J_{\text{TGV}}(p, \alpha_0, \alpha_1)$ and

$$g(\mathrm{x}) = \beta\Delta^2 p + \gamma p + \nabla^2 B^{-1}\mathrm{div}^2 p - \nabla^2 B^{-1} T^* f + \frac{1}{\epsilon_0} Q_\delta(p, \alpha_0) - \frac{1}{\epsilon_1}\nabla P_\delta(\mathrm{div} p, \alpha_1)$$

The adjoint variable $q\in V$ satisfies for an optimal triplet $(\overline{p}, \overline{\alpha}_0, \overline{\alpha}_1)$

$$\Big\langle (\mathrm{div}^2)^* J_0'(\mathrm{div}^2\overline{p}, p)\rangle_{V^*,V} + \langle \beta\Delta q + \gamma p + \nabla^2 B^{-1}\mathrm{div}^2 q + \frac{1}{\epsilon_0} D_1 Q_\delta(\overline{p}, \overline{\alpha}_0) q \\ - \frac{1}{\epsilon_1} D_1 \nabla P_\delta(\overline{p}, \overline{\alpha}_1) q, p\Big\rangle_{V^*,V} = 0, \tag{104}$$

$$\langle \lambda(-\Delta + I)\overline{\alpha}_1 - \frac{1}{\epsilon_1}(D_2\nabla P_\delta(\overline{p}, \overline{\alpha}_1))^* q, \alpha_1 - \overline{\alpha}_1\rangle_{H^1(\Omega)^*, H^1(\Omega)} \geq 0, \tag{105}$$

$$\Big\langle \frac{1}{\epsilon_0}(D_2 Q_\delta(\overline{p}, \overline{\alpha}_0))^* q, \alpha_0 - \overline{\alpha}_0 \Big\rangle_{\mathbb{R},\mathbb{R}} \geq 0, \tag{106}$$

for all $p\in V$, $\alpha_0 \in \mathcal{A}_{ad}^0$ and $\alpha_1\in\mathcal{A}_{ad}^1$. The derivative of the reduced objective is then computed by

$$\hat{J}'_{\text{TGV}}(\alpha_0, \alpha_1) = \lambda(-\Delta + I)\alpha_1 + \left(\frac{1}{\epsilon_0}(D_2 Q_\delta(p, \alpha_0)), -\frac{1}{\epsilon_1}(D_2\nabla P_\delta(p, \alpha_1))\right)^* q(\alpha_0, \alpha_1) \tag{107}$$

where again $q(\alpha_0, \alpha_1)$ solves (104) for $\overline{\alpha}_0 = \alpha_0$, $\overline{\alpha}_1 = \alpha_1$ and $\overline{p} = p(\alpha_0, \alpha_1)$.

For the reduced derivatives we have that $\hat{J}'_{\text{TV}}(\alpha) \in H^1(\Omega)^*$ and $\hat{J}'_{\text{TGV}}(\alpha_0, \alpha_1) \in (H^1(\Omega)\times\mathbb{R})^*$. In order to obtain the gradient of these functional which are essential for the design of gradient-based descent algorithms we can apply the inverse Riesz maps to the reduced gradients as follows

$$\nabla\hat{J}_{\text{TV}}(\alpha) := \mathcal{R}_{H^1}^{-1}\hat{J}'_{\text{TV}}(\alpha) \in H^1(\Omega), \tag{108}$$

$$\nabla\hat{J}_{\text{TGV}}(\alpha_0, \alpha_1) := \left(\mathcal{R}_{H^1}^{-1} P_1 \hat{J}'_{\text{TGV}}(\alpha_0, \alpha_1), P_2\hat{J}'_{\text{TGV}}(\alpha_0, \alpha_1)\right) \in H^1(\Omega)\times\mathbb{R}, \tag{109}$$

where for $(r_1, r_2) \in H^1 \times \mathbb{R}$ we have

$$\hat{J}'_{\text{TGV}}(\alpha_0, \alpha_1)[r_1, r_2] = P_1\hat{J}'_{\text{TGV}}(\alpha_0, \alpha_1)[r_1] + P_2\hat{J}'_{\text{TGV}}(\alpha_0, \alpha_1)[r_2].$$

The above representation of the derivative and gradients of the reduced functional together with the regularity results for the H^1-projection provide the basis for devising a function space-based projected gradient algorithm. Indeed such an algorithm is devised in Hintermüller et al. (2017) (compare *Algorithm 1* in that reference) for the case of bilevel TV and it can be similarly done for bilevel TGV.

5 Numerical examples

In the section we will provide some numerical examples for the bilevel TV and TGV algorithms which aim to solve the discretized versions of ($\mathbb{P}_{TV}$) and ($\mathbb{P}_{TGV}$), respectively. We will also discuss some discretization and algorithmic aspects.

In the discrete setting, (grey scale) images are considered as functions from $\Omega_h \to \mathbb{R}$ where $\Omega_h = \{1, 2, \ldots, n\} \times \{1, 2, \ldots, m\}$ is a discrete Cartesian grid that corresponds to the image pixels. We defined the mesh size, that is the distance between the grid points as $h = 1/\sqrt{nm}$. All the discrete differential operators will be computed on these points. However, we also emphasize that computing primal, dual variables as well as differential operators and their adjoints in different points has to be done with care. First, one has to devise discretizations which respect, on the discrete level, the relation between a discrete operator (including boundary conditions) and its associated adjoint. Moreover, so-called *inf-sup*-stability of the discretization is relevant. It takes care of a robust relation between the primal and dual problems under resolution refinements. For the predual of the TV problem, it has been found (Hintermüller et al., 2014) that a staggered grid discretization of the vector field p and the associated differential operators successfully eliminates some discretization artefacts. We depict such an example in Fig. 6 where the dual TV denoising problem (isotropic) is solved by using a standard discretization and a staggered grid as it is described in Hintermüller et al. (2014). It is known theoretically (Meyer, 2001) that applying the isotropic L^2–TV denoising problem with data f being a characteristic function of a disk yields as result a rescaled version of f (simple contrast lost). However, this is not the case with the standard discretization as a certain smoothing is observed on edges not aligned with the grid. By using a staggered grid this effect is essentially eliminated. We would also like to point out that examples like this demonstrate the necessity of studying variational imaging problems in function space setting in order to study the models independently of discretization artefacts. Further uses of staggered grid discretization methods can also be found in Hahn et al. (2012) and Tai et al. (2011b).

FIG. 6 Comparison of the standard vs staggered grid discretization see Hintermüller et al. (2014). The latter produces a sharper result which is closer to the one belonging to the original functional space problem.

5.1 Discrete operators for ($\mathbb{P}_{\text{TV}}$)

We will generally keep the same notation for the discrete differential operators (than for continuous ones). We set the following discrete spaces

$$U_h = \mathbb{R}^{\Omega_h}, \quad W_h = U_h \times U_h.$$

For an element $u \in U_h$ the discrete ℓ^2 norm is defined as

$$\| u \|^2_{\ell^2(\Omega_h)} = h^2 \sum_{(i,j)\in\Omega_h} |u_{i,j}|^2 \tag{110}$$

We define the discrete gradient $\nabla : U_h \to W_h$ and divergence operator $\text{div} : W_h \to U_h$ such that they satisfy the adjoint relation $\nabla = -\text{div}^\top$. For $p = (p^1, p^2) \in W_h$ the divergence is defined with standard backward differences as

$$(\text{div}p)_{i,j} = \frac{1}{h}\left(p^1_{i,j} - p^1_{i-1,j} + p^2_{i,j} - p^2_{i,j-1}\right), \quad (i,j) \in \Omega_h$$

setting $p^1_{\tilde{i},\tilde{j}} = p^2_{\tilde{i},\tilde{j}} = 0$ if $(\tilde{i},\tilde{j})$ falls outside the grid (ghost grid points). We also need to defined the discrete vectorial Laplacian $\Delta : W_h \to W_h$ where $\Delta p = (\Delta_D p^1, \Delta_D p^1)$ with $\Delta_D : U_h \to U_h$. In order to be consistent with the function space setting where $p \in H^1_0(\Omega)^2$ the discrete Laplacian must impose zero Dirichlet boundary conditions, hence (following good practice in numerical analysis) we avoid defining Δ_D as $\text{div}\nabla$ as the latter would result in (unsuitable) mixed boundary conditions. Thus, we define Δ_D using the discrete five-point Laplacian stencil

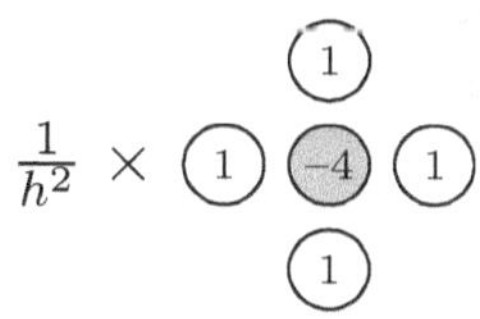

and again setting $p_{\tilde{i},\tilde{j}} = 0$ for ghost grid points. We will also make use of the discrete Laplacian with zero Neumann boundary conditions $\Delta_N : U_h \to U_h$ which will be used to act on the weight function α. These are the desired boundary conditions for α dictated by Proposition 7. For this purpose, we use the same stencil but we set the function value of ghost grid points to be the same as the function value of the nearest grid point in Ω_h. We follow a similar approach for the $\nabla\text{div} : W_h \to W_h$ operator for the denoising case where $T = B = Id$. We discretize ∇div directly as follows

$$\begin{aligned}(\nabla\text{div}p)_{i,j} = \frac{1}{h^2}\Big(&p^1_{i+1,j} - 2p^1_{i,j} + p^1_{i-1,j} + p^2_{i+1,j} - p^2_{i+1,j-1} - p^2_{i,j} + p^2_{i,j-1} \\ &p^2_{i,j+1} - 2p^2_{i,j} + p^2_{i,j-1} + p^1_{i,j+1} - p^1_{i-1,j+1} - p^1_{i,j} + p^1_{i-1,j}\Big)\end{aligned}$$

by setting again zero values outside the grid. Typically, in the implementations the variables are regarded as long vectors resulting from concatenating the columns of the pixel mask-based matrix representation.

We will also need discrete version of H^1-types of norms and norms of the corresponding duals spaces. For the discrete H^1-norm acting on the weight function we use

$$\|\alpha\|_{H^1(\Omega_h)} = h\sqrt{\alpha^\top (I-\Delta_N)\alpha} \tag{111}$$

while the dual norm is defined as

$$\|r\|_{H^1(\Omega_h)^*} = \|(I-\Delta_N)^{-1} r\|_{H^1(\Omega_h)} = h\sqrt{r^\top (I-\Delta_N)^{-1} r} \tag{112}$$

based on the $H^1(\Omega) \to H^1(\Omega)^*$ Riesz map $\alpha \mapsto r = (I - \Delta_N)\alpha$. We will also use the discrete dual $H_0^1(\mathrm{div}, \Omega_h)$-norm as

$$\|v\|^*_{H_0^1(\mathrm{div},\Omega_h)} = h\sqrt{v^\top (I - \nabla \mathrm{div})^{-1} v}. \tag{113}$$

For the discrete version of the averaging filter in the definition of the localized residuals (87) we use a spatially invariant averaging filter of size $n_w \times n_w$, that is, with entries of equal value whose sum is equal to one.

With these definitions the discrete version of $(\mathbb{P}_{\mathrm{TV}})$ is the following

$$\begin{cases} \min \frac{1}{2}\|(R(\mathrm{div}p) - \overline{\sigma}^2)^+\|^2_{\ell^2(\Omega_h)} + \frac{1}{2}\|(\underline{\sigma}^2 - R(\mathrm{div}p))^+\|^2_{\ell^2(\Omega_h)} + \frac{\lambda}{2}\|\alpha\|^2_{H^1(\Omega_h)} \\ \quad \text{over } (p,\alpha) \in W_h \times (\mathcal{A}_{ad})_h \\ \text{such that } \; -\beta\Delta_D p + \gamma p - \nabla B^{-1}\mathrm{div} p - \nabla B^{-1} T^* f + \frac{1}{\epsilon} P_\delta(p,\alpha) = 0. \end{cases} \tag{$\mathbb{P}^{\mathrm{h}}_{\mathrm{TV}}$}$$

Here we have

$$(\mathcal{A}_{ad})_h = \{\alpha \in U_h : \underline{\alpha} \le \alpha_{i,j} \le \overline{\alpha}, \text{ for all } (i,j) \in \Omega_h\}, \tag{114}$$

and $R(\mathrm{div}p)$ being a discrete convolution, with periodic boundary conditions. The penalty function $P_\delta : W_h \to W_h$ is also defined straightforwardly in the discrete setting by pointwise application of the function G_δ'.

Regarding the choice of the lower and upper bounds for the local variance $\underline{\sigma}^2$ and $\overline{\sigma}^2$, respectively, we follow here the rules

$$\overline{\sigma}^2 = \sigma^2\left(1 + \frac{\sqrt{2}}{n_w}\right), \quad \underline{\sigma}^2 = \sigma^2\left(1 - \frac{\sqrt{2}}{n_w}\right) \tag{115}$$

where σ^2 is the variance of Gaussian noise which is assumed to be known (or estimated from data). The derivation of the formulae (115) is done using statistics of the extremes; see Hintermüller et al. (2017, Section 4.2.1).

ALGORITHM 2 Discretized projected gradient method for the solution of the bilevel TV image reconstruction problem ($\mathbb{P}^h_{\mathrm{TGV}}$)

Input: f, $\underline{\alpha}$, $\overline{\alpha}$, $\overline{\sigma}$, $\underline{\sigma}$, λ, β, γ, ϵ, δ, n_w τ^0, $0 < c < 1$, $0 < \theta_- < 1 \leq \theta_+$
Initialise: $\alpha^0 \in \mathcal{A}^h_{ad}$ and set $k := 0$.
repeat
Use Algorithm 1 to compute the solution p^k of the lower level problem

$$g_{\mathrm{TV}}(p^k, \alpha^k) := -\beta\Delta_D p^k + \gamma p^k - \nabla B^{-1}\mathrm{div}p - \nabla B^{-1}T^* f + \frac{1}{\epsilon}P_\delta(p^k, \alpha^k) = 0$$

Solve the adjoint equation for q^k

$$-\nabla B^{-1}\mathrm{div}q^k - \beta\Delta q^k + \gamma q^k + \frac{1}{\epsilon}\left(G''_\delta(p^k - \alpha^k\mathbf{1}) + G''_\delta(-p^k - \alpha^k\mathbf{1})\right) q^k$$
$$= 2\nabla B^{-1}T^*\mathrm{div}p^k\left(w * \left((R(\mathrm{div}p^k) - \overline{\sigma}^2)^+ - (\underline{\sigma}^2 - R(\mathrm{div}p^k))^+\right)\right)$$

Compute the reduced derivative

$$\hat{J}'_{\mathrm{TV}}(\alpha^k) = \frac{1}{\epsilon}\mathbf{1}^\top\left(-G''_\delta(p^k - \alpha^k\mathbf{1}) + G''_\delta(-p^k - \alpha^k\mathbf{1})\right) q^k + \lambda(I - \Delta_N)\alpha^k$$

Compute the reduced gradient

$$\nabla\hat{J}_{\mathrm{TV}}(\alpha^k) = (I - \Delta_N)^{-1}\hat{J}'_{TV}(\alpha^k)$$

Compute the trial point $\alpha^{k+1} = P_{(\mathcal{A}_{ad})_h}(\alpha^k - \tau^k\nabla\hat{J}_{\mathrm{TV}}(\alpha^k))$
while $\hat{J}_{\mathrm{TV}}(\alpha^{k+1}) > \hat{J}_{\mathrm{TV}}(\alpha^k) + c\hat{J}'_{\mathrm{TV}}(\alpha^k)^\top(\alpha^{k+1} - \alpha^k)$ **do** (Armijo line search)
Set $\tau^k := \theta_-\tau^k$ and re-compute $\alpha^{k+1} = P_{(\mathcal{A}_{ad})_h}(\alpha^k - \tau^k\nabla\hat{J}_{\mathrm{TV}}(\alpha^k))$
end while
Update $\tau^{k+1} = \theta_+\tau^k$ and $k := k + 1$
until some stopping condition is satisfied

We now proceed with describing the algorithm for the numerical solution of ($\mathbb{P}^h_{\mathrm{TV}}$). In fact, we use a discretized projected gradient method with Armijo line search for globalization. The discretized gradient of the reduced functional is computed with the help of the adjoint equation which is the discrete version of (101). We summarize this in Algorithm 2. We note that in the algorithm above **1** denotes the matrix $[Id; Id]$ of size $nm \times 2nm$.

A few remarks on the solution of the lower level problem are in order. The target is to solve the problem for sufficiently small ϵ using Algorithm 1. In order to ensure robustness, we employ a path-following approach which, starting for a large ϵ^ℓ, successively solves $g_{\mathrm{TV},\epsilon}(p^\ell, \alpha) = 0$ for $\epsilon = \epsilon_\ell$ until a fixed tolerance

$$g_{\mathrm{TV},\epsilon^\ell}(p^{\ell+1}, \alpha) < \mathrm{tol}_{(l)}$$

is reached. Then $\epsilon^{\ell+1}$ is obtained by decreasing ϵ^ℓ by a factor $0 < \theta_\epsilon < 1$. This yields $\epsilon^{\ell+1} := \max(\theta_\epsilon\epsilon^\ell, \underline{\epsilon})$ for some $0 < \epsilon \ll 1$; compare also Hintermüller et al. (2017, Algorithm 3).

ALGORITHM 3 Discretized projected gradient method of the solution of the bilevel TGV image reconstruction problem ($\mathbb{P}^h_{\mathrm{TGV}}$).

Input: f, $\underline{\alpha}_0$, $\overline{\alpha}_0$, $\underline{\alpha}_1$, $\overline{\alpha}_1$, $\overline{\sigma}$, $\underline{\sigma}$, λ, β, γ, ϵ_0, ϵ_1, δ, n_w τ_0^0, τ_1^0, $0 < c < 1$, $0 < \theta_- < 1 \leq \theta_+$
Initialise: $\alpha_0^0 \in (\mathcal{A}_{ad}^0)_h$, $\alpha_1^0 \in (\mathcal{A}_{ad}^1)_h$ and set $k = 0$.
repeat

Use a Newton-type method to compute the solution p^k of the lower level problem

$$g_{\mathrm{TGV}}(p^k, \alpha_0^k, \alpha_1^k) := \beta\Delta^2 p^k + \gamma p^k + \nabla^2 B^{-1}\mathrm{div}^2 p^k - \nabla^2 B^{-1} T^* f + \frac{1}{\epsilon_0} Q_\delta(p^k, \alpha_0^k) - \frac{1}{\epsilon_1}\nabla P_\delta(\mathrm{div}p^k, \alpha_1^k) = 0$$

Solve the adjoint equation for q^k

$$\begin{aligned}\beta\Delta^2 q^k + \gamma q^k + \nabla^2 B^{-1}\mathrm{div}^2 q^k &+ \frac{1}{\epsilon_0}\left(G_\delta''(p^k - \alpha_0^k\mathbf{1}) + G_\delta''(-p^k - \alpha_0^k\mathbf{1})\right) q^k \\ &- \frac{1}{\epsilon_1}\nabla\left(G_\delta''(\mathrm{div}p^k - \alpha_1^k\mathbf{1}) + G_\delta''(-\mathrm{div}p^k - \alpha_1^k\mathbf{1})\right)\mathrm{div}q^k \\ &= -2\nabla B^{-1} T^*\mathrm{div}^2 p^k \left(w * \left((R(\mathrm{div}^2 p^k) - \overline{\sigma}^2)^+ - (\underline{\sigma}^2 - R(\mathrm{div}^2 p^k))^+\right)\right)\end{aligned}$$

Compute the reduced derivatives with respect to α_0 and α_1

$$\hat{J}'_{\mathrm{TGV},\alpha_0}(\alpha_0^k, \alpha_1^k) = \frac{1}{\epsilon_0}\mathbb{1}\left(-G_\delta''(p^k - \alpha_0^k\mathbf{1}) + G_\delta''(-p^k - \alpha_0^k\mathbf{1})\right) q^k$$

$$\hat{J}'_{\mathrm{TGV},\alpha_1}(\alpha_0^k, \alpha_1^k) = -\frac{1}{\epsilon_1}\nabla\left(-G_\delta''(\mathrm{div}p^k - \alpha_1^k\mathbf{1}) + G_\delta''(-\mathrm{div}p^k - \alpha_1^k\mathbf{1})\right) q^k + \lambda(I - \Delta_N)\alpha_1^k$$

Compute the reduced gradients

$$\nabla_{\alpha_0}\hat{J}_{\mathrm{TGV}}(\alpha_0^k, \alpha_1^k) = \hat{J}'_{\mathrm{TGV},\alpha_0}(\alpha_0^k, \alpha_1^k)$$
$$\nabla_{\alpha_1}\hat{J}_{\mathrm{TGV}}(\alpha_0^k, \alpha_1^k) = (I - \Delta_N)^{-1}\hat{J}'_{\mathrm{TV}}(\alpha_0^k, \alpha_1^k)$$

Compute the trial points

$$\alpha_0^{k+1} = P_{(\mathcal{A}_{ad}^0)_h}(\alpha_0^k - \tau_0^k\nabla_{\alpha_0}\hat{J}_{\mathrm{TGV}}(\alpha_0^k, \alpha_1^k))$$
$$\alpha_1^{k+1} = P_{(\mathcal{A}_{ad}^1)_h}(\alpha_1^k - \tau_1^k\nabla_{\alpha_1}\hat{J}_{\mathrm{TGV}}(\alpha_0^k, \alpha_1^k))$$

while

$$\hat{J}_{\mathrm{TGV}}(\alpha_0^{k+1}, \alpha_1^{k+1}) > \hat{J}_{\mathrm{TGV}}(\alpha_0^k, \alpha_1^k) + c\left(\hat{J}'_{\mathrm{TGV},\alpha_0}(\alpha_0^k, \alpha_1^k)^\top(\alpha_0^{k+1} - \alpha_0^k) + \hat{J}'_{\mathrm{TGV},\alpha_1}(\alpha_0^k, \alpha_1^k)^\top(\alpha_1^{k+1} - \alpha_1^k)\right)$$

do(Armijo line search)

Set $\tau_0^k := \theta_-\tau_0^k$, $\tau_1^k := \theta_-\tau_1^k$ and re-compute

$$\alpha_0^{k+1} = P_{(\mathcal{A}_{ad}^0)_h}(\alpha_0^k - \tau_0^k\nabla_{\alpha_0}\hat{J}_{\mathrm{TGV}}(\alpha_0^k, \alpha_1^k))$$
$$\alpha_1^{k+1} = P_{(\mathcal{A}_{ad}^1)_h}(\alpha_1^k - \tau_1^k\nabla_{\alpha_1}\hat{J}_{\mathrm{TGV}}(\alpha_0^k, \alpha_1^k))$$

end while

Update $\tau_0^{k+1} = \theta_+\tau_0^k$, $\tau_1^{k+1} = \theta_+\tau_1^k$ and $k := k + 1$

until some stopping condition is satisfied

The projection $P_{(\mathcal{A}_{Ad})_h}$ is computed by applying a semismooth Newton method to the associated minimization problem. It adapts the path-following method developed in Hintermüller and Kunisch (2006) to the projection problem; compare (Hintermüller et al., 2017, Algorithm 4). We just mention here that the original discretized H^1–projection problem $P_{(\mathcal{A}_{ad})_h}(\widetilde{\alpha})$

$$\begin{cases} \min \frac{1}{2} \| \alpha - \widetilde{\alpha} \|^2_{H^1(\Omega_h)} := \frac{h}{2} (\alpha - \widetilde{\alpha})^\top (I - \Delta_N)(\alpha - \widetilde{\alpha}) \\ \text{over } \alpha \in (\mathcal{A}_{ad})_h = \{\alpha \in U_h : \underline{\alpha} \leq \alpha_{i,j} \leq \overline{\alpha}\} \end{cases} \tag{116}$$

is substituted by the following version where the constraint is treated by a penalty term

$$\min_{\alpha \in U_h} \frac{1}{2} \| \alpha - \widetilde{\alpha} \|^2_{H^1(\Omega_h)} + \frac{1}{2\epsilon_\alpha} \left(\| (\alpha - \overline{\alpha})^+ \|^2_{\ell^2(\Omega_h)} + \| (\underline{\alpha} - \alpha)^+ \|^2_{\ell^2(\Omega_h)} \right) \tag{117}$$

for some small $\epsilon_\alpha > 0$.

5.2 Bilevel TV numerical experiments

For the test images shown in Fig. 7 with resolution $n = m = 256$, we now depict some examples where we compare some weighted TV reconstructions, produced by Algorithm 2, with scalar TV ones.

We first report the values of the different parameters that we used for these numerical examples. For the lower level problems parameters we used $\beta = 10^{-4}$, $\gamma = 10^{-4}$, $\delta = 10^{-3}$, $\epsilon = 10^{-8}$. Initially the lower level problem is solved for $\epsilon^0 = 10^{-3}$ and subsequently decreased by a factor of $\theta_\epsilon = 0.6$ down to a final value of $\underline{\epsilon} = 10^{-8}$. For every ϵ^l each problem is solved to an accuracy $\text{tol}_{(l)} = 10^{-10}$. The equation in Algorithm 1 corresponds to a linear system in finite dimensions. Here we simply used MATLAB's backslash command, even though preconditioned conjugate gradient methods can be used as well; see Hintermüler and Stadler (2006) and Kunisch and Hintermüller (2004). For the H^1-projection we used $\epsilon_\alpha = 10^{-8}$. For the constraint set $(\mathcal{A}_{ad})_h$ we used $\underline{\alpha} = 10^{-8}$ and $\overline{\alpha} = 10^{-2}$. For the upper level objective of the bilevel problem we chose $\lambda = 10^{-9}$ and w to be a $n_w \times n_w$ filter normalized (i.e., with entries $1/n_w^2$), with $n_w = 7$. The local variance barriers $\underline{\sigma}^2$ and $\overline{\sigma}^2$ were set according to (115). Since we are depicting examples with $\sigma^2 = 10^2$ and $\sigma^2 = 4 \times 10^{-4}$ the corresponding values for $(\underline{\sigma}, \overline{\sigma})$ are (0.00798, 0.01202) and (3.2×10^{-4}, 4.8×10^{-4}). For the Armijo line search parameters we set $\tau^0 = 10^{-3}$, $c = 10^{-8}$, $\theta_- = 0.25$ and $\theta_+ = 2$. The vector p^0 is chosen to be the zero vector.

FIG. 7 Test images, resolution 256 × 256.

Note that since we are confronted essentially with a nonconvex problem, i.e., the reduced objective is nonconvex, initialization of the algorithm with respect to α is crucial. Here we follow Hintermüller et al. (2017) and set the initial regularization weight sufficiently large $\alpha^0 = 2.5 \times 10^{-3}$ so that the associated initial image is cartoon-like.

We depict some first examples in Fig. 8. There we have corrupted the images in Fig. 7 with Gaussian noise of variance $\sigma^2 = 0.01$. In the second row we depict the best scalar TV examples, where the scalar parameter has been manually optimized via a bisection procedure with step 0.125×10^{-4}. We depict here the examples that correspond to the optimal parameters with respect to the Structural Similarity Index SSIM (Wang et al., 2004); see also

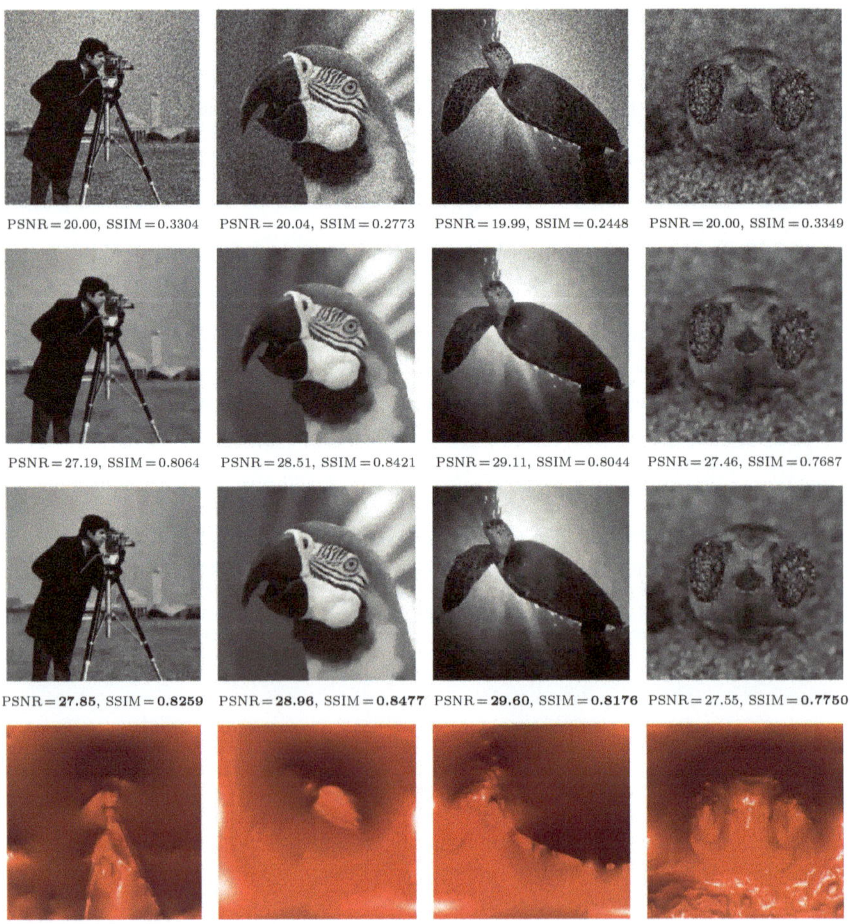

FIG. 8 *First row*: images corrupted with Gaussian noise of variance $\sigma^2 = 0.01$. *Second row*: Best scalar TV reconstructions in terms of SSIM. *Third row*: weighted TV reconstructions. *Fourth row*: the spatially adapted weight function α.

TABLE 1 PSNR and SSIM comparisons for the images of Figs. 8 and 9.

$\sigma^2 = 0.01$	Cameraman	Parrot	Turtle	Hatchling
Best scalar α for PSNR	27.54, 0.7857	28.88, 0.8119	29.27, 0.7924,	**27.57**, 0.7597
Best scalar α for SSIM	27.19, 0.8064	28.51, 0.8421	29.11, 0.8044	27.46, 0.7687
Bilevel TV	**27.85, 0.8259**	**28.96, 0.8477**	**29.60, 0.8176**	27.55, **0.7750**
$\sigma^2 = 4 \times 10^{-4}$	**Cameraman**	**Parrot**	**Turtle**	**Hatchling**
Best scalar α for PSNR	36.26, 0.9364	**37.29**, 0.9448	37.31, 0.9394	**36.31**, 0.9471
Best scalar α for SSIM	35.80, 0.9460	37.03, **0.9492**	37.06, 0.9443	36.02, **0.9521**
Bilevel TV	**36.30, 0.9486**	37.01, 0.9476	**37.50, 0.9481**	35.82, 0.9479

Every cell contains the corresponding PSNR and SSIM value. The highest PSNR and SSIM values for every image are highlighted in bold.

Table 1 for the optimal PSNR values as well. In the third row we depict the weighted TV denoising images u with the corresponding weight functions α in the fourth row. Note that the weighted TV reconstructions are better both visually and with respect to both, PSNR and SSIM. In fact we see that the weight functions get well adapted to the structure of the images, having lower values in areas of high details and larger in homogeneous areas. Thus, we obtained a structured nonsmooth prior.

We depict a second series of examples in Fig. 9, where we have used the same example images but with considerable lower Gaussian noise of variance $\sigma^2 = 4 \times 10^{-4}$. There we see that the low level of noise can make the weighting function α to adapt even more to the data. In the "cameraman" and "turtle" images, again the weighted TV result outperforms the best scalar TV one both in SSIM and PSNR; see again Table 1. In the "Parrot" the best scalar results are slightly better in terms of SSIM and PSNR, but perhaps not visually. The same holds for the "hatchling" image. We believe that this is due to the fact that in both of these images, the clean image itself already contains some noise – this is indeed the case in the background of the parrot image – which is still significant given the low level of artificial noise. In the "hatchling" image in fact, oscillatory features of different scale dominate. The bilevel TV also interferes with this type of noise which is naturally present in the images resulting in a lower PSNR and SSIM. Nevertheless the bilevel TV images are still visually more amenable than their best manually optimized scalar TV reconstructions.

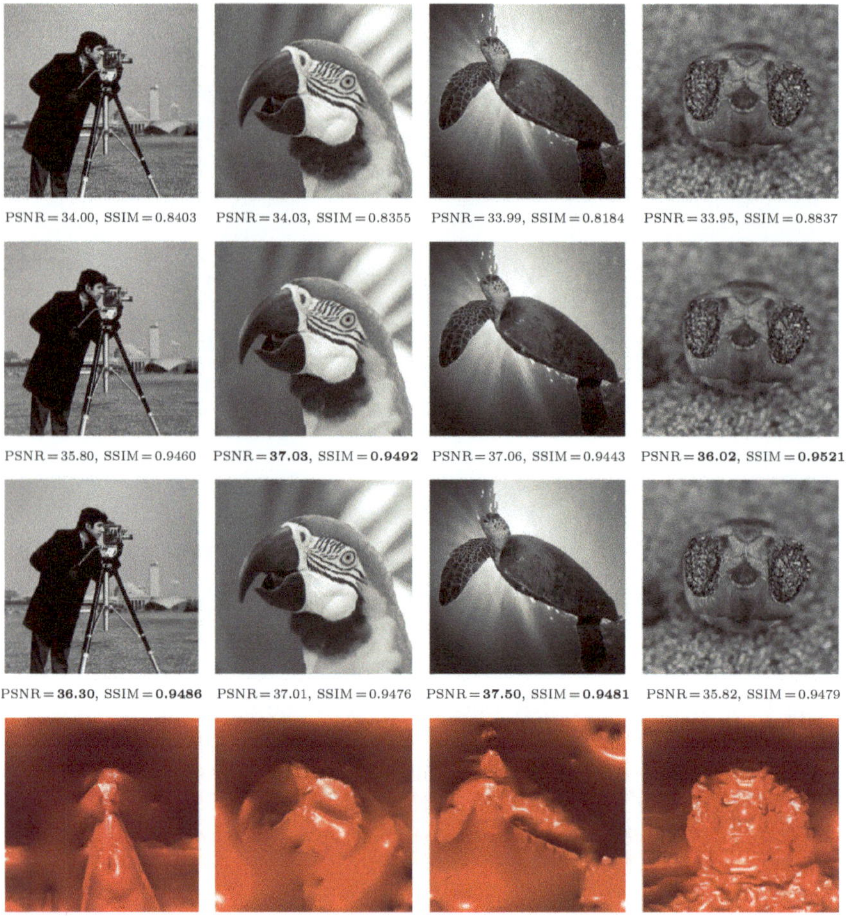

FIG. 9 *First row*: Images corrupted with Gaussian noise of variance $\sigma^2 = 4 \times 10^{-4}$. *Second row*: Best scalar TV reconstructions in terms of SSIM. *Third row*: Weighted TV reconstructions. *Fourth row*: The spatially adapted weight function α.

5.3 Discrete operators for ($\mathbb{P}_{\text{TGV}}$)

We now turn our attention to the discretized version of the bilevel TGV problem ($\mathbb{P}_{\text{TGV}}$). For that we need to defined the discrete space

$$V_h = U_h \times U_h \times U_h,$$

associated with the predual p, i.e., $p = (p^{11}, p^{12}, p^{22})$ with $p^{11}, p^{12}, p^{22} \in U_h$. In this problem, for the discrete gradient and divergence we have, $\nabla : W_h \to V_h$ and $\text{div} : V_h \to W_h$ satisfying again the adjoint relation $\nabla = -\text{div}^\top$. For $p \in V_h$, the divergence is defined again as

$$(\text{div}p)^1_{i,j} = \frac{1}{h}(p^{11}_{i,j} - p^{11}_{i-1,j} + p^{12}_{i,j} - p^{12}_{i,j-1}), \quad (i,j) \in \Omega_h,$$
$$(\text{div}p)^2_{i,j} = \frac{1}{h}(p^{12}_{i,j} - p^{12}_{i-1,j} + p^{22}_{i,j} - p^{22}_{i,j-1}), \quad (i,j) \in \Omega_h,$$

setting again zero values at the ghost nodes. For the second-order gradient $\nabla^2 : U_h \to V_h$ we have that $\nabla^2 = (D_{xx}u, D_{xy}u, D_{yy}u)$, where D_{xx}, D_{xy}, D_{xy} are operators $U_h \to V_h$ and are defined using the stencils as shown in Fig. 10 with zero values at ghost points. Note that the use of symmetric differences for the mixed derivative results to a symmetric matrix for the matrix representing Dxy. The resulting operators D_{xx}, D_{xy}, D_{yy} are self-adjoint. Hence for the discretized second divergence $\text{div}^2 : V_h \to U_h$, we have $\text{div}^2 p = D_{xx}p^{11} + 2D_{xy}p^{12} + D_{yy}p^{22}$. The vector bi-Laplacian is an operator $V_h \to V_h$ where $p \mapsto (\Delta^2 p^{11}, \Delta^2 p^{12}, \Delta^2 p^{22})$ with $\Delta^2 = D_{xxxx} + D_{yyyy} + D_{xxyy} + D_{yyxx}$. The resulting stencil for the Δ^2 is as shown in Fig. 11. In order to reflect the boundary conditions of $H^2_0(\Omega, \mathcal{S}^{2\times 2})$, the bi-Laplacian must be endowed with both, homogeneous Neumann and homogeneous Dirichlet boundary conditions. Again this is enforced by considering at any ghost points (up to two of them in the boundary) zero value. Finally we discuss the discretization of the operator $\nabla^2\text{div}^2 : V_h \to V_h$, which is equal to

$$(\nabla^2\text{div}^2 p)^{11} = D_{xxxx}p^{11} + 2D_{xxxy}p^{12} + D_{xxyy}p^{22}$$
$$(\nabla^2\text{div}^2 p)^{12} = D_{xyxx}p^{11} + 2D_{xyxy}p^{12} + D_{xyyy}p^{22}$$
$$(\nabla^2\text{div}^2 p)^{22} = D_{yyxx}p^{11} + 2D_{yyxy}p^{12} + D_{yyyy}p^{22}$$

where in fact it holds $D_{xxxy} = D_{xyxx}$, $D_{xxyy} = D_{xyxy} = D_{yyxx}$ and $D_{xyyy} = D_{yyxy}$. For these fourth-order discretized differential operators we use the stencils

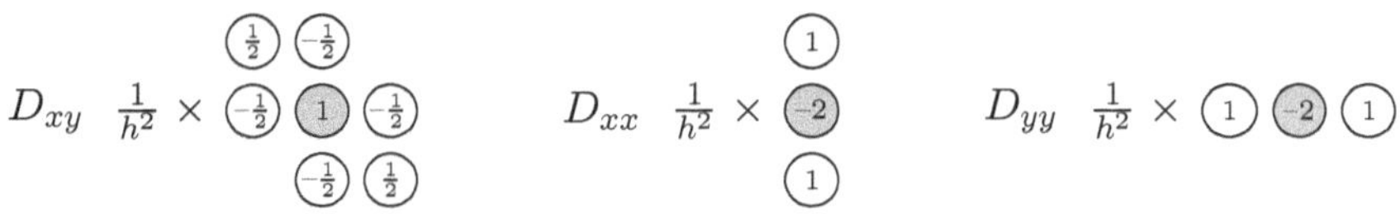

FIG. 10 Symmetric finite difference stencil for the second-order derivative operators.

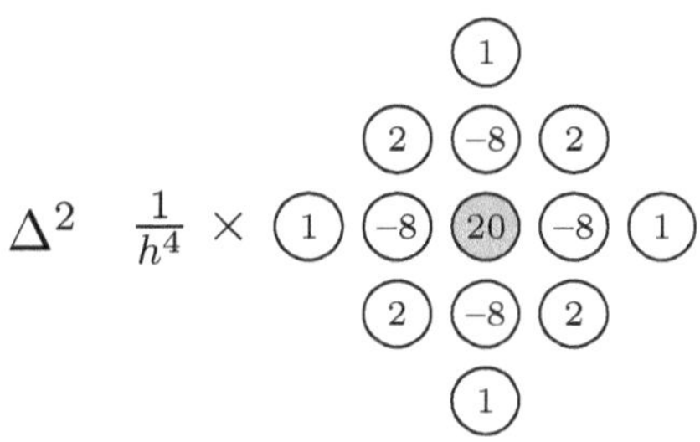

FIG. 11 Finite difference stencils that constitute the discrete bi-Laplacian Δ^2.

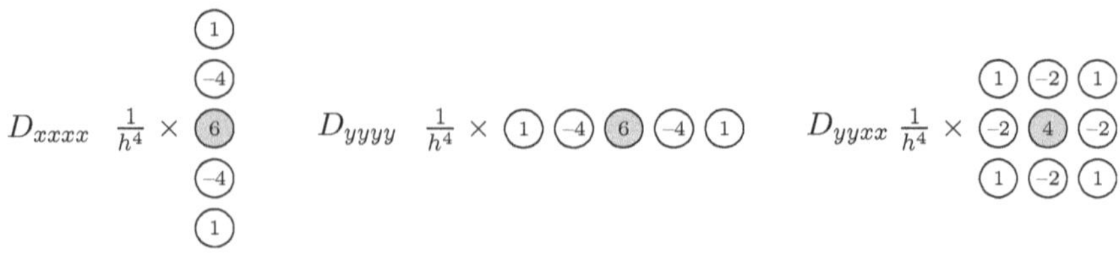

FIG. 12 Finite difference stencils that constitute the discrete bi-Laplacian Δ^2.

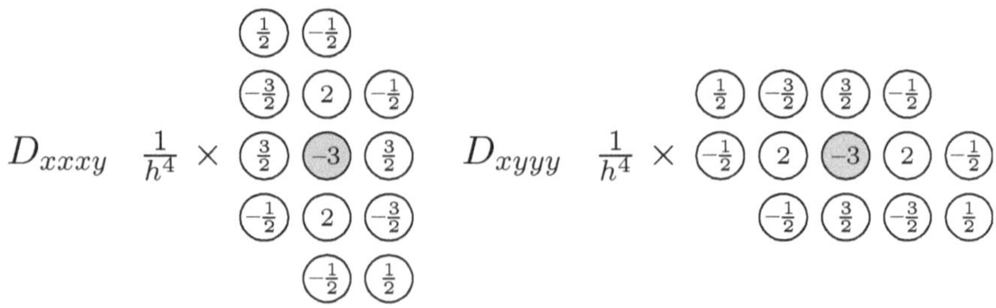

FIG. 13 Symmetric finite difference stencil for the fourth-order derivative operators.

shown in Figs. 12 and 13: and also using again the same rule to enforce homogeneous Neumann and homogeneous Dirichlet boundary conditions.

We remark that the matrix representing the operator $\nabla^2\mathrm{div}^2$ described above will not be symmetric due to the factor of 2 multiplying the terms concerning p^{12}. This leads to a nonsymmetric linear system when employing a Newton-type iteration. However, having a symmetric matrix is beneficial as this can lead to a more efficient and robust solution of the corresponding linear system via iterative solvers, such as conjugate gradients. A possible remedy is to solve for $(p^{11}, 2p^{12}, p^{22})$ rather than $(p^{11}, 2p^{12}, p^{22})$. This eliminates the factor 2 in the p^{12} part of the matrix that represents $\nabla^2\mathrm{div}^2$. In this case, however, other operators must be modified as well. For instance, the vector bi-Laplacian must take the form $(\Delta^2, \frac{1}{2}\Delta^2, \Delta^2)$, similarly for the other differential operators. The functions Q_δ and P_δ must be accordingly modified as well. The following version of the discrete dual $H_0^2(\Omega_h)^*$-norm is used for termination purposes:

$$\|v\|^*_{H_0^2(\Omega_h)} = h\sqrt{v^\top (I+\Delta^2)^{-1} v}. \tag{118}$$

We also use the discrete $H^1(\Omega_h)^*$-norm as in (112) along with the discrete Riesz map $\alpha_1 \mapsto (I - \Delta_B)\alpha_1$ as in the case of bilevel TV. We are now ready to write down the discrete version of $(\mathbb{P}_{\mathrm{TGV}})$:

$$\begin{cases} \min \frac{1}{2}\| (R(\mathrm{div}^2 p) - \overline{\sigma}^2)^+ \|^2_{\ell^2(\Omega_h)} + \frac{1}{2}\| (\underline{\sigma}^2 - R(\mathrm{div}^2 p))^+ \|^2_{\ell^2(\Omega_h)} + \frac{\lambda}{2}\| \alpha_1 \|^2_{H^1(\Omega_h)} \\ \quad \text{over } (p, \alpha_0, \alpha_1) \in V_h \times (\mathcal{A}^0_{ad})_h \times (\mathcal{A}^1_{ad})_h \\ \text{s.t. } \beta\Delta^2 p + \gamma p + \nabla^2 B^{-1}\mathrm{div}^2 p - \nabla^2 B^{-1} T^* f + \frac{1}{\epsilon_0} Q_\delta(p, \alpha_0) - \frac{1}{\epsilon_1}\nabla P_\delta(\mathrm{div}\, p, \alpha_1) = 0. \end{cases} \tag{$\mathbb{P}^h_{\mathrm{TGV}}$}$$

Here we have

$$(\mathcal{A}_{ad}^0)_h = \{\alpha_0 \in \mathbb{R} : \underline{\alpha}_0 \leq \alpha_0 \leq \overline{\alpha}_0\},$$

$$(\mathcal{A}_{ad}^1)_h = \{\alpha \in U_h : \underline{\alpha}_1 \leq (\alpha_1)_{i,j} \leq \overline{\alpha}_1, \text{ for all } (i,j) \in \Omega_h\}.$$

The discrete penalty functions $P_\delta : W_h \to W_h$ and $Q_\delta : V_h \to V_h$ are defined in the obvious way. We use the same rule as in (115) for the choice of $\underline{\sigma}^2, \overline{\sigma}^2$.

The corresponding projected gradient algorithm for the numerical solution of $(\mathbb{P}_{\text{TGV}}^h)$ is stated in Algorithm 3.

For the sake of notation, here **1** notes a matrix either of form $[Id; Id]$ or $[Id; Id; Id]$ of size $nm \times 2nm$ or $nm \times 3nm$, respectively, depending on whether it is applied to α_1 or α_0. On the other hand, 1 denotes a matrix of size $1 \times 3nm$ with all entries equal to one. The projection $P_{(\mathcal{A}_{ad}^1)_h}$ is computed as in the bilevel TV algorithm while $P_{(\mathcal{A}_{ad}^1)_h}(\alpha_0) = \max(\min(\alpha_0, \overline{\alpha}_0), \underline{\alpha}_0)$. As in the TV case, also here a path following scheme is used to solve $g_{\text{TGV},\ \epsilon_0,\ \epsilon 1}(p, \alpha_0, \alpha_1) = 0$. This is done successively for $\epsilon_0 = \epsilon_0^\ell$, $\epsilon_1 = \epsilon_1^\ell$ down to a tolerance

$$g_{\text{TGV},\epsilon_0^\ell,\epsilon_1^\ell}(p^{\ell+1}, \alpha_0, \alpha_1) \leq \text{tol}_{(l)}$$

and then setting $\epsilon_0^{\ell+1} := \max(\theta_\epsilon \epsilon_0^\ell, \epsilon_0)$, $\epsilon_1^{\ell+1} := \max(\theta_\epsilon \epsilon_1^\ell, \epsilon_1)$ for some $0 < \theta_\epsilon < 1$.

5.4 Bilevel TGV numerical experiments

Here we show a few numerical examples produced by Algorithm 3. The same test images as for the bilevel TV case were used; see Fig. 7. For the lower level TGV problem we used $\beta = 10^{-3}$, $\gamma = 0$, $\delta = 10^{-6}$, $\epsilon_0 = 10^{-12}$, $\epsilon_1 = 10^{-12}$. Initially the lower problem is solved for $\epsilon_0^0 = 10^3$, $\epsilon_1^0 = 10^3$ and each of these successively decreased by the same factor $\theta_\epsilon = 0.05$ down to finite values $\epsilon_0 = \epsilon_1 = 10^{-12}$.

We again used backslash for the solution of the linear systems, but sophisticated iterative solvers may be employed as well. We set $\underline{\alpha}_0 = 10^{-7}$, $\overline{\alpha}_0 = 10^{-2}$, while for the H^1-projection we set again $\epsilon_\alpha = 10^{-10}$, and $\underline{\alpha}_1 = 10^{-7}$, $\overline{\alpha}_1 = 10^{-2}$. The normalized filter w and the local variance barriers $\underline{\sigma}^2$ and $\overline{\sigma}^2$ were chosen as before. For the Armijo line search we set $\tau_0^0 = 1$, $\tau_1^0 = 10^{-12}$, while c, θ_-, θ_+ were chosen as in the TV case.

We note that extra attention must be paid to the initialization of the algorithm. As in the TV case α_0^0 and α_0^1 must be large enough in order to produce cartoon-like images, providing the local variance estimator with useful information. However, if α_0 is initially too large then there is a danger of falling into the regime of Theorem 3, in which the TGV functional and hence the solution map of (at least) the (nonregularized) lower level problem does not depend on α_0. This means that there is a danger that the derivative of the reduced functional with respect to α_0 will be close to zero, thus making only negligible progress towards optimality. This behaviour was confirmed by intensive numerical experimentation. Note that an analogous phenomenon can occur also in the case where α_0 is much smaller than α_1. Then the effect of α_1 becomes negligible. In fact this has been shown in Papafitsoros and Valkonen (2015, Proposition 2) for dimension one, but numerical experiments indicate that this can indeed be also a viable scenario in higher dimensions. In our examples we used $\alpha_0^0 =$ and $\alpha_1^0 = 9 \times 10^{-4}$ and 3.125×10^{-6}.

We show the results in Fig. 14. Once again we notice in general a clear improvement in image quality in comparison to the best scalar TGV results. The only exception is again the "hatchling" image where the scalar TGV result is slightly better than the weighted TGV one, even though the latter one is able to preserve better the detailed features in the eyes region. Finally in Fig. 15 we depict for the sake of comparison the regularization functions α and α_1 for the weighted TV and TGV results, respectively, that correspond to Figs. 8 and 14.

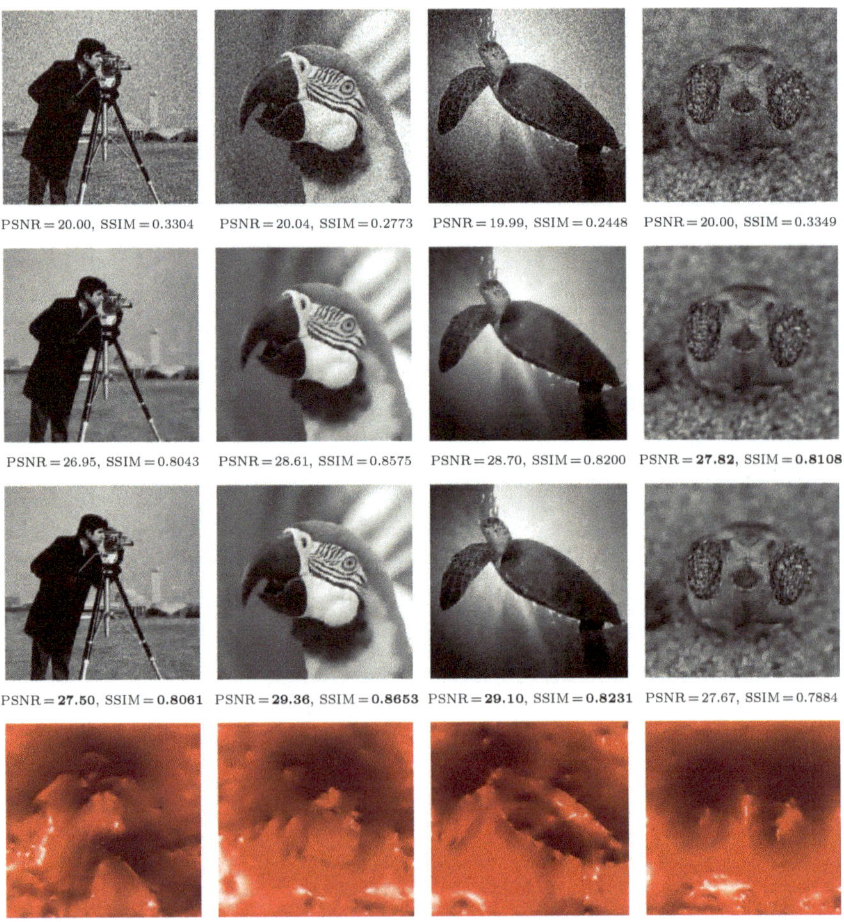

FIG. 14 *First row*: Images corrupted with Gaussian noise of variance $\sigma^2 = 0.01$. *Second row*: Best scalar TGV reconstructions in terms of SSIM. Third row: weighted TGV reconstructions. *Fourth row*: The spatially adapted weight function α_1 for weighted TGV.

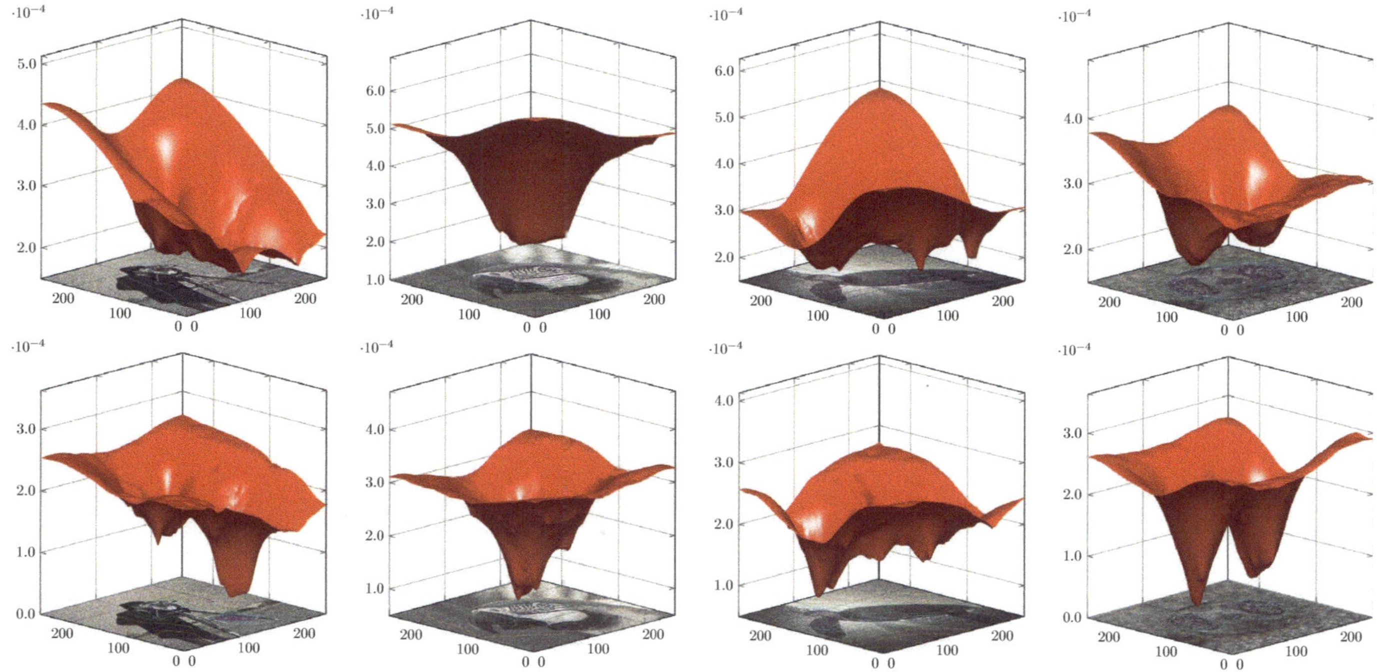

FIG. 15 *First row*: The regularization functions α for the weighted TV results of Fig. 8. *Second row*: The regularization functions α_1 for the weighted TGV results of Fig. 14.

References

Acar, R., Vogel, C.R., 1994. Analysis of bounded variation penalty methods for Ill-posed problems. Inverse Prob. 10 (6), 1217–1229. https://doi.org/10.1088/0266-5611/10/6/003.

Allard, W., 2008a. Total variation regularization for image denoising, I. Geometric theory. SIAM J. Math. Anal. 39 (4), 1150–1190. https://doi.org/10.1137/060662617.

Allard, W., 2008b. Total variation regularization for image denoising, II. Examples. SIAM J. Imaging Sci. 1 (4), 400–417. https://doi.org/10.1137/070698749.

Allard, W., 2009. Total variation regularization for image denoising, III. Examples. SIAM J. Imaging Sci. 2 (2), 532–568. https://doi.org/10.1137/070711128.

Almansa, A., Ballester, C., Caselles, V., Haro, G., 2008. A TV based restoration model with local constraints. J. Sci. Comput. 34 (3), 209–236. https://doi.org/10.1007/s10915-007-9160-x.

Alter, F., Durand, S., Froment, J., 2005. Adapted total variation for artifact free decompression of JPEG images. J. Math. Imaging Vis. 23, 199–211. https://doi.org/10.1007/s10851-005-6467-9.

Amar, M., De Cicco, V., Fusco, N., 2008. Lower semicontinuity and relaxation results in BV for integral functionals with BV integrants. ESAIM: Contr. Optim. Cal. Var. 14 (3), 456–477. https://doi.org/10.1051/cocv:2007061.

Ambrosio, L., Fusco, N., Pallara, D., 2000. Functions of Bounded Variation and Free Discontinuity Problems. Oxford University Press, USA. ISBN: 0198502451.

Anitescu, M., Tseng, P., Wright, S.J., 2007. Elastic-mode algorithms for mathematical programs with equilibrium constraints: global convergence and stationarity properties. Math. Program. 110 (2), 337–371. https://doi.org/10.1007/s10107-006-0005-4.

Attouch, H., Brezis, H., 1986. Duality for the sum of convex functions in general Banach spaces. North-Holland Math. Library 34, 125–133. https://doi.org/10.1016/S0924-6509(09)70252-1.

Attouch, H., Buttazzo, G., Michaille, G., 2014. Variational Analysis in Sobolev and BV Spaces: Applications to PDEs and Optimization. vol. 17. SIAM.

Babacan, S.D., Molina, R., Katsaggelos, A.K., 2008. Parameter estimation in TV image restoration using variational distribution approximation. IEEE Trans. Image Process. 17 (3), 326–339. https://doi.org/10.1109/TIP.2007.916051.

Barbu, V., 1984. Optimal control of variational inequalities. Res. Notes Math. 100.

Benning, M., Burger, M., 2011. Error estimates for general fidelities. Electron. Trans. Numer. Anal. 38, 44–68.

Benning, M., Burger, M., 2018. Modern regularization methods for inverse problems. Acta Numer. 27, 1–111. https://doi.org/10.1017/S0962492918000016.

Benning, M., Brune, C., Burger, M., Müller, J., 2013. Higher-order TV methods - enhancement via bregman iteration. J. Sci. Comput. 54 (2–3), 269–310. https://doi.org/10.1007/s10915-012-9650-3.

Benning, M., Gladden, L., Holland, D., Schönlieb, C.B., Valkonen, T., 2014. Phase reconstruction from velocity-encoded MRI measurements—a survey of sparsity-promoting variational approaches. J. Magn. Reson. 238, 26–43. https://doi.org/10.1016/j.jmr.2013.10.003.

Bertalmio, M., Caselles, V., Rougé, B., Solé, A., 2003. TV based image restoration with local constraints. J. Sci. Comput. 19 (1), 95–122. https://doi.org/10.1023/A:1025391506181.

Blomgren, P., Chan, T.F., 1998. Color TV: total variation methods for restoration of vector-valued images. IEEE Trans. Image Process. 7 (3), 304–309. https://doi.org/10.1109/83.661180.

Bredies, K., 2014. Recovering piecewise smooth multichannel images by minimization of convex functionals with total generalized variation penalty. In: Bruhn, A., Pock, T., Tai, X.C. (Eds.), Efficient Algorithms for Global Optimization Methods in Computer Vision, Lecture Notes in Computer Science, vol. 8293, Springer, Berlin, Heidelberg, pp. 44–77.

Bredies, K., Holler, M., 2012a. Artifact-free JPEG decompression with total generalized variation. In: VISAP 2012: Proceedings of the International Conference on Computer Vision and Applications.

Bredies, K., Holler, M., 2012b. A pointwise characterization of the subdifferential of the total variation functional. SFB-Report No. 2012-11. https://imsc.uni-graz.at/mobis/publications/SFB-Report-2012-011_Update2.pdf.

Bredies, K., Holler, M., 2012c. A total variation-based JPEG decompression model. SIAM J. Imaging Sci. 5 (1), 366–393. https://doi.org/10.1137/110833531.

Bredies, K., Holler, M., 2013. A TGV regularized wavelet based zooming model. In: Kuijper, A., Bredies, K., Pock, T., Bischof, H. (Eds.), Scale Space and Variational Methods in Computer Vision. SSVM 2013. Lecture Notes in Computer Science, vol. 7893. Springer, Berlin, Heidelberg, pp. 149–160.

Bredies, K., Holler, M., 2014. Regularization of linear inverse problems with total generalized variation. J. Inverse Ill-posed Prob. 22 (6), 871–913. https://doi.org/10.1515/jip-2013-0068.

Bredies, K., Holler, M., 2015. A TGV-based framework for variational image decompression, zooming, and reconstruction. Part I: Analytics. SIAM J. Imaging Sci. 8 (4), 2814–2850. https://doi.org/10.1137/15M1023865.

Bredies, K., Valkonen, T., 2011. Inverse problems with second-order total generalized variation constraints. In: Proceedings of SampTA 2011—9th International Conference on Sampling Theory and Applications, Singapore.

Bredies, K., Kunisch, K., Pock, T., 2010. Total generalized variation. SIAM J. Imaging Sci. 3 (3), 492–526. https://doi.org/10.1137/090769521.

Bredies, K., Dong, Y., Hintermüller, M., 2013a. Spatially dependent regularization parameter selection in total generalized variation models for image restoration. Int. J. Comput. Math. 90 (1), 109–123. https://doi.org/10.1080/00207160.2012.700400.

Bredies, K., Kunisch, K., Valkonen, T., 2013b. Properties of L^1-TGV^2: the one-dimensional case. J. Math. Anal. Appl. 398 (1), 438–454. https://doi.org/10.1016/j.jmaa.2012.08.053.

Brinkmann, E.-M., Burger, M., Rasch, J., Sutour, C., 2017. Bias reduction in variational regularization. J. Math. Imaging Vis. 59 (3), 534–566. https://doi.org/10.1007/s10851-017-0747-z.

Brune, C., Sawatzky, A., Wübbeling, F., Kösters, T., Burger, M., 2013. EM-TV methods for inverse problems with Poisson noise. In: Burger, M., Osher, S. (Eds.), Level Set and PDE Based Reconstruction Methods in Imaging. Lecture Notes in Mathematics, Springer International Publishing, pp. 71–142.

Burger, M., Müller, J., Papoutsellis, E., Schönlieb, C.B., 2014. Total variation regularization in measurement and image space for PET reconstruction. Inverse Prob. 30 (10), 105003. http://stacks.iop.org/0266-5611/30/i=10/a=105003.

Burger, M., Papafitsoros, K., Papoutsellis, E., Schönlieb, C.B., 2016. Infimal convolution regularisation functionals of BV and L^p spaces. Part I: the finite p case. J. Math. Imaging Vis. 55 (3), 343–369. https://doi.org/10.1007/s10851-015-0624-6.

Calatroni, L., Papafitsoros, K., 2019. Analysis and automatic parameter selection of a variational model for mixed Gaussian and Salt & Pepper noise removal. Inverse Prob. http://iopscience.iop.org/10.1088/1361-6420/ab291a

Calatroni, L., De Los Reyes, J.C., Schönlieb, C.-B., 2014. Dynamic sampling schemes for optimal noise learning under multiple nonsmooth constraints. In: Pötzsche, C., Heuberger, C., Kaltenbacher, B., Rendl, F. (Eds.), System Modeling and Optimization, IFIP Advances in Information and Communication Technology, vol. 443. Springer, Berlin, Heidelberg, pp. 85–95.

Calatroni, L., Chung, C., De Los Reyes, J.C., Schönlieb, C.-B., Valkonen, T., 2017a. Bilevel approaches for learning of variational imaging models. In: RADON Book Series on Computational and Applied Mathematics, vol. 18, De Gruyter, Berlin, Boston.

Calatroni, L., De Los Reyes, J., Schönlieb, C., 2017b. Infimal convolution of data discrepancies for mixed noise removal. SIAM J. Imaging Sci. 10 (3), 1196–1233. https://doi.org/10.1137/16M1101684.

Caselles, V., Chambolle, A., Novaga, M., 2007. The discontinuity set of solutions of the TV denoising problem and some extensions. Multiscale Model. Simul. 6 (3), 879–894. https://doi.org/10.1137/070683003.

Chambolle, A., 2004. An algorithm for total variation minimization and applications. J. Math. Imaging Vis. 20 (1), 89–97. https://doi.org/10.1023/B:JMIV.0000011325.36760.1e.

Chambolle, A., Lions, P.L., 1997. Image recovery via total variation minimization and related problems. Numer. Math. 76, 167–188. https://doi.org/10.1007/s002110050258.

Chambolle, A., Pock, T., 2011. A first-order primal-dual algorithm for convex problems with applications to imaging. J. Math. Imaging Vis. 40 (1), 120–145. https://doi.org/10.1007/s10851-010-0251-1.

Chambolle, A., Pock, T., 2016. An introduction to continuous optimization for imaging. Acta Numer. 25, 161–319. https://doi.org/10.1017/S096249291600009X.

Chambolle, A., Duval, V., Peyré, G., Poon, C., 2017. Geometric properties of solutions to the total variation denoising problem. Inverse Prob. 33 (1), 015002. http://stacks.iop.org/0266-5611/33/i=1/a=015002.

Chan, T.F., Esedoglu, S., 2005. Aspects of total variation regularized L^1 function approximation. SIAM J. Appl. Math. 1817–1837. https://doi.org/10.1137/040604297.

Chan, T.F., Shen, J., 2005. Variational image inpainting. Commun. Pure Appl. Math. 58 (5), 579–619. https://doi.org/10.1002/cpa.20075.

Chan, T., Golub, G., Mulet, P., 1999. A nonlinear primal-dual method for total variation-based image restoration. SIAM J. Sci. Comput. 20 (6), 1964–1977. https://doi.org/10.1137/S1064827596299767.

Chan, T., Marquina, A., Mulet, P., 2001. High-order total variation-based image restoration. SIAM J. Sci. Comput. 22 (2), 503–516. https://doi.org/10.1137/S1064827598344169.

Chan, T.F., Esedoglu, S., Park, F.E., 2007. Image decomposition combining staircase reduction and texture extraction. J. Vis. Commun. Image Represent. 18 (6), 464–486. https://doi.org/10.1016/j.jvcir.2006.12.004.

Chen, X., Nashed, Z., Qi, L., 2000. Smoothing methods and semismooth methods for nondifferentiable operator equations. SIAM J. Numer. Anal. 38, 1200–1216. https://doi.org/10.1137/S0036142999356719.

Chung, C.V., De los Reyes, J.C., Schönlieb, C.B., 2017. Learning optimal spatially-dependent regularization parameters in total variation image denoising. Inverse Prob. 33 (7), 074005. http://stacks.iop.org/0266-5611/33/i=7/a=074005.

Ciarlet, P.G., 1988. Mathematical Elasticity: Volume I: Three-Dimensional Elasticity. North-Holland.

Darbon, J., Sigelle, M., 2005. A fast and exact algorithm for total variation minimization. In: Marques, J.S., de la Blanca, N., Pérez, Pina, P. (Eds.), Pattern Recognition and Image Analysis, Springer, Berlin, Heidelberg, pp. 351–359.

De los Reyes, J.C., Schönlieb, C.B., 2013. Image denoising: learning the noise model via nonsmooth PDE-constrained optimization. Inverse Prob. Imaging 7 (4), 1183–1214. https://doi.org/10.3934/ipi.2013.7.1183.

De Los Reyes, J.C., Schönlieb, C.-B., Valkonen, T., 2016. The structure of optimal parameters for image restoration problems. J. Math. Anal. Appl. 434 (1), 464–500. https://doi.org/10.1016/j.jmaa.2015.09.023.

De Los Reyes, J.C., Schönlieb, C.-B., Valkonen, T., 2017. Bilevel parameter learning for higher-order total variation regularisation models. J. Math. Imaging Vis. 57 (1), 1–25. https://doi.org/10.1007/s10851-016-0662-8.

Deledalle, C., Vaiter, S., Fadili, J., Peyré, G., 2014. Stein unbiased gradient estimator of the risk (SUGAR) for multiple parameter selection. SIAM J. Imaging Sci. 7 (4), 2448–2487. https://doi.org/10.1137/140968045.

Demengel, F., Temam, R., 1984. Convex functions of a measure and applications. Indiana Univ. Math. J. 33, 673–709. ISSN 0022-2518.

Dempe, S., 2002. Foundations of Bilevel Programming. Springer, Berlin, Heidelberg.

Dong, G., Scherzer, O., 2017. Nonlinear flows for displacement correction and applications in tomography. In: Lauze, F., Dong, Y., Dahl Anders, B. (Eds.), Scale Space and Variational Methods in Computer Vision, Springer International Publishing, pp. 283–294.

Dong, Y., Hintermüller, M., Rincon-Camacho, M.M., 2010a. Automated regularization parameter selection in multi-scale total variation models for image restoration. J. Math. Imaging Vis. 40 (1), 82–104. https://doi.org/10.1007/s10851-010-0248-9.

Dong, Y., Hintermüller, M., Rincon-Camacho, M.M., 2010b. A multi-scale vectorial L^{τ}-TV framework for color image restoration. Int. J. Comput. Vis. 92 (3), 296–307. https://doi.org/10.1007/s11263-010-0359-1.

Dong, G., Patrone, A.R., Scherzer, O., Öktem, O., 2015. Infinite dimensional optimization models and PDEs for dejittering. In: Aujol, J.-F., Nikolova, M., Papadakis, N. (Eds.), Scale Space and Variational Methods in Computer Vision, Springer International Publishing, pp. 678–689.

Dong, G., Hintermüller, M., Papafitsoros, K., 2019. Quantitative magnetic resonance imaging: from fingerprinting to integrated physics-based models. SIAM J. Imaging Sci. 12 (2), 927–971. https://doi.org/10.1137/18M1222211.

Duval, V., Aujol, J.F., Gousseau, Y., 2009. The TVL1 model: a geometric point of view. SIAM J. Multiscale Model. Simul. 8 (1), 154–189. https://doi.org/10.1137/090757083.

Ekeland, I., Temam, R., 1976. Convex analysis and variational problems. vol. 1. Elsevier, North Holland.

Frick, K., Marnitz, P., Munk, A., 2012. Statistical multiresolution Dantzig estimation in imaging: fundamental concepts and algorithmic framework. Electron. J. Stat. 6, 231–268. https://doi.org/10.1214/12-EJS671.

Getreuer, P., 2012a. Rudin-Osher-Fatemi total variation denoising using split Bregman. Image Process. On Line 2, 74–95. https://doi.org/10.5201/ipol.2012.g-tvd.

Getreuer, P., 2012b. Total variation deconvolution using split Bregman. Image Process. On Line 2, 158–174. https://doi.org/10.5201/ipol.2012.g-tvdc.

Getreuer, P., 2012c. Total variation inpainting using split Bregman. Image Process. On Line 2, 147–157. https://doi.org/10.5201/ipol.2012.g-tvi.

Girault, V., Raviart, P.-A., 1986. Finite Element Method for Navier-Stokes Equation. Springer.

Goldstein, T., Osher, S., 2009. The split Bregman method for L1-regularized problems. SIAM J. Imaging Sci. 2 (2), 323–343. https://doi.org/10.1137/080725891.

Hahn, J., Wu, C., Tai, X.C., 2012. Augmented Lagrangian method for generalized TV-Stokes model. J. Sci. Comput. 50 (2), 235–264. https://doi.org/10.1007/s10915-011-9482-6.

Hinterberger, W., Scherzer, O., 2006. Variational methods on the space of functions of bounded Hessian for convexification and denoising. Computing 76 (1–2), 109–133. https://doi.org/10.1007/s00607-005-0119-1.

Hintermüler, M., Stadler, G., 2006. An infeasible primal-dual algorithm for total bounded variation-based inf-convolution-type image restoration. SIAM J. Sci. Comput. 28 (1), 1–23. https://doi.org/10.1137/040613263.

Hintermüller, M., 2007. Mesh-independence and fast local convergence of a primal-dual active-set method for mixed control-state constrained elliptic control problems. ANZIAM J. 49, 1–38. https://doi.org/10.1017/S1446181100012657.

Hintermüller, M., Kopacka, I., 2009. Mathematical programs with complementarity constraints in function space: C- and strong stationarity and a path-following algorithm. SIAM J. Optim. 20 (2), 868–902. https://doi.org/10.1137/080720681.

Hintermüller, M., Kunisch, K., 2006. Path-following methods for a class of constrained minimization problems in function space. SIAM J. Optim. 17 (1), 159–187. https://doi.org/10.1137/040611598.

Hintermüller, M., Rautenberg, C.N., 2015. On the density of classes of closed convex sets with pointwise constraints in Sobolev spaces. J. Math. Anal. Appl. 426 (1), 585–593. https://doi.org/10.1016/j.jmaa.2015.01.060.

Hintermüller, M., Rautenberg, C.N., 2017. Optimal selection of the regularization function in a weighted total variation model. Part I: Modelling and theory. J. Math. Imaging Vis. 59 (3), 498–514. https://doi.org/10.1007/s10851-017-0744-2.

Hintermüller, M., Rincon-Camacho, M.M., 2010. Expected absolute value estimators for a spatially adapted regularization parameter choice rule in L^1-TV-based image restoration. Inverse Prob. 26 (8), 085005. http://stacks.iop.org/0266-5611/26/i=8/a=085005.

Hintermüller, M., Rösel, S., 2016. A duality-based path-following semismooth Newton method for elasto-plastic contact problems. J. Comput. Appl. Math. 292, 150–173. https://doi.org/10.1016/j.cam.2015.06.010.

Hintermüller, M., Rösel, S., 2018. Duality results and regularization schemes for Prandtl-Reuss perfect plasticity. ESAIM: COCV, Forthcoming article. https://doi.org/10.1051/cocv/2018004.

Hintermüller, M., Surowiec, T.M., 2016. A bundle-free implicit programming approach for a class of elliptic MPECs in function space. Math. Program. 160 (1–2), 271–305. https://doi.org/10.1007/s10107-016-0983-9.

Hintermüller, M., Ulbrich, M., 2004. A mesh-independence result for semismooth Newton methods. Math. Program. 101 (1), 154–184. https://doi.org/10.1007/s10107-004-0540-9.

Hintermüller, M., Ito, K., Kunisch, K., 2002. The primal-dual active set strategy as a semismooth Newton method. SIAM J. Optim. 13 (3), 865–888. https://doi.org/10.1137/S1052623401383558.

Hintermüller, M., Mordukhovich, B.S., Surowiec, T.M., 2014. Several approaches for the derivation of stationarity conditions for elliptic MPECs with upper-level control constraints. Math. Program. 146 (1), 555–582. https://doi.org/10.1007/s10107-013-0704-6.

Hintermüller, M., Rautenberg, C.N., Hahn, J., 2014. Functional-analytic and numerical issues in splitting methods for total variation-based image reconstruction. Inverse Prob. 30 (5), 055014. http://stacks.iop.org/0266-5611/30/i=5/a=055014.

Hintermüller, M., Löbhard, C., Tber, H., 2015. An ℓ_1-penalty scheme for the optimal control of elliptic variational inequalities. In: Al-Baali, M., Grandinetti, L., Purnama, A. (Eds.), Numerical analysis and optimization, Springer Proceedings in Mathematics & Statistics, vol. 134. Springer, pp. 151–190.

Hintermüller, M., Papafitsoros, K., Rautenberg, C.N., 2017a. Analytical aspects of spatially adapted total variation regularisation. J. Math. Anal. Appl. 454 (2), 891–935. https://doi.org/10.1016/j.jmaa.2017.05.025.

Hintermüller, M., Rautenberg, C.N., Rösel, S., 2017b. Density of convex intersections and applications. Proc. R. Soc. London A Math. Phys. Eng. Sci. 473 (2205). https://doi.org/10.1098/rspa.2016.0919.

Hintermüller, M., Rautenberg, C.N., Wu, T., Langer, A., 2017. Optimal selection of the regularization function in a weighted total variation model. Part II: Algorithm, its analysis and numerical tests. J. Math. Imaging Vis. 59 (3), 515–533. https://doi.org/10.1007/s10851-017-0736-2.

Hintermüller, M., Holler, M., Papafitsoros, K., 2018. A function space framework for structural total variation regularization with applications in inverse problems. Inverse Prob. 34 (6), 064002. http://stacks.iop.org/0266-5611/34/i=6/a=064002.

Hinze, M., Pinnau, R., Ulbrich, M., Ulbrich, S., 2009. Optimization with PDE Constraints. vol. 23, Springer Science & Business Mediahttps://doi.org/10.1007/978-1-4020-8839-1.

Holler, M., Kunisch, K., 2014. On infimal convolution of TV-type functionals and applications to video and image reconstruction. SIAM J. Imaging Sci. 7 (4), 2258–2300. ISSN 1936-4954. https://doi.org/10.1137/130948793.

Hotz, Y., Marnitz, P., Stichtenoth, R., Davies, L., Kabluchko, Z., Munk, A., 2012. Locally adaptive image denoising by a statistical multiresolution criterion. Comput. Stat. Data Anal. 56 (3), 543–558. https://doi.org/10.1016/j.csda.2011.08.018.

Huber, P.J., 1964. Robust estimation of a location parameter. Annal. Stat. 53 (1), 73–101.

Ito, K., Kunisch, K., 2000. BV-type regularization methods for convoluted objects with edge, flat and grey scales. Inverse Prob. 16 (4), 909. http://stacks.iop.org/0266-5611/16/i=4/a=303.

Jalalzai, K., 2015. Some remarks on the staircasing phenomenon in total variation-based image denoising. J. Math. Imaging Vis. 54 (2), 256–268. https://doi.org/10.1007/s10851-015-0600-1.

Knoll, F., Bredies, K., Pock, T., Stollberger, R., 2011. Second order total generalized variation (TGV) for MRI. Magn. Reson. Med. 65 (2), 480–491. https://doi.org/10.1002/mrm.22595.

Kunisch, K., Hintermüller, M., 2004. Total bounded variation regularization as a bilaterally constrained optimization problem. SIAM J. Appl. Math. 64 (4), 1311–1333. https://doi.org/10.1137/S0036139903422784.

Kunisch, K., Pock, T., 2013. A bilevel optimization approach for parameter learning in variational models. SIAM J. Imaging Sci. 6 (2), 938–983. https://doi.org/10.1137/120882706.

Langer, A., 2017. Automated parameter selection for total variation minimization in image restoration. J. Math. Imaging Vis. 57 (2), 239–268. ISSN 1573-7683. https://doi.org/10.1007/s10851-016-0676-2.

Lefkimmiatis, S., Bourquard, A., Unser, M., 2012. Hessian-based norm regularization for image restoration with biomedical applications. IEEE Trans. Image Process. 21, 983–995. https://doi.org/10.1109/TIP.2011.2168232.

Lin, Y., Wohlberg, B., Guo, H., 2010. UPRE method for total variation parameter selection. Signal Process. 90 (8), 2546–2551. https://doi.org/10.1016/j.sigpro.2010.02.025.

Luo, T., Pang, J.S., Ralph, D., 1996. Mathematical programs with equilibrium constraints. Cambridge University Press, Cambridge.

Lustig, M., Donoho, D.L., Santos, J.M., Pauly, J.M., 2008. Compressed sensing MRI. IEEE Signal Process. Mag. 25 (2), 72–82. https://doi.org/10.1109/MSP.2007.914728.

Lysaker, M., Tai, X.C., 2006. Iterative image restoration combining total variation minimization and a second-order functional. Int. J. Comput. Vis. 66 (1), 5–18. https://doi.org/10.1007/s11263-005-3219-7.

Lysaker, M., Lundervold, A., Tai, X.C., 2003. Noise removal using fourth-order partial differential equation with applications to medical magnetic resonance images in space and time. IEEE Trans. Image Process. 12 (12), 1579–1590. https://doi.org/10.1109/TIP.2003.819229.

Malgouyres, F., Guichard, F., 2006. Edge direction preserving image zooming: a mathematical and numerical analysis. SIAM J. Numer. Anal. (1), 1–37. https://doi.org/10.1137/S0036142999362286.

Meyer, Y., 2001. Oscillating Patterns in Image Processing and Nonlinear Evolution Equations: The Fifteenth Dean Jacqueline B. Lewis Memorial Lectures. vol. 22, American Mathematical Society.

Mifflin, R., 1977. Semismooth and semiconvex functions in constrained optimization. SIAM J. Control Optim. 15, 959–972. https://doi.org/10.1137/0315061.

Natterer, F., 2001. The Mathematics of Computerized Tomography. Society for Industrial and Applied Mathematics. https://doi.org/10.1137/1.9780898719284.

Neittaanmaki, P., Sprekels, J., Tiba, D., 2006. Optimization of Elliptic Systems. Theory and Applications. Springer.

Nikolova, M., 2002. Minimizers of cost-functions involving nonsmooth data-fidelity terms. Application to the processing of outliers. SIAM J. Numer. Anal. 40 (3), 965–994. https://doi.org/10.1137/S0036142901389165.

Nikolova, M., 2004. A variational approach to remove outliers and impulse noise. J. Math. Imaging Vis. 20 (1), 99–120. https://doi.org/10.1023/B:JMIV.0000011326.88682.e5.

Osher, S., Burger, M., Goldfarb, D., Xu, J., Yin, W., 2005. An iterative regularization method for total variation-based image restoration. Multiscale Model. Simul. 4 (2), 460–489. https://doi.org/10.1137/040605412.

Outrata, J., Kocvara, M., Zowe, J., 1998. Nonsmooth approach to optimization problems with equilibrium constraints. In: Nonconvex Optimization and its Applications, vol. 28. Kluwer Academic Publishers, Dordrecht. https://doi.org/10.1007/978-1-4757-2825-5.

Papafitsoros, K., 2014. Novel Higher Order Regularisation Methods for Image Reconstruction (Ph.D. thesis). University of Cambridge. https://www.repository.cam.ac.uk/handle/1810/246692.

Papafitsoros, K., Bredies, K., 2015. A study of the one dimensional total generalised variation regularisation problem. Inverse Prob. Imaging 9 (2), 511–550. https://doi.org/10.3934/ipi.2015.9.511.

Papafitsoros, K., Schönlieb, C.B., 2014. A combined first and second order variational approach for image reconstruction. J. Math. Imaging Vis. 48 (2), 308–338. https://doi.org/10.1007/s10851-013-0445-4.

Papafitsoros, K., Valkonen, T., 2015. Asymptotic behaviour of total generalised variation. In: Aujol, J.-F., Nikolova, M., Papadakis, N. (Eds.), Scale Space and Variational Methods in Computer Vision: 5th International Conference, SSVM 2015, Proceedings. Springer International Publishing, pp. 702–714.

Pöschl, C., Scherzer, O., 2015. Exact solutions of one-dimensional total generalized variation. Commun. Math. Sci. 13 (1), 171–202. https://doi.org/10.4310/CMS.2015.v13.n1.a9.

Qi, L., Sun., J., 1993. A nonsmooth version of Newton's method. Math. Program. 58, 353–367. https://doi.org/10.1007/BF01581275.

Ring, W., 2000. Structural properties of solutions to total variation regularization problems. ESAIM: Math. Model. Numer. Anal. 34 (4), 799–810. https://doi.org/10.1051/m2an:2000104.

Rudin, L.I., Osher, S., Fatemi, E., 1992. Nonlinear total variation based noise removal algorithms. Physica D: Nonlinear Phenomena 60 (1–4), 259–268. https://doi.org/10.1016/0167-2789(92)90242-F.

Sawatzky, A., 2011. (Nonlocal) Total Variation in Medical Imaging (Ph.D. thesis). University of Münster.

Sawatzky, A., Brune, C., Müller, J., Burger, M., 2009. Total variation processing of images with Poisson statistics. In: Jiang, X., Petkov, N. (Eds.), Computer Analysis of Images and Patterns. CAIP 2009. Lecture Notes in Computer Science, vol. 5702. Springer, Berlin, Heidelberg, pp. 533–540.

Scheel, H., Scholtes, S., 2000. Mathematical Programs with Complementarity Constraints: Stationarity, Optimality, and Sensitivity. Math. Oper. Res. 25 (1), 1–22. https://doi.org/10.1287/moor.25.1.1.15213.

Scherzer, O., 1998. Denoising with higher order derivatives of bounded variation and an application to parameter estimation. Computing 60 (1), 1–27. https://doi.org/10.1007/BF02684327.

Setzer, S., 2011. Operator splittings, Bregman methods and frame shrinkage in image processing. Int. J. Comput. Vis. 92 (3), 265–280. https://doi.org/10.1007/s11263-010-0357-3.

Setzer, S., Steidl, G., 2008. Variational methods with higher order derivatives in image processing. In: Neamtu, M., Schumaker, L.L. (Eds.), Approximation XII. Nashboro Press, Brentwood, pp. 360–386.

Setzer, S., Steidl, G., Teuber, T., 2011. Infimal convolution regularizations with discrete ℓ_1-type functionals. Commun. Math. Sci. 9, 797–872. https://doi.org/10.4310/CMS.2011.v9.n3.a7.

Suquet, P.M., 1978. Existence et régularité des solutions de la placticité parfaite. Comptes Rendus de l'Académie des Sci. Ser. A. 286, 1201–1204.

Tai, X.C., Hahn, J., Chung, G.J., 2011a. A fast algorithm for Euler's elastica model using augmented Lagrangian method. SIAM J. Imaging Sci. 4 (1), 313–344. https://doi.org/10.1137/100803730.

Tai, X.C., Hahn, J., Chung, G.J., 2011b. A fast algorithm for Euler's elastica model using augmented Lagrangian method. SIAM J. Imaging Sci. 4 (1), 313–344. https://doi.org/10.1137/100803730.

Temam, R., 1985. Mathematical problems in plasticity. vol. 15, 5. Gauthier-Villars Paris.

Temam, R., Strang, G., 1980. Functions of bounded deformation. Arch. Rational Mech. Anal. 75 (1), 7–21. https://doi.org/10.1007/BF00284617.

Valkonen, T., 2015. The jump set under geometric regularization. Part 1: basic technique and first-order denoising. SIAM J. Math. Anal. 47 (4), 2587–2629. https://doi.org/10.1137/140976248.

Valkonen, T., 2017. The jump set under geometric regularisation. Part 2: Higher-order approaches. J. Math. Anal. Appl. 453 (2), 1044–1085. https://doi.org/10.1016/j.jmaa.2017.04.037.

Valkonen, T., Bredies, K., Knoll, F., 2013. Total generalized variation in diffusion tensor imaging. SIAM J. Imaging Sci. 6 (1), 487–525. https://doi.org/10.1137/120867172.

Vese, L., 2001. A study in the BV space of a denoising-deblurring variational problem. Appl. Math. Optim. 44 (2), 131–161. https://doi.org/10.1007/s00245-001-0017-7.

Vogel, C., Oman, M., 1996. Iterative methods for total variation denoising. SIAM J. Sci. Comput. 17 (1), 227–238. https://doi.org/10.1137/0917016.

Wang, Z., Bovik, A.C., Sheikh, H.R., Simoncelli, E.P., 2004. Image quality assessment: from error visibility to structural similarity. IEEE Trans. Image Process. 13 (4), 600–612. https://doi.org/10.1109/TIP.2003.819861.

Wotao, Y., Goldfarb, D., Osher, S., 2007. The total variation regularized L^1 model for multiscale decomposition. Multiscale Model. Simul. 6 (1), 190–211. https://doi.org/10.1137/060663027.

Wu, C., Tai, X.C., 2010. Augmented Lagrangian method, dual methods, and split Bregman iteration for ROF, vectorial TV, and high order models. SIAM J. Imaging Sci. 3 (3), 300–339. https://doi.org/10.1137/090767558.

Wu, C., Zhang, J., Tai, X.C., 2011. Augmented Lagrangian method for total variation restoration with non-quadratic fidelity. Inverse Prob. Imaging 5 (1), 237–261. https://doi.org/110.3934/ipi.2011.5.237.

Yin, W., Osher, S., Goldfarb, D., Darbon, J., 2008. Bregman iterative algorithms for ℓ_1-minimisation with applications to compressed sensing. SIAM J. Imaging Sci. 1, 142–168. https://doi.org/10.1137/070703983.

Zowe, J., Kurcyusz, S., 1979. Regularity and stability for the mathematical programming problem in Banach spaces. Appl. Math. Optim. 5 (1), 49–62. https://doi.org/10.1007/BF01442543.

Chapter 12

Graph-based optimization approaches for machine learning, uncertainty quantification and networks

Andrea L. Bertozzi[a,*] and Ekaterina Merkurjev[b]
[a]Department of Mathematics, University of California, Los Angeles, CA, United States
[b]Department of Mathematics and CMSE (Computational Mathematics, Science, and Engineering), Michigan State University, East Lansing, MI, United States
[*]Corresponding author: e-mail: bertozzi@math.ucla.edu

Chapter Outline

Abstract

Recently, the graphical framework, which provides information about the connections between pieces of data via a similarity graph, has become a popular setting for problems involving high dimensional data. Moreover, some of the best performing graph-based methods consist of solving variational problems on graphs. In this chapter, we describe several recent graph-based optimization algorithms for applications such as data classification, uncertainty quantification and the social sciences. In particular, the data classification procedures we consider are derived using optimization techniques, including

Handbook of Numerical Analysis, Vol. 20. https://doi.org/10.1016/bs.hna.2019.04.001
© 2019 Elsevier B.V. All rights reserved.

those involving the Ginzburg–Landau functional, the Merriman–Bence–Osher (MBO) scheme and total variation. Approaches to dealing with class size constraints are also included. Moreover, the uncertainty quantification algorithms we describe involve a Bayesian approach to uncertainty quantification for graph-based classification, using the posterior distribution on the classification labels. Finally, the last part of the chapter is dedicated to network analysis, which connects modularity optimization with graph cuts and total variation. Overall, the procedures detailed in this chapter form a powerful set of tools to be utilized for many applications and by many communities.

Keywords: Diffuse interfaces, Graphical models, Graph Laplacian, Machine learning, Uncertainty quantification, Social networks, Community detection, Data clustering, Modularity, MBO scheme

AMS Classifications Codes: Primary: 65K10, Secondary: 05C85, 62–09, 62H30, 91C20 68U10, 35R02, 91D30, 49M29, 35Q56, 94C15

1 Introduction

In recent years, algorithms based on the graphical framework have gained popularity as highly competitive for problems involving high dimensional data, with applications ranging from image processing to machine learning to the social sciences (Belkin et al., 2006; Chapelle et al., 2006; Digne, 2012; Elmoataz et al., 2015; Macdonald et al., 2013; Tian et al., 2009; Wang et al., 2008; Zhou and Schölkopf, 2004; Zhou et al., 2004; Zhu, 2006). The graphical setting has even been extended to deep learning techniques. In particular, Seo et al. (2018) introduce the concept of graph convolutional recurrent networks. Moreover, Bresson and Laurent (2017) compare graph recurrent neural networks with graph convolutional neural networks. Other works on deep learning methods on graphs include Levie et al. (2017), Monti et al. (2017) and Defferrard et al. (2016).

The success of such methods is partly due to the many advantages offered by using a graph-based approach. First, graphical techniques not only provide useful information about the overall structure of the data, but also about the connections between pieces of data, via a similarity graph, composed of nodes and edges, where the nodes correspond to elements of the data and the edge weights encode the degree of similarity between pairs of vertices. Second, as opposed to other algorithms, graph-based approaches reduce the high dimensionality of the problem by embedding the dimension of the features in the graphical structure. Further operations are performed on the graph itself. Moreover, we note that, for tasks such as data classification, the graphical framework serves as a way to handle nonlinear structure, thus allowing the user to process a variety of data. In addition, the graphical setting is not only the basic building block for network analysis, but in the case of image processing, is particularly effective in the capturing of texture due to its nonlocal nature. Overall, the graphical structure provides a general way to incorporate diverse types of data, such as hyperspectral data, text data, LIDAR data, images, video, etc.

In this chapter, we outline several recent graph-based optimization methods with a variety of applications, including machine learning, networks and uncertainty quantification. Section 2 provides a brief introduction to similarity graphs. Section 3 outlines recent methods for data classification using the Ginzburg–Landau (GL) functional on graphs and the Merriman–Bence–Osher (MBO) scheme, including approaches to handle volume penalties. Section 4 provides an introduction to recent total variation (TV) algorithms for machine learning, while Section 5 describes methods for uncertainty quantification (UQ) for graphical metrics. Section 6 involves network analysis, and we conclude in Section 7. Overall, this chapter is based on the material in Bertozzi (2018).

2 Graph theory

This section outlines the graphical framework, which is used in the formulations of the algorithms described in this chapter. The graphical setting consists of a graph $G = (V, E)$, where V and E are sets of vertices and edges, respectively.

The data set is embedded into the framework by considering each element of the data as a vertex, and connecting vertices i and j via an edge weighted by a weight function w_{ij}. The weight function is constructed in such a way that it assigns a large value to similar vertices, and a small value to dissimilar vertices. Several other important elements of the graphical setting are:

- W—the weight matrix of off-diagonal elements w_{ij}.
- D—the diagonal matrix of degrees, where the ith diagonal entry is the degree d_i, the sum of the weights connected to node i.
- L—the Laplacian matrix $D - W$.

Before computing the weight function, it is necessary to calculate a feature vector V_i for each vertex. Specifically, one must assign a vector to each node of the graph; the vector must contain important information about the vertex, which can be used to designate it to a class. For example, for a data set of thousands of images, each vertex in the graph represents an image. One way of forming a feature vector for this data is to perform principal component analysis to reduce each image to a r-dimensional vector, where r is much less than the number of pixels in each image. The vector is then assigned to be the feature vector of the image. For hyperspectral data sets, one can set the feature vector of each pixel to be the values at the channels of the pixel. For data sets consisting of a single image, each vertex of the graph represents a pixel in the image. Then, the values in a neighbourhood around each pixel, under a potential scaling and rotational invariance, can be used as a feature vector. Overall, the problem of computing the "correct" feature vector is a difficult task in itself, and is not the focus of this chapter.

After the feature vectors are computed, the next step is to calculate the weight function; some popular formulations include:

1. The Gaussian weight function

$$w_{ij} = \exp(-\|V_i - V_j\|^2/\tau), \tag{1}$$

where V_i is the feature vector of vertex i, and τ is a parameter. The Yaroslavsky filter (Yaroslavsky, 2012) and the nonlocal means filter (Buades et al., 2005) are different variants of the Gaussian weight function.

2. Gaussian weight function with cosine angle

$$w_{ij} = \exp - \frac{\left(1 - \frac{\langle V_i, V_j \rangle}{|V_i||V_j|}\right)^2}{2\sigma^2}, \tag{2}$$

where V_i is the feature vector of vertex i. This weight function is often used for hyperspectral data sets.

3. Zelnik-Manor and Perona weight function (Zelnik-Manor and Perona, 2004)

$$w_{ij} = \exp\left(-\frac{d(V_i, V_j)^2}{\sqrt{\tau(V_i)\tau(V_j)}}\right). \tag{3}$$

for the sparse case. Here, $d(V_i, V_j)$ is a metric between feature vectors, and $\sqrt{\tau(V_i)} = d(V_i, V_M)$ is a local parameter for each V_i, where V_M is the Mth closest vector to V_i. Often, the Zelnik-Manor and Perona weight function is used in combination with a nearest neighbour technique, where only k nearest neighbours are connected.

The weight function can then be used to compute the graph Laplacian matrix. In some cases, the operator is scaled to take into account the behaviour that arises as the sample size grows larger. Two popular normalization procedures for the graph Laplacian are:

- *The symmetric Laplacian*:

$$L_s = D^{-1/2}LD^{-1/2} = I - D^{-1/2}WD^{-1/2}, \tag{4}$$

where I is the identity matrix. This normalization is often used for the nonlocal means graph for images.

- *The random walk Laplacian*:

$$L_w = D^{-1}L = I - D^{-1}W, \tag{5}$$

which is related to discrete Markov processes.

A common property of the graph Laplacian, the symmetric Laplacian and the random walk Laplacian is that all three are positive semidefinite matrices and have n nonnegative real-values valued eigenvalues $0 = \lambda_1 \leq \lambda_2 \leq \lambda_3 \leq \cdots \leq \lambda_n$, where $n \times n$ is the dimension of the matrices.

There are several challenges to using the graphical framework. One difficulty is the $O(n^2)$ computational complexity of calculating the graph weights, where n is the number of data elements, although the graph can be sparsified to connect only certain pairs of vertices. A different strategy is to use the Nyström extension method (Belongie et al., 2002; Fowlkes et al., 2001, 2004), which computes the $O(n)$ low-rank approximation of the graph Laplacian matrix.

For data segmentation problems, another difficulty arises since the number of classes for segmentation may not be known; here, several strategies can be applied. One solution is to first approximate the number of classes, using methods such as the elbow algorithm and X-means clustering. The elbow method Ketchen and Shook (1996) plots the percentage of variance against the number of clusters to locate the "elbow criterion," which is then set to be the desired number of clusters, while X-means clustering Pelleg et al. (2000) proceeds similarly to k-means clustering, repeatedly alternating between subdividing clusters and advancing with the best split until a stopping criterion is reached. The final result possesses the desired number of clusters. Then, one can proceed with any of the classification methods mentioned in this chapter.

3 Recent methods for semisupervised and unsupervised data classification

Methods for segmenting data can be classified into several categories: unsupervised, semisupervised and supervised algorithms. In *unsupervised* learning, the algorithm is not provided with any training data, and it proceeds to obtain the structure and labels of the data without any prior information. In *semisupervised* learning, the method is equipped with a small set of labelled data, and the goal is to propagate the labels towards the unlabelled elements. *Supervised* approaches involve using a large training set, obtaining the parameters for the model, and testing the model on a smaller testing set. In this section, the focus is on recent semisupervised and unsupervised algorithms.

One of the most common and basic methods for segmenting data is spectral clustering, an unsupervised algorithm which clusters the data according to the k-means method applied to a small set of eigenfunctions of the graph Laplacian matrix (Von Luxburg, 2007). The procedure can be used along with the Nyström extension method to obtain a $O(n)$ low-rank approximation of the graph Laplacian matrix, where n is the number of elements in the data.

The recent decades have shown a surge in the development of algorithms for machine learning which are much more powerful than spectral clustering. In this section, we describe recent approaches, in particular those that are graph-based and optimization-based, for both semisupervised and unsupervised data classification. Specifically, recent algorithms derived from the Ginzburg–Landau functional (Section 3.1), the MBO scheme (Section 3.2), heat kernel pagerank (Section 3.3), Mumford–Shah model (Section 3.4) and auction techniques (Section 3.5) are outlined.

An optimization technique for data classification used in the context of machine learning often involves minimizing a general form of energy (or cost) functional:

$$E(u) = R(u) + \mu||u - \hat{u}||_b^p, \tag{6}$$

where u is the classification function, $R(u)$ is a regularization term and $||u - \hat{u}||_b^p$ is a fidelity term, incorporating most (supervised) or just a few (semisupervised) of the known values $\hat{u}$. In the case of unsupervised classification, the second term is replaced by a region homogeneity term, the first term is equipped with a penalty, or additional constraints are imposed.

The type of regularization term $R(u)$ has nontrivial consequences for the final classification accuracy. When choosing the regularization norm, it is important to conserve the sharp discontinuities that may appear in the boundaries between classes. The total variation seminorm has been particularly successful in the latter task.

Overall, Boyer et al. (2019) describe the structural properties of certain solutions of (6). Moreover, the authors of Unser et al. (2016) analyze l_1 vs l_2 regularization for the resolution of ill-posed linear inverse and compressed sensing problems. In addition, Unser et al. (2017) provide a definition and an analysis of a generalized total variation seminorm.

Recently, the Ginzburg–Landau (GL) energy has been explored as a regularization term for tasks such as semisupervised classification and image processing, since it serves as an approximation to total variation. Given a phase field variable u, the GL energy, introduced for the Euclidean space in the last century, involves a competition between the convex functional $\int (\nabla u)^2 dx$ that induces smoothing, with a double-well function $\int W(u)dx$, that separates its argument into phases. The next section describes recent algorithms derived using the GL functional.

Another multiphase image segmentation method, based on a phase transition model of Modica and Mortola in materials sciences, is described in Jung et al. (2007).

3.1 Semisupervised learning and the Ginzburg–Landau graph model

In Bertozzi and Flenner (2012), Bertozzi et al. formulate the first set of binary classifiers for semisupervised learning using a fully nonlinear model for similarity graphs. The algorithms require minimal training data and are derived from minimizing the GL functional as a smooth relaxation of graph total variation, used as a regularizer for semisupervised learning. For large datasets, efficient linear algorithms, such as the Nyström extension method (Belongie et al., 2002; Fowlkes et al., 2001, 2004), are integrated into the procedure. The SIGEST paper (Bertozzi and Flenner, 2016) is a republication of the above work, with more results.

The aforementioned *Ginzburg–Landau functional* takes the following form:

$$GL(u) = \frac{\varepsilon}{2}\int |\nabla u|^2 dx + \frac{1}{\varepsilon}\int W(u)dx, \tag{7}$$

where u is a scalar field representing the state of the phases in the system, ∇ is the spatial gradient operator, $W(u)$ denotes the double-well potential (i.e., $W(u) = u^2(1-u)^2$, where 0 and 1 are the two states of the system), and ε is a small positive constant. Energy (7) can be minimized in the L_2 sense using gradient descent, resulting in the *Allen–Cahn equation*

$$\frac{\partial u}{\partial t} = -\varepsilon \Delta u - \frac{1}{\varepsilon}W'(u). \tag{8}$$

The binary data segmentation model of Bertozzi and Flenner (2012) modifies the original Ginzburg–Landau functional (7) in two ways. First, the authors replace the first term of GL energy with the graph Dirichlet energy, $\sum_{ij} w_{ij}(u_i - u_j)^2$, equivalent to the inner product of Lu with u, where L indicates the graph Laplacian. Second, since the problem is semisupervised classification, the Ginzburg–Landau model is modified to include a semisupervised penalty term, such as an L^2 penalty for labelled data. In particular, the problem considered is minimizing

$$E(u) = \frac{1}{\varepsilon}\langle Lu, u\rangle + \varepsilon \sum_i (W(u_i)) + \frac{1}{2}\sum_{i \in V} \lambda_i (u_i - u_{0i})^2,$$

where the optimal u indicates the classification, u_{0i} indicates the initial data for node i, and λ_i is a constraint for semisupervision, which is equal to a constant or zero. For many methods described in this chapter, the accuracy of the model is competitive even when a small amount of training data, such as 5% or less of the data, is supplied to the algorithm.[a]

One advantage of using the graph-based Ginzburg–Landau functional, compared to total variation methods, is the fact that graph structure is contained only in the Dirichlet energy. Therefore, one can obtain a computationally efficient algorithm by exploiting efficient techniques, such as the Nyström extension method, to obtain approximations of the graph Laplacian. Moreover, the double-well structure of the functional simply reduces to thresholding during the minimization procedure. The second property of the Ginzburg–Landau functional is used to derive the graph MBO scheme algorithm, to be described in Section 3.2.

---

[a]In van Gennip and Bertozzi (2012), the authors show that a variant of the GL functional converges in the sense of Gamma convergence, and in the vanishing ε-limit, to the graph TV functional

$$E_{TV}(u) = \sum_{ij} w_{ij}|u_i - u_j|. \tag{9}$$

Extensions and variants of this result can be found in Thorpe and Theil (2017), Kohn and Sternberg (1989), Dobrosotskaya and Bertozzi (2008) and Dobrosotskaya and Bertozzi (2010).

The GL model can be solved very efficiently even for large datasets using spectral techniques and convex splitting methods (Schönlieb and Bertozzi, 2011). The spectral approach involves computing the eigenfunctions and eigenvalues of the graph Laplacian, which can be performed quickly using methods such as the Nyström extension algorithm (Belongie et al., 2002; Fowlkes et al., 2001, 2004). As noted earlier, the graph structure is only contained in the Dirichlet energy term of the model, which can be minimized using spectral approximations to the graph Laplacian. Since the Ginzburg–Landau functional serves as an approximation to total variation (TV), the TV minimization problem can be computed without calculating all the weights of the graph, which can be very computationally intensive for techniques such as nonlocal means in image processing (Buades et al., 2005; Gerhart et al., 2013; Gilboa and Osher, 2007a, 2008; Kang et al., 2014; Merkurjev et al., 2014; Peyré, 2008). The authors of Peyré (2008) provide an overview on image processing techniques with graphs.

The binary data segmentation model of Bertozzi and Flenner (2012) was extended to the multiclass case in Garcia-Cardona et al. (2014a). The extension was formulated by assigning each node a probability distribution over the classes. Specifically, for each node i, the vector $\mathbf{u}_i$ was required to be an element of the Gibbs simplex Σ^K, defined as

$$\Sigma^K := \left\{ (x_1,\ldots,x_K) \in [0,1]^K \,\middle|\, \sum_{k=1}^{K} x_k = 1 \right\}. \tag{10}$$

The final class of a vertex i is chosen to the argument of the minimum of the optimal vector $\mathbf{u}_i$.

A convex splitting scheme is used to minimize the GL functional in the phase field approach. The multiphase and graph-based Ginzburg energy model is decomposed into convex and concave parts as:

$$\begin{aligned}
E(\mathbf{U}) &= E_{\text{convex}}(\mathbf{U}) + E_{\text{concave}}(\mathbf{U})\\
E_{\text{convex}}(\mathbf{U}) &= \frac{\varepsilon}{2}\langle \mathbf{U}, L\mathbf{U}\rangle + \frac{C}{2}\langle \mathbf{U},\mathbf{U}\rangle\\
E_{\text{concave}}(\mathbf{U}) &= \frac{1}{2\varepsilon}\sum_{i\in V}\prod_{k=1}^{m}\frac{1}{4}\|\mathbf{u}_i - \boldsymbol{e}_k\|_{L_1}^2\\
&\quad + \sum_{i\in V}\frac{\mu_i}{2}\|\mathbf{u}_i - \hat{\mathbf{u}}_i\|_{L_2}^2 - \frac{C}{2}\langle \mathbf{U},\mathbf{U}\rangle,
\end{aligned}$$

where the matrix $\mathbf{U}$ is composed of the $\mathbf{u}_i's$ as rows, and $\hat{\mathbf{u}}$ indicates labelled data.

The two aforementioned papers have led to the development of many other new algorithms. In particular, more than 50 new papers have arisen from Bertozzi and Flenner (2012), including fast methods for nonlocal means image processing using the MBO scheme (Merkurjev et al., 2013), multiclass learning methods (Garcia-Cardona et al., 2014a; Iyer et al., 2017), parallel methods for exascale-ready platforms (Meng et al., 2016), hyperspectral video analysis (Hu et al., 2015; Meng et al., 2017; Merkurjev et al., 2014; Zhu et al., 2017), modularity optimization for network analysis (Boyd et al., 2018; Hu et al., 2013), measurement techniques in zoology (Calatroni et al., 2017), generalizations to hypergraphs (Bosch et al., 2018), Pagerank (Merkurjev et al., 2018) and Cheeger cut based methods (Merkurjev et al., 2017). This chapter reviews some of this literature.

3.2 The graph MBO scheme for data classification and image processing

A simplification of the algorithm in Bertozzi and Flenner (2012) and Bertozzi and Flenner (2016), described in the previous section, can be made using the MBO technique (Merriman et al., 1992), originally derived by Merriman, Bence and Osher to approximate motion by mean curvature, and later adapted to piecewise constant image segmentation (Esedog-lu and Tsai, 2006). The simple scheme consists of alternating between two steps: diffusion and thresholding.

Motivated by the aforementioned works, Merkurjev et al. (2013) extended the MBO scheme to the graphical setting for binary data classification and greyscale inpainting by solving a series of binary problems for each bit. The graph MBO scheme of Merkurjev et al. (2013) for semisupervised learning and image processing is comprised of alternating between the following two simple steps to obtain the next iterate:

1. Heat equation with forcing term. Propagate using

$$\frac{u^{(n+1/2)} - u^{(n)}}{dt} = -L_u^{(n)} - \lambda(i)(u^{(n)} - u_0)$$

2. Threshold.

$$u^{(n+1)} = \begin{cases} 1 & \text{if } u^{(n+1/2)} \geq 0, \\ 0 & \text{if } u^{(n+1/2)} < 0. \end{cases}$$

Here, u represents the classification, u_0 is the initial data, L is the graph Laplacian, and the $\lambda(i)(u^{(n)} - u_0)$ term is a fidelity term representing labelled data. As in the Ginzburg–Landau case, the heat equation with forcing term is solved using spectral methods and projections onto a small number of eigenfunctions of the graph Laplacian. Moreover, regarding the parameters, $\lambda = 0$ for unlabelled nodes and $\lambda = 1$ for labelled nodes. In the case of image inpainting, $\lambda = 0$ for the damaged region, and $\lambda = 1$ otherwise. The greyscale inpainting problem itself is approached by solving eight binary problems, one for each bit, and then obtaining the final classification from the eight different bit classifications.

The extension of the graph MBO scheme to the multiclass case (Garcia-Cardona et al., 2014a, b) can be performed using the Gibbs simplex (10) similarly to the procedure of the previous section. In this case, $u(x)$ is set to be a vector of probabilities over the classes for a data element x; the final class of x is the argument of the maximum of the vector. The only changes in the algorithm is first projecting to the Gibbs simplex after performing step 1, and then thresholding to the corners of the simplex during step 2.

The graph MBO scheme is remarkably efficient. In fact, its computational time is several times faster than that of the Split-Bregman method (Goldstein and Osher, 2009) applied to the Osher–Gilboa nonlocal means graph using the same datasets. Moreover, the speed-up is significant compared to the Ginzburg–Landau model described in the previous section. To demonstrate the efficiency of the procedure, one can consider the gas plume detection problem using the hyperspectral data set from Dugway Proving Ground, shown in Fig. 1 along with the results. The classification problem for hyperspectral data sets is especially difficult due to the high number of pixels and channels in the data. Computing the graph weights using the Nyström extension algorithm and spectral angle takes only two minutes on a regular laptop, while the two-step procedure finishes in seconds. To optimize the graph MBO scheme even further, Meng et al. (2016) have used an exascale-ready platform at the National Energy Research Supercomputing Center (Meng et al., 2016).

The MBO scheme has been analyzed rigorously in several recent works. For example, in Luo and Bertozzi (2017), the authors show that the convergence and stability of the procedure are independent not only of the size of the graph, but also of its sparseness. They also formulate theoretical convergence estimates (Luo and Bertozzi, 2017) for the Ginzburg–Landau model for semisupervised learning for different versions of the graph Laplacian. In particular, for the standard graph Laplacian, they derive maximum norm convergence results for minimizers of the GL energy using a combination of L^2-energy estimates and maximum principle results for the graph Laplacian

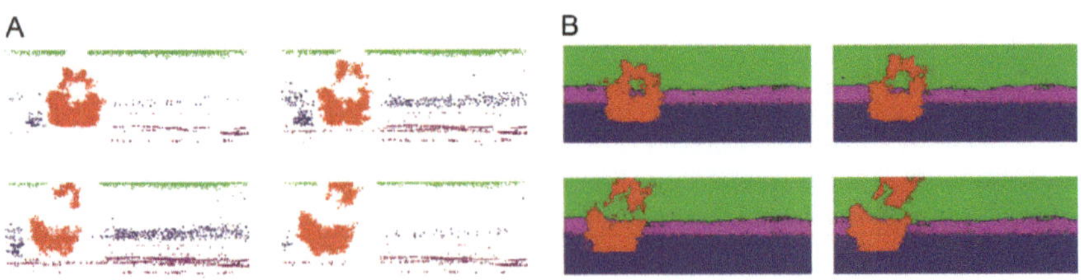

FIG. 1 The graph MBO scheme result for the hyperspectral video sequence from standoff detection of gas plume in the Dugway Proving Ground. Feature vectors are computed by taking the values in the channels for each pixel. Shown are four video frames with $N = 280{,}000$ graph nodes. The colours correspond to four classes: plume (*red*), sky (*green*), foreground (*blue*), mountain (*purple*). The *left side* of the figure shows the training data corresponding to only 8% of the data. The *right side* of the figure is the resulting classification—as in Merkurjev et al. (2014) and Meng et al. (2017). Code for this calculation is published online in Meng et al. (2017).

operator. Naturally, since the Ginzburg–Landau energy is nonconvex, the convergence is assumed to be to a local minimizer rather than a global one. Also, in Hu et al. (2015) and van Gennip et al. (2014), it is shown that if $\Gamma_\tau = e^{-\tau L}$ is the diffusion operator, where L is the graph Laplacian, and τ is the timestep of the MBO scheme, then the discrete energy

$$E_{\text{MBO}}(u) = \frac{1}{\tau}\langle 1 - u, \Gamma_\tau u\rangle$$

decreases at each timestep and approximates the graph total energy. Moreover, in van Gennip et al. (2014), the authors prove the first rigorous results connecting the graph counterparts of mean curvature flow, the Allen–Cahn equation and the MBO procedure. Overall, the MBO scheme finds the global minimum in most cases. For unusual problems that require a provably optimal solution, methods built around max-flow and ADMM techniques, such as those described in Section 4, are less efficient than the MBO algorithm, but they can guarantee a global optimal solution (Merkurjev et al., 2015).

3.3 Heat kernel pagerank method

In Merkurjev et al. (2018), heat kernel pagerank is used along with the MBO scheme techniques to solve the semisupervised classification problem. In addition to the main algorithm, the authors outline a semisupervised learning method that uses heat kernel pagerank *directly* as a classifier. This is the first application of heat kernel pagerank to semisupervised data classification. Moreover, all proposed methods of Merkurjev et al. (2018) avoid potentially computationally expensive linear algebra routines; for example, the calculation of eigenvalues and eigenvectors of large matrices.

Let $P = D^{-1}W$ be a transition probability matrix, where W and D denote the weight and degree matrices, respectively. We also denote the heat kernel of a graph G by the matrix $H_t = e^{-tL_{rw}}$, which is a solution to the heat equation $\frac{\partial u}{\partial t} = -L_{rw}u$. Using the aforementioned notation, the heat kernel pagerank is defined as $\rho_{t,f} = fH_t$ for some vector f, which can be written equivalently as

$$\rho_{t,f} = fH_t = \sum_{m=0}^{\infty} e^{-t}\frac{t^m}{m!} fP^m. \tag{11}$$

Since the random walk Laplacian is just a transition probability matrix for the random walk in which one moves from vertex i to vertex j with probability given by $(D^{-1}W)_{ij}$, one can see that the heat kernel pagerank is just the expected distribution of the random walk, where m random walk steps are taken with probability $e^{-t}\frac{t^m}{m!}$. Therefore, one can approximate the heat kernel pagerank vector by simulating r random walks (with the number of steps calculated according to the Poisson distribution) and then taking the average of the r final distributions. One can also compute an approximation to heat kernel pagerank using a completely different strategy by computing a finite number of terms in (11); we refer to this technique as the "series approach". The random walk technique of approximating heat kernel pagerank is more suited to the case of dense graphs, while the latter strategy can be applied more efficiently in the case of sparse graphs.

The solution to the following heat equation in a graphical setting:

$$\dot{u}(t) = -L_w u(t), \qquad u(0) \in R^n. \tag{12}$$

is given by $u(t) = D^{-1}\rho^{tr}_{t,f}$, where $f = u(0)^{tr}D$ and M^{tr} denote the transpose. Therefore, the heat equation in a graphical setting, defined using the random walk Laplacian, can be solved using heat kernel pagerank.

The classification procedure of Merkurjev et al. (2018) is derived by analyzing the dynamics of the graph MBO scheme, a procedure which consists of alternating between the application of the (modified) heat equation and a thresholding process. One variant of this process is to split the dynamics into three steps, with the first step consisting of solving the regular heat equation. This heat equation can be solved efficiently using heat kernel pagerank, which can be calculated approximately using two techniques: random walks and series. The time complexity of the algorithm in the case of a sparse graph is linear in the number of nodes in the data set. For a dense graph, the random walk approach requires $O(rnK)$ operations in the worst case, for a fixed number of classes m, where r is the number of random walks, n is the size of the data set, and K is the bound on the number of steps of each random walk. Of course, the more random walks are computed, and the longer they are, the better the approximation to heat kernel pagerank, with the exact error estimates stated in the paper.

3.4 Unsupervised learning and the Mumford–Shah model

In this section, we outline several recent graph-based unsupervised learning approaches using a generalization of the piecewise constant Mumford–Shah (MS) model (Chan and Vese, 2001; Esedog-lu and Tsai, 2006; Vese and Chan, 2002), which segments an image into m regions $\Omega_1, \ldots, \Omega_m$, according to a contour Φ. The contour is chosen so that the following energy is minimized:

$$E(\Phi, \{c_r\}_{r=1}^m) = |\Phi| + \lambda \sum_{r=1}^{m} \int_{\Omega_r} (u_0 - c_r)^2, \tag{13}$$

where c_r is a constant approximation of region Ω_r, for $r = 1, \ldots, m$, u_0 is the original image, and $|\Phi|$ designates the length of the contour Φ.

An extension of the MS model to the graphical framework is studied in Hu et al. (2015) and Meng et al. (2017). The authors investigate the following modification of the model:

$$E(u) = \frac{1}{2}E_{TV}(u) + \lambda\sum_{r=1}^{m}\sum_{i} u_r(n_i)\,\|u_0(n_i) - c_r\|^2. \tag{14}$$

Here, $|u|_{TV}$ represents the total variation norm in the graphical setting, and u is a function such that $u_r(n_i) = 1$ if n_i belongs to the rth class, and zero otherwise. Moreover, note that the first term of (14) corresponds to the first term of (13); the second term of (14) corresponds to the second term of (13).

The algorithms of Hu et al. (2015) and Meng et al. (2017) are developed using an MBO-like scheme. First, the total variation term of (14) is approximated using the Ginzburg–Landau energy, which allows one to leverage efficient techniques such as the Nyström extension procedure for the graph Laplacian operator. This allows one to formulate a three-step procedure for minimizing the graph-based Mumford–Shah functional (14) consisting of alternating between solving a modified heat equation with an extra term, thresholding and then updating the coefficients $c_1, \ldots, c_m$. Moreover, in Hu et al. (2015), theoretical analysis is developed and a Lyapunov functional is proven to decrease as the algorithm proceeds.

An important feature of the methods in Hu et al. (2015) and Meng et al. (2017) is that the image data are used both to compute the weights for the similarity graph and to solve the clustering problem, since the constants c_r, for $r = 1, \ldots, m$, in (14) are chosen to be averages of the feature vectors of elements in each class, and are recomputed each iteration. This procedure varies from that of the previous methods, for which the feature vectors were used only to calculate the graph weights that were then incorporated into the optimization model used to obtain the classification.

The algorithms in Hu et al. (2015) and Meng et al. (2017) are extremely efficient, partly due to the MBO-like scheme used to minimize (13) and the Nyström extension method. In fact, the techniques in Meng et al. (2017) take only a minute on a supercomputer to process 40 frames of a hyperspectral video sequence after the implementation of OpenMP directive-based parallelism.

3.5 Imposing volume constraints

For some real life data sets, information about the sizes of the classes, whether exact or approximate, is available in advance. For example, the distribution of digits in tax returns and postal codes is very well known. Incorporating class size information into the algorithm procedure improves classification accuracy, especially when the set of labelled nodes is small, which is a very realistic scenario, or when the problem is unsupervised.

In some cases, the sizes of the classes are expected to be reasonably distributed. For example, each digit 0–9 in a data set of images of handwritten digits can be assumed to be represented approximately equally. A well-studied way to incorporate this information is to minimize a graph cut energy with a penalty that forces the size of the classes to be of the same order. A penalty that can be used is a normalization of a graph cut; three commonly used versions are the ratio cut, the normalized cut and the Cheeger cut. For the three normalizations, the problem is to find a subset S of the graph that minimizes $cut(S,\overline{S})R(S)$, where

$$R = \begin{cases} 1/|S| + 1/|\overline{S}| & \text{for ratio cut,} \\ 1/vol(S) + 1/vol(\overline{S}) & \text{for normalized cut,} \\ R = min(|S|,|\overline{S}|)^{-1} & \text{for the Cheeger cut.} \end{cases}$$

By $vol(S)$, we denote the sum of the degrees of all the vertices in S. The three versions ensure a balanced cut since as the size of S or $\overline{S}$ goes to zero, the energy approaches infinity.

Several important papers related to clustering with volume penalties include Szlam and Bresson (2010a), Bresson et al. (2013b) and Bresson et al. (2012b) by Bresson, Szlam, Laurent and von Brecht, describing new procedures to solving the relaxed Cheeger cut problem. In Merkurjev et al. (2017), the authors present an extension of the graph-based Ginzburg–Landau functional to the volume constraints case. Their approach involves using the graph Laplacian to solve the fully nonlinear Cheeger cut and ratio cut problem, with the cut term approximated by the Ginzburg–Landau functional.

In Jacobs et al. (2018), Jacobs et al. have derived a novel technique for dealing with class size (volume) constraints. Specifically, the authors designed an efficient MBO-based method for solving the semisupervised data classification problem, which is able to incorporate the constraints. This work combines some of the best features of the MBO scheme in both Euclidean space and on graphs with a modification of a highly efficient algorithm of Bertsekas (1979) for the auction problem. Overall, the authors reinterpret the semisupervised classification problem using an auction dynamics framework (inspired by real life auctions) in which elements of the data set make bids to the class of their choice. This leads to a novel unconditionally stable forward and reverse auction method that readily incorporates class size constraints into an accurate and efficient algorithm requiring remarkably little labelled data.

In order to define the optimization problem, the authors give a notion of "smoothness" for partitions of V, which we denote by $\Sigma = \{\Sigma_1, \ldots, \Sigma_2\}$. First, note that the weight function assigns large values to similar vertices, and small values to dissimilar vertices. With this structure, one can consider the minimum cut problem

$$\text{argmin}_{\Sigma}\text{Cut}(\Sigma) = \frac{1}{2}\sum_{i=1}^{m}\sum_{x\in\Sigma_i}\sum_{y\notin\Sigma_i} w_{xy}, \tag{15}$$

which penalizes partitions that place strongly connected points in different classes. However, in order to avoid trivial solutions, the graph cut is usually either normalized or equipped with constraints. In the case of Jacobs et al. (2018), there are two types of constraints: labelled data and bounds on the sizes of the classes. We denote the labelled set of class i by F_i, and the upper and lower bounds of class i by U_i and L_i.

Combining the graph cut with the constraints, the authors of Jacobs et al. (2018) start deriving their classification method by considering

$$\operatorname{argmin}_{\Sigma} \operatorname{Cut}(\Sigma) = \frac{1}{2} \sum_{i=1}^{m} \sum_{x \in \Sigma_i} \sum_{y \notin \Sigma_i} w_{xy} \quad \text{s.t.} \quad F_i \subset \Sigma_i, \quad L_i \leq |\Sigma_i| \leq U_i. \tag{16}$$

This problem is difficult to solve due to the sheer number of partitions to consider. In this case, one technique is to examine a convex relaxation of the graph cut, the graph heat content (GHC):

$$\operatorname{GHC}(u, w) = \sum_{i=1}^{m} \sum_{x, y \in V} w_{xy} u_i(x)(1 - u_i(x)), \tag{17}$$

where $u : V \to \mathcal{S}_N = \{u : D \to [0, 1]^N : \sum_{i=1}^{N} u_i(x) = 1\}$ is an element of the convex relaxation of the space of m-phase partitions of V; $u_i(x)$ represents the probability that x belongs to class i.

Using GHC, one may obtain a graph Merriman–Bence–Osher (MBO) scheme by successively minimizing linearizations of GHC with the additional constraint that the size of each class must stay between user-specified upper and lower bounds. Under some simple transformations, minimizing linearizations of GHC under class size constraints is equivalent to the upper and lower volume bound assignment problem:

$$\operatorname{argmax}_{\mathbf{u}: \mathbf{V} \setminus \mathbf{F} \to \mathbb{R}, \mathbf{u} \geq 0} \sum_{i=1}^{m} \sum_{x \in V} a_i(x) u_i(x) \quad \text{s.t.} \ \sum_{i=1}^{m} u_i(x) = 1, \ L_i - |F_i| < \sum_{x \in V \setminus F} u_i(x) < U_i - |F_i|, \tag{18}$$

where $a_i(x) = 1 - \sum_{y \notin \Sigma_i} w_{xy}$, $V = \cup \Sigma_i$ is the set of elements of the data, $w_{x,y}$ is the weight between x and y, m is the number of classes, and $u_i(x)$ is the probability that x belongs to class i.

To interpret (18), suppose that each class is an institution that offers a limited number of memberships, and that the vertices $x \in V$ are people, and each person would like to become a member of a class. No person wants to have a membership in more than one class, and each class has a constraint on the number of memberships available. Finally, suppose that the coefficients $a_i(x)$ represent how much person x wants to be a member of class i. The solution to the assignment problem is the matching of people and classes that maximizes the total satisfaction of the population. Ideally, each person would like to become a member of their favourite phase. However, this is not possible if more than U_i people want to be members of some phase i. The main difficulty of the assignment problem is in understanding how to correctly handle these conflicts.

An interesting approach to the problem is to attempt to assign the memberships according to a market mechanism. Suppose that each phase i has a membership price p_i, and if x is a member of i, then they must pay p_i. It turns out that there exists an equilibrium price vector $\mathbf{p}_*$ such that assigning memberships, so that each person gets assigned to the class offering "the best value", gives a feasible matching in terms of class size constraints. In fact, $\mathbf{p}_*$ is the optimal solution to the *dual* of the assignment problem.

Motivated by Bertsekas (1979), Jacobs et al. (2018) propose to solve (18) with an intuitively economic approach. Due to the bounds on class sizes, the auction dynamics construction requires a more sophisticated procedure. The authors' technique is to run a *forward and reverse auction*:

- Step 1 (*Forward Auction*): The elements of the data compete for memberships to the class of their choice and place their bids, the value of which takes a particular form. The prices change each round according to demand; in particular, the prices of popular classes rise. Eventually, the prices of the most desired classes rise so high that it incentivizes unmatched elements to switch their bid to the class that might offer a better value. As the elements place their bids, each class accepts the nodes with the highest bids, but it cannot accept more than declared by the upper bound on its class size. Overall, the upper bounds fit nicely into this perspective, but the lower bounds might not be fulfilled. They are taken care of during Step 2.
- Step 2 (*Reverse Auction*): The classes compete for elements of the data set and provide incentives to attract the necessary number of customers, i.e., the number specified by the lower bound on the size of the class. Eventually, the incentives of deficient classes will rise high enough so that some matched people will change their membership to classes deficient in members. The bounds on the class sizes are fulfilled at the end of the reverse auction.

The proposed method is able to segment data with high accuracy even when the number of labelled nodes is very small.

4 Total variation methods for semisupervised and unsupervised data classification

This section focuses on describing recent methods formulated directly using total variation, without an approximation of the functional. Total variation (TV) has been increasingly more central to different applications, especially as a regularization term for data segmentation. In fact, TV has been often used to solve inverse problems in imaging (Lellmann et al., 2008). The classic paper by Rudin et al. (1992) sparked interest in total variation by applying it very successfully to denoising problems. Later, similar techniques were applied to areas such as inpainting, image restoration and blind deconvolution. For example, Chan and Wong (1998) present a blind deconvolution method using total variation (TV), in which they use an alternating minimization implicit iterative scheme.

In Bae and Merkurjev (2017) and Merkurjev et al. (2015), the authors develop a continuous-valued convex relaxation of a discrete-valued data classification problem involving graph-based total variation, class size constraints, supervised constraints and region homogeneity terms, converting the original NP-hard problem to a polynomial-time task which is approached using duality techniques. Under very mild assumptions on the dual problem, the authors prove that the minimizers of the original and relaxed problems are the same. Moreover, the relaxed problem easily accommodates all constraints. To solve the convex relaxation, an efficient duality-based algorithm, which incorporates all constraints implicitly, is proposed.

The graph-based data segmentation model considered, which includes the graph-based total variation operator (TV_w) is:

$$\min_{u \in B} \left\{ \sum_{i=1}^{m} TV_w(u_i) + \sum_{i=1}^{m} \sum_{x \in V} C_i(x) u_i(x) \right\}, \tag{19}$$

subject to class size constraints or penalty terms acting on class sizes that may be present. Here, the graph-based total variation operator is defined as

$$TV_w(g) = \frac{1}{2} \sum_{x,y \in V} w(x,y) |g(y) - g(x)|. \tag{20}$$

Also, $B = \{u : V \to \{0,1\}^m | \sum_{i=1}^{m} u_i(x) = 1, \forall x \in V\}$, m is the number of classes, and the optimal $u_i(x)$ represents the probability that x belongs to class i. The second term in the sum incorporates supervised information, such as labelled nodes and region homogeneity terms, which measure how well each element of the data set fits with each region. This type of problem, both in a graphical setting and using a different framework, has appeared in many recent papers. In fact, one of the papers considering a similar problem using the regular total variation operator is Pock et al. (2009).

Note that one encounters a few obstacles when solving (19). In particular, the problem is nonconvex, due to the binary constraint, and also NP-hard. Moreover, the constraints on class sizes might be difficult to impose.

The approach of Bae and Merkurjev (2017) and Merkurjev et al. (2015) in solving (19) is to use a convex relaxation of the original problem (19) by replacing B with $B' = \{u : V \to [0,1]^m | \sum_{i=1}^{m} u_i(x) = 1, \forall x \in V\}$. It turns out that the convex relaxation approximates the original problem extremely well; in particular, the main theorem of Merkurjev et al. (2015) is:

Theorem 1. *Let* (u^*, q^*, ρ^{1*}, ρ^{2*}) *be a primal-dual pair of the primal-dual problem P, derived from the convex relaxation of* (19), *where* q^*, ρ^{1*}, ρ^{2*} *are the dual variables. If the set*

$$I_{min}(x) = \text{argmin}_{i \in \{1,\ldots,m\}} \left(C_i(x) + (\text{div}_w q_i^*)(x) + \rho_i^{2*} - \rho_i^{1*} \right), \tag{21}$$

containing the argument(s) of the minimum of m numbers, consists of only one element at every point $x \in V$, *then the corresponding solution* u^* *is an exact global binary minimizer of the original nonconvex NP-hard problem* (19).

Intuitively, it is reasonable to assume that the conditions of Theorem 1 are satisfied since it is unlikely that two or more components of the vector $\vec{z}\,(x)$, composed of m elements, where m is the number of classes and $\vec{z}(x)_i = C_i(x) + (\text{div}_w q_i^*)(x) + \rho_i^{2*} - \rho_i^{1*}$, take exactly the same minimal value. In fact, the two smallest components of the vector are only expected to be close to each other for points near the interphase between the two classes. In the continuum limit, this is only expected to happen over a set of measure zero.

If there are no class size constraints, the result of Theorem 1 can be made even stronger. In this case, a binary global minimizer can be obtained if there are at most two minimal components to $\vec{z}(x)$ for all $x \in V$. Therefore, unless there are at least three minimal components to $\vec{z}(x)$ at some x, the global minimizer of the relaxed problem is binary. Note that the case of three minimal components corresponds exactly to triple junctions.

Therefore, the convex relaxation not only approximates the original problem very well, but is much easier to solve than the latter. By introducing dual and auxiliary variables, the authors specify the exact relationship:

Theorem 2. *The following problems are equivalent:*

- *Max-flow problem:*

$$E(p_s,p,q,\rho^1,\rho^2) = \sup_{p_s,p,q,\rho^1,\rho^2} \sum_{x \in V} p_s(x) + \sum_{i=1}^{m} \left(\rho_i^1 S_i^\ell - \rho_i^2 S_i^u\right) \tag{22}$$

subject to, for all i ∈{1, ..., m},

$$|q_i(x,y)| \leq 1, \qquad \forall (x,y) \in E, \tag{23}$$

$$p_i(x) \leq C_i(x), \qquad \forall x \in V, \tag{24}$$

$$(div_w q_i - p_s + p_i)(x) = \rho_i^1 - \rho_i^2, \quad \forall x \in V, \tag{25}$$

$$0 \leq \rho_i^1, \rho_i^2 \leq \gamma \qquad \forall x \in V. \tag{26}$$

- *The convex relaxed problem:*

$$\min_{u \in B'} E^P(u) = \sum_{i=1}^{m} TV_w(u_i) + \sum_{i=1}^{m} \sum_{x \in V} C_i(x) u_i(x),$$

where $B' = \{u : V \to [0,1]^m | \sum_{i=1}^m u_i(x) = 1, \forall x \in V\}$ *with possible class size constraints.*

Here, div_w is the graph-based divergence operator, and γ is a parameter that controls the type of class size constraints. Overall, the authors solve (22) with constraints (23)–(26) efficiently using the augmented Lagrangian technique. Note that the class size constraints are handled implicitly.

Experiments on semisupervised data classification indicate very high accuracies, and those on unsupervised classification of 3D point clouds acquired by a LaDAR in outdoor scenes demonstrate that the scenes can be accurately segmented into object classes such as vegetation, the ground plane and regular structures. An example is shown in Fig. 2.

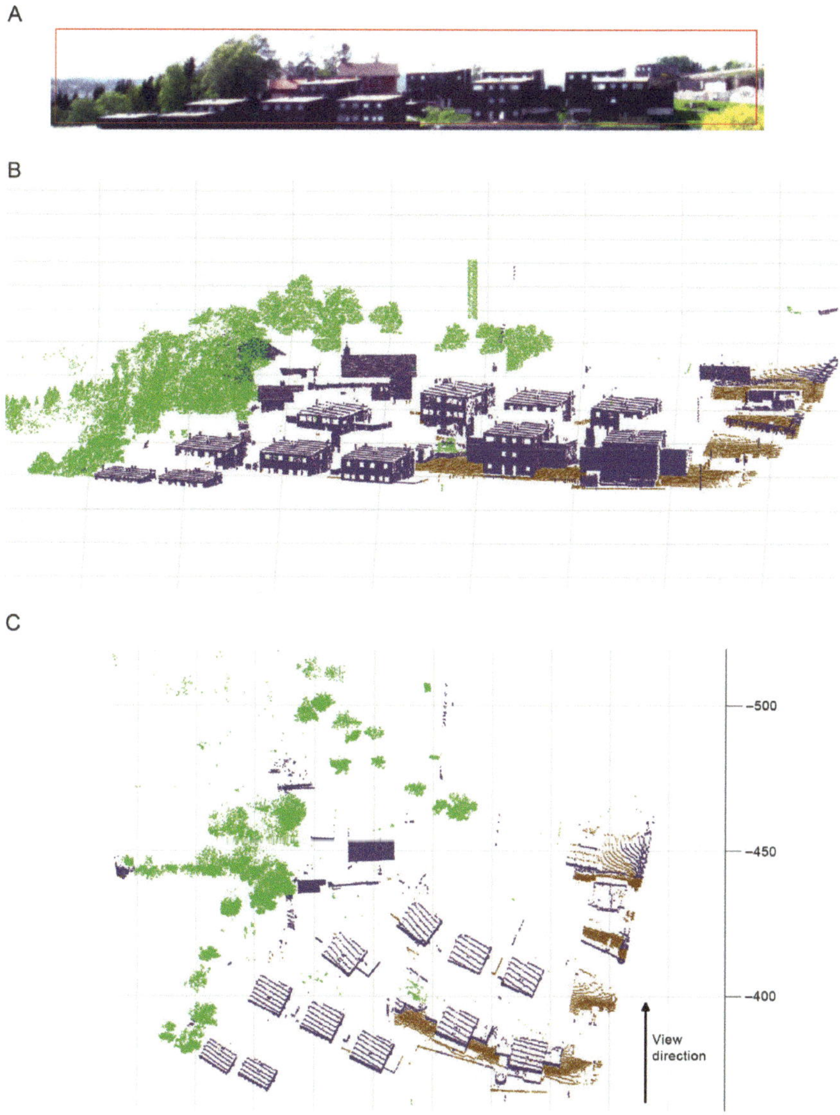

FIG. 2 (A) Scanning area. (B) and (C) Point cloud segmentation, consisting of 93,641 points, into ground plane (*brown*), vegetation (*green*) and human-made objects (*blue*). *Taken from Bae, E., Merkurjev, E., 2017. Convex variational methods on graphs for multiclass segmentation of high-dimensional data and point clouds. J. Math. Imag. Vis. 58 (3), 468–493. ISSN 1573-7683. https://doi.org10.1007/s10851-017-0713-9, which is distributed under the terms of the Creative Commons Attribution 4.0 International License (http://creativecommons.org/licenses/by/4.0/). No changes were made to the figures.*

5 Uncertainty quantification within the graphical framework

The applications of semisupervised learning, which combines labelled and unlabelled data, are wide-ranging. However, in many cases, there is either very little labelled data available, or none at all. For these applications, providing a measure of uncertainty, or uncertainty quantification, along with the classification, is crucial. Thus, obtaining the likelihood of the results for real-world applications, when ground truth is not available or when the existing ground truth is limited to a very small percentage of training data, is an important problem.

In Bertozzi et al. (2018), the authors introduce a Bayesian approach to uncertainty quantification for graph-based classification methods, using the posterior distribution on the classification labels, and formulate three different models. Moreover, they describe a unified framework within which one can view different methods formulated by various mathematical communities. They also develop algorithms to sample the posterior and compute MAP estimators, and describe the properties of the models.

In more detail, the three models investigated are probit classification (Williams and Rasmussen, 1996), the level set method for Bayesian inverse problems (Iglesias et al., 2016) and the Ginzburg–Landau optimization-based classifier (Bertozzi and Flenner, 2012; van Gennip and Bertozzi, 2012), the last two of which involve the Dirichlet energy and can thus be generalized to a Bayesian setting. All three models involve a common posterior and objective function:

$$J(w) = \frac{1}{2c}\langle w, Lw\rangle + \Phi(w), \quad P(u|y) \propto \exp(-J(u)). \tag{27}$$

Here, w represents a real-valued function on the nodes of the graph and is different for all three models, and L denotes the graph Laplacian. The classification problem can be formulated as an optimization of (27). To model the uncertainty, one can consider the probability distribution with density $P(w)$ proportional to $\exp(-J(w))$. One can extend the probit classification model to the graphical framework by defining a Gaussian measure through the graph Dirichlet energy and choosing a likelihood function involving thresholding a continuum latent variable plus noise. To compute the posterior, its PDF and the maximum a posteriori estimation, Bayes theorem can be used. In addition to studying the three models, Bertozzi et al. (2018) introduce efficient numerical methods, suited to large data-sets, for both MCMC-based sampling and gradient-based MAP estimation. The numerical methods are derived using the pCN method described in Cotter et al. (2013), Beskos et al. (2008) and Neal (1998). To make the algorithms scalable with the size of the graph, the authors introduce two variants of the pCN method, with spectral projection and with spectral approximation, which can sample the posterior distribution independently of the size of the graph. This useful

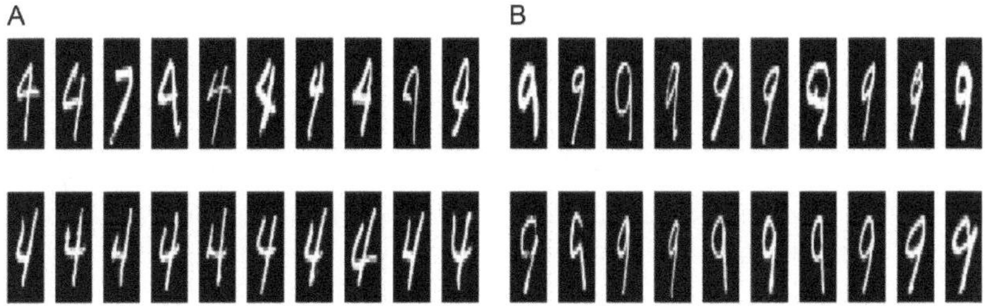

FIG. 3 "Hard to classify" vs "easy to classify" nodes in the MNIST (4, 9) dataset under the probit model. The digit "4" is labelled +1 and "9" is labelled −1. The top (bottom) row of the left column corresponds to images that have the lowest (highest) values of the node-wise posterior mean out of all the "4" digits. The right column is organized in the same way for "9" images except the top row now corresponds to the highest values of posterior mean. Higher posterior mean indicates higher confidence that the image is a 4 and not a "9", hence the top row could be interpreted as images that are "hard to classify" by the current model, and vice versa for the bottom row. (A) Fours in MNIST; (B) Nines in MNIST. *Taken from Bertozzi, A.L., Luo, X., Stuart, A.M., Zygalakis, K.C., 2018. Uncertainty quantification in the classification of high dimensional data. SIAM/ASA J. Uncertain. Quantif. 6 (2), 568–595. Copyright ©2018 Society for Industrial and Applied Mathematics and American Statistical Association. Reprinted with permission. All rights reserved.*

feature is achieved via a simple projection to the first ℓ eigenvectors and corresponding eigenvalues of L is performed.

One of the examples shown in Bertozzi et al. (2018) involves the MNIST database, consisting of 70,000 images of handwritten digits 0 through 9; see LeCun et al. (1998) for more information. The feature vectors are computed by taking the leading 50 principal components given by PCA. A k-nearest neighbour graph, with $k = 20$, is considered, along with a local rescaling as in (3). The labelled data are constructed by randomly selecting 4% of nodes from each class, and the experiment is conducted using the probit posterior. From the result in Fig. 3, one can see that nodes with posterior mean values closer to the binary ground truth labels $\pm$ 1 look visually more uniform than nodes with scores far from those labels, showing that the posterior mean can easily distinguish between outliers and inliers according to human perception.

6 Networks

Similarity graphs can arise in the form of networks, which capture pairwise interactions between discrete objects. For example, one can consider social networks which outline relationships between individuals, such as friendships on Facebook. Biological networks capture data in ecological, evolutionary and physiological contexts. Transportation networks describe structures which allow movement of vehicles. Other examples can be found in Newman (2010). This section outlines some of the recent methods for network analysis. The techniques of the algorithms can be applied to problems such as unsupervised clustering, even when the number of classes is not known.

Networks are often composed of cohesive groups, called communities, and their detection gives insight into the organization of the network. One popular approach to identifying communities is to maximize a quality function called modularity Q, which measures the fraction of total edge weight within communities vs what one might expect if edges were placed randomly according to some null model. The particular problem is to maximize over all possible partitions the function

$$Q = \frac{1}{2v} \sum_{ij} (w_{ij} - \gamma P_{ij}) \delta(g_i, g_j),$$

where g_i is the group assignment for node i. Here, P_{ij} is a probability null model (e.g., $P_{ij} = k_i k_j / 2v$), where $k_j = \sum_i w_{ij}$, $2v$ is the total volume of the graph ($\sum_i k_i$), and γ is a parameter.

The first connection of network modularity optimization to TV minimization on graphs was shown in Hu et al. (2013), derived using the null model introduced by Girvan and Newman (2004). Its main result is that maximizing modularity is equivalent to the graph cut problem, rewritten using the graph total variation functional, minus a second moment term:

$$\min_{u:G \rightarrow V^m} E(u) = |u|_{TV} - \gamma \| u - M_2(u) \|_{L_2}^2,$$

where m is the number of classes, V^m are the end nodes of the m-dimensional simplex, M_2 denotes a simple moment whose constraint can be introduced in a computationally tractable forcing term and u denotes the class assignment and takes values on the corners of the simplex. This fascinating connection allows one to leverage techniques from compressive sensing for modularity optimization.

In case the number of classes is unknown, one can minimize over a range of possible number of clusters. In particular, Hu et al. (2013) propose two implementations of their scheme. The first version is particularly suitable in the case when one expects a large number of communities. The second version is more appropriate when a small number of communities is more likely.

The algorithm in Hu et al. (2013) is very scalable, allowing the method to analyze large similarity graphs for machine learning problems. For example, for the MNIST dataset (LeCun et al., 1998), which contains tens of thousands of nodes, the procedure is 10–100 times faster computationally than the GenLouvain algorithm (Jutla et al., 2014) while generating results of comparable quality. The results also exceed those of basic fast greedy algorithms such as Newman (2006), Blondel et al. (2008) and other unsupervised clustering methods.

In the more recent work (Boyd et al., 2018), the authors show that modularity optimization is equivalent to minimizing a convex TV-based functional over a discrete domain. Therefore, it can also be regarded as a balanced cut problem over a discrete domain. Specifically, the authors have identified four different equivalent formulations of the modularity problem:

Theorem 3. (Equivalent forms of modularity (Boyd et al., 2018)).
For any subset S of the nodes of G, define $\mathrm{vol}S = \sum_{i \in S} k_i$. *Then the following optimization problems are all equivalent:*

$$\text{Standard formulation:} \quad \underset{m\in\mathbb{N},\{A_\ell\}_{\ell=1}^m \in \Pi(G)}{\operatorname{argmax}} \sum_{\ell=1}^{m} \sum_{ij\in A_\ell} w_{ij} - \gamma \frac{k_i k_j}{2v}, \tag{28}$$

$$\text{Balanced cut (I):} \quad \underset{m\in\mathbb{N},\{A_\ell\}_{\ell=1}^m \in \Pi(G)}{\operatorname{argmin}} \sum_{\ell=1}^{m} \left(\mathrm{Cut}(A_\ell, A_\ell^c) + \frac{\gamma}{2v} (\mathrm{vol}A_\ell)^2 \right), \tag{29}$$

$$\text{Balanced cut (II):} \quad \underset{m\in\mathbb{N},\{A_\ell\}_{\ell=1}^m \in \Pi(G)}{\operatorname{argmin}} \sum_{\ell=1}^{m} \left(\mathrm{Cut}(A_\ell, A_\ell^c) + \frac{\gamma}{2v} \left(\mathrm{vol}A_\ell - \frac{2v}{m} \right)^2 \right) + \gamma \frac{2v}{m}, \tag{30}$$

$$\text{Balanced TV (I):} \quad \underset{m\in\mathbb{N},u\in\Pi(G)}{\operatorname{argmin}} |u|_{TV} + \frac{\gamma}{2v} \left|\left| k^T u \right|\right|_2^2, \tag{31}$$

$$\text{Balanced TV (II):} \quad \underset{m\in\mathbb{N},u\in\Pi(G)}{\operatorname{argmin}} |u|_{TV} + \frac{\gamma}{2v} \left|\left| k^T u - \frac{2v}{m} \right|\right|_2^2 + \gamma \frac{2v}{m}. \tag{32}$$

Intuitively, the original formulation of modularity (28) regards communities as regions that are more connected than they would be if edges were totally random, while formulations (29) and (30) represent modularity as favouring sparsely interconnected regions with balanced volumes. The total variation formulations (31) and (32) seek a piecewise constant partition function u whose discontinuities have a small perimeter, under a balance-inducing quadratic penalty.

Note that in the formulations (29)–(32), the discrete constraint is the only source of nonconvexity. Therefore, modularity optimization can be equivalently considered as the minimization of a convex functional, but subject to a binary constraint. Such a reinterpretation of modularity serves as its connection to balance cut problems such as the Cheeger cut. Moreover, the ideas developed in Boyd et al. (2018) show that, while modularity optimization is inherently nonconvex, working with it as a constrained convex optimization problem produces results that are noticeably improved compared to prior methods that do not use such a formulation, such as Hu et al. (2013). In particular, the results from Boyd et al. (2018) on hyperspectral data, displayed in Fig. 4, show that the new approach is able to successfully segment the classes of a hyperspectral video sequence.

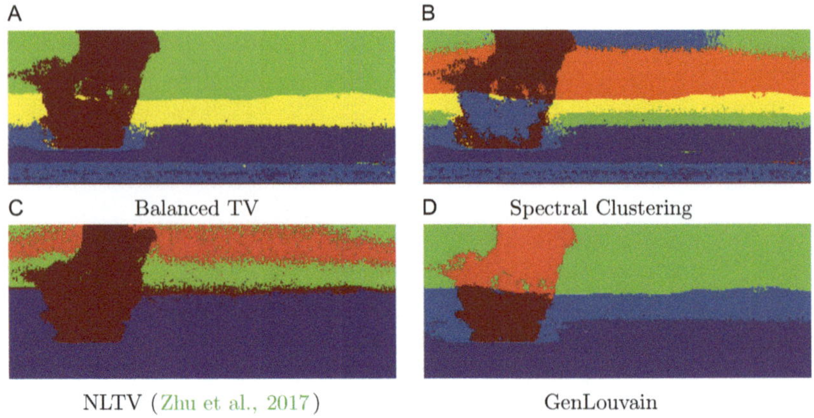

FIG. 4 Segmentations (Boyd et al., 2018) of the plume hyperspectral video using different methods. The Balanced TV is the only method that classifies the whole plume into a single class without any extraneous pixels. *Taken from Boyd, Z.M., Bae, E., Tai, X.C., Bertozzi, A.L., 2018. Simplified energy landscape for modularity using total variation. SIAM J. Appl. Math. 78 (5), 2439–2464. Copyright ©2018 Society for Industrial and Applied Mathematics. Reprinted with permission. All rights reserved.*

7 Conclusion

Mathematical models based on optimization techniques have had great success in applications such as machine learning, and especially in data classification problems (Bertozzi and Flenner, 2012; Bresson et al., 2012a, 2013a, 2014; Chapelle et al., 2006; Gilboa and Osher, 2007b; Hein and Bühler, 2010; Merkurjev et al., 2015; Szlam and Bresson, 2010b; Zhou et al., 2004). However, these models are often difficult to solve efficiently and accurately due to reasons including the high dimension and sheer amount of data, slow convergence, nonconvexity and diverse types of data. Moreover, the presence of constraints, such as bounds on the class size, make the problem even more challenging.

This chapter has described several recent, accurate and efficient graph-based optimization approaches for machine learning, network analysis and uncertainty quantification. The graphical framework used in the formulation of the algorithms provides several advantages such as the ability to capture data structure and connections between pieces of data, and the flexibility to incorporate diverse types of data, from text data to hyperspectral video sequences. The procedures themselves are formulated around optimizing objective functions, which allows one to inject powerful ideas from techniques such as duality. Moreover, additional techniques used in these methods, such as the Nyström extension procedure, play a big role in the efficiency of the algorithms. As part of a future direction, one can study the effect of sparsifying the graph on classification accuracy.

Overall, the procedures introduced in this chapter can be utilized for numerous practical applications such as email filtering, medical diagnosis, internet fraud detection, classifying DNA sequences, segmentation of speech signals, face and handwritten digits recognition, video tracking, document classification, financial predictions, etc. The methods can thus be used by many different disciplines and communities, and form a powerful tool to tackle a wide range of problems.

Acknowledgements

This work was supported by DARPA award number FA8750-18-2-0066 and NSF grants DMS-1737770 and DMS-1417674.

References

Bae, E., Merkurjev, E., 2017. Convex variational methods on graphs for multiclass segmentation of high-dimensional data and point clouds. J. Math. Imag. Vis. 58 (3), 468–493. ISSN 1573-7683. https://doi.org/10.1007/s10851-017-0713-9.

Belkin, M., Niyogi, P., Sindhwani, V., 2006. Manifold regularization: a geometric framework for learning from labeled and unlabeled examples. J. Mach. Learn. Res. 7, 2399–2434.

Belongie, S., Fowlkes, C., Chung, F., Malik, J., 2002. Spectral partitioning with indefinite kernels using the Nyström extension. In: European Conference on Computer Vision, pp. 531–542.

Bertozzi, A.L., 2018. Graphical models in machine learning, networks, and uncertainty quantification. In: Proceedings of ICM 2018.

Bertozzi, A.L., Flenner, A., 2012. Diffuse interface models on graphs for classification of high dimensional data. Multiscale Model. Simul. 10 (3), 1090–1118.

Bertozzi, A.L., Flenner, A., 2016. Diffuse interface models on graphs for classification of high dimensional data. SIAM Rev. 58 (2), 293–328.

Bertozzi, A.L., Luo, X., Stuart, A.M., Zygalakis, K.C., 2018. Uncertainty quantification in the classification of high dimensional data. SIAM/ASA J. Uncertain. Quantif. 6 (2), 568–595.

Bertsekas, D., 1979. A distributed algorithm for the assignment problem. Technical report, MIT.

Beskos, A., Roberts, G., Stuart, A.M., Voss, J., 2008. MCMC methods for diffusion bridges. Stoch. Dyn. 8 (03), 319–350.

Blondel, V.D., Guillaume, J.-L., Lambiotte, R., Lefebvre, E., 2008. Fast unfolding of communities in large networks. J. Stat. Mech. Theory Exp. 2008 (10), P10008.

Bosch, J., Klamt, S., Stoll, M., 2018. Generalizing diffuse interface methods on graphs: non-smooth potentials and hypergraphs. SIAM J. Appl. Math. 78 (3), 1350–1377.

Boyd, Z.M., Bae, E., Tai, X.C., Bertozzi, A.L., 2018. Simplified energy landscape for modularity using total variation. SIAM J. Appl. Math. 78 (5), 2439–2464.

Boyer, C., Chambolle, A., De Castro, Y., Duval, V., De Gournay, F., Weiss, P., 2019. On representer theorems and convex regularization. SIAM J. Optim. arxiv preprint. https://arxiv.org/abs/1806.09810.

Bresson, X., Laurent, T., 2017. Residual gated graph convnets. arXiv preprint arXiv:1711.07553.

Bresson, X., Laurent, T., Uminsky, D., von Brecht, J.H., 2012a. Convergence and energy landscape for Cheeger cut clustering. Adv. Neural Inf. Process. Syst. 1385–1393.

Bresson, X., Laurent, T., Uminsky, D., V. Brecht, J., 2012b. Convergence and energy landscape for Cheeger cut clustering. Adv. Neural Inf. Process. Syst. 1385–1393.

Bresson, X., Laurent, T., Uminsky, D., von Brecht, J., 2013a. Multiclass total variation clustering. Adv. Neural Inf. Process. Syst. 1421–1429.

Bresson, X., Laurent, T., Uminsky, D., von Brecht, J.H., 2013b. An adaptive total variation algorithm for computing the balanced cut of a graph. https://arxiv.org/abs/1302.2717.

Bresson, X., Tai, X.C., Chan, T.F., Szlam, A., 2014. Multi-class transductive learning based on ℓ_1 relaxations of Cheeger cut and Mumford-Shah-Potts model. J. Math. Imag. Vis. 49 (1), 191–201.

Buades, A., Coll, B., Morel, J.M., 2005. A non-local algorithm for image denoising. In: IEEE Computer Society Conference on Computer Vision and Pattern Recognition (CVPR'05), vol. 2, pp. 60–65.

Calatroni, L., van Gennip, Y., Schönlieb, C.-B., Rowland, H.M., Flenner, A., 2017. Graph clustering, variational image segmentation methods and hough transform scale detection for object measurement in images. J. Math. Imag. Vis. 57 (2), 269–291.

Chan, T., Vese, L.A., 2001. Active contours without edges. IEEE Trans. Image Process. 10, 266–277.

Chan, T.F., Wong, C.K., 1998. Total variation blind deconvolution. IEEE Trans. Imag. Proc. 7 (3), 370–375.

Chapelle, O., Schölkopf, B., Zien, A., 2006. Semi-Supervised Learning. vol. 2MIT Press.

Cotter, S.L., Roberts, G.O., Stuart, A.M., White, D., 2013. MCMC methods for functions: modifying old algorithms to make them faster. Stat. Sci. 28 (3), 424–446.

Defferrard, M., Bresson, X., Vandergheynst, P., 2016. Convolutional neural networks on graphs with fast localized spectral filtering. In: Adv. Neural Inf. Proces. Syst. 3844–3852. https://papers.nips.cc/paper/6081-convolutional-neural-networks-on-graphs-with-fast-localized-spectral-filtering.pdf.

Digne, J., 2012. Similarity based filtering of point clouds. In: IEEE Computer Society Conference on Computer Vision and Pattern Recognition Workshop, pp. 73–79.

Dobrosotskaya, J.A., Bertozzi, A.L., 2008. A wavelet-laplace variational technique for image deconvolution and inpainting. IEEE Trans. Imag. Proc. 17 (5), 657–663.

Dobrosotskaya, J.A., Bertozzi, A.L., 2010. Wavelet analogue of the Ginzburg-Landau energy and its γ-convergence. Int. Free Bound. 12 (4), 497–525.

Elmoataz, A., Touttain, M., Tenbrinck, D., 2015. On the p-Laplacian and ∞-Laplacian on graphs with application in image and data processing. SIAM J. Imaging Sci. 8, 2412–2451.

Esedog-lu, S., Tsai, Y.-H.R., 2006. Threshold dynamics for the piecewise constant Mumford-Shah functional. J. Comp. Phys. 211 (1), 367–384.

Fowlkes, C., Belongie, S., Malik, J., 2001. Efficient spatiotemporal grouping using the Nyström method. In: Proceedings of the IEEE Computer Society Conference on Computer Vision and Pattern Recognition, vol. 1, p. 231.

Fowlkes, C., Belongie, S., Chung, F., Malik, J., 2004. Spectral grouping using the Nyström method. IEEE Trans. Pattern Anal. Mach. Intell. 26 (2), 214–225.

Garcia-Cardona, C., Merkurjev, E., Bertozzi, A.L., Flenner, A., Percus, A.G., 2014a. Multiclass data segmentation using diffuse interface methods on graphs. IEEE Trans. Pattern Anal. Mach. Intell. 36 (8), 1600–1613.

Garcia-Cardona, C., Merkurjev, E., Bertozzi, A.L., Flenner, A., Percus, A.G., 2014b. Research announcement: diffuse interface methods for multiclass segmentation of high-dimensional data. Appl. Math. Lett. 33, 29–34.

Gerhart, T., Sunu, J., Lieu, L., Merkurjev, E., Chang, J.M., Gilles, J., Bertozzi, A.L., 2013. Detection and tracking of gas plumes in LWIR hyperspectral video sequence data. In: SPIE Conference on Defense Security and Sensing, vol. 8743, 87430J–87430J.

Gilboa, G., Osher, S., 2007a. Nonlocal linear image regularization and supervised segmentation. Multiscale Model. Simul. 6 (2), 595–630.

Gilboa, G., Osher, S., 2007b. Nonlocal linear image regularization and supervised segmentation. Multiscale Model. Simul. 6 (2), 595–630.

Gilboa, G., Osher, S., 2008. Nonlocal operators with applications to image processing. Multiscale Model. Simul. 7 (3), 1005–1028.

Girvan, M., Newman, M.E.J., 2004. Finding and evaluating community structure in networks. Phys. Rev. E. 69, 026113.

Goldstein, T., Osher, S., 2009. The split Bregman method for L1-regularized problems. SIAM J. Imaging Sci. 2 (2), 323–343.

Hein, M., Bühler, T., 2010. An inverse power method for nonlinear eigenproblems with applications in 1-spectral clustering and sparse PCA. Adv. Neural Inf. Process. Syst. 23, 847–855.

Hu, H., Laurent, T., Porter, M.A., Bertozzi, A.L., 2013. A method based on total variation for network modularity optimization using the MBO scheme. SIAM J. Appl. Math. 73 (6), 2224–2246.

Hu, H., Sunu, J., Bertozzi, A.L., 2015. Multi-class graph Mumford-Shah model for plume detection using the MBO scheme. In: International Workshop on Energy Minimization Methods in Computer Vision and Pattern Recognition, pp. 209–222.

Iglesias, M.A., Lu, Y., Stuart, A.M., 2016. A Bayesian level set method for geometric inverse problems. Interfaces Free Bound. 18, 181–217.

Iyer, G., Chanussot, J., Bertozzi, A.L., 2017. A graph-based approach for feature extraction and segmentation in multimodal images. In: Proceedings of the International Conference on Image Processing, Beijing, pp. 3320–3324.

Jacobs, M., Merkurjev, E., Esedog-lu, S., 2018. Auction dynamics: a volume constrained MBO scheme. J. Comp. Phys. 354, 288–310.

Jung, Y.M., Kang, S.H., Shen, J., 2007. Multiphase image segmentation via Modica-Mortola phase transition. SIAM J. Appl. Math. 67 (5), 1213–1232.

Jutla, I.S., Jeub, L.G.S., Mucha, P.J., 2014. A generalized Louvain method for community detection implemented in MATLAB (2011–2012). https://github.com/GenLouvain/GenLouvain. Accessed 1 May 2019.

Kang, S.H., Shafei, B., Steidl, G., 2014. Supervised and transductive multi-class segmentation using p-Laplacians and RKHS methods. J. Vis. Commun. Image Represent. 25 (5), 1136–1148.

Ketchen, D.J., Shook, C.L., 1996. The application of cluster analysis in strategic management research: an analysis and critique. Strateg. Manag. J. 17 (6), 441–458.

Kohn, R.V., Sternberg, P., 1989. Local minimisers and singular perturbations. Proc. Roy. Soc. Edinburgh Sect. A 111, 69–84.

LeCun, Y., Cortes, C., Burges, C.J., 1998. MNIST Handwritten Digit Database. AT&T Labs. http://yann.lecun.com/exdb/mnist.

Lellmann, J., Kappes, J. H., Yuan, J., Becker, F., Schnörr, C., 2008. Convex multi-class image labeling by simplex-constrained total variation. Technical Report IWR, University of Heidelberg.

Levie, R., Monti, F., Bresson, X., Bronstein, M.M., 2017. Cayleynets: graph convolutional neural networks with complex rational spectral filters. IEEE Trans. Signal Process. 67 (1), 97–109.

Luo, X., Bertozzi, A.L., 2017. Convergence analysis of the graph Allen-Cahn scheme. J. Stat. Phys. 167 (3), 934–958.

Macdonald, C.B., Merriman, B., Ruuth, S.J., 2013. Simple computation of reaction-diffusion processes on point clouds. Proc. Natl. Acad. Sci. 110, 3009–3012.

Meng, Z., Koniges, A., He, Y.H., Williams, S., Kurth, T., Cook, B., Deslippe, J., Bertozzi, A.L., 2016. OpenMP parallelization and optimization of graph-based machine learning algorithm. In: Proceedings of the 12th International Workshop on OpenMP (IWOMP).

Meng, Z., Merkurjev, E., Koniges, A., Bertozzi, A.L., 2017. Hyperspectral image classification using graph clustering methods. Ipol J. 7, 218–245. https://doi.org/10.5201/ipol.2017.204. (Published with code).

Merkurjev, E., Kostic, T., Bertozzi, A.L., 2013. An MBO scheme on graphs for classification and image processing. SIAM J. Imaging Sci. 6 (4), 1903–1930.

Merkurjev, E., Sunu, J., Bertozzi, A.L., 2014. Graph MBO method for multiclass segmentation of hyperspectral stand-off detection video. In: Proceedings of the International Conference on Image Processing, 689–693.

Merkurjev, E., Bae, E., Bertozzi, A.L., Tai, X.C., 2015. Global binary optimization on graphs for classification of high dimensional data. J. Math. Imag. Vis. 52 (3), 414–435.

Merkurjev, E., Bertozzi, A.L., Chung, F., 2018. A semi-supervised heat kernel pagerank MBO algorithm for data classification. Comm. Math. Sci. 16 (5), 1241–1265.

Merkurjev, E., Bertozzi, A., Yan, X., Lerman, K., 2017. Modified Cheeger and ratio cut methods using the Ginzburg-Landau functional for classification of high-dimensional data. Inverse Prob. 33 (7), 074003.

Merriman, B., Bence, J., Osher, S., 1992. Diffusion generated motion by mean curvature. In: Proc. Comput. Crystal Growers Workshop, 73–83.

Monti, F., Bronstein, M., Bresson, X., 2017. Geometric matrix completion with recurrent multi-graph neural networks. 3697–3707. https://papers.nips.cc/paper/6960-geometric-matrix-completion-with-recurrent-multi-graph-neural-networks.pdf.

Neal, R., 1998. Regression and classification using Gaussian process priors. Bayesian Stat. 6, 475.

Newman, M.E.J., 2006. Modularity and community structure in networks. Proc. Nat. Acad. Sci. 103 (23), 8577–8582.

Newman, M.E.J., 2010. Networks: An Introduction. Oxford University Press, Oxford, UK. ISBN: 9780199206650.

Pelleg, D., Moore, A.W., et al., 2000. X-means: extending K-means with efficient estimation of the number of clusters. In: ICML, vol. 1.

Peyré, G., 2008. Image processing with nonlocal spectral bases. Multiscale Model. Simul. 7 (2), 703–730.

Pock, T., Cremers, D., Bischof, H., Chambolle, A., 2009. An algorithm for minimizing the Mumford-Shah functional. In: 2009 IEEE 12th International Conference on Computer Vision, pp. 1133–1140.

Rudin, L.I., Osher, S., Fatemi, E., 1992. Nonlinear total variation based noise removal algorithms. Phys. D 60 (1), 259–268.

Schönlieb, C.-B., Bertozzi, A., 2011. Unconditionally stable schemes for higher order inpainting. Comm. Math. Sci. 9 (2), 413–457.

Seo, Y., Defferrard, M., Vandergheynst, P., Bresson, X., 2018. Structured sequence modeling with graph convolutional recurrent networks. In: International Conference on Neural Information Processing, pp. 362–373.

Szlam, A., Bresson, X., 2010a. Total variation and Cheeger cuts. In: Proceedings of the 27th International Conference on International Conference on Machine Learning, 1039–1046.

Szlam, A., Bresson, X., 2010b. A total variation-based graph clustering algorithm for Cheeger ratio cuts. In: Proceedings of the 27th International Conference on Machine Learning, 1039–1046.

Thorpe, M., Theil, F., 2017. Asymptotic analysis of the Ginzburg-Landau functional on point clouds. In: Proceedings of the Royal Society of Edinburgh Section A: Mathematics.

Tian, L., Macdonald, C.B., Ruuth, S.J., 2009. Segmentation on surfaces with the closest point method. In: Proceedings of the 16th IEEE International Conference on Image Processing, 3009–3012.

Unser, M., Fageot, J., Gupta, H., 2016. Representer theorems for sparsity-promoting ℓ_1 regularization. IEEE Trans. Inf. Theory 62 (9), 5167–5180.

Unser, M., Fageot, J., Ward, J.P., 2017. Splines are universal solutions of linear inverse problems with generalized TV regularization. SIAM Rev. 59 (4), 769–793.

van Gennip, Y., Bertozzi, A.L., 2012. γ-convergence of graph Ginzburg-Landau functionals. Adv. Diff. Equat. 17 (11–12), 1115–1180.

van Gennip, Y., Guillen, N., Osting, B., Bertozzi, A.L., 2014. Mean curvature, threshold dynamics, and phase field theory on finite graphs. Milan J. Math 82 (1), 3–65.

Vese, L.A., Chan, T.F., 2002. A multiphase level set framework for image segmentation using the Mumford-Shah model. Int. J. Comput. Vis. 50, 271–293.

Von Luxburg, U., 2007. A tutorial on spectral clustering. Stat. Comput. 17 (4), 395–416.

Wang, J., Jebara, T., Chang, S.F., 2008. Graph transduction via alternating minimization. In: Proceedings of the 25th International Conference on Machine Learning, 1144–1151.

Williams, C.K.I., Rasmussen, C.E., 1996. Gaussian processes for regression. MIT Press, 514–520.

Yaroslavsky, L.P., 2012. Digital Picture Processing: An Introduction. vol. 9. Springer Science & Business Media.

Zelnik-Manor, L., Perona, P., 2004. Self-tuning spectral clustering. In: Adv. Neural Inf. Process. Syst. 1601–1608.

Zhou, D., Schölkopf, B., 2004. A regularization framework for learning from graph data. In: Workshop on Statistical Relational Learning.

Zhou, D., Bousquet, O., Lal, T.N., Weston, J., Schölkopf, B., 2004. Learning with local and global consistency. Adv. Neural Inf. Proces. Syst. 16, 321–328.

Zhu, X., 2006. Semi-supervised Learning Literature Survey. Computer Science, University of Wisconsin-Madison.

Zhu, W., Chayes, V., Tiard, A., Sanchez, S., Dahlberg, D., Bertozzi, A.L., Osher, S., Zosso, D., Kuang, D., 2017. Unsupervised classification in hyperspectral imagery with nonlocal total variation and primal-dual hybrid gradient algorithm. IEEE Trans. Geo. Remote Sensing 55 (5), 2786–2798. ISSN 0196-2892.

Chapter 13

Survey of fast algorithms for Euler's elastica-based image segmentation

Sung Ha Kang[a,*], Xue-Cheng Tai[b,c] and Wei Zhu[d]
[a]*School of Mathematics, Georgia Institute of Technology, Atlanta, GA, United States*
[b]*Hong Kong Baptist University, Department of Mathematics, Kowloon Tong, Hong Kong*
[c]*University of Bergen, Department of Mathematics, Bergen, Norway*
[d]*Department of Mathematics, University of Alabama, Tuscaloosa, AL, United States*
[*]*Corresponding author: e-mail: kang@math.gatech.edu*

Chapter Outline

Abstract
Segmentation has been widely studied and is one of the fundamental and classical tasks of image processing. Starting from Mumford–Shah functional, we review multiphase segmentation using piecewise constant level set approach and relate to phase transition models for multiphase segmentation. In particular, we focus on Euler's elastica-based variational functionals, in the context of fast algorithms for image segmentation problem. Its relation to illusory contour is covered following the phase transition approach. Minimizing the functionals with curvature term helps to capture the geometry of the recovered shape, but introduces complexity due to their nonconvexity and high order derivatives. We review some of fast algorithms developed to efficiently compute the minimum of curvature-based segmentation functionals, such as Euler's elastica segmentation.

Keywords: Euler's elastica, Curvature, Segmentation augumented Lagrangian method, Fast algorithm, Interface model

AMS Classification Codes: 65R10, 94A08, 90C26

Handbook of Numerical Analysis, Vol. 20. https://doi.org/10.1016/bs.hna.2019.05.005
© 2019 Elsevier B.V. All rights reserved.

1 Introduction

The literature on segmentation has been the most wealthy and inspiring. From Geman and Geman's mixture random-field models (Geman and Geman, 1984), the data-driven Monte-Carlo Markov chain model (DDMCMC) of Tu and Zhu (2002), and the graph-cutting and spectral method of Shi and Malik (2000) to Mumford and Shah's piecewise smooth variational image models (Mumford and Shah, 1989), segmentation has been extensively studied by several major stochastic and deterministic machineries of modelling, analysis, and computation.

One of the most well-known variational segmentation models is by Mumford and Shah (1989):

$$E_{ms}[u,\Gamma|u_0]=\mathcal{H}^1(\Gamma)+\alpha\int_{\Omega\backslash\Gamma}|\nabla u|^2dx+\lambda\int_{\Omega}(u-u_0)^2dx.$$

Here Ω is a bounded Lipschitz domain, $u_0:\Omega\to\mathbb{R}_+\cup\{0\}$ is the given image, $\Gamma\subset\Omega$ denotes the edge set of the ideal image u, and $\mathcal{H}^1$ denotes the one-dimensional Hausdorff measure, extending the classical notion of length. This functional is well defined on the pair (u, Γ) with $u\in W^{1,2}(\Omega\setminus\Gamma)$, $\mathcal{H}^1(\Gamma)<\infty$, and Γ being relatively closed in Ω, provided that the given image u_0 belongs to $L^2(\Omega)$. The Chan–Vese model (2001) successfully implemented the piecewise constant segmentation using the celebrated level-set method (Osher and Sethian, 1988). For two phase segmentation, the corresponding energy becomes

$$E_{cv}[v,\Gamma|u_0]=\mathcal{H}^1(\Gamma)+\lambda_1\int_{\Omega_1}|u_0-c_1|^2dx+\lambda_2\int_{\Omega_2}|u_0-c_2|^2dx, \tag{1}$$

where Ω_i, $i=1,2$ denote the partitions of the whole domain Ω, i.e., Ω is divided into two sets Ω_1 and Ω_2, and c_1 and c_2 are the averages of u_0 on Ω_1 and Ω_2, respectively. The function $u:=c_1\chi_{\Omega_1}+c_2\chi_{\Omega_2}$ represents the segmented image.

In the multiphase context, this can be generalized to the following:

$$E[\Omega_i,c_i]=\frac{1}{2}\sum_{i,j=1}^{n}\mathcal{H}^1(\partial\Omega_i\cap\partial\Omega_j)+\lambda\sum_{i=1}^{n}\int_{\Omega_i}(c_i-u_0)^2dx. \tag{2}$$

A more general functional for the image segmentation problems is called the Potts model:

$$E_{Potts}(\{\Omega_i\}_{i=1}^n)=\sum_{i=1}^{n}\int_{\Omega_i}f_i(x)dx+\alpha R(\{\Omega_i\}_{i=1}^n), \tag{3}$$

where $f_i(x)$ is the region force term for the ith subregion and $R(\cdot)$ is the regularizer term. Since each subregion Ω_i can also be represented by its corresponding indicator function $\chi_i(x)$, (3) can also be written as

$$E_{Potts}(\chi)=\sum_{i=1}^{n}\int_{\Omega}f_i(x)\chi_i(x)dx+\sum_{i=1}^{n}\alpha\int_{\Omega}|\nabla\chi_i(x)|dx. \tag{4}$$

In this rich field of segmentation research: two phase vs multiphase models, continuous vs discrete settings, regularization of length minimization vs high order terms, we focus on curvature-based image segmentation approaches and its fast algorithm. We start with a short review of interface models to show the connection to the segmentation with depth problem and illusory contour, where the curvature constraint plays a visible role in the result. Then we present some details of fast algorithms for curvature-based segmentation, e.g., Euler's elastica segmentation.

2 Piecewise constant representation and interface problems to illusory contour with curvature term

In the seminal work of Chan–Esedougla–Nikolova (Chan et al., 2006), it was observed that binary segmentation models can be relaxed to get convex global minimization models and then get a binary label by threshold. Chambolle et al. (2012) extended this idea to more general problems by function-lifting, i.e., transfer the nonconvex minimization problem to a higher dimensional convex minimization problem. There are interesting extensions using graph-cut algorithms and global minimization in Bae and Tai (2009a,b), Bae et al. (2011), and Gu et al. (2012). One interesting observation in Bae et al. (2010) and Yuan et al. (2010a,b, 2014) is that the convex relaxation models of Chan et al. (2006) and Chambolle et al. (2012) are in fact the dual problem of a continuous min-cut problem which is also equivalent to a continuous max-flow problem. The dual problem and the max-flow problem are convex and thus have global minimizers. Using this interpretation, we can also cast the lifting technique of Chambolle et al. (2012), which is to solve problem (4) in the framework of continuous max-flow and min-cut. The detail explanations about this connection can be found in Bae et al. (2010, 2014).

There are also approaches using fuzzy region computation (Häuser and Steidl, 2013; He et al., 2012; Li et al., 2010a, b; Shafei and Steidl, 2012) which allows partial membership, other related work can be found at Chambolle et al. (2009), Gao and Bui (2005), Lellmann and Schnörr (2011), Li and Tai (2007), and Lie et al. (2006a). Brown et al. (2012) considered convex formulation of the Chan–Vese segmentation using lifting techniques. We also want to mention that a multilayer method was introduced by Chung and Vese (2003), where the authors used more than one levels of a level set to represent the discontinuity of the image, inspired by modelling island dynamics for epitaxial growth.

One of the typical ways to represent different regions is the piecewise constant level set method (Lie et al., 2005, 2006a,b):

$$\phi : \Omega \mapsto \{1, \ldots, n\} \text{ such that } \phi(x) = i \text{ if } x \in \Omega_i, \quad i = 1, \ldots, n.$$

A reason for the popularity of this approach of piecewise constant representation, is that nice convex relaxations often can be derived by relaxing the integrality constraints of the functions (Bae and Tai, 2015; Bae et al., 2011;

Chan et al., 2006; Lellmann et al., 2009; Pock et al., 2008, 2009; Zach et al., 2008). In Bae and Tai (2009b), algorithm is proposed for the piecewise constant level set method using graph cuts models. Using the discretized energy functional with new graph design, it becomes possible to design a fast algorithm. In Fig. 1, the graph for a 1D signal with 4-phase segmentation is shown. The edges in $\mathcal{E}_D$ are illustrated as the vertical arrows while the edges in $\mathcal{E}_R$ are illustrated as the horizontal arrows in Fig. 1. Then minimization is solved using the max-flow/min-cut algorithm. This algorithm, compared with traditional gradient decent methods, is normally 500 times faster for relatively large size 2D images Bae and Tai (2009b). Related work on generic convexification and graph cut method can be found at Liu et al. (2013).

This approach is closely related to interface problem and phase transition models. Ambrosio and Tortorelli (1990, 1992) first approximated the Mumford–Shah functional by means of quadratic, elliptic functionals using the theory of Γ-convergence. This method has been extensively studied and extended for segmentation, inpainting, and several other applications in image analysis and processing. For example, numerical implementations of such relaxed functionals were conducted in March (1992) and March and Dozio (1997), a relaxed version of (i.e., "soft") Mumford–Shah segmentation using stochastic-variational model was proposed in Shen (2006), Cahn–Hilliard model (1958) was applied to binary image inpainting in Bertozzi et al. (2006), a generalization of binary image inpainting to grey-value image inpainting was considered in Burger et al. (2009), Mumford–Shah–Euler inpainting model was proposed in Esedoglu and Shen (2002), and the

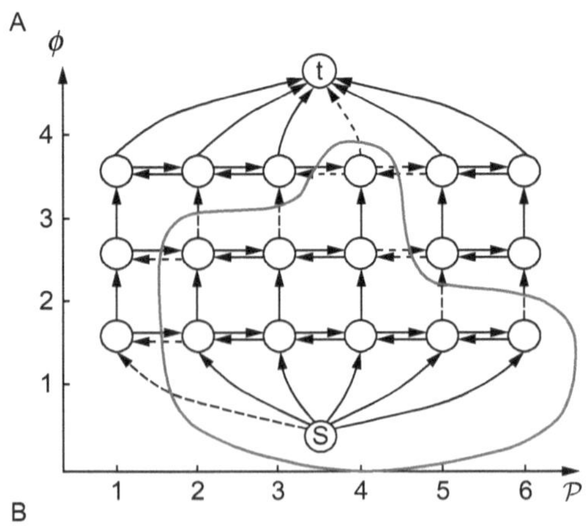

1	3	3	4	2	2

FIG. 1 Explanation of the multiphase graph cuts. (A) The graph corresponds to 1D signal of six grid points. A 3 level graph for this 4 phase segmentation problem is constructed. (B) The values of the level set function ϕ at each grid point corresponding to the cut in (A).

complex Ginzburg–Landau Equation was studied in Borzi et al. (2005) and Grossauer and Scherzer (2003). Moreover, in Aubert et al. (2005), the authors explored the Ginzburg–Landau model for detecting codimension, and in Dobrosotskaya and Bertozzi (2011), the authors developed a wavelet analogue of the Ginzburg–Landau energy and studied its Γ-convergence.

Adopting the model of Modica and Mortola (1977) with a sinusoidal potential, a model for multiphase segmentation was proposed in Jung et al. (2007). This is a relaxed version of the multiphase segmentation model (2) for a given number K of phases:

$$E_\epsilon[z|u_o] = \int_\Omega \left[\epsilon|\nabla z|^2 + \frac{1}{\epsilon}\sin^2\pi z\right] dx + \lambda \sum_{k=0}^{K-1} \int_\Omega |C_k - u_o|^2 \text{sinc}^2(z-k)\, dx. \quad (5)$$

Here the values C_k's have been conditionally optimized by the following equation

$$C_k = C_k[z] = \begin{cases} \dfrac{\int_\Omega u_o \text{sinc}^2(z-k)dx}{\int_\Omega \text{sinc}^2(z-k)\,dx} & \text{if } \int_\Omega \text{sinc}^2(z-k)dx > 0, \\ 0 & \text{otherwise.} \end{cases}$$

Fig. 2 shows a complex synthetic image that contains several generic visual structures, including an internal hole, occlusion and stacked objects, T-junctions, singular junctions where multiple objects (or phases) meet, as well as thin passages that reveal the bottleneck effect. With $K = 5$ phases, an initial condition is weakly supervised. By design, the model (5) naturally identifies multiple phases with only one scalar function z. However, due to the shape of

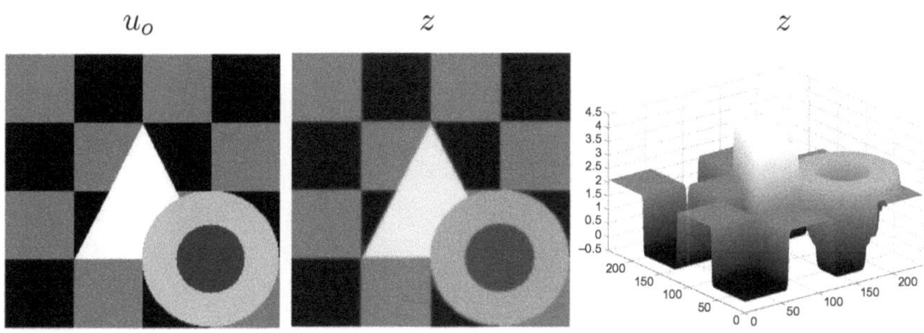

FIG. 2 A complex synthetic image u_o with multiple objects and several generic visual structures. From the given u_o, z is reconstructed using Eq. (5). *Copyright @2007 Society for Industrial and Applied Mathematics. Reprinted from Jung, Y. M., Kang, S. H., Shen, J., 2007. Multiphase image segmentation via Modica-Mortola phase transition. SIAM J. Appl. Math. 67 (5), 1213–1232. with permission. All rights reserved.*

sinusoidal function, the length is weighted by the phase distance between the two adjacent regions (for $K > 3$). More general model is considered in Kang and March (2014):

$$E_{\varphi}^{\epsilon}(\mathbf{u}) = \int_{\Omega}\left[\epsilon^{\alpha}\varphi(|\nabla\mathbf{u}|) + \frac{W(\mathbf{u})}{\epsilon}\right]dx + \lambda\int_{\Omega}(\langle\mathbf{c},\mathbf{u}\rangle - f)^2 dx, \tag{6}$$

where $\mathbf{u}:\Omega \to \mathbb{R}^N$ is a vector valued function, φ is a regularizing function, $\alpha \geq 0$ is a parameter, and $\mathbf{c} \in \mathbb{R}^N$ is a constant vector. Here the well potential $W(\mathbf{u}) = \prod_{i=1}^{N}|\mathbf{u} - \mathbf{e}_i|^2$ is defined using the standard basis of $\mathbb{R}^N$, $\{\mathbf{e}_1,\ldots,\mathbf{e}_N\}$, and the length between any two wells are all equal.

Four different variations of this functional (6) are explored: (i) $\varphi(t) = t^2$ ($p = 2$), a vectorial version of Modica–Mortola model (1987, 1977) where the regularization is quadratic, that is, the Baldo model; (ii) Superlinear model $\varphi(t) \approx t^p$ at infinity, with $1 < p < 2$, which gives the p-growth of the regularizing function φ, such as $\varphi(t) = \frac{t^2}{1+\mu\, t^{2-p}} \approx t^p$ for large t; (iii) Using the Perona–Malik (1990) regularization term, $\varphi(t) \approx t^p$ at infinity, with $0 \leq p < 1$, and (iv) the true length Total Variation model $\varphi(t) = |t|$ ($p = 1$), based on Rudin et al. (1992);

$$E_{tv}^{\epsilon}(\mathbf{u}) = \int_{\Omega}\left[|\nabla\mathbf{u}| + \frac{W(\mathbf{u})}{\epsilon}\right]dx + \lambda\int_{\Omega}(\langle\mathbf{c},\mathbf{u}\rangle - f)^2 dx, \tag{7}$$

if $\mathbf{u} \in \{\mathbf{u} \in W^{1,1}(\Omega;\mathbb{R}^N) : \mathbf{u}(x) \in \mathcal{C} \text{ a.e. in } \Omega\}$, otherwise $E_{tv}^{\epsilon}(\mathbf{u}) = +\infty$ in $L^1(\Omega;\mathbb{R}^N)$. Using standard numerical discretization method, the true length Total Variation model (7) is shown to be the simplest model, yet most stable and practical in practice. Fig. 3 shows the effect of true length (nonweighted),

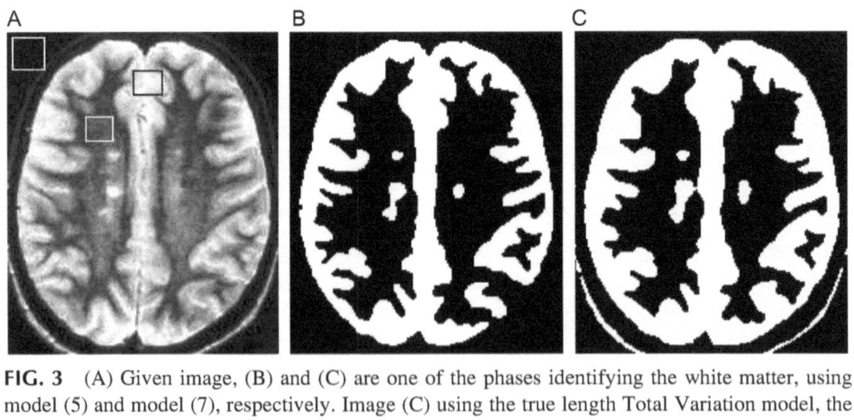

FIG. 3 (A) Given image, (B) and (C) are one of the phases identifying the white matter, using model (5) and model (7), respectively. Image (C) using the true length Total Variation model, the thin skull is also identified as the white region, since it is not biased by the height of the jumps in the transition model. *Reprinted from Kang, S. H., March, R., 2014. Multiphase image segmentation via equally distanced multiple well potential. J. Vis. Commun. Image Represent. 25 (6), 1446–1459 with permission.*

comparing results of Jung et al. (2007) and Kang and March (2014). From the given image (A), image (B) and (C) both show the third phase of each method. Notice the edge of the brain is correctly segmented as a part of the white matter using the true length Total Variation model in (C), since it is not biased by the height of the jumps in the transition model.

Interface model can also incorporate curvature term. One such example is finding illusory contour in Kang et al. (2014b), where curvature constraint is added to the reconstruction model. The objective of illusory contour is to identify an object, which only exists as an illusion, e.g., Fig. 4A. In the image of Kanizsa's Triangle, the actual triangle does not exists, but well-arranged three pac-man give a clear illusion of it. In Kang et al. (2014b), the authors used convex corners as a clue to find the illusory foreground and background shapes. With an introduction of corner bases, which expands 0-dimensional corner points to 2-dimensional micro structures, the corner information is extended to entire domain to find the illusory information. This extension is achieved by utilizing a curvature-based inpainting model via phase field approach;

$$E_\epsilon(u)=\int_{corner}\left\{a\left(\frac{\epsilon}{2}|\nabla u|^2+\frac{W(u)}{2\epsilon}\right)+\frac{b}{\epsilon}\left(\epsilon\Delta u-\frac{W'(u)}{2\epsilon}\right)^2\right\}dx+\int_{corner}\chi_S(u-\phi_S)^2dx, \tag{8}$$

where $W(u)$ denotes the double-well potential $W(u) = (u^2-1)^2$, and S is the union of all the supports of corner bases outside the given shape Q. As in the theory of Ginzburg and Landau (1950) or Cahn–Hilliard (1958), the chemical potential: $v=\epsilon\Delta u-\frac{W'(u)}{2\epsilon}$ is introduced, which could also be considered as the approximate curvature field. Then, the fourth-order equation can be rewritten as the a coupled system of second-order equations for (u, v), where fast Fourier transformation (FFT) is used inspired by Tai et al. (2011).

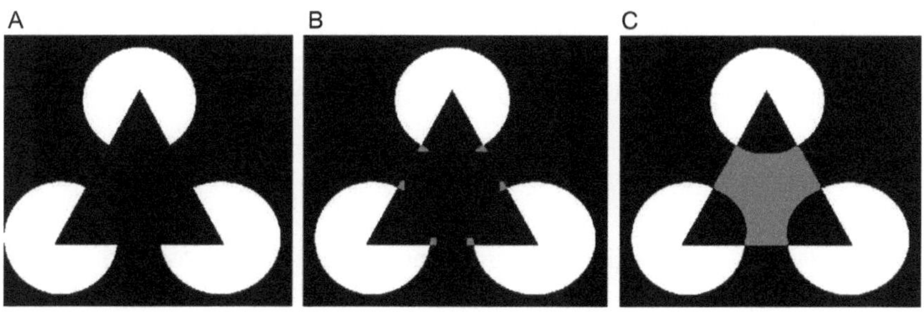

FIG. 4 (A) Kanizsa's Triangle. (B) Corner bases at the convex corners. (C) Elastica inpainting with interface model (8) constructs the *grey region*. In (C), the union of internal *black regions* and the *grey region* gives illusory contour, and the union of internal *black regions* and *white regions* gives the occluded shapes. *Copyright @2014 Society for Industrial and Applied Mathematics. Reprinted from Kang, S. H., Zhu, W., Shen, J., 2014b. Illusory shapes via corner fusion. SIAM J. Imaging Sci. 7 (4), 1907–1936 with permission. All rights reserved.*

Such case of large domain inpainting problem, the interface model and working with two second-order equations gave numerically stable and efficient results. Fig. 4 illustrates this procedure for the classical Kanizsa's Triangle. From the given configuration Q (white region), the corner bases are shown in (B), then the new merged region (grey) is shown in (C) using the model (8). With such partitions, the union of internal black regions and the grey region gives illusory contour, while the union of internal black regions and white regions leads to the occluded shapes.

Some related illusory contour work includes (Chan and Zhu, 2007; Jung and Shen, 2008; Sarti et al., 2002; Zhu and Chan, 2005, 2007). In Zhu and Chan (2007), the authors considered a model with Euler's elastica, using a similar set up of the Euler–Lagrange equation in Chan et al. (2003). The model preserves corners by incorporating curvature information of the given object boundaries. The choice of the second-order Euler's elastica used in the model is compared to the first-order Euler's elastica developed in Nitzberg, Mumford and Shiota's work on the problem of segmentation with depth (Nitzberg et al., 1993). Segmentation with depth is another interesting application where the curvature constraint plays an important role in the recovery of occluded objects. In Esedoglu and March (2003), the authors first applied the relaxed model of De Giorgi (1991), Modica (1987), and Modica and Mortola (1977) for an approximation of the 2.1D Sketch model proposed for segmentation with depth (Nitzberg et al., 1993).

3 Euler's elastica-based segmentation models and fast algorithms

Using Euler's elastica minimization model has a long history in image processing. It plays an important part in illusory contour (Kang et al., 2014b; Masnou and Morel, 2006), segmentation with depth (Esedoglu and March, 2003; Nitzberg et al., 1993; Zhu et al., 2006), and image inpainting problem. Assume that $u : \Omega \rightarrow \mathbb{R}$ denotes the image, where $\Omega \subset \mathbb{R}^2$ is an open bounded subset of $\mathbb{R}^2$ with Lipschitz continuous boundary, and $u(\mathbf{x})$ denotes the intensity of the grey level at point $\mathbf{x} \in \Omega$. The Euler's elastica inpainting model (Chan et al., 2003; Masnou and Morel, 1998) is defined by

$$\min_u \left\{ E_e(u) := \int_\Omega (a + b\kappa^2)|\nabla u| \, d\mathbf{x} + \frac{\mu}{2}\int_{\Omega \backslash D} (u - u_0)^2 d\mathbf{x} \right\},$$

where a, b, and λ are positive constant parameters, $D \subset \Omega$ denotes the missing region to be filled and is called the inpainting domain, u_0 is the available part of the image on $\Omega \backslash D$, and

$$\kappa = \kappa(\mathbf{x}) := \nabla \cdot \frac{\nabla u}{|\nabla u|}$$

is the curvature of the level lines of image at $\mathbf{x}$. The first term is the Euler's elastica regularizer capturing the geometrical feature of the image, and the second fidelity term enforces the known information on $\Omega \backslash D$ to be preserved. The parameter μ balances between the regularization term and the fidelity term. The parameter a is the weight for the total variation (TV), while b influences on the amount of curvature along the edges and on the connectivity of the separated level lines.

There are recent works on fast algorithms for Euler's elastica-based inpainting and denoising models. Tai et al. (2011) introduced a fast algorithm for inpainting problem, by reformulating the functional as a constrained minimization problem and using operator splitting and relaxation. The proposed functional is solved by augmented Lagrangian approach and shows a very successful results both in speed and result. A couple of methods with a less number of parameters are explored in Yashtini and Kang (2016): these methods are based on operator splitting and alternating direction method of multipliers, and subproblems can be solved efficiently by Fourier transforms and shrinkage operators. Zhang et al. (2017) proposed a linearized strategy to speed up augmented Lagrangian method (ALM) and applied it to the solution of image denoising problems. A new robust and stable algorithm, with essentially no parameters to choose is proposed in Deng et al. (2018) for image denoising. This algorithm is obtained by applying an operator-splitting-based time discretization scheme to an initial value problem, dynamical flow, associated with the optimality system. The subproblems have either closed form solutions or can be handled by fast dedicated solvers. This method is essentially parameter free, resulting in a very robust and stable algorithm. There are related new developments using approximated curvature-based models. For example, a convex lower semi-continuous approximation of Euler's elastica energy is proposed by Bredies et al. (2015), a direction of generalized TV for higher order method is explored in Papafitsoros et al. (2013), and mean curvature-based denoising is explore in Zhu et al. (2013b), and an overview on fast algorithms can be found in Wu and Tai (2010).

In the context of segmentation, we present some details of a couple of fast algorithms for Euler's elastica-based segmentation proposed in Bae et al. (2017), Tai and Duan (2017), and Zhu et al. (2013a). These approaches are based on ALM. ALM is shown to be very useful for the numerical treatment of high-order or/and nondifferentiable functionals. The main reason of ALM's fast performance lies in the following facts: by applying ALM, the original minimization can be reformulated as a constrained optimization problem, whose saddle points can be approached via solving several lower-order subproblems alternatingly and repeatedly; and these subproblems either have closed-form solutions or can be easily solved.

To illustrate the benefit of ALM approach, we present some details for Euler's elastica Chan–Vese segmentation (ECV-L^2) functional setup in Zhu et al. (2013a):

$$\begin{aligned}\mathbf{E}(\phi,c_1,c_2)=&\int_\Omega\left[a+b\left(\nabla\cdot\frac{\nabla\phi}{|\nabla\phi|}\right)^2\right]|\nabla H(\phi)|\\&+\int_\Omega[\mu(f-c_1)^2H(\phi)+(f-c_2)^2(1-H(\phi))]\end{aligned}\tag{9}$$

where ϕ is a level set function whose zero level curve presents the segmentation boundary, $H(\cdot)$ is the Heaviside function, c_1, c_2 are two scalars, and λ is a positive parameter. Here μ, a, b are positive parameters, which control the length and curvature of segmentation boundary. Here the parameter μ plays a more important role than in Chan–Vese model (1). It can relax the competition of the fitting term and the Euler's elastica term, aiming to complete missing boundaries.

The functional (9) involves the sign of the level set function ϕ. Following the idea of Chan et al. (2006) for finding the global minimizer of Chan–Vese model, a subsitution is made $u=H(\phi)$. As $\nabla\cdot\frac{\nabla H(\phi)}{|\nabla H(\phi)|}=\nabla\cdot\frac{\nabla\phi}{|\nabla\phi|}$, one can rewrite the functional (9) to be

$$\mathbf{E}(u,c_1,c_2)=\int_\Omega\mu(f-c_1)^2u+(f-c_2)^2(1-u)+\left[a+b\left(\nabla\cdot\frac{\nabla u}{|\nabla u|}\right)^2\right]|\nabla u|,$$

where u only takes on two values 0 and 1. We relax the restriction on u to be $0\le u\le 1$, then the functional becomes the following constrained minimization problem

$$\begin{aligned}&\min_{u,\mathbf{p},\mathbf{n}}\int_\Omega\mu(f-c_1)^2u+(f-c_2)^2(1-u)+\left[a+b(\nabla\cdot\mathbf{n})^2\right]|\mathbf{p}|,\\&\text{with } \mathbf{p}=\nabla u,\mathbf{n}=\frac{\nabla u}{|\nabla u|},u\in[0,1].\end{aligned}$$

Then, an augmented Lagrangian functional is formulated as follows:

$$\begin{aligned}\mathcal{L}(v,u,\mathbf{p},\mathbf{n},\mathbf{m};\lambda_1,\boldsymbol{\lambda}_2,\lambda_3,\boldsymbol{\lambda}_4)=&\int\mu(f-c_1)^2v+(f-c_2)^2(1-v)+\left[a+b(\nabla\cdot\mathbf{n})^2\right]|\mathbf{p}|\\&+r_1\int(|\mathbf{p}|-\mathbf{p}\cdot\mathbf{m})+\int\lambda_1(|\mathbf{p}|-\mathbf{p}\cdot\mathbf{m})\\&+\frac{r_2}{2}\int|\mathbf{p}-\nabla u|^2+\int\boldsymbol{\lambda}_2\cdot(\mathbf{p}-\nabla u)\\&+\frac{r_3}{2}\int(v-u)^2+\int\lambda_3(v-u)+\delta_{\mathcal{D}}(v)\\&+\frac{r_4}{2}\int|\mathbf{n}-\mathbf{m}|^2+\int\boldsymbol{\lambda}_4\cdot(\mathbf{n}-\mathbf{m})+\delta_{\mathcal{R}}(\mathbf{m}),\end{aligned}\tag{10}$$

where $\mathcal{D}=[0,1]$ and $\mathcal{R}=\{\mathbf{m}\in L^2(\Omega):|\mathbf{m}|\leq 1 \text{ a.e. in } \Omega\}$, and $\delta_{\mathcal{D}}(v)$ and $\delta_{\mathcal{R}}(\cdot)$ are the characteristic functions on the sets $\mathcal{D}$ and $\mathcal{R}$, respectively, which read as follows:

$$\delta_{\mathcal{D}}(v)=\begin{cases}0, & v\in\mathcal{D}\\ +\infty, & \text{otherwise,}\end{cases} \quad \text{and } \delta_{\mathcal{R}}(\mathbf{m})=\begin{cases}0, & \mathbf{m}\in\mathcal{R}\\ +\infty, & \text{otherwise.}\end{cases}$$

Here, $r_i's$, $i = 1, \ldots, 4$ are positive parameters, while λ_1, $\boldsymbol{\lambda}_2$, λ_3, $\boldsymbol{\lambda}_4$ are Lagrange multipliers, and a new variable $\mathbf{m}$ is introduced to simplify the associated subproblem on $\mathbf{p}$, just as in Tai et al. (2011). Indeed, as $\mathbf{m}$ is required to be inside $\mathcal{R}$, $|\mathbf{m}|\leq 1$, then $|\mathbf{p}|-\mathbf{p}\cdot\mathbf{m}\geq 0$ for any $\mathbf{p}$, and $|\mathbf{p}|-\mathbf{p}\cdot\mathbf{m}=0$ if and only if $\mathbf{m}=\frac{\mathbf{p}}{|\mathbf{p}|}$. This avoids the term $\int(|\mathbf{p}|-\mathbf{p}\cdot\mathbf{m})^2$, which results in a relatively complex functional on $\mathbf{p}$. Moreover, by using the new variable $\mathbf{m}$, the minimizer of the functional related to $\mathbf{p}$ can be obtained exactly and explicitly by using some appropriate shrinkage.

Based on the theory of optimization, the saddle point of the augmented Lagrangian functional (10) relates to a minimizer of the functional (9). Thus, one simply needs to find the saddle point of the augmented functional. An iterative algorithm is employed, that is, minimize the corresponding functional for each of the variable functions v, u, $\mathbf{p}$, $\mathbf{n}$, $\mathbf{m}$ by fixing the other ones, and after the advance of all the variable functions, update the Lagrange multipliers accordingly based on the optimization theory. One of the benefits of augmented Lagrangian functional in (10) is that each subproblem becomes very simple to solve, e.g., explicit solution form exists, shrinkage can be used, or fast methods such as FFT can be applied. The procedure is repeated until the variable functions are all convergent, which means a saddle point of the augmented Lagrangian functional is obtained.

The added curvature term, Euler's elastica in ECV-L^2, leads to a few new interesting features Zhu et al. (2013a): it can integrate missing or broken parts to form complete meaningful objects; it is able to selectively choose objects according to their sizes, in another words, the model only keeps objects of large sizes while omitting those ones of small sizes; moreover, it can also capture objects with tiny but elongated structures without preserving other negligible dots such as noise. Fig. 5 demonstrates the new features of the ECV-L^2 model when compared with the Chan–Vese model. This is a medical image with tubular vessels in an in-homogeneous background. This example demonstrates that the ECV-L^2 model well captures the elongated vessels, while also automatically interpolate the part of vessels that is not well defined by image grey intensity. Fig. 6 shows that the ECV-L^2 model is able to capture large objects while omitting small ones.

Further extension to L^1-Euler's elastica (Bae et al., 2017) is considered with the help of similar fast algorithms. The minimizing functional becomes

$$\mathbf{E}(\phi,c_1,c_2)=\int_\Omega[(f-c_1)^2H(\phi)+(f-c_2)^2(1-H(\phi))]+\int_\Omega\left[a+b\left|\nabla\cdot\frac{\nabla\phi}{|\nabla\phi|}\right|\right]|\nabla H(\phi)|. \tag{11}$$

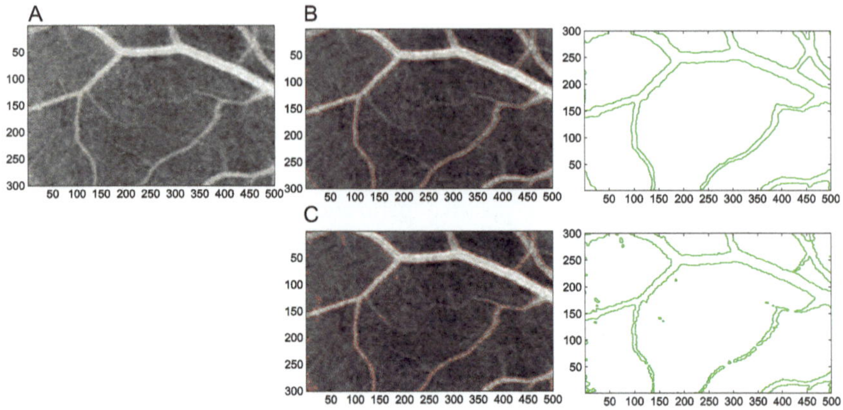

FIG. 5 (A) An image of tubular vessels and the segmentation results by the ECV-L^2 model and Chan–Vese's model. The parameters for the model (9) read $a = 10^{-3}$, $b = 3.0$, $\mu = 1.0$ while the parameters for ALM are chosen as follows: $r_1 = 0.08$, $r_2 = 1.0$, $r_3 = 0.5$, $r_4 = 1.0$. (B) ECV-L^2 model; (C) Chan–Vese's model. *Copyright @2013 Springer. Reprinted from Zhu, W., Tai, X.-C., Chan, T., 2013a. Image segmentation using Euler Os elastica as the regularization. J. Sci. Comput. 57 (2), 414–438 with permission.*

FIG. 6 (A) An image with two soaring eagles, (B) and (C) two different segmentation results of ECV-L^2 model using different curvature parameter $b = 0.1$, 2.0, respectively. The parameters for the model (9) read $a = 10^{-3}$, $\mu = 0.8$ while the parameters for ALM are chosen as follows: $r_1 = 0.6$, $r_2 = 2.0$, $r_3 = 0.1$, $r_4 = 1.0$. *Copyright @2013 Springer. Reprinted from Zhu, W., Tai, X.-C., Chan, T., 2013a. Image segmentation using Euler Os elastica as the regularization. J. Sci. Comput. 57 (2), 414–438 with permission.*

The most interesting feature of the model lies in that it prefers convex contours once the parameter b is sufficiently large. This regularity term also helps to single out objects of convex contours, which is among one of the first segmentation models that possess the shape driven property. This L^1-Euler's elastica-based regularizer was also adopted for curve denoising in Elsey and Esedoglu (2009).

To minimize the above functional (11), a different ALM-based algorithm is proposed. The most important difference lies in the treatment of the variables $\mathbf{p}$ and $\mathbf{n}$. In the above discussion, these two variables are connected through an intermediate variable $\mathbf{m}$ so that all the subproblems have nice forms. For the ECV-L^1 model, the variable $\mathbf{m}$ was removed, and the two variables are related directly by $|\mathbf{p}|\mathbf{n} - \mathbf{p} = 0$. This helps remove one penalty parameter, and thus considerably reduces the effort of choosing appropriate penalty parameters for the numerical implementation. Specifically, the new augmented Lagrangian functional for the ECV-L^1 model is given as follows

$$
\begin{aligned}
\mathcal{L}(\phi,q,\mathbf{p},\mathbf{n},c_1,c_2;\boldsymbol{\lambda}_1,\lambda_2,\boldsymbol{\lambda}_3) = & \int_\Omega \{(f-c_1)^2\left(\frac{1}{2}+\frac{1}{\pi}\arctan\left(\frac{\phi}{\epsilon}\right)\right) \\
& + (f-c_2)^2\left(\frac{1}{2}-\frac{1}{\pi}\arctan\left(\frac{\phi}{\epsilon}\right)\right)\} \\
& + \int_\Omega (a+b|q|)\frac{\epsilon}{\pi(\epsilon^2+\phi^2)}|\mathbf{p}| \\
& + \frac{r_1}{2}\int_\Omega |\mathbf{p}-\nabla\phi|^2 + \int_\Omega \boldsymbol{\lambda}_1\cdot(\mathbf{p}-\nabla\phi) \\
& + \frac{r_2}{2}\int_\Omega (q-\nabla\cdot\mathbf{n})^2 + \int_\Omega \lambda_2(q-\nabla\cdot\mathbf{n}) \\
& + \frac{r_3}{2}\int_\Omega ||\mathbf{p}|\mathbf{n}-\mathbf{p}|^2 + \int_\Omega \boldsymbol{\lambda}_3\cdot(|\mathbf{p}|\mathbf{n}-\mathbf{p}).
\end{aligned}
\tag{12}
$$

Since the variables $\mathbf{p}$ and $\mathbf{n}$ are connected through $|\mathbf{p}|\mathbf{n}-\mathbf{p} = 0$. This functional involves only three Lagrange multipliers. More careful numerical considerations have been made for efficient computation in Bae et al. (2017).

Fig. 7 shows effects of ECV-L^1 model for various different geometric shapes: a diamond, a staircase-like shape, and a square, two of which are convex shapes while the middle one is not convex. These three regions are designed to have the same area purposely. As the model parameter b increases while fixing all the other parameters, objects of smaller sizes will be omitted in the final segmentation contour. More importantly, for some special b, only the staircase-like shape is taken out from the final segmentation contour. This phenomenon also suggests a new shape driven property of the model, that is, it prefers objects of convex shapes. Fig. 8 shows that the ECV-L^1 model is able to segment the convex hat even though it is occluded by the boy's fact and neck.

Another model in Tai and Duan (2017) combines ideas of Zhu et al. (2013a) and Zhu et al. (2006) for more simplified segmentation model with curvature term. Utilizing the property of sign distance function and a relaxed representation of the length term, the numerical setup becomes quite simple. The functional is proposed to be

$$
\min_{\phi,u}\int_\Omega\left(\alpha+\beta\left|\nabla\cdot\frac{\nabla\phi}{|\nabla\phi|}\right|^2\right)|\nabla u|dx+\int_\Omega Q(c_1,c_2)u dx
$$

where ϕ is a continuous signed distance function, $u = H(\phi)$ is the Heaviside function of ϕ, and $Q(c_1, c_2) = (c_1-f)^2 - (c_2-f)^2$ for segmentation problem with a given input image f. This model uses the relaxed binary representation

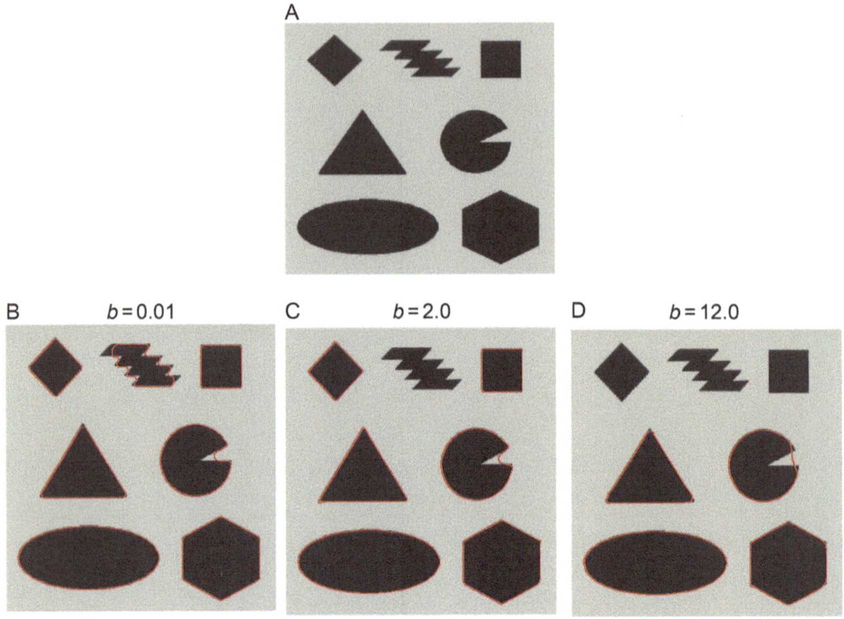

FIG. 7 The original image (A) and the segmentation results (B)–(D) with different curvature parameter b. The other parameters are $a = 10^{-4}$, $r_1 = 50$, $r_2 = 20$, $r_3 = 5$. *Copyright @2017 American Institute of Mathematical Sciences. Reprinted from Bae, E., Tai, X.-C., Zhu, W., 2017. Augmented Lagrangian method for an Eulers elastica based segmentation model that promotes convex contours. Inverse Prob. Imaging 11, 1–23 with permission.*

FIG. 8 The original image and the segmentation result. The result demonstrates that the invisible part of the hat can be restored by the proposed segmentation model with a relatively large curvature parameter b. In this experiment, the parameters are $a = 10^{-3}$, $b = 80$, $r_1 = 60$, $r_2 = 40$, $r_3 = 10$. *Copyright @2017 American Institute of Mathematical Sciences. Reprinted from Bae, E., Tai, X.-C., Zhu, W., 2017. Augmented Lagrangian method for an Eulers elastica based segmentation model that promotes convex contours. Inverse Prob. Imaging 11, 1–23 with permission.*

for the length term, and the signed distance representation for the calculation of curve curvature. Consequently, the original elastica problem becomes a TV type minimization problem and a redistance problem, which can be solved efficiently with recent fast approaches. The authors proposed alternating minimization of (i) the signed distance function computation $\phi^{k+1} = SDF(u^{k+1})$, and (ii) the indicator function u,

$$u^{k+1} = \arg \min_{u \in [0,1]} \int_\Omega \left(\alpha + \beta \left| \nabla \cdot \frac{\nabla \phi^k}{|\nabla \phi^k|} \right|^2 \right) |\nabla u| dx + \int_\Omega Q(c_1, c_2) u dx + \frac{\theta}{2} \int_\Omega (u - u^k)^2 dx.$$

The positive parameter θ controls the similarity between u^{k+1} and u^k and helps to reduce the speed of contour, thus contour evolutes slowly and steadily. The new algorithm only needs to solve a TV type minimization problem and a redistance problem, which results in fewer built-in parameters, and this model can be generalized to inpainting and illusory shape reconstruction with a simple modification of $Q(c_1, c_2)$.

There is recent related work on curvature-based model and fast algorithm, such as a new augmented Lagrangian primal dual algorithm for elastica regularization (Zhang and Chen, 2016), contrast driven elastica for image segmentation (El-Zehiry and Grady, 2016), finding global minimum for a Finsler elastica minimal path approach (Chen et al., 2017), and regularization with Euler's elastica using a discrete gradient scheme (Ringholm et al., 2018) just to name a few.

4 Discussion

Starting from the celebrated Mumford–Shah functional and Chan–Vese model, we considered a piecewise constant level set approach and interface models. The length minimization for multiple phases needs careful considerations, which can be addressed by an equal distance vectorial multiwell potential (6) as in Kang and March (2014). Multiphase segmentation is fundamentally a nonconvex problem, and related approaches such as Sandberg et al. (2010) or some supervised approach such as Kang et al. (2014a) maybe useful. We showed that interface approach is connected to illusory contour model, where curvature computation is achieved from a relaxed model (8) for an efficient computation.

In the context of image segmentation using Euler's elastica term, fast algorithms proposed in Bae et al. (2017), Tai and Duan (2017), and Zhu et al. (2013a) are reviewed. With recent developments in fast and efficient numerical methods, exploring new and interesting functionals with curvature terms became more feasible and practical. Such new models exhibit new interesting features, such as it can complete missing or broken parts to form meaningful objects, it is able to selectively choose objects according to their sizes, and it can also capture objects with tiny but elongated structures without preserving other negligible dots such as noise. In addition, ECV-L^1 results suggests a new shape driven property of the model that it prefers objects of convex shapes. Even with many new numerical developments, it is always challenging to design a good functional, which can induce efficient numerical schemes. This is an exciting field of research with a lot of possibilities.

References

Ambrosio, L., Tortorelli, V.M., 1990. Approximation of functionals depending on jumps by elliptic functionals via Γ-convergence. Comm. Pure Appl. Math. 43 (8), 999–1036.

Ambrosio, L., Tortorelli, V.M., 1992. On the approximation of free discontinuity problems. Boll. Un. Mat. Ital. 6-B, 105–123.

Aubert, G., Aujol, J.-F., Blanc-Feraud, L., 2005. Detecting codimension-two objects in an image with Ginzburg-Landau models. Int. J. Comput. Vision 65 (1/2), 29–42.

Bae, E., Tai, X.-C., 2009a. Efficient global minimization for the multiphase Chan-Vese model of image segmentation. In: Energy Minimization Methods in Computer Vision and Pattern Recognition, 7th International Conference, EMMCVPR 2009, Bonn, Germany, August 24–27, 2009, pp. 28–41.

Bae, E., Tai, X.-C., 2009b. Graph cut optimization for the piecewise constant level set method applied to multiphase image segmentation. In: Scale Space and Variational Methods in Computer Vision, Second International Conference, SSVM 2009, Voss, Norway, June 1–5, 2009, pp. 1–13.

Bae, E., Tai, X.-C., 2015. Efficient global minimization methods for image segmentation models with four regions. J. Math. Imaging Vision 51 (1), 71–97.

Bae, E., Yuan, J., Tai, X.-C., Boykov, Y., 2010. A Study on Continuous Max-Flow and Min-Cut Approaches. Part II: Multiple Linearly Ordered Labels. UCLA CAM Report 10-62.

Bae, E., Yuan, J., Tai, X.-C., 2011. Global minimization for continuous multiphase partitioning problems using a dual approach. Int. J. Comput. Vision 92 (1), 112–129.

Bae, E., Yuan, J., Tai, X.-C., Boykov, Y., 2014. A fast continuous max-flow approach to non-convex multi-labeling problems. In: Efficient Algorithms for Global Optimization Methods in Computer Vision, Springer, pp. 134–154.

Bae, E., Tai, X.-C., Zhu, W., 2017. Augmented Lagrangian method for an Euler's elastica based segmentation model that promotes convex contours. Inverse Prob. Imaging 11, 1–23.

Bertozzi, A., Esedoglu, S., Gilette, A., 2006. Inpainting of Binary Images Using the Cahn-Hilliard Equation. UCLA CAM Report 06-07.

Borzi, A., Grossauer, H., Scherzer, O., 2005. Analysis of iterative methods for solving a Ginzburg-Landau equation. Int. J. Comput. Vision 64 (2/3), 203–219.

Bredies, K., Pock, T., Wirth, B., 2015. A convex, lower semicontinuous approximation of Euler's elastica energy. SIAM J. Math. Anal. 47 (1), 566–613.

Brown, E.S., Chan, T.F., Bresson, X., 2012. Completely convex formulation of the Chan-Vese image segmentation model. Int. J. Comput. Vision 98 (1), 103–121.

Burger, M., He, L., Schönlieb, C., 2009. Cahn-Hilliard inpainting and a generalization for gray-value image. SIAM J. Imag. Sci. 2 (4), 1129–1167.

Cahn, J.W., Hilliard, J.E., 1958. Free energy of a nonuniform system. I. Interfacial free energy. J. Chem. Phys. 28 (2), 258–267.

Chambolle, A., Cremers, D., Bischof, H., Pock, T., 2009. A convex relaxation approach for computing minimal partitions. In: Computer Vision and Pattern Recognition, 2009. CVPR 2009. IEEE Conference on Date of Conference: 20–25 June 2009, pp. 810–817.

Chambolle, A., Cremers, D., Pock, T., 2012. A convex approach to minimal partitions. SIAM J. Imaging Sci. 5 (4), 1113–1158.

Chan, T.F., Vese, L., 2001. Active contours without edges. IEEE Trans. Image Process. 10 (2), 266–277.

Chan, T.F., Zhu, W., 2007. Capture illusory contours: a level set based approach. J. Math. Imaging Vis. 27, 29–40.

Chan, T.F., Kang, S.H., Shen, J., 2003. Euler's elastica and curvature-based inpainting. SIAM J. Appl. Math. 63 (2), 564–592.

Chan, T.F., Esedoglu, S., Nikolova, M., 2006. Algorithms for finding global minimizers of image segmentation and denoising models. SIAM J. Appl. Math. 66 (5), 1632–1648.

Chen, D., Mirebeau, J.-M., Cohen, L.D., 2017. Global minimum for a Finsler elastica minimal path approach. Int. J. Comput. Vision 122 (3), 458–483.

Chung, J.T., Vese, L.A., 2003. Image Segmentation Using a Multilayer Level-Set Approach. UCLA CAM Report 03-53.

De Giorgi, E., 1991. Some remarks on γ-convergence and least squares method. In: Composite Media and Homogenization Theory, Springer, pp. 135–142.

Deng, L.-J., Glowinski, R., Tai, X.-C., 2018. A new operator splitting method for Euler's elastica model. SIAM J. Imaging Sci. 12 (2), 1190–1230.

Dobrosotskaya, J.A., Bertozzi, A.L., 2011. Wavelet analogue of the Ginzburg-Landau energy and its Γ-convergence. Interfaces Free Bound. 12 (4), 497–525.

El-Zehiry, N.Y., Grady, L., 2016. Contrast driven elastica for image segmentation. IEEE Trans. Image Process. 25 (6), 2508–2518.

Elsey, M., Esedoglu, S., 2009. Analogue of the total variation denoising model in the context of geometry processing. Multiscale Model. Simul. 7 (4), 1549–1573.

Esedoglu, S., March, R., 2003. Segmentation with depth but without detecting junctions. J. Math. Imaging Vision 18 (1), 7–15.

Esedoglu, S., Shen, J., 2002. Digital inpainting based on the Mumford-Shah-Euler image model. European J. Appl. Math. 13, 353–370.

Gao, S., Bui, T.D., 2005. Image segmentation and selective smoothing by using Mumford-Shah model. IEEE Trans. Image Process. 14 (10), 1537–1549.

Geman, S., Geman, D., 1984. Stochastic relaxation, Gibbs distributions, and the Bayesian restoration of images. IEEE Trans. Pattern Anal. Machine Intell. 6, 721–741.

Ginzburg, V.L., Landau, L.D., 1950. On the theory of superconductivity. Soviet Phys. JETP 20, 1064–1082.

Grossauer, H., Scherzer, O., 2003. Using the complex Ginzburg-Landau equation for digital inpainting in 2D and 3D. In: Scale Space Methods in Computer Vision, Lecture Notes in Computer Science, 2695, pp. 225–236.

Gu, Y., Wang, L.-L., Tai, X.-C., 2012. A direct approach toward global minimization for multiphase labeling and segmentation problems. IEEE Trans. Image Process. 21 (5), 2399–2411.

Häuser, S., Steidl, G., 2013. Convex multiclass segmentation with shearlet regularization. Int. J. Comput. Math. 90 (1), 62–81.

He, Y., Shafei, B., Hussaini, M.Y., Ma, J., Steidl, G., 2012. A new fuzzy c-means method with total variation regularization for segmentation of images with noisy and incomplete data. Pattern Recogn. 45, 3436–3471.

Jung, Y.-M., Shen, J., 2008. First-order modeling and stability analysis of illusory contours. J. Visual Commun. Image Represent. 19, 42–55.

Jung, Y.M., Kang, S.H., Shen, J., 2007. Multiphase image segmentation via Modica-Mortola phase transition. SIAM J. Appl. Math. 67 (5), 1213–1232.

Kang, S.H., March, R., 2014. Multiphase image segmentation via equally distanced multiple well potential. J. Vis. Commun. Image Represent. 25 (6), 1446–1459.

Kang, S.H., Shafei, B., Steidl, G., 2014. Supervised and transductive multi-class segmentation using p-laplacians and RKHS methods. J. Visual Commun. Image Represent. 25 (5), 1136–1148.

Kang, S.H., Zhu, W., Shen, J., 2014. Illusory shapes via corner fusion. SIAM J. Imaging Sci. 7 (4), 1907–1936.

Lellmann, J., Schnörr, C., 2011. Continuous multiclass labeling approaches and algorithms. SIAM J. Imaging Sci. 4 (4), 1049–1096.

Lellmann, J., Kappes, J., Yuan, J., Becker, F., Schnörr, C., 2009. Convex multi-class image labeling by simplex-constrained total variation. In: International Conference on Scale Space and Variational Methods in Computer Vision, pp. 150–162.

Li, H., Tai, X.-C., 2007. Piecewise constant level set methods for multiphase motion. Int. J. Numer. Anal. Model. 4 (2), 291–305.

Li, F., Ng, M., Li, C., 2010. Variational fuzzy Mumford-Shah model for image segmentation. SIAM J. Appl. Math. 70 (7), 2750–2770.

Li, F., Ng, M.K., Zeng, T.Y., Shen, C., 2010. A multiphase image segmentation method based on fuzzy region competition. SIAM J. Imaging Sc. 3 (3), 277–299.

Lie, J., Lysaker, M., Tai, X.-C., 2005. Piecewise constant level set methods and image segmentation. In: International Conference on Scale-Space Theories in Computer Vision, Springer, pp. 573–584.

Lie, J., Lysaker, M., Tai, X.-C., 2006. A binary level set model and some applications to Mumford-Shah image segmentation. IEEE Trans. Image Process. 15 (5), 1171–1181.

Lie, J., Lysaker, M., Tai, X.-C., 2006. A variant of the level set method and applications to image segmentation. Math. Comput. 75 (255), 1155–1174.

Liu, J., Tai, X.-C., Leung, S., 2013. A generic convexification and graph cut method for multiphase image segmentation. In: International Workshop on Energy Minimization Methods in Computer Vision and Pattern Recognition, Springer, pp. 251–265.

March, R., 1992. Visual reconstruction with discontinuities using variational methods. Image Vision Comput. 10, 30–38.

March, R., Dozio, M., 1997. A variational method for the recovery of smooth boundaries. Image Vision Comput. 15, 705–712.

Masnou, S., Morel, J.M., 1998. Level lines based disocclusion. In: Proceedings 1998 International Conference on Image Processing. ICIP98 (Cat. No. 98CB36269). IEEE, pp. 259–263.

Masnou, S., Morel, J.-M., 2006. On a variational theory of image amodal completion. Rendiconti del Seminario Matematico della Universita di Padova 116, 211–252.

Modica, L., 1987. The gradient theory of phase transitions and the minimal interface criterion. Arch. Rational Mech. Anal. 98 (2), 123–142. ISSN 0003-9527.

Modica, L., Mortola, S., 1977. Un esempio di Γ^{-}-convergenza. Boll. Un. Mat. Ital. B (5) 14 (1), 285–299.

Mumford, D., Shah, J., 1989. Optimal approximations by piecewise smooth functions and associated variational problems. Commun. Pure Appl. Math. 42 (5), 577–685.

Nitzberg, M., Mumford, D., Shiota, T., 1993. Filtering, Segmentation and Depth. vol. 662, Springer.

Osher, S., Sethian, J.A., 1988. Fronts propagating with curvature-dependent speed: algorithms based on Hamilton-Jacobi formulations. J. Comput. Phys. 79 (1), 12–49.

Papafitsoros, K., Schoenlieb, C.B., Sengul, B., 2013. Combined first and second order total variation inpainting using split Bregman. Image Process. Line 3, 112–136.

Perona, P., Malik, J., 1990. Scale-space and edge detection using anisotropic diffusion. IEEE Trans. Pattern Anal. Machine Intell. 12, 629–639.

Pock, T., Schoenemann, T., Graber, G., Bischof, H., Cremers, D., 2008. A convex formulation of continuous multi-label problems. In: European Conference on Computer Vision, pp. 792–805.

Pock, T., Chambolle, A., Cremers, D., Bischof, H., 2009. A convex relaxation approach for computing minimal partitions. In: 2009 IEEE Conference on Computer Vision and Pattern Recognition, pp. 810–817.

Ringholm, T., Lazic, J., Schönlieb, C.-B., 2018. Variational image regularization with Euler's elastica using a discrete gradient scheme. SIAM J. Imaging Sc. 11 (4), 2665–2691.

Rudin, L., Osher, S., Fatemi, E., 1992. Nonlinear total variation based noise removal algorithms. Phys. D 60, 259–268.

Sandberg, B., Kang, S.H., Chan, T.F., 2010. Unsupervised multiphase segmentation: A phase balancing model. IEEE Trans. Image Process. 19 (1), 119–130.

Sarti, A., Malladi, R., Sethian, J.A., 2002. Subjective surfaces: a geometric model for boundary completion. Int'l J. Comp. Vision 46 (3), 201–221.

Shafei, B., Steidl, G., 2012. Segmentation of images with separating layers by fuzzy c-means and convex optimization. J. Vis. Commun. Image Represent. 23 (4), 611–621.

Shen, J., 2006. A stochastic-variational model for soft Mumford-Shah segmentation. Int. J. Biomed. Imaging 2006, 92329.

Shi, J., Malik, J., 2000. Normalized cuts and image segmentation. IEEE Trans. Pattern Anal. Machine Intell. 22 (8), 888–905.

Tai, X.-C., Duan, J., 2017. A simple fast algorithm for minimization of the elastica energy combining binary and level set representations. Int. J. Numer. Anal. Model. 14 (6), 809–821.

Tai, X.-C., Hahn, J., Chung, G.J., 2011. A fast algorithm for Euler's elastica model using augmented Lagrangian method. SIAM J. Imaging Sc. 4 (1), 313–344.

Tu, Z.W., Zhu, S.C., 2002. Image segmentation by data-driven Markov chain Monte Carlo. IEEE Trans. Pattern Anal. Machine Intell. 24 (5), 657–673.

Wu, C., Tai, X.-C., 2010. Augmented Lagrangian method, dual methods, and split Bregman iteration for ROF, vectorial TV, and high order models. SIAM J. Imaging Sci. 3 (3), 300–339.

Yashtini, M., Kang, S.H., 2016. A fast relaxed normal two split method and an effective weighted TV approach for Euler's elastica image inpainting. SIAM J. Imaging Sci. 9 (4), 1552–1581.

Yuan, J., Bae, E., Tai, X.-C., 2010a. A study on continuous max-flow and min-cut approaches. In: 2010 IEEE Computer Society Conference on Computer Vision and Pattern Recognition, IEEE, pp. 2217–2224.

Yuan, J., Bae, E., Tai, X.-C., Boykov, Y., 2010b. A continuous max-flow approach to Potts model. In: European Conference on Computer Vision, Springer, pp. 379–392.

Yuan, J., Bae, E., Tai, X.-C., Boykov, Y., 2014. A spatially continuous max-flow and min-cut framework for binary labeling problems. Numer. Math. 126 (3), 559–587.

Zach, C., Gallup, D., Frahm, J.-M., Niethammer, M., 2008. Fast global labeling for real-time stereo using multiple plane sweeps. VMV, pp. 243–252.

Zhang, J., Chen, K., 2016. A new augmented Lagrangian primal dual algorithm for elastica regularization. J. Algorithms Comput. Technol. 10 (4), 325–338.

Zhang, J., Chen, R., Deng, C., Wang, S., 2017. Fast linearized augmented Lagrangian method for Euler's elastica model. Numer. Math. Theory, Methods Appl. 10 (1), 98–115.

Zhu, W., Chan, T.F., 2005. Illusory Contours Using Shape Information. UCLA CAM Tech. Report 03-09.

Zhu, W., Chan, T.F., 2007. A variational model for capturing illusory contours using curvature. J. Math. Imaging Vision 27 (1), 29–40.

Zhu, W., Chan, T.F., Esedoglu, S., 2006. Segmentation with depth: a level set approach. SIAM J. Sci. Comput. 28 (5), 1957–1973.

Zhu, W., Tai, X.-C., Chan, T., 2013. Image segmentation using Euler Os elastica as the regularization. J. Sci. Comput. 57 (2), 414–438.

Zhu, W., Tai, X.-C., Chan, T.F., 2013. Augmented Lagrangian method for a mean curvature based image denoising model. Inverse Probl. Imaging 7 (4), 1409–1432.

Chapter 14

Recent advances in denoising of manifold-valued images

R. Bergmann[a], F. Laus[b], J. Persch[b] and G. Steidl[b,c,*]

[a]*Research Group Numerical Mathematics (Partial Differential Equations), Faculty of Mathematics, TU Chemnitz, Chemnitz, Germany*

[b]*Image Analysis Group, Department of Mathematics, TU Kaiserslautern, Kaiserslautern, Germany*

[c]*Fraunhofer ITWM, Kaiserslautern, Germany*

[*]*Corresponding author: e-mail: Steidl@mathematik.tu-kl.de*

Chapter Outline

Abstract

Modern signal and image acquisition systems are able to capture data that are no longer real-valued but may take values on a manifold. However, whenever measurements are taken, no matter whether manifold-valued or not, there occur tiny inaccuracies, which result in noisy data. In this chapter, we review recent advances in denoising of manifold-valued signals and images, where we restrict our attention to variational models and appropriate minimization algorithms. The algorithms are either classical as the subgradient algorithm or generalizations of the half-quadratic minimization method, the cyclic proximal point algorithm, and the Douglas–Rachford algorithm to manifolds. An important aspect when dealing with real-world data is the practical implementation. Here several groups provide software and toolboxes as the Manifold Optimization (Manopt) package and the manifold-valued image restoration toolbox (MVIRT).

Handbook of Numerical Analysis, Vol. 20. https://doi.org/10.1016/bs.hna.2019.07.005
© 2019 Elsevier B.V. All rights reserved.

Keywords: Manifold-valued images, Optimization on manifolds, Douglas–Rachford algorithm, TGV, Iteratively reweighted least squares, Hadamard manifolds

AMS Classification Codes: 49M37, 49Q20, 65K05, 65K10

1 Introduction

The mathematical notion of a manifold dates back to 1828, when Carl Friedrich Gauss established an important invariance property of surfaces while proving his Theorema Egregium. In his habilitation lecture in 1854, Bernhard Riemann intrinsically extended Gauss's theory making manifolds independent of their embedding in higher dimensional spaces. This is now called a Riemannian manifold. Nowadays modern signal and image acquisition methods are able to capture information that is no longer restricted to Euclidean spaces but can be manifold-valued. Sophisticated models for human colour perception involve non-Euclidean settings. Moreover, it is often advantageous to model information from large data as points on a certain manifold. Here are some typical examples.

Interferometric synthetic aperture radar (InSAR). An InSAR image is obtained by calculating the phase difference of two synthetic aperture radar (SAR) images of an area taken at different positions or times (Bürgmann et al., 2000). It can be used to measure elevation when the measurements are taken at the same time or to detect millimetre-scale deformations over days or years. It has applications in the geophysical monitoring of natural hazards, for example, earthquakes, volcanoes and landslides, and in structural engineering, in particular monitoring of subsidence and structural stability. InSAR produces phase-valued images, i.e., in each pixel the measurement lies on the circle $\mathbb{S}^1$, see Fig. 1 (left).

Image colour spaces. The usual RGB colour space has a vector space structure, but there exist models which are physically better suited for the human colour perception as they encode luminance independent of colour. Examples are the hue–saturation–value (HSV) colour space or the related Ich space, where the hue has its value on the circle $\mathbb{S}^1$, as well as the chromaticity-brightness (CB) colour space, where the chromaticity has values

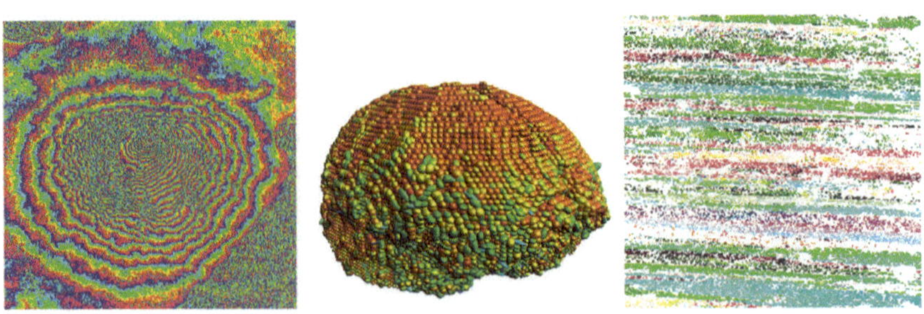

FIG. 1 *Left*: InSAR data from Mt. Vesuvius (Rocca et al., 1997). *Middle*: DT-MRI from the Camino project (Cook et al., 2006). *Right*: EBSD data of an aluminium sample. *Image courtesy: Institute of Materials Science and Engineering, TU Kaiserslautern.*

on the positive octant of the sphere $\mathbb{S}^2$. For a recent geometric model of brightness perception we refer to Batard and Bertalmío (2018).

Electron backscatter diffraction (EBSD). EBSD is a microstructural crystallography characterization technique used to analyze the microscopic structure of polycrystalline materials such as metals and minerals (Bachmann et al., 2011). Each point of a specimen is radiated by an electron beam and the diffraction pattern is measured, which gives information on the phase and the crystal orientation, a value on the rotation group SO(3). Regions of similar orientation are called grains. Material scientists are interested in the grain structure of the specimen, as it affects macroscopic properties such as ductility, electrical and lifetime properties. Since the atomic structure of a crystal is invariant under the symmetry of its atomic lattice, i.e., a symmetry group $S \subset \mathrm{SO}(3)$, the images obtained by EBSD have pixel values in $\mathrm{SO}(3)/S$, see Fig. 1 (right). The software MTEX (Bachmann and Hielscher, 2005-2016) is designed for processing EBSD data.

Diffusion tensor magnetic resonance imaging (DT-MRI). In DT-MRI the diffusion of water molecules perpendicular to a magnetic field is measured in biological tissue. Taking at least six different data sets measured with different magnetic fields, a DT-MRI image with values in the manifold $\mathcal{P}(3)$ of symmetric positive definite 3×3 matrices is computed at each pixel, see Fig. 1 (middle). The diffusion of water is influenced by the structure of the tissue. Hence, the knowledge about the diffusion can be used to distinguish between diseased and healthy parts. DT-MRI is a noninvasive and in vivo technique, which is extensively used in neurology, but can also be applied to detect defects in other tissue, like muscles.

Covariance matrices in texture analysis and brain computer interfaces. Textures form a special class of images and appear at the same time as an important image feature. There are different ways to model textures, one possibility is to encode the (local) dependence structure as a covariance matrix. This has been used, e.g., in Tuzel et al. (2006), where textures are characterized based on the empirical covariance of certain features, for example, colour intensities and derivatives of different orders. Under the assumption that the empirical covariance matrices are nondegenerated we are again faced with the manifold $\mathcal{P}(d)$ of symmetric positive definite $d \times d$ matrices, where d equals the number of features. In brain computer interfaces (BCI) approaches, EEG curves related to different human activities are measured at different brain positions at the same time. Their covariance matrices can be used to analyze the corresponding activity, see, e.g., Yair et al. (2018).

Beyond these applications, images with values in $\mathbb{S}^2$ appear when dealing with 3D directional information (Kimmel and Sochen, 2002), in foliage angle distribution (Helliwell et al., 2015) as well as in the analysis of liquid crystals (Alouges, 1997). The rotation group SO(3) also appears in cyro-electron microscopy (Hadani and Singer, 2011; Singer and Shkolnisky, 2012; Wang et al., 2013) and the special Euclidean group SE(3) is considered in tracking and (scene) motion analysis (Rosman et al., 2012; Tuzel et al., 2008).

Processing manifold-valued signals and images is a new challenge that affects classical tasks such as image restoration (denoising, inpainting), segmentation and clustering (Bergmann et al., 2017b), registration and large deformation diffeomorphic mapping (LDDMM) or metamorphosis between different images, see, e.g., Muñoz-Moreno et al. (2009), Neumayer et al. (2018) and Zhang et al. (2018).

This chapter focuses on variational denoising methods for manifold-valued images. The simplest idea is to embed the manifold into the Euclidean space and to apply the Euclidean models with the constraint that the image values have to lie in a manifold. Recall that by Whitney's theorem every smooth d-dimensional manifold can be smoothly embedded into an Euclidean space of dimension $2d$. Such an approach was given, e.g., in Rosman et al. (2014). The advantage is that optimization algorithms in Euclidean spaces can be applied, where the models are in general nonconvex due to the constraints. So-called lifting schemes were proposed for circular-valued data in Cremers and Strekalovskiy (2013) and for more general manifold-valued images in Lellmann et al. (2013). There, the problem is reformulated as a multilabel optimization problem which is approached using convex relaxation techniques. We also like to mention that current state-of-the-art methods for denoising of real-valued images include stochastic nonlocal patch-based approaches, as, e.g., the nonlocal Bayes' algorithm (Lebrun et al., 2013). A generalization of this minimum mean square estimator (MMSE)-based method to manifold-valued images was proposed in Laus et al. (2017).

This chapter deals with spatially discrete, intrinsic models and algorithms. Note that in Giaquinta et al. (1993) and Giaquinta and Mucci (2007), the notion of total variation of spatially continuous functions having their values on a manifold was investigated where the authors apply the theory of Cartesian currents. For a spatial continuous setting, the reader may also consult the recent paper (Ciak et al., 2019).

An important aspect when working with real data is the practical implementation of the developed methods and algorithms. In the spirit of reproducible research, several groups provide their software and toolboxes, e.g., the Manifold Optimization (Manopt) package (Boumal et al., 2014) and the manifold-valued image restoration toolbox (MVIRT Bergmann, 2017.

The outline of this chapter is as follows: starting with the necessary preliminaries in Section 2, we review several denoising models for manifold-valued images in Section 3. Appropriate minimization algorithms, namely

- half-quadratic minimization, in particular iteratively reweighted least squares,
- cyclic proximal point algorithm, and
- parallel Douglas–Rachford algorithm

are discussed in Section 4. Numerical examples in Section 5 illustrate the proposed methods.

2 Preliminaries on Riemannian manifolds

2.1 General notation

Throughout this chapter, let $\mathcal{M}$ be a connected, complete d-dimensional Riemannian manifold. By $T_x\mathcal{M}$ we denote the tangent space of $\mathcal{M}$ at $x \in \mathcal{M}$ with the Riemannian metric $\langle \cdot, \cdot \rangle_x$ and corresponding norm $\|\cdot\|_x$. Further, let $T\mathcal{M}$ be the tangent bundle of $\mathcal{M}$. By $\operatorname{dist}: \mathcal{M} \times \mathcal{M} \to \mathbb{R}_{\geq 0}$ we denote the geodesic distance on $\mathcal{M}$. Let $\mathcal{M}^n$ be the product or n-fold power manifold with product distance

$$\operatorname{dist}^2(x,y) := \left(\sum_{j=1}^{n} \operatorname{dist}^2(x_j, y_j) \right)^{\frac{1}{2}}.$$

Let $\gamma_{\widehat{x,y}}: [0,1] \to \mathcal{M}$ be a (not necessarily shortest) geodesic connecting $x, y \in \mathcal{M}$. Further, we apply the notation $\gamma_{x;\xi}$ to characterize the geodesics by its starting point $\gamma_{x;\xi}(0) = x$ and direction $\dot{\gamma}_{x;\xi}(0) = \xi \in T_x\mathcal{M}$. Note that the geodesic $\gamma_{\widehat{x,y}}$ is unique on manifolds with nonpositive curvature. The exponential map $\exp_x : T_x\mathcal{M} \to \mathcal{M}$ is defined by

$$\exp_x(\xi) := \gamma_{x;\xi}(1).$$

Since $\mathcal{M}$ is connected and complete, we know by the Hopf–Rinow theorem that the exponential map is indeed defined on the whole tangent space. The exponential map realizes a local diffeomorphism from a neighbourhood $\mathcal{D}_T(0_x)$ of the origin 0_x of $T_x\mathcal{M}$ into a neighbourhood of $x \in \mathcal{M}$. More precisely, extending the geodesic $\gamma_{x;\xi}$ from $t = 0$ to infinity is either minimizing $\operatorname{dist}(x, \gamma_{x;\xi}(t))$ all along or up to a finite time t_0 and not any longer afterwards. In the latter case, $\gamma_{x;\xi}(t_0)$ is called cut point and the set of all cut points of all geodesics starting from x is the cut locus $\mathcal{C}(x)$. This allows to define the inverse exponential map, also known as logarithmic map as

$$\log_x := \exp_x^{-1} : \mathcal{M} \backslash \mathcal{C}(x) \to T_x\mathcal{M}.$$

Then, the Riemannian distance between $x, y \in \mathcal{M}$, for $y \notin \mathcal{C}(x)$, can be written as

$$\operatorname{dist}(x,y) = \langle \log_x(y), \log_x(y) \rangle_x^{\frac{1}{2}} = \| \log_x(y) \|_x.$$

Let $F : \mathcal{M} \to \mathcal{N}$ be a smooth mapping between manifolds and $\xi \in T_x\mathcal{M}$. The linear mapping

$$DF(x) : T_x\mathcal{M} \to T_{F(x)}\mathcal{N}, \quad \xi \mapsto DF(x)[\xi],$$

is called differential of F at $x \in \mathcal{M}$. Let $F : \mathcal{M}_1 \to \mathcal{M}_2$ and $G : \mathcal{M}_2 \to \mathcal{M}_3$ be two smooth mappings. Then the differential of their concatenation $G \circ F$ applied to $\xi \in T_x\mathcal{M}_1$ is given by the chain rule

$$D(G \circ F)(x)[\xi] = DG(F(x))[DF(x)[\xi]].$$

For a function $f: \mathcal{M} \to \mathbb{R}$, the Riemannian gradient $\text{grad}_{\mathcal{M}}$ is defined by

$$\langle \text{grad}_{\mathcal{M}} f(x), \xi \rangle_x := Df(x)[\xi] \tag{1}$$

for all $\xi \in T_x\mathcal{M}$. A mapping $R_p: \mathcal{M} \to \mathcal{M}$ is called geodesic reflection at $x \in \mathcal{M}$, if

$$R_x(x) = x \text{ and } D(R_x(x)) = -I. \tag{2}$$

For $\mathcal{M} = \mathbb{R}^n$ we simply have $R_p(x) = 2p - x$. A connected Riemannian manifold $\mathcal{M}$ is (globally) symmetric if the geodesic reflection at any point $x \in \mathcal{M}$ is an isometry of $\mathcal{M}$, i.e., $\text{dist}(R_p(x), R_p(y)) = \text{dist}(x,y)$ for all $x, y \in \mathcal{M}$. All manifolds considered in this chapter are symmetric ones.

Let $\mathcal{X}(\mathcal{M})$ be the set of smooth vector fields on $\mathcal{M}$. Given a curve $\gamma: [0,1] \to \mathcal{M}$, we denote by $\mathcal{X}(\gamma)$ the set of smooth vector fields along γ, i.e., $X \in \mathcal{X}(\gamma)$ is a smooth mapping $X: [0,1] \to T\mathcal{M}$ with $X(t) \in T_{\gamma(t)}\mathcal{M}$. A vector field $X \in \mathcal{X}(\gamma)$ is called parallel to $\gamma: [0,1] \to \mathcal{M}$, if the covariant derivative along γ fulfils $\frac{D}{dt}X = 0$ for all $t \in [0, 1]$. We define the parallel transport of a tangent vector $\xi \in T_x\mathcal{M}$ to $T_y\mathcal{M}$ by

$$P_{x \to y}\xi := X(1),$$

where $X \in \mathcal{X}(\gamma_{\widehat{x,y}})$ is the vector field parallel to a minimizing geodesic $\gamma_{\widehat{x,y}}$ with $X(0) = \xi$. There exist analytical expressions of the parallel transport for few manifolds as spheres or positive definite matrices. However, the parallel transport can be locally approximated by Schild's ladder (Ehlers et al., 1972; Kheyfets et al., 2000) or by the pole ladder (Lorenzi and Pennec, 2014). Recently, it was shown that for connected, complete, symmetric Riemannian manifold, the pole ladder coincides with the parallel transport along geodesics (Pennec, 2018). Therefore, we prefer the pole ladder approach. Given $x, y \in \mathcal{M}$, the pole ladder transports $\xi \in T_x\mathcal{M}$ to $T_y\mathcal{M}$ by

$$P^{\mathrm{P}}_{x \to y}(\xi) := -\log_y\left(\gamma\left(\exp_x(\xi), \gamma\left(x, y; \frac{1}{2}\right); 2\right)\right) \in T_y\mathcal{M}, \tag{3}$$

where we use the notation $\gamma(x, y; t) := \gamma_{\widehat{x,y}}(t)$. For comparison, Schild's ladder transports as follows:

$$P^{\mathrm{S}}_{x \to y}(\xi) := \log_y\left(\gamma\left(x, \gamma\left(y, \exp_x(\xi); \frac{1}{2}\right); 2\right)\right) \in T_y\mathcal{M}.$$

Both transport schemes are illustrated in Fig. 2. In our minimization algorithms, we will need the Riemannian gradient of special functions, in particular of those appearing in the pole ladder (3). These gradients can be derived from differentials, see (1), and can be computed for symmetric Riemannian manifolds using the theory of Jacobi fields. Lemma 1 collects the final results which can be partially found in Bačák et al. (2016), Bredies et al. (2018) and do Carmo (1992). For the complete proof we refer to Persch (2018).

Lemma 1. *Let $\mathcal{M}$ be a symmetric Riemannian manifold and F one of the functions (i)–(v) below with parameter y, resp. u, together with the coefficient map $\alpha: \mathbb{R} \to \mathbb{R}$ and parameter T. Then the differential $DF(x)$ at $x \in \mathcal{M}$ is given for all $\xi \in T_x\mathcal{M}$ by*

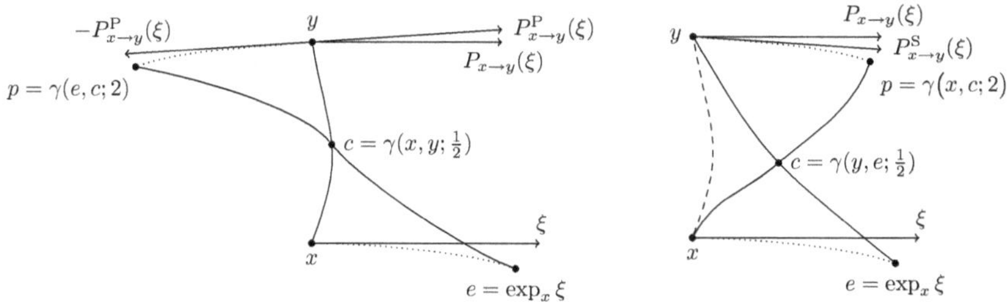

FIG. 2 Illustration of pole ladder (*left*) and Schild's ladder (*right*) for the approximation of $P_{x\to y}\xi$.

$$DF(x)[\xi] = \sum_{k=1}^{d} \langle \xi, \Xi_k(0)\rangle_x \alpha(\kappa_k)\Xi_k(T), \tag{4}$$

where $\{\Xi_k\}_{k=1}^d$ *denotes a parallel transported orthogonal frame along the geodesic* γ *with* $\gamma(0) = x$ *and* $\gamma(1) = y$, *and* $y := \exp_x(u)$ *if* F *depends on* u. *Further, the frame diagonalizes the Riemannian curvature tensor* $R(\,\cdot\,, \dot\gamma)\dot\gamma$ *with respective eigenvalues* κ_k, $k = 1, \ldots, d$. *The functions* F *and* $\alpha(\kappa)$ *are given as follows:*

F	T	$\alpha(\kappa)$
$\exp_{\cdot}(u)$	1	$\cosh(\sqrt{-\kappa})$ $\kappa<0$, 1 $\kappa=0$, $\cos(\sqrt{\kappa})$ $\kappa>0$
$\log_{\cdot}(y)$	0	$-\sqrt{-\kappa}\frac{\cosh(\sqrt{-\kappa})}{\sinh(\sqrt{-\kappa})}$ $\kappa<0$, -1 $\kappa=0$, $-\sqrt{\kappa}\frac{\cos(\sqrt{\kappa})}{\sin(\sqrt{\kappa})}$ $\kappa>0$
$\log_y(\cdot)$	1	$\frac{\sqrt{-\kappa}}{\sinh(\sqrt{-\kappa})}$ $\kappa<0$, 1 $\kappa=0$, $\frac{\sqrt{\kappa}}{\sin(\sqrt{\kappa})}$ $\kappa>0$
$\gamma_{\overset{\frown}{\cdot,y}}(\tau)$	τ	$\frac{\sinh(\sqrt{-\kappa}(1-\tau))}{\sinh(\sqrt{-\kappa})}$ $\kappa<0$, $1-\tau$ $\kappa=0$, $\frac{\sin(\sqrt{\kappa}(1-\tau))}{\sin(\sqrt{\kappa})}$ $\kappa>0$
$\gamma_{\overset{\frown}{y,\cdot}}(\tau)$	$1-\tau$	$\frac{\sinh(\sqrt{-\kappa}\tau)}{\sinh(\sqrt{-\kappa})}$ $\kappa<0$, τ $\kappa=0$, $\frac{\sin(\sqrt{\kappa}\tau)}{\sin(\sqrt{\kappa})}$ $\kappa>0$
$\exp_x(\cdot)$	1	$\frac{\sinh(\sqrt{-\kappa})}{(\sqrt{-\kappa})}$ $\kappa<0$, 1 $\kappa=0$, $\frac{\sin(\sqrt{\kappa})}{\sqrt{\kappa}}$ $\kappa>0$

In the last case, the differential $DF(u)$ of F at $u \in T_x\mathcal{M}$ is given by (4), where we have to replace $x \in \mathcal{M}$ by $u \in T_x\mathcal{M}$

The adjoint operator $(DF)^*(x): T_{F(x)}\mathcal{M} \to T_x\mathcal{M}$ of (4), which is also needed for the computation of the gradients, is given by

$$(DF)^*(x)[w] = \sum_{k=1}^{d} \langle w, \Xi_k \rangle_{F(x)} \alpha_k \xi_k, \quad w \in T_{F(x)}\mathcal{M}.$$

2.2 Convexity and Hadamard manifolds

A subset $C \subseteq \mathcal{M}$ is called weakly (strongly) convex if for all points $x, y \in C$ there exists a (unique) geodesic $\gamma_{\widehat{x,y}}$ of minimal length which is contained entirely in C. Let $\varphi: \mathcal{M} \supset C \to \mathbb{R} \cup \{\infty\}$ be a real-valued function on a weakly convex set C, then φ is called convex if

$$\varphi\left(\gamma_{\widehat{x,y}}(t)\right) \leq (1-t)\varphi(x) + t\varphi(y),$$

for all $x, y \in C$. The function φ is called strictly convex, if the above equation holds strictly for all $t \in (0, 1)$. A function $\varphi: \mathcal{M} \supseteq C \to \mathbb{R} \cup \{\infty\}$ is κ-strongly convex if

$$\varphi\left(\gamma_{\widehat{x,y}}(t)\right) \leq (1-t)\varphi(x) + t\varphi(y) - \kappa t(1-t)\mathrm{dist}^2(x,y).$$

Simply connected, complete Riemannian manifolds of nonpositive sectional curvature are called Hadamard manifolds. We denote them by $\mathcal{H}$. Examples are the manifold of positive definite matrices of fixed size with the affine invariant metric or hyperbolic spaces. An important property of Hadamard manifolds is, that the distance function $\mathrm{dist}(\cdot, \cdot)$ is jointly convex, which makes $\mathrm{dist}^2(\cdot, \cdot)$ strictly convex. Further, $\mathrm{dist}^2(\cdot, y)$ is a 1-strongly convex function on a Riemannian manifold if and only if $\mathcal{H}$ is a Hadamard manifold. The domain of a function $\varphi: \mathcal{H} \to \mathbb{R} \cup \{\infty\}$ is defined by

$$\mathrm{dom}(\varphi) := \{x \in \mathcal{H}: \varphi(x) < \infty\},$$

in general we work with proper functions, i.e., $\mathrm{dom}(\varphi) \neq \emptyset$. A function φ is called lower semicontinuous (lsc) if the set $\{x \in \mathcal{H}: \varphi(x) \leq c\}$ is closed for all $c \in \mathbb{R}$. Whenever $\mathrm{dist}(x, x_0) \to \infty$ for some $x_0 \in \mathcal{H}$, the function φ is called coercive if $\varphi(x) \to \infty$. Concerning minimizers of convex functions the following holds true: A proper, convex, lsc functions $\varphi: \mathcal{H} \to \mathbb{R} \cup \{\infty\}$ have a minimizer if it is coercive and unique minimizer if it is in addition strongly convex. For further information on more general Hadamard spaces and the basics of convex analysis therein, we refer to Bačák (2014b).

3 Intrinsic variational restoration models

We consider images as mappings from the image grid $\mathcal{G} = \{1,\ldots,n_1\} \times \{1,\ldots,n_2\}$ to a Riemannian manifold $\mathcal{M}$. Let $n := n_1 n_2$. Given a corrupted image $f : \mathcal{G} \to \mathcal{M}$, variational models generate a restored image $u : \mathcal{G} \to \mathcal{M}$ as a minimizer of a functional of the form

$$\mathcal{J}(u) := \mathcal{D}(u; f) + \alpha \mathcal{R}(u),$$

where $\mathcal{D}(\,\cdot\,; f)$ denotes the data-fitting term, $\alpha > 0$ the regularization parameter, and $\mathcal{R}$ the regularization term or prior.

For real-valued images typically a squared Euclidean distance is chosen as data-fitting term. For manifold-valued data, we can just use the squared distance on the manifold

$$\mathcal{D}(u; f) := \frac{1}{2} \sum_{i \in \mathcal{G}} \operatorname{dist}(u_i, f_i)^2. \tag{5}$$

Depending on the prior knowledge, several regularization terms were proposed in the Euclidean setting. The (discretized) total variation (TV) introduced by Rudin et al. (1992) is demonstrably a powerful, edge-preserving, nonsmooth, and convex regularizer. It sums up the norms of the gradients at the image points. A natural way to define the discrete gradient $\nabla := (\nabla_x, \nabla_y)^{\mathrm{T}}$ on manifold-valued images is as a vector field in the corresponding tangent spaces $(T_{u_i}\mathcal{M})^2$ with the difference operators

$$\nabla_x u_i := \begin{cases} \text{if } \log_{u_i} u_{i+(1,0)} & i + (1,0) \in \mathcal{G}, \\ 0 & \text{otherwise}, \end{cases}$$

and similarly for ∇_y. Now, the TV regularizer on manifold-valued images becomes

$$\begin{aligned} \mathrm{TV}(u) &:= \sum_{i \in \mathcal{G}} \left(\| \nabla_x u_i \|_{u_i}^p + \| \nabla_y u_i \|_{u_i}^p \right)^{\frac{1}{p}} \\ &= \sum_{i \in \mathcal{G}} \left(\sum_{j \in \mathcal{N}(i)} \operatorname{dist}(u_i, u_j)^p \right)^{\frac{1}{p}}, \quad p \in \{1, 2\}, \end{aligned} \tag{6}$$

with $\mathcal{N}(i) := \{i+(1,0), i+(0,1)\} \cap \mathcal{G}$. Here, $p = 1$ is used for the anisotropic and $p = 2$ for the isotropic model. This setting was considered in Lellmann et al. (2013) and Weinmann et al. (2014). If $\mathcal{M} = \mathcal{H}$ is an Hadamard manifold, then functional consisting of the data term (5) and the prior (6) has the same properties as its real-valued version, i.e., it is strongly convex and coercive and hence there exists a unique minimizer. The TV functional (6) is not differentiable. To apply minimization algorithms for differentiable functions we can recast it, using an even, differentiable function $\varphi : \mathbb{R}_{\geq 0} \to \mathbb{R}_{\geq 0}$, in the anisotropic case as

$$\mathrm{TV}_\varphi(u) := \sum_{i \in \mathcal{G}} \sum_{j \in \mathcal{N}(i)} \varphi\big(\mathrm{dist}(u_i, u_j)\big),$$

and in the isotropic one as

$$\mathrm{TV}_\varphi(u) := \sum_{i \in \mathcal{G}} \varphi\left(\left(\sum_{j \in \mathcal{N}(i)} \mathrm{dist}(u_i, u_j)^2\right)^{\frac{1}{2}}\right). \tag{7}$$

Typical functions φ are the Huber function and $\varphi(x) := \sqrt{x^2 + \varepsilon^2}$ with a small ε.

The minimizers of the TV regularized functionals prefer piecewise constant functions, a behaviour called staircasing. To avoid such artefacts second-order differences were incorporated into the regularizer. In the manifold-valued setting, we have to find a counterpart of such differences. A generalization of the anisotropic so-called second-order TV term was given in Bačák et al. (2016). Observing that in the Euclidean case the absolute second-order difference of $x, y, z \in \mathbb{R}^d$ can be rewritten as $|x - 2y + z| = 2|\frac{1}{2}(x+z) - y|$, a counterpart for $x, y, z \in \mathcal{M}$ is defined as

$$\mathrm{d}_2(x,y,z) := \min_{c \in \mathcal{C}_{x,z}} \mathrm{dist}(c,y),$$

where $\mathcal{C}_{x,z}$ denotes the set of midpoints $\gamma_{\widehat{x,z}}(\frac{1}{2})$ of all geodesics connecting x and z. Note that the geodesic $\gamma_{\widehat{x,z}}$ is unique on Hadamard manifolds. Similarly, second-order mixed differences were defined for $x, y, z, w \in \mathcal{M}$ as

$$\mathrm{d}_{1,1}(x,y,z,w) := \min_{c \in \mathcal{C}_{x,z}, \tilde{c} \in \mathcal{C}_{y,w}} \mathrm{dist}(c, \tilde{c}).$$

We emphasize that the absolute second-order difference d_2 is not convex in x and z on Hadamard manifolds. Now, we can introduce the absolute value of the second-order difference in x-direction as

$$\mathrm{d}_{xx} u_i := \begin{cases} \mathrm{d}_2(u_{i+(1,0)}, u_i, u_{i-(1,0)}) & \text{if } i \pm (1,0) \in \mathcal{G}, \\ 0 & \text{otherwise,} \end{cases}$$

and similarly in y-direction. As absolute value of the mixed differences we use

$$\mathrm{d}_{xy} u_i := \begin{cases} \mathrm{d}_{1,1}\big(u_i, u_{i+(0,-1)}, u_{i+(1,0)}, u_{i+(1,-1)}\big) & \text{if } \ i \pm (0,1) \wedge i + (1,0) \in \mathcal{G}, \\ 0 & \text{otherwise,} \end{cases}$$

and similarly for d_{yx}. Then we define

$$\mathrm{TV}_2(u) := \sum_{i\in\mathcal{G}} \left(\mathrm{d}_{xx}u_i^p + \mathrm{d}_{yy}u_i^p + \mathrm{d}_{xy}u_i^p + \mathrm{d}_{xy}u_i^p\right)^{\frac{1}{p}}, \quad p\in\{1,2\},$$

where $p = 1$ is used for the anisotropic model and $p = 2$ for the isotropic model. In the regularizer, the TV and TV_2 terms can appear separately or in a coupled way. Actually, their addition

$$\mathcal{R}(u) := \beta\mathrm{TV}(u) + (1-\beta)\mathrm{TV}_2(u), \quad \beta\in(0,1)$$

was considered in Bačák et al. (2016) and Bergmann and Weinmann (2015). Alternatively, couplings which generalize the infimal convolution approach Chambolle and Lions (1997) to the manifold-valued setting were proposed in Bergmann et al. (2017a) and Bergmann et al. (2018). In the Euclidean setting, the infimal convolution is related to the total generalized variation (TGV) approach of Bredies et al. (2010). Recently, Bredies et al. (2018) came also up with a TGV model for manifold-valued images, see also Valkonen et al. (2013) for DT-MRI. In the following we present a different TGV approach from Bergmann et al. (2018). For the relation between both model see Bergmann et al. (2018, Remark 5.1). In the Euclidean setting, the (discrete) TGV regularizer reads as

$$\min_{\xi\in(T_u\mathcal{M}^n)^2} \left\{ \sum_{i\in\mathcal{G}} \beta\, \|\nabla u_i - \xi_i\|_2 + (1-\beta)\, \|\widetilde{\nabla}\xi_i\|_2 \right\},$$

where $\widetilde{\nabla}$ denotes a certain symmetric backward difference operator. For a vector field $\xi = (\xi_i)_{i\in\mathcal{G}}$, $\xi_i \in (T_{u_i}\mathcal{M})^2$, the distance between ξ and ∇u in the first summand is given by

$$R_1(u,\xi) := \sum_{i\in\mathcal{G}} \left(\|\nabla_x u_i - \xi_{1,i}\|_{u_i}^p + \|\nabla_y u_i - \xi_{2,i}\|_{u_i}^p \right)^{\frac{1}{p}}, \quad p\in\{1,2\}.$$

To compute the backward differences $\widetilde{\nabla}\xi$ in the second summand, we need to compare tangent vectors from different tangent spaces. To this end, we apply the parallel transport between the tangent spaces via the pole ladder. Since the corresponding expression in (3) contains only exponential and logarithmic maps, the differentials required in the minimization procedure can be calculated using the chain rule and Lemma 1. More precisely, we define backward differences of a vector field $\zeta \in T_u\mathcal{M}^n$ in x-direction by

$$\widetilde{\nabla}_x^{\mathrm{P}}\zeta_i := \begin{cases} \zeta_i - P^{\mathrm{P}}_{u_{i-(1,0)}\to u_i}(\eta_{i-(1,0)}) & \text{if } i\pm(1,0)\in\mathcal{G}, \\ 0 & \text{otherwise,} \end{cases}$$

and similarly in y-direction. Then we define

$$R_2(\xi) := \sum_{i\in\mathcal{G}} \left(\|\widetilde{\nabla}_x^{\mathrm{P}}\xi_{1,i}\|_{u_i}^p + \|\widetilde{\nabla}_y^{\mathrm{P}}\xi_{1,i}\|_{u_i}^p + \|\widetilde{\nabla}_x^{\mathrm{P}}\xi_{2,i}\|_{u_i}^p + \|\widetilde{\nabla}_x^{\mathrm{P}}\xi_{2,i}\|_{u_i}^p \right)^{\frac{1}{p}}.$$

Note that due to simplified computations, this definition differs slightly from the symmetric arrangement of the backward differences in the Euclidean TGV setting. Now, we can define a TGV regularizer for manifold-valued images as

$$\mathrm{TGV}(u) := \inf_{\xi \in (T_u \mathcal{M}^n)^2} \{\beta R_1(u, \xi) + (1-\beta) R_2(\xi)\}, \quad \beta \in (0,1).$$

4 Minimization algorithms

To compute a minimizer of our functionals, Riemannian optimization methods can be applied. These intrinsic methods are often very efficient since they exploit the underlying geometric structure of the manifold, see, e.g., Absil et al. (2008) and Ring and Wirth (2012). For smooth functionals, various methods have been proposed, reaching from simple gradient descents on manifolds to more sophisticated trust region or (quasi) Newton methods.

We start by recalling the gradient decent algorithm or more precisely the subgradient algorithm which can also be applied for the minimization of nondifferentiable functions. This is a basic algorithm with convergence guarantee for manifolds with *nonnegative* sectional curvature. Then we turn to so-called half-quadratic minimization methods which can be applied if the energy functional is differentiable. A special case of half-quadratic minimization is the so-called iteratively reweighted least squares method.

Finally, we consider two algorithms having counterparts on vector spaces in convex analysis, namely the cyclic proximal point algorithm and the parallel Douglas–Rachford algorithm. These algorithms can also be applied for nonsmooth energy functionals. They make heavy use of so-called proximal mappings. Proximal mappings of distance functions in one or two arguments can be given analytically for Hadamard manifolds and for $\mathbb{S}^1$. Global convergence to a minimizer can be only proved for Hadamard manifolds, i.e., on manifolds having *nonpositive* sectional curvature.

4.1 Subgradient descent

The subdifferential of a convex function $\varphi : \mathcal{M} \to (-\infty, +\infty]$ at $x \in \operatorname{dom} f$ is defined by

$$\partial\varphi(x) := \left\{ v \in T_x\mathcal{M} : \varphi(y) \geq \varphi(x) + \langle v, \dot{\gamma}_{\widehat{x,y}}(0) \rangle \text{ for all } y \in \operatorname{dom} \varphi \right\},$$

see, e.g., Li et al. (2011) or Udrişte (1994) for finite functions φ. For any $x \in \operatorname{int}(\operatorname{dom} \varphi)$, the subdifferential is a nonempty convex and compact set in $T_x\mathcal{M}$. If the Riemannian gradient $\operatorname{grad}_{\mathcal{M}}\varphi(x)$ of φ in $x \in \mathcal{M}$ exists, then $\partial\varphi(x) = \{\operatorname{grad}_{\mathcal{M}}\varphi(x)\}$. Further, we see from the definition that $x \in \mathcal{M}$ is a global minimizer of φ if and only if $0 \in \partial\varphi(x)$.

Let $\varphi : \mathcal{M} \to \mathbb{R}$ be a convex function and $x^{(0)} \in \mathcal{M}$ the starting point. Given a sequence $(\tau_r)_{r\in\mathbb{N}}$ of nonegative numbers, the subgradient algorithm iterates

$$
\begin{aligned}
&\text{for} \quad r = 0, 1, \ldots \text{ until a stopping criterion is reached}\\
&\qquad s^{(r+1)} \in \partial\varphi(x^{(r)}),\\
&\qquad x^{(r+1)} = \exp_{x^{(r)}}\left(-\tau_r \frac{s^{(r+1)}}{\|s^{(r+1)}\|_{x^{(r)}}}\right).
\end{aligned}
$$

We have the following convergence result, see Ferreira and Oliveira (1998).

Theorem 1 (Convergence of subgradient algorithm)**.** *Let $\mathcal{M}$ be a Riemannian manifold with nonnegative sectional curvature, $\varphi : \mathcal{M} \to \mathbb{R} \cup \{\infty\}$ a convex function which has a minimizer, and $(\tau_r)_{r\in\mathbb{N}}$ a sequence of positive numbers in $\ell^2 \backslash \ell^1$. Then the sequence $\{x^{(r)}\}_{r\in\mathbb{N}}$ generated by subgradient algorithm converges to a minimizer of φ.*

For manifolds with curvature bounded from below the subgradient algorithm converges if the iterates stay in bounded sets, see Wang (2018) or Wang et al. (2015).

4.2 Half-quadratic minimization

Half-quadratic minimization methods belonging to the group of quasi-Newton methods are efficient minimization algorithms for the functionals with differentiable anisotropic or isotropic regularizers TV_φ. These methods, which cover iteratively reweighted least squares methods, were recently generalized to manifold-valued images (Bergmann et al., 2016a; Grohs and Sprecher, 2016). There exist additive and multiplicative versions of the method, see Geman and Reynolds (1992), Geman and Yang (1995) and Nikolova and Ng (2005). Here, we focus on the multiplicative one for the isotropic TV_φ regularizer, i.e., we want to minimize

$$
\mathcal{J}_\varphi(u) := \frac{1}{2}\sum_{i\in\mathcal{G}} \operatorname{dist}(u_i, f_i)^2 + \alpha \sum_{i\in\mathcal{G}} \varphi\left(\left(\sum_{j\in\mathcal{N}(i)^+} \operatorname{dist}(u_i, u_j)^2\right)^{\frac{1}{2}}\right).
$$

We consider the method based on the so-called c-transform. Given a function $c : \mathbb{R} \times \mathbb{R} \to \mathbb{R}$, the c-transform of a function $\varphi : \mathbb{R} \to \mathbb{R}$ is defined by

$$
\varphi^c(s) := \inf_{t\in\mathbb{R}} \{c(t, s) - \varphi(t)\}.
$$

We see immediately that $\varphi(t) + \varphi^c(s) \leq c(t, s)$. For $c(t, s) := -st$, the function $\varphi^c = -(-\varphi)^*$ is just the Fenchel transform of φ. We need the following proposition, see Bergmann et al. (2016a).

Proposition 1. *Let $\varphi : \mathbb{R} \to \mathbb{R}_{\geq 0}$ be an even, differentiable function and $c(t, s) := t^2 s$.*

(i) If the function $\Phi(t) := -\varphi(\sqrt{t})$ for $t \geq 0$ and $\Phi(t) := +\infty$ for $t < 0$ is convex, then $\varphi = \varphi^{cc}$, i.e., for $\psi(s) := \varphi^c(s)$ it holds

$$\varphi(t) = \inf_{s\in\mathbb{R}} \{c(t,s) - \psi(s)\}. \tag{8}$$

(ii) *If in addition* $\lim_{t\to\infty} \frac{\varphi(t)}{t^2} \to 0$, *and* $\varphi'(t) \geq 0$ *for* $t \geq 0$ *and* $\varphi''(0+) := \lim_{t\to 0+} \frac{\varphi'(t)}{t}$ *exists, then the infimum in* (8) *is attained for the tuple* $(t,s) = (t, s(t))$ *with*

$$s(t) := \begin{cases} \dfrac{\varphi'(t)}{2t} & \text{for } t > 0, \\ \dfrac{\varphi''(0+)}{2} & \text{for } t = 0. \end{cases} \tag{9}$$

These pairs fulfil $\varphi(t) + \psi(s) = c(t,s)$. *The choice is unique except for* $t = 0$, *where any* s *larger than* $\frac{\varphi''(0+)}{2}$ *is also a solution.*

(iii) *If* $\varphi'(t) > 0$ *for* $t > 0$ *and* $\varphi''(0+) > 0$, *then* $s(t) \in (0, \frac{\varphi''(0+)}{2}]$ *for all* $t > 0$.

Functions φ fulfilling the assumptions of Proposition 1 are listed in Table 1.

Then, replacing φ in TV_φ in the isotropic setting in (7) by the expression in (8), we can minimize instead of $\mathcal{J}_\varphi(u)$ the functional

$$J(u,v) := \frac{1}{2}\sum_{i\in\mathcal{G}} \mathrm{dist}^2(u_i, f_i) + \alpha \sum_{i\in\mathcal{G}} (c(\mathrm{d}_i, v_i) - \psi(v_i)),$$

where $\mathrm{d}_i := \left(\sum_{j\in\mathcal{N}(i)^+} \mathrm{dist}^2(u_i, u_j)\right)^{\frac{1}{2}}$. We apply alternating minimization over $v = (v_i)_{i\in\mathcal{G}} \in \mathbb{R}^n$ and $u \in \mathcal{M}^n$ and obtain together with (9) the following iterations:

$$\begin{aligned} &\text{for } \; r = 0, 1, \ldots \text{ until a stopping criterion is reached} \\ &\qquad v^{(r+1)} = \operatorname*{argmin}_v J(u^{(r)}, v) = s\left(\mathrm{d}^{(r)}\right), \\ &\qquad u^{(r+1)} \in \operatorname*{argmin}_u J(u, v^{(r+1)}). \end{aligned} \tag{10}$$

The minimization over u means to find a minimizer of

TABLE 1 Functions φ fulfilling the assumptions of Proposition 1.

	$\varphi(t)$	$s(t)$
$\varphi_1(t)$	$\sqrt{t^2+\varepsilon^2}$	$\dfrac{1}{2\sqrt{t^2+\varepsilon^2}}$
$\varphi_2(t)$	$\begin{cases} \frac{1}{2}t^2 & t<\varepsilon, \\ \varepsilon\lvert t\rvert - \frac{1}{2}\varepsilon^2 & t \geq \varepsilon \end{cases}$	$\begin{cases} \frac{1}{2} & t<\varepsilon, \\ \frac{\varepsilon}{2\lvert t\rvert} & t \geq \varepsilon \end{cases}$
$\varphi_3(t)$	$1-\exp(-\varepsilon^2 t^2)$	$\varepsilon^2 \exp(-\varepsilon^2 t^2)$

$$J(u,v^{(r)}) := \frac{1}{2}\sum_{i\in\mathcal{G}} \operatorname{dist}^2(u_i,f_i) + \alpha\sum_{i\in\mathcal{G}}\left(\sum_{j\in\mathcal{N}(i)^+} \operatorname{dist}^2(u_i,u_j)\right)v_i^{(r)}.$$

Here, we can apply, e.g., a gradient descent or a Riemann–Newton method, see Absil et al. (2008). Concerning the convergence of the algorithm we have the following theorem, see Bergmann et al. (2016a).

Theorem 2 (Convergence of half-quadratic minimization).
Let $\mathcal{H}$ be an Hadamard manifold and let $\varphi : \mathbb{R} \to \mathbb{R}_{\geq 0}$ fulfil the assumptions of Proposition 1. Then, the sequence $\{u^{(r)}\}_{r\in\mathbb{N}}$ generated by (10) converges to the uniquely determined minimizer of $\mathcal{J}_\varphi$.

4.3 Proximal point and Douglas–Rachford algorithm

In the Euclidean setting, tools from convex analysis, in particular powerful algorithms based on duality theory, were successfully applied to minimize the proposed functionals. A prominent example is the alternating directions method of multipliers (ADMM), which is equivalent to the Douglas–Rachford algorithm. A central ingredient of these algorithms are proximal mappings which can be efficiently computed for special regularization terms appearing in Euclidean image processing tasks, see Boyd et al. (2011) and Burger et al. (2017). Recently, several attempts have been made to translate these concepts to manifolds and it turns out that on Hadamard manifolds a certain theory of convex functions can be established. For example, the (inexact) cyclic proximal point algorithm can be introduced on these manifolds (Bačák, 2014b), and this method was also used to minimize the functional with first- and second-order TV regularizers in Bačák et al. (2016), Bergmann et al. (2014), Bredies et al. (2018) and Weinmann et al. (2014). Since the classical Douglas–Rachford algorithm relies on point reflections, it was natural to extend this algorithm to symmetric Hadamard manifolds (Bergmann et al., 2016b).

4.3.1 Proximal mapping

For $\lambda > 0$ and a proper, convex, lsc function $\varphi : \mathbb{R}^m \to (-\infty, +\infty]$, the proximal mapping $\operatorname{prox}_{\lambda\varphi} : \mathbb{R}^d \to \mathbb{R}^d$ defined by

$$\operatorname{prox}_{\lambda\varphi}(x) := \operatorname*{argmin}_{y\in\mathbb{R}^d}\left\{\frac{1}{2}\| x-y \|_2^2 + \lambda\varphi(y)\right\}$$

is uniquely determined. The counterpart on manifolds reads for $\varphi : \mathcal{M}^m \to [-\infty, +\infty]$ as

$$\operatorname{prox}_{\lambda\varphi}(x) := \operatorname*{argmin}_{y\in\mathcal{M}^d}\left\{\frac{1}{2}\operatorname{dist}(x,y)^2 + \lambda\varphi(y)\right\}.$$

Indeed, for proper, convex, lower semicontinuous functions on Hadamard manifolds $\mathcal{M} = \mathcal{H}$, the above minimizer is uniquely determined (Jost, 1997). Moreover, the proximal operator is nonexpansive. For the distance functions appearing in the sums of our models

$$\varphi := \mathrm{dist}(\,\cdot\,, y), \quad \psi := \mathrm{dist}(\,\cdot\,,\cdot\,),$$

the proximal mapping can be given analytically, see Ferreira and Oliveira (2002) and Weinmann et al. (2014). For our absolute second-order differences $d_2(\cdot, \cdot, \cdot)$ the proximal mapping can be computed numerically on certain manifolds by the (sub)gradient descent algorithm and Lemma 1 as outlined in Bačák et al. (2016). An analytical expression for the proximal mapping of $d_2(\cdot, \cdot, \cdot)$ on the sphere $\mathbb{S}^1$ was given in Bergmann et al. (2014).

The analytical expressions for the proximal mappings are summarized in the following lemmas.

Proposition 2. (Proximal mapping of distance functions).
Let $\mathcal{H}$ be an Hadamard manifold, $\lambda > 0$ and $y \in \mathcal{H}$. Then, for $p \in \{1, 2\}$, the proximal mappings of φ and ψ are given by

$$\mathrm{prox}_{\lambda\varphi^p}(x) = \gamma_{\widehat{x,y}}(\hat{t}), \quad \hat{t} := \begin{cases} \min\left\{\dfrac{\lambda}{\mathrm{dist}_{(x,y)}}, 1\right\} & \text{if } p = 1, \\ \dfrac{\lambda}{1+\lambda} & \text{if } p = 2, \end{cases}$$

$$\mathrm{prox}_{\lambda\psi^p}(x,y) = \left(\gamma_{\widehat{x,y}}(\hat{t}), \gamma_{\widehat{x,y}}(\hat{t})\right), \quad \hat{t} := \begin{cases} \min\left\{\dfrac{\lambda}{\mathrm{dist}_{(x,y)}}, \dfrac{1}{2}\right\} & \text{if } p = 1, \\ \dfrac{\lambda}{1+2\lambda} & \text{if } p = 2. \end{cases}$$

To give the analytical expressions for the proximal mappings on $\mathbb{S}^1$, we represent its elements by the angles in $[-\pi, \pi)$. For $a \in \mathbb{R}$, we denote by $(a)_{2\pi} \in [-\pi, \pi)$ those number for which there exists $k \in \mathbb{Z}$ such that $a + 2\pi k = (a)_{2\pi}$.

Proposition 3. (Proximal mapping of distance functions on $\mathbb{S}^1$).
Let $w_1 := (-1, 1)^T$, $w_2 := (1, -2, 1)^T$ and $\psi_1 = \psi = dist(\cdot, \cdot)$, $\psi_2 := d_2(\cdot, \cdot, \cdot)$. For x, $y \in [-\pi, \pi)^{\nu+1}$, $\nu \in \{1, 2\}$, set

$$s_\nu := \mathrm{sgn}(\langle x, w_\nu \rangle)_{2\pi}, \quad m_\nu := \min\left\{\lambda, \frac{|(\langle x, w_\nu \rangle)_{2\pi}|}{\|w_\nu\|_2^2}\right\}.$$

Then it holds for $\nu \in \{1, 2\}$ that

$$\mathrm{prox}_{\lambda\psi_\nu}(x) = \begin{cases} (x - s_\nu\, m_\nu\, w_\nu)_{2\pi} & \text{if } |(\langle x, w_\nu \rangle)_{2\pi}| < \pi, \\ (x \pm s_\nu\, m_\nu\, w_\nu)_{2\pi} & \text{if } |(\langle x, w_\nu \rangle)_{2\pi}| = \pi, \end{cases}$$

$$\mathrm{prox}_{\lambda\psi_\nu^2}(x) = \begin{cases} \left(x - \lambda \dfrac{(\langle x, w_\nu \rangle)_{2\pi}}{1 + \lambda\, \|w_\nu\|_2^2} w_\nu\right)_{2\pi} & \text{if } |(\langle x, w_\nu \rangle)_{2\pi}| < \pi \\ \left(x \pm \lambda \dfrac{\pi}{1 + \lambda\, \|w_\nu\|_2^2} w_\nu\right)_{2\pi} & \text{if } |(\langle x, w_\nu \rangle)_{2\pi}| = \pi. \end{cases},$$

$$\mathrm{prox}_{\lambda\varphi^2}(x) = \left(\frac{x + \lambda y}{1+\lambda} + \frac{\lambda}{1+\lambda} 2\pi\, v\right)_{2\pi},$$

where

$$v := \begin{cases} 0 & \text{if } |x-y| \le \pi, \\ \operatorname{sgn}(x-y) & \text{if } |x-y| > \pi. \end{cases}$$

4.3.2 Cyclic proximal point algorithm

Our functionals have the general form

$$\mathcal{J}(x) := \sum_{k=1}^{K} \varphi_k(x), \tag{11}$$

where appropriate splittings into the K summands must be determined. Given a starting point $x^{(0)} \in \mathcal{M}$ the cyclic proximal point algorithm (CPPA) iterates

$$\begin{aligned} &\text{for} \quad r = 0, 1, \ldots \text{ until a stopping criterion is reached} \\ &\qquad \text{for } k = 1, \ldots, K \\ &\qquad\quad x^{\left(r+\frac{k}{K}\right)} := \operatorname{prox}_{\tau_r \varphi_k}\left(x^{\left(r+\frac{k-1}{K}\right)}\right). \end{aligned}$$

We have the following convergence result from Bačák et al. (2016).
Theorem 3 (Convergence of cyclic PPA**).**
Let $\mathcal{H}$ be a Hadamard manifold and $\varphi_k : \mathcal{H} \to \mathbb{R}$, $k = 1, \ldots, K$, convex continuous functions such that $\mathcal{J}$ attains a (global) minimum. Assume that there exist $p \in \mathcal{H}$ and $C > 0$ such that for each $k = 1, \ldots, K$ and all $x, y \in \mathcal{H}$ we have

$$\varphi_k(x) - \varphi_k(y) \le C \operatorname{dist}(x, y)(1 + \operatorname{dist}(x, p)).$$

Then the sequence $\{x^{(r)}\}_{r\in\mathbb{N}}$ with nonnegative $\{\tau_r\}_{r\in\mathbb{N}} \in \ell^2 \backslash \ell_1$ converges for every starting point $x^{(0)}$ to a minimizer of $\mathcal{J}$.

The result can be generalized for the inexact cyclic PPA which iteratively generates the points $x^{\left(r+\frac{k}{K}\right)}$, $k = 1, \ldots, K$, $r \in \mathbb{N}_0$, fulfilling

$$\operatorname{dist}\left(x^{\left(r+\frac{k}{K}\right)}, \operatorname{prox}_{\tau_r \varphi_k}\left(x^{\left(r+\frac{k-1}{K}\right)}\right)\right) < \frac{\varepsilon_r}{K},$$

where $\{\varepsilon_r\}_{r\in\mathbb{N}_0}$ is a given sequence of positive reals with $\sum_{r=1}^{\infty} \varepsilon_r < \infty$, see Bačák et al. (2016).

4.3.3 Douglas–Rachford algorithm for symmetric Hadamard spaces

The Douglas–Rachford (DR) algorithm relies on reflections. In the Euclidean setting, the reflection of a proper, convex, lsc function $\varphi : \mathbb{R}^d \to (-\infty, +\infty]$ is defined as

$$R_\varphi(x) = 2 \operatorname{prox}_\varphi(x) - x.$$

It is a nonexpansive operator on $\mathbb{R}^d$ with respect to the Euclidean norm. Given two proper, convex, lsc functions $\varphi, \psi : \mathbb{R}^d \to (-\infty, +\infty]$, the DR algorithm aims to solve

$$\operatorname{argmin}_{x\in\mathbb{R}^d}\{\varphi(x)+\psi(x)\}$$

by iterating a starting point $t^{(0)}$ as follows:

$$\begin{aligned}
\text{for} \quad & r=0,1,\ldots \text{ until a stopping criterion is reached}\\
& t^{(r+1)} := \left((1-\tau_r)\,I+\tau_r R_{\eta\varphi}R_{\eta\psi}\right)\left(t^{(r)}\right).\\
& x^{(r+1)} := \operatorname{prox}_{\eta\psi}\left(t^{(r+1)}\right).
\end{aligned}$$

Note that $x^{(r)}$ must be only computed in the final step of the algorithm. It is known that the DR algorithm converges if $\operatorname{ri}(\operatorname{dom}\varphi)\cap\operatorname{ri}(\operatorname{dom}\psi)\neq\emptyset$, a minimizer exists, $\eta>0$ and $\sum_{r\in\mathbb{N}}\tau_r(1-\tau_r)=+\infty$. The DR algorithm can be considered a special case of the Krasnoselski–Mann iteration

$$t^{(r+1)} = \left((1-\lambda_r)\,I+\lambda_r T\right)\left(t^{(r)}\right),$$

with $T := R_{\eta\varphi}R_{\eta\psi}$. It is well known that the sequence of iterates to a fixed point of T if T is nonexpansive. This is clearly the case for our setting since reflections R_φ are nonexpansive and the concatenation of nonexpansive functions is nonexpansive again.

For two points $x, a\in\mathcal{H}$ on a symmetric Hadamard manifold, the geodesic reflection (2) can be written as $R_p(x)=\exp_p(-\log_p x)$. Then, the geodesic reflection of a proper, convex, lsc function $\varphi:\mathcal{H}^n\to(-\infty,+\infty]$ is the mapping

$$R_\varphi(x)=\exp_{\operatorname{prox}_\varphi(x)}\left(-\log_{\operatorname{prox}_\varphi(x)}(x)\right).$$

Now in order to minimize

$$\operatorname*{argmin}_{x\in\mathcal{H}^d}\{\varphi(x)+\psi(x)\}$$

the DR algorithm can be generalized as follows:

$$\begin{aligned}
\text{for} \quad & r=0,1,\ldots \text{ until a stopping criterion is reached}\\
& t^{(r+1)} := \gamma_{t^{(r)},\widehat{s^{(r)}}}(\tau_r), \quad s^{(r)} := R_{\eta\varphi}R_{\eta\psi}\left(t^{(r)}\right),\\
& x^{(r+1)} := \operatorname{prox}_{\eta\psi}\left(t^{(r+1)}\right).
\end{aligned}$$

Again, the algorithm can be seen as special case of the Krasnoselski–Mann iteration

$$t^{(r+1)} := \gamma\left(t^{(r)},\ T\left(t^{(r)}\right);\ \tau_r\right).$$

It was proved in Kakavandi (2013), see also Bačák (2014a, Theorem 6.2.1) that such iteration converges to a fixed point of T, if T is nonexpansive, has a nonempty fixed point set and $\sum_{r\in\mathbb{N}}\tau_r(1-\tau_r)=+\infty$. Unfortunately, reflections at proper, convex, lsc functions on Hadamard manifolds are in general *not nonexpansive*. However, for the distance functions involved in our functionals nonexpansiveness is guaranteed by the following theorem.

Theorem 4 (Reflections at distance functions).
Let $a \in \mathcal{H}$ and φ be one of the functions $\mathrm{dist}^p(a, \cdot)$, $\mathrm{dist}^p(\cdot, \cdot)$, $p \in \{1, 2\}$. Then the geodesic reflection $R_{\eta\varphi}$, $\eta > 0$ is nonexpansive.

In general we have more than two summands, i.e., we are interested in the minimization of (11) where $\varphi_k : \mathcal{H}^n \to (-\infty, +\infty]$, $k = 1, \ldots, K$, are proper, convex, lsc functions. Here the trick is to rewrite the functional as the sum of two special components

$$\operatorname*{argmin}_{x \in \mathcal{H}^{nK}} \{\Phi(x) + \iota_{\mathsf{D}}(x)\},$$

where $\Phi(x) := \sum_{k=1}^{K} \varphi_k(x_k)$, $x := (x_k)_{k=1}^{K}$, and

$$\mathsf{D} := \{x \in \mathcal{H}^{nK} : x_1 = \cdots = x_n \in \mathcal{H}^n\}.$$

Obviously, D is a nonempty, closed convex set so that its indicator function is proper, convex and lsc, see Bačák (2014a, p. 37). Now, the DR algorithm can be formulated as

$$\begin{aligned}
\text{for} \quad & r = 0, 1, \ldots \text{ until a stopping criterion is reached} \\
& t^{(r+1)} := \gamma_{\overrightarrow{t^{(r)}}, \overrightarrow{s^{(r)}}}(\lambda_r), \quad s^{(r)} := R_{\eta\Phi} R_{\iota_{\mathsf{D}}}\left(\vec{t}^{(r)}\right), \\
& x^{(r+1)} := \Pi_{\mathsf{D}}(t^{(r+1)}).
\end{aligned}$$

Note that the second step is indeed only necessary in the final iteration. Concerning the last step note that

$$\Pi_{\mathsf{D}}(x) = \left(\operatorname{argmin}_{x \in \mathcal{H}^n} \sum_{k=1}^{K} \operatorname{dist}(x_k, x)^2, \ldots, \operatorname{argmin}_{x \in \mathcal{H}^n} \sum_{k=1}^{K} \operatorname{dist}(x_k, x)^2 \right) \in \mathcal{H}^{nK}.$$

The minimizer of the sum is the so-called Karcher mean, which can be efficiently computed on Hadamard manifolds using the gradient descent algorithm or the cyclic proximal point algorithm.

Concerning the convergence of the parallel DR algorithm, for our setting, $R_{\eta\Phi}$ is nonexpansive since it contains only geodesic reflections of the distance functions in Theorem 4. Unfortunately, in symmetric Hadamard manifolds, geodesic reflections corresponding to orthogonal projections onto convex sets are in general not nonexpansive. This is also true for our special set D. The situation changes if we consider manifolds with constant curvature κ. Here those reflections are nonexpansive, see Fernández-León and Nicolae (2013) and Persch (2018). So, in summary, although the parallel DR algorithm showed very good numerical performance on general Hadamard manifolds in Bergmann et al. (2016b), theoretical convergence results remain up to now limited to manifolds with constant nonpositive curvature.

5 Numerical examples

In this section, we give some illustrative numerical examples. The experiments are carried out using MATLAB® 2017a and the MVIRT toolbox (Bergmann, 2017).[a] As a quality measure we use the mean squared error (MSE) defined by

$$\epsilon := \frac{1}{|\mathcal{G}|}\sum_{i\in\mathcal{G}} \mathrm{dist}^2(u_i, u_{0,i}),$$

where u_0 denotes the original image. The parameters in the models were obtained via a grid search with respect to the optimal ϵ and can be found in detail in the respective papers (Bergmann et al., 2016a, 2018). In Figs. 3 and 4 we compare the performance of our variational models with different regularizers, where TV-TV2 denotes the additive coupling of TV and TV_2. For comparison we added the results obtained with the patched-based methods from Bergmann et al. (2017c), namely with nonlocal means

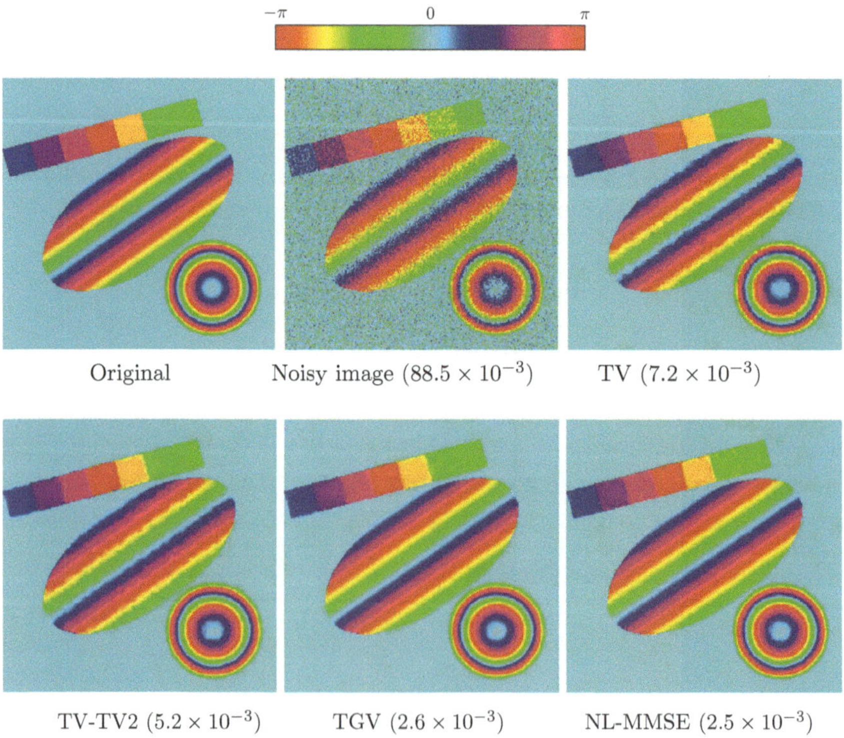

FIG. 3 Comparison of different variational models for an image with values on $\mathbb{S}^1$.

[a]Open source, available at https://ronnybergmann.net/mvirt/.

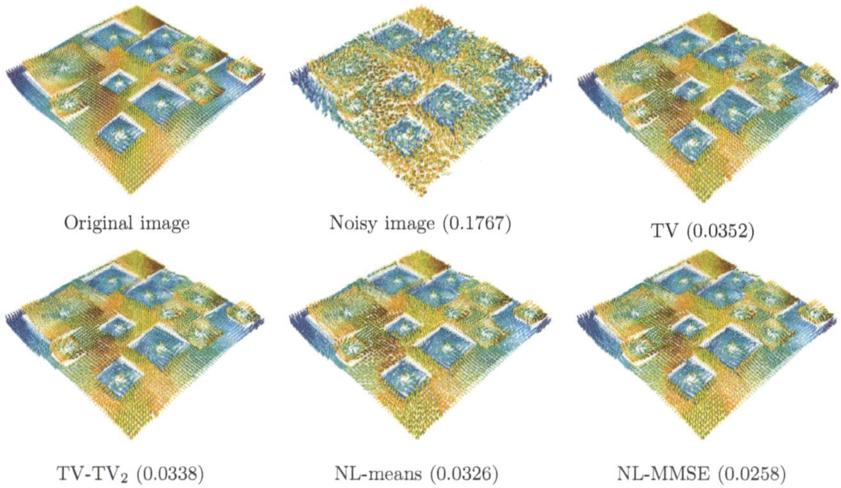

FIG. 4 Comparison of different variational models for an image with values on $\mathbb{S}^2$.

FIG. 5 Comparison of half-quadratic minimization applied to the anisotropic L_2-TV model with different functions φ_i, $i = 1, 2, 3$, on the chromaticity-brightness *colour model*, i.e., the manifold $\mathbb{R} \times \mathbb{S}^2$.

(NL-means) and nonlocal MMSE (NL-MMSE). In brackets we give the corresponding error value ϵ.

Figs. 5 and 6 show denoising results obtain by the half-quadratic minimization. Fig. 5 contains results for the different functions φ in Table 1 and for the nonsmooth TV regularizer. Here the parameters are optimized with respect to the PSNR of the RGB images and the numbers in brackets refer to the PSNR. In Fig. 6 we present a denoising result for the 3D DT-MRI image from Fig. 1.

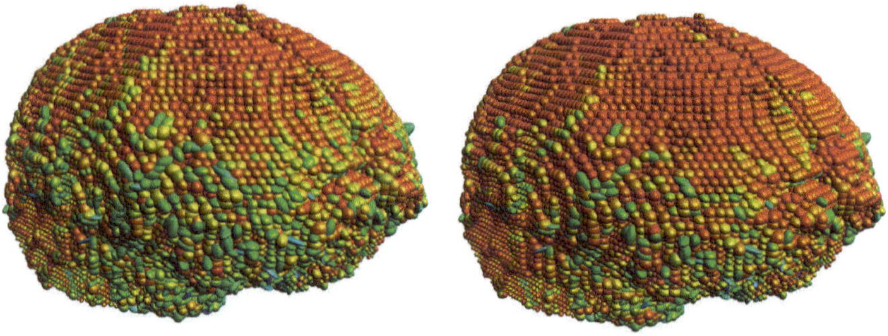

FIG. 6 Original "Camino" data set (*left*) and a denoised version by the smoothed TV regularizer with φ_1 (*right*).

6 Conclusions

The field of computational optimization on manifolds is an expanding area of research. Many problems in image and data processing can be modelled as optimization problems on finite or also infinite Riemannian manifolds, where the nonlinear structure of the underlying spaces better reflects problem setting. The absence of a vector space structure leads to new and significant challenges in the design, analysis and implementation of numerical algorithms. In particular, the monograph (Absil et al., 2008) led to a growing activity in the field, and recent developments as constrained optimization on manifolds and complexity bounds for Riemannian optimization (Boumal et al., 2019) as well as large number of real-world problems require further investigation.

References

Absil, P.A., Mahony, R., Sepulchre, R., 2008. Optimization Algorithms on Matrix Manifolds. Princeton University Press, Princeton.

Alouges, F., 1997. A new algorithm for computing liquid crystal stable configurations: the harmonic mapping case. SIAM J. Numer. Anal. 34 (5), 1708–1726.

Bačák, M., 2014a. Computing medians and means in Hadamard spaces. SIAM J. Optim. 24 (3), 1542–1566.

Bačák, M., 2014b. Convex Analysis and Optimization in Hadamard Spaces. De Gruyter Series in Nonlinear Analysis and Applications, vol. 22. De Gruyter, Berlin. https://doi.org/10.1515/9783110361629.

Bačák, M., Bergmann, R., Steidl, G., Weinmann, A., 2016. A second order non-smooth variational model for restoring manifold-valued images. SIAM J. Sci. Comput. 38 (1), A567–A597. https://doi.org/10.1137/15M101988X.

Bachmann, F., Hielscher, R., 2005-2016. MTEX - MATLAB toolbox for quantitative texture analysis. http:/mtex-toolbox.github.io/.

Bachmann, F., Hielscher, R., Schaeben, H., 2011. Grain detection from 2d and 3d EBSD data—specification of the MTEX algorithm. Ultramicroscopy 111 (12), 1720–1733.

Batard, T., Bertalmío, M., 2018. A geometric model of brightness perception and its application to color image correction. J. Math. Imaging Vision 60 (6), 849–881.

Bergmann, R., 2017. MVIRT, a toolbox for manifold-valued image restoration. In: IEEE International Conference on Image Processing, IEEE ICIP 2017, Beijing, China, September 17–20, 2017.

Bergmann, R., Weinmann, A., 2015. Inpainting of cyclic data using first and second order differences. In: Energy Minimization Methods in Computer Vision and Pattern Recognition. Springer, pp. 155–168.

Bergmann, R., Laus, F., Steidl, G., Weinmann, A., 2014. Second order differences of cyclic data and applications in variational denoising. SIAM J. Imag. Sci. 7 (4), 2916–2953. https://doi.org/10.1137/140969993.

Bergmann, R., Chan, R.H., Hielscher, R., Persch, J., Steidl, G., 2016a. Restoration of manifold-valued images by half-quadratic minimization. Inverse Prob. Imaging 10 (2), 281–304. https://doi.org/10.3934/ipi.2016001.

Bergmann, R., Persch, J., Steidl, G., 2016b. A parallel Douglas-Rachford algorithm for restoring images with values in symmetric Hadamard manifolds. SIAM J. Imaging Sci. 9 (3), 901–937. https://doi.org/10.1137/15M1052858.

Bergmann, R., Fitschen, J.H., Persch, J., Steidl, G., 2017a. Infimal convolution coupling of first and second order differences on manifold-valued images. In: Lauze, F., Dong, Y., Dahl, A.B. (Eds.), Scale Space and Variational Methods in Computer Vision 2017. Springer, Cham, pp. 447–459.

Bergmann, R., Fitschen, J.H., Persch, J., Steidl, G., 2017b. Iterative multiplicative filters for data labeling. Int. J. Comput. Vis. 123 (3), 435–453. https://doi.org/10.1007/s11263-017-0995-9.

Bergmann, R., Laus, F., Persch, J., Steidl, G., 2017c. Manifold-valued image processing. SIAM News 50 (8), 1–3. https:/sinews.siam.org/Details-Page/processing-manifold-valued-images.

Bergmann, R., Fitschen, J.H., Persch, J., Steidl, G., 2018. Priors with coupled first and second order differences for manifold-valued image processing. J. Math. Imaging Vision 60 (9), 1459–1481.

Boumal, N., Mishra, B., Absil, P.A., Sepulchre, R., 2014. Manopt, a MATLAB toolbox for optimization. J. Mach. Learn. Res. 15, 1455–1459.

Boumal, N., Absil, P.A., Cartis, C., 2019. Global rates of convergence for nonconvex optimization on manifolds. IMA J. Numer. Anal. 39 (1), 1–33.

Boyd, S., Parikh, N., Chu, E., Peleato, B., Eckstein, J., 2011. Distributed optimization and statistical learning via the alternating direction method of multipliers. Found. Trends Mach. Learning 3 (1), 101–122.

Bredies, K., Kunisch, K., Pock, T., 2010. Total generalized variation. SIAM J. Imaging Sci. 3 (3), 492–526. https://doi.org/10.1137/090769521.

Bredies, K., Holler, M., Storath, M., Weinmann, A., 2018. Total generalized variation for manifold-valued data. SIAM J. Imaging Sci. 11 (3), 1785–1848. https://doi.org/10.1137/17M1147597.

Burger, M., Sawatzky, A., Steidl, G., 2017. First order algorithms in variational image processing. In: Glowinski, R., Osher, S., Yin, W. (Eds.), Operator Splittings and Alternating Direction Methods. Springer.

Bürgmann, R., Rosen, P.A., Fielding, E.J., 2000. Synthetic aperture radar interferometry to measure earth's surface topography and its deformation. Annu. Rev. Earth Planet. Sci. 28 (1), 169–209. https://doi.org/10.1146/annurev.earth.28.1.169.

Chambolle, A., Lions, P.L., 1997. Image recovery via total variation minimization and related problems. Numer. Math. 76 (2), 167–188. https://doi.org/10.1007/s002110050258.

Ciak, R., Hirzmann, M., Scherzer, O., 2019. Regularization with metric double integrals of functions with values in a set of high-dimensional vectors. J. Math. Imaging Vision 61, 824–848.

Cook, P.A., Bai, Y., Nedjati-Gilani, S., Seunarine, K.K., Hall, M.G., Parker, G.J., Alexander, D.C., 2006. Camino: Open-source diffusion-MRI reconstruction and processing. In: 14th Scientific Meeting of the International Society for Magnetic Resonance in Medicine, p. 2759. Seattle, WA, USA.

Cremers, D., Strekalovskiy, E., 2013. Total cyclic variation and generalizations. J. Math. Imaging Vis. 47 (3), 258–277. https://doi.org/10.1007/s10851-012-0396-1.

do Carmo, M.P., 1992. Riemannian Geometry. vol. 115. Birkhäuser, Basel.

Ehlers, J., Pirani, F.A.E., Schild, A., 1972. The geometry of free fall and light propagation. In: O'Reifeartaigh, L. (Ed.), General Relativitiy. Oxford University Press, pp. 63–84.

Fernández-León, A., Nicolae, A., 2013. Averaged alternating reflections in geodesic spaces. J. Math. Anal. Appl. 402 (2), 558–566. https://doi.org/10.1016/j.jmaa.2013.01.060.

Ferreira, O.P., Oliveira, P.R., 1998. Subgradient algorithm on Riemannian manifolds. J. Optim. Theory Appl. 97 (1), 93–104. https://doi.org/10.1023/A:1022675100677.

Ferreira, O.P., Oliveira, P.R., 2002. Proximal point algorithm on Riemannian manifolds. Optimization 51 (2), 257–270.

Geman, D., Reynolds, G., 1992. Constrained restoration and the recovery of discontinuities. IEEE Trans. Pattern Anal. Mach. Intell. 14 (3), 367–383.

Geman, D., Yang, C., 1995. Nonlinear image recovery with half-quadratic regularization. IEEE Trans. Image Process. 4 (7), 932–946.

Giaquinta, M., Mucci, D., 2007. Maps of bounded variation with values into a manifold: total variation and relaxed energy. Pure Appl. Math. Q. 3 (2), 513–538.

Giaquinta, M., Modica, G., Souček, J., 1993. Variational problems for maps of bounded variation with values in S^1. Calc. Var. 1 (1), 87–121.

Grohs, P., Sprecher, M., 2016. Total variation regularization on Riemannian manifolds by iteratively reweighted minimization. Inf. Inference A J. IMA 5 (4), 353–378. https://doi.org/10.1093/imaiai/iaw011.

Hadani, R., Singer, A., 2011. Representation theoretic patterns in three dimensional Cryo-Electronmicroscopy I: the intrinsic reconstitution algorithm. Ann. Math. 174 (2), 1219–1241.

Helliwell, C.A., Anderssen, R.S., Robertson, M., Finnegan, E.J., 2015. How is FLC repression initiated by cold? Trends Plant Sci. 20, 76–82.

Jost, J., 1997. Nonpositive Curvature: Geometric and Analytic Aspects. Lectures in Mathematics ETH Zürich, Birkhäuser Verlag, Basel.

Kakavandi, B., 2013. Weak topologies in complete CAT(0) metric spaces. Proc. Am. Math. Soc. 141 (3), 1029–1039.

Kheyfets, A., Miller, W.A., Newton, G.A., 2000. Schild's ladder parallel transport procedure for an arbitrary connection. Int. J. Theor. Phys. 39 (12), 2891–2898. https://doi.org/10.1023/A:1026473418439.

Kimmel, R., Sochen, N., 2002. Orientation diffusion or how to comb a porcupine. J. Vis. Commun. Image Represent. 13 (1), 238–248. https://doi.org/10.1006/jvci.2001.0501.

Laus, F., Nikolova, M., Persch, J., Steidl, G., 2017. A nonlocal denoising algorithm for manifold-valued images using second order statistics. SIAM J. Imaging Sci. 10 (1), 416–448. https://doi.org/10.1137/16M1087114.

Lebrun, M., Buades, A., Morel, J.M., 2013. A nonlocal Bayesian image denoising algorithm. SIAM J. Imaging Sci. 6 (3), 1665–1688. https://doi.org/10.1137/120874989.

Lellmann, J., Strekalovskiy, E., Koetter, S., Cremers, D., 2013. Total variation regularization for functions with values in a manifold. IEEE International Conference on Computer Vision, pp. 2944–2951.

Li, C., Mordukhovich, B.S., Wang, J., Yao, J.C., 2011. Weak sharp minima on Riemannian manifolds. SIAM J. Optim. 21 (4), 1523–1560.

Lorenzi, M., Pennec, X., 2014. Efficient parallel transport of deformations in time series of images: from Schild's to pole ladder. J. Math. Imaging Vision 50 (1), 5–17. https://doi.org/10.1007/s10851-013-0470-3.

Muñoz-Moreno, E., Cárdenes-Almeida, R., Martín-Fernández, M., 2009. Review of techniques for registration of diffusion tensor imaging. In: Aja-Fernández, S., de Luis García, R., Tao, D., Li, X. (Eds.), Tensors in Image Processing and Computer Vision. Advances in Pattern Recognition, Springer, London.

Neumayer, S., Persch, J., Steidl, G., 2018. Morphing of manifold-valued images inspired by discrete geodesics in image spaces. SIAM J. Imaging Sci. 11 (3), 1898–1930.

Nikolova, M., Ng, M.K., 2005. Analysis of half-quadratic minimization methods for signal and image recovery. SIAM J. Sci. Comput. 27 (3), 937–966.

Pennec, X., 2018. Parallel transport with pole ladder: a third order scheme in affine connection spaces which is exact in affine symmetric spaces. ArXiv Preprint 1805.1143.

Persch, J., 2018. Recent Advances in Denoising of Manifold-Valued Images. TU Kaiserslautern.

Ring, W., Wirth, B., 2012. Optimization methods on Riemannian manifolds and their application to shape space. SIAM J. Optim. 22 (2), 596–627.

Rocca, F., Prati, C., Guarnieri, A.M., 1997. Possibilities and limits of SAR interferometry, vol. 407. ESA SP 15–26.

Rosman, G., Bronstein, M., Bronstein, A., Wolf, A., Kimmel, R., 2012. Group-valued regularization framework for motion segmentation of dynamic non-rigid shapes. Scale Space and Variational Methods in Computer Vision, Springer, pp. 725–736.

Rosman, G., Tai, X.-C., Kimmel, R., Bruckstein, A.M., 2014. Augmented-Lagrangian regularization of matrix-valued maps. Methods Appl. Anal. 21 (1), 121–138. https://doi.org/10.4310/MAA.2014.v21.n1.a5.

Rudin, L.I., Osher, S., Fatemi, E., 1992. Nonlinear total variation based noise removal algorithms. Phys. D 60 (1), 259–268.

Singer, A., Shkolnisky, Y., 2012. Viewing direction estimation in Cryo-EM using synchronization. SIAM J. Imaging Sci. 5 (3), 1088–1110.

Tuzel, O., Porikli, F., Meer, P., 2006. Region covariance: a fast descriptor for detection and classification. In: European Conference on Computer Vision, pp. 589–600.

Tuzel, O., Porikli, F., Meer, P., 2008. Learning on Lie groups for invariant detection and tracking. CVPR 2008. IEEE, pp. 1–8.

Udrişte, C., 1994. Convex Functions and Optimization Methods on Riemannian Manifolds. Mathematics and its Applications, vol. 297. Kluwer Academic Publishers Group, Dordrecht.

Valkonen, T., Bredies, K., Knoll, F., 2013. Total generalized variation in diffusion tensor imaging. SIAM J. Imaging Sci. 6 (1), 487–525. https://doi.org/10.1137/120867172.

Wang, X.M., 2018. Subgradient algorithms on Riemannian manifolds of lower bounded curvatures. Optimization 67 (1), 179–194. https://doi.org/10.1080/02331934.2017.1387548.

Wang, L., Singer, A., Wen, Z., 2013. Orientation determination of Cryo-EM images using least square deviations. SIAM J. Imaging Sci. 6 (4), 2450–2483.

Wang, X.M., Li, C., Yao, J.C., 2015. Subgradient projection algorithms for convex feasibility on Riemannian manifolds with lower bounded curvatures. J. Optim. Theory Appl. 164 (1), 202–217. https://doi.org/10.1007/s10957-014-0568-9.

Weinmann, A., Demaret, L., Storath, M., 2014. Total variation regularization for manifold-valued data. SIAM J. Imaging Sci. 7 (4), 2226–2257. https://doi.org/10.1137/130951075.

Yair, O., Ben-Chen, M., Talmon, R., 2018. Parallel transport on the cone manifold of SPD matrices for domain adaptation. ArXiv Preprint 1807.10479.

Zhang, P., Niethammer, M., Shen, D., Yapa, P.T., 2018. Large deformation diffeomorphic registration of diffusion-weighted imaging data. Med. Image Anal. 18 (8), 1290–1298.

Chapter 15

Image and surface registration

Ke Chen[a,*], Lok Ming Lui[b] and Jan Modersitzki[c]
[a]Department of Mathematical Sciences, University of Liverpool, Liverpool, United Kingdom
[b]Department of Mathematics, Chinese University of Hong Kong, Sha Tin, Hong Kong, China
[c]Institute of Mathematics and Image Computing, University of Lübeck, Lübeck, Germany
[*]Corresponding author: e-mail: k.chen@liverpool.ac.uk

Chapter Outline

Abstract

Image registration is one of the most useful and practical applications of image analysis. Among its many tasks are the tracking of changes between data from different time points or motion correction. Moreover, superimposing complementary information across image modalities is needed, as new imaging modalities emerge in the field. This chapter presents a review of the fundamental ideas and models in the field. The goal is to reflect the current state of the art of image registration (IR) to motivate the readers to refine these models, and to enable the tackling of new challenges as they arise. After

Handbook of Numerical Analysis, Vol. 20. https://doi.org/10.1016/bs.hna.2019.07.001
© 2019 Elsevier B.V. All rights reserved.

discussing the background (Section 2), we review main components of a variational model: a distance measure or data fidelity term (Section 3), regularization to ensure existence of solutions and constraints to further restrict the wanted transformation (Section 4). We also discuss diffeomorphic approaches which ensure local invertibility. We present the surface registration (SR) modelling in the same framework of variational models in Section 5 where the close relationship between IR and SR is also discussed, while we briefly discuss the numerical methods for IR and SR in Section 6. Finally in Section 7, we touch upon the main ideas in deep learning-based approaches for registration.

Keywords: Image registration, Surface registration, Variational models

AMS Classification Codes: 35A15, 37E30, 49J40, 49Q05, 62H35, 62M40, 65D18, 80M30

1 Introduction

Registration is the process of automatically establishing correspondences between geometrical data, such as images, surfaces or point clouds. Typically, registration aligns a pair, a stack or a sequence of such data. Of particular interest is the registration of two images. Often, one of these images is considered to be fixed, stationary or a reference and is denoted by R and the other one is considered a moving, floating or template image denoted by T. The necessity of registration can be found in diverse fields of sciences and engineering, including remote sensing, computer graphics, computer vision and medical imaging; see, e.g., Brown (1992), Sotiras et al. (2013), Glasbey and Mardia (1998), Maintz and Viergever (1998), Hill et al. (2001), Modersitzki (2009), Zitová and Flusser (2003), Scherzer (2008) and references therein. For example, in medical imaging, finding an accurate correspondence between biomedical images is crucial for statistical shape analysis of the anatomical structures. In computer graphics, a surface registration satisfying user-defined correspondences of landmark features is needed for the constrained texture mapping. In computer vision, image registration offers a crucial tool in combining, merging or fusing multiple data sources from multimodality images such as infrared images and satellite images. Despite its importance, the registration problem inhere many mathematical challenges: for various applications appropriate models are missing and the modelling is unclear, variational models are nonconvex and solutions nonunique (ill-posedness); see Weickert and Schnörr (2001), Droske and Rumpf (2004), Burger et al. (2013) and among others. Very often, a suitable choice of the regularization and a proper registration model is nontrivial to make. Robustness of a model remains a major challenge.

Different registration models have been recently developed. In terms of the type of information to be matched, existing approaches can mainly be divided into three categories, namely, feature-based registration, intensity-based registration and hybrid registration using both features and intensity information. Examples of such features are localized points a.k.a. landmarks, structures such as vessels or segmentations of organs.

Particularly landmark-based schemes are popular in computer sciences. One of the main advantages of landmarks-based methods is the straightforward incorporation of user-interaction during the registration. This provides an intuitive and user-validated assumption to achieve a meaningful deformation approximation. From the numerical point of view, a landmarks-based method is more computationally efficient when modelling large deformations, since more (initial) information about the deformation are provided through feature correspondences. The downside is that landmarks are generally difficult to determine, particularly in 3D medical data. Inaccurate, incorrect or noncorresponding landmarks may degrade the registration results considerably. Moreover, for many applications the number of required landmarks is unknown but registration results may depend heavily on this number. Intensity-based registration aims to match corresponding data without reducing the data to features such as landmarks. Registration is usually obtained by matching intensity functions, such as the image intensity for image registration or the surface curvature for surface geometric registration. The main advantage of the intensity-based registration is that more image information is taken into account and the derivation of features is not required. This is quite desirable in delivering an automatic algorithm. However, the lack of human supervision may cause inaccuracy in the registration result especially in cases where landmarks can be confidently identified with known knowledge. Hybrid registration that combines landmarks-based and intensity-based methods has gained increased attention. Hybrid approaches use both the landmarks and intensity information to guide the registration. This type of approaches can provide excellent initial guesses; see, e.g., Fischer and Modersitzki (2003) and Haber et al. (2009) for extended discussion.

A generic mathematical formulation of image registration (IR) takes a variational form. Here, a joint energy is to be minimized on a set of suitable transformations. This energy typically consists of an application conforming image distance or data fidelity, a regularization, and potentially additional penalty terms. The data fidelity term measures data mismatching, such as the landmark mismatching error or intensity mismatching error. The regularization term usually aims to enhance the smoothness of the registration map and may occasionally require an additional term to impose a geometric constraint on the map invertibility. An extra penalty might be added to emphasize particular aspects such as volume preservation or point to point correspondences. This term may either be phrased as a penalty or a hard constraint. We note that there also exists a connection to a Bayesian formulation of the registration problem. There, the probability of a transformation given the two images is maximized. However, a thorough discussion is beyond the scope of this chapter.

The regularization puts a bias on the set of transformation. Therefore, for different applications, different classes of regularizers have been proposed. For example, one of the common classes of mappings is the space of diffeomorphisms. Despite its computational costs, diffeomorphic registration has become more popular in recent years, especially in biomedical image registration. For many biomedical applications, a one-to-one transformation is a reasonable. Diffeomorphic mappings provide this feature but may result in a very smooth transformation fields. While this can be beneficial for some applications, it might be too restrictive for others. An example is the registration of lung images, as the lung slides along the rib cage. Other options to ensure bijectivity is to include a penalty or constraint on the Jacobian of the transformation. Other commonly used classes of transformations include rigid or affine transformations, conformal, quasi-conformal mappings as well as isometric mappings.

The ingredients of the variational model will be reviewed and discussed in this chapter, mostly for a pair of mono-modality images. However, the framework can be also be used for the registration of multimodal images. Here, the fidelity term has to be replaced appropriately. In fact, fusion of images with different modalities has challenged the medical imaging field very rapidly due to the presence of highly accessible patients' information in recent years. For example, cross platform nonrigid registration of CT with MRI images has found a significant role in different clinical application. There are many complementary modalities, such as MR images vs CT images, ultra-sound images vs PET images, CT images vs optical images, infrared images vs digital images and so on. In some instances labelling of anatomical features by medical experts is also involved to further improve the robustness, accuracy and authenticity of the registration. To tackle these challenges, various multimodality image registration models have been proposed. From a mathematical perspective, it is the same variational framework as for mono-modal images, simply with a different the data fidelity or distance measure. Note, however, that multimodal image registration is by far more complex than mono-modal registration. In practical applications issues such as prealignment, choice of parameters, local minima and robustness become much more severe.

The topic of surface registration (SR) is closely connected to IR. Though each problem class has own distinct features, there are common modelling ideas. In this chapter, we discuss both the similarities, commonalities as well as the differences. Some methods from one class can lend themselves to the other class; even mixed models are possible.

Both the theoretical and numerical analysis of IR and SR are very challenging because registration problems are typically nonconvex, the joint energy may have arbitrarily many global minimizers. Numerical approaches can easily end up in getting local minimizers and a proper initial guess can play a significant role. For example, the curvature regularizer discussed in Section 4 has an infinite dimensional kernel of harmonic functions. Moreover, not all models have been supplied with an existence theory. As a result, discretized functionals and related linear systems may not be positive definite and registration thus provides a rich source of challenging problems to design efficient and converging algorithms for. Note that current medical CT images can result in 512^3 voxels and a registration ends up in a nonconvex optimization problem with about $0.5 \cdot 10^9$ unknowns; we expect the registered results preferably in real time.

Finally we touch upon the artificial intelligence aspect of IR. In the era of big data, various learning-based approaches are being developed. Following the success of deep learning techniques in various imaging tasks, many recent studies have been carried out to apply deep learning techniques to improve various registration models. In particular, deep learning techniques have been applied to build prediction models of spatial transformations for achieving image registration under supervised learning framework. Landmark features can also be learnt for guiding the registration. Of course, the quality of the training data plays an important role for the accuracy of the obtained registration result. Exciting approaches also investigate to speedup optimization either by learning gradients or transformation manifolds.

In this chapter, our aim is to give a survey of different registration models and to motivate new research works in this challenging and yet existing field. Various variational models, statistical models and learning-based models will be explored.

2 Mathematical background

The goal of image registration is to find a suitable map y that maps a template T to a reference image R, such that $T(y)$ and R are aligned, that is corresponding points are placed at the same position. We also aim to provide a mathematical framework that is capable to cover as many image registration approaches as possible. This section will introduce this framework, modelling ideas and notation but leaving details to Sections 3–7.

2.1 Continuous and discrete images

From reading this volume of the handbook, the reader should be clear that mathematical imaging can offer powerful modelling tools (namely the variational framework) mainly because we can assume that the input images T, R are continuous. Then the full power of functional analysis can be applied to imaging and it is natural to define and discuss geometry (such as gradients, curvature, H_1 norm) of T, R. Also, a continuous interpretation of image data is much more convenient if geometrical transformations are to be applied.

In the continuous interpretation that is used in this chapter, a generic image is defined as a mapping $I : \Omega \rightarrow G$. Here, $\Omega \subset \mathbb{R}^d$ denotes the image domain and d is the image dimensionality, where typically Ω is a one-dimensional interval with $d = 1$, and a square with $d = 2$ or a cube with $d = 3$. The image range is denoted by G and may indicate binary images $G = \{0,1\}$,, grey scale images $G = \{0, \ldots, 255\}$, real images $G = \mathbb{R}$, or multispectral images such as colour images, mass spectroscopy, or tensors. In this chapter, we focus on real images mainly with $d = 2$. However, the discussed models extend to other images.

In applications, the image data I is typically given as a d-array of size $m_1 \times \cdots \times m_d$, which may be naturally viewed as sampled from a continuous image function I. Interpolation or approximation techniques must be used to embed the data into a function space; see, e.g., Modersitzki (2004) and Chen (2013). In this sense, the notation $I(y)$ implies that an interpolation must be implemented (unless y takes integer values only). This aspect is unique in IR and SR, different from other imaging problems. However, this chapter emphasizes the "existence" of a continuous image I (via interpretation) that is assumed available for discrete images.

If the continuous image I needs to be visualized, the domain Ω is partitioned into a number n of cells with cell-centres $x_j = (x_j^1, \ldots, x_j^d)$, and the discrete image $[I(x_j), j = 1, \ldots, n]$ is displayed. In imaging, a pixelized interpretation of image data is often used for processing; see Fig. 1 for an illustration.

Finally we remark that another popularly used notation for the deformed template $T(y)$ is $T \circ y$, where the transform y may also be written as $y = x + u(x)$ later when we apply regularization to the deformation field u.

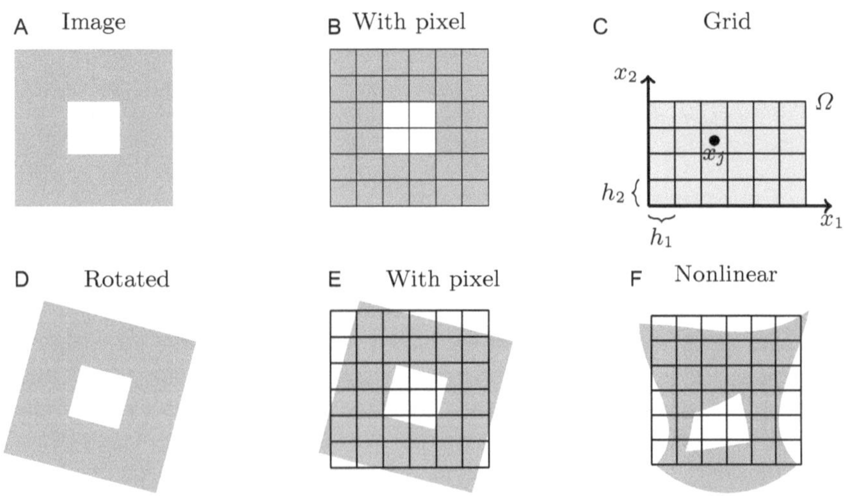

FIG. 1 An image (A) overlaid with a pixel grid (B), a rotated version of the image (D) overlaid with a pixel grid (E); a nonlinearly deformed version of the image overlaid with a pixel grid (F). Part (C) displays a partitioned image domain Ω with 6-by-4 cells with cell-centres $x_j = (x_j^1, x_j^2)$, $j = 1, \ldots, n = 24$.

2.2 A mathematical framework for image registration

As common for ill-posed problems, we use a variational framework that aims to minimize a joint energy functional constituting a data fidelity term or an image distance measure and a regularizer. Details on distance measures $\mathcal{D}$ and the regularizer $\mathcal{R}$ are provided in Sections 3 and 4.

The objective is to determine the wanted transformation $y : \mathbb{R}^d \to \mathbb{R}^d$ as a minimizer of a joint energy $\mathcal{J}$ over a feasible set of transformation $\mathcal{A}$, i.e.

$$\mathcal{J} : \mathcal{A} \to \mathbb{R}, \quad \mathcal{J}(y) := \mathcal{D}(T \circ y, R) + \mathcal{R}(y), \tag{1}$$

where the data fidelity term $\mathcal{D}$, the regularization term $\mathcal{R}$, and the feasible set $\mathcal{A}$ are discussed below.

We remark that, though the approach applies to both, mono- and multi-modal images, we focus on mono-modality for ease of presentation. We also remark that is often convenient to consider the deformation u rather than the transformation y, where the deformation or displacement denotes the change, $y(x) = x + u(x)$. Note, that in contrast to the differences of transformation and deformation the notations transformed image and deformed image are used synonymously in the literature.

3 Distance measures

An important piece of the puzzle in model (1) is the similarity or dissimilarity $\mathcal{D}$ of two given images. There are essentially two approaches. The first approach is to embed the data into a function space or a space of densities and to define similarity via norms or metrics in these spaces. The second approach is based on a projection of data onto a feature space and to base the similarity on feature similarity.

3.1 Volumetric differences

Probably the most intuitive volumetric distance measure is the energy of the difference image a.k.a. the sum of squared differences (SSD),

$$\mathcal{D}^{\text{SSD}}(T, R) := \|T - R\|_{L_2}^2 = \int_\Omega (T(x) - R(x))^2 \; dx$$

for a pair of mono-modality images. The underlying idea is that if image intensities match, i.e., $T(y(x)) \approx R(x)$, then the images correspond. Obviously, the difference can be replaced by any other meaningful residual function $r(T, R)$. Also, the L_2-norm $|r|^2$ may be replaced by another metric $\psi(r)$, and the integration may also include a weighted measure $d\omega$. We mention two particular options to replace the L_2-norm by an L_1-norm and a differential Huber-norm (Huber, 1964), respectively

$$|r|_{h,\varepsilon} = \sqrt{|r|^2 + \varepsilon^2} \quad \text{or} \quad |r|_{H,\varepsilon} = \begin{cases} 0.5|r|^2, & |r| \le \varepsilon, \\ \varepsilon(|r| - 0.5\varepsilon), & \text{otherwise.} \end{cases} \tag{2}$$

The least squares distance (LSD) approach of Heldmann (2010) relates the template to a projected version of the reference image. More precisely, using a transfer function $\gamma : \mathbb{R} \to \mathbb{R}$ similarly to a gamma-correction, the residual is phrased as $r(T, \gamma \circ R)$. Determining the optimal colour map γ might then be part of the overall optimization.

A weight function $\omega : \mathbb{R}^d \to \mathbb{R}$ might be use to either emphasize or deemphasize certain areas. For example, if $\Sigma \subset \Omega$ denotes a segmentation of a critical organ, then a weight $\omega := \chi_\Sigma$ would ignore points that are not in Σ. Here, χ_Σ denotes the characteristic function,

$$\chi_\Sigma = \begin{cases} 1, & x \in \Sigma, \\ 0, & x \notin \Sigma. \end{cases}$$

Note that ω is not necessarily restricted to $\{0, 1\}$ and may also be interpreted as a certainty or confidence. The weight may also depend on image intensities. For example, $\omega(x; T, R) := \exp(-(T(x) - R(x))^2)$ is suggested in Drobny et al. (2015) and Drobny (2015). Here, $\omega \approx 1$ in areas in which $T(x) \approx R(x)$, whereas $\omega \approx 0$ where intensity differences are huge. This weight therefore also yields an implicit soft segmentation into corresponding and noncorresponding structures. This concept can be beneficial if one images displays structures that are not contained in the other, e.g., tumours or air bubbles in the intestines; see also Chen et al. (2015) for similar strategies.

The above remarks also outline limitations of SSD-based distances. For multimodal images, i.e., images that are acquired with different sensors, it may holds that $T(x) \neq R(x)$ even if the images are perfectly aligned. A nasty example is illustrated in Fig. 2, mimicking the signal "coffee" sensed by nose (smell, smooth reference, grey) and by eye (vision, characteristic function for of the cup, black). Since the intensity of the smell does not relate to the size of

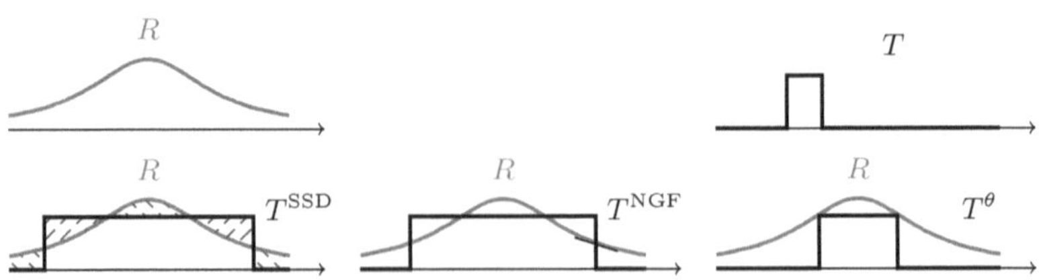

FIG. 2 "Coffee Problem:" Multimodal 1D registration problem aligning the smell of a cup of coffee (*top row left*: smooth reference R, *grey*) with the visual impression (*top row right*: characteristic function, template T, *solid black*). *Bottom row* shows solutions of L_2 minimization (*left*), correlation of maximum slope (*centre*), and thresholded reference; note that the scale of smell and vision may not relate.

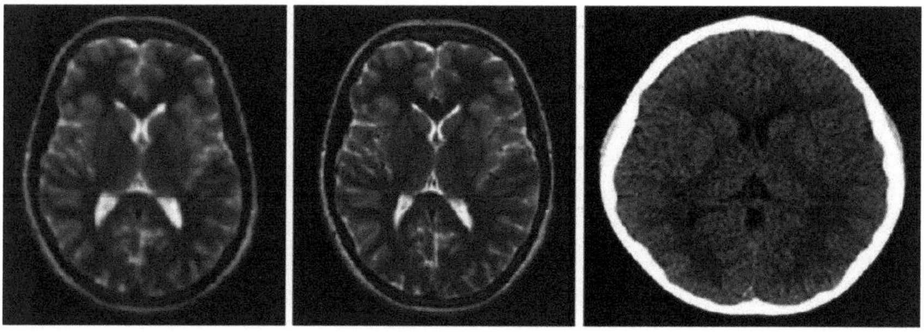

FIG. 3 Illustration of MR (*left, template*), CT data (*right, reference*) and the registered MR (*middle, T(y)*). The result is from the NGF model (Theljani and Chen, 2019) with $\alpha = 1/4$ and $N = 256$.

the cup, a minimization of $\|T \circ y - R\|_{L_2}$ is meaningless; see Fig. 2 for particular solutions. Several approaches for multimodal distance measures have been suggested. We note that to our best knowledge, none of these approaches is able to solve the "coffee problem".

The fusion of images with different modalities is a very important application field in image registration. For example, cross platform nonrigid registration of CT with MRI images has found a significant role in different clinical applications. Fig. 3 displays a typical example where the CT image shows structural (bone) and anatomical information and the MR image shows details on soft tissues. Registration or fusion of the two modalities thus locates complementary information in a normalized geometry.

In Haber and Modersitzki (2006a) it is suggested to use the collinearity of normalized gradient fields (NGF) of the images as a residual. More precisely, $r(x) := 1 - (g(T; x)^\top g(R, x))^2$, where $g(I; x) := \nabla I(x)/|\nabla I(x)|_{h,\varepsilon}$. The motivation is, that even if intensities of T and R do not match, intensity changes occur at corresponding positions. This assumption holds for many applications where structural changes appear at corresponding spatial positions such as the example in Fig. 3. However, it does not hold for the "coffee" example in Fig. 2.

The smoothing of the gradient fields is crucial. In the most likely advent of noise, the normalized gradient field is uninformative. The parameter ε is therefore introduced to discriminate between noise $|\nabla I(x)| \leq \varepsilon$ and signal change, i.e., $|\nabla I(x)| > \varepsilon$. We note that a proper embedding of the image data into a smooth function space as outlined in Section 2.1 can improve the performance significantly. Variants of NGF include to use individual noise parameters ε_T and ε_R. Additionally, one can work with the L_2 inner product for embedded gradients $\nabla_\varepsilon I := (\nabla I^\top, \varepsilon)^\top \in \mathbb{R}^{d+1}$. Exploration of the different variants is a topic of current research. For example, a recent improvement over NGF can be found in Theljani and Chen (2019) where a different measure of NGF is proposed.

Another commonly used similarity measure for multimodal images is based on the mutual information of the data (Collignon et al., 1995; Viola and Wells, 1997). Further approaches for multimodal image registration are based on so-called Wasserstein metrics (Oskarsson, 2015) or normalized cross correlation (Chelbi and Mekhmoukh, 2018; Sarvaiya et al., 2009).

In some applications such as histological serial sectioning, a sequence of $r > 2$ images I_j, $j = 1, \ldots, r$ needs not be registered. An obvious extension of the above framework is based on sequential concatenation of standard registration problems for two consecutive images I_{j-1} and I_j,

$$\mathcal{J}(y_1,\ldots,y_r) = \sum_{j=2}^{r} \{\mathcal{D}(I_{j-1} \circ y_{j-1}, I_j \circ y_j) + \mathcal{R}(y_j)\}, \quad y_1(x) = x.$$

This process can be very time consuming as the transformations y_j depend on the whole image sequence. Schatten-q-norms based approaches provide a promising alternative as it naturally combines all images of the sequence. Here, a Schatten-q-(quasi) norm Watrous (2018) of a data matrix A is defined as the q-norm of the vector containing the singular values of A: with a singular value decomposition $A = U\mathrm{diag}(\sigma)V^\top$ it holds $\|A\|_{S,q} := \|\sigma\|_q$. For more details and choices of the data matrix we refer to Arganda-Carreras et al. (2010), Nejadasl and Lindenbergh (2014), Tektonidis and Rohr (2017) and Brehmer et al. (2018).

3.2 Feature-based differences

A simple example for feature-based similarity is the location of markers, landmarks or keypoints. Markers are prominent spatial locations that have been attached to an object before imaging. Landmarks are usually outstanding points such as tips of fingers or eye centres that can be identified after imaging. Independent of whether these points have been identified a priori or a posteriori, the idea then is to determine the image transformation by matching the features. The transformation can be computed from an interpolation or approximation approach. To be more precise, let $(t_\ell)_{\ell=1}^L$ and $(r_\ell)_{\ell=1}^L$ denote two lists of corresponding landmarks, $t_\ell, r_\ell \in \mathbb{R}^d$, for the template and reference image, respectively. The goal is to determine the transformation y such that ideally $y(r_\ell) = t_\ell$ for $\ell = 1, \ldots, L$.

In case of a parametric approach, i.e., $y \in \mathcal{A}$ is parametrizable (by parameters $\{\alpha_j\}$), a least squares approach for the parameter α_j is typically used. Here, $\mathcal{A}$ denotes the set of feasible transformation, for example, rotations as in Fig. 1. More precisely, the optimal parameters are computed by minimizing the landmark distance $\mathcal{D}^{\mathrm{LM}}$ over the set $\mathcal{A}$, i.e.,

$$\min_{\alpha_1,\ldots,\alpha_N} \mathcal{D}^{\mathrm{LM}}(y) := \sum \|y(r_\ell) - t_\ell\|, \quad y \in \mathcal{A}. \tag{3}$$

Obviously, there is a huge degree of freedom in tuning the residuals. Mahalanobis distances (Mahalanobis, 1936) can be used if the location error is not isotropically distributed, weights can be incorporated to emphasize particular correspondences or to address further uncertainties. For a linear space $\mathcal{A}$, a solution can by obtained by solving the linear system $w = Q(r)^{\dagger}t$, where r and t denote the collection of landmarks, $Q(r)$ is the Vandermonde-matrix and superscript $\dagger$ denotes the pseudo-inverse; see Modersitzki (2004) for details.

Other approaches are to introduce a smoothness measure or regularizer $\mathcal{R}$ and to optimize with respect to the data fidelity, the regularization or a weighted compromise. For example, the thin-plate bending energy

$$\mathcal{R}^{\mathrm{TPS}}(y) := \sum_j \int_\Omega (\partial_{1,1}y^j)^2 + 2(\partial_{1,2}y^j)^2 + (\partial_{2,2}y^j)^2 \ dx \tag{4}$$

is used in thin-plate-spline registration (Rueckert et al., 1999). Variants of this techniques are based on different optimization approaches for the transformation:

1. $\mathcal{R}(y) \overset{!}{=} \min$ subject to $\mathcal{D}(y) \le \sigma$,
2. $\mathcal{D}(y) \overset{!}{=} \min$ subject to $\mathcal{R}(y) \le \sigma$,
3. $(1-\sigma)\mathcal{D}(y) + \sigma\mathcal{R}(y) \overset{!}{=} \min$,

where σ denotes a noise level.

For a variety of smoothness operators $\mathcal{R}$, solutions can be computed explicitly via radial basis functions or fundamental solutions ρ. For example, for $\mathcal{R}^{\mathrm{TPS}}$ as in (4) the radial basis function and thus the transformation of the first approach are given by

$$\rho(r) = \begin{cases} r^2 \log r, & d=2 \\ r, & d=1 \end{cases}, \quad y^j(x) = \sum_{k=1}^{L} c_k^j \rho(\|x - r_j\|) + w_0^j + \sum_{k=1}^{d} w_k^j x_k^j,$$

where the coefficients can be computed by solving the KKT systems

$$\begin{pmatrix} A & B \\ B^\top & 0 \end{pmatrix} \begin{pmatrix} c^j \\ w^j \end{pmatrix} = \begin{pmatrix} t^j \\ 0 \end{pmatrix}, \quad j = 1, \ldots, d,$$

with $A = [\rho(\|r_j - r_k\|)]_{j,k} \in \mathbb{R}^{L,L}$, $B = [1, r_k^1, \ldots, r_k^d]_k \in \mathbb{R}^{L,d+1}$ and $t^j = (t_1^j, \ldots, t_L^j)^\top \in \mathbb{R}^L$; see Modersitzki (2004) for details and discussions.

A major advantage of these approaches is that explicit solutions are available. If the number of landmarks is not too large (say less than 10,000), the above KKT system can generally be solved with reasonable effort and accuracy. Another advantage of landmark-based method is the straightforward incorporation of expertise knowledge during the registration process. This provides an intuitive and user-validated assumption to achieve a meaningful deformation approximation.

However, this approach also has a number of disadvantages: the impact of the number of landmarks and the distribution of landmarks is difficult to control. In practice, the determination of landmarks is cumbersome, difficult and erroneous. Also, when both the number of landmarks and the deformation of landmarks are large, parametric approaches to obtain landmark-matching registration usually fail to obtain a bijective deformation and overlaps can usually be observed in the obtained registration (unless extra constraint is imposed at the cost of not matching the landmarks accurately). To overcome these issues, various nonparametric approaches based on solving optimization problems have been proposed, such as the large deformation diffeomorphic metric matching (LDDMM) method (Beg et al., 2005; Joshi and Miller, 2000) and the quasi-conformal landmark registration (QCLR) method (Lam and Lui, 2014; Lui et al., 2012).

4 Regularization

As already outlined, image registration is an ill-posed problem and hence regularization becomes inevitable. In the registration context, the main goal of regularization is to ensure the existence of solutions. Uniqueness is a critical topic. For example, if one wants to register a square to a square, rotations of multiples of 90 degrees results in four meaningful global solutions. From an application point of view it can by questionable to declare one of these solutions to be best even if it makes the mathematical problem well-posed. The problem gets more exciting if one considers a circle rather than a square. We remark, that in some approaches the ambiguities are removed by adding boundary conditions, which may not comply the application demands. A smaller issue is related to the fact that many regularizers are based on derivatives and are thus blind with respect to translations. In practice, this is rarely a problem, since either the distance measure adds information, or alternative solutions may also be meaningful.

We remark that a regularizer is often imposed on the deformation rather than the underlying transform, expressed as a function of the increment, $\mathcal{R}(y-y_{\text{ref}})$, where the reference transformation y_{ref} may be obtained from a preregistration, an educated guess, or simply by setting $y_{\text{ref}}(x) = x$. Using this construction, it is also common to use zero Dirichlet boundary conditions for the update $u := y - y_{\text{ref}}$ (though Euler–Lagrange equations tend to impose Neumann's type boundary conditions). The boundary conditions result from the variational formulation and should not be phrased artificially.

4.1 Regularization by ansatz-spaces, parametric registration

In many applications the transformation is restricted to a certain set $\mathcal{A}$, such that the wanted transformation can be expressed as a finite combination of basis functions and parameters $w \in \mathbb{R}^p$. We call this class parametric registration schemes. Examples are the set of rigid transformations, the space of affine transformation or the space of spline transformations (Rueckert et al., 1999),

$$y^{\text{rigid}}(x,w) = \text{rotation}(w)\cdot x + \text{translation}(w),$$
$$y^{\text{affine}}(x,w) = \text{affine-map}(w)\cdot x + \text{translation}(w),$$
$$y_k^{\text{spline}}(x,w) = \sum w_j^k \cdot \text{spline}_j(x), \quad k = 1,\ldots,d.$$

The registration problem is to find the optimal $w \in \mathbb{R}^p$, where p denotes the degrees of freedom. In the 2D case for example, one may consider

$$y^{\text{rigid}}(x,w) = \begin{pmatrix} \cos w_1 & -\sin w_1 \\ \sin w_1 & \cos w_1 \end{pmatrix}\begin{pmatrix} x_1 \\ x_2 \end{pmatrix} + \begin{pmatrix} w_2 \\ w_3 \end{pmatrix}, \quad w \in \mathbb{R}^3 \text{ or}$$
$$y^{\text{affine}}(x,w) = \begin{pmatrix} w_1 & w_2 \\ w_4 & w_5 \end{pmatrix}\begin{pmatrix} x_1 \\ x_2 \end{pmatrix} + \begin{pmatrix} w_3 \\ w_6 \end{pmatrix}, \quad w \in \mathbb{R}^6.$$

In terms of basis functions, for the latter case, we can define them precisely: $\phi_1 = x_1 e_1$, $\phi_2 = x_2 e_1$, $\phi_3 = e_1$, $\phi_4 = x_1 e_2$, $\phi_5 = x_2 e_2$, $\phi_6 = e_2$. Then we have the equivalent form

$$y^{\text{affine}}(x,w) = \sum_{j=1}^{6} w_j \phi_j(x).$$

In many publications, regularization for the parametric case is often neglected, particularly for low degrees of freedom. Note, however, that registration is an ill-posed problem even in the parametric case and thus should be regularized; see also the above square or disk example. Particularly for the affine linear transformation $y^{\text{affine}}(x, w) = (I + A)x + b$ and the diffusion regularizer leads to $\mathcal{R}^{\text{diff}}(y^{\text{affine}}) = \|A\|_{\text{Fro}}^2$; see also Modersitzki (2009) and Chumchob and Chen (2009)

4.2 Quadratic regularizer

A huge class of regularizers is based on quadratic forms on the displacement u,

$$\mathcal{R}(y) = \int_\Omega (\mathcal{B}u)^\top (\mathcal{B}u)\ dx,$$

where $y(x) = x + u(x)$, $\mathcal{B}$ may be chosen differently but the Euler–Lagrange equations will contain the operator $\mathcal{B}^*\mathcal{B}$.

4.2.1 *Diffusion regularizer*

The classical optical flow regularizer (Horn and Schunck, 1981)

$$\mathcal{R}^{\text{diff}}(y) := \sum_j \int_\Omega \|\nabla u_j\|^2\ dx, \quad \mathcal{B}^*\mathcal{B} = -\Delta,$$

is used in registration as it enables fast algorithms due to its decoupled nature; see, e.g., Fischer and Modersitzki (2002), Bruhn et al. (2003), Haber and Modersitzki (2006b) and Chumchob and Chen (2011). However, the decoupling may also lead to unnatural and unpleasing transformations.

4.2.2 Elastic regularizer

A physically motivated extension of the diffusion regularizer is the elastic potential

$$\mathcal{R}^{\text{elas}}(y) := \int_\Omega \sum_{j,k} \frac{\mu}{4}(\partial_j u_k + \partial_k u_j)^2 + \frac{\lambda}{2}(\operatorname{div} u)^2 \; dx, \quad \mathcal{B}^*\mathcal{B} = -(\mu\Delta + (\lambda+\mu)\nabla \operatorname{div}),$$

with Lamé-constants λ, μ; see Fischler and Elschlager (1973), Broit (1981), Fischer and Modersitzki (1999) and Modersitzki (2004) for details on the elastic potential. Here, image features are considered to act like elastic bodies. We note that the results are often very pleasing, if displacements u are small, $|\partial_j u_k| \ll 1$.

4.2.3 Curvature regularizer

For some applications, smooth transformations might be desirable and a second-order regularizer can thus be beneficial. The linear curvature regularizer

$$\mathcal{R}^{\text{curv}}(y) := \sum_j \int_\Omega \Delta u_j \; dx, \quad \mathcal{B}^*\mathcal{B} = \Delta^2,$$

has been suggested in Fischer and Modersitzki (2003); for implementations see also Henn and Witsch (2005). This regularizer is of particular interest for registrations where volumetric distance measures and landmark conditions are combined,

$$\min \mathcal{J}(y) \quad \text{subject to} \quad \mathcal{D}^{\text{LM}}(y) = 0;$$

see Fischer and Modersitzki (2003) for details.

4.3 Nonquadratic regularizer

A disadvantage of the previous regularizers is that the regularization energy is always finite, even for unwanted transformation. Thus, an unwanted and nonregular transformation might be a minimizer if one reduce the distance measure sufficiently by adjusting the parameter. We therefore also discuss nonlinear, nonquadratic and nonconvex regularizers.

4.3.1 Mean curvature regularizer

An extension of the above curvature regularizer is to use the mean curvature as regularizer

$$\mathcal{R}^{\text{mean curv}}(y) := \sum_j \int_\Omega \kappa(u_j)^2 \; dx, \;\; \kappa(u_j) := \nabla \cdot \frac{\nabla u_j}{|\nabla u_j|}, \tag{5}$$

see Chumchob et al. (2011). The extended version may even be coupled by replacing κ with $\hat{\kappa}(u_j) = \nabla \cdot (\nabla u_j / |\nabla u|)$ in (5); see Zhang et al. (2017) for details.

4.3.2 Gaussian curvature regularizer

Gaussian curvature was investigated in Ibrahim et al. (2014). For 2D, κ in (5) is replaced by

$$\kappa_G(u_j) = \frac{u_{j;1,2}u_{j;2,1} - u_{j;1,1}u_{j;2,2}}{(u_{j;1}^2 + u_{j;2}^2 + 1)^2}.$$

It is fair to remark that both curvature models are nontrivial to solve efficiently; a related model could be built based on the elastica regulariser but again efficient solution is a topic of current research (Khan et al., 2018).

4.3.3 Fractional derivative based regularizer

Fractional derivatives are increasingly used in imaging to take advantage of nonlocal behaviour. We briefly review (Zhang and Chen, 2015), suggesting

$$\mathcal{R}^{\text{frac}}(y) := \sum_j \int_\Omega \|\nabla^\alpha u_j\|_2^2 \; dx, \;\; \nabla^\alpha u_j = \left(\partial_1^\alpha u_j, \ldots \partial_d^\alpha u_j\right)^\top, \tag{6}$$

$\partial_k^\alpha u_j$ denoting the fractional α-order derivative (Podlubny, 1998), $1 \leq \alpha \leq 2$ and the particular choice $\alpha = 1.6$. The fractional α-order derivative be based on the Riemann–Liouville, Grünwald–Letnikov or Caputo definition. For homogeneous zero boundaries condition and suitable smoothness assumption on u, all three definitions are equivalent.

We remark in passing that fractional derivatives, though nonlocal, generate structured matrices (Toeplitz form) so existing fast solvers can be utilized (Breiten et al., 2016).

4.3.4 Hyperelastic regularizer

We finally highlight the hyper elastic regularizer as proposed in Droske and Rumpf (2004) and Burger et al. (2013), to overcome the limitations of a quadratic regularizer and to also allow for large deformation:

$$\mathcal{R}^{\text{hyper}}(y) := \int_\Omega \alpha_l \text{length}(y) + \alpha_s \text{surface}(y) + \alpha_v \text{volume}(y) \; dx, \tag{7}$$

where $C := \text{cof}\nabla y$ denotes the cofactor matrix and $d := \det(\nabla y)$ the determinant of the matrix $F := \nabla y$, respectively. Here,

$$\begin{aligned} \text{length}(y) &= \psi_l(\nabla y), & \psi_l(F) &= \|F - \text{Id}\|_2^2, \\ \text{surface}(y) &= \psi_c(\text{cof}\nabla y), & \psi_c(C) &= \max\{\|C\|_2^2 - 3, 0\}^2, \\ \text{volume}(y) &= \psi_d(\det(\nabla y)); & \psi_d(d) &= ((d-1)^2/d)^2. \end{aligned}$$

For details, see Burger et al. (2013). In particular the penalty on the determinant ensures finite energies only for transformation with $\det \nabla y > 0$ and therefore enforces one-to-one transformations. Note that the penalty C also introduces a bias towards the identity transformation.

4.4 Registration penalties and constraints

The set of feasible transformation $\mathcal{A}$ can be used to restrict the wanted transformation. For example, by setting $\mathcal{A} = \{y: \mathcal{P}(y) = 0\}$, where the penalty could be a landmark match $\mathcal{P} = \mathcal{D}^{\text{LM}}(y)$ (cf. (3)), the restriction y being rigid on certain areas, or the determinant of the Jacobian to be positive.

In many applications, these restrictions or so-called hard constraints are replaced by so-called soft constraints or penalties. Here, the objective in (1) is augmented by a penalty term,

$$\mathcal{J}: \mathcal{A} \to \mathbb{R}, \quad \mathcal{J}(y) := \mathcal{D}(T \circ y, R) + \mathcal{R}(y) + \mathcal{P}(y). \tag{8}$$

We note, that the penalty approach comes with a number of drawbacks. Firstly, for a minimizer y^* of (8) generally $\mathcal{P}(y^*) \neq 0$. Secondly, even if $\mathcal{P}(y^*) \ll 1$, it might be big on small but potentially critical structures such as tumours or lessons. Thirdly, the penalty usually involves a weighting parameter that has to be tuned for practical applications. And fourthly, typically huge values are used for weighting, which then lead to ill-conditioned optimization problems. Despite its various conceptual disadvantages, penalty approach are very common in the registration community as they are generally easy to be implemented.

4.5 Penalties for locally invertible maps

The mapping $y = y(x) = x + u(x)$ is invertible in a neighbourhood of a point x, if the determinant of the Jacobian $d(x) := \det(\nabla y(x))$ is nonzero and the inverse function theorem applies. In order to also preserve orientation, one may constrain the transformation such that $d(x) > 0$. We remark that this constraint is mathematically challenging as the determinant may leave L_2 and solutions of the registration problem may not be evaluated pointwise; see, e.g., Haber and Modersitzki (2004) and Haber et al. (2010) and the extended discussion in Burger et al. (2013). In Haber and Modersitzki (2004), the poitnwise constraint $\mathcal{C}(y) = d(x) = 1$ has been introduced and extended to $k(x) \leq d(x) \leq K(x)$ in Haber and Modersitzki (2007). Here, the constraint applies to all points $x \in \Sigma \subset \Omega$, where Σ could be the whole domain or a subset indication critical organs or tumours; upper and lower bounds are denoted by k and K, respectively.

A commonly used idea in image registration is to add a penalty to the joint energy. In Rohlfing et al. (2003), the following penalty on the determinant of the Jacobian d is introduced

$$\mathcal{P}(y) = \int (\log(d))^2 \; dx,$$

Yanovsky et al. (2007) considered the penalty

$$\mathcal{P}(y) = \int |d-1| \cdot \log|d| \; dx.$$

Both approaches favour transformations with $d \approx 1$.

We remark that the hyperelastic approach of Burger et al. (2013) might be interpreted along the same lines as the volume term is

$$\text{volume}(y) = \int ((d-1)^2/d)^2 \; dx.$$

Note, however, that Burger et al. (2013) introduced a regularizer, i.e., the existence of a minimizing element with $d > 0$ is guaranteed, but the theory also requires the length and in particular the cofactor term; see Burger et al. (2013) for details and extended discussion.

The 2D case is simpler. In Zhang and Chen (2018), a penalty based on the Beltrami coefficient is proposed:

$$\mathcal{P}(y) = \int \frac{|\mu|^2}{(|\mu|^2-1)^2} \; dx, \quad |\mu|^2 = \frac{\| \nabla y \|_2^2 - 2\det(\nabla y)}{\| \nabla y \|_2^2 + 2\det(\nabla y)}. \tag{9}$$

However, the determinant is not easily accessible and it adds strong nonlinearity to a model. For $d = 3$ for example, the determinant is a polynomial of derivatives of degree 3 and thus may not even be in L_2. Furthermore, the determinate being nonnegative on certain grid points does not imply that the determinant is nonnegative everywhere. The algorithm from Burger et al. (2013) resolves these issues using a computational framework with finite elements.

4.6 Diffeomorphic registration

A diffeomorphic transform y is invertible, differentiable and smooth function (that offers one-to-one mapping) and its inverse is also smooth. For applications such as brain mapping and registration of MR images, diffeomorphic registration models are particularly suitable.

The above section discussed one-to-one maps and how to achieve them. Diffeomorphic registration models offer more than locally invertible maps. Here we review two such models.

In the first diffeomorphic registration model of large deformation metric mappings (LDMM) (Beg et al., 2005; Younes, 2010), an additional time component is introduced. Let $K := [0, 1]$ be a normalized time interval, $\Omega_K := \Omega \times K$, $I: \Omega_K \to \mathbb{R}$ and $y: \Omega_K \to \mathbb{R}^d$. For brevity, we write $z_t := z(\cdot, t)$ for functions depending on space and time. Moreover, we introduce the velocity v such that $\partial_t y_t(x) =: v(y_t(x))$. The objective is then to find a minimizing element (y, v) of an energy

$$\mathcal{J}^{\mathrm{LDDMM}}(y,v) := \mathcal{D}(T,R) + \mathcal{R}^{\mathrm{LDDMM}}(v), \tag{10}$$

constrained for all $(x, t) \in \Omega_K$ by

$$\begin{aligned} I_0(x) &= I(x,0) = R(x), & I_1(x) &= I(x,1) = T(x), \\ y_0(x) &= y(x,0) = x, & \partial_t y_t(x) &= v(y_t(x)). \end{aligned}$$

In the second diffeomorphic registration model of Lam and Lui (2014), the key idea is to look for y in the space of quasi-conformal maps, using

$$\mathcal{R}(u) = \int_\Omega |\nabla \mu|^2 dx, \quad \mathcal{C}(u) = \int_\Omega |\mu|^p dx + \mathcal{P}(y)$$

where $\mu = \mu(y)$ is the Beltrami coefficient (9), explicitly represented by $y = (y_1, y_2)$. More specifically, if one write $y = y_1 + \sqrt{-1} y_2$ in the complex form, then $\mu(y)$ is defined as $mu(y) = \frac{\partial y}{\partial \bar{z}} / \frac{\partial y}{\partial z}$, where $\frac{\partial}{\partial \bar{z}} = \frac{1}{2}(\frac{\partial}{\partial x} + \sqrt{-1}\frac{\partial}{\partial y})$ and $\frac{\partial}{\partial z} = \frac{1}{2}(\frac{\partial}{\partial x} - \sqrt{-1}\frac{\partial}{\partial y})$. $\mathcal{P}$ denotes the landmarks constraint. Note that the one-to-one mapping property can be seen from the equivalence of $\det(\nabla y) > 0$ and $|\mu| < 1$.

4.7 Registration by inverse consistent approach

This framework is used to enforce a directionally unbiased registration. Note that the optimal y for the images T and R may not relate to the optimal z for R and T, order matters. This is, among others, due to the fact that the distance as well as the regularizer are assigned to the reference image domain. A remedy suggested in Christensen and Johnson (2001) is to symmetrize the problem, i.e., to register R to T, as well as T to R, at the same time. Introducing the transformation $z : \mathbb{R}^d \to \mathbb{R}^d$, such that ideally $z = y^{-1}$, the objective now is to minimize

$$\mathcal{J}(y,z) = \mathcal{D}(T \circ y, R) + \mathcal{D}(R \circ z, T) + \mathcal{R}(y) + \mathcal{R}(z) + \mathcal{E}(y,z),$$

with

$$\mathcal{E}(y,z) = \|y \circ z^{-1} - \mathrm{id}\| + \|z \circ y^{-1} - \mathrm{id}\|;$$

for details, see, e.g., Christensen and Johnson (2001), Chen and Ye (2010), Ashburner (2007) and Thompson and Chen (2019).

5 Surface registration

In this section, we discuss state-of-the-art approaches for surface registration (SR); see, e.g., Lam and Lui (2014), Wong et al. (2012), Lui et al. (2010a), Ovsjanikov et al. (2016), Bronstein et al. (2010), Shi et al. (2008), Liang et al. (2012) and references therein.

SR is a special class of registration problem. Instead of warping between two images, SR aims to find a mapping between two curvilinear surfaces; see also Fig. 4. More precisely, the goal of SR is to establish a meaningful correspondences between two Riemann surfaces which are denoted by S^k, $k = 0, 1$. The correspondence should match the surface geometry as well as the data defined on the surfaces. As outlined in Section 3, SR can be interpreted as a special case of feature-based image registration, where surface are derived from the data by some kind of projection. However, a major difference between SR and the conventional IR is that surfaces have an intrinsic geometry. In IR, the domains of images are Euclidean spaces, whereas in SR, the domains of interest are Riemann surfaces embedded in $\mathbb{R}^d$. In this chapter we focus on $d = 3$ for ease of presentation. As such, the surface geometry is usually taken into consideration in SR. This is the reason why SR is usually also called shape matching (Bronstein et al., 2010).

5.1 Brief introduction to surface geometry

In practice, the surfaces to be registered often have different geometry. In SR, one main goal is to match two surfaces based on their geometry. For instance, in brain surface registration, the sulci and gyri on each brain surfaces should match with each other. To achieve this goal, the underlying idea of SR is that the similarity of two surfaces S^k can by measured by utilizing the associate Riemannian metrics g^k, $k = 0, 1$. More specifically, geometric distortion between S^0 and S^1 can be measured by comparing the so-called pull-back metric $f^*(g^1)$ with the original metric g^0.

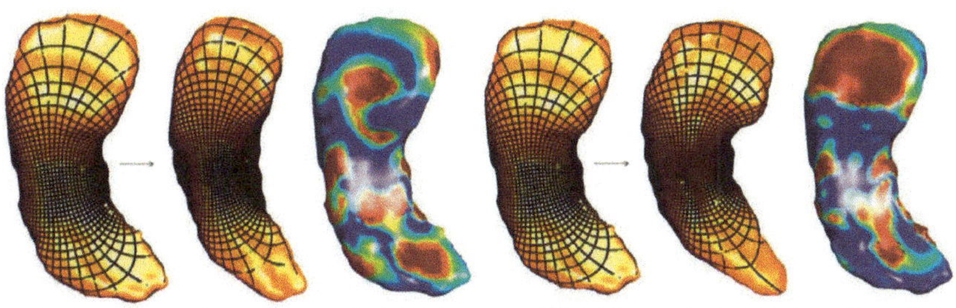

FIG. 4 Examples of surface registration between hippocampal surfaces for medical morphometry.

A Riemann surface is an arbitrarily smooth 2D manifold S equipped with an inner product g_p on the tangent space T_pS for each point $p \in S$; see, e.g., Lee (1997). The inner product g_p is also called a Riemannian metric and varies smoothly amongst points on S. More specifically, a Riemannian metric on S is a collection of inner product

$$g_p : T_pS \times T_pS \to \mathbb{R}, \quad p \in S,$$

such that for vector fields X and Y defined on S, the mapping $p \mapsto g_p(X(p), Y(p))$ is arbitrarily smooth.

The pull-back metric maps the metric of S^1 back to S^0. To this end let $f : S^0 \to S^1$ be one-to-one and differentiable. For any given point $p \in S^0$ and any given $u \in T_pS^0$, let γ be a smooth curve on S^0 with $\gamma(0) = p$ and $\gamma'(0) = u$. The differential df_p is defined by $df_p(u) := (f \circ \gamma)'(0)$ and the pull-back metric by

$$f_p^*(g^1) : T_pS^0 \times T_pS^0 \to \mathbb{R}, \quad f_p^*(g^1)(u,v) := g_p^1(df_p(u), df_p(v)).$$

Intuitively, a pull-back metric takes the inner product of two tangent vectors u and v of S^0 being pushed forward by f to two tangent vectors on S^1. The inner product differs from g^0 by the distortion introduced by f and the geometric distortion between the surfaces S^0 and S^1 can be measured by comparing the pull-back metric with the original metric g^0.

Over the last decades, different approaches for SR have been proposed. Below we will summarize some popular approaches from the literatures; see, e.g., Lam and Lui (2010), Wong et al. (2012), Bronstein et al. (2010), Liang et al. (2012). Specific examples include the isometric maps (Shi et al., 2008), the conformal maps (Lui et al., 2010a), and the quasi-conformal maps (Lam and Lui, 2014; Lui et al., 2012), which we discuss in more detail.

A map f is called isometric if $f^*(g^1) = g^0$. An isometric map captures shape deformations that preserve length, such as human poses. A map is called conformal if $f^*(g^1) = \lambda g^0$, where λ is the so-called conformal factor, a nonnegative smooth function $\lambda : S^0 \to \mathbb{R}_{>0}$. A conformal map preserves the local geometry under the map, although the local area is usually distorted. It captures deformations that does not change the local shape, such as biological deformations.

A map is called quasi-conformal, if $f^*(g^1) = |\frac{df}{dz}|^2 |dz + \mu d\bar{z}|^2$, where z is the local coordinate of S^0 and μ is a complex-valued function capturing the conformality distortion. It captures more general deformations with bounded amount of local geometric distortions. Depending on various applications, SR aims to look for optimal map in a suitable class of mappings that satisfies other prescribed constraints, such as the landmark constraints and curvature-matching condition.

5.2 Parameterization-based approaches

A commonly used approach for SR is based on the global surface parameterization (Lui et al., 2010a). The basic idea is to flatten the surface to a simple domain, such as a 2D unit disk or a 2D rectangle; see also Fig. 5. For instance, a surface can be interpreted as a 2D image, whose image intensity is given by some geometric quantities, if it is parameterized onto a 2D rectangle. SR can then be considered as finding an optimal mapping between two images that matches the corresponding geometric quantities, such as the curvatures. This problem can be solved by any registration approaches discussed in the previous sections. More specifically, suppose the surfaces S^0 and S^1 to be registered are parameterized onto the domain $\Omega \subset \mathbb{R}^2$. Denote the parameterizations of S^0 and S^1 by $\psi_0 : \Omega \to S^0$ and $\psi_1 : \Omega \to S^1$, respectively. Suppose the mean curvatures H_0 and H_1 of S^0 and S^1 are to be matched. The SR problem is reduced to an IR problem to match two 2D images given by $I^0 := H_0 \circ \psi_0 : \Omega \to \mathbb{R}$ and $I^1 := H_1 \circ \psi_1 : \Omega \to \mathbb{R}$. In other words, we aim for a transformation $T : \Omega \to \mathbb{R}^2$ such that

$$I^1 \circ T(x) = I^0(x) \text{ for all } x \in \Omega.$$

Sometimes, the constraints on landmark correspondences can be enforced. Suppose p_i on S^0 should be matched to q_i on S^1 for $i = 1, 2, \ldots, m$. Then, the transformation T should further satisfy the condition that $T\psi_0^{-1}(p_i) = \psi_1^{-1}(q_i)$ for $i = 1, 2, \ldots, m$. Similar to IR, a regularization on T is usually imposed to constrain the type of transformation. For instance, a common choice is to look for an optimal conformal parameterization that matches landmarks and the surface geometry (Lui et al., 2010a). Once an optimal T is obtained, the registration map between S^0 and S^1 can be obtained by the composition map $\psi_1 \circ T \circ \psi_0^{-1} : S^0 \to S^1$. Of course, the choice of parameterization is crucial in these approaches and may introduce distortions such as conformality distortion or metric distortion. To avoid theses distortions, conformal or isometric surface parameterizations are often used; cf., e.g., Lui et al. (2010a, 2012).

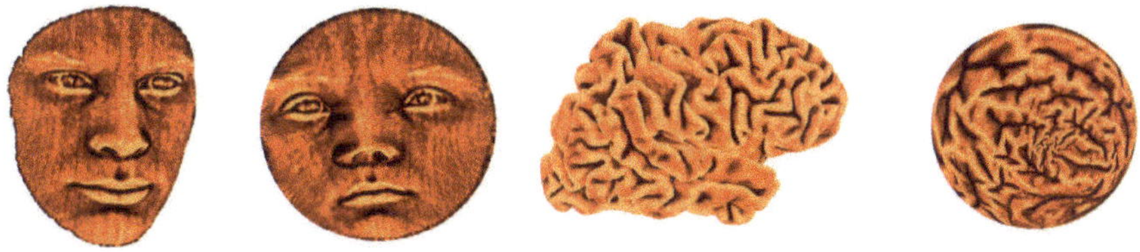

FIG. 5 Examples of surface conformal parameterizations of a human face (*left*) and a brain cortical surface onto a unit disk (*right*), respectively.

5.3 Laplace–Beltrami eigenmap approaches

In these approaches, one takes advantage of an eigensystem of the Laplace–Beltrami operator Δ_S on the surface S (Liang et al., 2012; Shi et al., 2008; Solomentsev and Shikin, 1994). Let (S, g) be a Riemann surface in $\mathbb{R}^3$, $g = (g_{i,j}) \in \mathbb{R}^{2,2}$ be the Riemann metric and $(g^{i,j})$ its inverse. For a smooth function $f: S \to \mathbb{R}$, the Laplace–Beltrami (L-B) operator Δ_S acting on f near a point p is defined by

$$\Delta_S f := (\det g)^{-\frac{1}{2}} \sum_{i,j=1}^{2} \partial_i [\det(g) \;\; g^{i,j} \partial_j f].$$

As the L-B operator Δ_S is self-adjoint and elliptic, it has a system of eigenvalue and corresponding eigenfunctions (λ_j, ϕ_j) with $-\Delta_S \phi_j = \lambda_j \phi_j$ and $\lambda_j \leq \lambda_{j+1}$ for all $j \in \mathbb{N}$. This L-B eigensystem can be used to define a shape signature for a surface up to isometry. Two isometric surfaces have the same L-B eigensystems (Shi et al., 2008). SR strategies have been investigated to match L-B eigensystems as good as possible for surfaces undergoing isometric deformations; see, e.g., Rustamov (2007).

The L-B approach is especially effective when registering shapes with different poses. For instance, for every point $p \in S$, the L-B eigensystem gives rise to a m-dimensional feature vector $s_m(p) := (\lambda_j^{-1/2} \phi_j(p))_{j=1}^{m}$, where the number m is user supplied. In a discrete scenario, we assume that the surfaces S^k are represented by a collection of vertices $\{p_j^k\}_{j=1}^{n}$, $k = 0, 1$. An L-B signature of S^k is then the $n \times m$ matrix B^k, where the jth row of B^k is given by $s_m^k(p_j^k)$. A match of the surfaces can be obtained by computing an optimal correspondence matrix C such that $||B^1 C - B^0||_{\text{Fro}}$ is minimized.

5.4 Metric approaches

In these approaches, a nonrigid shape is modelled as a metric space (X, dX), where X is a two-dimensional smooth compact connected and complete Riemannian surface (possibly with boundary) embedded into $\mathbb{R}^3$, and $dX: X \times X \to \mathbb{R}$ is a metric measuring distances between pairs of points on X. Two shapes (X, dX) and (Y, dY) are similar if the metrics between pairs of corresponding points on X and Y coincide, i.e., there exists a bijective map $\phi: X \to Y$ such that $dY \circ (\phi \times \phi) = dX$ (Bronstein et al., 2010). To simplify the computation of the shape correspondence ϕ, a low dimensional representation is often considered, by means of a minimum-distortion embedding $\phi: X \to Z$. The low dimensional representation is usually called the canonical form, which can be computed by solving the multidimensional scaling problem (Bronstein et al., 2010)

$$\min_{\phi: X \to Z} \max_{x, x' \in X} ||d_X(x, x') - d_Z(\phi(x), \phi(x'))||$$

where d_Z is the metric of Z. Once the embeddings $\phi(X)$ and $\psi(X)$ of X and Y are computed, an optimal correspondence can be obtained by minimizing the Hausdorff distance between $\phi(X)$ and $\psi(X)$. It is generally impossible to select a common metric space in which the geometry of any shape can be accurately represented. To solve this problem, the Gromov–Hausdorff distance, instead of the Hausdorff distance, is considered:

$$d_{GH}(X,Y) := \inf_{\phi:X\to Z,\psi:Y\to Z,Z} d_H^Z(\phi(X),\psi(Y)).$$

5.5 Functional map approaches

Another popular SR approach is based on a functional map representation (Ovsjanikov et al., 2016). For $k = 0, 1$, we consider the spaces $\mathcal{F}^k := \mathcal{F}(S^k, \mathbb{R})$ of mappings from S^k to $\mathbb{R}$. Then any one-to-one transformation $y : S^0 \to S^1$ defines the linear mapping $y_{\mathcal{F}} : \mathcal{F}^0 \to \mathcal{F}^1, \psi^0 \mapsto \psi^1 := \psi^0 \circ y^{-1}$.

For a given $y_{\mathcal{F}}$, the underlying transformation y can be reconstructed using point evaluation functionals δ_x (Ovsjanikov et al., 2016): for any $x^0 \in S^0$, $y_{\mathcal{F}}(\delta_{x_0}) =: \delta_{x^1}$ establishes the connection between $x^0 \in S^0$ and $x^1 \in S^1$.

Suppose now that for $k = 0, 1$, $\{b_j^k\}_{j=1}^{\infty}$ is a basis of $\mathcal{F}^k$. Thus we have the following representations, $f = \sum \zeta_\mu b_\mu^0$, $y_{\mathcal{F}}(b_\mu^0) = \sum \gamma_{\nu,\mu} b_\nu^1$, and

$$y_{\mathcal{F}}(f) = \sum \beta_\nu b_\nu^1 = y_{\mathcal{F}}\left(\sum \zeta_\mu b_\mu^0\right) = \sum \sum \gamma_{\nu,\mu} \zeta_\mu b_\nu^1.$$

With the column vectors (β_ν) and (ζ_μ) that are formed by the coefficients and the matrix representation $C = (\gamma_{\nu,\,\mu})$ of $y_{\mathcal{F}}$, we have $(\beta_\nu) = C(\zeta_\mu)$. With this setup, SR can be computed by different choices of basis functions as well as prescribing different conditions on $y_{\mathcal{F}}$.

For example, to obtain an isometric SR, one can choose $\{\phi_j^k\}_{j=1}^{\infty}$ to be the Laplace–Beltrami eigenfunctions on S^k and choose an optimal C that is closest to a diagonal matrix. To further enforce the landmark constraints $y_{\mathcal{F}}(\delta_{p_\ell^0}) = \delta_{p_\ell^1}$. Functional map method reduces the SR problem to optimizing a matrix C.

5.6 Relationship between SR and IR

SR and IR are closely related to each other. Many ideas from image registration can be applied to surface registration. In fact, both IR and SR can be formulated as an optimization problem over the space of admissible registration maps as follows:

$$E_{reg}(f) = \mathcal{D}(I_1, I_2 \circ f) + \mathcal{R}(f) \tag{11}$$

where $\mathcal{D}$ is the distance measure and $\mathcal{R}$ is a suitable regularization term. Common distance measures in IR are usually chosen as the L^1 or L^2 error between image intensities. These distance measures are often used in SR. The only challenge in SR, which is different from IR, is that meaningful "intensities" on surfaces are not readily defined. One important question in SR is to design suitable geometric quantities, such as curvatures, conformality distortions, Laplace–Beltrami eigenfunctions, to drive the registration process and obtain a geometric matching surface registration map.

In recent years, various "meaningful" intensities based on the surface geometry has been proposed to define the dissimilarity measure before surfaces. Another connection is the regularization terms in IR and SR. Most of the regularization terms applied in IR can be generalized to SR. Examples include the total variation (TV), harmonic energy and so on. The only difference is that on surfaces, instead of using the standard Euclidean differentiation, covariant derivatives defined according to the Riemannian metric have to be utilized. In other words, the problem of surface registration is indeed a combination of ideas from IR with Riemannian geometry.

More specifically, the connection between a surface S and an image I might be interpreted as follows. If the surface is parameterized as $S=\{(x_1,x_2,x_3)\in\mathbb{R}^3 : x_3=I(x_1,x_2)\}$, then the surface is characterized by the intensities of a 2D image I. For a comparison of the two surfaces, the distance measure can be chosen as the discrepancies of some geometric properties such as curvature. An example is the discrepancies of curvatures (Chan et al., 2016; Lui et al., 2010b):

$$\mathcal{D}(S^0,S^1):=\int_{S_1}\alpha(H^1-H^0)^2+\beta(K^1-K^0)^2\ ds,$$

where H^k denotes the mean curvatures and K^k the Gaussian curvatures on S^k, $k=0,1$. Here $\alpha,\beta>0$ are regularization weights. Another example is to match the Laplace–Beltrami eigenfunctions defined on each surface according to their Riemannian metrics.

As for the regularization term, various choices can be chosen. For example, in order the enhance the smoothness of the surface mapping, a harmonic energy can be used:

$$\mathcal{R}_{\text{harmonic}}\int_{S_1}|\nabla_{S_1}f|^2 ds,$$

where ∇_{S_1} is the gradient on the surface in term of covariant derivatives. If one aims for a biholomorphic mapping as regularization, the Riemann mapping theorem (Gray, 1994) suggest to add the Riemannian metric to the curvature mismatching term $\mathcal{D}(S^0,S^1)$,

$$\mathcal{D}(y):=\mathcal{D}(S^0,S^1\circ y)+\gamma|\mu(T)|^2.$$

Here, $\mu(y)$ is the Beltrami coefficient measuring the metric distortion under the transformation y. This gives rise to various geometric measurements. Interestingly, if α, β, $\gamma>0$, this gives a full geometric measurement.

This is that $\mathcal{D}(y)=0$ when two surfaces are the same up to a rigid motion. For $\beta=0$, $\mathcal{D}(S^0,S^1)=0$ when the two surfaces are isometric to each other (Lui et al., 2010b).

6 Numerical methods

Image registration is phrased as a variational problem and the objective is to find a minimizing element of $\mathcal{J}$ in an admissible set $\mathcal{A}$. Typically, analytic solutions are not at hand and numerical schemes are to be applied. As the data to be process can range to 3×1024^3 and time is valuable, the choice of a numerical solution strategy is nontrivial. In this chapter, we only briefly outline some of the basic concepts as the topic is too big; for details see, e.g., Modersitzki (2009).

There are mainly two directions to go. The first is called optimize-then-discretize. Here, one derives the so-called Euler–Lagrange equations (ELE), which are conditions for a minimizer of the functional $\mathcal{J}$ in (1). One then discretizes the ELE which yields a nonlinear set of equation. Finally, theses equations are solved numerically. However, efficient solvers are nontrivial to design, mainly to the fidelity term and depending on the regularisers used. For example, the ELE for $\mathcal{J}$ based on the SSD distance measure and the diffusion regularizer and $y_{\text{ref}}=0$ yields

$$\begin{aligned}\mathcal{J}(y) &:= \mathcal{D}^{\text{SSD}}(T\circ y,R)+\alpha\mathcal{R}^{\text{diff}}(y)=\frac{1}{2}\| T\circ y-R\|^2_{L^2(\Omega)}=+\frac{\lambda}{2}\|\mathcal{B}y\|^2_{L^2(\Omega)}\\ 0 &= (T\circ y-R)^\top\nabla T\circ y-\Delta y,\end{aligned}$$

see (Modersitzki, 2004) for details. The transformation can be discretized using a finite differences or finite element approach; see, e.g., Modersitzki (2004, 2009), Lam and Lui (2014), Zhang and Chen (2018) and the references therein.

This approach has a number of drawbacks. First of all, the operator is derived from partial integration and the approach therefore comes with additional assumptions on z. For example, for the diffusion regularizer, the Gâteaux derivative for a suitable perturbation z, one gets

$$\begin{aligned}\lim_{\varepsilon\to 0}\frac{1}{\varepsilon}(\mathcal{R}(y+\varepsilon z)-\mathcal{R}(y)) &= \sum_j\int_\Omega(\nabla y_j)^\top(\nabla z_j)\ dx\\ &= \sum_j\left([z_j\cdot n^\top\nabla y_j]_{\partial\Omega}-\int_\Omega\Delta y_j\cdot z_j\ dx\right).\end{aligned}$$

With the artificial arguments that with appropriate boundary conditions the boundary term vanishes and that y is twice differentiable, the above ELE follows. Secondly, it is neither clear nor obvious that discretization of the ELE yields a symmetric Hessian system.

The so-called discretize-then-optimize approach offers a better alternative (Haber and Modersitzki, 2006b). Here, the objective $\mathcal{J}$ and the wanted transformation y are discretized, and then a sequence of finite dimensional optimization problems is to be solved. For the above example, one ends up with

$$J^h(y^h) = \frac{h}{2}\|T(y^h) - R(x^h)\|^2_{\mathbb{R}^m} + \lambda\frac{h}{2}\| B^h y^h \|^2$$

where x^h is a collection of m grid points, Section 2.1, $R(x^h) \in \mathbb{R}^m$ are image values on the grid, $y^h = y(x^h)$ is the representation of the transformation on the grid, $T(y^h)$ are image values of the transformed template, h is the volume element. For $d = 1$, the matrix B^h is given by

$$B^h = \frac{1}{h}\begin{pmatrix} -1 & 1 & & \\ & \ddots & \ddots & \\ & & -1 & 1 \end{pmatrix} \in \mathbb{R}^{m-1,m}. \tag{12}$$

In this approach, the problem due to data fidelity takes the form of a highly nonconvex term and also a nonpositive definite Hessian.

Advantages of the discretize-then-optimize approach are that no additional assumptions on the solutions are used, boundary conditions are derived from the optimization problem and are automatically taken care of, and the Hessian on the discrete problem is naturally symmetric. Finally, stopping criteria and step-length control follow standard optimization rules; see, e.g., Nocedal and Wright (1999). The optimization is typically based on first-order or second-order schemes, mainly quasi-Newton schemes. In particular in a multilevel framework, where a numerical solution of a coarse grid representation is used as a starting guess for a finer grid, one expects an excellent starting point and hence fast convergence due to a higher order scheme.

Solving the problem numerically is still challenging. Memory and time restriction prohibits the use of direct solvers for the Hessian systems. Highly efficient multigrid solvers for B^*B or $M + B^*B$ are used, where M is some approximation to the Hessian of the data-fit; see Modersitzki (2009) and Chumchob and Chen (2011) for details.

In particular for the SSD measures and similarly to optical flow problems, some authors use a Gauss–Newton type approximation of the data-fit. With $y^{k+1} = y^k + u$, $T^k := T \circ y^k$ and $dT := \nabla T \circ y^k$,

$$\mathcal{D}(T \circ y^{k+1}, R) = \|T \circ (y^k + u) - R\|^2_\Omega \approx \| T + dT \cdot u - R \|^2_\Omega =: q(u)$$

the objective becomes a convex quadratic function.

Development of new and faster algorithms with optimal computational complexity remains an interesting topic.

7 Deep learning-based registration

Machine learning techniques have advanced rapidly over the last decade, especially in image classification and segmentation. The main idea of deep learning-based algorithms is to aggregate information of various complexities in the datasets of training images to drive the registration process. In recent years, supervised and unsupervised deep learning techniques have been successfully employed for image and surface registration (Cao et al., 2017; Jaderberg et al., 2017; Krebs et al., 2017; Liao et al., 2017; Miao et al., 2016; Sokooti et al., 2017; Yang et al., 2017) (in particular the latter five references on convolutional neural networks for image registration). Convolutional neural networks of various architectures have been used to learn meaningful feature descriptors (Balntas et al., 2016; Rocco et al., 2017; Zagoruyko and Komodakis, 2015).

Though possessing many similarities in the variation framework to other imaging models, registration is quite different in the context of deep learning. The major difference lies in the nonreadily available sets of training data, e.g., in medical imaging, few clinicians can accurately draw the deformation to a good accuracy. This sets the scene of the state of the art in deep learning-based registration which we briefly review; refer to Litjens et al. (2017), Maier et al. (2019) and Lundervold and Lundervold (2019).

Firstly, we discuss the supervised learning. To deal with the problem of getting enough training data, one could generate simpler transforms for various images so that training tuples of the form (T, R, y) are available, instead of the usual registration pair (T, R) being given as in Sections 1–6. Deep learning is also employed to estimate the momentum in the framework of large deformation diffeomorphic metric matching (LDDMM) (Yang et al., 2017) where the same idea of generating a space of priori deformations is used to design training data. Of course, once training data are given or rather generated, various deep learning architectures are applicable as with solving other imaging problems by learning. Refer to recent surveys Litjens et al. (2017) and Maier et al. (2019). Clearly the supervised model is good for problems where the types of deformations can be predicted a priori, though not robust.

Secondly, we mention the unsupervised learning where no training data are required so that more generality is offered. Then as with nonlearning models, intensity-based registration models drive the registration process through matching image intensities. The key step is to replace the loss function (usually measuring the distance between the given deformation and the iterate) by an energy functional that underlines a variational registration model (see Sections 1–6). In this idea, various deep neural networks have been proposed to transform images to match image intensities. For instance, the spatial transformer network (Jaderberg et al., 2017) transforms an input image through a neural network such that the classification task can be simplified. Convolutional neural networks with reinforcement learning are trained to predict descent directions of the transformation towards optimal alignment (Liao et al., 2017; Miao et al., 2016). Convolutional neural network has also been applied to predict parameters in the thin-plate spline registration model (Cao et al., 2017). General dense displacement vector field can also be predicted by training a deep convolutional neural network, without applying a fixed parametric registration model (Sokooti et al., 2017). Other ideas of merging small sets of training data and unsupervised learning may also be considered, to design new learning models.

8 Conclusions

Registration between corresponding data is a crucial process and has important applications in various fields. In this chapter, we give an overview of the registration problem. The basic mathematical background and the mathematical framework for registration problems are firstly introduced. State-of-the-art registration models are then described. In particular, we carefully examine various existing distance measures and regularization terms. Their advantages and disadvantages are also discussed. Furthermore, mathematical models for surface registration, which aims to find meaningful one-to-one correspondence between two surfaces embedded in 3D space, are also reviewed. The connection between surface registration and image registration is also studied. Numerical methods for different registration models are also briefly explained. As a popular research direction, various deep learning-based registration models invented recently are also described. In the future, we expect registration models, which combine mathematical models and machine learning techniques, will define a trend for solving more challenging registration problems.

References

Arganda-Carreras, I., Sorzano, C.O.S., Thévenaz, P., Muñoz-Barrutia, A., Kybic, J., Marabini, R., Carazo, J.M., de Solorzano, C.O., 2010. Non-rigid consistent registration of 2D image sequences. Phys. Med. Biol. 55 20, 6215–6242.

Ashburner, J., 2007. A fast diffeomorphic image registration algorithm. NeuroImage 38 (1), 95–113.

Balntas, V., Johns, E., Tang, L., Mikolajczyk, K., 2016. Pn-net: conjoined triple deep network for learning local image descriptors. Preprint arXiv:1601.05030.

Beg, M.F., Miller, M.I., Trouve, A., Younes, L., 2005. Computing large deformation metric mappings via geodesic flows of diffeomorphisms. Int. J. Comput. Vis. 61 (2), 139–157. https://doi.org/10.1023/B:VISI.0000043755.93987.aa.

Brehmer, K., Wacker, B., Modersitzki, J., 2018. Simultaneous registration of image sequences—a novel singular value based images similarity measure. In: Proceedings in Applied Mathematics and Mechanics—89th GAMM Annual Meeting. Wiley.

Breiten, T., Simoncini, V., Stoll, M., 2016. Low-rank solvers for fractional differential equations. Electron. Trans. Numer. Anal. 45, 107–132.

Broit, C., 1981. Optimal Registration of Deformed Images (Ph.D. thesis). Computer and Information Science, University of Pennsylvania, USA.

Bronstein, A.M., Bronstein, M.M., Kimmel, R., Mahmoudi, M., Sapiro, G., 2010. A Gromov-Hausdorff framework with diffusion geometry for topologically-robust non-rigid shape matching. Int. J. Comput. Vis. 89 (2–3), 266–286.

Brown, L.G., 1992. A survey of image registration techniques. ACM Comput. Surv. 24 (4), 325–376.

Bruhn, A., Weickert, J., Feddern, C., Kohlberger, T., Schnörr, C., 2003. Real-time optic flow computation with variational methods. Comput. Anal. Images Patterns 2756, 222–229.

Burger, M., Modersitzki, J., Ruthotto, L., 2013. A hyperelastic regularization energy for image registration. SIAM J. Sci. Comput. 35 (1), B132–B148.

Cao, X., Yang, J., Zhang, J., Nie, D., Kim, M., Wang, Q., Shen, D., 2017. Deformable image registration based on similarity-steered CNN regression. In: Medical Image Computing and Computer Assisted Intervention—MICCAI 2017, vol. 28, pp. 300–308.

Chan, H.L., Li, H.F., Lui, L.M., 2016. Quasi-conformal statistical shape analysis of hippocampal surfaces for Alzheimer disease analysis. J. Neurocomput. 175 (A), 177–187.

Chelbi, S., Mekhmoukh, A., 2018. Features based image registration using cross correlation and radon transform. Alex. Eng. J. 57 (4), 2313–2318. ISSN 1110-0168. https://doi.org/10.1016/j.aej.2017.07.013.

Chen, L., 2013. Digital Functions and Data Reconstruction. Springer.

Chen, Y.M., Ye, X.J., 2010. Inverse consistent deformable image registration. In: The Legacy of Alladi Ramakrishnan in the Mathematical Sciences. Springer, New York, pp. 419–440.

Chen, K.L., Derksen, A., Heldmann, S., Hallmann, M., Berkels, B., 2015. Deformable image registration with automatic non-correspondence detection. In: Aujol, J.F., Nikolova, M., Papadakis, N. (Eds.), Scale Space and Variational Methods in Computer Vision: 5th International Conference, SSVM 2015, Lège-Cap Ferret, France, May 31–June 4, 2015, Proceedings. Springer International Publishing, Cham, pp. 360–371.

Christensen, G.E., Johnson, H.J., 2001. Consistent image registration. IEEE Trans. Med. Imaging 20 (7), 568–582.

Chumchob, N., Chen, K., 2009. A robust affine image registration method. Int. J. Numer. Anal. Model. 6 (2), 311–334.

Chumchob, N., Chen, K., 2011. A robust multigrid approach for variational image registration models. J. Comput. Appl. Math. 236 (5), 653–674.

Chumchob, N., Chen, K., Brito-Loeza, C., 2011. A fourth order variational image registration model and its fast multigrid algorithm. SIAM J. Multiscale Model. Simul. 9 (1), 89–128.

Collignon, A., Maes, F., Delaere, D., Vandermeulen, D., Suetens, P., Marchal, G., 1995. Automated multi-modality image registration based on information theory. In: IPMI, pp. 263–274.

Drobny, D., 2015. Automatic Weighting of Non-corresponding Regions in Image Registration (Ph.D. thesis). University of Lubeck, Lubeck, Germany.

Drobny, D., Carolus, H., Kabus, S., Modersitzki, J., 2015. Handling non-corresponding regions in image registration. In: Handels, H., Deserno, T.M., Meinzer, H.P., Tolxdorff, T. (Eds.), Bildverarbeitung Fur die Edizin 2015. Springer, Berlin, Heidelberg, pp. 107–112.

Droske, M., Rumpf, M., 2004. A variational approach to non-rigid morphological image registration. SIAM Appl. Math. 64 (2), 668–687.

Fischer, B., Modersitzki, J., 1999. Fast inversion of matrices arising in image processing. Numer. Algorithms 22, 1–11. ISSN 1017-1398 (Print), 1572-9265 (Online). https://doi.org/10.1023/A:1019194421221.

Fischer, B., Modersitzki, J., 2002. Fast diffusion registration. Contemp. Math. 313, 117–129.
Fischer, B., Modersitzki, J., 2003. Combining landmark and intensity driven registrations. PAMM 3, 32–35.
Fischler, M.A., Elschlager, R.A., 1973. The representation and matching of pictorial structures. IEEE Trans. Comput. 22 (1), 67–92.
Glasbey, C.A., Mardia, K.V., 1998. A review of image-warping methods. J. Appl. Stat. 25 (2), 155–171.
Gray, J., 1994. On the history of the Riemann mapping theorem. Rend. Circ. Mat. Palermo 2 (Suppl. 34), 47–94.
Haber, E., Modersitzki, J., 2004. Numerical methods for volume preserving image registration. Inverse Prob. 20, 1621–1638.
Haber, E., Modersitzki, J., 2006. Intensity gradient based registration and fusion of multi-modal images. In: Barillot, C., Haynor, D.R., Hellier, P. (Eds.), Medical Image Computing and Computer-Assisted Intervention—MICCAI 2006, vol. 3216. Springer LNCS, pp. 591–598.
Haber, E., Modersitzki, J., 2006. A multilevel method for image registration. SIAM J. Sci. Comput. 27 (5), 15941607.
Haber, E., Modersitzki, J., 2007. Image registration with a guaranteed displacement regularity. Int. J. Comput. Vis. 71, 361–372. https://doi.org/10.1007/s11263-006-8984-4.
Haber, E., Heldmann, S., Modersitzki, J., 2009. A scale-space approach to landmark constrained image registration. In: Proceedings of the Second International Conference on Scale Space Methods and Variational Methods in Computer Vision (SSVM), Springer LNCS, pp. 1–12.
Haber, E., Horesh, R., Modersitzki, J., 2010. Numerical methods for constrained image registration. Numer. Linear Algebra Appl. 17 (2–3), 343–359.
Heldmann, S., 2010. Multimodal registration of MR images with a novel least-squares distance measure. In: Medical Imaging 2010: Image Processing, SPIE, San Diego.
Henn, S., Witsch, K., 2005. A variational image registration approach based on curvature scale space. In: Scale Space and PDE Methods in Computer Vision, vol. 3459, pp. 143–154.
Hill, D., Batchelor, P.G., Holden, M., Hawkes, D., 2001. Medical image registration. Phys. Med. Biol. 46, R1–R45.
Horn, B.K.P., Schunck, B.G., 1981. Determining optical flow. Artif. Intell. 17, 185–203.
Huber, P.J., 1964. Robust estimation of a location parameter. Ann. Stat. 53 (1), 73–101.
Ibrahim, M., Chen, K., Brito, C., 2014. A novel variational model for image registration using Gaussian curvature. J. Geome. Imaging Comput. 1 (4), 417–446.
Jaderberg, M., Simonyan, K., Zisserman, A., Kavukcuoglu, K., 2017. Spatial transformer networks. Adv. Neural Inf. Proces. Syst. 28, 2017–2025.
Joshi, S.C., Miller, M.I., 2000. Landmark matching via large deformation diffeomorphisms. IEEE Trans. Image Process. 9 (8), 1357–1370.
Khan, A.M., Chen, W., Khalil, A., Lin, J., 2018. Euler's elastica and curvature based model for image restoration. PLOS ONE 13, e0202464. https://doi.org/10.1371/journal.pone.0202464.
Krebs, J., Mansi, T., Delingette, H., Zhang, L., Ghesu, F.C., Miao, S., Maier, A.K., Ayache, N., Liao, R., Kamen, A., 2017. Robust nonrigid registration through agentbased action learning. In: Medical Image Computing and Computer Assisted Intervention—MICCAI 2017, vol. 28, pp. 344–352.
Lam, K.C., Lui, L.M., 2010. Optimized conformal parameterization with controllable area distortions. Commun. Math. Sci. 3 (1), 52–78.
Lam, K.C., Lui, L.M., 2014. Landmark- and intensity-based registration with large deformations via quasi-conformal maps. SIAM J. Imaging Sci. 7 (4), 2364–2392.
Lee, J., 1997. Riemannian Manifolds: An Introduction to Curvature. Springer.

Liang, J., Lai, R., Wong, T.W., Zhao, H.K., 2012. Geometric understanding of point clouds using Laplace-Beltrami operator. In: Computer Vision and Pattern Recognition (CVPR), pp. 214–221.

Liao, R., Miao, S., de Tournemire, P., Grbic, S., Kamen, A., Mansi, T., Comaniciu, D., 2017. An artificial agent for robust image registration. In: Proceedings of the Thirty-First AAAI Conference on Artificial Intelligence, vol. 28.

Litjens, G., Kooi, T., Bejnordi, B.E., Setio, A.A.A., Ciompi, F., Ghafoorian, M., van der Laak, J.A.W.M., van Ginneken, B., Snchez, C.I., 2017. A survey on deep learning in medical image analysis. Med. Image Anal. 42, 60–88. ISSN 1361-8415. https://doi.org/10.1016/j.media.2017.07.005.

Lui, L.M., Thiruvenkadam, S., Wang, Y., Thompson, P.M., Chan, T.F., 2010. Optimized conformal surface registration with shape-based landmark matching. SIAM J. Imaging Sci. 3 (1), 52–78.

Lui, L.M., Wong, T.W., Gu, X.F., Thompson, P.M., Chan, T.F., Yau, S.T., 2010. Hippocampal shape registration using Beltrami holomorphic flow. 6362, 323–330.

Lui, L.M., Wong, T.W., Zeng, W., Gu, X.F., Thompson, P.M., Chan, T.F., Yau, S.T., 2012. Optimization of surface registrations using Beltrami holomorphic flow. J. Sci. Comput. 50 (3), 557–585.

Lundervold, A.S., Lundervold, A., 2019. An overview of deep learning in medical imaging focusing on MRI. Z. Med. Phys. 29 (2), 102–127. ISSN 0939-3889. https://doi.org/10.1016/j.zemedi.2018.11.002.

Maier, A., Syben, C., Lasser, T., Riess, C., 2019. A gentle introduction to deep learning in medical image processing. Z. Med. Phys. 29 (2), 86–101. ISSN 0939-3889. https://doi.org/10.1016/j.zemedi.2018.12.003.

Mahalanobis, P.C., 1936. On the generalised distance in statistics. In: Proceedings of the National Institute of Science of India, vol. 2, pp. 49–55.

Maintz, J.B., Viergever, M.A., 1998. A survey of medical image registration. Med. Image Anal. 2 (1), 1–36.

Miao, S., Wang, Z.J., Liao, R., 2016. A CNN regression approach for realtime 2D/3D registration. IEEE Trans. Med. Imaging 35 (5), 1352–1363.

Modersitzki, J., 2004. Numerical Methods for Image Registration. Oxford University Press, New York.

Modersitzki, J., 2009. FAIR: Flexible Algorithms for Image Registration. SIAM, Philadelphia.

Nejadasl, F.K., Lindenbergh, R., 2014. Sequential and automatic image-sequence registration of road areas monitored from a hovering helicopter. Sensors 14 (9), 16630–16650. ISSN 1424-8220. https://doi.org/10.3390/s140916630. http://www.mdpi.com/1424-8220/14/9/16630.

Nocedal, J., Wright, S.J., 1999. Numerical Optimization. Springer, New York.

Oskarsson, M., 2015. Regularizing image intensity transformations using the Wasserstein metric. In: Paulsen, R.R., Pedersen, K.S. (Eds.), Image Analysis, Springer, Cham, pp. 275–286.

Ovsjanikov, M., Corman, E., Bronstein, M.M., Rodolà, E., Ben-Chen, M., Guibas, L., Chazal, F., Bronstein, A.M., 2016. Computing and processing correspondences with functional maps. In: Proc. SIGGRAPH Courses (SIGGRAPH 2016).

Podlubny, I., 1998. Fractional Differential Equations. Academic Press.

Rocco, I., Arandjelović, R., Sivic, J., 2017. Convolutional neural network architecture for geometric matching. In: Proceedings of the IEEE Conference on Computer Vision and Pattern Recognition.

Rohlfing, T., Maurer, C.R., Bluemke, D.A., Jacobs, M.A., 2003. Volume-preserving nonrigid registration of MR breast images using free-form deformation with an incompressibility constraint. IEEE Trans. Med. Imaging 22 (6), 730–741.

Rueckert, D., Sonoda, L.I., Hayes, C., Hill, D.L.G., Leach, M.O., Hawkes, D.J., 1999. Nonrigid registration using free-form deformations: application to breast MR images. IEEE Trans. Med. Imaging 18 (8), 712–721.

Rustamov, R.M., 2007. Laplace-Beltrami eigenfunctions for deformation invariant shape representation. In: Symposium on Geometry Processing, pp. 225–233.

Sarvaiya, J.N., Patnaik, S., Bombaywala, S., 2009. Image registration by template matching using normalized cross-correlation. In: 2009 International Conference on Advances in Computing, Control, and Telecommunication Technologies, pp. 819–822.

Scherzer, O., 2008. Variational Methods in Imaging. Springer.

Shi, Y., Lai, R., Kern, K., Sicotte, N., Dinov, I., Toga, A.W., 2008. Harmonic surface mapping with Laplace-Beltrami eigenmaps. In: Medical Image Computing and Computer Assisted Intervention (MICCAI)147–154.

Sokooti, H., de Vos, B., Berendsen, F., Lelieveldt, B. P. F., Isgum, I., Staring, M., 2017. Nonrigid image registration using multiscale 3D convolutional neural networks. In: Medical Image Computing and Computer Assisted Intervention—MICCAI 2017, pp. 232–239.

Solomentsev, E.D., Shikin, E.V., 1994. LaplaceBeltrami equation. In: Hazewinkel, M. (Ed.), Encyclopedia of Mathematics, Springer/Kluwer Academic Publishers. https://www.encyclopediaofmath.org/index.php/Laplace-Beltrami_equation.

Sotiras, A., Davatzikos, C., Paragios, N., 2013. Deformable medical image registration: a survey. IEEE Trans. Med. Imaging 32 (7), 1153–1190.

Tektonidis, M., Rohr, K., 2017. Diffeomorphic multi-frame non-rigid registration of cell nuclei in 2D and 3D live cell images. IEEE Trans. Image Process. 26 (3), 1405–1417. ISSN 1057-7149. https://doi.org/10.1109/TIP.2017.2653360.

Theljani, A., Chen, K., 2019. An augmented Lagrangian method for solving a new variational model based on gradients similarity measures and high order regularization for multimodality registration. Inverse Prob. Imaging 13 (2), 309–335.

Thompson, T., Chen, K., 2019. An effective diffeomorphic model and its fast multigrid algorithm for registration of lung CT images. Comput. Methods Appl. Math. (to appear). https://doi.org/10.1515/cmam-2018-0126.

Viola, P., Wells, W.M., 1997. Alignment by maximization of mutual information. Int. J. Comput. Vis. 24 (2), 137–154. https://doi.org/10.1023/A:1007958904918.

Watrous, J., 2018. Theory of Quantum Information. Cambridge University Press, UK.

Weickert, J., Schnörr, C., 2001. A theoretical framework for convex regularizers in PDE-based computation of image motion. Int. J. Comput. Vis. 45 (3), 245–264.

Wong, T.W., Lui, L.M., Thompson, P.M., Chan, T.F., 2012. Instrinic feature extraction and hippocampal surface registration using harmonic eigenmap. SIAM J. Imaging Sci. 5 (2), 746–768.

Yang, X., Kwitt, R., Styner, M., Niethammer, M., 2017. Quicksilver: fast predictive image registration—a deep learning approach. NeuroImage 158, 378–396.

Yanovsky, I., Thompson, P., Osher, S., Leow, A., 2007. Large deformation unbiased diffeomorphic nonlinear image registration: theory and implementation. In: IEEE conference CVPR' 2007 (see also UCLA CAM Report 06-71), vol. 71.

Younes, L., 2010. Shapes and Diffeomorphisms. vol. 171. Springer-Verlag. (see also 2019 edition).

Zagoruyko, S., Komodakis, N., 2015. Learning to compare image patches via convolutional neural networks. In: Proceedings of the IEEE Conference on Computer Vision and Pattern Recognition.

Zhang, J.P., Chen, K., 2015. Variational image registration by a total fractional-order variation model. J. Comput. Phys. 293, 442–461. ISSN 0021-9991. https://doi.org/10.1016/j.jcp.2015.02.021.

Zhang, D.P., Chen, K., 2018. A novel diffeomorphic model for image registration and its algorithm. J. Math. Imaging Vis. 60 (8), 1261–1283.

Zhang, J., Chen, K., Yu, B., 2017. An efficient numerical method for mean curvature-based image registration model. East Asian J. Appl. Math. 7 (01), 125–142.

Zitová, B., Flusser, J., 2003. Image registration methods: a survey. Image Vis. Comput. 21 (11), 977–1000.

Chapter 16

Metric registration of curves and surfaces using optimal control

Martin Bauer[a], Nicolas Charon[b] and Laurent Younes[b,*]
[a]*Department of Mathematics, Florida State University, Tallahassee, FL, United States*
[b]*Department of Applied Mathematics and Statistics, Johns Hopkins University, Baltimore, MD, United States*
[*]*Corresponding author: e-mail: laurent.younes@jhu.edu*

Chapter Outline

Abstract

This chapter presents an overview of recent developments in the analysis of shapes such as curves and surfaces through Riemannian metrics. We show that several constructions of metrics on spaces of submanifolds can be unified through the prism of Riemannian submersions, with shape space metrics being induced from metrics defined on the top spaces. Computing the resulting Riemannian distances involves solving geodesic matching problems with boundary conditions. To deal efficiently with such variational problems, one can rely on an auxiliary family of "chordal" distances to simplify the

 https://doi.org/10.1016/bs.hna.2019.03.001
© 2019 Elsevier B.V. All rights reserved.

treatment of boundary conditions, which we use to come up with a relaxed inexact formulation of the matching problem. This also allows to turn shape matching into optimal control problems and give a common framework to address them in practice. We then specify our analysis to the cases of intrinsic shape metrics defined using invariant Sobolev metrics on parametrized immersions, outer shape metrics induced from metrics on diffeomorphism groups of the ambient space and finally a recent hybrid model that combines those two approaches.

Keywords: Shape analysis, Curve matching, Intrinsic metrics, Varifolds

AMS Classification Codes: 68Q25, 68R10, 68U05

1 Introduction

Curves and surfaces are natural representations of shapes in two or three dimensions and are mathematically special cases of submanifolds of the Euclidean space $\mathbb{R}^d$ for $d = 2$ or 3. While such submanifolds are conveniently described through parametrizations, or atlases, which are relatively simple mathematical objects (forming special classes of functions), shapes, as geometric objects, form equivalence classes of such functions modulo reparametrization, leading to quotient spaces that can be significantly more complex to study, especially in infinite dimensions. We will refer to these equivalence classes of parametrized submanifolds modulo reparametrization, as "unparametrized submanifolds", mainly for clarity since these objects are what is commonly referred to as, simply, submanifolds.

Constructing tractable comparison tools between such unparametrized sets is at the core of shape analysis and its many applications. Among the various approaches that were proposed (Bauer et al., 2014b; Bronstein et al., 2010a; Feiszli and Mumford, 2007; Glaunès et al., 2008; Gu and Yau, 2008; Klassen et al., 2004; Mémoli, 2011; Mennucci et al., 2008; Michor and Mumford, 2007; Mio and Srivastava, 2004; Ovsjanikov et al., 2016; Sharon and Mumford, 2006a; Srivastava and Klassen, 2016; Younes, 1996, 2010; Zeng and Gu, 2011), our goal is here to focus the review on a class of metrics defined through the introduction of Riemannian (or sub-Riemannian) metrics on spaces of parametrized submanifolds with suitable invariance properties, allowing these metrics, through mechanisms that will be described below, to "project" onto distances modulo reparametrization. The computation of these distances will always require estimating a minimizing path between the two endpoints, resulting in a class of algorithms that we will refer to as "metric registration".

In our setting, parametrized manifolds will be represented as functions $q: M \to \mathbb{R}^d$, where M, the parameter space, is a p-dimensional oriented compact manifold. Typical examples are $M = \mathbb{S}^p$, the p-dimensional unit sphere, or a regular closed subset of $\mathbb{R}^p$ with a smooth boundary (such as a closed interval, or a closed disc). Situations in which M has no boundary are usually more amenable to theoretical analysis, but the with-boundary case also has important practical applications (e.g., for open curves).

In order for q to parametrize a p-dimensional submanifold of $\mathbb{R}^d$, one needs to assume that it provides an embedding, i.e., a one-to-one C^1 mapping from M into $\mathbb{R}^d$, such that $q^{-1} : q(M) \to M$ is continuous and $dq(m)$ is a one-to-one linear mapping for all $m \in M$ (where dq denotes the differential of q). The set of embeddings will be denoted $\mathrm{Emb}(M,\mathbb{R}^d)$. However, for several metrics described in this paper, it will be useful to relax the embedding constraint and assume that q is an immersion instead, where an immersion is a C^1 mapping from M into $\mathbb{R}^d$ such that $dq(m)$ is a one-to-one linear mapping for all $m \in M$ (so that an embedding is an immersion which is, in addition, a homeomorphism onto its image). We let $\mathrm{Imm}(M,\mathbb{R}^d)$ denote the space of immersions from M to $\mathbb{R}^d$. The main reason for considering immersions rather than embeddings is that several of the metrics that will be introduced have geodesic paths that remain immersions at all times, but not necessarily embeddings (even when looking for shortest paths between two embeddings).

If X is a space of C^1 functions from M to $\mathbb{R}^d$, we will use the notation Emb_X, Imm_X for the set of embeddings or immersions $q \in X$. We will typically use $X = C^r = C^r(M,\mathbb{R}^d)$ (functions with at least r continuous derivatives) with $r \geq 1$, or $X = H^s = H^s(M,\mathbb{R}^d)$ (functions with at least s generalized derivatives which are all square integrable) with $s > p/2 + 1$, or $X = C^\infty = C^\infty(M,\mathbb{R}^d)$ (smooth functions).

Let $\mathrm{Diff}(M)$ denote the space of orientation-preserving diffeomorphisms of M, i.e., the space of invertible C^1 mappings τ from M onto itself with a C^1 inverse such that, for all $m \in M$, $d\tau(m)$ maps positively oriented bases of T_mM onto positively oriented bases of $T_{\tau(m)}M$. Given $q \in \mathrm{Imm}(M,\mathbb{R}^d)$, we let $S_q = q(M)$ denote the image of M by q. If $\tau \in \mathrm{Diff}(M)$, the mapping $q \cdot \tau = q \circ \tau$ is also an immersion and $S_{q\cdot\tau} = S_q$. One says that τ is a change of parameter, or reparametrization, and the unparametrized surface associated with q is the equivalent class

$$[q] = q \cdot \mathrm{Diff}(M) = \{q \circ \tau, \tau \in \mathrm{Diff}(M)\}.$$

Notice that, when considering immersions that belong to a given class X, the space $\mathrm{Diff}(M)$ must also be restricted to diffeomorphisms that preserve X, i.e., such that $q \circ \tau \in X$ for all $q \in X$.

As indicated earlier, it will be relatively "easy" to equip $\mathrm{Imm}_X(M,\mathbb{R}^d)$ with a Riemannian metric and various approaches will be described later in this chapter. We will then use a standard mechanism—Riemannian submersions, described in the next section—to define, through the mapping $q \mapsto [q]$, a distance between unparametrized shapes. More precisely, this chapter is organized as follows. Following this introduction, Section 2 describes the general concept of Riemannian submersion and how it can be utilized to define metrics in spaces acted upon by deformation groups. Section 3 summarizes basic results in optimal control that will be instrumental in the computation of the geodesic distances. Section 4 reviews a special class of metrics in which shapes spaces are embedded in suitably chosen Hilbert spaces and compared using the associated Hilbert norm, yielding "chordal metrics" between shapes, to distinguish them from Riemannian metrics directly defined on shape spaces, which are our main focus. Chordal metrics will here mainly serve as auxiliary terms in order to relax variational problems associated to the computation of geodesics into computationally feasible optimal control problems. Section 5 introduces a first class of metrics, called intrinsic, because they derive from standard Sobolev metrics on the space of immersions. Section 6 describes "outer deformation metrics" that are derived from the action of groups of diffeomorphisms of the ambient space on immersions. Finally, Section 7 describes some attempts at combining the two approaches, leading to "hybrid metrics". These last three sections will also include some experimental results and a discussion of numerical methods. We will conclude the chapter by a brief discussion in Section 8.

2 Building metrics via submersions

The following discussion describes some metric constructions involving Riemannian submersions. They can be rigorously defined in finite dimensions, but also serve as motivational foundations for similar constructions in infinite dimensions (which will be our main focus), even though not all of the properties listed below may generalize, or they may require additional assumptions to hold.

If $\mathcal{F}$ and $\mathcal{B}$ are differential manifolds, with $\dim(\mathcal{B}) \leq \dim(\mathcal{F})$, a mapping π from $\mathcal{F}$ onto $\mathcal{B}$ is a submersion if $d\pi(f)$ is onto from $T_f\mathcal{F}$ to $T_{\pi(f)}\mathcal{B}$ for all $f \in \mathcal{F}$. For such a submersion, one defines

$$V_f = \mathrm{Null}(d\pi(f)) = \{\eta \in T_f\mathcal{F} : d\pi(f)\eta = 0\},$$

the vertical space at f for the submersion π, which is also the tangent space at f to the submanifold $\pi^{-1}(b)$, where $b = \pi(f)$. If $\mathcal{F}$ is a Riemannian manifold, one can also define, for $f \in \mathcal{F}$, the horizontal space

$$H_f = V_f^{\perp} = \{\eta \in T_f\mathcal{F} : g_f^{\mathcal{F}}(\eta, \zeta) = 0 \text{ for all } \zeta \in V_f\},$$

where $g^{\mathcal{F}}$ denotes the Riemannian metric on $\mathcal{F}$, so that $g_f^{\mathcal{F}}(\eta, \zeta)$ is the Riemannian inner product between η and ζ in $T_f\mathcal{F}$. Moreover, the restriction of $d\pi(f)$ to H_f is an isomorphism between H_f and $T_b\mathcal{B}$ where $b = \pi(f)$. If $\mathcal{B}$ is also Riemannian, then one says that π is a *Riemannian submersion* if and only if $d\pi(f)$ is an isometry from H_f to $T_b\mathcal{B}$, i.e., for all η, $\eta' \in H_f$:

$$g_b^{\mathcal{B}}(d\pi(f)\eta, d\pi(f)\eta') = g_f^{\mathcal{F}}(\eta, \eta'). \tag{1}$$

If $\pi(f) = \pi(f') = b$, this implies that the horizontal spaces H_f and $H_{f'}$ are isometric, via the transformations $\psi_{f,f'} = d\pi(f')^{-1} \circ d\pi(f) : H_f \to H_{f'}$. Conversely, if the metric on $\mathcal{F}$ is such that all transformations $\psi_{f,f'}$ are isometries whenever $\pi(f) = \pi(f')$, then one can equip $\mathcal{B}$ with a uniquely defined Riemannian metric, through (1), such that π is a Riemannian submersion. This is the construction that we will use in Sections 3–7, in which $\mathcal{F}$ will be a set on which Riemannian metrics can be easily defined, with properties ensuring that horizontal spaces are isometric, resulting in a metric on $\mathcal{B}$.

Let π be a Riemannian submersion and $h \in T_b\mathcal{B}$ for some $b \in \mathcal{B}$. If $f \in \pi^{-1}(b)$, denote by h^f the unique vector in H_f such that $d\pi(f)h^f = h$, so that $g_b^{\mathcal{B}}(h,h) = g_b^{\mathcal{F}}(h^f, h^f)$. Any vector $\eta \in d\pi(f)^{-1}h$ can be uniquely decomposed as $\eta = h^f + \zeta$, where $\zeta \in V_f$, so that the metric on $\mathcal{B}$ can also be derived from the one on $\mathcal{F}$ with

$$g_b^{\mathcal{B}}(h,h) = \min\{g_f^{\mathcal{F}}(\eta,\eta) : d\pi(f)\eta = h\} \tag{2}$$

for any f such that $\pi(f) = b$. This immediately implies that minimizing geodesics in $\mathcal{B}$, which are functions $t \mapsto b(t)$ such that $b(0) = b_0$ and $b(1) = b_1$ for given b_0, b_1 and that minimize

$$\int_0^1 g_{b(t)}^{\mathcal{B}}(\dot{b}(t), \dot{b}(t))dt$$

can also be defined as $b(t) = \pi(f(t))$ where $f(\cdot)$ minimizes

$$\int_0^1 g_{f(t)}^{\mathcal{F}}(\dot{f}(t), \dot{f}(t))dt \tag{3}$$

subject to the constraints $\pi(f(0)) = b_0$ and $\pi(f(1)) = b_1$. One can also show that any optimal $f(\cdot)$ must be horizontal (i.e., $\dot{f}(t) \in H_{f(t)}$ for all t). Moreover, because of the isometry between horizontal spaces, one can simplify the first constraint by choosing $f(0) = f_0$ arbitrarily in $\pi^{-1}(b_0)$, while $f(1)$ still needs to be optimized over $\pi^{-1}(b_1)$. This remark will be used in our algorithms, because optimization over trajectories in $\mathcal{F}$ will generally lead to simpler formulations. Moreover, we will often relax the constraint at $t = 1$ and replace it with a penalty, minimizing

$$\int_0^1 g_{f(t)}^{\mathcal{F}}(\dot{f}(t), \dot{f}(t))dt + U(f(1), f_1) \tag{4}$$

where f_1 is a fixed element in $\pi^{-1}(b_1)$ and U is a nonnegative function blind to "fibre" variations, such that $U(f, f') = 0$ if and only if $\pi(f) = \pi(f')$. The definition of such functions when comparing parametrized curves or surfaces will be addressed in Section 4.

We will, in particular, consider two specific situations, both associated with group actions, for which we introduce some notation. Let $\mathcal{G}$ be a Lie group, i.e., both a group and a manifold, in which the group operation is smooth. The Lie algebra of $\mathcal{G}$, denoted $\mathfrak{g}$, is the tangent space to $\mathcal{G}$ at the identity. The left or right action of a Lie group $\mathcal{G}$ on a manifold $\mathcal{M}$ will generally be denoted with a dot: $(g, m) \mapsto g \cdot m$ (left action) and $(g, m) \mapsto m \cdot g$ (right action). This action is also assumed to be smooth as a mapping from $\mathcal{G} \times \mathcal{M}$ to $\mathcal{M}$. Assuming a left action, and fixing m, we let $A_m : g \mapsto g \cdot m$ and one defines the infinitesimal action of $\mathfrak{g}$ on $\mathcal{M}$ by $v \cdot m = dA_m(\mathrm{id})v$, $v \in \mathfrak{g}$. (A similar definition can be made for right actions, with the infinitesimal action denoted $m \cdot v$.) We will also use the notation $\xi_m = dA_m(\mathrm{id})$, which is a linear mapping between $\mathfrak{g}$ and $T_m\mathcal{M}$.

We here describe three constructions involving Riemannian submersions that have been introduced for metric shape registration. The first two of them will serve as blueprints for the constructions of Sections 5–7.

(1) Assume that $\mathcal{G}$ is a group acting on $\mathcal{F}$ (on the right) and $\mathcal{B}$ is the quotient space

$$\mathcal{F}/\mathcal{G} = \{[f] = f \cdot \mathcal{G} : f \in \mathcal{F}\}.$$

Assume that $\mathcal{F}$ and $\mathcal{G}$ are such that this quotient space is also a manifold. Sufficient conditions for this to hold (see, e.g., Marsden and Ratiu, 2013, Chapter 9) typically require that $\mathcal{G}$ acts freely on $\mathcal{F}$, i.e., that, for all $f \in \mathcal{F}$, the identity $f \cdot g = f$ only holds when $g = \mathrm{id}$, in which case $\pi : f \mapsto f \cdot \mathcal{G}$ is a submersion. Assume that the metric on $\mathcal{F}$ is such that the action $\theta_g : f \mapsto f \cdot g$ is an isometry for all $g \in \mathcal{G}$. Then one can build a metric on $\mathcal{B}$ such that π is a Riemannian submersion. Indeed, $\pi(f) = \pi(f')$ implies that $f' = T_g f$ for some g, so that $T_f\mathcal{F}$ and $T_{f'}\mathcal{F} = d\theta_g(f)T_f\mathcal{F}$ are isometric, and so are V_f and $V_{f'}$, because $\pi \circ \theta_g = \pi$ implies that $d\pi(f')d\theta_g(f) = d\pi(f)$. This implies that $d\theta_g(f)$ also is an isometry between H_f and $H_{f'}$, and that it is equal to $\psi_{f,f'}$.

Under this assumption, (3) must be minimized subject to $f(0) = f_0$ and $f(1) = f_1 \cdot g$ for some $g \in \mathcal{G}$ (so that g can be considered as an additional variable in the optimization). If one uses (4), the function U must be such that $U(f, f') = 0$ if and only if $f' = f \cdot g$.

(2) For the second example, assume that $\mathcal{F}$ is a Lie group (with Lie algebra $\mathfrak{f}$) acting transitively on the left on $\mathcal{B}$, so that, for all $b, b' \in \mathcal{B}$, there exists $f \in \mathcal{F}$ such that $f \cdot b = b'$. Fix an element $b_0 \in B$ and define $\pi : \mathcal{F} \to \mathcal{B}$ by $\pi(f) = f \cdot b_0$, which is a submersion. If one assumes that the metric on $\mathcal{F}$ is right invariant, so that transformations $\theta_f : f' \mapsto f'f$ are isometries, then horizontal spaces are isometric. Indeed, the right-invariance assumption ensures that the tangent spaces to $\mathcal{F}$ are isometric, with $T_f\mathcal{F} = d\theta_f(\mathrm{id})\mathfrak{f}$. Moreover, it is easy to check that vertical spaces are right translations of the spaces

$$\mathfrak{v}_b = \{h \in \mathfrak{f} : h \cdot b = 0\}$$

through $V_f = d\theta_f(\mathrm{id})\mathfrak{v}_b$ when $\pi(f) = b$. This in turn implies that $H_f = d\theta_f(\mathrm{id})\mathfrak{h}_b$ with $\mathfrak{h}_b = \mathfrak{v}_b^{\perp}$, proving that horizontal spaces are all isometric. Moreover, if $\pi(f) = \pi(f')$, then $\pi \circ \theta_f = \pi \circ \theta_{f'}$ and $d\pi(f)d\theta_f(\mathrm{id}) = d\pi(q')d\theta_{f'}(\mathrm{id})$ so that $\psi_{f,f'}d\theta_f(\mathrm{id}) = d\theta_{f'}(\mathrm{id})$ and $\psi_{f,f'}$ is an isometry.

The right-invariance property of the group metric provides an alternate formulation of (3) or (4). Indeed, this property implies that

$$g_f^{\mathcal{F}}(\dot{f},\dot{f}) = g_{\mathrm{id}}^{\mathcal{F}}(d\theta_{f^{-1}}(f)\dot{f}, d\theta_{f^{-1}}(f)\dot{f})$$

and introducing the new variable $v(t) = d\theta_{f(t)^{-1}}(f)(\dot{f}(t)) \in \mathfrak{f}$, (3) can be expressed in the form of an optimal control problem, namely minimizing

$$\int_0^1 g_{\mathrm{id}}^{\mathcal{F}}(v(t), v(t))dt \tag{5}$$

subject to $f(0) = f_0, \dot{f} = \theta_f v$ and $\pi(f(1)) = b_1$ (where f_0 is chosen so that $\pi(f_0) = b_0$). A similar transformation can be done with (4), in which it suffices to take the function U in the form $U(f, f') = \tilde{U}(f \cdot b_0, f' \cdot b_0)$, where $\tilde{U}$ is defined on $\mathcal{B} \times \mathcal{B}$, to obtain the required invariance. This leads to the formulation that is used in Section 6 for the introduction of outer deformation metrics.

(3) In our third example, we assume that $\mathcal{G}$ is a Lie group with a left action on $\mathcal{B}$, and take $\mathcal{F} = \mathcal{G} \times \mathcal{B}$. (We do not require the action to be transitive.) We consider the mapping $\pi : \mathcal{F} \to \mathcal{B}$ such that $\pi(g, \beta) = g \cdot \beta$, which is a submersion. If one equips $\mathcal{F}$ with a Riemannian metric that is invariant by the right action of $\mathcal{G}$ on $\mathcal{F}$ defined by $(h, \beta) \cdot g = (hg, g^{-1} \cdot \beta)$, which leaves fibres invariant since $\pi((h, \beta) \cdot g) = \pi(h, \beta)$, then it is easy to show that horizontal spaces over a given point $b \in \mathcal{B}$ are isometric, therefore providing a new Riemannian metric. This construction corresponds to the metamorphosis metric introduced in Miller and Younes (2001) and Trouvé and Younes (2005), which can be expressed as

$$g_b^{\mathcal{B}}(\eta, \eta) = \min\{G_{(\mathrm{id}, b)}^{\mathcal{F}}((v, h), (v, h)) : \eta = v \cdot b + h\}.$$

Notice that one can use (1) after (2) in sequence where $\mathcal{F}$ acts transitively on $\mathcal{B}$ and $\mathcal{G}$ has another (free) action on $\mathcal{B}$ also. It is then necessary that the norm obtained after (1) be invariant by $\mathcal{G}$ in order for (2) to apply. We will review examples in Section 6.

3 Optimal control framework

As we will evidence in the sections to follow, there is a particular advantage in interpreting the variational problems appearing in metric shape registration as optimal control problems, which gives a natural and convenient framework to derive optimality equations or design algorithms to numerically solve them. In this section, we give a brief overview of the main principles of optimal control, which will be then applied to the cases of intrinsic, outer and hybrid metric matching in Sections 5–7. We will, however, restrict our presentation to systems governed by finite-dimensional ODEs. We should yet point out that, unless discretized, the optimal control problems appearing in shape analysis are generally infinite dimensional, which adds specific difficulties. We refer readers to Arguillère et al. (2015) for a partial extension of the results mentioned below to the infinite-dimensional case.

We assume that the state of the system at any time t is modelled by a variable $x(t) \in \mathbb{R}^n$ that evolves according to an ODE of the form $\dot{x}(t) = f(x(t), u(t))$ with $u(t) \in \Omega \subset \mathbb{R}^k$ being the control variable, Ω its constraint set, and $f : \mathbb{R}^n \times \Omega \to \mathbb{R}^n$ a function satisfying the adequate regularity assumptions for solutions of the ODE to exist. The class of optimal control problems we consider consists in determining a control $t \mapsto u(t)$ that brings the system from its initial state $x(0) = x_0$ to a final state $x(1) = x_1$ with x_1 belonging to a certain target set $\mathcal{C}_1$ (e.g., a single point or a submanifold of $\mathbb{R}^n$) while minimizing a cost functional of the form:

$$C(u) = \int_0^1 L(x(t), u(t))dt + U(x(1)) \tag{6}$$

in which $L : \mathbb{R}^n \times \Omega$ is called the Lagrangian functional of the system while $U(x(1))$ is a cost term on the end-time state. The two typical situations we encounter in this chapter are the cases where $\mathcal{C}_1 = \{x_1\}$ with $U = 0$ which corresponds to a boundary value (or Lagrange) problem or $\mathcal{C}_1 = \mathbb{R}^n$ with $U(x(1))$ measuring the distance of $x(1)$ to the target set which corresponds to a relaxed (or Mayer–Lagrange) problem.

There are several questions that optimal control theory attempts to address, the first of which being the existence of solutions. This existence problem usually involves first the existence of admissible trajectories between the initial state x_0 and the target set $\mathcal{C}_1$: this is the issue of *controllability* and is particularly important in the case of boundary value problems. It is, however, a difficult question to address in general and often only local controllability can be established. Conditions of global controllability exist nonetheless in the cases of linear or sub-Riemannian control systems (c.f., for instance, the presentation in Coron (2007)). Next, one needs to show that the minimum of C over all admissible controls is achieved. This typically relies on a compactness and lower semi-continuity argument on the cost functional, for which some conditions have to be carefully checked with each specific problem at hand.

The second point of optimal control is to provide some characterization of the solutions. To that end, a very powerful result is the *Pontryagin maximum principle* (PMP), first introduced in Pontryagin et al. (1962). For the control system and cost functional considered here, it can be written as follows. We first define the *Hamiltonian* $H_\lambda : \mathbb{R}^n \times \mathbb{R}^n \times \Omega \to \mathbb{R}$ as:

$$H_\lambda(x, p, u) = p^T f(x, u) - \lambda L(x, u) \tag{7}$$

where p denotes the co-state variable that we will often call *momentum*. The PMP states that if u_* is an optimal control then there must exist $\lambda \in \{0, 1\}$ and a time-dependent momentum $p : [0, 1] \to \mathbb{R}^n$ such that $(\lambda, p) \neq (0, 0)$ and the following Hamiltonian system of equations is satisfied:

$$\begin{cases} \dot{x}(t) = \partial_p H_\lambda(x(t), p(t), u_*(t)), \\ \dot{p}(t) = -\partial_x H_\lambda(x(t), p(t), u_*(t)) \\ u_*(t) = \mathrm{argmax}_{u \in \Omega} H_\lambda(x(t), p(t), u). \end{cases} \tag{8}$$

Moreover, the following transversality condition holds: $p(1) + \lambda \nabla U(x(1))$ is orthogonal to the tangent space $T_{x(1)}\mathcal{C}_1$ of the constraint set at $x(1)$.

There are thus two possible types of solutions for such optimal control problems. When $\lambda = 0$, u_* and the optimal trajectory are independent of the Lagrangian: these are called *abnormal solutions* of the system. Note that such solutions are rather singular and exist only for special systems and/or choice of boundary conditions, and we generally not consider such solutions in practice. For $\lambda = 1$, we get *normal solutions*. The transversality condition is directly related to the target set $\mathcal{C}_1$: in particular, in the unconstrained case $\mathcal{C}_1 = \mathbb{R}^n$, it reduces to $p(1) = -\nabla U(x(1))$ whereas in the boundary problem $\mathcal{C}_1 = \{x_1\}$, we get no conditions on $p(1)$ (but instead $x(1) = x_1$).

The third equation in (8) may, in some cases, be used to express the optimal control in feedback form, namely as a function of the state and the co-state, i.e., as $u_*(t) = \eta(x(t), p(t))$. For instance, if Ω is an open subset, one can write first order optimality equations $\partial_u H_\lambda(x(t), p(t), u_*(t)) = 0$ and attempt to make u_* explicit from the implicit equations. In this case, the Hamiltonian system can be rewritten in more compact form. Indeed, if we denote by $\widetilde{H}(x,p) = p^T f(x, \eta(x,p)) - L(x, \eta(x,p))$ the *reduced Hamiltonian*, (8) becomes:

$$\begin{cases} \dot{x}(t) = \partial_p \widetilde{H}(x(t), p(t)), \\ \dot{p}(t) = -\partial_x \widetilde{H}(x(t), p(t)) \end{cases} \tag{9}$$

These reduced Hamiltonian equations eventually show that (normal) optimal solutions are fully characterized by the initial momentum $p_0 \in \mathbb{R}^n$. This leads to a considerable reduction of the search space which can be exploited in numerical approaches.

In fact, several different algorithmic strategies can be implemented to approximate solutions of the previous problems. We briefly discuss two among the main ones. A first approach, commonly known as *trajectory optimization*, is to start from a discretization in time of all the problem variables. Considering time samples $0 = t^{(0)} < t^{(1)} < \cdots < t^{(N)} = 1$ and the associated collocation points for the state $x^{(0)}$, $x^{(1)}$, …, $x^{(N)}$ and control $u^{(0)}$, $u^{(1)}$, …, $u^{(N)}$, the principle is to define a numerical scheme (such as Euler and Runge–Kutta) for the ODE $\dot{x}(t) = f(x(t), u(t))$ and a quadrature rule to approximate the cost functional (6) in order to rewrite both with respect to the $x^{(i)}$ and $u^{(i)}$. This turns the optimal control problem into a nonlinear programming one, which can be tackled with standard constrained optimization solvers such as interior point method or sequential quadratic programming. Trajectory optimization methods have the advantage of being easy to design and relatively robust to the initialization of the state and control variables. However, in order to get a good accuracy for the solution, they can often result in very high-dimensional optimization problems.

A second possible approach is to exploit, when available, the dimensionality reduction given by the PMP and in particular the reduced Hamiltonian equations (9) by searching directly for an optimal initial momentum p_0. This is the principle of *shooting* methods. To keep this section short, we will only say that those basically amount in optimizing the cost functional with respect to p_0 with first or second order descent schemes, the main difficulty being to compute the gradient of (6) as a function of p_0. Numerical or automatic differentiation techniques can be very often used for that purpose. Another common strategy is to obtain the gradient by numerically solving the adjoint Hamiltonian system of equations. The optimization problem to solve is typically of much lower dimension than with trajectory optimization, although shooting methods may be more sensitive to the initialization of p_0 and the presence of local minima. The interested reader can refer to Betts (2010), Bryson and Ho (1975) and Trélat (2008) for more details and examples of those methods' practical uses.

4 Chordal metrics on shapes

4.1 Motivation

The Riemannian metric framework introduced in Section 2 leads to geodesic distances between two given shapes b_0 and b_1 that are formally expressed through a minimization over paths connecting $q_0 \in \pi^{-1}(b_0)$ to $q(1) \in \pi^{-1}(b_1)$. As already mentioned, it is often necessary to relax the end-time constraint for multiple reasons.

Indeed, in certain situations, there may not exist paths connecting q_0 to an element of the fibre $\pi^{-1}(b_1)$. Moreover, one could argue more generally that many variations between shapes may be the result of undesirable perturbations like acquisition noise, segmentation issues or boundary effects. Imposing exact matching constraint is then very likely to result in distances with poor robustness properties to such perturbations and thus of little interest in problems such as statistical shape classification on real datasets.

A second issue is the complexity of optimizing (3) under the constraint that $q(1)$ belongs to $\pi^{-1}(b_1)$, which requires an additional exploration of vertical motions in the fibre of b_1. Those consist essentially in reparametrizations of the shape b_1. While numerical frameworks for discretizing and optimizing over Diff(M) have been quite commonly used for closed and open curves (i.e., when $M = \mathbb{S}^1$ or $M = [0, 1]$) in works like Klassen et al. (2004), Srivastava et al. (2011), Bauer et al. (2017), and Srivastava and Klassen (2016), these are harder to extend to cases like closed or open surfaces and with the general class of metrics we are considering here. In this direction, only work for a particular metric of order one on the space of sphere-like surfaces has been done in Kurtek et al. (2013) and Laga et al. (2017).

In this section, we thus seek an auxiliary family of shape distances to serve as a relaxation of the constraint $q(1) \in \pi^{-1}(b_1)$, which would be (1) defined explicity, (2) reparametrization-blind by construction and (3) easy/fast to compute numerically. These shape distances are based on the idea of "embedding" shape spaces into certain functional spaces, in contrast with the previous submersion setting, and we thus refer to them as chordal metrics.

4.2 General principle

Chordal metrics are constructed by mapping $\mathrm{Imm}(M,\mathbb{R}^n)$ into a certain Banach space $(\mathcal{H}, \|\cdot\|_{\mathcal{H}})$ (typically a space of distributions or measures over a certain feature set as we detail below). The norm on $\mathcal{H}$ can be then pulled back to $\mathrm{Imm}(M,\mathbb{R}^d)$ based on this mapping.

More specifically let $\mu : \mathrm{Imm}(M,\mathbb{R}^d) \to \mathcal{H}$ be a mapping such that the representation of any $q \in \mathrm{Imm}(M,\mathbb{R}^d)$ as $\mu_q \in \mathcal{H}$ is *independent of parametrization*, that is for all $\tau \in \mathrm{Diff}(M)$ we have $\mu_{q\circ\tau} = \mu_q$. Then, one has a well-defined quotient mapping $[\mu]$ on $\mathcal{S} = \mathrm{Imm}(M,\mathbb{R}^d)/\mathrm{Diff}(M)$ and we can introduce for $q_1, q_2 \in \mathrm{Imm}(M,\mathbb{R}^d)$:

$$d_{\mathrm{chor}}([q_1],[q_2]) = d_{\mathrm{chor}}(q_1,q_2) = \| \mu_{q_1} - \mu_{q_2} \|_{\mathcal{H}} \tag{10}$$

which is precisely the chordal distance in $\mathcal{H}$ on the image of $\mathrm{Imm}(M,\mathbb{R}^d)$ by μ, as illustrated in Fig. 1. However, (10) is in general only a pseudo-distance on $\mathcal{S}$. To obtain a true distance, one needs the additional assumption that $[\mu]$ is injective, in other words that the space of shapes is embedded in $\mathcal{H}$.

Note that the resulting (pseudo-)distances are defined between any two elements of $\mathcal{S}$ but are fundamentally different from the Riemannian or sub-Riemannian metrics of Section 2 since, for chordal distances, there are no corresponding geodesics on the shape space. On the other hand, with adequate choices of μ and of the norm $\|\cdot\|_{\mathcal{H}}$, such distances can still provide a reasonable proximity criterion which can be computed in a much more direct way.

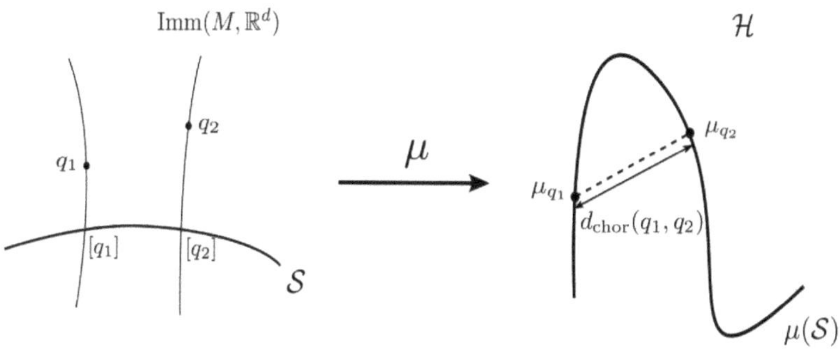

FIG. 1 Shape space embedding and chordal distance.

4.3 Oriented varifold distances

There are multiple possible constructions of a mapping μ satisfying the conditions above. A popular approach, inspired from the field of geometric measure theory, is to represent shapes as measures or distributions and compare them with the corresponding dual metrics. This is the basic idea behind such frameworks as measures (Glaunès et al., 2004), currents (Glaunès et al., 2008), varifolds (Charon and Trouvé, 2013), and recently the works of Roussillon and Glaunès (2016) on normal cycles and (Feydy et al., 2017) on optimal transport discrepancies. In the rest of this section, we focus on the setting of *oriented varifold* for curves and surfaces presented in Kaltenmark et al. (2017), which has the advantage of encompassing several of these distance families in one single model. We will also restrict the presentation to the case where shapes are either curves or hypersurfaces in $\mathbb{R}^d$, i.e., $\dim(M) = 1$ or $\dim(M) = d - 1$.

We let $\mathcal{H} = W^*$ with W being a Hilbert space of test functions on $\mathbb{R}^d \times \mathbb{S}^{d-1}$, with the continuous embedding $W \hookrightarrow C_0(\mathbb{R}^d \times \mathbb{S}^{d-1})$, and define μ as follows. For any $q \in \mathrm{Imm}(M, \mathbb{R}^d)$, μ_q is the distribution in W^* such that for all $\omega \in W$:

$$\mu_q(\omega) = \int_M \omega(q(m), \vec{t}(m)) d\,\mathrm{vol}_q(m) \tag{11}$$

where $\vec{t}\,(m) \in \mathbb{S}^{d-1}$ denotes the unit oriented tangent (resp. normal) vector to the curve (resp. hypersurface) at $q(m)$, and vol_q is the intrinsic volume density of q (in the case of curves, this would correspond to the integration with respect to arc length). A simple change of variable shows that μ_q is independent of the parametrization. Intuitively, (11) amounts in representing a shape as a distribution of its tangent spaces attached to the different positions of its points.

We can then introduce the associated chordal pseudo-distance in the dual Hilbert space W^* which we shall write:

$$d_{\mathrm{var}}([q_1],[q_2])^2 = \| \mu_{q_1} - \mu_{q_2} \|_{W^*}^2 = \| \mu_{q_1} \|_{W^*}^2 - 2\langle \mu_{q_1}, \mu_{q_2} \rangle_{W^*} + \| \mu_{q_2} \|_{W^*}^2 . \tag{12}$$

Such a distance depends obviously of the choice of the test space W and its metric. Yet the main interest of this model is that d_{var} can be in fact expressed explicitly in the right setting. Indeed, given the assumption $W \hookrightarrow C_0(\mathbb{R}^d \times \mathbb{S}^{d-1})$, W is a reproducing kernel Hilbert space (RKHS) on $\mathbb{R}^d \times \mathbb{S}^{d-1}$ which implies that W has a unique reproducing kernel $K_W : (\mathbb{R}^d \times \mathbb{S}^{d-1})^2 \to \mathbb{R}$. The reproducing kernel property allows one to write $\| \cdot \|_{W^*}$ in terms of K_W and one can show in particular that

$$\| \mu_q \|_{W^*}^2 = \iint_{M \times M} K_W((q(m), \vec{t}(m)), (q(m'), \vec{t}(m'))) d\,\mathrm{vol}_q(m) d\,\mathrm{vol}_q(m').$$

In practice, we proceed in the reverse way and start from a positive kernel on $\mathbb{R}^d \times \mathbb{S}^{d-1}$ that implicitly specifies a space W. We generally restrict to separable kernels, such that K_W is the product of a positive kernel on $\mathbb{R}^d$ and of a positive kernel on $\mathbb{S}^{d-1}$. Furthermore, it is often relevant to seek metrics that are invariant to the action of rotations and translations. This can be achieved with kernels of the following form:

$$K_W((x,\vec{t}),(x',\vec{t}'))=\rho(|x-x'|^2)\gamma(\langle\vec{t},\vec{t}'\rangle),$$

which are functions only of the distance between positions x and x' and of the angle between oriented unit vectors $\vec{t}$ and $\vec{t}'$, with $\rho:\mathbb{R}_+\to\mathbb{R}$ and $\gamma:[-1,1]\to\mathbb{R}$. Given two such functions ρ and γ that define positive kernels on $\mathbb{R}^d$ and $\mathbb{S}^{d-1}$, respectively, the oriented varifold metric becomes fully explicit:

$$\|\mu_q\|_{W^*}^2=\iint_{M\times M}\rho(|q(m)-q(m')|^2)\gamma\left(\left\langle\vec{t}(m),\vec{t}(m')\right\rangle\right)d\,\mathrm{vol}_q(m)d\,\mathrm{vol}_q(m'). \tag{13}$$

As mentioned earlier, d_{Var} defined by (12) and (13) is only a pseudo-distance on $\mathcal{S}$ unless the RKHS W associated with the kernel is such that the mapping $[\mu]$ is injective. Unfortunately, it can be seen that, in the case of oriented varifold representation (11), this is actually impossible for the full space of unparametrized immersed curves, no matter the choice of W (see, for example, Bauer et al. (2019) for a counter-example with curves). However, sufficient conditions exist for d_{var} to be a distance on the subset of *embedded* submanifolds $\mathrm{Emb}(M,\mathbb{R}^d)/\mathrm{Diff}(M)$.

Theorem 1. *Assume that ρ and γ are C^1-functions, that the kernel defined by ρ is C_0-universal on $\mathbb{R}^d$ and that $\gamma(1)>0$, $\gamma(-u)\neq\gamma(u)$ for all $u\in[-1,1]$. Then d_{var}, defined by (12) and (13), is a distance on the set of embedded shapes.*

The C_0-universality assumption is essentially a density property of the RKHS associated with ρ (c.f., Carmeli et al. (2010) for details). It is satisfied by families of kernels such as Gaussian $e^{-\frac{|x-x'|^2}{\sigma^2}}$, Cauchy $1/\left(1+\frac{|x-x'|^2}{\sigma^2}\right)$, Wendland, and so on. The assumptions on γ are much milder: typical choices include for instance the linear kernel $\gamma(\langle\vec{t},\vec{t}'\rangle)=\langle\vec{t},\vec{t}'\rangle$ or the restricted Gaussian kernel on $\mathbb{S}^{d-1}\gamma(\langle\vec{t},\vec{t}'\rangle)=e^{\frac{2}{\sigma^2}(1-\langle\vec{t},\vec{t}'\rangle)}$.

Although we do not obtain, with this framework, a true distance on $\mathcal{S}$, we point out that the result of Theorem 1 may in fact still hold with the same assumptions for a larger class than embedded shapes: for instance, it was extended to immersed curves with transverse self-intersections in Bauer et al. (2019). In all cases, and for the purpose of this work, d_{var} gives in practice a simple and explicit notion of shape proximity to be used as a relaxation of the exact matching constraint.

4.4 Numerical aspects

In practical cases, shapes are discretized and oriented varifold metrics are computed based on approximations of the integrals appearing in (13). Although there exist different possible discretization schemes for curves and surfaces, a common situation is to deal with vertices and a mesh structure such as lists of segments or triangles. In such cases, we can view the parameter space M as a reunion of simplices $M = \bigcup_{i=1}^{N_S} M_i$ where M_i are disjoint faces delimited by vertices $m_{i,1}, \ldots, m_{i,p}$ ($p = 2$ for segments, $p = 3$ for triangles). Then (13) may be rewritten as:

$$\| \mu_q \|_{W^*}^2 = \sum_{i,j=1}^{N_S} \iint_{M_i \times M_j} \rho(|q(m) - q(m')|^2)\gamma(\langle \vec{t}(m), \vec{t}(m') \rangle) d\, \mathrm{vol}_q(m) d\, \mathrm{vol}_q(m')$$

and we can approximate every integral on the product simplex $M_i \times M_j$ using a certain quadrature rule. For example, the approach proposed in Charon and Trouvé (2013) and Kaltenmark et al. (2017) consists in simply approximating the integrals by their interpolated value at the centre of simplices. Specifically, denoting $\overline{q}_i$ the barycentre of the vertices $(q(m_{i,1}), \ldots, q(m_{i,p}))$, $\vec{t}_i$ the tangent or normal orientation vector to this simplex and $\mathrm{vol}_q(M_i)$ the length or area of the simplex $q(M_i)$, the previous norm is approximated as:

$$\| \mu_q \|_{W^*}^2 \approx \sum_{i,j=1}^{N_S} \rho(|\overline{q}_i - \overline{q}_j|^2)\gamma\left(\left\langle \vec{t}_i, \vec{t}_j \right\rangle\right) \mathrm{vol}_q(M_i) \mathrm{vol}_q(M_j). \quad (14)$$

It can be shown that the error in (14) is controlled by the maximum diameter of the simplices. In addition, under certain assumptions, convergence of discrete to continuous varifold shape representations has been established. From a numerical standpoint, the computational cost of (14) or of its gradient with respect to vertex positions is typically quadratic in the number of simplices. It is, however, highly parallelizable and many recent implementations in CUDA take advantage of GPU architectures. We refer the interested reader to Arguillère et al. (2016), Kaltenmark et al. (2017), and Charon et al. (2018) for more detailed discussions on discrete varifold approximations and computations.

5 Intrinsic metrics

In this section we will study reparametrization-invariant Sobolev metrics on the space of parametrized shapes $\mathrm{Imm}(M, \mathbb{R}^d)$. The reparametrization invariance of the metric will allow us to consider induced metrics on the space of unparametrized shapes as described in Section 2 paragraph (1). We call these metrics *intrinsic* as they will be defined solely in terms of the shape (immersion, resp.), which is in contrast to the *outer deformation metric models* studied in Section 6 that are defined using the geometry of the ambient space of the (parametrized) shapes. In order to simplify the presentation we will first introduce the class of intrinsic metrics for the space of infinitely smooth immersions and will later show that they can be extended to smooth Riemannian metrics on the bigger space of all immersions of (sufficiently high) Sobolev class.

5.1 Reparametrization-invariant metrics on parametrized shapes

To view $\mathrm{Imm}_{C^\infty}(M,\mathbb{R}^d)$ as an (infinite-dimensional) manifold we observe that it is an open subset of the vector space of all smooth functions $C^\infty(M,\mathbb{R}^d)$. Thus tangent vectors to an immersion q—vector fields along the immersion—can be identified with smooth functions $h\in C^\infty(M,\mathbb{R}^d)$, see Cervera et al. (1991). To define a Riemannian metric on this infinite-dimensional manifold we have to specify a family of inner products

$$\mathrm{Imm}_{C^\infty}(M,\mathbb{R}^d)\times C^\infty(M,\mathbb{R}^d)\times C^\infty(M,\mathbb{R}^d)\ni(q,h,k)\mapsto G_q(h,k)\in\mathbb{R}, \tag{15}$$

that depends smoothly on the footpoint $q\in\mathrm{Imm}_{C^\infty}(M,\mathbb{R}^d)$. As our main goal is the study of the space of unparametrized shapes we will always require the metric to be invariant under the action of the group of reparametrizations, i.e.,

$$G_q(h,k)=G_{q\circ\tau}(h\circ\tau,k\circ\tau), \tag{16}$$

for all $\tau\in\mathrm{Diff}_{C^\infty}(M)$. This assumption is a necessary condition for a metric G to induce a Riemannian metric on the quotient space $\mathcal{S}$ such that the projection π is a Riemannian submersion. In this infinite-dimensional situation one also has to prove the existence of the horizontal space, as defined in Section 2. For all the metrics studied in this section this second condition will be satisfied, which is in contrast to metrics on $\mathrm{Imm}_{C^\infty}(M,\mathbb{R}^d)$, that are induced via outer deformation metric models, where the horizontal bundle only exists in some Sobolev completion of the tangent bundle (Micheli et al., 2013).

We are now ready to define the specific class of metrics that we will study in this section. Therefore we let Δ_q be the Laplace operator[a] induced by the submanifold q. A reparametrization-invariant Sobolev metric of order s is given by:

$$G_q(h,k)=\int_M\langle(1+\Delta_q)^{s/2}h,(1+\Delta_q)^{s/2}k\rangle_{\mathbb{R}^d}\,\mathrm{vol}_q, \tag{17}$$

[a]To formally define the Laplace operator let $g=q^*\langle .,.\rangle$ be the induced surface metric with corresponding covariant derivative ∇. Then the induced Laplacian of a function $h\in C^\infty(M,\mathbb{R}^d)$ is given by $\Delta_q h=\mathrm{tr}(g^{-1}\nabla^2 h)$. It will be important for us, that the Laplacian is equivariant under the action of the diffeomorphism group, i.e., $\Delta_{q\circ\tau}(h\circ\tau)=(\Delta_q h)\circ\tau$ for all $\tau\in\mathrm{Diff}_{C^\infty}(M)$.

where vol_q denotes integration with respect to the intrinsic volume density of the submanifold q. For odd (or non-integer) s we define the fractional Laplacian using analytic functional calculus, see, e.g., Bauer et al. (2018). For closed manifolds M we can use integration by parts to rewrite the formula of the metric as

$$G_q^s(h,k) = \int_M \langle (1+\Delta_q)^s h, k\rangle_{\mathbb{R}^d} \mathrm{vol}_q. \tag{18}$$

This class of metrics has been first introduced in the case of planar curves in Michor and Mumford (2007), Sundaramoorthi et al. (2011), Mennucci et al. (2008) and Sundaramoorthi et al. (2008) and has been later generalized to immersions between general Riemannian manifolds in Bauer et al. (2011, 2012b, 2014b).

The invariance of the metric under the action of the reparametrization group follows directly from the equivariance of the Laplacian and the transformation formula for multidimensional integration.

Remark 1. It is worth to note that the metric is in addition invariant under the Euclidean motion group (translations and rotations) and thus induces a metric on the space of shapes modulo Euclidean motions. It is, however, not invariant under the action of the scaling group. To achieve a scale invariant version one can add weights depending on the total volume of the surface into the definition of the metric, see Bauer et al., 2012b; Mennucci et al., 2008; Younes et al., 2008.

More generally, one can consider metrics defined in terms of fields of (pseudo)-differential operators L_q:

$$G_q^L(h,k) = \int_M \langle L_q h, L_q k\rangle_{\mathbb{R}^d} \mathrm{vol}_q, \tag{19}$$

where L_q is for each $q \in \mathrm{Imm}(M,\mathbb{R}^d)$ a positive, (pseudo) differential operator, that is equivariant under the action of the reparametrization group. The equivariance property is necessary to ensure the required invariance of the induced metric.

The following result states that Sobolev metrics as defined above can be extended to smooth Riemannian metrics on spaces of immersions of finite regularity:

Theorem 2. *For any $s \in \mathbb{N}$, $\mathbb{R} \ni r \geq s$ and $r > \frac{p}{2} + 1$ the Sobolev metric of order s extends to a smooth Riemannian metric on the space of Sobolev immersions $\mathrm{Imm}_{H^r}(M,\mathbb{R}^d)$ of order r.*

Remark 2 (Weak and strong Riemannian metrics). In infinite dimensions there exist two types of Riemannian metrics, which are referred to as weak and strong metrics: a Riemannian metric on an infinite-dimensional manifold $\mathcal{F}$ is called strong if the induced topology of the geodesic distance function corresponds with the original manifold topology (Lang, 1999). As a consequence one can view the metric as a bijective mapping $G: T\mathcal{F} \to T^*\mathcal{F}$. For weak Riemannian metrics the topology of the metric is strictly weaker than the standard topology and the mapping $G: T\mathcal{F} \to T^*\mathcal{F}$ is only injective but not surjective. As a consequence many standard results of finite-dimensional geometry (in which all metrics are strong) may fail in this situation, e.g., the geodesic spray might not exist and the induced geodesic distance can vanish identically on the manifold. We will see an example of such an ill-behaviour later in Section 5.3 in the context of the L^2-metric. Sobolev metrics on spaces of immersions are usually only weak Riemannian metrics. The only exception being the metric of order $s > \frac{p}{2} + 1$ on the space of Sobolev immersions $\mathrm{Imm}_{H^s}(M, \mathbb{R}^d)$ of the same order.

Example 1. To illustrate the above defined metrics, we want to present the metric of order one in the special situation $\mathrm{Imm}(\mathbb{S}^1, \mathbb{R}^2)$, i.e., for the case of metrics on the space of planar curves. In that case the induced volume density of a curve c is given by arc-length integration $\mathrm{vol}_c = |\partial_\theta c| d\theta$ and the Laplacian of the curve c is given in terms of arc-length differentiation $\Delta_c = -\frac{1}{|\partial_\theta c|}\partial_\theta \frac{1}{|\partial_\theta c|}\partial_\theta$. Thus the metric can be written as:

$$G_c^1(h,k) = \int_M \left\langle \left(1 - \frac{1}{|\partial_\theta c|}\partial_\theta \frac{1}{|\partial_\theta c|}\partial_\theta\right) h, k \right\rangle_{\mathbb{R}^2} |\partial_\theta c|\, d\theta$$

$$= \int_M \left(\langle h, k\rangle_{\mathbb{R}^2} |\partial_\theta c| + \frac{1}{|\partial_\theta c|} \langle \partial_\theta h, \partial_\theta k\rangle_{\mathbb{R}^2} \right) d\theta.$$

Using the Sobolev multiplication theorem it is easy to see that this metric extends to the Sobolev completion $\mathrm{Imm}_{H^r}(M, \mathbb{R}^d)$ for $r > \frac{p}{2} + 1$.

5.2 The metric on the space of unparametrized shapes

As described in Section 2 the metric on the quotient space $\mathcal{S}$ corresponds to the metric on the space of immersions restricted to horizontal tangent vectors. To determine the horizontal space we first have to calculate the vertical space of the submersion $\pi: \mathrm{Imm}_{C^\infty}(M, \mathbb{R}^d) \to \mathcal{S}$: therefore we recall that elements in $\mathrm{Imm}_{C^\infty}(M, \mathbb{R}^d)$ that differ only by their parametrization are identified on the space of unparametrized shapes. Tangent vectors h that only change the parametrization of q are vector fields that are always tangent to the surface, i.e.,

$$V_q = \{h = dq.X : X \in C^\infty(M, TM)\}.$$

To determine the horizontal space we have to calculate the orthogonal complement of V with respect to the metric G. It is tempting to believe that horizontal tangent vectors are all vectors that are pointwise normal to the surface. However, it turns out that this is not true in general. Instead one has to solve a partial differential equation to calculate the horizontal path of a tangent vector:

Lemma 1 (Bauer et al., 2011). *Let M be a closed manifold. Then the horizontal bundle of the projection* $\pi : \mathrm{Imm}_{C^\infty}(M, \mathbb{R}^d) \to \mathcal{S}$ *with respect to the Sobolev metric G of order s is given by*

$$H_q = \left\{ h \in T_q \mathrm{Imm}_{C^\infty}(M, \mathbb{R}^d) : \langle (1+\Delta)^{2s} h,\, dq.X \rangle = 0,\ \forall X \in C^\infty(M, TM) \right\}.$$

In cases where the horizontal projection of tangent vectors is easy to compute this can give a useful alternative approach for the calculation of the geodesic distance between two unparametrized shapes $[q_0]$ and $[q_1]$: instead of minimizing in addition over the endpoint in the fibre of $[q_1]$ as we will describe in Section 5.5, one can alternatively minimize the horizontal energy between two fixed parametrized shapes. However, since, for the class of Sobolev metrics, calculating the horizontal projection is rather expensive, we will not follow this approach.

5.3 The induced geodesic distance

Given a Riemannian metric one can consider the induced geodesic distance, which is defined as the infimum of the Riemannian length of all paths that connect two given points (shapes), i.e.,

$$d_{\mathrm{Imm}}(q_0, q_1) = \inf_q L(q) := \inf_q \int_0^1 \sqrt{G_{q(t)}(\partial_t q(t), \partial_t q(t))}dt, \tag{20}$$

where this infimum is calculated over all paths of immersions

$$q : [0,1] \to \mathrm{Imm}_{C^\infty}(M, \mathbb{R}^d) \tag{21}$$

with $q(0) = q_0$ and $q(1) = q_1$. We also obtain an induced geodesic distance function on the space of unparametrized shapes:

$$d_{\mathcal{S}}([q_0],[q_1]) = \inf_{\tau \in \mathrm{Diff}_{C^\infty}(M)} d_{\mathrm{Imm}}(q_0, q_1 \circ \tau). \tag{22}$$

In a similar way, one can obtain a geodesic distance function on spaces of shapes modulo translations and rotations. To factor out the group of scalings, one has to define a scale invariant metric on the space of parametrized shapes, c.f. Remark 1.

On a finite-dimensional manifold the geodesic distance always defines a "true" metric, i.e., it is symmetric, satisfies the triangle inequality and non-degenerate. In infinite dimensions this result might not be true anymore as the geodesic distance can degenerate or even vanish identically. The first instance of this phenomenon has been found by Eliashberg and Polterovich for the H^{-1}-metric on the symplectomorphism group in Eliashberg and Polterovich

(1993). In the context of shape analysis of submanifolds a similar result has been found by Michor and Mumford for the L^2-metric on the space of unparametrized submanifolds and on the diffeomorphism group, see Michor and Mumford (2005). These results have been later extended to spaces of parametrized submanifolds and to fractional-order metrics on diffeomorphism groups, see Bauer et al. (2012a, 2013). For Sobolev metrics of order $s \geq 1$ on spaces of submanifolds the following theorem shows that this ill behaviour cannot appear, which renders this class of metrics relevant for applications in shape analysis.

Theorem 3 (Bauer et al., 2011; Michor and Mumford, 2007). *The geodesic distance of the Sobolev metric* (17) *of order $s \geq 1$ is non-degenerate both on the space of parametrized and unparametrized shapes.*

This result has only been formally proven for the space of unparametrized shapes. For parametrized shapes it follows from the existence of simplifying transformation for H^1-metrics (Kurtek et al., 2010, 2012; Srivastava et al., 2011; Younes et al., 2008) and the fact that H^s-metrics can be bounded from below by the H^1-metric for $s \geq 1$.

5.4 The geodesic equation

A classical result of Riemannian geometry states that the squared geodesic distance is also the infimum over all paths of the Riemannian energy,

$$d_{\text{Imm}}(q_0, q_1)^2 = \inf_q E(q) = \inf_q \int_0^1 G_{q(t)}(\partial_t q(t), \partial_t q(t)) dt. \tag{23}$$

Thus we can find critical points of the length functional (i.e., locally shortest paths) by looking for critical points of the Riemannian energy functional. These critical points of the energy functional are called *geodesics* and the first order condition for critical points, $dE(c) = 0$ is the *geodesic equation*. For our class of metrics the geodesic equation is a nonlinear partial differential equation for the function $q = q(t, x)$ that is of order two in t and of order $2s$ in $x \in M$. We refrain from presenting this equation here as it is rather complicated and not very insightful. Instead we refer the interested reader to the articles (Bauer et al., 2011; Michor and Mumford, 2007).

Since we are working in infinite dimensions the existence of geodesics is a highly non-trivial question and analytic solution formulas are only available in scarce special cases. The following theorem summarizes local and global well-posedness results that are known for the class of reparametrization-invariant metrics on the space of parametrized submanifolds.

Theorem 4. *Let M be a closed manifold and let G be the reparametrization-invariant Sobolev metric of order $s \in \mathbb{N}$, $s \geq 1$, as defined in (17) on the space of parametrized submanifolds. Then for any $r > \frac{p}{2} + 1$ the initial value problem for the geodesic equation has unique local solutions in the manifold*

$\mathrm{Imm}_{H^{r+2s}}(M,\mathbb{R}^d)$. *The solutions depend smoothly on t and on the initial conditions* $q(0,\cdot)$ *and* $q_t(0,\cdot)$ *and the domain of existence (in t) is uniform in r and thus this also holds in the smooth category* $\mathrm{Imm}_{C^\infty}(M,\mathbb{R}^d)$

Longtime existence of geodesics on the manifold $\mathrm{Imm}(M,\mathbb{R}^d)$ remains an open question for general parameter manifolds M. For the case $M=\mathbb{S}^1$, i.e., the space of curves in $\mathbb{R}^d$, the completeness of metrics of order $s\geq 2$ has been shown by Bruveris et al. (2014) and Bruveris (2015), see also Nardi et al. (2016).

Theorem 5. *Let G be the Sobolev metric of order* $s\in\mathbb{N}$, $s\geq 2$, *on the space of closed curves. We have:*

(1) *The Riemannian manifold* $(\mathrm{Imm}_{C^\infty}(\mathbb{S}^1,\mathbb{R}^d),G)$ *is geodesically complete, i.e., given any initial conditions* $(c_0,u_0)\in T\mathrm{Imm}_{C^\infty}(\mathbb{S}^1,\mathbb{R}^d)$ *the solution of the geodesic equation for the metric G with initial values* (c_0,u_0) *exists for all time.*

(2) *The metric completion of the space* $\mathrm{Imm}_{C^\infty}(\mathbb{S}^1,\mathbb{R}^d)$ *equipped with the geodesic distance* dist^G *is the space* $\mathrm{Imm}_{H^s}(\mathbb{S}^1,\mathbb{R}^d)$ *of immersions of Sobolev class* H^s. *Furthermore, any two curves in the same connected component of the space* $\mathrm{Imm}_{H^s}(\mathbb{S}^1,\mathbb{R}^d)$ *can be joined by a minimizing geodesic.*

There is some evidence that suggests that a similar result should be valid for general M if the order s of the metric satisfies $s>\frac{p}{2}+1$. A proof of this statement is, however, largely missing.

5.5 An optimal control formulation of the geodesic problem on the space of unparametrized shapes

In the following we will describe the optimal control formulation for the geodesic boundary value problem, as introduced in Section 3, in the specific situation studied in this section. We recall, that a geodesic between $b_0=\pi(q_0)=[q_0]$ and $b_1=\pi(q_1)=[q_1]$ minimizes $\int_0^1 G_{q(t)}(h(t),h(t))dt$ subject to $\dot{q}(t)=h(t)\in T_{q(t)}$ $\mathrm{Imm}(M,\mathbb{R}^d)$ and the boundary constraints $q(0)=q_0$ and $q(1)\in\pi^{-1}(b_1)$. The latter constraint can be replaced by jointly optimizing over variations in the fibre $q(1)\circ\tau$ for $\psi\in\mathrm{Diff}(M)$. This leads to the following formulation of the boundary value problem on the space of unparametrized shapes:

$$\min_{h,\tau}\int_0^1 G_{q(t)}(h(t),h(t))dt \tag{24}$$

over $h\in L^2([0,1],C^\infty(M,\mathbb{R}^d))$, $\tau\in\mathrm{Diff}(M)$ with $q(0)=q_0$, $q(1)=q_1\circ\tau$ and $\dot{q}(t)=h(t)$. It is often more practical to relax the end-time constraint using a fidelity term blind to fibre variations, in other words to reparametrizations of $q(1)$ and q_1. This is where the chordal distances defined in Section 4 fit in. In particular, using oriented varifold distances, the inexact matching problem may be formulated as:

$$\min_h \int_0^1 G_{q(t)}(h(t),h(t)) \;\; dt + \lambda d_{\text{var}}(q(1),q_1)^2 \tag{25}$$

with $\dot{q}(t) = h(t)$, $q(0) = q_0$ and $\lambda > 0$ being a weight parameter. Interestingly, (25) turns into an infinite-dimensional *optimal control* problem in which, at each time t, the state of the system is the immersion $q(t)$ and the control is the vector field $h(t)$.

5.6 Numerical aspects

We will now describe a discretization of the relaxed geodesic estimation problem (25). We will focus on the discretization of the intrinsic metric part of the cost, the discretization of the varifold metric having been discussed in Section 4. In the language of Section 3, our approach will be based on a trajectory optimization approach. For the space of curves and Sobolev metrics of first order, there exist simplifying transformations—related to the so-called square root velocity mapping—that map the manifold of shapes into a subset of a flat space. This in turn allows one to obtain (almost) explicit formulas for geodesics and geodesic distance, which makes the numerical computations on the space of parametrized shapes trivial (Bauer et al., 2014a; Kurtek and Needham, 2018; Srivastava et al., 2011; Younes, 1998; Younes et al., 2008). Since we aim to describe a numerical framework for a more general class of metrics, we will not describe this approach, but focus on the discretization of the energy functional. Furthermore, we will focus on the case of open and closed curves. For a four-parameter family of second order Sobolev metrics, the presented framework is available under an open source license.[b]

Theoretically the described approach will directly generalize to higher dimensional objects. However, so far only the case of open and closed curves has been implemented for this general class of metrics and is available in open access. It is an ongoing project of the authors to obtain a similar numerical framework for the space of surfaces in $\mathbb{R}^3$. (Note, however, that, based on a generalization of the square root velocity transform, a numerical scheme for a particular Sobolev metric of first order on the space of surfaces has been developed in Laga et al. (2017) and Kurtek et al. (2013).)

To discretize the energy of a path of curves with respect to the intrinsic metric G it is convenient to use a B-splines representation of the involved elements, i.e., a path of curves $c(t, \theta)$ is represented by a tensor-product

[b]The code can be downloaded at github: https://www.github.com/h2metrics/h2metrics, see Bauer et al. (2019) for a detailed description of the implementation.

B-spline on knot sequences of orders n_t in time and n_θ in space. Note that the degree of the spline in space n_θ should be greater than the order of the metric s. Using this discretization one obtains $N_t N_\theta$ basis splines $B_i(t)C_j(\theta)$, where $i \in\{1, \ldots, N_t\}$ and $j \in\{1, \ldots, N_\theta\}$ and N_t and N_θ are the number of control points in each variable, respectively. Note that $B_i(t)$ are B-splines defined by an equidistant simple knot sequence on [0, 1] with full multiplicity at the boundary knots, and $C_j(\theta)$ are defined by an equidistant simple knot sequence on $[0, 2\pi]$ with either periodic boundary conditions or full multiplicity at the boundary for closed or open curves, respectively. Now, paths of curves $c(t, \theta)$ are parametrized as

$$c(t,\theta)=\sum_{i=1}^{N_t}\sum_{j=1}^{N_\theta} c_{i,j}B_i(t)C_j(\theta). \tag{26}$$

More details on this approach can be found in Bauer et al. (2017, 2019). The full multiplicity of the boundary knots in t simplify the boundary constraints, since we have

$$c(0,\theta) \;=\sum_{j=1}^{N_\theta} c_{1,j}C_j(\theta),\quad c(1,\theta)=\sum_{j=1}^{N_\theta} c_{N_t,j}C_j(\theta).$$

Thus the initial curve $c(0)$ is given by the control points $c_{1,\,j}$ only. Using this B-spline representation, it is then straightforward to discretize the Riemannian energy: we specifically use Gaussian quadrature with quadrature nodes placed between knots where the curves are smooth. In the article (Bauer et al., 2017) we presented a careful analysis of the robustness of the B-spline discretization including a Γ-convergence result on the resulting discrete energy.

This discretization then yields a trajectory optimization approach on the spline control points, which in that case boils down to an unconstrained minimization problem—the free variables being the intermediate control points $c_{i,j}$ for $i = 2,\ldots, N_{t-1}$—which can be tackled with standard methods of finite-dimensional optimization such as the L-BFGS algorithm from the HANSO library (Overton, 2016) that we use. The complexity of the algorithm at each optimization step is approximately $O(N_\theta N_t)$ (evaluation of splines and their derivatives to compute the energy and its gradient for which the required number of basic operations is linear in the total number of spline control points). To solve the geodesic matching problem, one also needs to compute at each optimization step the varifold distance and its gradient which have complexity $O(N_\theta^2)$.

We show examples of geodesics on the space of unparametrized curves in Fig. 2. One can clearly see the influence of the choice of metric constants on the resulting geodesics. As a comparison we will present the same example for the outer and hybrid deformation metric models (that will be described in the next sections) in Fig. 3.

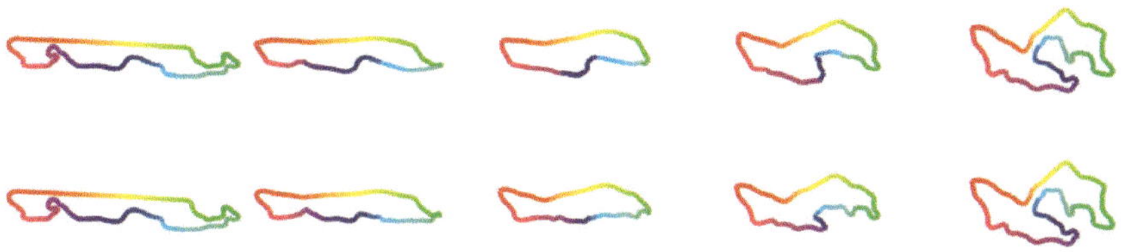

FIG. 2 Optimal deformations between two unparametrized shapes with respect to an intrinsic H^1-metric at selected time steps ($t = 0, 0.25, 0.5, 0.75, 1$). These boundary shapes represent the outlines of Florida 5th congressional district and Maryland 7th congressional district. In the second row the amount of H^1-energy is increased by a factor of 100. One can observe that the higher curvature regions are easier resolved in the first row as compared to the second row.

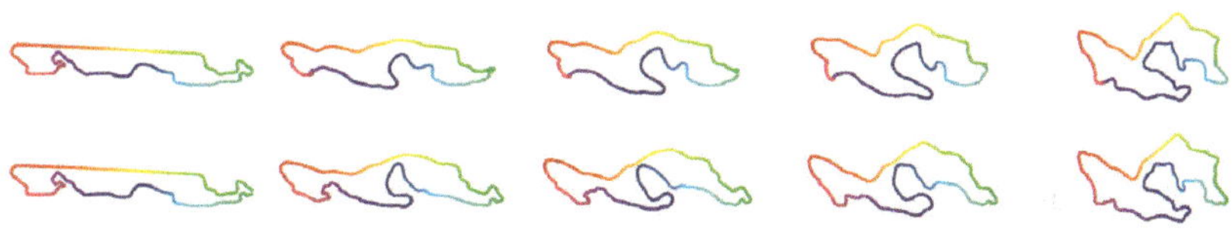

FIG. 3 Geodesics for outer and hybrid metrics on closed curves at selected time steps ($t = 0, 0.3, 0.5, 0.7, 1$). *Top row* (Outer metric): a few portions of the template curve are dramatically compressed to align the two shapes (in the *green region* on the right and in the *purple/pink* transition), at a point making them almost invisible at the image resolution. *Bottom row* (hybrid metric): the regions that were compressed in the outer metric geodesic now properly unwrap since compression along the curve is now more strongly penalized.

6 Outer deformation metric models

We now switch to a construction based on the transitive group action model described in Section 2 paragraph (2), in which $\mathrm{Imm}(M,\mathbb{R}^d)$ becomes the base space and the top space is now provided by the diffeomorphism group of $\mathbb{R}^d$ ($\mathrm{Diff}(\mathbb{R}^d)$) with action $\phi \cdot q = \phi \circ q$.

If $\boldsymbol{B}$ is a Banach space of vector fields over $\mathbb{R}^d$, we let $\mathrm{Diff}_{\boldsymbol{B}}$ denote the space of diffeomorphisms ϕ on $\mathbb{R}^d$ such that $\phi - \mathrm{id}$ and $\phi^{-1} - \mathrm{id}$ both belong to $\boldsymbol{B}$. We will in particular use $\mathrm{Diff}_{C_0^r}(\mathbb{R}^d)$ to denote the space corresponding to $\boldsymbol{B} = C_0^r(\mathbb{R}^d,\mathbb{R}^d)$, the space of vector fields v on $\mathbb{R}^d$ that are r times continuously differentiable (with $r \geq 1$) and tend to 0 (together with their first r derivatives) uniformly at infinity. Such spaces are Banach manifolds with $\boldsymbol{B} = T_\phi$ $\mathrm{Diff}_{\boldsymbol{B}}$ for all ϕ (and in particular at $\phi = \mathrm{id}$). We will always assume that $\boldsymbol{B}$ is continuously embedded in $C_0^1(\mathbb{R}^d,\mathbb{R}^d)$.

In this infinite-dimensional setting, the Riemannian framework needs to be weakened into a sub-Riemannian one, in which the metric will only be defined on subspaces of the tangent spaces, inducing the notion of admissible trajectories. More precisely, let $\mathbf{V}$ be a Hilbert space of vector fields, continuously embedded in $\boldsymbol{B}$, and let $\langle \cdot , \cdot \rangle_{\mathbf{V}}$ and $\|\cdot\|_{\mathbf{V}}$ denote the inner product and norm. For $\phi \in \mathrm{Diff}_{\boldsymbol{B}}$, we define

$$\mathbf{V}_\phi = \mathbf{V} \circ \phi = \{v \circ \phi, v \in \mathbf{V}\}$$

(so that $\mathbf{V} = \mathbf{V}_{\mathrm{id}}$), and the sub-Riemannian metric

$$G_\phi(w,w') = \langle w \circ \phi^{-1}, w' \circ \phi^{-1} \rangle_{\mathbf{V}}, \text{ for } w,w' \in \mathbf{V}_\phi.$$

(This metric is called sub-Riemannian because it is only defined on a subset of the tangent bundle of $\mathrm{Diff}_{\boldsymbol{B}}$.)

A differentiable path $\psi(\cdot)$ on $\mathrm{Diff}_{\boldsymbol{B}}$ is then called admissible if $\dot{\psi}(t) \in V_{\psi(t)}$ for all t (such paths are often called horizontal in the sub-Riemannian geometry literature, but we are using horizontal with another meaning here). The energy of an admissible path (defined for $t \in [0, 1]$) is given by

$$E(\psi) = \int_0^1 G_{\psi(t)}(\dot{\psi}(t), \dot{\psi}(t))\, dt$$

A diffeomorphism $\phi \in \mathrm{Diff}_{\boldsymbol{B}}$ is called attainable (from the identity) is there exists an admissible path with finite energy such that $\psi(0) = \mathrm{id}$ and $\psi(1) = \phi$. The set of attainable diffeomorphisms will be called $\mathrm{Diff}_{\mathbf{V}}^{att}$. One also defines a right-invariant sub-Riemannian distance d_V on $\mathrm{Diff}_{\mathbf{V}}^{att}$ such that $d_{\mathbf{V}}(\phi, \phi') = d_{\mathbf{V}}(\mathrm{id}, \phi' \circ \phi^{-1})$ is the square root of the minimal energy among all admissible paths between id and $\phi' \circ \phi^{-1}$.

We then have the following result (see Younes (2010) for a proof).

Theorem 6 (Trouvé). *If $\boldsymbol{B}$ is embedded in $C_0^r(\mathbb{R}^d, \mathbb{R}^d)$, then $\mathrm{Diff}_{\mathbf{V}}^{att}$ is a subgroup of $\mathrm{Diff}_{C_0^r}(\mathbb{R}^d)$ and a complete space for the sub-Riemannian distance d_V. Moreover, for any $\phi \in \mathrm{Diff}_{\mathbf{V}}^{att}$, there exists an optimal path $\psi(\cdot)$ such that $E(\psi)$ is minimal over all admissible paths between id and ϕ.*

Notice that $\mathrm{Diff}_{\mathbf{V}}^{att}$ is in general distinct from $\mathrm{Diff}_{\mathbf{V}}$. Exceptions in which the two sets coincide are provided when $\mathbf{V}$ equals a Sobolev space of vector fields of high enough order (see Bruveris and Vialard, 2014).

$\mathrm{Diff}_{\boldsymbol{B}}$ acts on both $\mathrm{Imm}_{C^s}(M, \mathbb{R}^d)$ and $\mathrm{Emb}_{C^s}(M, \mathbb{R}^d)$ as soon as $r \geq s$ through the action $\phi \cdot q = \phi \circ q$ (because, if q is in any of these two spaces, $\phi \cdot q$ also belongs to the same space). The infinitesimal action is $v \cdot q = v \circ q$. The optimal control problems described in Eqs. (3) and (4) can then be rewritten as follows. The square distance on the base space between q_0 and q_1 is obtained by minimizing

$$\int_0^1 \| v(t) \|_{\mathbf{V}}^2\, dt \tag{27}$$

subject to $q(0) = q_0$, $q(1) = q_1$ and $\dot{q} = v \circ q$. The relaxed version removes the constraint $q(1) = q_1$ and replaces it by a penalty $U(q(1), q_1)$ added to the integral.

Horizontal and vertical subspaces can be defined within this sub-Riemannian framework as subsets of $\mathbf{V}$ or $\mathbf{V}_\phi$, letting, for a given q, $\mathfrak{v}_q = \{v \in \mathbf{V} : v \circ q = 0\}$, $\mathfrak{h}_q = \mathfrak{v}_q^\perp$ with $V_\phi = \mathfrak{v}_{\phi \circ q} \circ \phi$, $H_\phi = \mathfrak{h}_{\phi \circ q} \circ \phi$. The specific instances taken by these spaces can, however, be surprising in some cases. It is for instance possible to find examples for which $\mathfrak{v}_q = \{0\}$ so that $\mathfrak{h}_q = \mathbf{V}$, which is counter-intuitive compared to the finite-dimensional setting in which $\mathcal{B}$ and $\mathfrak{h}_q$ should have the same dimension.

As described in Section 2, the induced (sub-Riemannian) metric on $\mathrm{Imm}_{C^s}(M, \mathbb{R}^d)$ is defined by

$$G_q^{\mathbf{V}}(h,h) = \min\{\| v \|_{\mathbf{V}}^2 : v \in \mathbf{V}, v \circ q = h\} \tag{28}$$

for all $h \in \mathbf{V}_q = \mathbf{V} \circ q \subset C^s(M, \mathbb{R}^d)$. If $h \in \mathbf{V}_q$, then there is a unique $v = h^q \in \mathfrak{h}^q$ such that $G_q^{\mathbf{V}}(h,h) = \| h^q \|_{\mathbf{V}}$, and the minimization in (27) (or its relaxed version) can be reduced with the additional constraint $v(t) \in \mathfrak{h}_{q(t)}$ without loss of generality. This remark will be useful, in particular, for the discrete version of the problem that is described below.

In order to allow this metric to formally descend to unparametrized curves $[q] \in \mathcal{S}$, we need to check that it is right invariant by diffeomorphisms $\tau \in \mathrm{Diff}_{C^s}(M)$, i.e., that

$$G_{q \circ \tau}^{\mathbf{V}}(h \circ \tau, h \circ \tau) = G_q^{\mathbf{V}}(h,h)$$

with $\mathbf{V}_{q \circ \tau} = \mathbf{V}_q \circ \tau$, which is clear from the definitions. So, if U is any function such that $U(q, q') = 0$ if and only if q' is a reparametrization of q (which holds for the chordal metrics introduced in Section 4 for instance), the minimization of

$$\int_0^1 \| v(t) \|_{\mathbf{V}}^2 \, dt + U(q(1), q_1)$$

subject to $q(0) = q_0$ and $\dot{q}(t) = v(t) \circ q(t)$ provides a metric registration method between unparametrized immersions, inducing a method called large deformation diffeomorphic metric mapping, or "LDDMM" (Beg et al., 2005; Glaunès et al., 2008; Joshi and Miller, 2000; Vaillant and Glaunès, 2006; Younes, 2010). This problem can also be expressed in sub-Riemannian form, minimizing

$$\int_0^1 G_{q(t)}^{\mathbf{V}}(\dot{q}(t), \dot{q}(t)) \, dt + U(q(1), q_1)$$

subject to $q(0) = 0$, $\dot{q} \in \mathbf{V}_{q(t)}$ for all t. Notice that any admissible path starting from $\mathrm{Emb}_{C^s}(M, \mathbb{R}^d)$ will remain in that space at all times, and that, when restricted to that space, this optimal control problem becomes one between embedded submanifold of $\mathbb{R}^d$ (diffeomorphic to M).

Unsurprisingly, the complications that result from the infinite-dimensional nature of the shape space disappear after discretization, in which case the variational framework is a direct application of the formulation described in Section 2. Using discrete curves and triangulated surfaces, the manifold M can be replaced by a simplicial complex, which, in the surface case, is provided by a family of singletons, pairs and triples of integers. In the special case of a triangulated surface, the complex can be described as a finite set of indices, say $\mathcal{I}$ together with a set $\mathcal{T}$ consisting of triples (i, j, k) where triples that differ with a circular permutation are identified, i.e., $(i, j, k) = (j, k, i)$. Edges are deduced from $\mathcal{I}$ and $\mathcal{T}$ as the set containing all pairs that are subsets of triples in $\mathcal{T}$. Letting $\mathcal{M} = (\mathcal{I}, \mathcal{T})$, We can then denote by $\mathrm{Imm}(\mathcal{M}, \mathbb{R}^3)$ the set of all one-to-one mappings $q = (q(i), i \in \mathcal{I})$ from $\mathcal{I}$ to $\mathbb{R}^3$, and $(q, \mathcal{T})$ then forms a triangulated surface. (Typically, $\mathcal{M}$ is obtained though a triangulation of a continuous surface M.) With the action of diffeomorphisms still defined by $\phi \cdot q = \phi \circ q$, and the infinitesimal action $v \cdot q = v \circ q$, the vertical space $\mathfrak{v}_q$ is given by

$$\mathfrak{v}_q = \{v \in \mathbf{V} : v(q(i)) = 0, i \in \mathcal{I}\}.$$

The space $\mathfrak{h}_q = \mathfrak{v}_q^\perp$ can be described using the reproducing kernel of $\mathbf{V}$, which is an RKHS of vector fields since it is embedded in $C_0^1(\mathbb{R}^r, \mathbb{R}^d)$. Denoting this kernel by K, which is therefore a matrix-valued function $(x, y) \mapsto K(x, y)$, the space $\mathfrak{h}_q$ is simply

$$\mathfrak{h}_q = \left\{ v = \sum_{i \in \mathcal{I}} K(\cdot, q(i)) a(i) : a(i) \in \mathbb{R}^3, i \in \mathcal{I} \right\}.$$

Moreover, if $v = \sum_{i \in \mathcal{I}} K(\cdot, q(i)) a(i) \in \mathfrak{h}_q$, then

$$\| v \|_{\mathbf{V}}^2 = \sum_{i, j \in \mathcal{I}} a(i)^T K(q(i), q(i)) a(i)$$

is also explicit. This implies that the problem in (27) with the additional constraint $v \in \mathfrak{h}_q$ can now be rephrased as a finite-dimensional optimal control problem, minimizing

$$\int_0^1 L(q(t), a(t))\, dt$$

subject to $q(0, \cdot) = q_0$, $q(1, \cdot) = q_1$ and $\dot{q}(t, i) = \sum_{j \in \mathcal{I}} K(q(t, i), q(t, j)) a(t, j)$, where

$$L(q, a) = \sum_{i, j \in \mathcal{I}} a(i)^T K(q(i), q(j)) a(j).$$

The corresponding metric on $\mathrm{Imm}(\mathcal{M}, \mathbb{R}^3)$ is actually Riemannian (not just sub-Riemannian), with

$$G_q^{\mathbf{V}}(h, h) = h^T \mathcal{K}(q)^{-1} h$$

where $h = (h(i), i \in \mathcal{I})$ with $h(i) \in \mathbb{R}^3$ is identified with a column vector and $\mathcal{K}(q)$ is the matrix formed with blocks $K(q(i), q(j))$, $i, j \in \mathcal{I}$.

Finally, the relaxed version of this problem requires using data attachment terms $U(q, q')$ adapted to triangulated surfaces, such as those described in Section 4. Several algorithms and open-access numerical implementations of curve/surface LDDMM have been developed. In the vocabulary of Section 3, shooting-based implementations include codes such as Deformetrica,[c] fshapesTk,[d] catchine[e] while other implementations rely instead on trajectory optimization over the time-varying costates $a(t)$, such as py-lddmm.[f] Matching results based on the latter code are shown in the first rows of Fig. 3 for curves and of Fig. 5 for surfaces.

7 A hybrid metric model

The outer metric framework of the previous section has the advantage of enforcing a diffeomorphic transformation of the template shape and prevents geodesics from changing topology along optimal paths, which is often an important constraint in applications. Yet, penalizing an extrinsic deformation of the ambient space as measured by a global metric G_q^V can also end up being limited. This is the case, most notably, when each shape is itself a complex of multiple curves or surfaces in close proximity to one another (Arguillère et al., 2016; Durrleman et al., 2014). Indeed, while one expects smooth deformations of each individual component, it is usually critical to also estimate more local changes of alignment and sliding motions between them: these two conditions are generally hard to jointly satisfy in the LDDMM model, even with multiscale approaches. In this section, we take the advantages of both intrinsic and outer metrics and combine them into an hybrid model. The following is a condensed version of the more complete presentation of Younes (2018).

The construction of hybrid metrics follows a similar setting as Section 6. The key difference consists in replacing the cost function in (27) by:

$$\int_0^1 \| v(t) \|_{q(t)}^2 \, dt \tag{29}$$

where, for $q \in \mathrm{Imm}_{C^s}(M, \mathbb{R}^d)$, $\| v \|_q^2 = \| v \|_V^2 + G_q(v \circ q, v \circ q)$, G_q being an intrinsic metric of order at most s like the Sobolev metric of (17). Note that the LDDMM framework corresponds to the particular case $G_q = 0$. Now the induced metric on $\mathrm{Imm}_{C^s}(M, \mathbb{R}^d)$ in (28) becomes

[c] http://www.deformetrica.org/.
[d] https://github.com/fshapes/fshapesTk.
[e] www.mi.parisdescartes.fr/~glaunes/catchine.zip.
[f] https://bitbucket.org/laurent_younes/registration.git.

$$G_q^{\mathbf{V}}(h,h) = \min\{\| v \|_{\mathbf{V}}^2 + G_q(h,h) : v \in \mathbf{V}, v \circ q = h\} \tag{30}$$

for all $h \in \mathbf{V}_q = \mathbf{V} \circ q \subset C^s(M, \mathbb{R}^d)$. Note that the two terms in this hybrid model play a different and complementary role. The intrinsic term constrains the regularity of the deformation of the shape $q(M)$ itself while the LDDMM term is mostly here to enforce the global transformation resulting from the full vector field v to be diffeomorphic. In particular, small scale kernels for the norm on $\mathbf{V}$ can be typically used in this context, in order, for example, to allow more local transformations of the background space.

The inexact matching problem under the hybrid metrics then consists in the minimization of:

$$\int_0^1 \left[\| v(t) \|_{\mathbf{V}}^2 + G_{q(t)}(v(t) \circ q(t), v(t) \circ q(t))\right] dt + F(q(1), q_1)$$

over $v \in L^2([0, 1], \mathbf{V})$ and subject to $q(0) = q_0$, $\dot{q}(t) = v(t) \circ q(t)$. As in Section 6, if one considers discrete simplicial shapes, this problem can be turned into a finite-dimensional optimal control problem. Specifically, using the same notations as previously, the discrete Lagrangian is now:

$$L(q,a) = a^T \mathcal{K}(q) a + (\mathcal{K}(q)a)^T \Lambda(q)(\mathcal{K}(q)a)$$

where $\mathcal{K}(q)$ is again the matrix of the kernel of V evaluated at the vertices, $\mathcal{K}(q)a \in \mathbb{R}^{d|\mathcal{I}|}$ the vectorized deformation field at all vertices, and $\Lambda(q) \in \mathbb{R}^{(d|\mathcal{I}|)\times(d|\mathcal{I}|)}$ is the matrix that corresponds to the discretization of the differential operator defining the intrinsic part of the metric. $\Lambda(q)$ is typically a sparse matrix obtained from either a finite difference or a finite element scheme. For instance, Younes (2018) uses an H^1 metric for G_q which is discretized based on a finite difference scheme on each simplex of the shape. In that case, the resulting approximations $(\mathcal{K}(q)a)^T \Lambda(q)(\mathcal{K}(q)a)$ become:

$$\sum_{(i,j) \in \mathcal{T}} \frac{|h(j) - h(i)|^2}{|q(j) - q(i)|} \quad \text{(curve)} \tag{31}$$

$$\sum_{(i,j,k) \in \mathcal{T}} \frac{1}{4A_{ijk}} |h(i)(q(k) - q(j)) + h(j)(q(i) - q(k)) + h(k)(q(j) - q(i))|^2 \quad \text{(surface)} \tag{32}$$

where $h \in \mathbb{R}^{|\mathcal{I}| \times d}$ are the values of v at the different vertices $q(i)$ and, in the case of triangulated surfaces, for each triangle (i, j, k) in $\mathcal{T}$, A_{ijk} denotes its area which is given by the norm of $(q(j) - q(i)) \wedge (q(k) - q(i))$.

The eventual optimization procedure to solve the discrete problem can be then derived from one of the numerical methods for optimal control that were discussed in Section 3. Fig. 3 (second row) shows the geodesic obtained with a hybrid LDDMM + H^1 metric on the same curve example as in Fig. 2. In this example, we used an H^1 norm penalizing only the tangential part of the derivative of the vector field along the curve (a special case of the two-parameter model introduced in Mio and Srivastava (2004)). We finally perform a similar experiment with the example of the incomplete ring surfaces shown in Fig. 4, where the hybrid model uses the standard H1 norm discretized according to (32). The resulting geodesics are shown in Fig. 5.

FIG. 4 Deformed incomplete rings. *Left*: template; *Right*: target.

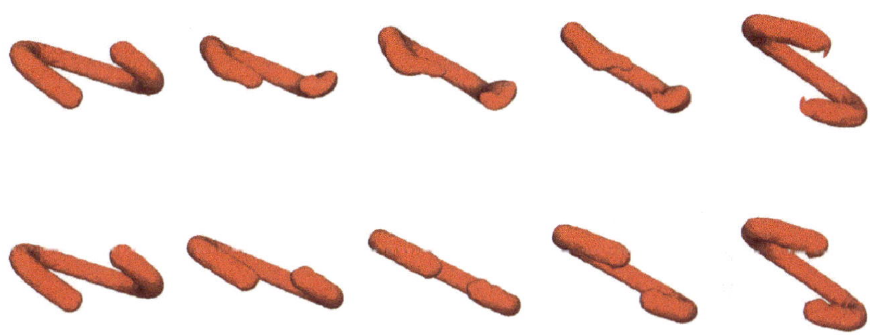

FIG. 5 Geodesics between surfaces for outer and hybrid metrics at selected time steps ($t = 0$, 0.3, 0.5, 0.7, 1). Top row: outer metric, bottom row: hybrid metric. The hybrid-metric geodesic exhibit significantly less distortion in order to allow the arms of the open ring to pass nearby during the transition from template to target.

8 Conclusion

We have summarized in this chapter some of the most recent developments in metric shape analysis and registration. While this presentation covered a large range of works by our group and others, it certainly does not account for the whole spectrum of literature that involves shape spaces. For example, we only briefly mentioned metamorphosis, which leads to interesting metrics that are still investigated, including recent work such as Holm et al. (2009), Richardson and Younes (2013), Richardson and Younes (2016), Charon et al. (2016), and Effland et al. (2018). We also made a short account of the large body of work on the H^1 metric on curves or on the LDDMM algorithms and applications of these two methods, the relevant literature being too large to be listed in this chapter (but the reader may refer to books such as Grenander et al. (2007); Srivastava and Klassen (2016); Younes (2010) and the references they contain). Recent developments on the numerical analysis of this class of problems, include variational time discretization of geodesic calculus (Rumpf and Wirth, 2015), conformal mapping approaches for curve comparison (with the Weil–Petersson metric introduced in Sharon and Mumford (2006b) and Feiszli and Mumford (2007)) and for surface registration (Gu and Yau, 2008). One can also cite computational work around the Gromov–Hausdorff (or Gromov–Wasserstein) distance between surfaces (Bronstein et al., 2010b; Memoli, 2007; Mémoli, 2011) or recent methods relaxing of the registration problem through the introduction of functional maps (Ovsjanikov et al., 2012).

The framework discussed in this chapter offers multiple open theoretical or computational problems. In addition to those that were already mentioned in the main text, there is the important issue of generalizing the models beyond the framework of manifolds, including more general objects such as stratified spaces, or trees and graphs, on which very little has been achieved so far (Duncan et al., 2016). Also, an important recent and current body of work is focusing on stochastic processes on shape spaces (Arnaudon et al., 2017a, b; Holm, 2015; Sommer et al., 2017; Staneva and Younes, 2017; Trouvé and Vialard, 2012), leading to new avenues in the statistical analysis of shapes. The extensive activity within the field is certainly justified by the originality of the mathematical problems that are raised, combined with the diversity and importance of its applications.

Acknowledgements

N.C. is partially supported by the National Science Foundation under Grant No 1819131. L.Y. is partially supported by the National Institutes of Health under Grant No R01-DC016784.

References

Arguillère, S., Trélat, E., Trouvé, A., Younes, L., 2015. Shape deformation analysis from the optimal control viewpoint. J. Math. Pures. Appl. 104 (1), 139–178.

Arguillère, S., Miller, M., Younes, L., 2016. Diffeomorphic surface registration with atrophy constraints. SIAM J. Imaging Sci. 9 (3), 975–1003.

Arguillère, S., Trélat, E., Trouvé, A., Younes, L., 2016. Registration of multiple shapes using constrained optimal control. SIAM J. Imaging Sci. 9 (1), 344–385.

Arnaudon, A., Holm, D.D., Pai, A., Sommer, S., 2017a. A stochastic large deformation model for computational anatomy. In: International Conference on Information Processing in Medical Imaging, pp. 571–582.

Arnaudon, A., Holm, D.D., Sommer, S., 2017b. A geometric framework for stochastic shape analysis. arXiv preprint arXiv:1703.09971.

Bauer, M., Harms, P., Michor, P.W., 2011. Sobolev metrics on shape space of surfaces. J. Geom. Mech. 3 (4), 389–438. ISSN 1941-4889.

Bauer, M., Bruveris, M., Harms, P., Michor, P.W., 2012a. Vanishing geodesic distance for the Riemannian metric with geodesic equation the KdV-equation. Ann. Global Anal. Geom. 41 (4), 461–472. ISSN 0232-704X. https://doi.org/10.1007/s10455-011-9294-9.

Bauer, M., Harms, P., Michor, P.W., 2012b. Sobolev metrics on shape spaces, II: weighted Sobolev metrics and almost local metrics. J. Geom. Mech. 4 (4), 365–383.

Bauer, M., Bruveris, M., Michor, P.W., 2013. Geodesic distance for right invariant Sobolev metrics of fractional order on the diffeomorphism group. II. Ann. Global Anal. Geom. 44 (4), 361–368. ISSN 0232-704X. https://doi.org/10.1007/s10455-013-9370-4.

Bauer, M., Bruveris, M., Marsland, S., Michor, P.W., 2014a. Constructing reparameterization invariant metrics on spaces of plane curves. Differential Geom. Appl. 34, 139–165. ISSN 0926-2245. https://doi.org/10.1016/j.difgeo.2014.04.008.

Bauer, M., Bruveris, M., Michor, P.W., 2014b. Overview of the geometries of shape spaces and diffeomorphism groups. J. Math. Imaging. Vis. 50, 60–97. ISSN 0924-9907. https://doi.org/10.1007/s10851-013-0490-z.

Bauer, M., Bruveris, M., Harms, P., Møller-Andersen, J., 2017. A numerical framework for Sobolev metrics on the space of curves. SIAM J. Imaging Sci. 10 (1), 47–73. ISSN 1936-4954.

Bauer, M., Bruveris, M., Harms, P., Michor, P.W., 2018. Smoothness of the Fractional Laplacian on the Space of Riemannian Metrics of Sobolev Order. University of Vienna.

Bauer, M., Bruveris, M., Charon, N., Møller-Andersen, J., 2019. A relaxed approach for curve matching with elastic metrics. ESAIM COCV. ArXiv:1803.10893 (Forthcoming).

Beg, M.F., Miller, M.I., Trouvé, A., Younes, L., 2005. Computing large deformation metric mappings via geodesic flows of diffeomorphisms. Int. J. Comput. Vis. 61, 139–157.

Betts, J., 2010. Practical Methods for Optimal Control and Estimation Using Nonlinear Programming, second ed. Society for Industrial and Applied Mathematics.

Bronstein, A.M., Bronstein, M., Kimmel, R., Mahmoudi, M., Sapiro, G., 2010a. A Gromov-Hausdorff framework with diffusion geometry for topologically-robust non-rigid shape matching. Int. J. Comput. Vis. 89 (2), 266–286. ISSN 0920-5691. http://www.springerlink.com/index/b2714k181406618q.pdf.

Bronstein, A.M., Bronstein, M.M., Kimmel, R., Mahmoudi, M., Sapiro, G., 2010b. A Gromov-Hausdorff framework with diffusion geometry for topologically-robust non-rigid shape matching. Int. J. Comput. Vis. 89 (2–3), 266–286.

Bruveris, M., 2015. Completeness properties of Sobolev metrics on the space of curves. J. Geom. Mech. 7 (2), 125–150. ISSN 1941-4889.

Bruveris, M., Vialard, F.-X., 2014. On completeness of groups of diffeomorphisms. arXiv preprint arXiv:1403.2089.

Bruveris, M., Michor, P.W., Mumford, D., 2014. Geodesic completeness for Sobolev metrics on the space of immersed plane curves. Forum Math. Sigma 2, e19. ISSN 2050-5094.

Bryson, A.E., Ho, Y., 1975. Applied Optimal Control: Optimization, Estimation, and Control. John Wiley & Sons.

Carmeli, C., De Vito, E., Toigo, A., Umanita, V., 2010. Vector valued reproducing kernel Hilbert spaces and universality. Anal. Appl. 8 (01), 19–61.

Cervera, V., Mascaró, F., Michor, P.W., 1991. The action of the diffeomorphism group on the space of immersions. Differential Geom. Appl. 1 (4), 391–401.

Charon, N., Trouvé, A., 2013. The varifold representation of non-oriented shapes for diffeomorphic registration. SIAM J. Imag. Sci. 6 (4), 2547–2580.

Charon, N., Charlier, B., Trouvé, A., 2016. Metamorphoses of functional shapes in Sobolev spaces. Found. Comput. Math. 1–62.

Charon, N., Charlier, B., Glaunès, J., Roussillon, P., Gori, P., 2018. Fidelity metrics between curves and surfaces: currents, varifolds and normal cycles. In: Riemannian Geometric Statistics in Medical Image Analysis.

Coron, J.M., 2007. Control and Nonlinearity. American Mathematical Society.

Duncan, A., Zhang, Z., Srivastava, A., 2016. An elastic Riemannian framework for shape analysis shape analysis of curves and tree-like structures. In: Algorithmic Advances in Riemannian Geometry and Applications. Springer, pp. 187–205.

Durrleman, S., Prastawa, M., Charon, N., Korenberg, J.R., Joshi, S., Gerig, G., Trouvé, A., 2014. Deformetrics: morphometry of shape complexes with space deformations. Neuroimage 101, 35–49.

Effland, A., Rumpf, M., Schäfer, F., 2018. Image extrapolation for the time discrete metamorphosis model: existence and applications. SIAM J. Imaging Sci. 11 (1), 834–862.

Eliashberg, Y., Polterovich, L., 1993. Biinvariant metrics on the group of Hamiltonian diffeomorphisms. Int. J. Math. 4 (05), 727–738.

Feiszli, M., Mumford, D., 2007. Shape representation via conformal mapping. In: Computational Imaging V6498, pp. 64980G.

Feydy, J., Charlier, B., Vialard, F.X., Peyré, G., 2017. Optimal transport for diffeomorphic registration. In: Medical Image Computing and Computer Assisted Intervention—MICCAI 2017, pp. 291–299.

Glaunès, J., Trouvé, A., Younes, L., 2004. Diffeomorphic matching of distributions: a new approach for unlabelled point-sets and sub-manifolds matching. In: IEEE Computer Society Conference on Computer Vision and Pattern Recognition, vol. 2. IEEE Computer Society, Los Alamitos, CA, USA, pp. 712–718.

Glaunès, J., Qiu, A., Miller, M.I., Younes, L., 2008. Large deformation diffeomorphic metric curve mapping. Int. J. Comput. Vis. 80 (3), 317–336. ISSN 0920-5691.

Grenander, U., Miller, M.I., et al., 2007. Pattern Theory: From Representation to Inference. Oxford University Press.

Gu, X.D., Yau, S.-T., 2008. Computational Conformal Geometry. International Press Somerville.

Holm, D.D., 2015. Variational principles for stochastic fluid dynamics. Proc. R. Soc. A 471 (2176), 20140963.

Holm, D., Trouvé, A., Younes, L., 2009. The Euler-Poincaré theory of metamorphosis. Q. Appl. Math. 67 (4), 661–685.

Joshi, S.C., Miller, M.I., 2000. Landmark matching via large deformation diffeomorphisms. IEEE Trans. Image Process. 9 (8), 1357–1370.

Kaltenmark, I., Charlier, B., Charon, N., 2017. A general framework for curve and surface comparison and registration with oriented varifolds. In: Computer Vision and Pattern Recognition (CVPR).

Klassen, E., Srivastava, A., Mio, W., Joshi, S.H., 2004. Analysis of planar shapes using geodesic paths on shape spaces. IEEE Trans. Pattern Anal. Mach. Intell. 26 (3), 372–383. ISSN 0162-8828. https://doi.org/10.1109/TPAMI.2004.1262333.

Kurtek, S., Needham, T., 2018. Simplifying transforms for general elastic metrics on the space of plane curves. arXiv preprint arXiv:1803.10894.

Kurtek, S., Klassen, E., Ding, Z., Srivastava, A., 2010. A novel Riemannian framework for shape analysis of 3D objects. In: 2010 IEEE Conference on Computer Vision and Pattern Recognition (CVPR), pp. 1625–1632.

Kurtek, S., Klassen, E., Gore, J.C., Ding, Z., Srivastava, A., 2012. Elastic geodesic paths in shape space of parameterized surfaces. IEEE Trans. Pattern Anal. Mach. Intell. 34 (9), 1717–1730.

Kurtek, S., Srivastava, A., Klassen, E., Laga, H., 2013. Landmark-guided elastic shape analysis of spherically-parameterized surfaces. In: Computer Graphics Forum, vol. 32, pp. 429–438.

Laga, H., Xie, Q., Jermyn, I.H., Srivastava, A., et al., 2017. Numerical inversion of SRNF maps for elastic shape analysis of genus-zero surfaces. IEEE Trans. Pattern Anal. Mach. Intell. 39 (12), 2451–2464.

Lang, S., 1999. Fundamentals of Differential Geometry. Graduate Texts in Mathematics, vol. 191. Springer-Verlag, New York, ISBN: 0-387-98593-X, xviii+535. https://doi.org/10.1007/978-1-4612-0541-8.

Marsden, J.E., Ratiu, T.S., 2013. Introduction to mechanics and symmetry: a basic exposition of classical mechanical systems, vol. 17. Springer Science & Business Media.

Memoli, F., 2007. On the use of Gromov-Hausdorff distances for shape comparison. In: Botsch, M., Pajarola, R., Chen, B., Zwicker, M. (Eds.), Eurographics Symposium on Point-Based Graphics. The Eurographics Association.

Mémoli, F., 2011. Gromov-Wasserstein distances and the metric approach to object matching. Found. Comput. Math. 11 (4), 417–487.

Mennucci, A.C., Yezzi, A., Sundaramoorthi, G., 2008. Properties of Sobolev-type metrics in the space of curves. Interfaces Free Bound. 10 (4), 423–445.

Micheli, M., Michor, P.W., Mumford, D., 2013. Sobolev metrics on diffeomorphism groups and the derived geometry of spaces of submanifolds. Izv. Math. 77 (3), 541.

Michor, P.W., Mumford, D., 2005. Vanishing geodesic distance on spaces of submanifolds and diffeomorphisms. Doc. Math. 10, 217–245.

Michor, P.W., Mumford, D., 2007. An overview of the Riemannian metrics on spaces of curves using the Hamiltonian approach. Appl. Comput. Harmon. Anal. 23 (1), 74–113.

Miller, M.I., Younes, L., 2001. Group actions, homeomorphisms, and matching: a general framework. Int. J. Comput. Vis. 41 (1–2), 61–84.

Mio, W., Srivastava, A., 2004. In: Proceedings of the 2004 IEEE Computer Society Conference on Computer Vision and Pattern Recognition, 2004. CVPR 2004, pp. II–10-II-15. ISSN 1063-6919. https://doi.org/10.1109/CVPR.2004.1315138.

Nardi, G., Peyré, G., Vialard, F.-X., 2016. Geodesics on shape spaces with bounded variation and Sobolev metrics. SIAM J. Imag. Sci. 9 (1), 238–274.

Overton, M. L., 2016. HANSO: hybrid algorithm for non-smooth optimization 2.2. https://cs.nyu.edu/overton/software/hanso/.

Ovsjanikov, M., Ben-Chen, M., Solomon, J., Butscher, A., Guibas, L., 2012. Functional maps: a flexible representation of maps between shapes. ACM Trans. Graph. 31 (4), 30.

Ovsjanikov, M., Corman, E., Bronstein, M., Rodolà, E., Ben-Chen, M., Guibas, L., Chazal, F., Bronstein, A., 2016. Computing and processing correspondences with functional maps. In: SIGGRAPH ASIA 2016 Courses on—SA '16. ACM Press, New York, New York, USA, pp. 1–60. http://dl.acm.org/citation.cfm?doid=2988458.2988494.

Pontryagin, L., Boltyanskii, V., Gamkrelidze, R., Mishchenko, E., 1962. The Mathematical Theory of Optimal Processes. John Wiley & Sons.

Richardson, C.L., Younes, L., 2013. Computing metamorphoses between discrete measures. J. Geom. Mech. 5 (1), 131–150.

Richardson, C.L., Younes, L., 2016. Metamorphosis of images in reproducing kernel Hilbert spaces. Adv. Comput. Math. 42 (3), 573–603.

Roussillon, P., Glaunès, J., 2016. Kernel metrics on normal cycles and application to curve matching. SIAM J. Imag. Sci. 9 (4), 1991–2038. https://doi.org/10.1137/16M1070529.

Rumpf, M., Wirth, B., 2015. Variational time discretization of geodesic calculus. IMA J. Numer. Anal. 35 (3), 1011–1046.

Sharon, E., Mumford, D., 2006a. 2D-Shape analysis using conformal mapping. Int. J. Comput. Vis. 70 (1), 55–75. ISSN 0920-5691. https://doi.org/10.1007/s11263-006-6121-z. http://www.springerlink.com/index/10.1007/s11263-006-6121-z.

Sharon, E., Mumford, D., 2006b. 2D-shape analysis using conformal mapping. Int. J. Comput. Vis. 70 (1), 55–75.

Sommer, S., Arnaudon, A., Kuhnel, L., Joshi, S., 2017. Bridge simulation and metric estimation on landmark manifolds. In: Graphs in Biomedical Image Analysis, Computational Anatomy and Imaging Genetics. Springer, pp. 79–91.

Srivastava, A., Klassen, E., 2016. Functional and Shape Data Analysis. Springer Series in Statistics.

Srivastava, A., Klassen, E., Joshi, S.H., Jermyn, I.H., 2011. Shape analysis of elastic curves in Euclidean spaces. IEEE T. Pattern Anal. 33 (7), 1415–1428. https://doi.org/10.1109/TPAMI.2010.184.

Staneva, V., Younes, L., 2017. Learning shape trends: parameter estimation in diffusions on shape manifolds. In: 2017 IEEE Conference on Computer Vision and Pattern Recognition Workshops (CVPRW), pp. 717–725.

Sundaramoorthi, G., Yezzi, A., Mennucci, A.C., 2008. Coarse-to-fine segmentation and tracking using Sobolev active contours. IEEE Trans. Pattern Anal. Mach. Intell. 30 (5), 851–864. ISSN 0162-8828. https://doi.org/10.1109/TPAMI.2007.70751.

Sundaramoorthi, G., Mennucci, A.C., Soatto, S., Yezzi, A., 2011. A new geometric metric in the space of curves, and applications to tracking deforming objects by prediction and filtering. SIAM J. Imaging Sci. 4 (1), 109–145. ISSN 1936-4954. https://doi.org/10.1137/090781139.

Trélat, E., 2008. Contrôle Optimal. Vuibert.

Trouvé, A., Vialard, F.-X., 2012. Shape splines and stochastic shape evolutions: a second order point of view. Q. Appl. Math. 219–251.

Trouvé, A., Younes, L., 2005. Metamorphoses through Lie group action. Foundations of Computational Mathematics 5 (2), 173–198.

Vaillant, M., Glaunès, J., 2006. Surface matching via currents. Proc. Inform. Process. Med. Imaging (IPMI), Lect. Notes Comput. Sci 3565 (381–392).

Younes, L., 1996. A distance for elastic matching in object recognition. Comptes Rendus de l'Académie des sciences. Série 1, Mathématique 322 (2), 197–202.

Younes, L., 1998. Computable elastic distances between shapes. SIAM J. Appl. Math. 58 (2), 565–586.

Younes, L., 2010. Shapes and Diffeomorphisms, vol. 171. Springer Science & Business Media.

Younes, L., 2018. Hybrid Riemannian metrics for diffeomorphic shape registration. Ann. Math. Sci. Appl. 3 (1), 189–210.

Younes, L., Michor, P.W., Shah, J., Mumford, D., 2008. A metric on shape space with explicit geodesics. Atti Accad. Naz. Lincei Cl. Sci. Fis. Mat. Natur. 19 (1), 25–57.

Zeng, W., Gu, X.D., 2011. Registration for 3D surfaces with large deformations using quasi-conformal curvature flow. In: 2011 IEEE Conference on Computer Vision and Pattern Recognition (CVPR), pp. 2457–2464.

Chapter 17

Efficient and accurate structure preserving schemes for complex nonlinear systems

Jie Shen*
Department of Mathematics, Purdue University, West Lafayette, IN, United States
Corresponding author: e-mail: shen7@purdue.edu

Chapter Outline

Abstract

The scalar auxiliary variable (SAV) approach was recently proposed in Shen et al. (2018) to construct efficient and accurate unconditionally energy stable schemes for gradient flows. We present the SAV approach in a more general setting, and consider a few interesting extensions that enable us to use the SAV approach to deal with a large class of complex dissipative/conservative systems.

Keywords: Scalar auxiliary variable, Gradient flows, Conservative system, Structure preserving, Energy stability, Phase-field model

AMS Classification Codes: 65M12, 35K20, 35K35, 35K55, 65Z05

1 Introduction

The ability of accurate simulation of complex phenomena governed by highly complex dissipative/conservative systems is central to our understanding of

Handbook of Numerical Analysis, Vol. 20. https://doi.org/10.1016/bs.hna.2019.06.004
© 2019 Elsevier B.V. All rights reserved.

many important issues, such as advanced materials, quantum mechanics, semiconductors, imaging, optimal transport, nonconvex optimization, etc. Examples of these dissipative systems include Allen–Cahn and Cahn–Hilliard equations, phase-field models (Anderson et al., 1998; Gurtin et al., 1996; Liu and Shen, 2003; Lowengrub and Truskinovsky, 1998; Yue et al., 2004a), thin film models (Giacomelli and Otto, 2001; Otto, 1998), liquid crystal models (Doi and Edwards, 1988; Forest et al., 2004a, b; Larson, 1990; Leslie, 1979; Yu et al., 2010), and examples of conservative systems include nonlinear Schrödinger equations, Sine-Gordon equations, etc.

However, it is a tremendous challenge to numerically solving highly complex nonlinear systems which preserve energy dissipation/conservation, and/or physical constraints such as positivity, mass conservation, etc. Simple explicit or semi-implicit approaches may induce severe stability conditions on the time step so they are not efficient in practice. On the other hand, unconditionally energy stable schemes may be constructed with a fully implicit approach (Du and Nicolaides, 1991) but it leads to nonlinear systems which may not be easy to solve and whose unique solvability may still dictate a time step constraint.

It is in general very challenging to develop efficient and accurate unconditionally structure preserving schemes which preserve essential physical quantities for these stiff systems with highly complex nonlinearities. Many efforts have been made in the last three decades, we refer to Shen et al. (2019) for a brief review on the existing approaches for gradient flows which is a particular class of such problems, and to Du and Feng (2019) for an extensive review. In particular, the recently proposed scalar auxiliary variable (SAV) approach (Shen et al., 2018) enjoys several remarkable advantages (cf. Shen et al., 2019). Thanks to its simplicity, efficiency and generality, the SAV approach can be applied to a large class of problems with energy dissipation/conservation. In addition to the various applications presented in Shen et al. (2019), it has been applied to, among others, the epitaxial thin film growth models in Cheng et al. (2019), the multicomponent two-phase compressible flow in Kou et al. (2018), a phase-field surfactant model in Zhu et al. (2018), a phase-field vesicle membrane model in Cheng and Shen (2018), a phase-field model for solid-state dewetting problems (He et al., 2018), the incompressible Navier–Stokes equations in Yang and Dong (2018), to strongly anisotropic Cahn–Hilliard equations in Yang (2018a), Sine-Gordon equations in Cai et al. (2018), and to compute the stationary solutions of one- and multicomponent Bose–Einstein condensates in Zhuang and Shen (2019).

The aim of this paper is to introduce the recently proposed scalar auxiliary variable (SAV) approach (Shen et al., 2018) in a more general context, discuss how it can be made more effective for a large class of problems, and present a few interesting extensions.

The rest of the paper is organized as follows. In Section 2, we describe the SAV approach in a general context, present its properties. In Section 3, we present several interesting extensions of the SAV approach.

2 The SAV approach

We describe below the scalar auxiliary variable (SAV) approach introduced in Shen et al. (2018). Given a free energy in the form

$$\mathcal{E}(\phi)=\int_\Omega\left[\frac{1}{2}\phi\,\mathcal{L}\phi+F(\phi)\right]dx, \tag{1}$$

where $\mathcal{L}$ is a linear positive definite operator and $F(\phi)$ is a nonlinear free energy, we consider a general gradient flow given by:

$$\begin{aligned}\phi_t&=-\mathcal{G}\mu,\\ \mu:&=\frac{\delta\mathcal{E}}{\delta\phi}=\mathcal{L}\phi+F'(\phi),\end{aligned} \tag{2}$$

where $\mathcal{G}$ is a positive definite operator describing the relaxation process of the gradient flow, e.g., $\mathcal{G}=I$ is the usual L^2-gradient flow while $\mathcal{G}=-\Delta$ is the so-called H^{-1}-gradient flow, μ is the so-called chemical potential. The above system satisfies the energy law:

$$\frac{\partial}{\partial t}\mathcal{E}(\phi)=\frac{\partial}{\partial t}\int_\Omega\left[\frac{1}{2}\phi\,\mathcal{L}\phi+F(\phi)\right]dx=-(\mathcal{G}\mu,\mu)\le 0. \tag{3}$$

Assume that

$$E_1(\phi):=\int_\Omega F(\phi)dx \text{ is bounded from below, i.e., } E_1(\phi)>-C_0 \text{ for some } C_0>0. \tag{4}$$

We introduce one scalar auxiliary variable (SAV), $r(t)=\sqrt{E_1(\phi)+C_0}$, and rewrite the original gradient flow (2) as:

$$\begin{aligned}\frac{\partial\phi}{\partial t}&=-\mathcal{G}\mu\\ \mu&=\mathcal{L}\phi+\frac{r(t)}{\sqrt{E_1(\phi)+C_0}}F'(\phi)\\ r_t&=\frac{1}{2\sqrt{E_1(\phi)+C_0}}\int_\Omega F'(\phi)\phi_t dx.\end{aligned} \tag{5}$$

The above system is equivalent to (2) under the initial condition $r(0)=\sqrt{E_1(\phi|_{t=0})+C_0}$. However, it allows more flexibility in time discretization. For example, with the principle of treating linear terms implicitly and all nonlinear terms explicitly, a second-order scheme for the above reformulated system can be constructed as follows:

$$\begin{aligned}
\frac{\phi^{n+1}-\phi^n}{\Delta t} &= -\mathcal{G}\mu^{n+1/2},\\
\mu^{n+1/2} &= \mathcal{L}\phi^{n+1/2} + \frac{r^{n+1/2}}{\sqrt{E_1\left(\tilde{\phi}^{n+1/2}\right)+C_0}}F'\left(\tilde{\phi}^{n+1/2}\right),\\
\frac{r^{n+1}-r^n}{\Delta t} &= \int_\Omega \frac{F'\left(\tilde{\phi}^{n+1/2}\right)}{2\sqrt{E_1\left(\tilde{\phi}^{n+1/2}\right)+C_0}}\frac{\phi^{n+1}-\phi^n}{\Delta t}dx,
\end{aligned} \tag{6}$$

where $h^{n+1/2}:=\frac{1}{2}(h^{n+1}+h^n)$ for any sequence $\{h^k\}$, and $\tilde{\phi}^{n+1/2}:=\frac{3}{2}\phi^n-\frac{1}{2}\phi^{n-1}$.

It is clear that the above scheme is second-order accurate. Furthermore, by taking the inner products of three equations in (6) with $\mu^{n+1/2}$, $\phi^{n+1}-\phi^n$, and $2r^{n+1/2}$, respectively, notice that the two terms with $F'(\phi)$ cancel each other regardless the form of $F(\phi)$, so we obtain immediately the following result:

Theorem 1. *The scheme* (6) *is unconditionally energy stable in the sense that*

$$\bar{\mathcal{E}}^{n+1}(\phi,r)-\bar{\mathcal{E}}^n(\phi,r)=-\Delta t(\mathcal{G}\mu^{n+1/2},\mu^{n+1/2}),$$

where $\bar{\mathcal{E}}^k(\phi,r):=\int_\Omega\frac{1}{2}\phi^k\,\mathcal{L}\phi^k dx+r^k$ *is the modified free energy at* t^k.

Note that the modified energy $\bar{\mathcal{E}}^n(\phi,r)$ is an approximation of the original energy $\mathcal{E}(\phi^n)$ provided that $(r^n)^2$ is an approximation of $E_1(\phi^n)+C_0$. We observe from the above theorem that not only the scheme (6) is unconditionally energy stable, it also preserve the energy dissipation rate. In particular, it is energy conservative if $\mathcal{G}$ is skew-symmetric.

Another important fact is that the SAV scheme (6) is easy to implement. To this end, we write (6) in the following matrix form:

$$\begin{pmatrix} \frac{1}{\Delta t}I & \frac{1}{2}\mathcal{G} & 0\\ \mathcal{L} & -I & *\\ * & 0 & \frac{1}{\Delta t}\end{pmatrix}\begin{pmatrix}\phi^{n+1}\\ \mu^{n+1}\\ r^{n+1}\end{pmatrix}=\bar{b}^n, \tag{7}$$

where $\bar{b}^n$ is the vector with known quantities, and * represents known vectors with variable coefficients. Hence, we can solve r^{n+1} with a block Gaussian elimination, which requires solving a system with constant coefficients of the form

$$\begin{pmatrix} \frac{1}{\Delta t}I & \frac{1}{2}\mathcal{G} \\ \mathcal{L} & -I \end{pmatrix} \begin{pmatrix} \phi \\ \mu \end{pmatrix} = \overline{b}. \tag{8}$$

Once r^{n+1} is known, we can obtain (ϕ^{n+1}, μ^{n+1}) by solving one more equation in the above form.

Remark 1. Several remarks are in order:

- It should be noted that the SAV approach is inspired by the IEQ/EQ methods (Yang, 2016; Zhao et al., 2017). However, with the IEQ/EQ approach, one has to solve a coupled linear system with time dependent variable coefficients, so in this sense the SAV approach is computationally more efficient.
- While, for the sake of simplicity, we described the scheme and its stability in semidiscrete form, the above stability result carries over to any consistent Galerkin type approximations, e.g. finite-elements or finite-differences, which respect proper integration by parts.
- The linear system (8) can be reduced to

$$\left(\frac{1}{\Delta t}I + \frac{1}{2}\mathcal{G}\mathcal{L}\right)\phi = \psi, \tag{9}$$

  with some known ψ. With any consistent Galerkin approximations of $\mathcal{G}$ and $\mathcal{L}$, the corresponding linear system for (9) is positive definite so can be efficiently solved by one's favourite method, e.g., preconditioned conjugate gradient method. In many common situations such as $\mathcal{L} = -\Delta$ and $\mathcal{G} = I$ or $-\Delta$ with suitable boundary conditions, the corresponding linear system for (9) can even be solved by fast direct methods such as fast Poisson solvers. Hence, the scheme (6) is very efficient.
- Note that in a linearly implicit scheme with or without stabilization, one also has to solve (8) at each time step. Therefore, the cost of the SAV scheme is essentially twice of linearly implicit schemes which are usually conditionally stable, and the benefit is that the scheme becomes unconditionally stable so an adaptive time stepping can be employed to further reduce the computational cost.

For dissipative systems, it is usually better to use BDF type schemes rather than the Crank–Nicolson scheme. For example, a semi-implicit second-order scheme based on BDF2 formula is

$$\frac{3\phi^{n+1} - 4\phi^n + \phi^{n-1}}{2\Delta t} = \mathcal{G}\mu^{n+1}, \tag{10a}$$

$$\mu^{n+1} = \mathcal{L}\phi^{n+1} + \frac{r^{n+1}}{\sqrt{E_1(\overline{\phi}^{n+1})}} F'(\overline{\phi}^{n+1}), \tag{10b}$$

$$\frac{3r^{n+1} - 4r^n + r^{n-1}}{2\Delta t} = \int_\Omega \frac{F'(\overline{\phi}^{n+1})}{2\sqrt{E_1(\overline{\phi}^{n+1})}} \frac{3\phi^{n+1} - 4\phi^n + \phi^{n-1}}{2\Delta t} d\boldsymbol{x}, \tag{10c}$$

where $\overline{\phi}^{n+1} := 2\phi^n - \phi^{n-1}$.

Taking the inner products of the three equations with μ^{n+1}, $(3\phi^{n+1} - 4\phi^n + \phi^{n-1})/(2\Delta t)$, $2r^{n+1}$, respectively, and using the identity:

$$2(a^{k+1}, 3a^{k+1} - 4a^k + a^{k-1}) = |a^{k+1}|^2 + |2a^{k+1} - a^k|^2 + |a^{k+1} - 2a^k + a^{k-1}|^2 - |a^k|^2 - |2a^k - a^{k-1}|^2, \tag{11}$$

we can obtain the following result:

Theorem 2. *The scheme (10) is second-order accurate, and unconditionally energy stable in the sense that*

$$\begin{aligned}\widetilde{\mathcal{E}}^{n+1}(\phi,r) - \widetilde{\mathcal{E}}^n(\phi,r) = &- \Delta t(\mathcal{G}\mu^{n+1}, \mu^{n+1}) \\ &- \left\{\frac{1}{4}(\phi^{n+1} - 2\phi^n + \phi^{n-1}, \mathcal{L}(\phi^{n+1} - 2\phi^n + \phi^{n-1})) \right. \\ &\left. + \frac{1}{2}(r^{n+1} - 2r^n + r^{n-1})^2\right\},\end{aligned}$$

where the modified discrete energy at t^{n+1} is defined as

$$\begin{aligned}\widetilde{\mathcal{E}}^{n+1}(\phi,r) = &\frac{1}{4}((\phi^{n+1}, \mathcal{L}\phi^{n+1}) + (2\phi^{n+1} - \phi^n, \mathcal{L}(2\phi^{n+1} - \phi^n))) \\ &+ \frac{1}{2}\left((r^{n+1})^2 + (2r^{n+1} - r^n)^2\right).\end{aligned}$$

Note that the modified discrete energy $\mathcal{E}^{n+1}(\phi,r)$ Comparing with (6), the scheme (10) introduces extra dissipation terms so the dissipation rate is not exactly preserved as in (6). On the other hand, we can easily construct higher-order SAV schemes by replacing BDF2 approximation with BDF-k ($k \leq 6$) approximation. Numerical results (Shen et al., 2019) indicate that these BDF-k ($k \leq 6$) schemes are also unconditionally energy stable although a rigorous proof is still elusive.

We note that while the SAV schemes such as (6) and (10) are unconditionally energy stable for any type of free energies as long as we can split the free energy in the form of (1) satisfying (4). However, for the SAV schemes to be most effective, the following two considerations are very important.

2.1 Suitable energy splitting

The unconditional energy stability does not necessarily mean that the corresponding SAV schemes such as (6) and (10) will be effective. For example, if we take $\mathcal{L}=0$, the SAV schemes (6) and (10) are still unconditionally energy stable, but they are fully explicit so exceedingly small time steps are required to obtain reasonable accurate approximations. Therefore, the performance of the SAV schemes depend intrinsically on how to split the free energy. In principle, the splitting should be such that the linear term $\frac{1}{2}\phi\mathcal{L}\phi$ should "dominate" the other term $F(\phi)$. Otherwise, accurate approximate solutions can only be obtained with very small time steps. Hence, it is important to find suitable splittings such that good approximations can be obtained with reasonably large time steps.

Consider, for instance, a typical example with free energy

$$\mathcal{E}[\phi]=\int\frac{1}{2}|\nabla\phi|^2+\frac{1}{4\epsilon^2}(1-\phi^2)^2d\boldsymbol{x}, \tag{12}$$

where ϵ is a small parameter. This type of free energy is often used in Allen–Cahn (Allen and Cahn, 1979) and Cahn–Hilliard equations (Cahn and Hilliard, 1958, 1959), and in the study of interfacial dynamics (Ainsworth and Mao, 2017; Allen and Cahn, 1979; Anderson et al., 1998; Gurtin et al., 1996; Kim, 2012; Liu and Shen, 2003; Lowengrub and Truskinovsky, 1998; Rogers and Desai, 1989; Yue et al., 2004a). In typical situation, ϵ, representing the interfacial width, is usually small so that the nonlinear term is not "small" compared with the linear term. The consequence is that very small time steps are needed for the approximate solution to be accurate. However, this situation can be easily dealt with by using a suitable splitting of the free energy as follows:

$$\mathcal{E}(\phi)=\int\left[\frac{1}{2}|\nabla\phi|^2+\frac{\beta}{\epsilon^2}\phi^2\right]+\left[\frac{1}{4\epsilon^2}(1-\phi^2)^2-\frac{\beta}{\epsilon^2}\phi^2\right]d\boldsymbol{x}, \tag{13}$$

where β is a parameter of our choice. It can be easily shown that $E_1(\phi)=\int[\frac{1}{4\epsilon^2}(1-\phi^2)^2-\frac{\beta}{\epsilon^2}\phi^2]\,d\boldsymbol{x}$ has a lower bound for any β so SAV approach can be applied.

As an illustrative example, we implemented the SAV scheme (10) for the Cahn–Hilliard equation, which is the gradient flow with the above free energy and $\mathcal{L}=\mathcal{G}=-\Delta$. We assume the periodic boundary conditions in the domain $[0, 2\pi)$, and use the Fourier-spectral method in space with initial condition $\phi(x,0)=0.2\sin x$. In the right of Fig. 1, we plot the numerical solutions of the scheme with $\beta=0$ (no stabilization) and $\beta=1$ (with stabilization) using a small $\Delta t=10^{-4}$. We observe that both solutions have the correct profile. However, with a relatively large $\Delta t=4\times10^{-3}$, see the left of Fig. 1, the solution with no stabilization exhibits oscillations while that with stabilization is still accurate.

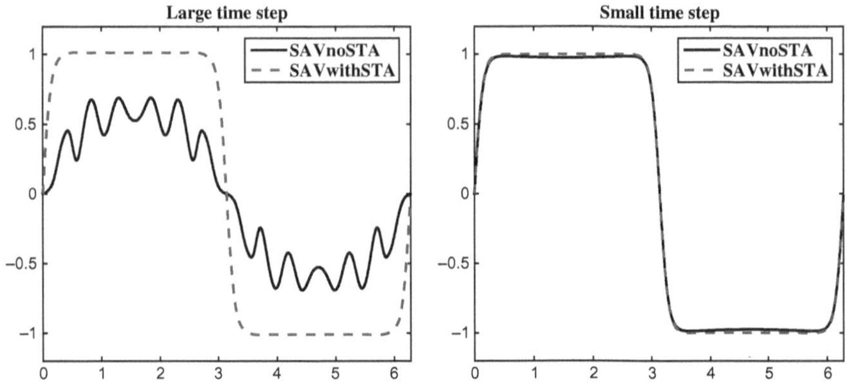

FIG. 1 The solution at $T = 0.1$ with $\epsilon = 0.1$. *Left*: $\Delta t = 4 \times 10^{-3}$; *Right*: $\Delta t = 10^{-4}$. The *red dashed lines* represent solutions with stabilization, while the *black solid lines* represent solutions without stabilization.

2.2 Adaptive time stepping

A main advantage of unconditional energy stable schemes is that they can be easily implemented with an adaptive time stepping strategy so that the time step is only dictated by accuracy rather than by stability as with conditionally stable schemes. This is particularly important for those situations where the solution may vary drastically in certain time intervals, but changes very little elsewhere.

It should be noted that variation of energy alone does not provide enough information to determine appropriate time steps. A suitable time adaptive strategy similar to that employed in Gomez and Hughes (2011) is proposed in Shen et al. (2016) (see also Shen et al. (2019)). In particular, the formula for updating the time step size is given by

$$A_{dp}(e,\tau) = \rho \left(\frac{tol}{e} \right)^{1/2} \tau, \tag{14}$$

along with prescribed minimum and maximum time steps. In the above, ρ is a default safety coefficient, *tol* is a reference tolerance, and e is the relative error at each time level. As an illustrative example, we consider the 2D Cahn–Hilliard equation with $\mathcal{L} = \mathcal{G} = -\Delta$ on $[0, 2\pi) \times [0, 2\pi)$ with periodic boundary conditions and random initial data. We take $\epsilon = 0.1$, and use the Fourier-spectral method with $N_x = N_y = 256$. We choose $\rho = 0.9$ and $tol = 10^{-3}$. The minimum and maximum time steps are taken as $\tau_{min} = 10^{-5}$ and $\tau_{max} = 10^{-2}$, respectively. The initial time step is taken as τ_{min}. Numerical results are shown in Fig. 2. We observe that the adaptive time stepping leads to correct energy evolution with the largest time step being several hundred times larger than the smallest time step. This indicates that savings up to several orders of magnitude times

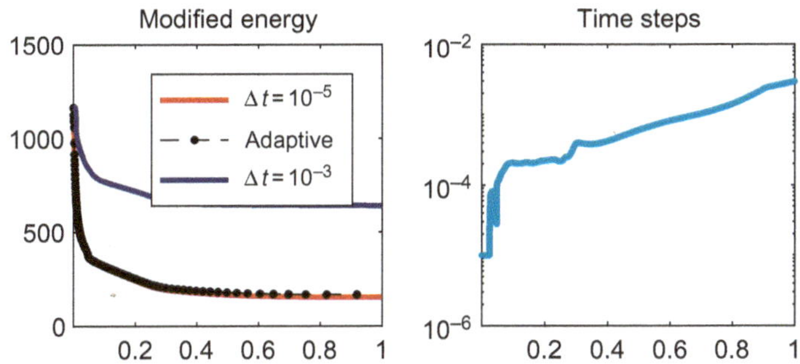

FIG. 2 Adaptive time stepping. *Left*: Energy evolution with small time steps, adaptive time steps, and large time steps; *Right*: step sizes of adaptive time stepping.

can be achieved by combining SAV schemes with an adaptive time stepping compared with semi-implicit schemes whose time steps are dictated by stability constraint.

In summary, the SAV schemes possess some amazing advantages such as: (i) it only requires solving elliptic systems with constant coefficients at each time step; (ii) it is second-order accurate, unconditionally energy stable and can be extended to higher-order. Ample numerical results in Shen et al. (2019, 2018) have shown its advantages compared with other schemes. Furthermore, it has been shown in Shen and Xu (2018) that the SAV schemes converge under essentially the same conditions as are required for the well posedness of the system (2), thanks to the unconditionally energy stability. On the other hand, convergence of linearly implicit schemes (Condette et al., 2011; Kessler et al., 2004; Shen and Yang, 2010b) usually requires that the derivative of nonlinear free energy, $F'(\phi)$ is uniformed bounded, a condition which is not even satisfied by the double well potential. The convergence results and error estimates in Shen and Xu (2018) for the semidiscrete case have also been extended to fully discrete cases with finite-differences in Li et al. (2019) and with finite-elements in Chen and Shen (2019).

3 Several extensions of the SAV approach

In the following, we present a few interesting extensions which enable us to use the SAV approach for a larger class of complex nonlinear systems.

3.1 Problems with global constraints

Many complex systems not only obey energy dissipation/conservation but also are constrained with other conservations such as mass, volume, surface area, etc. A popular and effective approach is to introduce penalty terms in the free energy so as to enforce, approximately, these constraints. Then, we can apply the SAV approach.

Consider, for example, the phase-field vesicle membrane model in Du et al. (2004) with the elastic bending energy

$$
\begin{aligned}
E_b(\phi) &= \frac{\epsilon}{2}\int_\Omega \left(-\Delta\phi + \frac{1}{\epsilon^2}G(\phi)\right)^2 dx \\
&= \frac{\epsilon}{2}\int_\Omega \left(|\Delta\phi|^2 - \frac{2}{\epsilon^2}|\nabla\phi|^2 + \frac{6}{\epsilon^2}\phi^2|\nabla\phi|^2 + \frac{1}{\epsilon^4}(G(\phi))^2\right) d\boldsymbol{x},
\end{aligned}
\tag{15}
$$

where $G(\phi) = F'(\phi)$ with $F(\phi) = \frac{1}{4}(\phi^2 - 1)^2$, and subject to volume and surface area conservation:

$$
A(\phi) = \frac{1}{2}\int_\Omega (\phi + 1) d\boldsymbol{x} \equiv \alpha, \quad \text{and} \quad B(\phi) = \int_\Omega \left(\frac{\epsilon}{2}|\nabla\phi|^2 + \frac{1}{\epsilon}F(\phi)\right) d\boldsymbol{x} \equiv \beta, \tag{16}
$$

where α, β represent the initial volume and surface area.

There is no mechanism in the SAV approach to enforce these constraints directly. However, we can penalize the free energy to approximately enforce them. More precisely, we consider the following free energy with additional penalization terms:

$$
E_{tot}(\phi) = E_b(\phi) + \frac{1}{2\gamma}(A(\phi) - \alpha)^2 + \frac{1}{2\eta}(B(\phi) - \beta)^2, \tag{17}
$$

where γ and η are two small parameters. In order to apply the SAV approach, we split $E_{tot}(\phi)$ as follows

$$
\begin{aligned}
E_{tot}(\phi) &= \frac{\epsilon}{2}\int_\Omega \left\{\frac{\epsilon}{2}\left(|\Delta\phi|^2 - \frac{2}{\epsilon^2}|\nabla\phi|^2\right) + \frac{1}{2\gamma}(A(\phi) - \alpha)^2\right\} \\
&\quad + \int_\Omega \left\{\frac{\epsilon}{2}\left(\frac{6}{\epsilon^2}\phi^2|\nabla\phi|^2 + \frac{1}{\epsilon^4}(G(\phi))^2\right) + \frac{1}{2\eta}(B(\phi) - \beta)^2\right\}.
\end{aligned}
\tag{18}
$$

Note that the volume constraint is linear, so it is included in the first part. Then, we can apply the SAV approach to the above energy to construct unconditional energy stable SAV schemes. Usually such schemes would be effective, but due to the three small parameters in the free energy, it is observed in Cheng and Shen (2018) that such SAV schemes, while unconditionally energy stable, require exceedingly small time steps to obtain reasonable approximate solutions.

To remedy this situation, a multiple SAV approach is introduced in Cheng and Shen (2018). The basic strategy is to introduce one SAV for each small parameter in the nonlinear part of the free energy. So we are led to introduce two SAVs:

$$
U = B(\phi) - \beta, \quad V = \sqrt{\int_\Omega \left(\frac{6}{\epsilon^2}\phi^2|\nabla\phi|^2 + \frac{1}{\epsilon^4}(G(\phi))^2\right) d\boldsymbol{x} + C}, \tag{19}
$$

where C is a positive constant, so the total energy becomes

$$E_{tot}(\phi,U,V)=\frac{\epsilon}{2}\int_\Omega\left(|\Delta\phi|^2-\frac{2}{\epsilon^2}|\nabla\phi|^2\right)d\boldsymbol{x}+\frac{1}{2\gamma}(A(\phi)-\alpha)^2+\frac{U^2}{2\eta}+\frac{\epsilon}{2}(V^2-C).$$

Then, the L^2 gradient flow with a mobility M can be written as:

$$\phi_t=-M\mu, \tag{20}$$

$$\mu=\frac{\delta E_{tot}}{\delta\phi}=\epsilon\Delta^2\phi+\frac{2}{\epsilon}\Delta\phi+\frac{1}{\gamma}(A(\phi)-\alpha)+\frac{1}{\eta}U\frac{\delta U}{\delta\phi}+\epsilon V\frac{\delta V}{\delta\phi}, \tag{21}$$

$$U_t=\int_\Omega\frac{\delta U}{\delta\phi}\phi_t d\boldsymbol{x},\quad V_t=\int_\Omega\frac{\delta V}{\delta\phi}\phi_t d\boldsymbol{x}. \tag{22}$$

We can then apply the SAV approach to construct efficient energy stable schemes. For example, a second-order SAV scheme based on Crank–Nicolson is

$$\frac{\phi^{n+1}-\phi^n}{\Delta t}=-M\frac{\mu^{n+1}+\mu^n}{2}, \tag{23}$$

$$\frac{\mu^{n+1}+\mu^n}{2}=\epsilon\Delta^2\frac{\phi^{n+1}+\phi^n}{2}+\frac{2}{\epsilon}\Delta\frac{\phi^{n+1}+\phi^n}{2}+\frac{1}{\gamma}\left(A\left(\frac{\phi^{n+1}+\phi^n}{2}\right)-\alpha\right) \tag{24}$$

$$+\frac{1}{\eta}\frac{U^{n+1}+U^n}{2}\frac{\delta U}{\delta\phi}\left(\phi^{\star,n+\frac{1}{2}}\right)+\epsilon\frac{V^{n+1}+V^n}{2}\frac{\delta V}{\delta\phi}\left(\phi^{\star,n+\frac{1}{2}}\right),$$

$$\frac{U^{n+1}-U^n}{\Delta t}=\int_\Omega\frac{\delta U}{\delta\phi}\left(\phi^{\star,n+\frac{1}{2}}\right)\frac{\phi^{n+1}-\phi^n}{\Delta t}d\boldsymbol{x}, \tag{25}$$

$$\frac{V^{n+1}-V^n}{\Delta t}=\int_\Omega\frac{\delta V}{\delta\phi}\left(\phi^{\star,n+\frac{1}{2}}\right)\phi^{n+1}-\phi^n\Delta t d\boldsymbol{x}, \tag{26}$$

where $\phi^{\star,n+\frac{1}{2}}=\frac{3}{2}\phi^n-\frac{1}{2}\phi^{n-1}$. Taking the inner products of (23) with $\frac{\mu^{n+1}+\mu^n}{2}$, (24) with $\frac{\phi^{n+1}-\phi^n}{\Delta t}$, (25) with $\frac{U^{n+1}+U^n}{2}$ and (26) with $\frac{V^{n+1}+V^n}{2}$, respectively, and summing up the results, we obtain the following results:

Theorem 3. *The scheme* (23)–(26) *is unconditionally energy stable in the sense that*

$$E^{n+1}(\phi,U,V)-E^n(\phi,U,V)=-\Delta tM\parallel\mu^{n+\frac{1}{2}}\parallel^2,$$

where

$$E^{n+1}(\phi,U,V)=\frac{\epsilon}{2}\parallel\Delta\phi^{n+1}\parallel^2-\frac{1}{\epsilon}\parallel\nabla\phi^{n+1}\parallel^2+\frac{1}{2\eta}(U^{n+1})^2+\frac{\epsilon}{2}(V^{n+1})^2+\frac{1}{2\gamma}(A(\phi^{n+1})-\alpha)^2.$$

The above scheme can also be implemented as in the case with one SAV. Indeed, we can write it as a matrix system

$$\begin{pmatrix} \frac{1}{\Delta t}I & \frac{M}{2} & 0 \\ -\frac{\epsilon}{2}\Delta^2 - \epsilon\Delta - \frac{1}{4\gamma}I & \frac{1}{2}I & * \\ * & 0 & I_2 \end{pmatrix} \begin{pmatrix} \phi^{n+1} \\ \mu^{n+1} \\ \overline{r}^{n+1} \end{pmatrix} = \overline{b}^n,$$

where $\overline{r}^{n+1} = (U^{n+1}, V^{n+1})^t$, I is the identity operator, I_2 is the identity matrix of order 2, $*$ and $\overline{b}^n$ include only the terms from previous time steps. Therefore, we can first solve $\overline{r}^{n+1}$ using a block Gaussian elimination, which requires solving two systems with constant coefficients of the form

$$\begin{pmatrix} \frac{1}{\Delta t}I & \frac{M}{2} \\ -\frac{\epsilon}{2}\Delta^2 - \epsilon\Delta - \frac{1}{4\gamma}I & \frac{1}{2}I \end{pmatrix} \begin{pmatrix} \phi \\ \mu \end{pmatrix} = \overline{b},$$

which is a fourth-order equation with constant coefficients. With $\overline{r}^{n+1}$ known, we can obtain (ϕ^{n+1}, μ^{n+1}) by solving one more system in the above form. We refer to Cheng and Shen (2018) for more details with numerical validations.

3.2 L^1 minimization via hyper regularization

In imaging processing and many other applications, one often considers minimization of the free energy (Rudin et al., 1992)

$$E(\phi) = \int_\Omega \left[|\nabla\phi| + \frac{\lambda}{2}(\phi - g)^2 \right] dx, \tag{27}$$

where g is a given function. Oftentimes, one attempts to find the minimizer by finding the steady state solution of its gradient flow:

$$\frac{\partial\phi}{\partial t} = \nabla \cdot \frac{1}{|\nabla\phi|}\nabla\phi - \lambda(\phi - g). \tag{28}$$

However, there is no suitable energy splitting which would make the SAV approach effective. Therefore, we consider a hyper regularized free energy

$$E_\epsilon(\phi) = \int_\Omega \left\{ \frac{\epsilon}{2}|(-\Delta)^\beta\phi|^2 + \frac{\lambda}{2}(\phi - g)^2 + F_\epsilon(\phi) - \epsilon^2 \right\} dx, \tag{29}$$

where $\epsilon \ll 1$ and $\beta \geq 1$ are two parameters of our choice, and $F_\epsilon(\phi) = \sqrt{|\nabla\phi|^2 + \epsilon^2}$. Its gradient flow takes the form

$$\frac{\partial\phi}{\partial t} = -\epsilon(-\Delta)^{2\beta}\phi - \lambda(\phi - g) - F_\epsilon'(\phi). \tag{30}$$

We can then introduce a SAV $r(t)=\sqrt{E_{1,\epsilon}(\phi)}$ with $E_{1,\epsilon}(\phi)=\int_\Omega F_\epsilon(\phi)$, and rewrite the hyper regularized gradient flow (30) as

$$\begin{aligned}\frac{\partial\phi}{\partial t}&=-\epsilon(-\Delta)^{2\beta}\phi-\lambda(\phi-g)-\frac{r(t)}{\sqrt{E_{1,\epsilon}(\phi)}}F'_\epsilon(\phi),\\ r_t&=\frac{1}{2\sqrt{E_{1,\epsilon}(\phi)}}\int_\Omega F'_\epsilon(\phi)\phi_t dx.\end{aligned}\tag{31}$$

Then, a first-order scheme based on the above regularized SAV formulation is:

$$\begin{aligned}\frac{\phi^{n+1}-\phi^n}{\Delta t}&=-\epsilon(-\Delta)^{2\beta}\phi^{n+1}-\lambda(\phi^{n+1}-g)-\frac{r^{n+1}}{\sqrt{E_{1,\epsilon}(\phi^n)}}F'_\epsilon(\phi^n),\\ \frac{r^{n+1}-r^n}{\Delta t}&=\frac{1}{\sqrt{E_{1,\epsilon}(\phi^n)}}\int_\Omega F'_\epsilon(\phi^n)\frac{\phi^{n+1}-\phi^n}{\Delta t}dx.\end{aligned}\tag{32}$$

Second- and higher-order SAV schemes can be constructed as usual.

Taking the inner products of two equations above with $\frac{\phi^{n+1}-\phi^n}{\Delta t}$ and $2r^{n+1}$, respectively, one obtains:

Theorem 4. *The scheme* (32) *is unconditionally energy diminishing in the sense that*

$$E^{n+1}(\phi,r)-E^n(\phi,r)=-\Delta t\,\|\frac{\phi^{n+1}-\phi^n}{\Delta t}\|^2-\frac{\epsilon}{2}\|(-\Delta)^\beta(\phi^{n+1}-\phi^n)\|^2-|r^{n+1}-r^n|^2,$$

where $E^{n+1}(\phi,r)=\frac{\epsilon}{2}\|(-\Delta)^\beta\phi^{n+1}\|^2+\frac{\lambda}{2}\|\phi^{n+1}-g\|^2+|r^{n+1}|^2$.

Note that at each time step, one only need to solve, twice, a system of the form

$$\alpha\phi+\epsilon(-\Delta)^{2\beta}\phi=h,$$

with suitable boundary conditions. For imaging processing, periodic boundary condition can be used so the above equation can be solved fast a Fourier spectral method. One can then tune ϵ and β to achieve desired results.

3.3 Free energies with highly nonlinear terms

For problems with nonlinear terms in the highest derivative terms, such as in anisotropic Cahn–Hilliard equations (Torabi et al., 2009), it may not be easy or even possible to find a suitable splitting. In these cases, one may combine the SAV approach with a suitable regularization (Yang, 2018b). Consider for example an free energy with a nonlinear operator $\mathcal{L}$:

$$E(\phi)=\int_\Omega[\gamma(\phi)|\nabla\phi|^2+F(\phi)]dx,\tag{33}$$

where $\gamma(\phi) \geq 0$ is certain nonlinear function of ϕ. The corresponding anisotropic Cahn–Hilliard equation is

$$\phi_t = \Delta\mu; \quad \mu = \frac{\delta E}{\delta\phi}. \tag{34}$$

We shall approximate the above equation with the following regularized equation:

$$\phi_t = \Delta\mu; \quad \mu = -\sum_{k=0}^{2} \epsilon_k (-1)^k \Delta^k \phi_{tt} + \frac{\delta E}{\delta\phi}, \tag{35}$$

where $\epsilon_k \ll 1$ are free parameters and their choices depend on $\gamma(\phi)$ and $F(\phi)$. Introducing a SAV $r(t) = \sqrt{E(\phi) + C_0}$ where C_0 is a suitable constant such that $E(\phi) + C_0 > 0$, we can rewrite (35) as

$$\begin{aligned} \phi_t &= \Delta\mu, \\ \mu &= -\sum_{k=0}^{2} \epsilon_k (-1)^k \Delta^k \phi_{tt} + \frac{r(t)}{\sqrt{E(\phi) + C_0}} \frac{\delta E}{\delta\phi}, \\ r_t &= \frac{1}{2\sqrt{E(\phi) + C_0}} \int_\Omega \frac{\delta E}{\delta\phi} \phi_t dx. \end{aligned} \tag{36}$$

Then, a second-order SAV Crank–Nicolson scheme is

$$\begin{aligned} \frac{\phi^{n+1} - \phi^n}{\Delta t} &= \Delta\mu^{n+1/2}, \\ \mu^{n+1/2} &= -\sum_{k=0}^{2} \frac{\epsilon_k}{\Delta t^2} (-1)^k \Delta^k (\phi^{n+1} - 2\phi^n + \phi^{n-1}) + \frac{r^{n+1/2}}{\sqrt{E\left(\tilde{\phi}^{n+1/2}\right) + C_0}} \frac{\delta E}{\delta\phi}\left(\tilde{\phi}^{n+1/2}\right), \\ \frac{r^{n+1} - r^n}{\Delta t} &= \frac{1}{2\sqrt{E(\tilde{\phi}^{n+1/2}) + C_0}} \int_\Omega \frac{\delta E}{\delta\phi}\left(\tilde{\phi}^{n+1/2}\right) \frac{\phi^{n+1} - \phi^n}{\Delta t} dx, \end{aligned} \tag{37}$$

where $h^{n+1/2} := \frac{1}{2}(h^{n+1} + h^n)$ for any sequence $\{h^k\}$, and $\tilde{\phi}^{n+1/2} = \frac{3}{2}\phi^n - \frac{1}{2}\phi^{n-1}$.

Taking the inner products of the first equation with $\mu^{n+1/2}$, the second equation with $\frac{\phi^{n+1} - \phi^n}{\Delta t}$ and the third equation with $2r^{n+1/2}$, summing up the results, we obtain immediately the following results:

Theorem 5. *The scheme* (37) *is unconditionally energy stable in the sense that*

$$\tilde{\mathcal{E}}^{n+1}(\phi, r) - \tilde{\mathcal{E}}^n(\phi, r) = -\Delta t \| \nabla\mu^{n+1/2} \|^2 - \sum_{k=0}^{2} \frac{\epsilon_k}{2\Delta t} \| \Delta^{k/2} (\phi^{n+1} - 2\phi^n + \phi^{n-1}) \|^2,$$

where $\tilde{\mathcal{E}}^{j+1}(\phi, r) := \sum_{k=0}^{2} \frac{\epsilon_k}{2\Delta t^2} \| \Delta^{k/2} (\phi^{j+1} - \phi^j) \|^2 + (r^{j+1})^2$ *is the modified free energy at* t^{j+1}.

The above approach was successfully used in Yang (2018b) to approximate the strongly anisotropic Cahn–Hilliard equation proposed in Torabi et al. (2009). We refer to Yang (2018b) for details on the choice of ϵ_k and their effects on the accuracy.

3.4 Coupling with other physical conservation laws

In complex thermodynamic systems, e.g., phase-field models of multiphase complex fluids (Liu and Shen, 2003; Yue et al., 2004b), only the order parameters such as the phase variable (Liu and Shen, 2003) and/or director field (Yue et al., 2004b), are governed by gradient flows which are coupled to other physical conservation laws such as momentum, mass, and energy conservations. The SAV approach can be used to deal with the nonlinear terms in gradient flow part to construct efficient numerical schemes for the coupled thermodynamic systems (see also Zhu et al. (2018) and Kou et al. (2018)).

Consider for example a phase-field model for the mixture of two incompressible, immiscible fluids. Let ϕ be a labelling function to identify the two fluids, i.e.,

$$\phi(x,t)=\begin{cases}1 & x\in \text{in fluid } 1,\\ -1 & x\in \text{in fluid } 2,\end{cases} \tag{38}$$

with a smooth interfacial layer of thickness η. Consider a mixing free energy

$$E_{mix}(\phi)=\lambda\int_\Omega\left(\frac{1}{2}|\nabla\phi|^2+F(\phi)\right)d\boldsymbol{x},$$

where $F(\phi)=\frac{1}{4\eta^2}(\phi^2-1)^2$ and λ is a mixing coefficient. For the sake of simplicity, we consider the two fluids having the same density $\rho_0=1$. Then, the Navier–Stokes Cahn–Hilliard phase-field model for the two-phase incompressible flow is as follows (cf., for instance, Anderson et al., 1998; Liu and Shen, 2003):

$$\frac{\partial\phi}{\partial t}+\nabla\cdot(u\phi)=\nabla\cdot(\gamma\nabla\mu), \tag{39}$$

$$\mu=\frac{\delta E_{mix}}{\delta\phi}=-\lambda\Delta\phi+\lambda F'(\phi); \tag{40}$$

$$\frac{\partial u}{\partial t}+(u\cdot\nabla)u=\nu\Delta u-\nabla p-\phi\nabla\mu; \tag{41}$$

$$\nabla\cdot u=0; \tag{42}$$

subject to suitable boundary conditions for ϕ, μ, u. In the above, γ is a relaxation coefficients and ν is the viscosity coefficient; the unknown are ϕ, μ, u,

p with u being the velocity and p the pressure. Taking the inner products of (39), (40) and (53) with μ, $\frac{\partial\phi}{\partial t}$ and u, respectively, we obtain the following energy dissipation law:

$$\frac{d}{dt}\int_\Omega \left\{\frac{1}{2}|u|^2 + \frac{\lambda}{2}|\nabla\phi|^2 + \lambda F(\phi)\right\} d\boldsymbol{x} = -\int_\Omega \{\nu|\nabla u|^2 + \gamma|\nabla\mu|^2\}\, d\boldsymbol{x}.$$

To apply the SAV approach, we set $E_1(\phi) = \int_\Omega F(\phi)$, introduce $r(t) = \sqrt{E_1(\phi)+\delta}$ with $\delta > 0$, and replace (40) by

$$\begin{aligned}\mu &= -\lambda\Delta\phi + \lambda\frac{r(t)}{\sqrt{E_1(\phi)+\delta}}F'(\phi),\\ \frac{dr}{dt} &= \frac{1}{2\sqrt{E_1(\phi)+\delta}}\int_\Omega F'(\phi)\frac{\partial\phi}{\partial t}\, d\boldsymbol{x}.\end{aligned} \tag{43}$$

Then, we can combine the scheme (3.9) in Shen and Yang (2010a) and the SAV approach to construct the following scheme for (39)–(42):

$$\begin{aligned}&\frac{3\phi^{n+1} - 4\phi^n + \phi^{n-1}}{2\Delta t} + \hat{u}^{n+1}\cdot\nabla\overline{\phi}^{n+1} = \gamma\Delta\mu^{n+1},\\ &\mu^{n+1} = -\lambda\Delta\phi^{n+1} + \frac{\lambda r^{n+1}}{\sqrt{E_1(\overline{\phi}^{n+1})+\delta}}F'(\overline{\phi}^{n+1}),\\ &\frac{3r^{n+1} - 4r^n - r^{n-1}}{2\Delta t} = \int_\Omega \frac{F'(\overline{\phi}^{n+1})}{2\sqrt{E_1(\overline{\phi}^{n+1})+\delta}}\frac{3\phi^{n+1} - 4\phi^n + \phi^{n-1}}{2\Delta t}\, dx;\end{aligned} \tag{44}$$

$$\begin{aligned}&\frac{3\tilde{u}^{n+1} - 4u^n + u^{n-1}}{2\Delta t} + \overline{u}^{n+1}\cdot\nabla\tilde{u}^{n+1}\\ &-\nu\Delta\tilde{u}^{n+1} + \nabla p^n - \mu^{n+1}\nabla\overline{\phi}^{n+1} = 0;\end{aligned} \tag{45}$$

$$\begin{aligned}&\frac{3}{2\Delta t}(u^{n+1} - \tilde{u}^{n+1}) + \nabla(p^{n+1} - p^n) = 0,\\ &\nabla\cdot u^{n+1} = 0;\ \ u^{n+1}\cdot\boldsymbol{n}|_{\partial\Omega} = 0.\end{aligned} \tag{46}$$

In the above, $\overline{\psi}^{n+1} := 2\psi^n - \psi^{n-1}$ for any sequence $\{\psi^k\}$. As for $\hat{u}^{n+1}$, we can choose $\hat{u}^{n+1} = \tilde{u}^{n+1}$ or $\hat{u}^{n+1} = 2u^n - u^{n-1}$.

If $\hat{u}^{n+1} = \tilde{u}^{n+1}$, we can show that the scheme is unconditionally energy stable. More precisely, we have the following result:

Theorem 6. *The scheme* (44)–(46) *with* $\hat{u}^{n+1} = \tilde{u}^{n+1}$ *is unconditionally energy stable in the sense that*

$$\mathcal{E}^{n+1}(\phi,u,p) - \mathcal{E}^n(\phi,u,p) \le -\Delta t[\gamma\parallel\nabla\mu^{n+1}\parallel^2 + \nu\parallel\nabla\tilde{u}^{n+1}\parallel^2],$$

where $\mathcal{E}^{n+1}(\phi,u,p)$ *is the modified total energy at* t^{n+1} *given by*

$$\mathcal{E}^{n+1}(\phi,\mu,u,p)=\frac{\lambda}{4}(\|\nabla\phi^{n+1}\|^2+\|\nabla(2\phi^{n+1}-\phi^n)\|^2)+\frac{\lambda}{2}(|r^{n+1}|^2+|2r^{n+1}-r^n|^2)$$
$$+\frac{1}{4}(\|u^{n+1}\|^2+\|2u^{n+1}-u^n\|^2)+\frac{\Delta t^2}{3}\|\nabla p^{n+1}\|^2.$$

Proof. Taking the inner products of three equations in (44) with $\Delta t\mu^{n+1}$, $\frac{1}{2}(3\phi^{n+1}-4\phi^n+\phi^{n-1})$ and $2\Delta t\lambda r^{n+1}$, respectively, using the identity (11), we obtain

$$\left[\frac{\lambda}{4}(\|\nabla\phi^{n+1}\|^2+\|\nabla(2\phi^{n+1}-\phi^n)\|^2)+\frac{\lambda}{2}(|r^{n+1}|^2+|2r^{n+1}-r^n|^2)\right]$$
$$-\left[\frac{\lambda}{4}(\|\nabla\phi^n\|^2+\|\nabla(2\phi^n-\phi^{n-1})\|^2)+\frac{\lambda}{2}(|r^n|^2+|2r^n-r^{n-1}|^2)\right]+\Delta t(\tilde{u}^{n+1}\nabla\overline{\phi}^{n+1},\mu^{n+1})$$
$$=-\Delta t\gamma\|\mu^{n+1}\|^2-\frac{\lambda}{4}\|\nabla(\phi^{n+1}-2\phi^n+\phi^{n-1})\|^2-\frac{\lambda}{2}|r^{n+1}-2r^n+r^{n-1}|^2. \tag{47}$$

Next, taking the inner product of (45) with Δtu^{n+1}, and using the property

$$(u\cdot\nabla v,v)=0,\ \forall u\in H,\ v\in(H_0^1(\Omega))^d, \tag{48}$$

where $H=\{u\in(L^2(\Omega))^d\colon\ \nabla\cdot u=0,\ u\cdot n|_{\partial\Omega}=0\}$, we obtain

$$\frac{1}{2}(3\tilde{u}^{n+1}-4u^n+u^{n-1},\tilde{u}^{n+1})+\Delta t(\nabla p^n,\tilde{u}^{n+1})-\Delta t(\mu^{n+1}\nabla\overline{\phi}^{n+1},\tilde{u}^{n+1})=-\mu\Delta t\|\nabla\tilde{u}^{n+1}\|^2. \tag{49}$$

Then, we rearrange (46) as

$$\frac{\sqrt{3}}{2}u^{n+1}+\frac{\Delta t}{\sqrt{3}}\nabla p^{n+1}=\frac{\sqrt{3}}{2}\tilde{u}^{n+1}+\frac{\Delta t}{\sqrt{3}}\nabla p^n, \tag{50}$$

and take the inner products of both side with itself to get

$$\frac{3}{4}\|u^{n+1}\|^2+\frac{\Delta t^2}{3}\|p^{n+1}\|^2=\frac{3}{4}\|\tilde{u}^{n+1}\|^2+\frac{\Delta t^2}{3}\|p^n\|^2+\Delta t(\nabla p^n,\tilde{u}^{n+1}). \tag{51}$$

Since the term $\Delta t(\nabla p^n,\tilde{u}^{n+1})$ in (51) and (49) cancel each other, and the term $\Delta t(\mu^{n+1}\nabla\overline{\phi}^{n+1},\tilde{u}^{n+1})$ in (49) and (47) cancel each other, it remains to deal with the first term in (49), which we write as

$$\frac{1}{2}(3\tilde{u}^{n+1}-4u^n+u^{n-1},\tilde{u}^{n+1})=\frac{3}{2}(\tilde{u}^{n+1}-u^{n+1},\tilde{u}^{n+1})$$
$$+\frac{1}{2}(3u^{n+1}-4u^n+u^{n-1},\tilde{u}^{n+1}-u^{n+1}) \tag{52}$$
$$+\frac{1}{2}(3u^{n+1}-4u^n+u^{n-1},u^{n+1}):=I_1+I_2+I_3.$$

We have $I_1 = \frac{3}{4}(\| \tilde{u}^{n+1}\|^2 - \| u^{n+1}\|^2 + \| \tilde{u}^{n+1} - u^{n+1}\|^2)$, and for I_3 we use (11). On the other hand, by using (46), we derive that $I_2 = 0$. Finally, collect all the results in the above identities, we obtain the desired result. □

Several remarks are in order:

- The pressure-correction step (46) is decoupled from the rest. More precisely, if $\tilde{u}^{n+1}$ is known, one can determine (p^{n+1}, u^{n+1}) from (46) by solving a Poisson equation for $p^{n+1} - p^n$ with a homogeneous Neumann boundary condition.
- If we take $\hat{u}^{n+1} = 2u^n - u^{n-1}$, one can eliminate r^{n+1} from (44) as before, so the scheme is linear, decoupled, second-order, and only requires solving a sequence of elliptic type equations at each time step, but not unconditionally energy stable.
- On the other hand, if we take $\hat{u}^{n+1} = \tilde{u}^{n+1}$, the scheme is unconditionally energy stable, linear and second-order. However, it is weakly coupled between $(\phi^{n+1}, w^{n+1}, \tilde{u}^{n+1})$ by the term $\tilde{u}^{n+1} \cdot \nabla\overline{\phi}^{n+1}$. The weakly coupled linear system is positive definite, and one can use the decoupled scheme with $\hat{u}^{n+1} = 2u^n - u^{n-1}$ as a preconditioner for the coupled scheme with $\hat{u}^{n+1} = \tilde{u}^{n+1}$.

3.5 Dissipative/conservative systems which are not driven by free energy

In a recent work (Lin and Dong (2018)), the authors presented an interesting extension of the SAV approach to the following incompressible Navier–Stokes equations (NSE):

$$\begin{aligned} u_t + (u \cdot \nabla)u &= \nu\Delta u - \nabla p, \\ \nabla \cdot u = 0, &\quad u|_{\partial\Omega} = 0, \end{aligned} \tag{53}$$

which is not a gradient flow but consists of only physical conservation laws and satisfies an energy dissipation law:

$$\frac{1}{2}\frac{d}{dt}\int |u|^2 dx = -\nu \int_\Omega |\nabla u|^2. \tag{54}$$

A usual semi-implicit scheme for (53) is

$$\begin{aligned} \frac{u^{n+1} - u^n}{\Delta t} + (u^n \cdot \nabla)u^n &= \nu\Delta u^{n+1} - \nabla p^{n+1}, \\ \nabla \cdot u^{n+1} = 0, &\quad u^{n+1}|_{\partial\Omega} = 0. \end{aligned} \tag{55}$$

However, it is not unconditionally energy stable.

Let $E(t) = \int_\Omega \frac{1}{2}|u|^2 dx + \delta$ with any $\delta > 0$, and introduce a SAV variable $R(t) = \sqrt{E(t)}$. We can rewrite the NSE (53) as

$$\begin{aligned} &u_t + \frac{R(t)}{\sqrt{E(t)}}(u\cdot\nabla)u = \nu\Delta u - \nabla p, \quad u|_{\partial\Omega} = 0;\\ &\nabla\cdot u = 0;\\ &2R(t)R'(t) = (u_t, u) = \left(u_t + \frac{R(t)}{\sqrt{E(t)}}(u\cdot\nabla)u, u\right). \end{aligned} \tag{56}$$

In the last equation, we have used the property $((u\cdot\nabla)u, u) = 0$ for all u such that $\nabla\cdot u = 0$ and $u|_{\partial\Omega} = 0$. With $R(0) = \sqrt{E(0)}$, the above system is equivalent to the original NSE (53).

The following semi-implicit SAV scheme based on the above reformulation is proposed in Lin and Dong (2018):

$$\begin{aligned} &\frac{u^{n+1}-u^n}{\Delta t} + \frac{R^{n+1}}{\sqrt{E(t^{n+1})}}(u^n\cdot\nabla)u^n = \nu\Delta u^{n+1} - \nabla p^{n+1},\\ &\nabla\cdot u^{n+1} = 0, \quad u^{n+1}|_{\partial\Omega} = 0;\\ &2R^{n+1}\frac{R^{n+1}-R^n}{\Delta t} = \left(\frac{u^{n+1}-u^n}{\Delta t} + \frac{R^{n+1}}{\sqrt{E(t^{n+1})}}(u^n\cdot\nabla)u^n, u^{n+1}\right). \end{aligned} \tag{57}$$

Taking the inner product of the first equation with u^{n+1} and sum up with the third equation, we obtain:

Theorem 7. *The scheme* (57) *is unconditionally energy stable in the sense that all its solutions satisfy*

$$\frac{1}{\Delta t}(|R^{n+1}|^2 - |R^n|^2 + |R^{n+1} - R^n|^2) + \nu\,\|\nabla u^{n+1}\|^2 = 0.$$

The above result is very surprising since the nonlinear term is treated explicitly.

The above scheme can also be implemented very efficiently. Indeed, denote $S^{n+1} = \frac{R^{n+1}}{\sqrt{E(t^{n+1})}}$ and set

$$u^{n+1} = u_1^{n+1} + S^{n+1}u_2^{n+1}, \; p^{n+1} = p_1^{n+1} + S^{n+1}p_2^{n+1}.$$

Plugging the above in (57), we find that (u_i^{n+1}, p_i^{n+1}) $(i = 1,2)$ can be determined separately from:

$$\begin{aligned} &\frac{u_1^{n+1}-u^n}{\Delta t} = \nu\Delta u_1^{n+1} - \nabla p_1^{n+1}, \quad u_1^{n+1}|_{\partial\Omega} = 0;\\ &\nabla\cdot u_1^{n+1} = 0; \end{aligned} \tag{58}$$

and

$$\frac{u_2^{n+1}}{\Delta t} + (u^n \cdot \nabla)u^n = \nu\Delta u_2^{n+1} - \nabla p_2^{n+1}, \quad u_2^{n+1}|_{\partial\Omega} = 0; \qquad (59)$$
$$\nabla \cdot u_2^{n+1} = 0.$$

Once (u_i^{n+1}, p_i^{n+1}) $(i = 1,2)$ are known, the last equation in (57) becomes a nonlinear algebraic equation for S^{n+1} which can be solved by a Newton iteration.

Several remarks are in order:

- The cost of solving the nonlinear algebraic equation for S^{n+1} is negligible. The main cost at each time step is to solve (58) and (59). So the computational cost is about twice of that for the usual semi-implicit scheme (55) but it has the advantage of being unconditionally energy stable.
- Existence of a positive solution for S^{n+1} is still elusive. But in practice, this is always true as long as Δt is not "too" large. Since S^{n+1} is designed to be an approximation to 1, the difference $|S^{n+1} - 1|$ provides a "free" estimator for adaptive time stepping: if $|S^{n+1} - 1|$ is not "sufficiently small", then, one needs to reduce Δt.
- Ample numerical results in Lin and Dong (2018) show that the SAV approach is more efficient and robust than the usual semi-implicit schemes.
- This approach is not restricted to the Navier–Stokes equations and can be applied to other nonlinear systems. For example, we can couple this approach with the scheme (44)–(46) to construct new efficient schemes for two-phase phase-field models, we refer to Yang and Dong (2018) for an attempt in this direction.

4 Conclusion

We provided an updated account for the recently proposed SAV approach for complex dissipative/conservative systems. The SAV approach was originally developed for dealing with gradient flows. But as shown in the last section, the idea can be extended to deal with a large class of complex dissipative/conservative systems which are not gradient flows, and still lead to efficient and accurate numerical schemes.

Acknowledgements

This work is partially supported by NSF DMS-1620262 & DMS-1720442 and AFOSR FA9550-16-1-0102.

References

Ainsworth, M., Mao, Z., 2017. Analysis and approximation of a fractional Cahn-Hilliard equation. SIAM J. Numer. Anal. 55 (4), 1689–1718. ISSN 0036-1429.

Allen, S.M., Cahn, J.W., 1979. A microscopic theory for antiphase boundary motion and its application to antiphase domain coarsening. Acta Metall. 27 (6), 1085–1095.

Anderson, D.M., McFadden, G.B., Wheeler, A.A., 1998. Diffuse-interface methods in fluid mechanics. Annu. Rev. Fluid Mech. 30 (1), 139–165.

Cahn, J.W., Hilliard, J.E., 1958. Free energy of a nonuniform system. I. Interfacial free energy. J. Chem. Phys. 28 (2), 258–267.

Cahn, J.W., Hilliard, J.E., 1959. Free energy of a nonuniform system. III. Nucleation in a two-component incompressible fluid. J. Chem. Phys. 31 (3), 688–699.

Cai, W., Jiang, C., Wang, Y., 2018. Structure-preserving algorithms for the two-dimensional sine-Gordon equation with Neumann boundary conditions. arXiv preprint arXiv:1809.02704.

Chen, H., Shen, J., 2019. Optimal error estimates for SAV finite-element schemes for gradient flows. Numer. Math. (submitted for publication).

Cheng, Q., Shen, J., 2018. Multiple scalar auxiliary variable (MSAV) approach and its application to the phase-field vesicle membrane model. SIAM J. Sci. Comput. 40 (6), A3982–A4006.

Cheng, Q., Shen, J., Yang, X., 2019. Highly efficient and accurate numerical schemes for the epitaxial thin film growth models by using the SAV approach. J. Sci. Comput. 78 (3), 1467–1487.

Condette, N., Melcher, C., Süli, E., 2011. Spectral approximation of pattern-forming nonlinear evolution equations with double-well potentials of quadratic growth. Math. Comp. 80 (273), 205–223. ISSN 0025-5718. https://doi.org/10.1090/S0025-5718-10-02365-3.

Doi, M., Edwards, S.F., 1988. The Theory of Polymer Dynamics. vol. 73. Oxford University Press.

Du, Q., Feng, X., 2019. The phase field method for geometric moving interfaces and their numerical approximations. arXiv preprint arXiv:1902.04924.

Du, Q., Nicolaides, R.A., 1991. Numerical analysis of a continuum model of phase transition. SIAM J. Numer. Anal. 28 (5), 1310–1322. ISSN 0036-1429. https://doi.org/10.1137/0728069.

Du, Q., Liu, C., Wang, X., 2004. A phase field approach in the numerical study of the elastic bending energy for vesicle membranes. J. Comput. Phys. 198 (2), 450–468.

Forest, M.G., Wang, Q., Zhou, R., 2004. The flow-phase diagram of Doi-Hess theory for sheared nematic polymers II: finite shear rates. Rheol. Acta 44 (1), 80–93.

Forest, M.G., Wang, Q., Zhou, R., 2004. The weak shear kinetic phase diagram for nematic polymers. Rheol. Acta 43 (1), 17–37.

Giacomelli, L., Otto, F., 2001. Variatonal formulation for the lubrication approximation of the Hele-Shaw flow. Calc. Var. 13 (3), 377–403.

Gomez, H., Hughes, T.J.R., 2011. Provably unconditionally stable, second-order time-accurate, mixed variational methods for phase-field models. J. Comput. Phys. 230 (13), 5310–5327. ISSN 0021-9991. https://doi.org/10.1016/j.jcp.2011.03.033.

Gurtin, M.E., Polignone, D., Vinals, J., 1996. Two-phase binary fluids and immiscible fluids described by an order parameter. Math. Models Methods Appl. Sci. 6 (06), 815–831.

He, Z., Chen, J., Chen, Z., 2018. Efficient linearly and unconditionally energy stable schemes for the phase field model of solid-state dewetting problems. In: International Conference on Computational Science, pp. 102–112.

Kessler, D., Nochetto, R.H., Schmidt, A., 2004. A posteriori error control for the Allen-Cahn problem: circumventing Gronwall's inequality. M2AN Math. Model. Numer. Anal. 38 (1), 129–142. ISSN 0764-583X. https://doi.org/10.1051/m2an:2004006.

Kim, J., 2012. Phase-field models for multi-component fluid flows. Commun. Comput. Phys. 12 (03), 613–661.

Kou, J., Sun, S., Wang, X., 2018. Linearly decoupled energy-stable numerical methods for multi-component two-phase compressible flow. SIAM J. Numer. Anal. 56 (6), 3219–3248.

Larson, R.G., 1990. Arrested tumbling in shearing flows of liquid crystal polymers. Macromolecules 23 (17), 3983–3992.

Leslie, F.M., 1979. Theory of flow phenomena in liquid crystals. Adv. Liq. Cryst. 4, 1–81.

Li, X., Shen, J., Rui, H., 2019. Energy stability and convergence of SAV block-centered finite difference method for gradient flows. Math. Comp. 88 (319), 2047–2068.

Lin, L., Dong, S., 2018. Numerical approximation of incompressible Navier-Stokes equations based on an auxiliary energy variable. arXiv:1804.10859.

Liu, C., Shen, J., 2003. A phase field model for the mixture of two incompressible fluids and its approximation by a Fourier-spectral method. Physica D 179 (3), 211–228.

Lowengrub, J., Truskinovsky, L., 1998. Quasi-incompressible Cahn-Hilliard fluids and topological transitions. In: Proceedings of the Royal Society of London A: Mathematical, Physical and Engineering Sciences, vol. 454. The Royal Society, pp. 2617–2654.

Otto, F., 1998. Lubrication approximation with prescribed nonzero contact angle. Commun. Partial Differ. Equ. 23 (11–12), 2077–2164.

Rogers, T.M., Desai, R.C., 1989. Numerical study of late-stage coarsening for off-critical quenches in the Cahn-Hilliard equation of phase separation. Phys. Rev. B 39 (16), 11956.

Rudin, L.I., Osher, S., Fatemi, E., 1992. Nonlinear total variation based noise removal algorithms. Physica D 60 (1–4), 259–268.

Shen, J., Xu, J., 2018. Convergence and error analysis for the scalar auxiliary variable (SAV) schemes to gradient flows. SIAM J. Numer. Anal. 56 (5), 2895–2912. ISSN 0036-1429. https://doi.org/10.1137/17M1159968.

Shen, J., Yang, X., 2010. Energy stable schemes for Cahn-Hilliard phase-field model of two phase incompressible flows equations. Chin. Ann. Math. B 31, 743–758.

Shen, J., Yang, X., 2010. Numerical approximations of Allen-Cahn and Cahn-Hilliard equations. Discrete Contin. Dynam. Systems. A 28, 1669–1691.

Shen, J., Tang, T., Yang, J., 2016. On the maximum principle preserving schemes for the generalized Allen-Cahn equation. Comm. Math. Sci. (14), 1517–1534.

Shen, J., Xu, J., Yang, J., 2019. A new class of efficient and robust energy stable schemes for gradient flows. SIAM Rev. (submitted for publication).

Shen, J., Xu, J., Yang, J., 2018. The scalar auxiliary variable (SAV) approach for gradient flows. J. Comput. Phys. 353, 407–416. ISSN 0021-9991. https://doi.org/10.1016/j.jcp.2017.10.021.

Torabi, S., Lowengrub, J., Voigt, A., Wise, S., 2009. A new phase-field model for strongly anisotropic systems. In: Proceedings of the Royal Society of London A: Mathematical, Physical and Engineering Sciences. The Royal Society. pp. rspa–2008.

Yang, X., 2016. Linear, first and second-order, unconditionally energy stable numerical schemes for the phase field model of homopolymer blends. J. Comput. Phys. 327, 294–316.

Yang, X., 2018. Efficient schemes with unconditionally energy stability for the anisotropic Cahn-Hilliard equation using the stabilized-scalar augmented variable (S-SAV) approach. arXiv preprint arXiv:1804.02619.

Yang, X., 2018. Efficient schemes with unconditionally energy stability for the anisotropic Cahn-Hilliard equation using the stabilized-scalar augmented variable (S-SAV) approach. arXiv preprint arXiv:1804.02619.

Yang, Z., Dong, S., 2018. An unconditionally energy-stable scheme based on an implicit auxiliary energy variable for incompressible two-phase flows with different densities involving only precomputable coefficient matrices. arXiv preprint arXiv:1811.07888.

Yu, H., Ji, G., Zhang, P., 2010. A nonhomogeneous kinetic model of liquid crystal polymers and its thermodynamic closure approximation. Commun. Comput. Phys. 7 (2), 383.

Yue, P., Feng, J.J., Liu, C., Shen, J., 2004. A diffuse-interface method for simulating two-phase flows of complex fluids. J. Fluid Mech. 515, 293–317.
Yue, P., Feng, J.J., Liu, C., Shen, J., 2004. A diffuse-interface method for simulating two-phase flows of complex fluids. J. Fluid Mech. 515, 293–317.
Zhao, J., Wang, Q., Yang, X., 2017. Numerical approximations for a phase field dendritic crystal growth model based on the invariant energy quadratization approach. Int. J. Numer. Methods Eng. 110 (3), 279–300.

Zhu, G., Kou, J., Sun, S., Yao, J., Li, A., 2018. Decoupled, energy stable schemes for a phase-field surfactant model. Comput. Phys. Commun. 233, 67–77.

Zhuang, Q., Shen, J., 2019. Efficient SAV approach for imaginary time gradient flows with applications to one- and multi-component Bose-Einstein Condensates. J. Comput. Phys. (submitted for publication).

Index

Note: Page numbers followed by "*f*" indicate figures, "*t*" indicate tables, and "*b*" indicate boxes.

G

H

I

J

K

L

M

N

O

P

FSC
www.fsc.org
MIX
Papier | Fördert
gute Waldnutzung
FSC® C083411

Kolibri 360 GmbH
Ferdinand-Jühlke-Straße 7
99095 Erfurt, Deutschland
produktsicherheit@kolibri360.de